AF819192

Solution Guide

Non-Commutative Harmonic Analysis

by

Raymond Fabec

DP

Drexville Publishing, St. Gabriel, LA

TeX and $\mathcal{A}_{\mathcal{M}}\mathcal{S}$-LaTeX are trademarks of the American Mathematical Society.
This book was prepared in $\mathcal{A}_{\mathcal{M}}\mathcal{S}$-LaTeX by the author,
Raymond C. Fabec.

A Solution Guide to Non-Commutative Harmonic Analysis
ISBN: 978-0-9913266-1-7 (paperback)

Copyright

©2022 Raymond Fabec

All rights reserved.

No part of this book may be reproduced in any form or by any means without permission in writing from the author.

Printed in the United States of America

Contents

Introduction

It is the intent of this text to provide a full account of the exercises and their solutions in the text Non-Commutative Harmonic Analysis. These extend the material covered and provide students with results that could have made up an enlarged volume. The easier problems provide a stepping stone for students new to the subject on how proofs are obtained and argued. The more difficult problems broaden the scope of the text and present readers with results connected to the overall presentation.

Also some problems were either already part of the main text or in wrong sections. These are addressed when appropriate. A few problems had to be slightly restated to overcome difficulty in their proofs. The author feels the arguments in themselves make a fine and useful addition to the original text.

There is no bibliography. For this one should reference the original text. The index and the original text index may be used to locate topics of interest to the reader.

Chapter 1

Fourier Series

Exercise Set 1.1

1. Prove equation (1.9): $\sum_{n=-\infty}^{\infty} r^{|n|}e^{in\theta} = \frac{1-r^2}{1-2r\cos(\theta)+r^2}$ for $0 \leqslant r < 1$.
Solution: We assume $|r| < 1$. Then using geometric series,

$$\begin{aligned}
\sum r^{|n|}e^{in\theta} &= 1 + 2\mathrm{Re}\sum_{n=1}^{\infty} r^n e^{in\theta} \\
&= 1 + 2\mathrm{Re}\frac{re^{i\theta}}{1-re^{i\theta}} \\
&= 1 + 2\mathrm{Re}\frac{re^{i\theta}(1-re^{-i\theta})}{(1-r\cos\theta)^2 + r^2\sin^2\theta} \\
&= 1 + 2\frac{r\cos\theta - r^2}{1+r^2-2r\cos\theta} \\
&= \frac{1+r^2-2r\cos\theta+2r\cos\theta-2r^2}{1+r^2-2r\cos\theta} \\
&= \frac{1-r^2}{1-2r\cos\theta+r^2}.
\end{aligned}$$

□

2. Prove the following:

(a) $P(r,\theta) \geqslant 0$ and $\frac{1}{2\pi}\int_0^{2\pi} P(r,\theta)\,d\theta = 1$ for all $1 > r \geqslant 0$.

(b) The maximum of $\theta \mapsto P(r,\theta)$ occurs at $\theta = 0$ and $\max_\theta P(r,\theta) = \frac{1+r}{1-r}$. In particular $P(r,0) \to \infty$ as $r \to 1-$.

(c) The function $\theta \mapsto P(r,\theta)$ takes its minimum at $\theta = \pi$. Evaluate $P(r,\pi)$.

Solution: First note $P(r,\theta) = \frac{1-r^2}{1-2r\cos\theta+r^2} \geqslant \frac{1-r^2}{1+2r+r^2} = \frac{1-r}{1+r} > 0$ for $0 \leqslant r < 1$. Also the maximum of $P(r,\theta)$ occurs when $\cos\theta = 1$ or $\theta = 2\pi n$ for this is when the denominator is minimum and this maximum is $\frac{1-r^2}{(1-r)^2} = \frac{1+r}{1-r}$. The minimum occurs when $\cos\theta = -1$ for this gives the largest denominator. Thus $\theta = \pi + 2\pi n$. The minimum is $\frac{1-r}{1+r}$.

Since the series converges uniformly to $P(r,\theta)$ for fixed r in $[0,1)$, we see

$$\begin{aligned}\int_{-\pi}^{\pi} P(r,\theta)\,d\theta &= \sum_n \int_{-\pi}^{\pi} r^{|n|}e^{in\theta}\,d\theta \\ &= 2\pi + \sum_{n\neq 0} r^{|n|}\frac{e^{in\pi} - e^{-in\pi}}{in} \\ &= 2\pi.\end{aligned}$$

□

3. Suppose that f is 2π-periodic and piecewise continuous. Show that

$$\lim_{r\to 1-} \frac{1}{2\pi}\int_0^{2\pi} f(\phi)P(r,\theta-\phi)\,d\phi = f(\theta)$$

if f is continuous at θ.

Solution: First note integrals of 2π periodic functions from 0 to 2π are the same as integrals from $-\pi$ to π. By translating f we may suppose f is piecewise continuous, 2π periodic, and continuous at 0. Next note if $|\theta| \geqslant \delta > 0$ where $\delta < \pi$, then the maximum of $P(r,\theta)$ for $\delta \leqslant |\theta| \leqslant \pi$ occurs when $1-2r\cos\theta+r^2$ has a minimum or when $|\theta| = \delta$. So $|P(r,\theta)| \leqslant \frac{1-r^2}{1-2r\cos\delta+r^2}$. Thus$|P(r,\theta)|$ converges uniformly to 0 for $\delta \leqslant |\theta| \leqslant \pi$ as $r \to 1-$. Hence using $\int_{-\pi}^{\pi} P(r,\theta)\,d\theta = 2\pi$ from the previous exercise, one has using the positivity of $P(r,\theta)$ and continuity of f at 0 that

$$\begin{aligned}|f(0) - \frac{1}{2\pi}\int_{-\pi}^{\pi} f(\theta)P(r,\theta)\,d\theta| &= \frac{1}{2\pi}|\int_{-\pi}^{\pi}(f(0)-f(\theta))P(r,\theta)\,d\theta| \\ &\leqslant \frac{1}{2\pi}\int_{\delta\leqslant|\theta|\leqslant\pi} P(r,\theta)|f(0)-f(\theta)|\,d\theta + \frac{\epsilon}{4\pi}\int_{|\theta|\leqslant\delta} P(r,\theta)\,d\theta\end{aligned}$$

if δ is chosen so that $|f(\theta) - f(0)| \leqslant \frac{\epsilon}{4\pi}$ for $|\theta| \leqslant \delta$. Since f is piecewise continuous, $|f(\theta) - f(0)|$ has an upperbound M and consequently if r_0 is chosen so that $|P(r,\theta)| \leqslant \frac{\epsilon}{2M+1}$ for $\delta \leqslant |\theta| \leqslant \pi$ and $r_0 \leqslant r < 1$, then

$$\begin{aligned}|f(0) - \frac{1}{2\pi}\int_{-\pi}^{\pi} f(\theta)P(r,\theta)\,d\theta| &\leqslant \frac{\epsilon}{2\pi(2M+1)}M(2\pi) + \frac{\epsilon}{2} \\ &< \epsilon.\end{aligned}$$

□

4. Write $u(x,y) = u(z)$ where $z = x+iy$. Suppose that f is continuous. Show that u is holomorphic on D if and only if $a_n = 0$ for all $n < 0$.

Solution: We have $u(r,\theta) = u(z) = \sum a_n r^{|n|}e^{in\theta} = a_0 + \sum_{n>0}(a_n z^n + a_{-n}\bar{z}^n)$ where $z = re^{i\theta}$ and $a_n = \frac{1}{2\pi}\int_0^{2\pi} f(\theta)e^{-in\theta}\,d\theta$. Since f is continuous, f has a maximum M on $[0, 2\pi]$ and thus each $|a_n| \leqslant M$ for all n. Thus the series $\sum_{n>0} a_n z^n$ and $\sum_{n>0} a_{-n}\bar{z}^n$ converge uniformly on the balls $|z| \leqslant R$ when $R < 1$. Thus $\sum_{n\geqslant 0} a_n z^n$ is holomorphic on each ball $|z| < R < 1$ and thus is holomorphic on $|z| < 1$. Now by a similar argument, if $g(z) = \sum_{n>0} a_{-n}\bar{z}^n$, then $\overline{g(z)}$ is holomorphic on $|z| < 1$. Hence $u(z)$ is holomorphic on the disk if and only if $u(z) - \sum_{n\geqslant 0} a_n z^n = g(z)$ is holomorphic on the disk. Consequently this occurs if and only if $g(z)$ and $\overline{g(z)}$ are both holomorphic. But if both are holomorphic $g(z) + \overline{g(z)} = 2\text{Re}g(z)$ is holomorphic. But the only real valued holomorphic functions are constants. Consequently, $g(z) = 0$ precisely when u is holomorphic. □

Exercise Set 1.2

1. Show if f is a function on $\mathbb{T}$ such that $f \circ \kappa$ is a simple measurable function on $\mathbb{R}$, then

$$\int f(z)\, d\mu(z) = \int_0^1 f(e^{2\pi i x})\, dx.$$

Solution: Suppose $f \circ \kappa$ is a simple measurable function on $\mathbb{R}$. Thus $f \circ \kappa = \sum_{m=1}^n a_m \chi_{E_k}$ where E_k are Lebesgue measurable subsets of $\mathbb{R}$. By definition of the measure μ, we have $\mu(F) = m(\kappa^{-1}(F) \cap [0,1))$. We note $f(e^{2\pi i x}) = \sum_{j=1}^n a_j \chi_{E_j}(x) = \sum_{j=1}^n a_j g_j(e^{2\pi i x})$ where $g_j(e^{2\pi i x}) = \chi_{E_j}(x)$ for $x \in [0,1)$. Hence by additivity, we may assume $f(e^{2\pi i x}) = a\chi_E(x)$ for $x \in [0,1)$. Thus $f(z) = a$ if and only if $e^{2\pi i x} = z$ for some $x \in [0,1) \cap E$. So $f = a\chi_W$ where $W = e^{2\pi i (E \cap [0,1))}$. Consequently,

$$\begin{aligned}
\int_T f(z)\, d\mu(z) &= a\mu(W) \\
&= am(\kappa^{-1}(W) \cap [0,1)) \\
&= a\int_0^1 \chi_W(e^{2\pi i x})\, dx \\
&= \int_0^1 f(e^{2\pi i x})\, dx.
\end{aligned}$$

□

2. Show $L^p(\mathbb{T}) \subseteq L^q(\mathbb{T})$ and $|f|_p \geqslant |f|_q$ for $p \geqslant q$.
Solution: Note $1 \in L^p(\mathbb{T})$ for all $p \in [1, \infty]$, one has by Hölder's inequality, $\int |f|\, d\mu \leqslant ||f||_p ||1||_q = ||f||_p$ if $\frac{1}{p} + \frac{1}{q} = 1$. Thus if $p \geqslant q \geqslant 1$, then $\frac{p}{q} \geqslant 1$ and thus $\int |f|^q\, d\mu \leqslant || \, |f|^q ||_{\frac{p}{q}} = \left(\int |f|^p\, d\mu\right)^{q/p}$. Thus $||f||_q^q \leqslant ||f||_p^q$ and we see $||f||_q \leqslant ||f||_p$. □

3. Consider $C(\mathbb{T})$ with norm $|\cdot|_\infty$. For $w \in \mathbb{T}$, define $\lambda(w)f(z) = f(w^{-1}z)$ for $f \in C(\mathbb{T})$.

(a) Show $\lambda(w)$ is a linear isometry for each w.

(b) Show λ is a homomorphism of $\mathbb{T}$ into the isometry group of $C(\mathbb{T})$.

(c) Show λ is not continuous from $\mathbb{T}$ into the Banach space of bounded linear operators on $C(\mathbb{T})$.

(d) Show for each $f \in C(\mathbb{T})$, the mapping $w \mapsto \lambda(w)f$ is continuous from $\mathbb{T}$ into $C(\mathbb{T})$

Solution: For (a), note $\lambda(w)$ is a linear mapping. Moreover, if $|f|$ has a maximum at z_0, then $|\lambda(w)f|$ has the same maximum at wz_0. So $|\lambda(w)f|_\infty = |f|_\infty$. (b) is a direct calculation. For (c), let $\epsilon = \frac{1}{2}$. Suppose $\delta > 0$ satisfies $|\lambda(w)f - f| < \epsilon$ if $|w-1| \leqslant \delta$ and $|f|_\infty \leqslant 1$. Choose $f \in C(\mathbb{T})$ so that $f(e^{i\theta}) = 1$ if $-\frac{\delta}{4} \leqslant \theta \leqslant \frac{\delta}{4}$ and $f(e^{i\theta}) = 0$ if $\frac{\delta}{2} \leqslant |\theta| \leqslant \pi$ and $0 \leqslant f \leqslant 1$. Then $|e^{i\delta} - 1| \leqslant \delta$ but $|\lambda(e^{i\delta})f - f|_\infty = 1$. To see (d), note $\theta \mapsto f(e^{i\theta})$ is uniformly continuous on $[-2\pi, 2\pi]$ and thus if $\epsilon > 0$, there is a $\delta > 0$ with $|f(e^{i\theta}) - f(e^{i\phi})| < \epsilon$ if $|\phi - \theta| < \delta$. Thus $|f(e^{-i\theta}w) - f(w)| < \epsilon$ if $|\theta| < \delta$. So $|\lambda(e^{i\theta})f - f|_\infty < \epsilon$ if $|\theta| < \delta$. This gives $|\lambda(e^{i\theta})\lambda(w)f - \lambda(w)f|_\infty = |\lambda(w)\lambda(e^{i\theta})f - \lambda(w)f|_\infty = |\lambda(e^{i\theta})f - f|_\infty < \epsilon$ if $|\theta| < \delta$. So $|\lambda(z)f - \lambda(w)f|_\infty < \epsilon$ when $|z - w| < \delta$ for if $|z - w| < \delta$, there is a θ with $|\theta| < \delta$ and $z = e^{i\theta}w$. □

4. Let $1 \leqslant p < \infty$. Let $\lambda(z)f(w) = f(z^{-1}w)$ for $f \in L^p(\mathbb{T})$. Show for each f and each $\epsilon > 0$, there is a $\delta > 0$ such that $|\lambda(z)f - f|_p < \epsilon$ if $|z - 1| < \delta$.

Solution: Let $1 \leqslant p < \infty$. Since $C(\mathbb{T})$ is dense in $L^p(\mathbb{T})$, we can find a $g \in C(T)$ with $|g - f|_p < \frac{\epsilon}{3}$. Now g is uniformly continuous. Hence we can find a $\frac{\pi}{4} > \delta > 0$ such that $|g(e^{i\theta}w) - g(w)| < \epsilon/3$ for all w when $|\theta| < \delta$. So $|\lambda(e^{i\theta})g - g|_p^p = \int |g(e^{-i\theta}w) - g(w)|^p d\mu(w) \leqslant \frac{\epsilon^p}{3^p}$. This gives $|\lambda(z)g - g|_p < \frac{\epsilon}{3}$ if $z = e^{i\theta}$ and $|\theta| < \delta$. Now note $0 \leqslant \frac{\theta}{2} \leqslant \sin\theta$ for $0 \leqslant \theta < \frac{\pi}{4}$. Hence if $|z - 1| \leqslant \frac{\delta}{2}$ and $z = e^{i\theta} = \cos\theta + i\sin\theta$ where $|\theta| < \frac{\pi}{4}$, then since $|\cos\theta + i\sin\theta - 1| = |\cos\theta||1 + i\tan\theta - \sec\theta| \geqslant |\cos\theta\tan\theta| = |\sin\theta|$, we see $|\sin\theta| < \frac{\delta}{2}$ and so $|\theta| < \delta$. Thus $|\lambda(z)g - g|_p < \frac{\epsilon}{2}$ if $|z - 1| < \frac{\delta}{2}$. Hence $|z - 1| < \frac{\delta}{2}$ implies

$$\begin{aligned}
|\lambda(z)f - f|_p &\leqslant |\lambda(z)f - \lambda(z)g|_p + |\lambda(z)g - g|_p + |g - f|_p \\
&= 2|g - f|_p + |\lambda(z)g - g|_p \\
&< 2\left(\frac{\epsilon}{3}\right) + \frac{\epsilon}{3} = \epsilon.
\end{aligned}$$

□

5. For $1 \leqslant p < q$, find $f \in L^p(\mathbb{T})$ such that $f \notin L^q(\mathbb{T})$.

Solution: Let $f_s(e^{i\theta}) = \frac{1}{\theta^s}$ for $0 < \theta < 1$ and 0 otherwise. Then $f_s \in L^p(\mathbb{T})$ if and only if $\frac{1}{\theta^{sp}}$ is integrable on $(0, 1)$ if and only if $1 - sp > 0$ if and only if $p < \frac{1}{s}$. Thus if $1 \leqslant p < q$ and $\frac{1}{s} \in (p, q)$, we see $f_s \in L^P(\mathbb{T})$ and $f_s \notin L^q(\mathbb{T})$. □

6. Let $h \in L^2(\mathbb{T})$. Show $|h * g|_2 \leqslant |h|_1\,|g|_2$ for all $g \in L^2(\mathbb{T})$. Then show if $h \in L^1(\mathbb{T})$ and $h_n \in L^2(\mathbb{T})$ converges in $L^1(\mathbb{T})$ to h, then $h * g$ defined to be the L^1 limit of $h_n * g$ is well defined and satisfies $|h * g|_2 \leqslant |h|_1|g|_2$. (**Hint:** Let $f \in L^2(\mathbb{T})$. Then $f\overline{h * g}$ is integrable and $f \mapsto \int f(z)\overline{h * g(z)}\, d\mu(z)$ is a continuous linear functional on $L^2(\mathbb{T})$.)

Solution: Let $h \in L^2(\mathbb{T})$ and g be in $L^2(\mathbb{T})$. We first note by Proposition 1.15 that $h * g(z) = \int h(w)g(w^{-1}z)\,d\mu(w)$ is a continuous function. Moreover, if $f \in L^2(\mathbb{T})$, then

$$\begin{aligned}
\left|\int f(z)\overline{(h*g)(z)}\,d\mu(z)\right| &\leqslant \left|\int\int f(z)\overline{h(zw)g(w^{-1})}\,d\mu(w)\,d\mu(z)\right| \\
&\leqslant \left|\int\int f(z)\overline{h(zw)g(w^{-1})}\,d\mu(z)\,d\mu(w)\right| \\
&\leqslant \left|\int\int \bar{h}(z)f(zw^{-1})\bar{g}(w^{-1})\,d\mu(z)\,d\mu(w)\right| \\
&\leqslant \int |h(z)|\left|\int f(zw)\bar{g}(w)\,d\mu(w)\right|\,d\mu(z) \\
&\leqslant \int |h(z)|\,|f|_2|g|_2\,d\mu(z) = |h|_1|f|_2|g|_2.
\end{aligned}$$

Thus $h * g$ is in L^2 and $|h * g|_2 \leqslant |h|_1|g|_2$.

Now assume $h \in L^1(\mathbb{T}) \supseteq L^2(\mathbb{T})$. Choose a sequence $h_n \in L^2(\mathbb{T})$ with $h_n \to h$ in $L^1(\mathbb{T})$. Then $|(h_n - h_m)*g|_2 \leqslant |h_n - h_m|_1|g|_2$ and since $|h_n - h_m|_1 \to 0$ as $m, n \to \infty$, we see $h_n * g$ is Cauchy in $L^2(\mathbb{T})$. By completeness of $L^2(\mathbb{T})$, $h_n * g$ converges in $L^2(\mathbb{T})$ and so $h * g = \lim_n h_n * g$ exists as a limit in $L^2(\mathbb{T})$. Since $|h_n * g|_2 \leqslant |h_n|_1|g|_2$ for all n, one has $|h * g|_2 \leqslant |h|_1|g|_2$. □

Exercise Set 1.3

1. Suppose σ_k are seminorms on vector space X and ρ_k are seminorms on vector space Y. Give X and Y the topological vector space topologies defined by these seminorms. Show a linear transformation $T : X \to Y$ is continuous if and only if $\rho_k \circ T$ is continuous at 0 for each k.

(Hint: Recall a subset U of X will be open in the topology defined by the seminorms σ_k if for each $p \in U$, there is an $\epsilon > 0$ and finitely many seminorms $\sigma_{k_1}, \sigma_{k_2}, \ldots, \sigma_{k_n}$ so that if $\sigma_{k_i}(q - p) < \epsilon$ for $i = 1, 2, \ldots, n$, then $q \in U$.)

Solution: Assume $T : X \to Y$ is continuous and ρ_k is a seminorm on Y. Then ρ_k is continuous. Thus $\rho_k \circ T$ is continuous being a composition of continuous functions. Now assume $\rho_k \circ T$ is continuous for each ρ_k. Let U be a nonempty open subset of Y and let $x_0 \in T^{-1}(U)$. Since $Tx_0 \in U$ and U is open, there is an $\epsilon > 0$ and seminorms $\rho_{k_1}, \rho_{k_2}, \ldots, \rho_{k_n}$ such that $\{y \in Y \mid \rho_{k_j}(y, Tx_0) < \epsilon,\ j = 1, 2, \ldots, n\} \subseteq U$.

Since $\rho_{k_j} \circ T$ is continuous for each j and the set of j is finite, we can find finitely many seminorms $\sigma_{k_1}, \sigma_{k_2}, \ldots, \sigma_{k_m}$ and a δ (just put all the needed σ together and take a smallest positive necessary δ) such that if $\sigma_{k_i}(x_0, x) < \delta$ for $i = 1, 2, \ldots, m$, then $\rho_{k_j}(Tx, Tx_0) < \epsilon$. Thus $T^{-1}U$ contains the open neighborhood $\{x \mid \sigma_{k_i}(x, x_0) < \delta \text{ for } i = 1, 2, \ldots, m\}$. □

2. Show that $f \in L^2([0, 2\pi], dx)$ can be written in the form

$$f(x) = a_0 + \sum_{n=1}^{\infty} a_n \cos(nx) + \sum_{n=1}^{\infty} b_n \sin(nx).$$

Find an expression for a_n and b_n.

Solution: We know if $f \in L^2(\mathbb{T})$, then $f = \sum c_n e^{in\theta}$ where this converges in $L^2[0, 2\pi]$ and each c_n is given by $c_n = \frac{1}{2\pi}\int_0^{2\pi} f(\theta)e^{-in\theta}\, d\theta$. Now one can rearrange this series so that $f = c_0 + \sum_{n\geqslant 1}(c_n e^{in\theta} + c_{-n}e^{in\theta})$ and this converges in L^2. But

$$\begin{aligned} c_n e^{in\theta} + c_{-n}e^{-in\theta} &= (c_n + c_{-n})\cos n\theta + (ic_n - ic_{-n})\sin n\theta \\ &= a_n \cos n\theta + b_n \sin n\theta. \end{aligned}$$

Here

$$a_n = \frac{1}{2\pi}\int_0^{2\pi} f(\theta)\,(\cos n\theta + \cos n\theta)\, d\theta = \frac{1}{\pi}\int_0^{2\pi} f(\theta)\, \cos n\theta\, d\theta$$

and

$$b_n = \frac{i}{2\pi}\int_0^{2\pi} f(\theta)\,(i\sin(-n\theta) - i\sin(n\theta))\, d\theta = \frac{1}{\pi}\int_0^{2\pi} f(\theta)\sin n\theta\, d\theta.$$

Thus if $a_0 = c_0 = \frac{1}{2\pi}\int_0^{2\pi} f(\theta)\, d\theta$, we see

$$f(\theta) = a_0 + \sum_{n=1}^{\infty}(a_n \cos n\theta + b_n \sin n\theta)$$

in L^2. But the functions $\cos n\theta$ and $\sin m\theta$ are orthogonal in L^2 and this implies one can take any rearrangement and still have L^2 convergence. Thus $f(\theta) = a_0 + \sum_{n=1}^{\infty} a_n \cos n\theta + \sum_{n=1}^{\infty} b_n \sin n\theta$. $\square$

3. Let g be the function on the torus given by $g(e^{i\theta}) = |\theta|$ for $\theta \in [-\pi, \pi)$.

(a) Find $\hat{g}(n)$.

(b) Show that the Fourier series converges uniformly.

Solution: We have $g(\theta) = |\theta|$. Thus using the evenness of $|\theta|\cos n\theta$ and the oddness of $|\theta|\sin n\theta$, we see for $n \neq 0$,

$$\begin{aligned} \hat{g}(n) &= \frac{1}{2\pi}\int_{-\pi}^{\pi} |\theta|\, e^{-in\theta}\, d\theta = \frac{1}{2\pi}\int_{-\pi}^{\pi} |\theta| \cos n\theta\, d\theta - \frac{i}{2\pi}\int_{-\pi}^{\pi} |\theta| \sin n\theta\, d\theta \\ &= \frac{1}{\pi}\int_0^{\pi} \theta\cos n\theta\, d\theta = \frac{1}{\pi}\left(\theta\frac{\sin n\theta}{n}\Big|_0^{\pi} + \frac{\cos n\theta}{n^2}\Big|_0^{\pi}\right) = \frac{(-1)^n - 1}{n^2\pi}. \end{aligned}$$

Now $\hat{g}(0) = \frac{1}{2\pi}\int_{-\pi}^{\pi} |\theta|\, d\theta = \frac{1}{\pi}\frac{\theta^2}{2}\Big|_0^{\pi} = \frac{\pi}{2}$.

Thus in L^2, one has

$$|\theta| = \frac{\pi}{2} + \sum_{n \text{ odd}} \frac{-2}{n^2\pi}(\cos n\theta + i \sin n\theta).$$

In this sum dealing with $+n$ and $-n$ for $n > 0$ gives the cancellation of the sine terms and doubles up the cosine terms. Thus

$$|\theta| = \frac{\pi}{2} - \sum_{n \text{ odd,pos}} \frac{4}{n^2\pi} \cos n\theta.$$

We could also use $f(\theta) = a_0 + \sum_{n=1}^{\infty}(a_n \cos n\theta + b_n \sin n\theta)$ where $a_0 = \frac{1}{2\pi}\int_{-\pi}^{\pi} f(\theta)\, d\theta = \frac{1}{\pi}\int_0^{\pi} \theta\, d\theta = \frac{\pi}{2}$ and $a_n = \frac{1}{\pi}\int_{-\pi}^{\pi} |\theta| \cos n\theta\, d\theta = \frac{2}{\pi}\int_0^{\pi} \theta \cos n\theta\, d\theta = \left(\frac{2}{n\pi}\theta \sin n\theta - \frac{2}{n^2\pi}\cos n\theta\right)\Big|_0^{\pi}$. Thus $a_n = 0$ if n is even and $a_n = \frac{-4}{n^2\pi}$ if n is odd. One obtains $b_n = 0$ for all n for $|\theta| \sin n\theta$ is odd. Thus

$$|\theta| = \frac{\pi}{2} - \frac{4}{\pi}\cos\theta - \frac{4}{9\pi}\cos 3\theta - \frac{4}{25\pi}\cos 5\theta - \frac{4}{49\pi}\cos 7\theta - \cdots.$$

This series converges uniformly for $\sum \frac{1}{(2k+1)^2} < \infty$. □

4. Use the Fourier transform to evaluate the sum $\sum_{n=1}^{\infty} \frac{1}{n^2}$.
Solution: Let $f(\theta) = \theta$ for $-\pi < \theta < \pi$. Note $\hat{f}(0) = 0$. If $n \neq 0$,

$$\begin{aligned}
\hat{f}(n) &= \frac{1}{2\pi}\int_{-\pi}^{\pi} \theta e^{-in\theta}\, d\theta \\
&= \frac{1}{2\pi}\frac{e^{-in\theta}}{-in}\theta\Big|_{-\pi}^{\pi} - \frac{1}{2\pi}\int_{-\pi}^{\pi} \frac{e^{-in\theta}}{-in}\, d\theta \\
&= \frac{i}{2\pi n}((-1)^n\pi - (-1)^n(-\pi)) - \frac{1}{2\pi}\frac{e^{-in\theta}}{(in)^2}\Big|_{-\pi}^{\pi} \\
&= \frac{(-1)^n i}{n}.
\end{aligned}$$

By the Plancherel Theorem, $\sum_{-\infty}^{\infty} |\hat{f}(n)|^2 = \frac{1}{2\pi}\int_{-\pi}^{\pi} |f(\theta)|^2\, d\theta$. So $2\sum_{n=1}^{\infty}\frac{1}{n^2} = \frac{1}{\pi}\int_0^{\pi} \theta^2\, d\theta = \frac{1}{\pi}\cdot\frac{\pi^3}{3} = \frac{\pi^2}{3}$. Hence

$$\sum \frac{1}{n^2} = \frac{\pi^2}{6}.$$

□

5. Suppose $f \in L^2[0,1]$ has Fourier series

$$f(x) = \sum_n c_n e^{2\pi i n x}.$$

Show

$$\int_0^x f(t)\, dt = i \sum_{n=-\infty}^{\infty} c_n \frac{1 - e^{2\pi i n x}}{2\pi n}$$

for $0 \leqslant x \leqslant 1$ where $\frac{1-e^{2\pi i 0x}}{2\pi 0}$ is defined appropriately.
Solution: Since $|f|_1 \leqslant |f|_2$ for $f \in L^2[0,1]$ and one has L^2 convergence for L^2 functions, we have $f\chi_{[0,x]} = \sum c_n e^{2\pi int}\chi_{[0,x]}(t)$ in L^2 and hence in L^1. Thus

$$\begin{aligned}
\int_0^x f(t)\,dt &= \sum_n c_n \int_0^x e^{2\pi int}\,dt \\
&= c_0 x + \sum_{n\neq 0} c_n \frac{e^{2\pi int}}{2\pi in}\bigg|_{t=0}^{t=x} \\
&= c_0 x + \sum_{n\neq 0} c_n \frac{e^{2\pi inx}-1}{2\pi in} \\
&= c_0 x + i\sum_{n\neq 0} c_n \frac{1-e^{2\pi inx}}{2\pi n} \\
&= i\sum_n c_n \frac{1-e^{2\pi inx}}{2\pi n}
\end{aligned}$$

where here we are taking $\frac{1-e^{2\pi i0x}}{2\pi 0} = \lim_{y\to 0}\frac{1-e^{2\pi ixy}}{2\pi y} = -ix$. ☐

6. Let f be the periodic function with period 1 corresponding to $\chi_{[-\frac{1}{2},0)} - \chi_{[0,\frac{1}{2})}$. Evaluate $\hat{f}(n)$.
Solution: We have

$$\begin{aligned}
\hat{f}(n) &= \int_{-1/2}^{1/2} (\chi_{[-\frac{1}{2},0)}(t) - \chi_{[0,\frac{1}{2})}(t))e^{-2\pi int}\,dt \\
&= \int_{-\frac{1}{2}}^{0} e^{-2\pi int}\,dt - \int_0^{\frac{1}{2}} e^{-2\pi int}\,dt \\
&= \frac{-1}{2\pi in}\left(1 - e^{\pi in} - e^{-\pi in} + 1\right) \\
&= \frac{i}{\pi n}(1-\cos(n\pi)) \\
&= \frac{i}{\pi n}(1-(-1)^n)
\end{aligned}$$

when $n \neq 0$ and $c_0 = 0$. Thus $c_{2n} = 0$ and $c_{2n+1} = \frac{2i}{(2n+1)\pi}$ for all $n \in \mathbb{Z}$. ☐

7. Let f be a C^∞ function of compact support on $\mathbb{R}$. Define F by

$$F(x) = \sum_{n\in\mathbb{Z}} f(x+n).$$

(a) Show F is a C^∞ function of period 1.

(b) Show

$$\sum_n f(n) = \sum_n \hat{F}(n).$$

Solution: For (a), one has $F(x+1) = F(x)$ for $\sum_n f(x+n+1) = \sum_n f(x+n)$ and each of these sums are finite for f vanishes off some bounded interval. To see F is C^∞, fix x. Take $m \in \mathbb{N}$ so that $f(y) = 0$ if $|y| \geqslant m-1$. Let W be the finite set of integers k with $|x+k| < m$. Then if $|y-x| < \frac{1}{m}$,

$$\begin{aligned} F(y) &= \sum_k f(y+k) \\ &= \sum_k f(x+y-x+k) \\ &= \sum_{k \in W} f(y+k) \end{aligned}$$

for if $k \notin W$, then $|y+k| = |x+y-x+k| \geqslant |x+k| - |y-x| \geqslant m - \frac{1}{m} \geqslant m-1$. Consequently, if $|y-x| < \frac{1}{m}$, $F(y) = \sum_{k \in W} f(y+k)$ which is a finite sum of C^∞ functions on the interval $x - \frac{1}{m} < y < x + \frac{1}{m}$. Thus F is C^∞ on this interval.

By (a), F is C^∞ and periodic with period 1. Using Corollary 1.23 one has $\sum \hat{F}(n)e^{2\pi inx}$ converges uniformly to F. Hence $F(0) = \sum \hat{F}(n)$. But $F(0) = \sum f(n)$. □

8. A function f on the torus is even if $f(z) = f(z^{-1})$ and odd if $f(z) = -f(z^{-1})$. Suppose that $f \in C^2(\mathbb{T})$. Show the following:

(a) If f is even, then $f(e^{2\pi i\theta}) = \sum \hat{f}(n)\cos(2\pi n\theta)$;

(b) If f is odd, then $f(e^{2\pi i\theta}) = i\sum \hat{f}(n)\sin(2\pi n\theta)$.

Solution: We have if f is even, then

$$\begin{aligned} \hat{f}(-n) &= \frac{1}{2\pi}\int_{-\pi}^{\pi} f(e^{i\theta})e^{in\theta}\,d\theta = \frac{1}{2\pi}\int_{-\pi}^{\pi} f(e^{-i\theta})e^{in\theta}\,d\theta \\ &= \frac{1}{2\pi}\int_{-\pi}^{\pi} f(e^{i\theta})e^{-in\theta}\,d\theta \\ &= \hat{f}(n). \end{aligned}$$

Similarly, if f is odd, $\hat{f}(-n) = -\hat{f}(n)$. So if f is even, one has $\hat{f}(n)i\sin n\theta + \hat{f}(-n)i\sin(-n\theta) = 0$. This implies $\hat{f}(n)e^{in\theta} + \hat{f}(-n)e^{in\theta} = \hat{f}(n)\cos n\theta + \hat{f}(-n)\cos(-n\theta)$ and similarly if f is odd, $\hat{f}(n)e^{in\theta} + \hat{f}(-n)e^{in\theta} = \hat{f}(n)i\sin n\theta + \hat{f}(-n)i\sin(-n\theta)$. Putting these together gives the result. □

9. Let $g \in C^1(\mathbb{T})$ and $f \in L^1(\mathbb{T})$. Show $f * g \in C^1(\mathbb{T})$ and $D(f*g) = f * Dg$.

Solution: Let $g \in C^1(\mathbb{T})$. Thus $\theta \mapsto g(e^{i\theta})$ is differentiable. Thus if f is in $L^1(\mathbb{T})$, then $f * g(e^{i\theta}) = \frac{1}{2\pi}\int_{-\pi}^{\pi} f(e^{i\gamma})g(e^{i(\theta-\gamma)})\,d\gamma$ exists for the integrand is integrable being bounded by $|g|_\infty|f|$. Moreover, if θ and θ_0 are given, then by the Mean Value Theorem and the Chain Rule,

$$g(e^{i(\theta-\gamma)}) - g(e^{i(\theta_0-\gamma)}) = h'(\theta(\gamma)-\gamma)(\theta-\theta_0)$$

where $h(\theta) = g(e^{i\theta})$ and $\theta(\gamma)$ is between θ_0 and θ. Thus

$$\frac{f * g(e^{i\theta}) - f * g(e^{i\theta_0})}{\theta - \theta_0} = \frac{1}{2\pi}\int_{-\pi}^{\pi} f(e^{i\gamma})h'(\theta(\gamma) - \gamma)\,d\gamma$$

$$= \frac{1}{2\pi}\int_{-\pi}^{\pi} f(e^{i\gamma})h'(\theta_0 - \gamma)\,d\gamma + \frac{1}{2\pi}\int_{-\pi}^{\pi} f(e^{i\gamma})(h'(\theta(\gamma) - \gamma) - h'(\theta_0 - \gamma))\,d\gamma$$

$$\to \frac{1}{2\pi}\int_{-\pi}^{\pi} f(e^{i\gamma})h'(\theta_0 - \gamma)\,d\gamma = f * Dg(e^{i\theta_0})$$

as $\theta \to \theta_0$ for h' is uniformly continuous and thus for $\epsilon > 0$, $|h'(\theta(\gamma) - \gamma) - h'(\theta_0 - \gamma)| < \frac{\epsilon}{|f|_1+1}$ for all γ if θ is sufficiently near θ_0. Hence for these θ, one has

$$\frac{1}{2\pi}\int_{-\pi}^{\pi} |f(e^{i\gamma})(h'(\theta(\gamma) - \gamma) - h'(\theta_0 - \gamma))|\,d\gamma < \epsilon.$$

□

10. Let $L > 0$. Let f be a L-periodic function such that $\int_0^L |f(t)|^2\,dt < \infty$. Show that there are constants $a_n \in \mathbb{C}$ such that in $L^2([0, L))$ we have

$$f(t) = \sum_{n=-\infty}^{\infty} a_n e^{2\pi int/L}.$$

Find an expression for a_n.
Solution: We define a unitary mapping from $L^2[0, 1]$ onto $L^2[0, L]$ by

$$Uf(t) = \frac{1}{\sqrt{L}} f(\frac{t}{L}).$$

Thus U is one-to-one, satisfies

$$\begin{aligned}|Uf|_2^2 &= \int_0^L \frac{1}{L}|f(\frac{t}{L})|^2\,dt \\ &= \frac{1}{L}\int_{-\infty}^{\infty} \chi_{[0,L]}(t)|f(\frac{t}{L})|^2\,dt \\ &= \int_{-\infty}^{\infty} L\chi_{[0,L]}(Lt)|f(t)|^2\,dt \\ &= \int_0^1 |f(t)|^2\,dt\end{aligned}$$

and has inverse $U^{-1}F(s) = \sqrt{L}F(sL)$. Thus if $f \in L^2[0, L]$, one has by the Plancherel Theorem 1.20,

$$U^{-1}f(s) = \sum_{n=-\infty}^{\infty} c_n e_n(s)$$

where

$$c_n = \int_0^1 U^{-1} f(s) e^{-2\pi i n s}\, ds = \int_0^1 \sqrt{L} f(sL) e^{-2\pi i n s}\, ds = \frac{1}{\sqrt{L}} \int_0^L f(t) e^{-2\pi n i t/L}\, dt.$$

So

$$\begin{aligned} f(t) &= \sum c_n U e_n(t) \\ &= \sum c_n \frac{1}{\sqrt{L}} e_n(\frac{t}{L}) \\ &= \sum c_n \frac{1}{\sqrt{L}} e^{2\pi i n t/L} \end{aligned}$$

in $L^2[0, L]$. Take $a_n = \frac{1}{\sqrt{L}} c_n = \frac{1}{L} \int_0^L f(t) e^{-2\pi i n t/L}\, dt$ and we have the result. □

11. Show that the seminorms ρ_k, ρ'_k, and ρ''_k all define the same topology on $\mathcal{S}(\mathbb{Z})$.

Solution: We have $\rho_k(a) = \sup_n (1 + |n|)^k |a_n|$, $\rho'_k(a) = \sup_n |n|^k |a_k|$, and $\rho''_k(a) = \sum_n |n|^k |a_n|$ where k is a positive integer.

We first show the topologies defined by ρ_k and ρ'_k are the same. We need only show they have the same neighborhoods of 0. Since $\rho_0 \leqslant \rho_1 \leqslant \cdots \leqslant \rho_k \leqslant \cdots$, U is a neighborhood of 0 for the ρ_k's if and only if there is a $k > 1$ and a $\epsilon > 0$ such that $U_k = \{a \mid \rho_k(a) < \epsilon\} \subseteq U$. Now $\rho'_k(a) \leqslant \rho_k(a) \leqslant \rho'_0(a) + 2^k \rho'_k(a)$. Hence for this U, the set V consisting of the a with $\rho'_0(a) < \frac{\epsilon}{2}$ and $\rho'_k(a) < \frac{\epsilon}{2^{k+1}}$ is a neighborhood of 0 for the ρ'_k and is contained in U. Thus U is a neighborhood of 0 for the ρ'_k's. Also if W is a neighborhood of 0 for the ρ'_k's, then since ρ'_k are increasing for $k \geqslant 1$, there is a $k \geqslant 1$ and an $\epsilon > 0$ such that if $\rho'_0(a) < \epsilon$ and $\rho'_k(a) < \epsilon$, then $a \in W$. Thus since $\rho_0 = \rho'_0$ and $\rho'_k \leqslant \rho_k$, we see if $\rho_0(a) < \epsilon$ and $\rho_k(a) < \epsilon$, then $a \in W$. So W is a neighborhood of 0 for the ρ_k's.

Similarly we note $\rho'_k \leqslant \rho''_k$ for all $k \geqslant 0$ and

$$\begin{aligned} \rho''_k(a) &= \sum_{n \neq 0} |n|^{k+2} |a_n| \frac{1}{|n|^2} + 0^k |a_0| \\ &\leqslant C \rho'_{k+2}(a) + \rho'_0(a) \end{aligned}$$

where $C = \sum_{n \neq 0} |n|^{-2} < \infty$. So we have $\rho'_k \leqslant \rho''_k \leqslant C\rho'_{k+2} + \rho'_0$ and a similar argument to the one above shows the neighborhoods of 0 for the ρ' and the ρ'' are the same. □

12. Let $\mathbb{T}^k := \{\mathbf{z} = (z_1, \ldots, z_k) \mid z_j \in \mathbb{T}\}$ with the product topology. For $\mathbf{z}, \mathbf{w} \in \mathbb{T}^k$, let $\mathbf{zw} = (z_1 w_1, \ldots, z_k w_k)$. For $\mathbf{n} = (n_1, \ldots, n_k) \in \mathbb{Z}^k$, let $e_{\mathbf{n}}(\mathbf{z}) = z_1^{n_1} \cdots z_k^{n_k}$. Show the following:

(a) $\mathbb{T}^k$ is a topological group, i.e., the map $\mathbb{T}^k \times \mathbb{T}^k \ni (\mathbf{z}, \mathbf{w}) \mapsto \mathbf{z}^{-1}\mathbf{w} \in \mathbb{T}^k$ is continuous.

(b) $e_{\mathbf{n}} : \mathbb{T}^k \to \mathbb{T}$ is a continuous homomorphism.

(c) If $\chi : \mathbb{T}^k \to \mathbb{T}$ is a continuous homomorphism, then there exists a $\mathbf{n} \in \mathbb{Z}^k$ such that $\chi = e_{\mathbf{n}}$.

Solution: For (a), note if U is an open subset containing $z_0^{-1}w_0$ in $\mathbb{T}^k$, then there is an $\epsilon > 0$ such that if $|z(j) - z_0^{-1}(j)w_0(j)| < \epsilon$ for $j = 1, 2, \ldots, k$, then $z \in U$. Now note

$$\begin{aligned}
|z(j)^{-1}w(j) - z_0(j)^{-1}w_0(j)| &\leqslant |z(j)^{-1}w(j) - z_0(j)^{-1}w(j)| \\
&\quad + |z_0(j)^{-1}w(j) - z_0(j)^{-1}w_0(j)| \\
&= |\overline{z(j)} - \overline{z_0(j)}| + |w(j) - w_0(j)| \\
&= |z(j) - z_0(j)| + |w(j) - w_0(j)| \\
&< \epsilon
\end{aligned}$$

if $|z(j) - z_0(j)| < \frac{\epsilon}{2}$ and $|w(j) - w_0(j)| < \frac{\epsilon}{2}$ for $j = 1, 2, \ldots, k$. So $(z, w) \mapsto z^{-1}w$ is continuous.

To see (b), note first that $z \mapsto z(j)^{n_j}$ is continuous for each j. Since $\mathbb{T}$ is a topological group and compositions of continuous mappings are continuous, we have

$$z \mapsto (z(1)^{n_1}, z(2)^{n_2}, \ldots, z(k)^{n_k}) \mapsto z(1)^{n_1} z(2)^{n_2} \cdots z(k)^{n_k}$$

is continuous. Also

$$\begin{aligned}
e_{\mathbf{n}}(z^{-1}w) &= (z^{-1}w)(1)^{n_1} \cdots (z^{-1}w)(k)^{n_k} \\
&= (z^{-1})(1)^{n_1} \cdots (z^{-1})(k)^{n_k} w(1)^{n_1} \cdots w(k)^{n_k} \\
&= (z(1)^{n_1} z(2)^{n_2} \cdots z(k)^{n_k})^{-1} (w(1)^{n_1} w(2)^{n_2} \cdots w(k)^{n_k}) \\
&= e_{\mathbf{n}}(z)^{-1} e_{\mathbf{n}}(w).
\end{aligned}$$

So $e_{\mathbf{n}}$ is a homomorphism.

Now for (c), note if χ is a continuous homomorphism of $\mathbb{T}^k$ into $\mathbb{T}$, then $z(j) \mapsto \chi(1, 1, \ldots, z(j), \ldots, 1)$ is a continuous homomorphism of $\mathbb{T}$ into $\mathbb{T}$ and thus there is an n_j with $x(1, 1, \ldots, z(j), \ldots, 1) = z(j)^{n_j}$. Thus

$$\begin{aligned}
\chi(z(1), z(2), \ldots, z(k)) &= \chi(\prod_{j=1}^{k}(1, 1, \ldots, z(j), 1, \ldots, 1)) \\
&= \prod_{j=1}^{n} \chi(1, 1, \ldots, 1, z(j), 1, \ldots, 1) \\
&= \prod z(j)^{n_j} \\
&= e_{\mathbf{n}}(z).
\end{aligned}$$

□

13. Let $\mu_k = \mu \times \cdots \times \mu$ be the product measure on $\mathbb{T}^k$. For $f \in L^1(\mathbb{T}^k)$, define $\hat{f} : \mathbb{Z}^k \to \mathbb{C}$ by

$$\hat{f}(\mathbf{n}) := \int_{\mathbb{T}^k} f(\mathbf{z})\, e_{-\mathbf{n}}(\mathbf{z})\, d\mu_k(\mathbf{z}) .$$

Show the following:

(a) If $f \in L^2(\mathbb{T}^k)$, then $\|f\|_2 = \sqrt{\sum_{\mathbf{n}\in\mathbb{Z}^k} \left|\hat{f}(\mathbf{n})\right|^2}$.

(b) If $f \in L^2(\mathbb{T}^k)$, then $f = \sum_{\mathbf{n}\in\mathbb{Z}^k} \hat{f}(\mathbf{n})\, e_{\mathbf{n}}$ in $L^2(\mathbb{T}^k)$.

Solution: For (a), note

$$
\begin{aligned}
\|f\|_2^2 &= \int\cdots\int |f(z_1, z_2, \ldots, z_k)|^2 \, d\mu(z_1)\, d\mu(z_2) \cdots d\mu(z_k) \\
&= \int\cdots\int\int \sum_{n_1} \left|\int f(z_1, z_2, \ldots, z_n) e_{-n_1}(z_1)\, d\mu(z_1)\right|^2 d\mu(z_2)\cdots d\mu(z_k) \\
&= \int\cdots\int \sum_{n_1,n_2} \left|\int\int f(z_1, z_2, \ldots, z_n) e_{-n_1}(z_1) e_{-n_2}(z_2)\, d\mu(z_1)\, d\mu(z_2)\right|^2 \prod_{j=3}^{k} d\mu(z_j) \\
\cdot &= \cdot \\
\cdot &= \cdot \\
\cdot &= \cdot \\
&= \sum_{n_1,\ldots,n_k} \left|\int\int\cdots\int f(z_1, z_2, \ldots, z_k) e_{-n_1}(z_1) e_{-n_2}(z_2) \cdots e_{-n_k}(z_k) \prod_{j=1}^{k} d\mu(z_j)\right|^2 \\
&= \sum_{\mathbf{n}} \left|\int f(\mathbf{z}) e_{-\mathbf{n}}(z)\, d\mu(z)\right|^2
\end{aligned}
$$

where we have used Fubini's Theorem repeatedly and the inside sums are giving us repeatedly partial L^2 integrals, e.g. the third inside sum gives

$$\int\int |f(z_1, z_2, \ldots, z_k)|^2 \, d\mu(z_1)\, d\mu(z_2)$$

for almost all $z_3, z_4, \ldots, z_k$.

To see (b), we first note the $e_{\mathbf{n}}(z)$ are orthonormal functions in $L^2(\mathbb{T}^k)$ which follows by Fubini's Theorem and then using the fact that any orthonormal set is an orthonormal basis if it satisfies (a).

One can also mimic the proof of the Plancherel Theorem for $L^2(\mathbb{T})$, the linear span of the $e_{\mathbf{n}}$ form an algebra in which the Stone-Weierstrass Theorem applies and then again use orthogonality and the density of $C(\mathbb{T}^k)$ in $L^2(\mathbb{T}^k)$ to obtain the result. □

14. Generalize Corollary 1.23 to $\mathbb{R}^k$. Namely let f be a C^∞ function on $\mathbb{R}^k$ satisfying $f(\mathbf{x} + \mathbf{n}) = f(\mathbf{x})$ for all $\mathbf{x} \in \mathbb{R}^k$ and $\mathbf{n} \in \mathbb{Z}^k$. Define $\hat{f}(\mathbf{n}) = \int_{[0,1]^k} f(\mathbf{x}) e^{-2\pi i \mathbf{x}\cdot\mathbf{n}}\, d\mathbf{x}$. Show the series $\sum_{\mathbf{n}\in\mathbb{Z}^k} \hat{f}(\mathbf{n}) e^{2\pi i \mathbf{n}\cdot\mathbf{x}}$ and its derivatives converge uniformly to f and f's derivatives on $\mathbb{R}^k$.

Solution: Let f be C^∞ and satisfy $f(\mathbf{x} + \mathbf{n}) = f(\mathbf{x})$ for all $\mathbf{x}$ and $\mathbf{n}$. Let $\alpha = (\alpha_1, \ldots, \alpha_k)$ be a k-tuple of nonnegative integers and consider $D^\alpha f$. We note using integration by parts and periodicity,

$$\begin{aligned}\widehat{D^\alpha f}(\mathbf{n}) &= \frac{1}{(2\pi)^k}\int_0^1\cdots\int_0^1 D_1^{\alpha_1}D_2^{\alpha_2}\cdots D_k^{\alpha_k} f(x_1,x_2,\ldots,x_k)\,e^{-2\pi i n_1 x_1}dx_1\\ &\qquad\times e^{-2\pi i n_2 x_2}dx_2\cdots e^{-2\pi i n_k x_k}dx_k\\ &= \frac{(2\pi i n_1)^{\alpha_1}(2\pi i n_2)^{\alpha_2}\cdots(2\pi i n_k)^{\alpha_k}}{(2\pi)^k}\int_0^1\cdots\int_0^1 f(x_1,x_2,\ldots,x_k)\\ &\qquad\times\, e^{-2\pi i n_1 x_1}\,dx_1\,e^{-2\pi i n_2 x_2}dx_2\cdots e^{-2\pi i n_k x_k}dx_k\\ &= (2\pi i n_1)^{\alpha_1}\cdots(2\pi i n_k)^{\alpha_k}\hat{f}(\mathbf{n}).\end{aligned}$$

Since $D^\alpha f$ is L^2, we know by Exercise 1.3.13, that

$$(2\pi)^{2\sum\alpha_j}\sum_{\mathbf{n}} n_1^{2\alpha_1}n_2^{2\alpha_2}\cdots n_k^{2\alpha_k}|\hat{f}(\mathbf{n})|^2 = |D^\alpha f|_2^2 < \infty$$

for each α. This implies

$$\sum_{\mathbf{n}}\widehat{D^\alpha f}(\mathbf{n})e^{2\pi i\mathbf{x}\cdot\mathbf{n}}$$

converges uniformly and in L^2 to $D^\alpha f$.

Indeed, uniform convergence follows since

$$\begin{aligned}\sum_{\mathbf{n}}|\widehat{D^\alpha f}(\mathbf{n})| &= (2\pi)^{\sum\alpha_j}\sum_{\mathbf{n}}\prod_{j=1}^k\left((1+n_j^2)n_j^{\alpha_j}\right)|\hat{f}(\mathbf{n})|\left(\prod_{j=1}^k(1+n_j^2)\right)^{-1}\\ &\leqslant (2\pi)^{\sum\alpha_j}\left(\sum_{\mathbf{n}}\prod_{j=1}^k(1+n_j^2)^2 n_j^{2\alpha_j}|\hat{f}(\mathbf{n})|^2\right)^{1/2}\left(\sum_{\mathbf{n}}\prod_{j=1}^k(1+n_j^2)^{-2}\right)^{1/2}\\ &\leqslant (2\pi)^{\sum\alpha_j}M\left(\sum_{\mathbf{n}}\prod_{j=1}^k(1+n_j^2)^{-2}\right)^{1/2}\\ &= (2\pi)^{\sum\alpha_j}M\left(\sum_{n_1}(1+n_1^2)^{-2}\sum_{n_2,\ldots,n_k}(1+n_2^2)^{-2}\cdots(1+n_k^2)^{-2}\right)^{1/2}\\ &= \cdots\\ &= (2\pi)^{\sum\alpha_j}M\left(\sum_{j=1}^\infty\frac{1}{(1+j^2)^2}\right)^{k/2} < \infty\end{aligned}$$

since by the above

$$\begin{aligned}M^2 &= \sum_{\mathbf{n}}\prod_{j=1}^k(1+n_j^2)^2 n_j^{2\alpha_j}|\hat{f}(\mathbf{n})|^2\\ &= \sum_{\text{finitely many }\beta}\ \sum_{\mathbf{n}} n_1^{\beta_1}n_2^{\beta_2}\cdots n_k^{\beta_k}|\hat{\mathbf{f}}(\mathbf{n})|^2\\ &< \infty\,.\end{aligned}$$

We thus have the Fourier series for $D^\alpha f$ converges uniformly. Moreover, since $D^\alpha f$ is in L^2, we know by Exercise 1.3.13 that the Fourier series converges in L^2 to $D^\alpha f$. Since uniform convergence implies L^2 convergence, one has the Fourier series for $D^\alpha f$ converges uniformly to $D^\alpha f$. $\square$

15. Show if $f \in C^\infty(\mathbb{R}^k)$ and $f(\mathbf{x}+\mathbf{n}) = f(\mathbf{x})$ for all $\mathbf{n} \in \mathbb{Z}^n$, then $\hat{f} : \mathbb{Z}^k \to \mathbb{C}$ is rapidly decreasing; i.e.,

$$\sup_{\mathbf{n}\in\mathbb{Z}^k} |p(\mathbf{n})\hat{f}(\mathbf{n})|$$

is finite for every polynomial p. Use this to show the absolute uniform convergence of the series $\sum_{\mathbf{n}} \hat{f}(\mathbf{n}) D^\alpha(e^{2\pi i \mathbf{x}\cdot\mathbf{n}})$ for each $\alpha \in \mathbb{N}_0^n$.

Solution: The same argument as in Exercise 1.3.14 shows for each k-tuple α,

$$\sum_{\mathbf{n}} n_1^{2\alpha_1} n_2^{2\alpha_2} \cdots n_k^{2\alpha_k} |\hat{f}(\mathbf{n})|^2 < \infty .$$

Hence for each α, one has

$$\sup_{\mathbf{n}} n_1^{\alpha_1} n_2^{\alpha_2} \cdots n_k^{\alpha_k} |\hat{f}(\mathbf{n})| < \infty .$$

Since $p(\mathbf{n})$ is a finite linear combination of terms of form $n_1^{\alpha_1} n_2^{\alpha_2} \cdots n_k^{\alpha_k}$, the result follows. $\square$

16. Show that there is no differentiable function f on $\mathbb{T}$ such that $Df = 1$.

Solution: Suppose f is a C^1 function on $\mathbb{T}$ such that $Df = 1$. Then $F(t) = f \circ \kappa(t) = f(e^{it})$ is C^1 on $\mathbb{R}$ and $F'(t) = 1$ for all t. This implies $F(t) = t + c$ for some constant c. So $F(0) = c$ and $F(2\pi) = 2\pi + c$. But $F(2\pi) = f(e^{2\pi i}) = f(1) = f(e^{i0}) = F(0) = c$, a contradiction. $\square$

17. Let $p(z) = \sum_{n=0}^k a_n z^n$ be a polynomial. Define $p(D) : C^\infty(\mathbb{T}) \to C^\infty(\mathbb{T})$ by

$$p(D)f = \sum_{n=0}^{k} a_n D^n f .$$

Show that if $g \in C^\infty(\mathbb{T})$ is such that $\hat{g}(n) = 0$ if $p(2\pi i n) = 0$, then the differential equation $p(D)f = g$ has a solution.

Solution: Let $g \in C^\infty(\mathbb{T})$. By Corollary 1.23, we know $g(z) = \sum_n c_n z^n$ where $\{c_n\}_{n=1}^\infty$ is rapidly decreasing. Now $Dz^n = 2\pi i n z^n$ since $\frac{d}{dt}(e^{2\pi i n t}) = 2\pi i n e^{2\pi i n t}$.

Thus if $f(z) = \sum_n b_n z^n$ where $\{b_n\}$ is rapidly decreasing, then

$$\begin{aligned} p(D)f(z) &= \sum_n b_n \sum_{m=0}^k a_m (2\pi i n)^m z^n \\ &= \sum_n b_n p(2\pi i n) z^n \\ &= \sum_n c_n z^n \end{aligned}$$

which has a solution precisely when for each $c_n \neq 0$, $p(2\pi i n) \neq 0$ and then the solution is $f(z) = \sum_{c_n \neq 0} \frac{c_n}{p(2\pi i n)} z^n$. $\square$

18. Let $f \in L^2(\mathbb{T})$. Show there exists a unique $g \in L^2(\mathbb{T})$ such that

$$\hat{g}(n) = -i\,\mathrm{sgn}(n)\,\hat{f}(n)\,, \quad \forall n \in \mathbb{Z}\,.$$

Define $Hf = g$. Then $H : L^2(\mathbb{T}) \to L^2(\mathbb{T})$ is linear. Prove the following:

(a) $|Hf|_2 = |f|_2$ if and only if $\int f\,d\mu = 0$.

(b) If $f \in C^\infty(\mathbb{T})$, then $Hf \in C^\infty(\mathbb{T})$.

(c) If $f \in C^\infty(\mathbb{T})$, then $Hf(1) = -\frac{1}{\pi}\sum_{n=1}^\infty \int_0^{2\pi} f(e^{i\theta})\sin(n\theta)\,d\theta\,.$

Solution: To see g exists, it suffices to note by Theorem 1.22 the mapping f to $\hat{f}$ is a linear isometry of $L^2(\mathbb{T})$ onto l^2. Now $\sum_n |\hat{g}(n)|^2 = \sum_{n\neq 0}|\hat{f}(n)|^2 = |f|_2^2 - |\hat{f}(0)|^2 < \infty$. Moreover, we then also see $|g|_2^2 = |\hat{g}|_2^2 = |f|_2^2 - |\hat{f}(0)|^2$. Thus $|Hf|_2 = |f|_2$ if and only if $\hat{f}(0) = 0$. The map H is linear for the mapping $\hat{f}$ to $\hat{g}$ is linear. For (b), note the sequence $\{\hat{f}(n)\}_n$ is rapidly decreasing if and only if the sequence $\{\hat{g}(n)\}_n$ is rapidly decreasing. Finally to see (c), note since $f \in C^\infty(\mathbb{T})$ that the Fourier series for Hf converges pointwise uniformly to Hf. Moreover, since cos is even and sin is odd, we see

$$\begin{aligned}
Hf(1) &= \sum_{n\neq 0} -i\mathrm{sgn}(n)\hat{f}(n)1^n \\
&= \sum_{n\neq 0} \frac{-i\mathrm{sgn}(n)}{2\pi}\int_0^{2\pi} f(e^{i\theta})e^{-in\theta}\,d\theta \\
&= \sum_{n=1}^\infty \frac{-i}{2\pi}\int_0^{2\pi} f(e^{i\theta})(e^{-in\theta} - e^{in\theta})\,d\theta \\
&= \sum_{n=1}^\infty \frac{-i}{2\pi}\int_0^{2\pi} f(e^{i\theta})(-2i\sin n\theta)\,d\theta \\
&= -\frac{1}{\pi}\sum_{n=1}^\infty \int_0^{2\pi} f(e^{i\theta})\sin n\theta\,d\theta.
\end{aligned}$$

□

Exercise Set 1.4

1. Show that the sequence $a_n = (-1)^n$ is Abel and Cesáro summable to 1/2.
Solution: To see $a_n = (-1)^n$ is Cesáro summable, note $s_n = \sum_{k=0}^n a_k = 0$ if n is odd and is 1 if n is even. Thus $\sum_{n=0}^N s_n = \sum_{n \text{ even}} 1 = [\frac{N+2}{2}]$ where $[\cdot]$ is the greatest integer function. Thus $\frac{1}{N+1}\sum_{n=0}^N s_n = \frac{1}{N+1}[\frac{N+2}{2}] \to \frac{1}{2}$ as $N \to \infty$.

For Abel summability, note $\sum_{n=0}^\infty (-1)^n r^n = \frac{1}{1+r}$ for $|r| < 1$ and thus

$$\lim_{r\to 1-}\sum_{n=0}^\infty (-1)^n r^n = \frac{1}{2}$$

and so we see the Abel sum is $\frac{1}{2}$. □

2. Show that if $\sum a_n$ converges to L, then $\{a_n\}$ is Abel and Cesáro summable to L. (**Hint:** To show $\sum a_k$ is Abel summable, use Abel's summation formula $\sum_{k=1}^n a_k b_k = (\sum_{k=1}^n a_k)b_n + \sum_{k=1}^{n-1}(\sum_{l=1}^k a_l)(b_k - b_{k+1})$ to show the series $\sum a_k r^k$ converges uniformly for $0 \leqslant r \leqslant 1$.)

Solution: To see Césaro summability of a sequence $\{a_n\}$ with $\sum a_n$ existing, it suffices to show if b_n converges to L, then $\frac{1}{n+1}\sum_{k=0}^n b_k$ converges to L, for then we apply this to $s_n = \sum_{k=0}^n a_k$ which converges.

Thus suppose $a_n \to L$ as $n \to \infty$. Let $\epsilon > 0$. Choose N such that $k > N$ implies $|a_k - L| < \frac{\epsilon}{3}$. Then choose $N_0 > N$ so that $\frac{1}{n+1}\sum_{k=0}^N |a_k| < \frac{\epsilon}{3}$ and $\frac{1+N}{n+1}|L| < \frac{\epsilon}{3}$ for $n > N_0$. Then

$$
\begin{aligned}
|\frac{1}{n+1}\sum_{k=0}^n a_k - L| &\leqslant \frac{1}{n+1}\left|\sum_{k=0}^N a_k\right| + \left|\frac{1}{n+1}\sum_{k=N+1}^n a_k - L\right| \\
&\leqslant \frac{1}{n+1}\sum_{k=0}^N |a_k| + \left|\frac{1}{n+1}\sum_{k=N+1}^n (a_k - L) + \frac{n-N}{n+1}L - L\right| \\
&\leqslant \frac{1}{n+1}\sum_{k=0}^N |a_k| + \frac{1}{n+1}\sum_{k=N+1}^n |a_k - L| + |\frac{n-N}{n+1} - 1|\,|L| \\
&\leqslant \frac{1}{n+1}\sum_{k=0}^N |a_k| + \frac{1}{n+1}\sum_{k=N+1}^n |a_k - L| + \frac{1+N}{n+1}|L| \\
&< \frac{\epsilon}{3} + \frac{\epsilon}{3} + \frac{\epsilon}{3} < \epsilon
\end{aligned}
$$

for $n > N_0$ and we see the result.

For Abel summability, we use Abel's formula

$$
\sum_{k=0}^n a_k b_k = \left(\sum_{k=0}^n a_k\right) b_n + \sum_{k=0}^{n-1}\left(\sum_{l=0}^k a_l\right)(b_k - b_{k+1}).
$$

Note this formula can be verified by reversing summation in the second sum;

i.e.,

$$\begin{aligned}
\left(\sum_{k=0}^{n} a_k\right) b_n + \sum_{k=0}^{n-1}\left(\sum_{l=0}^{k} a_l\right)(b_k - b_{k+1}) &= \left(\sum_{k=0}^{n} a_k\right) b_n + \sum_{l=0}^{n-1}\sum_{k=l}^{n-1} a_l(b_k - b_{k+1}) \\
&= \left(\sum_{k=0}^{n} a_k\right) b_n + \sum_{l=0}^{n-1} a_l \sum_{k=l}^{n-1}(b_k - b_{k+1}) \\
&= \left(\sum_{k=0}^{n} a_k\right) b_n + \sum_{l=0}^{n-1} a_l(b_l - b_n) \\
&= \left(\sum_{k=0}^{n} a_k\right) b_n - \sum_{l=0}^{n-1} a_l b_n + \sum_{l=0}^{n-1} a_l b_l \\
&= a_n b_n + \sum_{l=0}^{n-1} a_l b_l \\
&= \sum_{l=0}^{n} a_l b_l.
\end{aligned}$$

Now if the series $\sum a_n$ converges to L, there is an M such that $|\sum_{l=0}^{n} a_l| < M$ for all n and thus for $0 \leqslant r \leqslant 1$, one has

$$\begin{aligned}
\sum_{k=0}^{n} a_k r^k &= \left(\sum_{k=0}^{n} a_k\right) r^n + \sum_{k=0}^{n-1}(\sum_{l=0}^{k} a_l)(r^k - r^{k+1}) \\
&\to Lr^n + \sum_{k=0}^{\infty}(\sum_{l=0}^{k} a_l)(r^k - r^{k+1})
\end{aligned}$$

uniformly for $\sum_{k=0}^{n-1} |\left(\sum_{l=0}^{k} a_l\right)(r^k - r^{k+1})| < \sum_{k=0}^{n-1} M(r^k - r^{k+1}) < M(1-r^n)$ and thus is absolutely uniformly convergent.

Therefore $\sum_n a_n r^n$ converges uniformly in r for $r \in [0,1]$ and the resulting function $f(r)$ is continuous. Thus $\lim_{r\to 1-} f(r) = f(1)$ and we see

$$\lim_{r\to 1-} \sum_{k=0}^{\infty} a_k r^k = \sum_{k=0}^{\infty} a_k.$$

□

3. Show if the series $\sum a_n$ is Cesáro summable, then $\lim_{n\to\infty} \frac{a_n}{n} = 0$.

Solution: Let $S_n = a_1 + a_2 + \cdots + a_n$. Note $S_n - S_{n-1} = a_n$. We show $\frac{S_n}{n} \to 0$. Let $C_n = \frac{1}{n}(S_1 + \cdots + S_n)$. Note $C_n - \frac{n-1}{n}C_{n-1} \to 0$. But this is $\frac{a_n}{n}$. □

4. Show the series $\sum_{n=0}^{\infty}(-1)^n \frac{(n+k)!}{n!}$ is Abel summable and find its sum. (**Hint:** Consider $\left(\frac{d}{dr}\right)^k (1+r)^{-1}$ expressed as a series.)

Solution: $\frac{1}{1+r} = \sum_{n=0}^{\infty}(-1)^n r^n$ and this converges uniformly on intervals $[0, R]$ where $0 \leqslant R < 1$. Now the term by term derivative of the series is given bys $\sum_{n=k}^{\infty}(-1)^n n(n-1)\cdots(n-k+1)r^{n-k} = \sum_{n=0}^{\infty}(-1)^{n+k}(n+k)(n+k-1)\cdots(n+1)\,r^n$. For each k, the series converges uniformly on $[0, R]$ for $0 < R < 1$ since $\sum_{n=0}^{\infty}\frac{(n+k)!}{n!}R^n$ converges by the ratio test. This implies $(-1)^k\left(\frac{d}{dr}\right)^k(1+r)^{-1} = k!(1+r)^{-k-1} = \sum_{n=0}^{\infty}(-1)^n\frac{(n+k)!}{n!}r^n$ for each k. Thus the Abel sum is $\frac{k!}{2^{k+1}}$. □

5. Let $\delta_0 \geqslant \delta_1 \geqslant \cdots \geqslant 0$ be a decreasing sequence with $\lim \delta_n = 0$. Define $a_0 := \delta_0$ and $a_n := \delta_{n-1} - \delta_n$ for $n \geqslant 1$. Let $g(z) := \sum a_n z^n$ for $z \in \mathbb{T}$. Show that g is continuous, $g * D_N$ converges uniformly to g, and $|g * D_N - g|_\infty \leqslant \delta_N$.
Solution: We start by showing

$$\delta_0 + \sum_{n=1}^{\infty}(\delta_{n-1} - \delta_n)z^n$$

converges uniformly for $z \in \mathbb{T}$. Indeed, $\sum_{n=N}^{M}|(\delta_{n-1} - \delta_n)z^n| \leqslant \sum_{n=N}^{M}(\delta_{n-1} - \delta_n) = \delta_{N-1} - \delta_M \to 0$ as $M > N \to \infty$. Thus the series $\sum a_n z^n$ converges uniformly. So $g(z) = \sum_{n=0}^{\infty} a_n z^n$ is continuous and thus is in $L^2(\mathbb{T})$. Now by Lemma 1.30, $s_N(g)(z) = D_N * g(z) = \sum_{n=0}^{N} a_n z^n$. So $g * D_N$ converges uniformly to g as $N \to \infty$. Also $|D_N * g(z) - g(z)| = |\sum_{n=N+1}^{\infty}|a_n z^n| \leqslant \sum_{n=N+1}^{\infty}(\delta_{n-1} - \delta_n) = \delta_N$. So $|g * D_N - g|_\infty \leqslant \delta_N$. □

6. Suppose $f \in L^1(\mathbb{T})$. Show if $\hat{f}(n) = 0$ for all n, then $f(z) = 0$ a.e. z.
Solution: Suppose $f \in L^1(\mathbb{T})$ and $\hat{f}(n) = 0$ for all n. We know $\sigma_N(f) = f * \Sigma_N = 0$ where Σ_N is the Fejer kernel. But $\sigma_N(f)(z) = \frac{1}{N+1}\sum_{n=0}^{N} s_n(f)(z) = 0$ for $s_N(f) = \sum_{n=-N}^{N}\hat{f}(n)e_n$. But by Theorem 1.40, $f * \Sigma_N \to f$ in L^1 as $N \to \infty$. So $f = 0$ in $L^1(\mathbb{T})$. □

7. Let F be the function on $\mathbb{T}$ defined by $F(e^{i\theta}) = \theta$ for $-\pi \leqslant \theta < \pi$.

(a) Evaluate $\hat{F}(n)$.

(b) Show the series $\sum \hat{F}(n)e^{in\pi}$ diverges.

(c) Show for each N, $\sum_{k=-N}^{N}\hat{F}(n)e^{in\pi} = 0$.

(d) Define f by $f(x) = F(e^{2\pi ix})$. Show f satisfies the conditions of Theorem 1.35 at $x = \frac{1}{2}$.

Solution: For $n \neq 0$,

$$\begin{aligned}
\hat{F}(n) &= \frac{1}{2\pi}\int_{-\pi}^{\pi} F(e^{i\theta})e^{-in\theta}\,d\theta \\
&= \frac{1}{2\pi}\int_{-\pi}^{\pi}\theta e^{-in\theta}\,d\theta \\
&= \frac{1}{2\pi}\left(\frac{\theta e^{-in\theta}}{-in} - \int\frac{e^{-in\theta}}{-in}\,d\theta\right)\Bigg|_{-\pi}^{\pi}
\end{aligned}$$

$$
\begin{aligned}
&= \frac{1}{2\pi}\left(\frac{\pi e^{-in\pi}}{-in} - \frac{-\pi e^{in\pi}}{-in}\right) - \frac{1}{2\pi}\left.\frac{e^{-in\theta}}{(in)^2}\right|_{-\pi}^{\pi} \\
&= \frac{1}{2\pi}\left(\frac{\pi(-1)^n}{-in} - \frac{-\pi(-1)^n}{-in}\right) - \frac{1}{2\pi}\left(\frac{(-1)^n}{(in)^2} - \frac{(-1)^n}{(in)^2}\right) \\
&= \frac{1}{2\pi}\frac{2\pi(-1)^n}{(-in)} \\
&= \frac{(-1)^n i}{n}.
\end{aligned}
$$

Also $\hat{F}(0) = \frac{1}{2\pi}\int_{-\pi}^{\pi} \theta \, d\theta = 0$. Thus the series $\sum_n \hat{F}(n)e^{in\pi} = \sum_n \frac{(-1)^{2n} i}{n} = \sum_{n\neq 0} \frac{i}{n}$ diverges.

But $\sum_{n=-N}^{N} \hat{F}(n)e^{in\pi} = \sum_{n=1}^{N}\left(\frac{(-1)^n i}{n} - \frac{(-1)^n i}{n}\right)(-1)^n = 0$. Now let $f(t) = F(e^{2\pi i t})$. Note f is discontinuous at $\frac{1}{2}$, but $f(\frac{1}{2}-) = \pi$ and $f(\frac{1}{2}+) = -\pi$. Also $f(t) = 2t\pi$ for $0 < t < \frac{1}{2}$ and $f(t) = f(t-1) = 2(t-1)\pi$ for $\frac{1}{2} < t < 1$. Therefore $f(t+\frac{1}{2}) - f(\frac{1}{2}+) = 2(t-\frac{1}{2})\pi + \pi = 2t\pi$ and $f(\frac{1}{2}-t) - f(\frac{1}{2}-) = (1-2t)\pi - \pi = -2t\pi$. Thus Theorem 1.35 is satisfied with $K = 2\pi$ and $\alpha = 1$. □

8. Show the series $\sum_{n=1}^{\infty} \frac{\sin nx}{n}$ converges pointwise everywhere and determine its limit. Does this series converge uniformly? (**Hint:** Consider the odd function of period 2π equal to $\pi - x$ on the interval $0 < x \leqslant \pi$.)

Solution: Let $f(x)$ be the odd function of period 2π satisfying $f(x) = \pi - x$ for $0 < x \leqslant \pi$. Note $f(0+) = \pi$ and $f(0-) = -\pi$. Set $c_n = \frac{1}{2\pi}\int_{-\pi}^{\pi} f(x)e^{-inx}\,dx$. Write $c_n = a_n + ib_n$. Since f is odd, $a_n = 0$ and $b_n = -\frac{1}{2\pi}\int_{-\pi}^{\pi} f(x)\sin nx\,dx = -\frac{1}{\pi}\int_0^{\pi}(\pi - x)\sin nx\,dx$. Hence

$$
\begin{aligned}
b_n &= \frac{1}{\pi}\left[\left.\frac{\pi\cos nx}{n}\right|_0^{\pi} - \left.\frac{x\cos nx}{n}\right|_0^{\pi} + \int_0^{\pi}\frac{\cos nx}{n}\,dx\right] \\
&= \frac{1}{\pi}\left[\frac{(-1)^n\pi}{n} - \frac{\pi}{n} + \frac{(-1)^{n+1}\pi}{n}\right] \\
&= -\frac{1}{n}.
\end{aligned}
$$

By Corollary 1.36, $\sum_{n=-N}^{N} c_n e^{inx} = \sum_{n=-N}^{N} -b_n \sin nx = 2\sum_{n=1}^{N}\frac{1}{n}\sin nx$ converges as $N \to \infty$ to $\frac{1}{2}(f(x+) + f(x-))$. Hence

$$
\sum_{n=1}^{\infty}\frac{\sin nx}{n} = \begin{cases} \frac{1}{2}(\pi - x) \text{ if } 0 < x < 2\pi \\ 0 \text{ if } x = 0 \text{ for } x = 2\pi \end{cases}
$$

and thus the series converges for all x. □

9. Show

$$
\frac{\pi}{4} = \sum_{k=0}^{\infty}\frac{\sin(2k+1)x}{2k+1}
$$

for $0 < x < \pi$.

Solution: Consider the odd periodic function $f(x) = 1$ for $0 < x < \pi$ and $f(x) = -1$ for $-\pi < x < 0$.

Then $\hat{f}(0) = 0$ and

$$\hat{f}(n) = \frac{1}{2\pi}\int_{-\pi}^{\pi} f(x)(\cos nx - i\sin nx)\,dx = -\frac{i}{\pi}\int_0^{\pi} \sin nx\,dx$$
$$= -\frac{i}{\pi}\,\frac{-\cos nx}{n}\Big|_0^{\pi} = \frac{-i}{n\pi}\left(1-(-1)^n\right).$$

Thus $\hat{f}(2n) = 0$ and $\hat{f}(2n+1) = \frac{-2i}{(2n+1)\pi}$.

By Corollary 1.36, $\sum_{k=-N}^{N} \hat{f}(k)e^{ikx} \to f(x)$ as $N \to \infty$ for $0 < x < \pi$. Thus

$$\frac{-i}{\pi}\sum_{k=0}^{n}\left(\frac{2}{2k+1}e^{i(2k+1)x} - \frac{2}{2k+1}e^{-i(2k+1)x}\right) \to 1$$

as $n \to \infty$.

So

$$\sum_{k=0}^{n} \frac{1}{2k+1}\sin(2k+1)x \to \frac{\pi}{4} \text{ as } n \to \infty.$$

□

10. Suppose $f \in L^1(\mathbb{T})$ and let $a_n = \hat{f}(n)$ for $n \in \mathbb{Z}$. Set $u(r,\theta) = \frac{1}{2\pi}\int_0^{2\pi} P(r,\theta-\phi)f(e^{i\theta})\,d\theta$ for $0 < r < 1$. Show $u(r,\theta) = \sum_{n=-\infty}^{\infty} a_n r^{|n|}e^{in\theta}$.

Solution: Indeed, we know $P(r,\theta) = \sum_{n=-\infty}^{\infty} r^{|n|}e^{in\theta}$. Set $u_r(\theta) = u(r,\theta)$. Then since $\sum r^{|m|}e^{im\theta}$ converges in $L^\infty(\mathbb{T})$ for $0 < r < 1$, we see

$$\hat{u}_r(n) = \frac{1}{2\pi}\int_0^{2\pi} u_r(e^{i\theta})e^{-in\theta}\,d\theta$$
$$= \frac{1}{2\pi}\int_0^{2\pi}\frac{1}{2\pi}\int_0^{2\pi} P(r,\theta-\phi)f(\phi)\,d\phi\, e^{-in\theta}\,d\theta$$
$$= \frac{1}{2\pi}\int_0^{2\pi}\frac{1}{2\pi}\int_0^{2\pi}\sum_m r^{|m|}e^{im(\theta-\phi)}f(\phi)\,d\phi\, e^{-in\theta}\,d\theta$$
$$= \frac{1}{2\pi}\int_0^{2\pi}\sum_m \frac{1}{2\pi}\int_0^{2\pi} r^{|m|}e^{-im\phi}f(\phi)\,d\phi\, e^{-i(n-m)\theta}\,d\theta$$
$$= \frac{1}{2\pi}\int_0^{2\pi}\sum_m \hat{f}(m)r^{|m|}e^{-i(n-m)\theta}\,d\theta$$
$$= \sum_m r^{|m|}\hat{f}(m)\frac{1}{2\pi}\int_0^{2\pi} e^{-i(n-m)\theta}\,d\theta$$
$$= r^{|n|}\hat{f}(n).$$

Moreover, $u_r = P_r * f \in L^\infty(\mathbb{T})$ for $P_r \in L^\infty(\mathbb{T})$ and $f \in L^1(\mathbb{T})$. Now the series $\sum r^{|n|}\hat{f}(n)e^{in\theta}$ converges uniformly for $f \in L^1(\mathbb{T})$ since $\hat{f}$ is bounded. We know since u_r is also in L^2, one has $\sum_m r^{|n|}\hat{f}(n)e^{in\theta} = u_r$ in $L^2(\mathbb{T})$. So it must be u_r in $C(\mathbb{T})$. □

11. Let $u(x+iy)$ be a continuous function defined on the disc $|z| \leqslant 1$ which is harmonic on $|z| < 1$; i.e., u is C^2 on $|z| < 1$ and $u_{xx} + u_{yy} = 0$. Show if $u = 0$ on $|z| = 1$, then $u = 0$. (**Hint:** Use the divergence theorem (Stoke's Theorem) on this disk and its boundary applied to the vector field $u\nabla u$.)

Now suppose $u(x+iy)$ is harmonic on a nonempty open subset U of $\mathbb{C}$. Show:

(a) If $z_0 \in U$ and the disc $|z - z_0| \leqslant a$ is contained in U, then if $0 \leqslant r < a$,

$$u(z_0 + re^{i\theta}) = \frac{1}{2\pi}\int_0^{2\pi} P(\frac{r}{a}, \theta - \phi)u(z_0 + ae^{i\phi})\, d\phi.$$

(**Hint:** Consider Theorem 1.42.)

(b) Conclude

$$u(z_0) = \frac{1}{2\pi}\int_0^{2\pi} u(z_0 + ae^{i\phi})\, d\phi = \frac{1}{2\pi a}\oint_{|z-z_0|=a} u(z)\, |dz|.$$

(c) **(Maximum Principle)** Use (c) to show u cannot assume a local maximum or minimum inside U.

Solution: For the first part, we have $\int\int_{|z|\leqslant 1} \nabla\cdot(u\nabla u)\, dA = \int_{|z|=1}(u\nabla u)\vec{n}\, ds = 0$ for u is zero on the circle. But $\nabla \cdot (u\nabla u) = \nabla u \cdot \nabla u + u\nabla^2 u = |\nabla u|^2$. Thus $\iint_{|z|\leqslant 1} |\nabla u|^2\, dA = 0$. So $\nabla u = 0$. Hence u is constant inside the circle. But $u = 0$ on the boundary. So $u = 0$.

(a) Let $|z - z_0| \leqslant a$ be a disc which is a subset of U. Define $v(r, \theta) = u(z_0 + are^{i\theta})$ for $0 \leqslant r \leqslant 1$. Then using $z = re^{i\theta} = x + iy$, one has v is harmonic on $|z| < 1$ and continuous on $|z| \leqslant 1$. Moreover, if we define $\tilde{v}(r, \theta) = \frac{1}{2\pi}\int_0^{2\pi} P(r, \theta - \phi)v(1, \phi)\, d\phi$ for $0 \leqslant r < 1$, we have by Theorem 1.42 that $\tilde{v}$ is harmonic on $|z| < 1$ and $\tilde{v}(r, \theta) \to v(1, \theta)$ uniformly as $r \to 1-$. Thus if we define $\tilde{v}(z) = v(z)$ if $|z| = 1$, then $\tilde{v}$ is continuous on the disc $|z| \leqslant 1$ and harmonic on $|z| < 1$. Consequently, since $\tilde{v} - v$ is 0 on $|z| = 1$, we see by the first part that $\tilde{v} = v$ on $|z| \leqslant 1$. So for $0 \leqslant r < a$,

$$\begin{aligned} u(z_0 + re^{i\theta}) &= v(\frac{r}{a}, \theta) \\ &= \tilde{v}(\frac{r}{a}, \theta) \\ &= \frac{1}{2\pi}\int_0^{2\pi} P(\frac{r}{a}, \theta - \phi)u(z_0 + ae^{i\phi})\, d\phi\,. \end{aligned}$$

(b) is immediate. For (c) note if u has a local maximum at z_0, then wolog $u(z_0) = \frac{1}{2\pi}\int u(z_0 + ae^{i\phi})\, d\phi$ where $u(z_0) \geqslant u(z_0 + ae^{i\phi})$ for all ϕ. Since we are dealing with continuous functions $u(z_0 + ae^{i\phi}) = u(z_0)$ for all ϕ. By (a) u is constant on the disk $|z - z_0| \leqslant a$. Hence u would not have a local maximum at z_0. □

Chapter 2

Function Spaces on $\mathbf{R}^n$

Exercise Set 2.1

1. Show a topological vector space X is separated if and only if it is Hausdorff.

Solution: Assume $x \neq y$. Then $x - y \neq 0$. Hence there is an open neighborhood V of 0 with $x - y \notin V$. Choose an open neighborhood U of 0 with $U - U \subseteq V$. Then $(x + U) \cap (y + U) = y + (x - y + U) \cap U = \emptyset$. $\square$

2. Let $||\cdot||$ be a seminorm on a vector space X. Show $U = \{x \mid ||x|| < 1\}$ is balanced, saturating, and convex and each point in U is internal.

Solution: To see U is balanced, note if $|\alpha| \leqslant 1$ and $x \in U$, then $||\alpha x|| = |\alpha|\,||x|| < 1$. Now let $||x|| \neq 0$. Then $\frac{1}{2||x||}x \in U$ and so $x \in 2||x||U$. Thus U is saturating. Finally to see every point $x \in U$ is internal, note if $y \in X$, then $x + \alpha y \in U$ if $||x|| + |\alpha|\,||y|| < 1$, which is true for all α if $||y|| = 0$ and for $|\alpha| < \frac{1-||x||}{||y||}$ if $||y|| \neq 0$. $\square$

3. Let X be a topological vector space. Show every interior point of a subset U of X is internal.

Solution: Let x be an interior point of U. If $y \in X$ and V is an open subset of X containing x contained in U, then since $\alpha \mapsto x + \alpha y$ is continuous and $x + 0y \in V$, there is a $\delta > 0$ with $x + \alpha y \in V$ for $|\alpha| < \delta$. Hence x is internal in U. $\square$

4. Let X be a topological vector space. Show the interior of any balanced subset of X containing the zero vector as an interior point is balanced.

Solution: Let U be a balanced set containing 0 as an interior point. Let $x \in U^\circ$. Since $0 \in U^\circ$, we see $\alpha x \in U^\circ$ if $\alpha = 0$. Suppose $0 < |\alpha| \leqslant 1$. Then since x is interior, there is an open set V with $x \in V \subseteq U$. Hence $\alpha x \in \alpha V \subseteq \alpha U \subseteq U$. Since αV is open, αx is interior to U. So $\alpha x \in U^\circ$. $\square$

5. Let X be a topological vector space. Show the interior of any convex set is convex.

Solution: Suppose W is a convex subset of X and U is its interior. Let $x, y \in U$. Choose open neighborhoods U_x and U_y of x and y which are subsets of W. Suppose $0 < \alpha < 1$. Then $\alpha x + (1-\alpha)y \in \alpha U_x + (1-\alpha)U_y \subseteq W$. Since α is nonzero and $z \mapsto \alpha z$ is a homeomorphism when $\alpha \neq 0$, we see αU_x is an open subset containing αx. Since translation is a homeomorphism, $\alpha U_x + (1-\alpha)U_y = \bigcup_{z \in (1-\alpha)U_y}(\alpha U_x + z)$ is open. Thus since $\alpha U_x + (1-\alpha)U_y$ is an open subset of W containing $\alpha x + (1-\alpha)y$, U° is convex. □

6. Give an example of an internal point of a subset of a normed space X which is not interior.

Solution: Let $X = C[0,1]$ with norm $||\cdot||_2$. Set $U = \{f \mid ||f||_\infty < 1\}$. We claim 0 is not an interior point. Indeed, we need a $\delta > 0$ such that if $||f||_2 < \delta$, then $||f||_\infty < 1$; but clearly no such δ exists. We claim 0 is internal. Indeed, if $f \in X$, take $\delta_f = \frac{1}{||f||_\infty + 1}$. Then if $0 \leqslant |\alpha| \leqslant \delta_f$, then $||0 + \alpha f||_\infty < \frac{1}{||f||_\infty + 1}||f||_\infty < 1$. □

7. Let W be a convex, saturating, and balanced set in a vector space X with corresponding seminorm $||\cdot||$ where

$$||x|| = \inf\{\lambda > 0 \mid x \in \lambda W\}.$$

Show $\{x \mid ||x|| < 1\}$ equals the set of internal points in W. In particular, the set of internal points of W has only internal points and is convex, saturating, and balanced.

Solution: Let $E = \{x \mid ||x|| < 1\}$. By Lemma 2.4, $|\cdot|$ is a seminorm and thus E is convex, balanced, and saturating. We need to show x is in E if and only if x is internal in W. Suppose x is in E, then there is a $1 > \lambda > 0$ with $x \in \lambda W$. So $\lambda W \subseteq W$. Moreover if $y \in X$, we have $y \in \mu W$ for some $\mu \neq 0$. Take $\delta_y = \frac{1-\lambda}{|\mu|}$. Then if $|\alpha| < \delta_y$,

$$\begin{aligned} x + \alpha y &\in \lambda W + \alpha(1-\lambda)\frac{\mu}{(1-\lambda)}\frac{1}{\mu}y \\ &\in \lambda W + (1-\lambda)\frac{\alpha\mu}{(1-\lambda)}W \\ &\subseteq \lambda W + (1-\lambda)W \\ &\subseteq W \end{aligned}$$

for $\left|\frac{\alpha\mu}{1-\lambda}\right| < 1$ and W is a balanced and convex set. So x is an internal point of W.

Conversely, suppose x is an internal point of W. If $|x| = 0$, then $x \in E$; so we may suppose $|x| > 0$. Now there is a $\delta > 0$ with $x + \alpha x \in W$ if $|\alpha| \leqslant \delta$. Thus $x = \frac{1}{1+\delta}(x + \delta x) \in \frac{1}{1+\delta}W$. This implies $|x| \leqslant \frac{1}{1+\delta} < 1$. So $x \in E$. □

8. Show there is a unique Hausdorff vector space topology on $\mathbb{R}^n$. (**Hint:** Show all compact sets in the standard topology are compact and hence closed in any other topology. Then show if V is open and contains the zero vector and $\alpha V \subseteq \{x \mid ||x|| \neq 1\}$ where $|\alpha| \leqslant \delta$, then V is bounded.)

Solution:

Let $\mathcal{T}$ be a Hausdorff vector space topology on $\mathbb{R}^n$. Let $U \in \mathcal{T}$ be an open neighborhood of 0 and let $e_1, e_2, \ldots, e_n$ be the usual basis of $\mathbb{R}^n$. Choose $r > 0$ such that $\alpha e_j \in V$ if $|\alpha| < r$ where V is an open neighborhood of 0 in the $\mathcal{T}$ topology so that $V + V + \cdots + V \subseteq U$. We then have $\alpha_1 e_1 + \alpha_2 e_2 + \cdots + \alpha_n e_n \in U$ if the $|\alpha_i| < r$. Thus if $\sum \alpha_i^2 < r^2$, then $|\alpha_i| < r$ for $i = 1, 2, \ldots, n$ and thus

$$\alpha_1 e_1 + \alpha_2 e_2 + \cdots + \alpha_n e_n \in U.$$

Thus U contains an open neighborhood of 0 in the standard topology. Hence every set in $\mathcal{T}$ is open in the standard topology.

Let W be a compact set in the standard topology. Let U_α be open sets in $\mathcal{T}$ which cover W. Since the U_α are open in the standard topology, finitely many of the U_α cover W. Thus W is compact in the $\mathcal{T}$ topology. Since $\mathcal{T}$ is Hausdorff, W is closed.

Since $\{x \mid ||x|| = 1\}$ is compact, the set $\{||x|| \neq 1\}$ is open. Thus there is a $\delta > 0$ and an open neighborhood V of 0 in $\mathcal{T}$ such that $\alpha V \subseteq \{||x|| \neq 1\}$ for $|\alpha| \leqslant \delta$. Suppose there is an α_0 and a $v \in V$ with $|\alpha_0| \leqslant \delta$ and $|\alpha_0 v| > 1$. Then $\left|\alpha_0 \frac{1}{\alpha_0 |v|} v\right| = 1$ and $\left|\alpha_0 \frac{1}{\alpha_0 |v|}\right| \leqslant |\alpha_0| \leqslant \delta$ which is impossible. So $|\alpha v| < 1$ for $|\alpha| \leqslant \delta$ and $v \in V$. So $|\delta v| < 1$ if $v \in V$ and we see $||v|| < \frac{1}{\delta}$ for $v \in V$. Thus V is bounded. Thus if U is a standard open neighborhood of 0, there is a $r > 0$ with $\{||x|| < r\} \subseteq U$. But then $r\delta V \subseteq U$ and thus U is a neighborhood of 0 in $\mathcal{T}$. This implies every standard open set is in $\mathcal{T}$.

A slightly different argument:

Consider the usual topology on $\mathbb{R}^n$. Call this topology $\mathcal{U}$ and let $\mathcal{T}$ be another Hausdorff vector space topology on $\mathbb{R}^n$. We claim $\mathcal{T} \subseteq \mathcal{U}$. Indeed for each i, the mapping $x \mapsto xe_i$ is continuous. It follows that $v \mapsto v$ continuous from $(\mathbb{R}^n, \mathcal{U})$ into $(\mathbb{R}^n, \mathcal{T})$. Thus $\mathcal{T} \subseteq \mathcal{U}$. Consider any compact set in $(\mathbb{R}^n, \mathcal{U})$. Since $\mathcal{T} \subseteq \mathcal{U}$,it is compact in $\mathcal{T}$. Since $\mathcal{T}$ is Hausdorff, it is closed in $\mathcal{T}$. Thus the sphere S of x with $||x|| = 1$ is closed and compact in $\mathcal{T}$. Consider $\mathbb{R}^n - S$. It is open in $\mathcal{T}$. So there is an open neighborhood V of 0 in $\mathcal{T}$ and a $\delta > 0$ such that $\alpha V \subseteq \mathbb{R}^n - S$ for $|\alpha| \leqslant \delta$. Replacing V by δV, we have $\alpha V \subseteq \mathbb{R}^n - S$ for $|\alpha| \leqslant 1$. This implies V is bounded; in fact $||v|| < 1$ for $v \in V$. Indeed, if not choose v with $||v|| \geqslant 1$. Consider $g(t) = tv$ for $0 \leqslant t \leqslant 1$. Note g is a path from 0 to v. Thus there would be a t with $||g(t)|| = 1$, a contradiction. Now let U be a neighborhood of 0 in $\mathcal{U}$. Choose $r > 0$ with $B_r(0) \subseteq U$. Then $rV \subseteq B_r(0) \subseteq U$. Thus every neighborhood of 0 in $\mathcal{U}$ is a neighborhood of 0 in $\mathcal{T}$.

□

9. Let X be a locally convex topological vector space. Let $\mathcal{F}$ be the family of all continuous seminorms on X. Show the topology on X is the locally convex vector space topology on X defined by using all the seminorms in $\mathcal{F}$.

Solution: Let $\mathcal{T}_\mathcal{F}$ be the locally convex topology on X defined by the seminorms in $\mathcal{F}$. To show this is the locally convex topology on X, it suffices to show there is a common neighborhood subbase at 0 for both the topology $\mathcal{T}_\mathcal{F}$ and the topology on X.

But a subbase at 0 for the topology $\mathcal{T}_{\mathcal{F}}$ consists of all the set $\{x \mid n(x) < r\}$ where $r > 0$ and $n \in \mathcal{F}$. But these sets are open in X for each $n \in \mathcal{F}$ is continuous. Moreover, by Proposition 2.9, these sets also form a subbase at 0 for the locally convex topology on X. □

10. Let X be a locally convex topological vector space.

(a) Show $\frac{t}{1+t} \leqslant \frac{r}{1+r} + \frac{s}{1+s}$ whenever $r, s, t \geqslant 0$ and $t \leqslant r + s$.

(b) Suppose the locally convex topology on X is defined by a countable family $\{||\cdot||_i \mid i \in \mathbb{N}\}$ of seminorms. Define d on $X \times X$ by

$$d(x,y) = \sum_{i=1}^{\infty} \frac{1}{2^i}\left(\frac{||x-y||_i}{1+||x-y||_i}\right).$$

Show d is a pseudometric on X; i.e., $d(x,x) = 0$, $d(x,y) = d(y,x)$, and $d(x,z) \leqslant d(x,y) + d(y,z)$ for all x, y and z in X.

(c) Show the topology on X defined by the pseudometric d is the topology on X.

(d) Show X is metrizable if and only if X is Hausdorff and the locally convex topology on X can be defined by countably many seminorms.

Solution: Set $f(x) = \frac{x}{1+x}$. Note $f'(x) = \frac{1}{(1+x)^2}$ and thus f is increasing. Thus $f(t) \leqslant f(r+s)$ if $0 \leqslant t \leqslant r+s$. Hence

$$\frac{t}{1+t} \leqslant \frac{r+s}{1+r+s} \leqslant \frac{r}{1+r+s} + \frac{s}{1+r+s} < \frac{r}{1+r} + \frac{s}{1+s}$$

when in addition $r \geqslant 0$ and $s \geqslant 0$. This gives (a).

For (b), $d(x,y) = \sum_i \frac{1}{2^i}\left(\frac{||x-y||_i}{1+||x-y||_i}\right)$ clearly satisfies $d(x,x) = 0$, $d(x,y) = d(y,x)$. For the triangle inequality we note using (a) since $||x-z||_i \leqslant ||x-y||_i + ||y-z||_i$ that

$$\begin{aligned} d(x,z) &= \sum_i \frac{1}{2^i}\left(\frac{||x-y+y-z||_i}{1+||x-y+y-z||_i}\right) \\ &\leqslant \sum_i \frac{1}{2^i}\left(\frac{||x-y||_i}{1+||x-y||_i}\right) + \sum_i \frac{1}{2^i}\left(\frac{||y-z||_i}{1+||y-z||_i}\right) \\ &= d(x,y) + d(y,z)\,. \end{aligned}$$

Now for (c). We show the pseudometric d and the topology defined by the $||\cdot||_i$ have the same neighborhoods. Fix x. Let V be a neighborhood of x in the topology defined by the seminorms. Then there are an ϵ and $i_1, i_2, \ldots, i_k$ with $1 > \epsilon > 0$ and $i_1 < i_2 < \ldots < i_k$ such that

$$\{y \mid ||y-x||_{i_j} < \epsilon, j = 1, 2, \ldots, k\} \subseteq V\,.$$

Let $W = \{y \mid d(y,x) < \frac{\epsilon}{2^{i_k+1}}\}$. Note if $y \in W$, then $\frac{1}{2^{i_j}}\left(\frac{||y-x||_{i_j}}{1+||y-x||_{i_j}}\right) < \frac{\epsilon}{2^{i_k+1}}$. Thus $\frac{||y-x||_{i_j}}{1+||y-x||_{i_j}} < 2^{i_j-i_k-1}\epsilon \leqslant \frac{\epsilon}{2} < \frac{1}{2}$. Now $\frac{t}{1+t}$ is increasing in t and at $t=1$ we have $\frac{1}{2}$. So $||y-x||_{i_j} < 1$. Thus $||y-x||_{i_j} < (1+||y-x||_{i_j})2^{i_j-i_k-1} < 2^{i_j-i_k}\epsilon \leqslant \epsilon$ for $j = 1, 2, \ldots, k$. So $y \in V$ and $W \subseteq V$. So V is a d neighborhood of x.

Conversely, let W be a d neighborhood of x. Then there is an $\epsilon > 0$ with $d(y,x) < \epsilon$ implying $y \in W$. Choose k_0 such that $\sum_{k_0+1}^{\infty} 2^{-k} < \frac{\epsilon}{2}$. Set V to be the set of y with $||y-x||_i < \frac{\epsilon}{2}$ for $i = 1, 2, \ldots, k_0$. Then for $y \in V$,

$$\begin{aligned}
d(y,x) &= \sum_{i=1}^{k_0} \frac{1}{2^i}\left(\frac{||x-y||_i}{1+||x-y||_i}\right) + \sum_{i=k_0+1}^{\infty} \frac{1}{2^i}\left(\frac{||x-y||_i}{1+||x-y||_i}\right) \\
&< \sum_{i=1}^{k_0} \frac{\epsilon}{2^{i+1}} + \sum_{i=k_0+1} 2^{-i} \leqslant \frac{\epsilon}{2} + \frac{\epsilon}{2} = \epsilon\,.
\end{aligned}$$

Thus $V \subseteq W$. This shows the topologies are the same.

For (d), suppose that X is metrizable. Thus X is Hausdorff. Moreover, the sets $V_n = \{y \mid d(y,0) < \frac{1}{n}\}$ for $n > 0$ form a neighborhood base at 0. Now for each n, there exist a finite set F_n of continuous seminorms and an $\epsilon_n > 0$ such that if $||y|| < \epsilon_n$ for $||\cdot|| \in F_n$, then $y \in V_n$. Let $S = \cup F_n$. Then S is countable and by the above, every neighborhood V_n is a neighborhood of the topology on X defined by the seminorms in S. Thus X has a topology defined using countable many seminorms.

Conversely, suppose a Hausdorff topology on X is defined by countably many seminorms $||\cdot||_i$. By (c), this topology has a pseudometric d. To see it is a metric, we need only show if $d(y,x) = 0$, then $y = x$. But $d(y,x) = 0$ implies $||y-x||_i = 0$ for all i. Thus $y - x$ is in every neighborhood of 0 in X. Since X is Hausdorff, $y - x = 0$ and so $y = x$. □

11. Show if X is a Fréchet space, there is a complete metric ρ on X satisfying $\rho(x,y) = \rho(x-y,0)$ for all $x, y \in X$.

Solution: By Exercise 2.1.10, there is a countable family $||\cdot||_i$ of seminorms such that $\rho(x,y) = \sum \frac{1}{2^i}\left(\frac{||x-y||_i}{1+||x-y||_i}\right)$ is a metric which defines the topology on X. Clearly $\rho(x,y) = \rho(x-y,0)$. We need to show ρ is complete. Let x_n be Cauchy for ρ. This implies x_n is Cauchy in X and thus there is an x with $||x_n - x||_i \to 0$ as $n \to \infty$ for each i. Thus if $\epsilon > 0$ and $\sum_{k+1}^{\infty} \frac{1}{2^i} < \frac{\epsilon}{2}$ and we choose N so that $||x_n - x||_i < \frac{\epsilon}{2}$ for $n > N$ and $i = 1, 2, \ldots, k$, then

$$\begin{aligned}
\rho(x_n, x) &< \sum_{i=1}^{k} \frac{1}{2^i}\left(\frac{||x-y||_i}{1+||x-y||_i}\right) + \sum_{k+1}^{\infty} \frac{1}{2^i} \\
&< \sum_{i=1}^{k} \frac{1}{2^i}\left(\frac{\epsilon}{2}\right) + \frac{\epsilon}{2} < \epsilon
\end{aligned}$$

for $n > N$. □

12. Let $X = C[0,1]$ be the vector space of real valued continuous functions on the closed interval $[0,1]$. For each $x \in [0,1]$, define $||f||_x = |f(x)|$.

(a) Show the family of seminorms $||\cdot||_x$ define a locally convex vector space topology on X.

(b) Show this topology is not metrizable; i.e., show there is no metric whose topology defines the same topology as the seminorms $||\cdot||_x$.

(c) Show the space is not complete.

(d) Find a completion for X; i.e., a complete locally convex topological vector space Y such that X is a linear subspace of Y, X has the relative topology of Y, and X is dense in Y.

Solution: Note it is easy to verify the functions $||\cdot||_x$ are seminorms; so (a) holds.

We do (b). We show the topology on X is not first countable. Suppose there is a countable base U_n at 0. Hence there is a countable set C of points $x \in [0,1]$ such that for each n, there is a finite set $F_n \subseteq C$ and a $\delta_n > 0$ so that if $||f||_x < \delta_n$ for $x \in F_n$, then $f \in U_n$. Choose $y \notin C$ and take $V = \{f \mid ||f||_y < 1\}$. Choose n so that $U_n \subseteq V$. Clearly we can find an $f \in X$ with $f(y) > 1$ but $|f(x)| < \delta_n$ for $x \in F_n$. Thus $f \in U_n$ and $f \notin V$, a contradiction.

For (c) let $f_n(x) = 0$ for $0 \leqslant x \leqslant 1 - \frac{1}{n}$ and $f_n(x) = n(x - 1 + \frac{1}{n})$ for $1 - \frac{1}{n} \leqslant x \leqslant 1$. Then each f_n is continuous and for $0 \leqslant x < 1$, one has $||f_k - f_m||_x = 0$ if $m, k > n$ where $1 - \frac{1}{n} > x$. For $x = 1$, one has $||f_k - f_m||_x = 0$ for all k and m. So $\{f_k\}_k$ is Cauchy. But there is no $f \in C[0,1]$ with $f_k \to f$. Indeed, then $f_k(1) \to 1$ and $f_k(x) \to 0$ for $0 \leqslant x < 1$. So f would have to satisfy $f(x) = 0$ for $0 \leqslant x < 1$ and $f(1) = 1$. But this f is discontinuous.

For (d), consider the space Y of all real valued functions on $[0,1]$ with seminorms $||f||_x = |f(x)|$. Clearly $X \subseteq Y$ and X has the relative topology. We claim X is dense in Y. Let $f \in Y$ and suppose $\epsilon > 0$ and $x_1, x_2, \ldots, x_n$ are in $[0,1]$. We need to show there is a $g \in C[0,1]$ with $|g(x_j) - f(x_j)| < \epsilon$ for $j = 1, 2, \ldots, n$. But indeed, one can easily find a g with $g(x_j) = f(x_j)$ for $j = 1, 2, \ldots, n$. □

13. Let X be a locally convex Hausdorff topological vector space whose topology is defined by the seminorms $||\cdot||_\alpha$ for $\alpha \in A$.

(a) Show a subset E of X is bounded if and only if for each $\alpha \in A$, there is a $B_\alpha \geqslant 0$ such that
$$||x||_\alpha \leqslant B_\alpha \text{ for } x \in E.$$

(b) Show every compact subset of X is closed and bounded.

(c) Give an example of an X which has a closed bounded subset which is not compact.

(d) Show a subset E of X is bounded if and only if for each continuous seminorm $|\cdot|$ on X there is a constant $M \geqslant 0$ such that

$$|x| \leqslant M \text{ for all } x \in E.$$

Solution: Recall a set E is bounded if for each open neighborhood U of 0, there is an $r > 0$ such that $E \subseteq rU$. Now each set $U_\alpha = \{x \mid ||x||_\alpha < 1\}$ is open. So there is a $B_\alpha > 0$ such that $E \subseteq B_\alpha U_\alpha$. Thus if $y \in E$, $y = B_\alpha x$ for some x with $||x||_\alpha < 1$. Thus $||y||_\alpha = ||B_\alpha x||_\alpha = B_\alpha ||x||_\alpha < B_\alpha$.

Conversely, let there be a B_α such that $||x||_\alpha \leqslant B_\alpha$ for $x \in E$. If U is a neighborhood of 0, there is a finite set F of α and an $\epsilon > 0$ with $||x||_\alpha < \epsilon$ for $\alpha \in F$ implying $x \in U$. Hence if $x \in E$ and $x \neq 0$, then if $B > \max_{\alpha \in F} B_\alpha$, one has $||\frac{\epsilon}{B}x||_\alpha < \frac{\epsilon}{B_\alpha}||x||_\alpha = \epsilon$ for $\alpha \in F$. Thus $\frac{\epsilon}{B}x \in U$ and we see $E \subseteq \frac{B}{\epsilon}U$. So (a) holds.

To see (b), let K be a compact subset of X. Let U be an open neighborhood of 0. Then there is a finite set F of α and an $\epsilon > 0$ such that the set N_ϵ of x with $||x||_\alpha < \epsilon$ for $\alpha \in F$ is a subset of U. But the sets rN_ϵ for $r > 0$ cover X and hence K. Hence $K \subseteq r_1 N_\epsilon \cup \cdots \cup r_k N_\epsilon$ for some finite sequence $0 < r_1 < r_2 < \cdots < r_k$. But then $K \subseteq r_k N_\epsilon \subseteq r_k U$ and we see K is bounded. Since X is Hausdorff, every compact subset is closed.

Now let $X = C[0,1]$ with maximum norm. So $||f|| = \max_x |f(x)|$. Set $C = \{f \in X \mid -1 \leqslant f(x) \leqslant 1 \text{ for } 0 \leqslant x \leqslant 1\}$. Note C is bounded and if $||g|| > 1$, then $\{h \mid ||h - g|| < \frac{||g||-1}{2}\}$ is a neighborhood of g which misses C. So C is closed. We need to show C is not compact. This follows from the Ascoli Theorem, for a subset of X is compact if and only if it is closed, bounded, and equicontinuous.

For (d), note if $|\cdot|$ is continuous, then $V = \{x \mid |x| < 1\}$ is open. So if E is bounded, $E \subseteq rV$ for some $r > 0$ and thus $|x| < r$ for $x \in E$. The converse follows by (a). □

14. Show every Cauchy sequence in a locally convex topological vector space is bounded.

Solution: Let $\{x_n\}$ be a Cauchy sequence and let $||\cdot||$ be a continuous seminorm on X. There is an $N > 0$ with $||x_m - x_n|| < 1$ for $m, n > N$. Thus $||x_n|| \leqslant \max\{||x_1||, ||x_2||, \ldots, ||x_N||, 1 + ||x_{N+1}||\}$ for all n for

$$||x_n|| \leqslant ||x_n - x_{N+1}|| + ||x_{N+1}|| < 1 + ||x_{N+1}||$$

for $n > N$. Hence the sequence $\{x_n\}$ is bounded. □

15. Let X be a locally convex topological vector space. Show the closure of any bounded subset of X is bounded.

Solution: Let S be a bounded set and let $||\cdot||$ be a continuous seminorm. Then there is a B with $||x|| \leqslant B$ for $x \in S$. Now let y be in the closure of S. The set $\{z \mid ||z - y|| < 1\}$ is an open neighborhood of y and thus meets S. So there is a $z \in S$ in this set. So $||y|| \leqslant ||y - z|| + ||z|| < 1 + B$. Hence the continuous seminorms are bounded on the closure of S. □

16. Let X and Y be locally convex topological vector spaces. Suppose T is a continuous linear transformation from X into Y. Show if B is a bounded subset of X, then $T(B)$ is a bounded subset of Y.

Solution: Let S be a bounded set and let $||\cdot||$ be a continuous seminorm on Y. Let $N = \{y \in Y \mid ||y|| < 1\}$. This is an open set in Y containing 0. So $T^{-1}(N)$ is an open set in X containing 0. But then there is a continuous seminorm $|\cdot|$ on X with $|x| < 1$ implying $x \in T^{-1}(N)$. Indeed, we know $T^{-1}(N)$ contains an open balanced set U and if $|\cdot|$ is the seminorm on X defined using U, then $U = \{x \mid |x| < 1\} \subseteq T^{-1}(N)$. Since S is bounded, there is a $B > 0$ such that $|x| < B$ if $x \in S$. Thus $\frac{x}{B} \in U$. Thus $\frac{1}{B}T(x) \in T(U) \subseteq N$. This implies $\frac{1}{B}||T(x)|| < 1$ and consequently $||T(x)|| < B$ for $x \in S$. Hence $T(S)$ is bounded. □

17. Let V be a complex locally convex topological Hausdorff topological space. Let V^* be the dual space of V. Thus V^* is the vector space whose elements are the continuous complex valued linear functionals on V. The topology on V defined by the seminorms $|\cdot|_f$ where $|v|_f = |f(v)|$ for $f \in V^*$ is called the weak topology on V while the topology on V^* defined by seminorms $|\cdot|_v$ on V^* where $|f|_v = |f(v)|$ for $f \in V^*$ and $v \in V$ is called the weak-$*$ topology on V^*.

(a) Show V^* is Hausdorff.

(b) Assume V is a Hilbert space. Show the conjugate linear isomorphism of V onto V^* given by $w \mapsto f_w$ where $f_w(v) = (v, w)$ is a homeomorphism of V with the weak topology onto V^* with the weak-$*$ topology.

Solution: For (a), note if $f_1 \neq f_2$, then there is a v with $f_1(v) \neq f_2(v)$. Set $\epsilon = \frac{|f_1(v)-f_2(v)|}{2}$ and $N_1 = \{f \mid |f - f_1|_v < \epsilon\}$, $N_2 = \{f \mid |f - f_2|_v < \epsilon\}$. Then N_1 and N_2 are open neighborhoods of f_1 and f_2, respectively and $N_1 \cap N_2 = \varnothing$; for otherwise there is an f with $|f_1(v) - f(v)| < \epsilon$ and $|f(v) - f_2(v)| < \epsilon$. Consequently,

$$2\epsilon = |f_1(v) - f_2(v)| \leqslant |f_1(v) - f(v)| + |f(v) - f_2(v)| < 2\epsilon$$

which is impossible.

Now if V is a Hilbert space, we know by the Riesz Theorem that every f in V^* is given uniquely by a $w \in V$ with $f(v) = (v, w)$. It is easy to verify the mapping $w \mapsto f_w$ where $f_w(v) = (w, v)$ is conjugate linear. Thus to finish part (b), it suffices to show a preimage under this map of a neighborhood subbase at 0 for the weak-$*$ topology on V^* is a neighborhood subbase at 0 for V which defines the weak topology on V. Now the sets $\{f \mid |f|_v < \epsilon\}$ where $\epsilon > 0$ and $v \in V$ is a neighborhood subbase for V^*. Its preimage consists of the sets $\{w \mid |f_w|_v < \epsilon\}$ over v and ϵ. These are the sets $\{w \mid |f_v|_w < \epsilon\}$ or the sets $\{w \in V \mid |w|_f < \epsilon\}$ for $f \in V^*$ and $\epsilon > 0$. These sets are a subbase defining the weak topology on V. □

18. Let X be a vector space and suppose X_i for $i \in I$ form a family of locally convex topological vector subspaces of X whose union is X. Give X the inductive limit topology. Show the relative inductive limit topology on each X_i is the topology on X_i for all i if and only if there is a family $|\cdot|_\alpha$, $\alpha \in A$ of continuous seminorms on X such that the restrictions of these seminorms to each X_i define the topology on X_i. An example is $\mathcal{D}(\Omega)$ and the seminorms

$$||\phi||_\alpha = \max_{x \in \text{supp } \phi} |D^\alpha \phi(x)| \text{ for } \alpha \in \mathbb{N}_0^n.$$

Solution: Suppose X has the inductive limit topology and the relative topology on each X_i is the topology of X_i. Let $|\cdot|_{\alpha \in A}$ be a family of continuous seminorms on X which define the locally convex topology on X. Since X_i has the relative topology, each of the sets $N_\alpha(\epsilon) = \{x \in X_i \mid |x|_\alpha < \epsilon\} = X_i \cap \{x \in X \mid |x|_\alpha < \epsilon\}$ is open in X_i and these sets form a neighborhood subbase for relative topology of X on X_i. But this relative topology is the topology of X_i. Since these neighborhoods are the neighborhoods of X_i defined by the restrictions of the continuous seminorms $|\cdot|_\alpha$ to X_i, we see their restrictions define the topology on X_i.

Conversely, suppose there is a family of continuous seminorms $|\cdot|_{\alpha \in A}$ for X whose restrictions to each X_i define the topology for X_i. By definition the relative topology of the inductive limit topology for X on X_i is weaker than the topology of X_i. But the sets $\{x \in X_i \mid |x|_\alpha < \epsilon\}_{\alpha \in A, \epsilon > 0}$ are open in the relative topology and these define the topology on X_i. So the relative topology from X on X_i is also as strong as the topology on X_i. So the relative topology on each X_i is the topology on X_i. □

19. Let X_n be an increasing sequence of locally convex topological vector subspaces of a vector space X whose union is X. Let X have the inductive limit topology. Suppose M is a vector subspace of X. Give M the inductive limit topology of the relative topologies of X_n on $X_n \cap M$.

(a) Show the relative topology of X on M is weaker than the topology on M.

(b) Give an example where the relative topology is strictly weaker.

Solution: For (a), we note that X has a neighborhood base consisting of the balanced convex sets U such that $U \cap X_n$ is open in X_n for all n. This implies the sets $U \cap M$ for such sets U form a neighborhood base for the relative topology of X on M. Consequently, the sets $U \cap M \cap X_n$ form a neighborhood base for the relative topology of the inductive limit topology on M to the subspace X_n. But the sets $U \cap X_n$ are open in X_n and so the sets $U \cap M \cap X_n$ are open in the relative topology of X_n on M. Since these sets are balanced and convex, Corollary 2.17 implies they are open in the inductive limit topology obtained by taking the inductive limit topology of the relative topologies on $X_n \cap M$.

For (b), let $X_n = \{f \in L^2(\mathbb{R}) \mid f\chi_{[n,\infty)} = 0 \text{ for some } n \in \mathbb{N}\}$ with norm $|\cdot|_2$. Then $X = \cup_n X_n$ is a topological vector space and has an inductive limit topology. Set Y to be the vector space of L^2 functions on $\mathbb{R}$ vanishing a.e. off some interval $[-n, n]$ where $n \in \mathbb{N}$ and let $Y_n = X_n \cap Y$. We claim the inductive limit topology of the Y_n on Y is strictly stronger than the relative topology of X on Y.

To do this we show there is a continuous seminorm on Y which is not continuous in the relative topology from X. In fact, the L^1 norm $|f|_1 = \int |f(x)|\,dx$ exists for $f \in Y$ for if $f \in Y_n$, then $|f|_1 = \int_{-n}^{n} |f(x)|\,dx = (\chi_{[-n,n]}, |f|)_2 \leqslant |\chi_{[-n,n]}|_2 |f|_2 = \sqrt{2n}\,|f|_2$. In particular, we see $|\cdot|_1$ is continuous on Y_n and by Proposition 2.16, one has $|\cdot|_1$ is continuous on Y.

Now consider $V = \{f \in Y \mid |f|_1 < 1\}$. This is an open balanced convex neighborhood of 0 in the inductive limit topology on Y. If Y has the relative topology from X, then V contains a set of form $U \cap Y$ where U is an open balanced convex neighborhood of 0 in X. Now $\frac{1}{x}\chi_{(-\infty,-1]} \in X$ and thus $\frac{a}{x}\chi_{(-\infty,-1]} \in U$ for some $a > 0$. Also $\frac{a}{x}\chi_{(-\infty,-1]} - \frac{a}{x}\chi_{(-\infty,-m]} = \frac{a}{x}\chi_{(-m,-1)}$ in L^2 and $\frac{a}{x}\chi_{(-\infty,-m]} \to 0$ in X as $m \to \infty$. To see this, we need only show if $|\cdot|$ is a seminorm on X with $|\cdot|$ restricted to each X_n continuous, then $|\frac{a}{x}\chi_{(-\infty,-m]}| \to 0$ as $m \to \infty$. But then there is a $\delta > 0$ for X_1 so that if $|f|_2 < \delta$, then $|f| < 1$ whenever $f \in X_1$. But $|\frac{a}{x}\chi_{(-\infty,-m]}|_2 = a\sqrt{\int_m^\infty \frac{1}{x^2}\,dx} = \frac{a}{m} \to 0$ as $m \to \infty$ and consequently, $|\frac{a}{x}\chi_{(-\infty,-m]}| \to 0$ as $m \to \infty$. Since U is open, we conclude $\frac{a}{x}\chi_{(-m.-1)} \in Y \cap U \subseteq V$ for large m. Thus $|\frac{a}{x}\chi_{(-m,-1)}|_1 < 1$ for all large $m \in \mathbb{N}$. This cannot hold for $\int_1^m \frac{a}{x}\,dx = a \ln m \to \infty$ as $m \to \infty$. $\square$

Exercise Set 2.2

1. Let Ω be an open subset of $\mathbb{R}^n$ and let $F = (f_1, f_2, \ldots, f_m)$ be a function from Ω into $\mathbb{R}^m$. Suppose F is differentiable at a point x in Ω. Show the partial derivatives $D_i f_j(x)$ exist and the linear transformation D is defined by

$$D(a_1, a_2, \ldots, a_n) = (\sum_{i=1}^{n} a_i D_i f_1(x), \sum_{i=1}^{n} a_i D_i f_2(x), \ldots, \sum_{i=1}^{n} a_i D_i f_m(x)).$$

Solution: Let $F = (f_1, f_2, \ldots, f_m)$ be differentiable at a point x_0 in an open subset Ω of $\mathbb{R}^n$ and suppose $D : \mathbb{R}^n \to \mathbb{R}^m$ is a linear operator satisfying

$$F(x) = F(x_0) + D(x - x_0) + o(x - x_0).$$

Then taking $D(e_j) = (D_{1,j}, D_{2,j}, \dots, D_{m,j})$, one has

$$\begin{aligned}
D(a_1, \dots, a_n) &= \sum_{j=1}^{n} a_j D(e_j) \\
&= \sum_{j=1}^{n} a_j (D_{1,j}, D_{2,j}, \dots, D_{m,j}) \\
&= (\sum a_j D_{1,j}, \sum a_j D_{2,j}, \dots \sum a_j D_{m,j}) \\
&= (a_1, a_2, \dots, a_n) \begin{bmatrix} D_{1,1} & D_{2,1} & \dots & D_{m,1} \\ D_{1,2} & D_{2,2} & \dots & D_{m,2} \\ \cdot & \cdot & \cdots & \cdot \\ \cdot & \cdot & \cdots & \cdot \\ \cdot & \cdot & \cdots & \cdot \\ D_{1,n} & D_{2,n} & \cdots & D_{m,n} \end{bmatrix}.
\end{aligned}$$

Moreover,

$$f_j(x) = f_j(x_0) + \sum_{i=1}^{n} D_{j,i}(x_i - x_{0,i}) + o_j(x - x_0).$$

Thus for $x = x_0 + he_i$ and h near 0 so that x lies in Ω, one has

$$\frac{f_j(x_0 + he_i) - f_j(x_0)}{h} = D_{j,i} + \frac{o_j(he_i)}{h}.$$

Taking the limit as $h \to 0$ gives

$$D_i f_j(x_0) = D_{j,i}.$$

□

2. Let $F : \Omega \to \mathbb{R}^m$ be a function on an open subset Ω of $\mathbb{R}^n$. Show F is continuous at each point where it is differentiable.

Solution: Let $F = (f_1, f_2, \dots, f_m)$ be differentiable at a point x_0 in an open subset Ω of $\mathbb{R}^n$ and suppose $D : \mathbb{R}^n \to \mathbb{R}^m$ is a linear operator satisfying

$$F(x) = F(x_0) + D(x - x_0) + o(x - x_0)$$

for $x \in \Omega$. Clearly one has

$$\begin{aligned}
\lim_{x \to x_0} F(x) &= F(x_0) + \lim_{x \to x_0} D(x - x_0) + \lim_{x \to x_0} \frac{|o(x - x_0)|}{|x - x_0|} \lim_{x \to x_0} |x - x_0| \\
&= F(x_0)
\end{aligned}$$

since $x \mapsto D(x)$ is continuous and $D(0) = 0$ and each factor in the product of the limits is 0. □

3. Give an example of a function $f : \mathbb{R}^2 \to \mathbb{R}$ where $D_1 f(0,0)$ and $D_2 f(0,0)$ exist but f is not differentiable.

Solution: Define f on $\mathbb{R}^2$ by

$$f(x,y) = \begin{cases} \frac{xy}{x^2+y^2} & \text{if } x^2+y^2>0 \\ 0 & \text{if } (x,y)=(0,0) \end{cases}.$$

Then $D_1 f(0,0) = 0$ and $D_2 f(0,0) = 0$, but since $f(x,x) = \frac{1}{2}$ for $x \neq 0$, we see f is not continuous at $(0,0)$. By Exercise 2.2.2, f cannot be differentiable at $(0,0)$. □

4. Show that $F = (f_1, f_2, \ldots, f_m)$ mapping an open set $\Omega \subseteq \mathbb{R}^n$ into $\mathbb{R}^m$ is differentiable at a point x in Ω if and only if each f_i is differentiable at x.
Solution: Assume $F = (f_1, f_2, \ldots, f_m)$ is differentiable at x. Then there exists an a little o with

$$F(y) = F(x) + (y-x)\,[D_{i,j}] + o(y-x) \tag{A}$$

This implies

$$f_j(y) = f_j(x) + \sum_{i=1}^{n} (y_i - x_i) D_{i,j} + o_j(y-x) \tag{B}$$

where $o(z) = (o_1(z), o_2(z), \ldots, o_m(z))$. Since $\frac{|o_j(z)|}{|z|} \leqslant \frac{|o(z)|}{|z|} \to 0$ as $z \to 0$ in $\mathbb{R}^n$, we see each f_j is differentiable at x.

Conversely working the above statements backwards, we see if each f_j is differentiable at x as in (B), then if $o(z) = (o_1(z), o_2(z), \ldots, o_m(z))$, the statement (A) holds. Moreover,

$$\begin{aligned}\frac{|o(z)|}{|z|} &= \frac{\sqrt{\sum o_j(z)^2}}{|z|} \\ &= \sqrt{\sum \frac{o_j(z)^2}{|z|^2}} \to 0\end{aligned}$$

as $z \to 0$ shows o is a little o function. Hence F is differentiable. □

5. Let Ω be a nonempty open subset of $\mathbb{R}^n$. Assume $f : \Omega \to \mathbb{R}$ is a function such that the partial derivatives $D_i f$ exist and are continuous on Ω. Show f is continuously differentiable on Ω.
Solution: Let $f : \Omega \to \mathbb{R}$ have continuous partial derivatives. Then by the mean value theorem, one has if x and y are in a ball inside Ω, there is a z in this ball with each z_i between x_i and y_i such that

$$f(y) - f(x) = \sum_{i=1}^{n} D_i f(z)(y_i - x_i).$$

Thus

$$\begin{aligned} f(y) &= f(x) + \sum_i (y_i - x_i) D_i f(x) + \sum_i (y_i - x_i)(D_i f(z) - D_i f(x)) \\ &= f(x) + \sum_i (y_i - x_i) D_i f(x) + o(y-x) \end{aligned}$$

where

$$o(y-x)=\sum_i (y_i-x_i)(D_if(z)-D_if(x)).$$

Now

$$\begin{aligned}\lim_{y\to x}\frac{|o(y-x)|}{|y-x|} &\leqslant \lim_{y\to x}\sum_i \frac{|y_i-x_i|}{|y-x|}|D_if(z)-D_if(x)| \\ &\leqslant \lim_{y\to x}\sum_{i=1}^n |D_if(z)-D_if(x)| = 0\end{aligned}$$

by continuity of the D_if. □

6. Let $\alpha\in\mathbb{N}_0^n$ and suppose $f_1, f_2, \ldots, f_m$ are C^∞ functions on $\mathbb{R}^n$. Show

$$D^\alpha(f_1f_2\cdots f_m)=\sum_{\beta_1+\beta_2+\cdots+\beta_m=\alpha}\binom{\alpha}{\beta_1,\ \beta_2,\ \cdots,\ \beta_m}D^{\beta_1}f_1D^{\beta_2}f_2\cdots D^{\beta_m}f_m$$

where

$$\binom{\alpha}{\beta_1,\ \beta_2,\ \cdots,\ \beta_m}=\frac{\alpha!}{\beta_1!\beta_2!\cdots\beta_m!}.$$

Solution: First we note $D_i^k(f_1f_2)=\sum_{l=0}^k\binom{k}{l}(D_i^lf_1D_i^{k-l}f_2)$. This is proved by induction. We do the induction step. Assume it is true for k. Then

$$\begin{aligned}D_i^{k+1}(f_1f_2) &= \sum_{l=0}^k\binom{k}{l}(D_i^lf_1D_i^{k+1-l}f_2+D_i^{l+1}f_iD_i^{k-l}f_2)\\ &= f_1D_i^{k+1}f_2+\sum_{l=1}^k\binom{k}{l}D_i^lf_1D_i^{k+1-l}f_2+\sum_{l=0}^{k-1}\binom{k}{l}D_i^{l+1}f_1D_i^{k-l}f_2+D_i^{k+1}f_1f_2\\ &= f_1D_i^{k+1}f_2+\sum_{l=1}^k\left(\binom{k}{l}+\binom{k}{l-1}\right)D_i^lf_1D_i^{k+1-l}f_2+D_i^{k+1}f_1f_2\\ &= f_1D_i^{k+1}f_2+\sum_{l=1}^k\binom{k+1}{l}D_i^lf_1D_i^{k+1-l}f_2+D_i^{k+1}f_1f_2\\ &= \sum_{l=0}^{k+1}\binom{k+1}{l}D_i^lf_1D_i^{k+1-l}f.\end{aligned}$$

From this we argue

$$D_i^k(f_1f_2\cdots f_m)=\sum_{k_1+k_2+\cdots+k_m=k}\frac{k!}{k_1!k_2!\cdots k_m!}D_i^{k_1}f_1D_i^{k_2}f_2\cdots D_i^{k_m}f_m.$$

Indeed, we have done this for $m = 2$. Assume true for m. Then

$$\begin{aligned} D_i^k(f_1 f_2 \cdots f_m f_{m+1}) &= D_i^k\left((f_1 f_2 \cdots f_m) f_{m+1}\right) \\ &= \sum_{l=0}^{k} \binom{k}{l} D_i^l (f_1 f_2 \cdots f_m) D_i^{k-l} f_{m+1} \\ = \sum_{l=0}^{k} \binom{k}{l} & \sum_{l_1+l_2+\cdots+l_m=l} \frac{l!}{l_1! l_2! \cdots l_m!} D_i^{l_1} f_1 D_i^{l_2} f_2 \cdots D_i^{l_m} f_m D_i^{k-l} f_{m+1}. \end{aligned}$$

Letting $k - l = l_{m+1}$, we see

$$\begin{aligned} &D_i^k(f_1 f_2 \cdots f_m f_{m+1}) \\ &= \sum_{l_1+l_2+\cdots+l_{m+1}=k} \frac{k!}{(k-l_1-l_2-\cdots-l_m)! l!} \cdot \frac{l!}{l_1! l_2! \cdots l_m!} D_i^{l_1} f_1 D_i^{l_2} f_2 \cdots D_i^{l_m} f_m D_i^{l_{m+1}} f_{m+1} \\ &\qquad = \sum_{l_1+l_2+\cdots+l_{m+1}=k} \frac{k!}{l_1! l_2! \cdots l_m! l_{m+1}!} D_i^{l_1} f_1 D_i^{l_2} f_2 \cdots D_i^{l_m} f_m D_i^{l_{m+1}} f_{m+1}. \end{aligned}$$

Finally,

$$\begin{aligned} &D^{\alpha}(f_1 f_2 \cdots f_m) = D_1^{\alpha_1} D_2^{\alpha_2} \cdots D_n^{\alpha_n} (f_1 f_2 \cdots f_m) \\ &= (D_1^{\alpha_1} D_2^{\alpha_2} \cdots D_{n-1}^{\alpha_{n-1}}) \sum_{l_{n,1}+l_{n,2}+\cdots+l_{n,m}=\alpha_n} \frac{\alpha_n!}{l_{n,1}! l_{n,2}! \cdots l_{n,m}!} (D_n^{l_{n,1}} f_1 D_n^{l_{n,2}} f_2 \cdots D_n^{l_{n,m}} f_m) \\ &\qquad = (D_1^{\alpha_1} D_2^{\alpha_2} \cdots D_{n-2}^{\alpha_{n-2}}) \sum \sum_{l_{n-1,1}+l_{n-1,2}+\cdots+l_{n-1,m}=\alpha_{n-1}} \frac{\alpha_{(n-1)}!}{l_{n-1,1}! l_{n-1,2}! \cdots l_{n-1,m}!} \\ &\sum_{l_{n,1}+l_{n,2}+\cdots+l_{n,m}=\alpha_n} \frac{\alpha_n!}{l_{n,1}! l_{n,2}! \cdots l_{n,m}!} (D_{n-1}^{l_{n-1,1}} D_n^{l_{n,1}} f_1 \cdot D_{n-1}^{l_{n-1,2}} D_n^{l_{n,2}} f_2 \cdots D_{n-1}^{l_{n-1,m}} D_n^{l_{n,m}} f_m). \end{aligned}$$

Repeating this the remaining $n - 2$ times gives

$$\begin{aligned} &D^{\alpha}(f_1 f_2 \cdots f_m) = \sum_{l_{1,1}+l_{1,2}+\cdots+l_{1,m}=\alpha_1} \sum_{l_{2,1}+l_{2,2}+\cdots+l_{2,m}=\alpha_2} \cdots \sum_{l_{n,1}+l_{n,2}+\cdots+l_{n,m}=\alpha_n} \\ &\frac{\alpha_1! \alpha_2! \cdots \alpha_n!}{l_{1,1}! l_{1,2}! \cdots l_{1,m}! l_{2,1}! l_{2,2}! \cdots l_{2,m}! \cdots l_{n,1}! l_{n,2}! \cdots l_{n,m}!} D_1^{l_{1,1}} D_2^{l_{2,1}} \cdots D_n^{l_{n,1}} f_1 \cdot D_1^{l_{1,2}} D_2^{l_{2,2}} \cdots D_n^{l_{n,2}} f_2 \\ &\qquad \cdots D_1^{l_{1,m}} D_2^{l_{2,m}} \cdots D_n^{l_{n,m}} f_m = \sum_{\beta_1+\beta_2+\cdots+\beta_m=\alpha} \frac{\alpha!}{\beta_1! \beta_2! \cdots \beta_m!} D^{\beta_1} f_1 D^{\beta_2} f_2 \cdots D^{\beta_m} f_m \end{aligned}$$

and we have the result. □

7. In $\mathbb{R}^n$, one has coordinates

$$\begin{aligned} x_1 &= r \cos\theta_1 \\ x_2 &= r \sin\theta_1 \cos\theta_2 \\ x_3 &= r \sin\theta_1 \sin\theta_2 \cos\theta_3 \\ &\cdots \\ x_{n-1} &= r \sin\theta_1 \sin\theta_2 \cdots \sin\theta_{n-2} \cos\theta_{n-1} \\ x_n &= r \sin\theta_1 \sin\theta_2 \cdots \sin\theta_{n-2} \sin\theta_{n-1} \end{aligned}$$

where $0 \leqslant \theta_j \leqslant \pi$ for $j = 1, \cdots, n-2$ and $0 \leqslant \theta_{n-1} \leqslant 2\pi$. Show in terms of these coordinates, one has

$$dx_1 dx_2 \cdots dx_n = r^{n-1} \sin^{n-2}\theta_1 \sin^{n-3}\theta_2 \cdots \sin\theta_{n-2} dr d\theta_1 \cdots d\theta_{n-1}.$$

Solution: We do a proof by induction. Let $y_1, y_2, \ldots, y_{n-1}$ be the coordinates for $\mathbb{R}^{n-1}$ given by

$$\begin{aligned}
y_1 =& r\cos\theta_1 \\
y_2 =& r\sin\theta_1\cos\theta_2 \\
y_3 =& r\sin\theta_1\sin\theta_2\cos\theta_3 \\
&\cdots \\
y_{n-2} =& r\sin\theta_1\sin\theta_2\cdots\sin\theta_{n-3}\cos\theta_{n-2} \\
y_{n-1} =& r\sin\theta_1\sin\theta_2\cdots\sin\theta_{n-3}\sin\theta_{n-2}.
\end{aligned}$$

Then the coordinates $x_1, x_2, \ldots, x_n$ on $\mathbb{R}^n$ are given by

$$x_1 = y_1, x_2 = y_2, \ldots, x_{n-2} = y_{n-2}, x_{n-1} = y_{n-1}\cos\theta_{n-1}, x_n = y_{n-1}\sin\theta_{n-1}.$$

Thus $(x_1, x_2, \cdots, x_n)$ is a composition of the transformation T defined by

$$T(y_1, y_2, \ldots, y_{n-1}, \theta_{n-1}) = (y_1, y_2, \ldots, y_{n-1}\cos\theta_{n-1}, y_{n-1}\sin\theta_{n-1})$$

with the transformation

$$\begin{aligned}
S(r, \theta_1, \theta_2, \ldots, \theta_{n-1}) = (&r\cos\theta_1, r\sin\theta_1\cos\theta_2, \ldots, r\sin\theta_1\cdots \\
&\sin\theta_{n-3}\cos\theta_{n-2}, r\sin\theta_1\cdots\sin\theta_{n-3}\cos\theta_{n-2}, \theta_{n-1}).
\end{aligned}$$

Hence the Jacobian is given by $\mathrm{Jac}(x_1, x_2, \ldots, x_n) = \mathrm{Jac}(T)\ \mathrm{Jac}(S)$. But

$$\mathrm{Jac}(T) = \det\begin{bmatrix} I_{n-2} & 0 & \\ 0 & \cos\theta_{n-1} & \sin\theta_{n-1} \\ & -y_{n-1}\sin\theta_{n-1} & y_{n-1}\cos\theta_{n-1} \end{bmatrix} = y_{n-1}$$

and $\mathrm{Jac}(S) = \mathrm{Jac}(y_1, y_2, \ldots, y_{n-2})$. Thus one sees

$$\mathrm{Jac}(x_1, x_2, \ldots, x_n) = (r\sin\theta_1\sin\theta_2\cdots\sin\theta_{n-3}\sin\theta_{n-2})\,\mathrm{Jac}(y_1, \ldots, y_{n-2}).$$

Setting J_n to be the Jacobian for the coordinate system on $\mathbf{R}^n$, we see $J_n = (r\sin\theta_1\sin\theta_2\cdots\sin\theta_{n-3}\sin\theta_{n-2})J_{n-1}$. Using this recursion formula working down to J_2 gives

$$J_n = r^{n-2}\sin^{n-2}\theta_1\sin^{n-3}\theta_2\cdots\sin\theta_{n-2}J_2.$$

But $J_2 = \det\begin{bmatrix} \cos\theta_1 & \sin\theta_1 \\ -r\sin\theta_1 & r\cos\theta_1 \end{bmatrix} = r$. Hence

$$J_n = r^{n-1}\sin^{n-2}\theta_1\sin^{n-3}\theta_2\cdots\sin\theta_{n-2}.$$

This gives $dx = r^{n-1}\sin^{n-2}\theta_1\sin^{n-3}\theta_2\cdots\sin\theta_{n-2}\, dr\, d\theta_1\, d\theta_2\cdots d\theta_{n-1}$. □

8. Let $B_n = \{x \in \mathbb{R}^n \mid |x| \leqslant 1\}$. Show

$$\lambda_n(B_n) = \frac{\pi^{n/2}}{\Gamma(n/2+1)}.$$

(**Hint:** Give an inductive argument and use the Beta function $B(r,s) = \int_0^1 t^{r-1}(1-t)^{s-1}\,dt = \frac{\Gamma(r)\Gamma(s)}{\Gamma(r+s)}$.)

Solution: We will use the Beta function $B(r,s) = \int_0^1 t^{r-1}(1-t)^{s-1}\,dt = \frac{\Gamma(r)\Gamma(s)}{\Gamma(r+s)}$ for $r,s>0$ and $\Gamma(\frac{1}{2}) = \sqrt{\pi}$.

Note $\lambda_1(B_1) = 2$ and $\lambda_2(B_2) = \pi$. Now

$$\lambda_{n+1}(B_{n+1}) = \int\int \chi_{B_{n+1}}(x, x_{n+1})\,d\lambda_n(x)\,d\lambda_1(x_{n+1}).$$

Thus

$$\begin{aligned}\lambda_{n+1}(B_{n+1}) &= \int_{-1}^{1} \lambda_n\{x \mid |x|^2 + x_{n+1}^2 \leqslant 1\}\,dx_{n+1}\\ &= \int_{-1}^{1} \lambda_n(\sqrt{1-x_{n+1}^2}B_n)\,dx_{n+1}\,.\end{aligned}$$

Hence $\lambda_{n+1}(B_{n+1}) = \lambda_n(B_n)\int_{-1}^{1}\left(1-x_{n+1}^2\right)^{n/2}\,dx_{n+1}$. But

$$\begin{aligned}\int_{-1}^{1}(1-x^2)^{n/2}\,dx &= 2\int_0^1(1-x^2)^{n/2}\,dx = 2\int_0^1(1-t)^{n/2}\frac{1}{2}t^{-1/2}\,dt\\ &= \int_0^1 t^{1/2-1}(1-t)^{\frac{n+2}{2}-1}\,dt = B(\frac{1}{2},\frac{n+2}{2}) = \frac{\Gamma(1/2)\Gamma(\frac{n+2}{2})}{\Gamma(\frac{n}{2}+\frac{3}{2})}.\end{aligned}$$

Hence

$$\begin{aligned}\lambda_n(B_n) &= \frac{\Gamma(.5)\Gamma(.5(n+1))}{\Gamma(.5(n+2))}\lambda_{n-1}(B_{n-1})\\ &= \frac{\Gamma(.5)\Gamma(.5(n+1))}{\Gamma(.5(n+2))}\frac{\Gamma(.5)\Gamma(.5(n))}{\Gamma(.5(n+1))}\lambda_{n-2}(B_{n-2})\\ &= \frac{\Gamma(.5)\Gamma(.5(n+1))}{\Gamma(.5(n+2))}\frac{\Gamma(.5)\Gamma(.5(n))}{\Gamma(.5(n+1))}\frac{\Gamma(.5)\Gamma(.5(n-1))}{\Gamma(.5(n))}\lambda_{n-3}(B_{n-3})\\ &= \cdots\\ &= \frac{\Gamma(.5)^k\Gamma(.5(n+2-k))}{\Gamma(.5(n+2))}\lambda_{n-k}(B_{n-k}).\end{aligned}$$

Thus $\lambda_{2n}(B_{2n}) = \frac{\Gamma(.5)^{2n-2}\Gamma(2)}{\Gamma(n+1)}\lambda_2(B_2) = \frac{\pi^{n-1}\pi}{n!} = \frac{\pi^n}{n!}$.

Also

$$\begin{aligned}
\lambda_{2n+1}(B_{2n+1}) &= \frac{\Gamma(.5)^{2n}\Gamma(.5(2n+1+2-2n))}{\Gamma(.5(2n+1+2))}\lambda_1(B_1) \\
&= \frac{2\pi^n\Gamma(3/2)}{\Gamma(n+\frac{1}{2}+1)} \\
&= \frac{\pi^n\Gamma(.5)}{\Gamma(n+\frac{1}{2}+1)} \\
&= \frac{\pi^{n+1/2}}{\Gamma(n+\frac{1}{2}+1)}
\end{aligned}$$

As a second method, we use Exercise 2.2.7 and spherical coordinates. First we find $A_k = \int_0^\pi \sin^k\theta\,d\theta$. We note if $x = \sin\theta$, then $A_k = 2\int_0^{\pi/2}\sin^k\theta\,d\theta = 2\int_0^1 x^k \frac{1}{\sqrt{1-x^2}}\,dx$. Thus taking $t = x^2$, one sees

$$\begin{aligned}
A_k &= 2\int_0^1 x^k(1-x^2)^{-1/2}\,dx \\
&= \int_0^1 x^{k-1}(1-x^2)^{-1/2}\,(2x\,dx) \\
&= \int_0^1 t^{(k+1)/2-1}(1-t)^{1/2-1}\,dt \\
&= B(\frac{k+1}{2},\frac{1}{2}).
\end{aligned}$$

So $A_k = B(\frac{k+1}{2},\frac{1}{2}) = \frac{\Gamma(\frac{k+1}{2})\Gamma(\frac{1}{2})}{\Gamma(k/2+1)}$. Thus for $a > 0$, one has

$$\begin{aligned}
\lambda_n(aB_n) &= \int_0^a r^{n-1}\,dr \int_0^{2\pi} d\theta_{n-1} \prod_{k=1}^{n-2}\int_0^\pi \sin^k\theta_{n-1-k}\,d\theta_{n-1-k} \\
&= \frac{2\pi a^n}{n}\prod_{k=1}^{n-2}\frac{\Gamma(\frac{k+1}{2})\sqrt{\pi}}{\Gamma(k/2+1)} \\
&= \frac{2\pi^{1+(n-2)/2}a^n}{n}\,\frac{\Gamma(1)\Gamma(3/2)\Gamma(4/2)\cdots\Gamma((n-1)/2)}{\Gamma(3/2)\Gamma(4/2)\Gamma(5/2)\cdots\Gamma(n/2)} \\
&= \frac{2\pi^{n/2}a^n}{n\Gamma(n/2)} = \frac{\pi^{n/2}a^n}{(n/2)\Gamma(n/2)} \\
&= \frac{\pi^{n/2}a^n}{\Gamma(n/2+1)}.
\end{aligned}$$

□

9. Show that $\sigma(S^{n-1})$ is the usual surface measure of S^{n-1}; i.e.,

$$\sigma(S^{n-1}) = \lim_{h\to 0+}\frac{1}{h}\lambda((1+h)B_n - B_n)).$$

Solution: By Lemma 2.24,

$$\begin{aligned}\lambda((1+h)B_n) &= \lambda(\psi((0,1+h)\times S^{n-1}))\\ &= \sigma(S^{n-1})\int_0^{1+h} r^{n-1}dr\\ &= \frac{(1+h)^n}{n}\sigma(S^{n-1}).\end{aligned}$$

Thus

$$\begin{aligned}\frac{\lambda((1+h)B_n)-\lambda(B_n)}{h} &= \frac{1}{h}\left(\frac{(1+h)^n}{n}-\frac{1}{n}\right)\sigma(S^{n-1})\\ &= \frac{1}{n}\sum_{k=1}^{n}\binom{n}{k}h^{k-1}\sigma(S^{n-1}) \to \sigma(S^{n-1}) \text{ as } h\to 0\,.\end{aligned}$$

□

10. Show the measure σ is invariant under rotations.

Solution: We note since orthogonal matrices of determinant 1 give rotations, Corollary 2.23 shows if Q is a Borel subset of $\mathbb{R}^n$ and T is a rotation, TQ and Q have the same volume. In fact,

$$\begin{aligned}\lambda(QT) &= \int 1_{QT}(x)\,d\lambda(x)\\ &= \int 1_Q(xT^{-1})\,d\lambda(x)\\ &= \det(T)\int 1_Q(x)\,d\lambda(x)\\ &= \lambda(Q)\,.\end{aligned}$$

By Lemma 2.24, if W is a Borel subset of S^{n-1}, then since $(\psi((0,1]\times W))T = \psi((0,1]\times(WT))$, we see

$$\frac{1}{n}\sigma(WT) = \frac{1}{n}\sigma(W)\,.$$

□

11. Let $\alpha > 0$ and $1 \leqslant p < \infty$. Show that the Chebyshev's inequality

$$\lambda_n\left(\{x\in\mathbb{R}^n \mid |f(x)| > \alpha\}\right) \leqslant \left(\frac{|f|_p}{\alpha}\right)^p$$

holds.

Solution: For $\alpha > 0$, we have

$$\begin{aligned}|f|_p^p &= \int_{|f(x)|>\alpha}|f(x)|^p\,d\lambda_n(x) + \int_{|f(x)|\leqslant\alpha}|f(x)|^p\,d\lambda_n(x)\\ &\geqslant \int_{|f(x)|>\alpha}\alpha^p\,d\lambda_n(x)\\ &= \alpha^p\lambda_n(\{x \mid |f(x)|>\alpha\})\,.\end{aligned}$$

So

$$\lambda_n(\{x \mid |f(x)| > \alpha\}) \leqslant \left(\frac{|f|_p}{\alpha}\right)^p.$$

□

12. Show Corollary 2.26 follows from Proposition 2.25.
Solution: From Proposition 2.25, one can easily check

$$\int \psi(x)\, d\lambda_n(x) = \int_0^\infty \int_{S^{n-1}} r^{n-1}\psi(rs)\, d\sigma(s)\, dr$$

for any nonnegative simple function ψ on $\mathbb{R}^n$.

Now Corollary 2.26 holds if it holds for any positive Lebesgue integrable function f on $\mathbb{R}^n$. But for such a function f, there is a nondecreasing sequence $\{\psi_k\}$ of nonnegative simple functions satisfying $\psi_k(x) \to f(x)$ for each $x \in \mathbb{R}^n$. By the monotone convergence theorem,

$$\begin{aligned}\int f(x)\, d\lambda_n(x) &= \lim_{k\to\infty} \int \psi_k(x)\, d\lambda_n(x)\\ &= \lim_{k\to\infty} \int_0^\infty \int_{S^{n-1}} r^{n-1}\psi_k(rs)\, d\sigma(s)\, dr\\ &= \int_0^\infty \int_{S^{n-1}} r^{n-1} f(rs)\, d\sigma(s)\, dr.\end{aligned}$$

□

13. Let $\mathcal{H}$ be a Hilbert space and $T \in \mathcal{B}(\mathcal{H})$. Show $\ker(T^*) = T(\mathcal{H})^\perp$.
Solution:

Note $v \in \ker T^*$ if and only if $T^*v = 0$ if and only if $(T^*v, w) = 0$ for all $w \in \mathcal{H}$ if and only if $(v, Tw) = 0$ for all $w \in \mathcal{H}$ if and only if $v \in T(\mathcal{H})^\perp$.

□

14. Let $T \in \mathcal{B}(\mathcal{H}, \mathcal{K})$ where $\mathcal{H}$ and $\mathcal{K}$ are Hilbert spaces. Show $\mathcal{H} = \ker T \oplus \overline{T^*(\mathcal{K})}$ is an orthogonal decomposition of $\mathcal{H}$.
Solution:

First we note, $\ker T = T^*(\mathcal{K})^\perp$. In fact $v \in \ker T$ if and only if $Tv = 0$ if and only if $(Tv, w) = 0$ for all $w \in \mathcal{K}$ if and only if $(v, T^*w) = 0$ for all $w \in \mathcal{K}$ if and only if $v \in T^*(\mathcal{K})^\perp$. Thus the sum $\ker T + \overline{T^*(\mathcal{K})}$ is orthogonal and direct. To finish we show this sum is $\mathcal{H}$. First note it is a closed subspace of $\mathcal{H}$. Indeed, if $x_n = v_n + v_n'$ gives a Cauchy sequence with $v_n \in \ker T$ and $v_n' \in \overline{T^*(\mathcal{K})}$, then since $||x_n||_{\mathcal{H}}^2 = ||v_n||_{\mathcal{H}}^2 + ||v_n'||^2$, both $\{v_n\}$ and $\{v_n'\}$ are Cauchy. Since the spaces $\ker T$ and $\overline{T^*(\mathcal{K})}$ are closed, the limits v and v' are in $\ker T$ and $\overline{T^*(\mathcal{K})}$. Thus $x_n \to x \equiv v + v' \in \ker T \oplus \overline{T^*(\mathcal{K})}$. To see this subspace is all of $\mathcal{H}$, note if y is perpendicular to this subspace, then $y \in T^*(\mathcal{K})^\perp$ and so $y \in \ker T \cap (\ker T)^\perp = \{0\}$. So $y = 0$.

□

15. Suppose $\{e_\alpha\}_{\alpha\in A}$ is an orthonormal basis of $\mathcal{H}$ and w_α for $\alpha \in A$ are vectors in a Hilbert space $\mathcal{K}$. Show if $\sum_{\alpha\in A} ||w_\alpha||^2 < \infty$, the sums $T = \sum_\alpha w_\alpha \otimes \bar{e}_\alpha$ and $S = \sum e_\alpha \otimes \bar{w}_\alpha$ converge in operator norm and define Hilbert-Schmidt operators T and S with $T^* = S$ and $Te_\alpha = w_\alpha$ for all $\alpha \in A$. In particular, T is a Hilbert-Schmidt operator if and only if $T = \sum Te_\alpha \otimes \bar{e}_\alpha$ where $\sum ||Te_\alpha||^2_{\mathcal{K}} < \infty$.
Solution:

We first show the $\sum_\alpha w_\alpha \otimes \bar{e}_\alpha$ converges in norm to a bounded operator T. Let F and F' be finite subsets of A with $F \subseteq F'$. Then for a unit vector u in $\mathcal{H}$, by Parseval's and Cauchy-Schwarz inequalities, we have

$$\begin{aligned} || \sum_{\alpha\in F'} (w_\alpha \otimes \bar{e}_\alpha)(u) - \sum_{\alpha\in F} (w_\alpha \otimes \bar{e}_\alpha)(u)|| &= || \sum_{\alpha\in F'\setminus F} (u, e_\alpha) w_\alpha|| \\ &\leqslant \sum_{\alpha\notin F} |(u, e_\alpha)| \, ||w_\alpha||_{\mathcal{K}} \\ &\leqslant (\sum_\alpha |(u, e_\alpha)|^2)^{1/2} (\sum_{\alpha\notin F} ||w_\alpha||^2)^{1/2} \\ &\leqslant ||u||_{\mathcal{H}} (\sum_{\alpha\notin F} ||w_\alpha||^2)^{1/2} . \end{aligned}$$

This converges to 0 as $F \nearrow$ and defines an operator on $\mathcal{H}$ which satisfies

$$|| \sum (w_\alpha \otimes \bar{e}_\alpha)(u)|| \leqslant ||u||_{\mathcal{H}} \left(\sum ||w_\alpha||^2_{\mathcal{K}}\right)^{1/2} .$$

Thus the operator is bounded. Clearly this operator T satisfies $Te_\alpha = w_\alpha$. Similarly, since by Cauchy-Schwarz,

$$\begin{aligned} || \sum_\alpha (e_\alpha \otimes \bar{w}_\alpha)(u)||_{\mathcal{H}} &= (\sum_\alpha |(u, w_\alpha)|^2)^{1/2} \\ &\leqslant (\sum_\alpha ||w_\alpha||^2)^{1/2} , \end{aligned}$$

the operator $S = \sum e_\alpha \otimes \bar{w}_\alpha$ is bounded. Finally

$$\begin{aligned} (T^* w, v) &= (w, Tv) \\ &= (w, \sum (v, e_\alpha) w_\alpha) \\ &= \sum (e_\alpha, v)(w, w_\alpha) \\ &= \sum ((w, w_\alpha) e_\alpha, v) \\ &= (\sum_\alpha (w, w_\alpha) e_\alpha, v) \\ &= (\sum (e_\alpha \otimes \bar{w}_\alpha) w, v) \\ &= (Sw, v) . \end{aligned}$$

So $T^* = S$.

□

16. Let K and L be Hilbert-Schmidt operators on $L^2(\mathbb{R}^n)$ with L^2 kernels R and S. Show $\int R(x,z)S(z,y)\,dz$ is an L^2 kernel for KL.

Solution:

Let M be the operator KL. Thus for $f \in L^2(\mathbb{R}^n)$,

$$\begin{aligned} Mf(z) &= K(Lf)(z) \\ &= \int R(z,y)Lf(y)\,dy \\ &= \int R(z,y) \int S(y,x)f(x)\,dx\,dy \end{aligned}$$

for a.e. z. Now we already know that $y \mapsto \int S(y,x)f(x)\,dx$ is in L^2 and $x \mapsto S(y,x)f(x)$ is integrable for a.e. y. Next note for a.e. pair (z,x), the function $y \mapsto R(z,y)S(y,x)$ is integrable. Indeed, by Fubini, since for a.e. z, $y \mapsto R(z,y)$ is L^2 and for a.e. x, $y \mapsto S(y,x)$ is L^2, we see by Cauchy-Schwarz that for a.e. (z,x), $y \mapsto R(z,y)S(y,x)$ is integrable. Furthermore, the function T given by

$$T(z,x) = \int R(z,y)S(y,x)\,dy$$

is an L^2 function, for

$$\begin{aligned} \iint |T(z,x)|^2\,dz\,dx &= \iint \left| \int R(z,y)S(y,x)\,dy \right|^2 dz\,dx \\ &\leqslant \iint \left(\int |R(z,y)|^2\,dy \right) \left(\int |S(y,x)|^2\,dy \right) dz\,dx \\ &= \iint |R(z,y)|^2\,dz\,dy \iint |S(y,x)|^2\,dy\,dx \\ &= |R|_2^2\,|S|_2^2 < \infty\,. \end{aligned}$$

So T defines a Hilbert-Schmidt operator N. We show it is M.

We do this by showing first it is true if R is continuous and has compact support. Note $R(z,y)S(y,x)f(x)$ is integrable in x and y for a.e. z. Indeed, $\int |S(y,x)f(x)|\,dx$ is an L^2 function in y and so for all z, $R(z,y)\int |S(y,x)f(x)|\,dx$ is integrable. So $\int |R(z,y)| \int |S(y,x)f(x)|\,dx\,dy$ is finite for all z. Hence by Fubini,

$$\begin{aligned} Nf(z) &= \int T(z,x)f(x)\,dx \\ &= \int\int R(z,y)\,S(y,x)\,dy\,f(x)\,dx \\ &= \int\int R(z,y)S(y,x)f(x)\,dx\,dy \\ &= \int R(z,y)Lf(y)\,dy \\ &= KLf(z) \\ &= Mf(z) \end{aligned}$$

for a.e. z. Now for general R, we know if $R \in L^2(\mathbb{R}^n \times \mathbb{R}^n)$, there is a sequence $\{R_j\}$ in $C_c(\mathbb{R}^n \times \mathbb{R}^n)$ which converges in L^2 to R. Now if K_j is the Hilbert-Schmidt operator with kernel R_j, we have

$$||K_jL - KL|| = ||K_j - K||\,||L|| \leqslant ||K_j - K||_2\,||L|| = |R_j - R|_2\,||L|| \to 0 \text{ as } j \to \infty.$$

Thus as L^2 limits,

$$\begin{aligned} Mf(z) &= KLf(z) \\ &= \lim_j K_jLf(z) \\ &= \lim_j \int T_j(z,x)f(x)\,dx \\ &= \int T(z,x)f(z)\,dx \\ &= Nf(z). \end{aligned}$$

Note here we have used $T_j \to T$ in L^2. This follows easily:

$$\begin{aligned} |T_j - T|_2^2 &= \int\int |T_j(z,x) - T(z,x)|^2\,dz\,dx \\ &\leqslant \iint \left(\int |(R_j(z,y) - R(z,y))S(y,x)|\,dy \right)^2 dz\,dx \\ &\leqslant \iint \left(\int |R_j(z,y) - R(z,y)|^2\,dy \right) \left(\int |S(y,x)|^2\,dy \right) dz\,dx \\ &= |R_j - R|_2^2\,|S|_2^2 \to 0. \end{aligned}$$

□

17. Let (X,μ) and (Y,ν) be σ-finite measure spaces. Suppose T is a Hilbert-Schmidt operator from $L^2(Y,\nu)$ into $L^2(X,\mu)$ with kernel K. Show T^* is Hilbert-Schmidt and has kernel K^* where $K^*(y,x) = \overline{K(x,y)}$.

Solution:

Let T be the Hilbert-Schmidt operator with $L^2(X \times Y)$ kernel $K(x,y)$. Thus $Tf(x) = \int K(x,y)f(y)\,d\nu(y)$. Now if $g \in L^2(X,\mu)$, one has since we have already seen $K(x,y)f(y)\bar{g}(x)$ is integrable that

$$\begin{aligned} (Tf,g)_{L^2(\mu)} &= \int \left(\int K(x,y)f(y)\,d\nu(y) \right) \bar{g}(x)\,d\mu(x) \\ &= \iint K(x,y)f(y)\bar{g}(x)\,d\nu(y)\,d\mu(x) \\ &= \iint f(y)\overline{\bar{K}(x,y)g(x)}\,d\mu(x)\,d\nu(y) \\ &= \int f(y)\overline{\int \overline{K(x,y)}g(x)\,d\mu(x)}\,d\nu(y) \\ &= (f,Sg)_{L^2(\nu)} \end{aligned}$$

where S is the Hilbert-Schmidt operator with kernel $K^*(y,x) = \overline{K(x,y)}$.

□

18. Let (X,μ) and (Y,ν) be σ-finite measure spaces. Let $e_\alpha(\cdot)$, $\alpha \in A$ and $f_\beta(\cdot)$, $\beta \in B$ be complete orthonormal bases of $L^2(X)$ and $L^2(Y)$, respectively. Show the functions $e_\alpha \times f_\beta$ defined by $e_\alpha \times f_\beta(x,y) = e_\alpha(x) f_\beta(y)$ form a complete orthonormal basis of $L^2(X \times Y, \mu \times \nu)$.

Solution: We have μ is σ finite. We show $e_\alpha \times f_\beta$ is an o.n. basis of $L^2(X \times Y)$. It is clearly orthonormal. The closure of its linear span contains all $F(x)G(y)$ where $F \in L^2(X)$, $G \in L^2(Y)$. So suppose $f(x,y)$ perpendicular to all $F \times G$. Since μ is σ-finite, we can work on $X_0 \times Y_0$ where $\mu(X_0) < \infty$ and $\nu(Y_0) < \infty$. Wolog μ and ν are finite. We thus have $f \in L^1(X \times Y)$ and $\int_{E \times F} f(x,y)\, d\mu \times \nu = 0$ for all measurable E and F. Thus $\int_A f d\mu \times \nu = 0$ for all A in the algebra $\mathcal{A}$ of finite disjoint unions of measurable rectangles. By the Lebesgue Dominated Convergence Theorem, $\int_W f d\mu \times \nu = 0$ for all $W \in \mathcal{A}_\sigma$. Again we get by the Lebesgue Dominated Convergence Theorem, $\int_W f\, d\mu \times \nu = 0$ for all $W \in \mathcal{A}_{\sigma\delta}$. But if E is measurable in the product space, there is a $W \in \mathcal{A}_{\sigma\delta}$ with $E \subseteq W$ $\mu \times \mu(W - E) = 0$. Thus $\int_E f\, d\mu \times \nu = 0$ for all measurable E. Thus $f = 0$. □

19. Let v and w be vectors in a Hilbert space $\mathcal{H}$ and let A and B be bounded linear operators on $\mathcal{H}$. Show: $||v \otimes \bar{w}|| = ||v||\,||w|| = ||v \otimes \bar{w}||_2$, $(v \otimes \bar{w})^* = w \otimes \bar{v}$, and $A(v \otimes \bar{w})B^* = Av \otimes \overline{Bw}$.

Solution: The results are clearly true if v or w is 0. So assume $v \neq 0$ and $w \neq 0$. Recall $v \otimes \bar{w}(u) = (u,w)v$. Thus $||v \otimes \bar{w}(u)|| = |(u,w)|\,||v|| \leqslant ||u||\,||v||\,||w||$. Thus $||v \otimes \bar{w}|| \leqslant ||v||\,||w||$. Since $||v \otimes \bar{w}(w)|| = ||w||^2\,||v|| \leqslant ||v \otimes \bar{w}||\,||w||$, we also have $||v \otimes \bar{w}|| \geqslant ||v||\,||w||$. Hence $||v \otimes \bar{w}|| = ||v||\,||w||$.

Since

$$
\begin{aligned}
(v \otimes \bar{w}(u), u') &= ((u,w)v, u') \\
&= (v, u')(u, w) \\
&= (u, (u', v)w) \\
&= (u, w \otimes \bar{v}(u')),
\end{aligned}
$$

we have

$$(v \otimes \bar{w})^* = w \otimes \bar{v}\,.$$

Now let A and B be bounded linear operators. One has

$$
\begin{aligned}
A(v \otimes \bar{w})B^* u &= (B^* u, w)Av \\
&= (u, Bw)Av \\
&= (Av \otimes \overline{Bw}\,)u\,.
\end{aligned}
$$

Finally, the Hilbert-Schmidt norm can be calculated by taking an orthonormal basis $\{e_\alpha\}$ with one of the $e_\alpha = \frac{w}{||w||}$. Then

$$\begin{aligned} ||v \otimes \bar{w}||_2^2 &= \sum_\alpha ||(v \otimes \bar{w})(e_\alpha)||^2 \\ &= \sum_\alpha ||(e_\alpha, w)v||^2 \\ &= \frac{(w,w)^2}{||w||^2} ||v||^2 \\ &= ||v||^2 \, ||w||^2 . \end{aligned}$$

□

20. Recall if $f,\, h \in L^2(\mathbb{R}^n)$, then $f \otimes h$ is the rank one operator given by

$$(f \otimes h)(g)(x) = \left[\int g(y)h(y)\, dy \right] f(x).$$

Let $\{h_j\}$ be an orthonormal basis. Show T is a Hilbert-Schmidt on $L^2(\mathbb{R}^n)$ if and only if $T = \sum_{j,k} \lambda_{j,k} h_j \otimes h_k$ where $\sum |\lambda_{j,k}|^2 < \infty$.
Solution:

Let $\{h_j\}$ be an orthonormal basis of $L^2(\mathbb{R}^n)$. By Proposition 2.42, the vectors $h_j \otimes \bar{\bar{h}}_k = h_j \otimes h_k$ form an orthonormal basis of $\mathcal{B}_2(L^2(\mathbb{R}^n))$. So every Hilbert-Schmidt operator T on $L^2(\mathbb{R}^n)$ has form $T = \sum_{j,k} \lambda_{j,k} h_j \otimes h_k$ where $\sum |\lambda_{j,k}|^2 = ||T||_2^2 \geqslant ||T||^2$.

□

21. Show T is a Hilbert-Schmidt operator on $L^2(\mathbb{R}^n)$ if and only if T has form $T = \sum f_j \otimes h_j$ where $f_j,\, h_j \in L^2(\mathbb{R}^n)$, the sequence $\{h_j\}_{j=1}^\infty$ is orthogonal, and $\sum ||f_j||^2 ||h_j||^2 < \infty$. In this case $||T||_2^2 = \sum ||f_j||^2 ||h_j||^2$.
Solution: Note first we may assume all h_j are not 0. We will take a complete orthonormal base $\{e_i\}$ of $L^2(\mathbb{R}^n)$ such that the orthonormal vectors $\frac{h_j}{||h_j||_2}$ lie amongst the e_i. Now if T is Hilbert-Schmidt, we know $\sum ||T\bar{e}_i||_2^2 < \infty$ and thus by Exercise 2.2.15, $\sum T\bar{e}_i \otimes e_i$ is Hilbert-Schmidt and

$$\sum T\bar{e}_i \otimes e_i(\bar{e}_j) = \sum (\bar{e}_j, \bar{e}_i)_2 T\bar{e}_i = \sum_i (e_i, e_j)_2 T\bar{e}_i = T\bar{e}_j.$$

Since they agree on the orthonormal basis $\{\bar{e}_i\}$, $T = \sum_i T\bar{e}_i \otimes e_i$.

Conversely, if $T = \sum f_j \otimes h_j$ where $\sum ||f_j||_2^2 \, ||h_j||_2^2 < \infty$, then

$$T = \sum_j ||h_j||_2 f_j \otimes \frac{h_j}{||h_j||_2}$$

and in this case $\sum_i ||T\bar{e}_i||^2 = \sum_j ||h_j||_2^2 ||f_j||_2^2 < \infty$. So T is Hilbert-Schmidt.

□

22. Find a bounded linear transformation T that cannot be written in form $\sum v_i \otimes \bar{w}_i$ where this series converges in the norm topology.

Solution:

Let $\mathcal{H}$ be a separable infinite dimensional Hilbert space. Then there is a countable orthonormal basis $\{e_i\}_{i=1}^{\infty}$. Let I be the identity operator on $\mathcal{H}$. It is clearly bounded. Assume $I = \sum_{j=1}^{\infty} v_j \otimes \bar{w}_j$ where this sum converges in the operator topology. Define $T_n = \sum_{j=1}^{n} v_j \otimes \bar{w}_j$. Then T_n is a bounded operator with finite dimensional range. Thus by Theorem 2.46 and Proposition 2.48, the norm limit I would be a compact operator. But I is not compact for the sequence $\{Ie_i\}_{i=1}^{\infty}$ has no convergent subsequence. So I cannot be written as $\sum v_j \otimes \bar{w}_j$ in the norm topology.

□

23. Let $\mathcal{H}$ be a Hilbert space and let $\mathcal{B}(\mathcal{H})$ be the Banach space of bounded linear operators on $\mathcal{H}$. For each $v \in \mathcal{H}$, define a seminorm $|\cdot|_v$ on $\mathcal{B}(\mathcal{H})$ by

$$|T|_v = ||Tv||_{\mathcal{H}}.$$

The locally convex vector space topology defined on $\mathcal{B}(\mathcal{H})$ by the seminorms $|\cdot|_v$ is called the strong operator topology on $\mathcal{B}(\mathcal{H})$.

(a) Show the strong operator topology on $\mathcal{B}(\mathcal{H})$ is weaker than the norm topology on $\mathcal{B}(\mathcal{H})$.

(b) Show if T is a bounded linear operator on $\mathcal{H}$ and $\{e_\alpha\}_{\alpha\in A}$ is a complete orthonormal basis, then

$$T = \sum_{\alpha} Te_\alpha \otimes \bar{e}_\alpha$$

in the strong operator topology.

Solution:

For (a), it suffices to show every neighborhood of 0 in the strong operator topology is a neighborhood of 0 in the norm topology. So let $v_1, v_2, \ldots, v_k$ be vectors in $\mathcal{H}$ and consider

$$N = \{T \mid ||Tv_i|| < \epsilon,\ i = 1, 2, \ldots, k\}.$$

Choose $\delta > 0$ so that $||T|| < \delta$ implies $||Tv_i|| < \epsilon$ for $i = 1, 2, \ldots, k$; i.e. take $\delta = \min_i \frac{\epsilon}{||v_i||}$. Then if $||T|| < \delta$, $T \in N$.

For (b) it suffices to show $||T - \sum_{\alpha\in F} Te_\alpha \otimes \bar{e}_\alpha||_v \to 0$ as $F \nearrow$ where the F is a finite subset of A. But

$$\begin{aligned} ||T - \sum_{\alpha\in F} Te_\alpha \otimes \bar{e}_\alpha||_v &= ||Tv - \sum_{\alpha\in F} (v, e_\alpha)Te_\alpha|| \\ &= ||T(v - \sum_{\alpha\in F} (v, e_\alpha)e_\alpha)|| \\ &\leqslant ||T||\,||v - \sum_{\alpha\in F} (v, e_\alpha)e_\alpha|| \end{aligned}$$

which converges to 0 for $\sum_{\alpha\in F}(v, e_\alpha)e_\alpha$ converges to v in $\mathcal{H}$.

□

24. Show the set of trace class operators form an ideal in $\mathcal{B}(\mathcal{H})$.
Solution:

By Corollary 2.41, the left or right product of a trace class operator with any operator in $\mathcal{B}(\mathcal{H})$ is trace class operator and by Definition 2.37, the sum of two trace class operators is trace class and the product of a trace class operator by a scalar is trace class.

□

25. Show T^* is trace class if T is trace class and $\mathrm{Tr}(T^*) = \overline{\mathrm{Tr}(T)}$.
Solution:

By Proposition 2.40, T is trace class if and only if $T = RS$ for some Hilbert-Schmidt operators R and S. Thus $T^* = S^*R^*$. So since by Corollary 2.33, R^* and S^* are both Hilbert-Schmidt, the operator T^* is trace class. Now by Proposition 2.39,

$$\mathrm{Tr}(T^*) = \sum_\alpha (T^*e_\alpha, e_\alpha) = \sum_\alpha (e_\alpha, Te_\alpha) = \overline{\sum_\alpha (Te_\alpha, e_\alpha)} = \overline{\mathrm{Tr}(T)}$$

where $\{e_\alpha\}$ is an orthonormal basis of $\mathcal{H}$.

□

26. Give another proof of Theorem 2.46 by showing that if $\{T_n\}$ is a sequence of compact operators converging in norm to T, then the closure of the set $\{Tv \mid ||v|| = 1\}$ is totally bounded. (Recall a subset of a metric space is compact if and only if it is complete and totally bounded.)
Solution:

Let T_n be a sequence of compact operators converging in norm to T. We show the closure of $S = \{Tv \mid ||v|| = 1\}$ is totally bounded. If not there is an $\epsilon > 0$ such that no finite set of neighborhoods of radius ϵ cover S. Pick v_1 with length 1. Since $S \nsubseteq B_\epsilon(Tv_1)$, there is a v_2 with $||v_2|| = 1$ and $Tv_2 \notin B_\epsilon(Tv_1)$. Since $S \nsubseteq B_\epsilon(Tv_1) \cup B_\epsilon(Tv_2)$, we can choose a v_3 of length 1 with $Tv_3 \notin B_\epsilon(Tv_1) \cup B_\epsilon(Tv_2)$. Repeating inductively, we can find a sequence $\{v_n\}$ of unit vectors with $Tv_k \notin \cup_{j=1}^{k-1} B_\epsilon(Tv_j)$. This implies that whenever $k < l$, $||Tv_l - Tv_k|| \geqslant \epsilon$ and thus the sequence $\{Tv_n\}_{n=1}^\infty$ has no convergent subsequence, a contradiction of the compactness of the operator T.

□

27. Let $\mathcal{H}$ be a Hilbert space. Show the adjoint of any finite rank operator on $\mathcal{H}$ is a finite rank operator.
Solution:

We first note that a finite rank operator T has form $\sum_{k=1}^m v_k \otimes \bar{w}_k$ for some finite set of vectors. Indeed, since T is continuous, the kernel $\ker T$ is a closed linear subspace of $\mathcal{H}$ and thus T maps $(\ker T)^\perp$ onto the range of T for $T(\mathcal{H}) = T(\ker T \oplus (\ker T)^\perp) = T((\ker T)^\perp)$. Hence there is an orthonormal basis $e_1, e_2, \ldots, e_m$ of $(\ker T)^\perp$ such that $w_1 = Te_1$, $w_2 = Te_2, \ldots, Te_m$ form a basis of the range of T. This implies

$$T = \sum_{k=1}^{m} w_k \otimes \bar{e}_k$$

for the operator $\sum_{k=1}^{m} w_k \otimes \bar{e}_k$ is zero on $\ker T$ and agrees with T on $(\ker T)^{\perp}$ for

$$\sum_{k=1}^{m} w_k \otimes \bar{e}_k(e_j) = \sum_{k=1}^{m} (e_j, e_k) w_k = w_j = Te_j.$$

Now by Exercise 2.2.19,

$$T^* = \left(\sum_{k=1}^{m} w_k \otimes \bar{e}_k \right)^* = \sum_{k=1}^{m} e_k \otimes \bar{w}_k$$

which is a finite rank operator.

□

28. Let $\mathcal{H}$ be a Hilbert space. Show the finite rank operators on $\mathcal{H}$ form a norm dense subspace of the space of compact operators.
Solution:

By Theorem 2.52, we know if a compact operator T does not have finite dimensional range, there is a decreasing sequence λ_n of positive numbers converging to 0 and orthonormal sequences $\{e_n\}$ and $\{f_n\}$ in $\mathcal{H}$ such that $T = \sum_{k=1}^{\infty} \lambda_k e_k \otimes \bar{f}_k$. Set $T_n = \sum_{k=1}^{n} \lambda_k e_k \otimes \bar{f}_k$. Then $T - T_n = \sum_{k=n+1}^{\infty} \lambda_k e_k \otimes \bar{f}_k$. Thus by Bessel's inequality,

$$\begin{aligned} ||(T - T_n)v||^2 &= || \sum_{k=n+1}^{\infty} \lambda_k (v, f_k) e_k ||^2 \\ &= \sum_{k=n+1}^{\infty} |\lambda_k|^2 |(v, f_k)|^2 \\ &\leqslant |\lambda_{n+1}|^2 \sum_{k=n+1}^{\infty} |(v, f_k)|^2 \\ &\leqslant |\lambda_{n+1}|^2 \, ||v||^2 \end{aligned}$$

and hence $||T - T_n|| \leqslant |\lambda_{n+1}| \to 0$ as $n \to \infty$.

□

29. Let T be a bounded linear operator on a Hilbert space $\mathcal{H}$. Show:

(a) The operator T is a Hilbert-Schmidt operator if and only if T has form

$$T = \sum \lambda_{i,j} e_i \otimes \bar{f}_j$$

where $\sum |\lambda_{i,j}|^2 < \infty$ and the sequences $\{e_i\}$ and $\{f_j\}$ are orthonormal.

(b) The operator T is a trace class operator if and only if T has form

$$T = \sum \lambda_{i,j} e_i \otimes \bar{f}_j$$

where $\sum |\lambda_{i,j}| < \infty$ and the sequences $\{e_i\}$ and $\{f_i\}$ are orthonormal.

Solution:

By Corollary 2.49, we know every Hilbert-Schmidt operator is compact. By the structure theorem 2.52 these operators have form $\sum_k \lambda_k e_k \otimes \bar{f}_k$ where $|\lambda_k| \searrow 0$ and the sequences $\{e_k\}$ and $\{f_k\}$ are orthonormal. If T is a Hilbert-Schmidt operator of this form, then

$$||T||_2^2 \geqslant \sum_n ||Tf_n||^2 = \sum_n |\lambda_n|^2 .$$

Thus T has the desired form. For the reverse argument, assume $T = \sum_{i,j} \lambda_{i,j} e_i \otimes \bar{f}_j$ where $\sum_{i,j} |\lambda_{i,j}|^2 < \infty$. By taking $\lambda_{i,j} = 0$ for an extended set of f_j's, we may assume $\{f_j\}$ is an orthonormal basis. Now

$$\begin{aligned}
\sum_k ||Tf_k||_2^2 &= \sum_k || \sum_{i,j} \lambda_{i,j} e_i \otimes \bar{f}_j(f_k)||^2 \\
&= \sum_k || \sum_i \lambda_{i,j} e_i ||^2 \\
&= \sum_k \sum_i |\lambda_{i,j}|^2 < \infty.
\end{aligned}$$

So T is Hilbert-Schmidt.

Now by Proposition 2.40, if T is trace class, $T = RS$ where R and S are Hilbert-Schmidt. Moreover, as seen in the proof of this proposition, we may assume $R = \sum v_i \otimes \bar{e}_i$, $S = \sum e_i \otimes \bar{w}_i$, the sequence $\{e_i\}$ is orthonormal and $\sum ||v_i||^2 < \infty$, $\sum ||w_i||^2 < \infty$. Also by increasing the size of the collection $\{e_i\}$ and taking the additional v_i and w_i to be 0, we may assume all the vectors v_i and w_j are in the closure of the linear span of the e_i's. Thus if $\mu_{i,j} = (v_j, e_i)$ and $\nu_{i,j} = (w_j, e_i)$, we have by Parseval's equality, $\sum_i |\mu_{i,j}|^2 = ||v_j||^2$ and $\sum_i |\nu_{i,j}|^2 = ||w_j||^2$. Moreover,

$$R = \sum_j \left(\sum_i (v_j, e_i) e_i \right) \otimes \bar{e}_j = \sum_{i,j} \mu_{i,j} e_i \otimes \bar{e}_j$$

and

$$S = \sum_j e_j \otimes \overline{\sum_i (w_j, e_i) e_i} = \sum_{i,j} \bar{\nu}_{j,i} e_j \otimes \bar{e}_i .$$

These sum interchanges can be verified using the square summability of the sets

$\{|\mu_{i,j}|\}$ and $\{|\nu_{i,j}|\}$. Thus

$$\begin{aligned} T = RS &= \left(\sum_{i,m} \mu_{i,m} e_i \otimes \bar{e}_m\right)\left(\sum_{k,j} \bar{\nu}_{k,j} e_k \otimes \bar{e}_j\right) \\ &= \sum_i \sum_j \sum_m \mu_{i,m}\bar{\nu}_{m,j} e_i \otimes \bar{e}_j \\ &= \sum_{i,j} \lambda_{i,j} e_i \otimes \bar{e}_j \end{aligned}$$

where $\lambda_{i,j} = \sum_m \mu_{i,m}\bar{\nu}_{m,j}$. Now note

$$\begin{aligned} \sum_{i,l} |\lambda_{i,j}| &\leqslant \sum_{i,j} \sum_m |\mu_{i,m}\bar{\nu}_{m,j}| \\ &\leqslant \sum_{i,j} \left(\sum_m |\mu_{i,m}|^2\right)^{1/2} \left(\sum_m |\nu_{m,j}|^2\right)^{1/2} \\ &= \sum_{i,j} ||v_i||\,||w_j|| \\ &\leqslant \left(\sum ||v_i||^2\right)^{1/2} \left(\sum_j ||w_j||^2\right)^{1/2} < \infty . \end{aligned}$$

Thus T has the desired form.

Conversely, suppose $T = \sum_{i,j} \lambda_{i,j} e_i \otimes \bar{f}_j$ where $\sum |\lambda_{i,j}| < \infty$ and the sequences $\{e_i\}$ and $\{f_j\}$ are orthonormal. Define vector $v_{i,j} = \frac{\lambda_{i,j}}{\sqrt{|\lambda_{i,j}|}} e_i$ and vector $w_{i,j} = \sqrt{|\lambda_{i,j}|} f_j$. Then $T = \sum_{i,j} v_{i,j} \otimes \bar{w}_{i,j}$. We also have

$$\sum_{i,j} ||v_{i,j}||^2 = \sum_{i,j} ||w_{i,j}||^2 = \sum_{i,j} |\lambda_{i,j}| < \infty .$$

So by definition, T is trace class.

□

30. Let F be a measurable subset of $\mathbb{R}^n$. Let E be the set of x in F such that there is a $\delta > 0$ with $\lambda(B_\delta(x) \cap F) = 0$. Show that E has measure 0. (**Hint:** Take a countable base for the topology of $\mathbb{R}^n$.)

Solution:

Let F be a measurable subset of $\mathbb{R}^n$ and let E be the points x in F such that $\lambda(B_\delta(x) \cap F) = 0$ for some $\delta > 0$. Consider a countable base for the topology of $\mathbb{R}^n$ consisting of the sets $B_{r_n}(y_n)$ where $n = 1, 2, 3, \ldots$. For each $x \in E$, pick a $n(x) \in \mathbb{N}$ with $x \in B_{r_{n(x)}}(y_{n(x)}) \subseteq B_\delta(x)$. Then $E \subseteq \cup_{x \in E} B_{r_{n(x)}}(y_{n(x)})$ and since $B_{r_{n(x)}}(y_{n(x)}) \cap F \subseteq B_\delta(x) \cap F$, $\lambda(B_{r_n(x)}(y_{n(x)}) \cap F) = 0$ for all $x \in E$. But the collection $\{B_{r_{n(x)}}(y_{n(x)})) \mid x \in E\}$ is countable and thus has form $\{B_{r_i}(y_i) \mid i = 1, 2, \ldots\}$. This implies $\lambda(E) \leqslant \sum_{i=1}^\infty \lambda(B_{r_i}(y_i) \cap F) = 0$.

□

Exercise Set 2.3

1. Show the function f defined by

$$f(t) = \begin{cases} 0 & \text{if } t \leqslant 0 \\ e^{-\frac{1}{t}} & \text{if } t > 0 \end{cases}$$

is C^∞ on $\mathbb{R}$.

Solution:

The only place that needs to be checked is $t = 0$. Indeed, by induction $f^{(n)}(t) = 0$ for all n and all $t < 0$ and $f^{(n)}(t) = P_n(\frac{1}{t})e^{-\frac{1}{t}}$ for some polynomial of degree n for $t > 0$.

We next note if Q is a polynomial $Q(\frac{1}{t})e^{-\frac{1}{t}} \to 0$ as $t \to 0+$. In fact, it suffices to note

$$\begin{aligned} \frac{1}{t^m}e^{-\frac{1}{t}} &= \frac{1}{t^m e^{\frac{1}{t}}} \\ &< \frac{1}{t^m\left(\frac{(m+1)!}{t^{m+1}}\right)} \\ &= \frac{t}{(m+1)!} \to 0 \text{ as } t \to 0+ \,. \end{aligned}$$

We now claim $f^{(n)}(0) = 0$ for all n. We have this for $n = 0$ and suppose true for n. Then since $\frac{f^{(n)}(h)-0}{h} = 0$ for $h < 0$, we need only note

$$\begin{aligned} \lim_{h\to 0+} \frac{f^{(n)}(h) - 0}{h} &= \lim_{h\to 0+} \frac{P_n(\frac{1}{h})e^{-\frac{1}{h}} - 0}{h} \\ &= \lim_{h\to 0+} \frac{1}{h} P_n(\frac{1}{h})e^{-\frac{1}{h}} = 0\,. \end{aligned}$$

So $f^{(n+1)}(0) = 0$ and we have all derivatives of f at 0 are 0. Finally

$$\lim_{t\to 0+} f^{(n)}(t) = \lim_{t\to 0+} P_n(\frac{1}{t})e^{-\frac{1}{t}} = 0 = f^{(n)}(0)\,.$$

Since the limit from the left is also 0, we have $f^{(n)}$ is continuous at 0.

□

2. Let $p(x)$ be a polynomial function and let $\lambda > 0$. Show the function $x \mapsto p(x)e^{-\lambda|x|^2}$ is in $\mathcal{S}(\mathbb{R}^n)$.

Solution:

A simple inductive argument shows $(D^\alpha)^{\text{th}}$ derivative of the function $f(x) = p(x)e^{-\lambda|x|^2}$ has form $f^{(\alpha)}(x) = Q_\alpha(x)e^{-\lambda|x|^2}$ for some polynomial Q_α of degree $\deg p + |\alpha|$. Thus f is C^∞. To see it is Schwartz, since $(1+|x|^2)^m$ is a polynomial, it suffices to check $x^\alpha e^{-\lambda|x|^2} \to 0$ as $|x| \to \infty$ for any monomial x^α. But

$|x^\alpha| = |x_1^{\alpha_1} x_2^{\alpha_2} \cdots x_n^{\alpha_n}| \leqslant \left(\sqrt{x_1^2 + x_2^2 + \cdots + x_n^2}\right)^{|\alpha|} = |x|^{|\alpha|}$. Thus it suffices to show $t^m e^{-\lambda t^2} \to 0$ as $t \to \infty$ for any $m \in \mathbb{N}$. This is a direct consequence of

$$t^m e^{-\lambda t^2} = \frac{t^m}{e^{\lambda t^2}} \leqslant \frac{(m+1)!t^{m+1}}{(\lambda t^2)^{m+1}} = \frac{(m+1)!}{\lambda^{m+1} t^{m+1}}$$

for $m \geqslant 0$, $\lambda > 0$, and $t > 1$.

□

3. Show that

$$\int_{\mathbb{R}^n} e^{-\pi|x|^2}\, dx = 1\,.$$

(Hint: Do the case $n = 2$ by using polar coordinates. Then use Fubini's Theorem to do the other cases.)

Solution:

We do $n = 2$. Then

$$\iint\limits_{\mathbb{R}^2} e^{-\pi(x^2+y^2)}\, dx\, dy = \int_0^\infty \int_0^{2\pi} e^{-\pi r^2} r\, d\theta\, dr$$
$$= 2\pi \left(\frac{-e^{-\pi r^2}}{2\pi}\right)\Bigg|_0^\infty$$
$$= 1.$$

Thus by Fubini,

$$\left(\int_{\mathbb{R}} e^{-\pi x^2}\, dx\right)^2 = \int_{\mathbb{R}} e^{-\pi x^2} \int_{\mathbb{R}} e^{-\pi y^2}\, dy\, dx$$
$$= \iint\limits_{\mathbb{R}^2} e^{-\pi(x^2+y^2)}\, dx\, dy$$
$$= 1\,.$$

So $\int_{\mathbb{R}} e^{-\pi x^2}\, dx = 1$. Hence by Fubini,

$$\int_{\mathbb{R}^n} e^{-\pi|x|^2}\, dx = \int_{\mathbb{R}} e^{-\pi x_1^2} dx_1 \int_{\mathbb{R}} e^{-\pi x_2^2}\, dx_2 \cdots \int_{\mathbb{R}} e^{-\pi x_n^2}\, dx_n = 1\,.$$

□

4. Show $\frac{1+|x+y|^2}{(1+|x|^2)(1+|y|^2)} \leqslant 2$.

Solution:

$$\frac{1+|x+y|^2}{(1+|x|^2)(1+|y|^2)} \leqslant \frac{1+|x|^2+2|x|\,|y|+|y|^2}{1+|x|^2+|y|^2}$$
$$\leqslant \frac{2+2|x|^2+2|y|^2}{1+|x|^2+|y|^2}$$
$$= 2$$

for $|x|^2 + 2|x|\,|y| + |y|^2 \leqslant |x|^2 + 2|x|\,|y| + |y|^2 + (|x| - |y|)^2 = 2|x|^2 + 2|y|^2$. □

5. Show that the following are equivalent:

- $f \in \mathcal{S}(\mathbb{R}^n)$;
- For all $\alpha, \beta \in \mathbb{N}_0^n$, $|f|_{\alpha,\beta} := \sup \left|D^\alpha (x^\beta f)(x)\right| < \infty$;
- For all $\alpha, \beta \in \mathbb{N}_0^n$, $|f|'_{\alpha,\beta} := \sup \left|x^\alpha D^\beta f(x)\right| < \infty$;
- For all polynomials p and all $\alpha \in \mathbb{N}_0^n$,

$$|f|_{p,\alpha} = \sup |p(x) D^\alpha f(x)| < \infty;$$

- For all polynomials p and all $\alpha \in \mathbb{N}_0^n$,

$$|f|_{\alpha,p} = \sup |D^\alpha (p(x) f(x))| < \infty;$$

- For all $N \in \mathbb{N}_0$ and $\alpha \in \mathbb{N}_0^n$, $|f|'_{N,\alpha} = \sup (1+|x|)^N |D^\alpha f(x)| < \infty$.

Furthermore show each family of seminorms $\{|\cdot|_{N,\alpha}\}$, $\{|\cdot|_{\alpha,\beta}\}$, $\{|\cdot|'_{\alpha,\beta}\}$, $\{|\cdot|_{p,\alpha}\}$, $\{|\cdot|_{\alpha,p}\}$, and $\{|\cdot|'_{N,\alpha}\}$ defines the same topology on $\mathcal{S}(\mathbb{R}^n)$.

Solution:

We show the locally convex topologies defined by each family of seminorms are all the same. To do this is equivalent to showing the neighborhoods of 0 defined by any one seminorm in one family is a neighborhood of 0 in the topology defined by the other family of seminorms.

Case 1. The seminorms $|\cdot|_{\alpha,\beta}$ and $|\cdot|_{\alpha,p}$

Since each $|\cdot|_{\alpha,\beta}$ is a $|\cdot|_{\alpha,p_\beta}$ where $p_\beta(x) = x^\beta$, every neighborhood of 0 for $|\cdot|_{\alpha,\beta}$ is a neighborhood of 0 for the $|\cdot|_{\alpha,p}$.

Now consider a polynomial, $p(x) = \sum_{|\beta|<k} c_\beta x^\beta$. Then

$$D^\alpha (p(x) f(x)) = \sum_{|\beta|<k} c_\beta D^\alpha (x^\beta f(x)).$$

Thus

$$\begin{aligned} |f|_{\alpha,p} &\leqslant \sum_{\beta<k} |c_\beta| \sup_x |D^\alpha (x^\beta f(x))| \\ &\leqslant \sum_{|\beta|<k} |c_\beta| \, |f|_{\alpha,\beta}. \end{aligned}$$

So $|f|_{\alpha,p} < \epsilon$ if $|f|_{\alpha,\beta} < \delta$ where $\sum_{|\beta|<k} |c_\beta| \delta < \epsilon$.

Case 2. The seminorms $|\cdot|'_{\alpha,\beta}$ and $|\cdot|_{p,\alpha}$

Clearly every seminorm $|\cdot|'_{\alpha,\beta}$ is one of the $|\cdot|_{p,\alpha}$. So we need only show a neighborhood of 0 for $|\cdot|_{p,\alpha}$ is a neighborhood of 0 in the topology defined by the seminorms $|\cdot|'_{\alpha,\beta}$. But if $p(x) = \sum_{|\beta|<k} c_\beta x^\beta$, then

$$|p(x) D^\alpha f(x)| \leqslant \sum_{|\beta|<k} |c_\beta| \, |x^\beta D^\alpha f(x)| \leqslant \sum_{|\beta|<k} |c_\beta| \, |f|'_{\beta,\alpha}.$$

Thus $|f|_{p,\alpha} < \epsilon$ if $|f|'_{\beta,\alpha} < \delta$ for $|\beta| < k$ where δ satisfies $\sum |c_\beta|\delta < \epsilon$.

Case 3. The seminorms $|\cdot|'_{\alpha,\beta}$ and the Schwarz topology using seminorms $|\cdot|_{N,\alpha}$.
Note

$$\begin{aligned}(1+|x|^2)^N|D^\alpha f(x)| &= \sum_{k=0}^{N}\binom{N}{k}|x|^{2k}|D^\alpha f(x)|\\ &= \sum_{k=0}^{N}\binom{N}{k}(x_1^2+x_2^2+\cdots+x_n^2)^k|D^\alpha f(x)|\\ &= \sum_{k=0}^{N}\binom{N}{k}\sum_{|\beta|=k}\frac{k!}{\beta!}x^{2\beta}|D^\alpha f(x)|\\ &\leqslant \sum_{k=0}^{N}\sum_{|\beta|=k}\binom{N}{k}\frac{k!}{\beta!}|f|'_{2\beta,\alpha}\,.\end{aligned}$$

So again if $\epsilon > 0$ and $\delta > 0$ is chosen so that $\delta\sum_{k=0}^{N}\sum_{|\beta|=k}\binom{N}{k}\frac{k!}{\beta!} < \epsilon$, then if $|f|'_{2\beta,\alpha} < \epsilon$ for $|\beta| = k$, then $|f|_{N,\alpha} < \epsilon$.

Conversely, consider the neighborhood $\{f \mid |f|'_{\alpha,\beta} < \epsilon\}$. By Propositions 2.57 and 2.58, the mapping $f \mapsto x^\alpha D^\beta f(x)$ is continuous from $\mathcal{S}(\mathbb{R}^n)$ into $\mathcal{S}(\mathbb{R}^n)$. Thus if $\epsilon > 0$, there is a $\delta > 0$ and seminorms $|\cdot|_{N_i,\alpha_i}$, $i = 1,\ldots,k$ such that $|f|'_{\alpha,\beta} = \sup_x(1+|x|^2)^0|D^{(0,\cdots,0)}x^\alpha D^\beta f(x)| < \epsilon$ if $|f|_{N_i,\alpha_i} < \delta$ for $i = 1,2,\ldots,k$. Thus $\{f \mid |f|'_{\alpha,\beta} < \epsilon\}$ is a neighborhood of 0 in the Schwartz topology.

Case 4. The seminorms $|\cdot|_{\alpha,\beta}$ and $|\cdot|'_{\beta,\alpha}$
First note using Leibniz's rule, Lemma 2.21

$$\begin{aligned}D^\alpha(x^\beta f(x)) &= \sum_{\gamma\leqslant\alpha}\binom{\alpha}{\gamma}D^\gamma x^\beta D^{\alpha-\gamma}f(x)\\ &= \sum_{\gamma\leqslant\alpha,\,\gamma\leqslant\beta}\binom{\alpha}{\gamma}\frac{\beta!}{(\beta-\gamma)!}x^{\beta-\gamma}D^{\alpha-\gamma}f(x)\,.\end{aligned}$$

Thus $|f|_{\alpha,\beta} \leqslant \sum_{\gamma\leqslant\alpha,\,\gamma\leqslant\beta}\binom{\alpha}{\gamma}\frac{\beta!}{(\beta-\gamma)!}|f|'_{\beta-\gamma,\alpha-\gamma} < \epsilon$ if $\delta > 0$ is chosen so that $\delta\sum_{\gamma\leqslant\alpha,\,\gamma\leqslant\beta}\binom{\alpha}{\gamma}\frac{\beta!}{(\beta-\gamma)!} < \epsilon$ and each $|f|'_{\beta-\gamma,\alpha-\gamma} < \delta$.

Now from the above by solving for $x^\beta D^\alpha f(x)$, we have

$$x^\beta D^\alpha f(x) = D^\alpha(x^\beta f(x)) - \sum_{0\neq\gamma\leqslant\alpha,\gamma\leqslant\beta}\binom{\alpha}{\gamma}\frac{\beta!}{(\beta-\gamma)!}x^{\beta-\gamma}D^{\alpha-\gamma}f(x).$$

Note if $\alpha = \beta = 0$, we have $|f|'_{\alpha,\beta} = |f|_{\beta,\alpha}$ and thus $|f|'_{\alpha,\beta} < \epsilon$ if $|f|_{\beta,\alpha} < \epsilon$. Now assume for all strictly smaller $\beta' < \beta$ and $\alpha' < \alpha$, there is for each $\epsilon > 0$, a $\delta > 0$ such that $|f|'_{\beta',\alpha'} < \epsilon$ if $|f|_{\alpha,\beta} < \delta$ and all $|f|_{\beta'',\alpha''} < \delta$ for $\alpha'' \leqslant \alpha'$ and $\beta'' \leqslant \beta'$.

Hence for $\epsilon > 0$, choose a $\delta > 0$ so that $\delta < \frac{\epsilon}{2}$ and if $|f|_{\alpha'',\beta''} < \delta$ for $\alpha'' < \alpha$ and $\beta'' < \beta$, then $\delta \sum_{0\neq\gamma\leqslant\alpha,\gamma\leqslant\beta} \binom{\alpha}{\gamma} \frac{\beta!}{(\beta-\gamma)!} |f|'_{\beta-\gamma,\alpha-\gamma} < \frac{\epsilon}{2}$. This implies

$$|x^\beta D^\alpha f(x)| < \frac{\epsilon}{2} + \sum_{0\neq\gamma\leqslant\alpha,\, \gamma\leqslant\beta} \binom{\alpha}{\gamma} \frac{\beta!}{(\beta-\gamma)!} |f|'_{\beta-\gamma,\alpha-\gamma} < \epsilon$$

whenever $|f|_{\beta,\alpha} < \delta$ and $|f|_{\alpha',\beta'} < \delta$ for $\alpha' < \alpha$ and $\beta' < \beta$. Thus $|f|'_{\beta,\alpha} < \epsilon$ if all $|f|_{\alpha',\beta'} < \delta$ for $\alpha' \leqslant \alpha$ and $\beta' \leqslant \beta$.

Case 5. The seminorms $|\cdot|'_{N,\alpha}$ and $|\cdot|_{N,\alpha}$ for the Schwarz topology. First note since $|x| \leqslant |x|^2$ for $|x| \geqslant 1$ that

$$(1+|x|)^N |D^\alpha f(x)| \leqslant 2^N |D^\alpha f(x)| + (1+|x|^2)^N |D^\alpha f(x)|$$

for all x. Thus

$$|f|'_{N,\alpha} \leqslant 2^N |f|_{0,\alpha} + |f|_{N,\alpha}.$$

Thus if $\epsilon > 0$, $\{f \mid |f|'_{N,\alpha} < \epsilon\}$ contains the f with $|f|_{0,\alpha} < \frac{\epsilon}{2}$ and $|f|_{N,\alpha} < \frac{\epsilon}{2}$.

Similarly

$$(1+|x|^2)^N |D^\alpha f(x)| \leqslant 2^N |D^\alpha f(x)| + (1+|x|)^{2N} |D^\alpha f(x)|$$

and thus $|f|_{N,\alpha} \leqslant 2^N |f|'_{0,\alpha} + |f|'_{2N,\alpha}$ and so the neighborhood $\{f \mid |f|_{N,0} < \epsilon\}$ contains the f with $|f|'_{0,\alpha} < \frac{\epsilon}{2}$ and $|f|'_{2N,\alpha} < \frac{\epsilon}{2}$.

□

6. Let g be a tempered measurable complex valued function on $\mathbb{R}^n$. Show that the map

$$\mathcal{S}(\mathbb{R}^n) \ni f \mapsto T_g(f) := \int_{\mathbb{R}^n} f(x)g(x)\,dx \in \mathbb{C}$$

is a well defined continuous linear functional on $\mathcal{S}(\mathbb{R}^n)$.

Solution:

Let g be a tempered complex measurable function on $\mathbb{R}^n$. Then there is an N such that $|g(x)| \leqslant K(1+|x|^2)^N$ for some constant $K > 0$. Thus by Remark 2.27, $(1+|x|^2)^{N-M}$ is integrable if and only if $M - N > \frac{n}{2}$. Take $M \in \mathbb{N}$ with this property. One has

$$\begin{aligned}
|T_g(f)| &= |\int_{\mathbb{R}^n} f(x)g(x)\,dx| \\
&\leqslant K \int |f(x)|(1+|x|^2)^N\,dx \\
&\leqslant K \int |f(x)|(1+|x|^2)^M (1+|x|^2)^{-(M-N)}\,dx \\
&\leqslant K|f|_{M,0} \int (1+|x|^2)^{-(M-N)}\,dx \\
&\leqslant KL|f|_{M,0}
\end{aligned}$$

where

$$L = \int_{\mathbb{R}^n} (1 + |x|^2)^{-(M-N)}\, dx < \infty.$$

Thus T is continuous at 0 and so is continuous.

$\square$

7. Let A be a linear transformation of $\mathbb{R}^n$ given by

$$A(x_1, \ldots, x_n) = (\sum a_{1,j}x_j, \sum a_{2,j}x_2, \ldots, \sum a_{n,j}x_j).$$

Suppose $f \in C^\infty(\mathbb{R}^n)$. Show

$$\left(\frac{\partial}{\partial x_k}\right)^m (f \circ A) = \sum_{|\beta|=m} (a_{1,k}, a_{2,k}, \ldots, a_{n,k})^\beta (D^\beta f) \circ A$$

Solution:

We show this by induction using the chain rule. For $m = 1$,

$$\begin{aligned}\left(\frac{\partial}{\partial x_k}\right)(f \circ A)(x_1, x_2, \ldots, x_n) &= \frac{\partial}{\partial x_k}(f(\sum a_{1,j}x_j, \sum a_{2,j}x_j, \ldots, \sum a_{n,j}x_j)) \\ &= \sum_{i=1}^{n} (D_i f)(A(x)) \frac{\partial}{\partial x_k}(\sum a_{i,j}x_j) \\ &= \sum_{i=1}^{n} a_{i,k}(D_i f)(A(x)) \\ &= \sum_{|\beta|=1} (a_{1,k}, a_{2,k}, \ldots, a_{n,k})^\beta (D^\beta f) \circ A(x)\,.\end{aligned}$$

We assume the result is true for m. Then for $m+1$, we have

$$\left(\frac{\partial}{\partial x_k}\right)^{m+1}(f\circ A)(x_1,x_2,\ldots,x_n) =$$

$$\sum_{|\beta|=m}(a_{1,k},a_{2,k},\ldots,a_{n,k})^{\beta}\times$$

$$\frac{\partial}{\partial x_k}((D^{\beta}f)(\sum a_{1,j}x_j,\sum a_{2,j}x_j,\ldots,\sum a_{n,j}x_j))$$

$$=\sum_{|\beta|=m}(a_{1,k},a_{2,k},\ldots,a_{n,k})^{\beta}\times$$

$$\sum_{i=1}^{n}(D_i(D^{\beta}f))(\sum a_{1,j}x_j,\sum a_{2,j}x_j,\ldots,\sum a_{n,j}x_j)\frac{\partial}{\partial x_k}(\sum_{j=1}^{n}a_{i,j}x_j)$$

$$=\sum_{|\beta|=m}(a_{1,k},a_{2,k},\ldots,a_{n,k})^{\beta}\sum_{i=1}^{n}a_{i,k}(D_i(D^{\beta}f))(\sum a_{1,j}x_j,\sum a_{2,j}x_j,\ldots,\sum a_{n,j}x_j)$$

$$=\sum_{|\beta|=m}\sum_{i=1}^{n}a_{i,k}(a_{1,k},a_{2,k},\ldots,a_{n,k})^{\beta}(D_i(D^{\beta}f))(\sum a_{1,j}x_j,\sum a_{2,j}x_j,\ldots,\sum a_{n,j}x_j)$$

$$=\sum_{|\beta|=m}\sum_{i=1}^{n}(a_{1,k},a_{2,k},\ldots,a_{n,k})^{\beta+\epsilon_i}(D^{\beta+\epsilon_i}f)(\sum a_{1,j}x_j,\sum a_{2,j}x_j,\ldots,\sum a_{n,j}x_j)$$

$$=\sum_{|\gamma|=m+1}(a_{1,k},a_{2,k},\ldots,a_{n,k})^{\gamma}((D^{\gamma}f)\circ A)(x)$$

□

8. Let $\Phi=(\phi_1,\ldots,\phi_n)$ be C^{∞} mapping of $\mathbb{R}^n$ onto $\mathbb{R}^n$ with C^{∞} inverse $\Psi=(\psi_1,\ldots,\psi_n)$. Assume all the ϕ_i and ψ_j are in $\mathcal{S}'_{\infty}(\mathbb{R}^n)$; i.e., all their derivatives have polynomial growth. Show

$$f\mapsto f\circ\Phi$$

is a linear homeomorphism of $\mathcal{S}_n(\mathbb{R}^n)$.
Solution: We estimate $p(x)D^{\alpha}(f\circ\Phi)$. The chain rule gives

$$p(x)D^{\alpha}(f\circ\Phi)(x)=p(x)\sum_{|\beta|\leqslant|\alpha|}P_{\beta}([D^{\gamma}\phi_j(x)\mid 1\leqslant j\leqslant n,|\gamma|\leqslant|\alpha|])(D^{\beta}f)\circ\Phi(x)$$

where P_{β} is polynomial in the list of entries provided.

It therefore suffices to show each term

$$p(x)\prod_{\gamma,j\in F}(D^{\gamma}\phi_j)^{m_{\gamma,j}}(x)D^{\beta}f\circ\Phi(x)$$

can be controlled by a finite number of seminorms for $\mathcal{S}_n$.

Choose $P_0(x)$ and $P_{\gamma,j}(x)$ such that $|D^{\gamma}\phi_j(x)|\leqslant P_{\gamma,j}(x)$ and $|p(x)|\leqslant P_0(x)$. We may assume each of these polynomials are everywhere larger than 1. Take $M=\max\{m_{\gamma,j}\}$. Then

$$|p(x)\prod_{\gamma,j\in F}(D^{\gamma}\phi_j)^{m_{\gamma,j}}(x)D^{\beta}f\circ\Phi(x)|\leqslant P(x)^M|D^{\beta}f(\Phi(x))|$$

where $P(x) = P_0(x)\prod P_{\gamma,j}(x)$. Now $|P(x)| = |P(\Psi(\Phi(x))|$. Since P is a polynomial and there are polynomials Q_j such that $|\psi_j(y)| \leqslant Q_j(y)$, we have $|P(\Psi(y))| \leqslant Q(y)$ for some polynomial Q. Hence $|P(x)| \leqslant Q(\Phi(x))$. Thus we have

$$P(x)^M|D^\beta f(\Phi(x))| \leqslant Q(\Phi(x))^M|D^\beta f(\Phi(x))| \leqslant |f|_{Q^M,\beta}.$$

So if $\epsilon > 0$ and we make all the $|f|_{Q^M,\beta}$ small for each of the terms occurring in $p(x)D^\alpha(f \circ \Phi)$, we can insure $|f \circ \Phi|_{p,\alpha} < \epsilon$. □

9. Suppose $\phi \in \mathcal{S}(\mathbb{R})$ and $\int_{-\infty}^{\infty} \phi(t)\,dt = 0$. Show

$$F(x) = \int_{-\infty}^{x} \phi(t)\,dt$$

is in $\mathcal{S}(\mathbb{R})$. (**Hint:** Consider $\int_x^\infty \phi(t)\,dt$.)

Solution: Since $F'(x) = \phi(x)$, we need only show $(1+x^2)^n F(x) \to 0$ as $x \to \pm\infty$. We note if $x \to -\infty$, we have $F(x) = \int_{-\infty}^{x} \phi(t)\,dt = \int_{-x}^{\infty} \phi(-t)\,dt$ and $-x \mapsto \infty$. We will show $(1+x^2)^n \int_x^\infty \psi(t)\,dt \to 0$ as $x \to \infty$ for all $\psi \in \mathcal{S}$. This then shows $(1+x^2)^n F(x) \to 0$ as $x \to -\infty$. It will also show

$$(1+x^2)^n F(x) \to 0 \text{ as } x \to +\infty$$

for

$$F(x) + \int_x^\infty \phi(t)\,dt = \int_{-\infty}^{\infty} \phi(t)\,dt = 0.$$

So $F(x) = -\int_x^\infty \phi(t)\,dt$.

Thus suppose ψ is Schwartz. Then

$$\begin{aligned}(1+x^2)^n \int_x^\infty |\psi(t)|\,dt &= \frac{(1+x^2)^n}{(1+x^2)^n}\int_x^\infty (1+x^2)^n(1+t^2)^{-m}(1+t^2)^m|\psi(t)|\,dt\\ &\leqslant |\psi|_{m,0}\int_x^\infty \frac{(1+x^2)^n}{(1+t^2)^m}\,dt\\ &= |\psi|_{m,0}\int_x^\infty \frac{(1+x^2)^n}{(1+t^2)^n}\frac{1}{(1+t^2)^{m-n}}\,dt\\ &\leqslant |\psi|_{m,0}\int_x^\infty \frac{1}{(1+t^2)^{m-n}}\,dt\\ &\to 0 \text{ as } x \to \infty\end{aligned}$$

for $m > n$. □

10. Let f be Schwartz on $\mathbb{R}^m \times \mathbb{R}^n$. Show $y \mapsto f(x,y) = f_x(y)$ is Schwartz on $\mathbb{R}^n$ for each $x \in \mathbb{R}^m$.

Solution: Clearly f_x is C^∞. Now

$$\begin{aligned}(1+|y|^2)^N|D^{0,\beta}f(x,y)| &\leqslant (1+|y|^2)^N\frac{(1+|x|^2+|y|^2)^N}{(1+|x|^2+|y|^2)^N}|D^{0,\beta}f(x,y)|\\ &\leqslant |f|_{N,(0,\beta)}.\end{aligned}$$

□

11. Show $x \mapsto f_x$ where $f_x(y) = f(x,y)$ is continuous from $\mathbb{R}^m$ into $\mathcal{S}(\mathbb{R}^n)$ for each $f \in \mathcal{S}(\mathbb{R}^m \times \mathbb{R}^n)$.

Solution: We consider $|f_x - f_{x_0}|_{p,\beta}$. We have

$$|p(y)D^{(0,\beta)}f(x,y) - p(y)D^{(0,\beta)}f(x_0,y)| = |p(y)G(x,y) - p(y)G(x_0,y)|$$

where $G(x,y) = D^{(0,\beta)}f(x,y)$. So by the Mean Value Theorem, there is an s with

$$\begin{aligned}
|p(y)(G(x,y) - G(x_0,y))| &= |p(y)\nabla G(sx_0 + (1-s)x, y) \cdot (x - x_0)| \\
&= \left|p(y)\left(\sum D^{(e_j,0)}G(sx_0 + (1-s)x), y)(x - x_0)_j\right)\right| \\
&\leqslant \sum_{j=1}^{n} |(x - x_0)^{e_j} p(y) D^{(e_j,0)} G(sx_0 + (1-s)x, y)| \\
&\leqslant \sum_{j=1}^{n} \left|\frac{(x - x_0)^{e_j}}{1 + |(sx_0 + (1-s)x, y)|^2}\right| |G|_{q(x,y),(e_j,0)}
\end{aligned}$$

where $q(x,y) = (1 + |(x,y)|^2)p(y)$.

This clearly goes to 0 as $x \to x_0$ for any $p(y)$. □

12. Let I be a continuous linear functional on $\mathcal{S}(\mathbb{R}^n)$ and suppose $f \in \mathcal{S}(\mathbb{R}^m \times \mathbb{R}^n)$. Define $I(f)(x) = I(f_x)$ where $f_x(y) = f(x,y)$. Show I is a continuous linear transformation of $\mathcal{S}(\mathbb{R}^m \times \mathbb{R}^n)$ into $\mathcal{S}(\mathbb{R}^m)$.

Solution: We know $x \mapsto f_x$ is continuous from $\mathbb{R}^m$ into $\mathcal{S}(\mathbb{R}^n)$ by Exercise 2.3.11. Thus $I(f)$ is continuous. To see $I(f)$ is C^∞, we show $D^\alpha I(f) = I(D^{(\alpha,0)}f)$ for $\alpha \in \mathbb{N}_0^n$. This can be done by induction. It is clear if $\alpha = 0$. Suppose true for $|\alpha| = n$. Then since

$$D^{\alpha+e_i}I(f) = D^{e_i}I(D^{(\alpha,0)}f)$$

it suffices to show $D^{e_i}I(f) = I(D^{(e_i,0)}f)$ for $f \in \mathcal{S}(\mathbb{R}^m \times \mathbb{R}^n)$. Now

$$\begin{aligned}
D^{e_i}I(f)(x) &= \lim_{h\to 0} \frac{I(f)(x + he_i) - I(f)(x)}{h} \\
&= \lim_{h\to 0} \frac{I(f_{x+he_i}) - I(f_x)}{h} \\
&= \lim_{h\to 0} I\left(\frac{f_{x+he_i} - f_x}{h}\right) \\
&= I(D^{(e_i,0)}f)(x)
\end{aligned}$$

for if we use the Mean Value Theorem twice, one has

$$
\begin{aligned}
&(1+|y|^2)^N \left| \frac{D^\beta f_{x+he_i}(y) - D^\beta f_x(y)}{h} - D^\beta (D^{(e_i,0)} f)_x(y) \right| \\
&\quad = (1+|y|^2)^N \left| \frac{D^{(0,\beta)} f(x+he_i,y) - D^{(0,\beta)} f(x,y)}{h} - D^{(e_i,\beta)} f(x,y) \right| \\
&\quad = (1+|y|^2)^N \left| D^{(e_i,0)} D^{(0,\beta)} f(x+u(h,y)e_i,y) - D^{(e_i,\beta)} f(x,y) \right| \\
&\quad = (1+|y|^2)^N \left| D^{(e_i,\beta)} f(x+u(h,y)e_i) - D^{(e_i,\beta)} f(x,y) \right| \\
&\quad = (1+|y|^2)^N \frac{|u(h,y)|}{(1+|(x+v(h,y)e_i,y)|^2)^N} \\
&\qquad \times (1+|(x+v(h,y)e_i y)|^2)^N |D^{(2e_i,\beta)} f(x+v(h,y)e_i,y)| \\
&\quad = (1+|y|^2)^N \frac{|u(h,y)|}{(1+|(x+v(h,y)e_i,y)|^2)^N} |f|_{N,(2e_i,\beta)} \\
&\quad \leqslant |h|\,|f|_{N,(2e_i,\beta)}
\end{aligned}
$$

where $|u(h,y)| < h$ and $|v(h,y)| < |u(h,y)|$. Thus

$$
\left| \frac{f_{x+he_i} - f_x}{h} - (D^{(e_i,0)} f)_x \right|_{N,\beta} \leqslant |h|\,|f|_{N,(2e_i,\beta)} \to 0
$$

as $h \to 0$. So $x \mapsto I(f_x)$ is C^∞.

We now show $I(f)$ is Schwartz and $f \mapsto I(f)$ is continuous. First note if $k \in \mathbb{N}_0$, then

$$
(1+|x|^2)^k If(x) = I(\tilde{f})(x)
$$

where $\tilde{f}(x,y) = (1+|x|^2)^k f(x,y)$. But the continuity of I implies there is an N and an M so that $|I(g)| \leqslant M$ if $|g|_{N,\beta} \leqslant 1$ for $|\beta| \leqslant N$. But $(1+|y|^2)^N |D^\beta \tilde{f}_x(y)| \leqslant (1+|(x,y)|^2)^N |D^{(0,\beta)} f(x,y)|$. Thus if $|f|_{(N,\beta)} \leqslant 1$ for $|\beta| \leqslant N$, then $|\tilde{f}_x|_{N,\beta} \leqslant 1$ for $|\beta| \leqslant N$ and thus $|I(\tilde{f}_x)| \leqslant M$. Consequently, $|(1+|x|^2)^k If(x)| \leqslant M$ if $|f|_{N,\beta} \leqslant 1$ for $|\beta| \leqslant N$. Applying this to $D^{(\alpha,0)} f$ for any $\alpha \in \mathbb{N}_0^m$ shows one can make $(1+|x|^2)^k |D^\alpha If(x)| = (1+|x|^2)^k |I(D^{(\alpha,0)} f)(x)| \leqslant M_\alpha$ if $|f|_{N_\alpha,\beta} \leqslant 1$ for $|\beta| \leqslant N_\alpha$. This implies the continuity of $f \mapsto I(f)$ from $\mathcal{S}(\mathbb{R}^m \times \mathbb{R}^n)$ into $\mathcal{S}(\mathbb{R}^m)$.

□

EXERCISE SET 2.4

1. Show if Ω is a nonempty open subset of $\mathbb{R}^n$, then the inclusion mapping from $\mathcal{D}(\Omega)$ into $\mathcal{E}(\Omega)$ is continuous.

Solution: Let K^* be an arbitrary compact subset of Ω. To show i, the inclusion mapping is continuous it suffices to show $i|_{\mathcal{D}_{K^*}(\Omega)}$ is continuous. Now take one of the seminorms $|\cdot|_{K,\alpha}$ which define the topology on $\mathcal{E}(\Omega)$. So $K \subseteq \Omega$ is compact and $\alpha \in \mathbb{N}_0^n$. Then $|i(\phi)|_{K,\alpha} = \max_{x\in K} |D^\alpha \phi(x)| \leqslant \max_{x\in K^*} |D^\alpha \phi(x)| = |\phi|_{K^*,\alpha}$ for $\phi \in \mathcal{D}_{K^*}(\Omega)$. So i is continuous. □

2. Show if Ω_1 and Ω_2 are nonempty open subsets of $\mathbb{R}^n$ and $\Omega_1 \subseteq \Omega_2$, then the restriction mapping from $\mathcal{E}(\Omega_2)$ into $\mathcal{E}(\Omega_1)$ is continuous.

Solution: Let $|\cdot|_{K,\alpha}$ be a defining seminorm for $\mathcal{E}(\Omega_1)$. So $K \subseteq \Omega_1$ is compact and $\alpha \in \mathbb{N}_0^n$. Let $\phi \in \mathcal{E}(\Omega_2)$, then

$$|\phi_{|\Omega_1}|_{K,\alpha} = \max_{x\in K\subseteq\Omega_1} |D^\alpha\phi(x)| = \max_{x\in K\subseteq\Omega_2} |\phi(x)| = |\phi|_{K,\alpha}.$$

Thus the restriction mapping is continuous. □

3. Give a proof for Proposition 2.69.

Solution:

By Proposition 2.73, a linear transformation T of $\mathcal{D}(\mathbb{R}^n)$ into $\mathcal{D}(\mathbb{R}^n)$ is continuous if and only if for any sequence ϕ_k converging to ϕ in $\mathcal{D}(\mathbb{R}^n)$, the sequence $T(\phi_k)$ converges to $T(\phi)$ in $\mathcal{D}(\mathbb{R}^n)$. But by Corollary 2.72, we need only show if ϕ_k is a sequence in $\mathcal{D}_K(\mathbb{R}^n)$ for some compact subset K with $\phi_k \to \phi$ in $\mathcal{D}_K(\mathbb{R}^n)$, then $T(\phi_k) \to T(\phi)$ in $\mathcal{D}(\mathbb{R}^n)$. But again by Corollary 2.72, this occurs if and only if there is a compact subset K' such that $T(\phi_k) \to T(\phi)$ in $\mathcal{D}_{K'}(\mathbb{R}^n)$.

For D^α we would then have $|D_\alpha\phi_k - D_\alpha\phi|_{K,\beta} = |\phi_k - \phi|_{K,\alpha+\beta} \to 0$ as $k \to \infty$ and thus with $K' = K$ we see $D_\alpha\phi_k \to D_\alpha\phi$ in $\mathcal{D}_{K'}(\mathbb{R}^n)$.

For $\lambda(y)$, take $K' = K+y$. Then note $\lambda(y)f(x) \neq 0$ if and only if $f(x-y) \neq 0$ if and only if $x - y \in \{z \mid f(z) \neq 0\}$ if and only if $x \in \{z \mid f(z) \neq 0\} + y$. Thus supp $\lambda(y)f = \text{supp } f + y$. So $\lambda(y)\phi_k \in \mathcal{D}_{K'}(\mathbb{R}^n)$, $\lambda(y)\phi \in D_{K'}(\mathbb{R}^n)$ and

$$\begin{aligned}
|\lambda(y)\phi_k - \lambda(y)\phi|_{K',\alpha} &= \max_{x\in K} |D^\alpha\lambda(y)\phi_k(x+y) - D^\alpha\lambda(y)\phi(x+y)| \\
&= \max_{x\in K} |\lambda(y)(D^\alpha\phi_k)(x+y) - \lambda(y)(D^\alpha\phi)(x+y)| \\
&= \max_{x\in K} |D^\alpha\phi_k(x) - D^\alpha\phi(x)| \\
&= |\phi_k - \phi|_{K,\alpha} \to 0
\end{aligned}$$

as $k \to \infty$. So $\lambda(y)$ is continuous.

For $\delta(a)$ where $a > 0$ note $\text{supp}(\delta(a)f) = a\,\text{supp}(f)$ for C^∞ functions of compact support for $f(a^{-1}y) \neq 0$ if and only if $y \in a\{z \mid f(z) \neq 0\}$. Thus if $K' = aK$, we have

$$\begin{aligned}
|\delta(a)\phi_k - \delta(a)\phi|_{K',\alpha} &= \max_{x\in K} |D^\alpha(\delta(a)\phi_k)(ax) - D^\alpha(\delta(a)\phi)(ax)| \\
&= a^{-|\alpha|} \max_{x\in K} |\delta(a)(D^\alpha\phi_k)(ax) - \delta(a)(D^\alpha\phi)(ax)| \\
&= a^{-n/2-|\alpha|} \max_{x\in K} |D^\alpha\phi_k(x) - D^\alpha\phi(x)| \\
&= a^{-n/2-|\alpha|}|\phi_k - \phi|_{K,\alpha} \to 0
\end{aligned}$$

as $k \to \infty$ by either the chain rule or the direct calculation

$$\begin{aligned} D_i(\delta(a)f)(y) &= \lim_{h\to 0} \frac{\delta(a)f(y+he_i) - \delta(a)f(y)}{h} \\ &= a^{-1} \lim_{h\to 0} \frac{a^{-n/2}f(a^{-1}y + a^{-1}he_i) - a^{-n/2}f(a^{-1}y)}{a^{-1}h} \\ &= a^{-1}a^{-n/2}D_i f(a^{-1}y) \\ &= a^{-1}\delta(a)(D_i f)(y). \end{aligned}$$

For conjugation, one takes $K' = \bar{K}$ and notes since $D^\alpha \bar{f} = \overline{D^\alpha f}$, that $|\bar{\phi}_k - \bar{\phi}|_{K',\alpha} = |\phi_k - \phi|_{K,\alpha}$.

For check one takes $K' = \check{K}$ and notes $|\check{\phi}_k - \check{\phi}|_{\check{K},\alpha} = |\phi_k - \phi|_{K,\alpha}$. Indeed, $D^\alpha(\check{f})(z) = (-1)^{|\alpha|}\widecheck{D^\alpha f}(z)$. Similarly for adjoints, one takes $K' = \bar{\check{K}}$ and then notes $D^\alpha(f^*)(z) = (-1)^{|\alpha|}(D^\alpha f)^*(z)$. In this case one has

$$\begin{aligned} |\phi_k^* - \phi^*|_{K',\alpha} &= \max_{x\in K} |D^\alpha(\phi_k^*)(-\bar{x}) - D^\alpha(\phi^*)(-\bar{x})| \\ &= \max_{x\in K} |(-1)^{|\alpha|}(D^\alpha\phi_k)^*(-\bar{x}) - (-1)^{|\alpha|}(D^\alpha\phi)^*(-\bar{x})| \\ &= \max_{x\in K} |D^\alpha\phi_k(x) - D^\alpha\phi(x)| \\ &= |\phi_k - \phi|_{K,\alpha} \to 0 \end{aligned}$$

as $k \to \infty$.

□

4. Let Ω be an open subset of $\mathbb{R}^n$.

(a) Show there is a countable increasing sequence $K_1 \subseteq K_2 \subseteq K_3 \subseteq \cdots$ of compact subsets of Ω such that $\Omega = \cup K_i$ and $K_i \subseteq K_{i+1}^\circ$ for all i. Conclude $\mathcal{D}(\Omega) = \cup \mathcal{D}_{K_i}(\Omega)$.

(b) Show the inductive limit topology defined by the subspaces $\mathcal{D}_{K_i}(\Omega)$ is the same as the inductive limit topology defined by all the subspaces $\mathcal{D}_K(\Omega)$ where K is a compact subset of Ω.

Solution:

(a) Let Ω be a nonempty subset of $\mathbb{R}^n$. Since Ω is second countable and locally compact, we can find a countable base U_i for the topology so that each $\bar{U}_i$ is compact and is a subset of Ω. Take $K_1 = \bar{U}_1$. Since the sets U_j cover K_1, take a finite subcover, $U_{1,1}, \ldots, U_{1,m_1}$ and set $K_2 = \cup_{j=1}^{m_1}\bar{U}_{1,j}$. Clearly $K_1 \subseteq K_2^\circ$. Suppose we have $K_1 \subseteq K_2^\circ \subseteq K_2 \subseteq K_3^\circ \subseteq \cdots \subseteq K_{l-1}^\circ \subseteq K_l$ where all are subsets of Ω and K_l contains U_l. Let $U_{l,1}, U_{l,2}, \ldots, U_{l,m_l}$ be a cover of $K_1 \cup K_2 \cup \cdots \cup K_l$. Take $K_{l+1} = \bar{U}_l \cup \left(\cup_{j=1}^{m_l}\bar{U}_{l,j}\right)$. Then K_{l+1} is a compact subset of Ω and $K_l \subseteq K_l^\circ$. Repeat inductively. We obtain the needed sequence for $\cup K_l$ contains $\cup U_l = \Omega$.

Since any $\phi \in \mathcal{D}(\Omega)$ is an element in $\mathcal{D}_K(\Omega)$ where $K = \operatorname{supp}\phi$ and since $K \subseteq \cup_i K_i^\circ$, $K \subseteq K_i^\circ$ for some i; there being a finite subcover. So $\phi \in \mathcal{D}_{K_i}(\Omega)$.

(b) Let $\mathcal{T}$ be the strongest locally convex topology on $\mathcal{D}(\Omega)$ such that its relative topology on each $\mathcal{D}_K(\Omega)$ is weaker than the topology on $\mathcal{D}_K(\Omega)$ and let $\mathcal{T}'$ be the strongest locally convex topology such that its relative topology on each $\mathcal{D}_{K_i}(\Omega)$ is weaker than the topology on $\mathcal{D}_{K_i}(\Omega)$ for each i. Then $\mathcal{T} \subseteq \mathcal{T}'$. Now consider the relative topology of $\mathcal{T}'$ on $\mathcal{D}_K(\Omega)$. Choose K_i with $K \subseteq K_i$. This is possible since K is compact and $K \subseteq \cup_i K_i^\circ$. Now the relative topology of $\mathcal{T}'$ on $\mathcal{D}_{K_i}(\Omega)$ is weaker than the topology on $\mathcal{D}_{K_i}(\Omega)$. But the relative topology of $\mathcal{D}_{K_i}(\Omega)$ on $\mathcal{D}_K(\Omega)$ is the topology of $\mathcal{D}_K(\Omega)$ for the seminorms $|\cdot|_{K_i,\alpha}$ restricted to $\mathcal{D}_K(\Omega)$ are the seminorms $|\cdot|_{K,\alpha}$ which define define the topology on $\mathcal{D}_K(\Omega)$. Thus the relative topology of $\mathcal{T}'$ on $\mathcal{D}_K(\Omega)$ is weaker than the $\mathcal{D}_K(\Omega)$ topology. Since this is true for all K, $\mathcal{T}' \subseteq \mathcal{T}$.

□

5. Prove that $\mathcal{E}(\Omega)$ is complete.
Solution:

The topology on $\mathcal{E}(\Omega)$ is defined using the seminorms $|\cdot|_{K,\alpha}$. Thus if $\{\phi_k\}$ is a Cauchy sequence, one has $\max_{x\in K} |D^\alpha\phi_k(x) - D_l^\alpha\phi_k(x)| \to 0$ as $k, l \to \infty$. This implies $D^\alpha\phi_k$ converges uniformly on each compact subset of Ω to a function ψ_α and so this function is defined on all of Ω. Moreover $\max_{x\in K} |D^\alpha\phi_k(x) - \psi_\alpha(x)| \to 0$ as $k \to \infty$ for each compact set K contained in Ω. As seen in Exercise 2.4.4 there is an increasing sequence K_i of compact sets with $\cup_i K_i^\circ = \Omega$. Since the convergence of $D^\alpha\phi_k$ is uniform on K_i, the uniform limit ψ_α is continuous on K_i°. Moreover, having all derivatives $D^\alpha\phi_k$ converge uniformly on K_i° implies the limit function ϕ is C^∞ and the derivatives of $D^\alpha\phi$ are the uniform limits of the $D^\alpha\phi_k$. So $D^\alpha\phi = \psi_\alpha$ on K_i°. This gives $D^\alpha\phi(x) = \psi_\alpha(x)$ for all $x \in \Omega$ and consequently, $|\phi_k - \phi|_{K,\alpha} = \max_{x\in K} |D^\alpha\phi_k(x) - D^\alpha\phi(x)| \to 0$ for each K and α. So $\phi_k \to \phi$ as $k \to \infty$ and we see $\mathcal{E}(\Omega)$ is complete.

□

6. Let Ω be an open subset of $\mathbb{R}^n$ and suppose K is a nonempty compact subset.

(a) Show if B is a bounded subset of $\mathcal{D}_K(\Omega)$, then the collection of functions inside B is equicontinuous.

(b) Show every bounded subset of $\mathcal{D}_K(\Omega)$ has compact closure. Conclude that the space $\mathcal{D}_K(\Omega)$ has the Heine-Borel property; i.e., a subset of $\mathcal{D}_K(\Omega)$ is compact if and only if it is closed and bounded.

Solution: Assume B is bounded. Then $\bar{B}$ is bounded. So all for all α, $|D^\alpha f|_\infty \leqslant M_\alpha < \infty$ for $f \in B$. This implies the collection $D^\alpha f$ for $f \in B$ is equicontinuous and uniformly bounded.

Enumerate $\mathbb{N}_0^n$ into a sequence α_k, $k = 1, 2, \ldots$ and let f_j be a sequence in $\bar{B}$. Then by the Ascoli Theorem, there is a subsequence $f_{j,1}$ such that the sequence $D^{\alpha_1} f_{j,1}$ converges uniformly.

Again by Ascoli, there is a subsequence $f_{j,2}$ of $f_{j,1}$ such that $D^{\alpha_2} f_{j,2}$ converges uniformly. Repeat.

Obtain $f_{j,k}$ where $\{f_{j,k}\}_j$ is a subsequence of $\{f_{j,k-1}\}_j$ for $k = 2, 3, \ldots$ and $D^{\alpha_k} f_{j,k} \to F_{\alpha_k}$ uniformly on K as $j \to \infty$. Consider the Cantor diagonal $f_{j,j}$. We then have $D^\alpha f_{j,j}$ converges uniformly on K for all α to a function F_α. Set $f = F_0$. Assume we know $D^\alpha f$ is exists and is continuous and $D^\alpha f = F_\alpha$. Then if $x \in K$ and the line from x to $x + \epsilon x$ lies in Ω, we have

$$D^\alpha f_{j,j}(x + \epsilon e_i) = D^\alpha f_{j,j}(x) + \int_0^\epsilon D^{\alpha+e_i} f_{j,j}(x + te_i)\, dt.$$

Taking a limit, we see

$$D^\alpha f(x + \epsilon e_i) = D^\alpha f(x) + \int_0^\epsilon F_{\alpha+e_i}(x + te_i)\, dt.$$

Replacing ϵ by h, this gives

$$\frac{D^\alpha f(x + he_i) - D^\alpha f(x)}{h} = \frac{1}{h}\int_0^h F_{\alpha+e_i}(x + te_i)\, dt.$$

Taking $h \to 0$, we obtain

$$D^{\alpha+e_i} f(x) = F_{\alpha+e_i}(x).$$

Thus by induction, $D^\beta f$ exists and equal F_β for all $\beta \in \mathbb{N}_0$.

As $|f_{j,j} - f|_{K,\beta} = |D^\beta f_{j,j} - D^\beta f|_\infty = |D^\beta f_{j,j} - F_\beta|_\infty \to 0$ as $j \to \infty$, we see $f_{j,j}$ converges to f in $\mathcal{D}_K(\Omega)$. □

7. Show $C_c^\infty(\mathbb{R})$ with the relative topology from $\mathcal{S}_1$ is not complete in the sense that there is a sequence $\{f_n\}$ in $C_c^\infty(\mathbb{R})$ such that for each open neighborhood V of 0 in $\mathcal{S}_1$, there is a $K \in \mathbb{N}$ with $f_m - f_n \in V$ for $m, n \geqslant K$, but $\{f_n\}$ does not have a limit in $\mathcal{D}_1$. (**Hint:** Use Lemma 2.53 to find a ϕ which is C^∞, is nonzero, and vanishes for $x \notin [0, 1]$. Then take $f_n = \phi + \frac{1}{2}\lambda(1)\phi + \frac{1}{4}\lambda(2)\phi + \cdots \frac{1}{2^{n-1}}\lambda(n)\phi$.)
Solution: Take $f_n = \phi + \frac{1}{2}\phi_1 + \frac{1}{4}\phi_2 + \cdots \frac{1}{2^n}\phi_n$ where ϕ is C^∞ and which is one on $[\frac{1}{4}, \frac{3}{4}]$, $\phi = 0$ off $[0, 1]$ and $0 \leqslant \phi \leqslant 1$. (Here $\phi_k(x) = \phi(x - k)$.) Since $0 \leqslant |\frac{1}{2^k} D^l(\phi_k)|_\infty < \frac{1}{2^k} \max |D^l \phi(x)|$, we see f_n and all of its derivatives converge uniformly on $\mathbb{R}$. Thus $f = \lim f_n$ is C^∞. We claim $f_n \to f$ in the Schwartz topology. Indeed, $|f_n - f|_{N,k} = \sup_x (1 + |x|^2)^N |D^k(f - f_n)(x)|$. But $f - f_n = \sum_{l=n+1}^\infty \frac{1}{2^l}\phi_l$. Hence $D^k(f - f_n) = \sum_{l=n+1}^\infty \frac{1}{2^l}(D^k\phi)_l$. Hence $D^k(f - f_n)(x) = 0$ if $x \leqslant n + 1$. Moreover, if $l \leqslant x \leqslant l + 1$, since the supports of ϕ_l are disjoint, one has $|D^k(f - f_n)(x)| \leqslant \frac{1}{2^l} \max |D^k\phi(x)|$. Hence $|D^k(f - f_n)(x)| \leqslant \frac{M_k}{2^l}$ if $l \leqslant x \leqslant l + 1$ and $l \geqslant n + 1$ where $M_k = \max |D^k\phi(x)|$ while $D^k(f - f_n)(x) = 0$ for $x \leqslant n + 1$. Thus

$$\begin{aligned}
|f - f_n|_{N,k} &= \sup_{x \geqslant n+1} (1 + |x|^2)^N |D^k(f - f_n)(x)| \\
&= \sup_{l \geqslant n+1} \sup_{l \leqslant x \leqslant l+1} (1 + |x|^2)^N |D^k(f - f_n)(x)| \\
&\leqslant \sup_{l \geqslant n+1} \frac{M_k}{2^l} \sup_{l \leqslant x \leqslant l+1} (1 + |x|^2)^N \\
&\leqslant M_k \sup_{l \geqslant n+1} \frac{1}{2^l}(1 + (l+1)^2)^N.
\end{aligned}$$

Since $\frac{(1+(l+1)^2)^N}{2^l} \to 0$ as $l \to \infty$, we see $|f - f_n|_{N,k} \to 0$ as $n \to \infty$. Thus $f_n \to f$ in $\mathcal{S}_n$. Clearly $f \notin \mathcal{D}_n$. This implies $\mathcal{D}_n$ is not complete in the relative Schwartz topology. □

8. Let Ω_1 and Ω_2 be nonempty open subsets of $\mathbb{R}^n$ with $\Omega_1 \subseteq \Omega_2$. We can redefine $\mathcal{D}(\Omega)$ to be all C^∞ functions on $\mathbb{R}^n$ which have compact support inside Ω.

(a) Show the inclusion mapping of $\mathcal{D}(\Omega_1)$ into $\mathcal{D}(\Omega_2)$ is continuous.

(b) Show the relative topology of $\mathcal{D}(\mathbb{R})$ on $\mathcal{D}(-1,1)$ is strictly weaker than the topology of $\mathcal{D}(-1,1)$.

Solution: Let $|\cdot|$ be a continuous seminorm on $\mathcal{D}(\Omega_2)$; this is equivalent to $|\cdot|_{\mathcal{D}_K(\Omega_2)}$ being continuous on $\mathcal{D}_K(\Omega_2)$ for all $K \subseteq \Omega_2$. Thus $|\cdot|_{\mathcal{D}_K(\Omega_1)}$ is continuous for all $K \subseteq \Omega_1$. So $|\cdot|$ is continuous on $\mathcal{D}(\Omega_1)$. Thus the inclusion mapping $\mathcal{D}(\Omega_1)$ into $\mathcal{D}(\Omega_2)$ is continuous. So the relative topology is weaker.

Let ϕ be C^∞ function with compact support $[-1,1]$. Define $\phi_k(x) = a_k\phi(a_k x)$ where $a_k \searrow 1$. Note $\phi_k(x)$ has support $[-\frac{1}{a_k}, \frac{1}{a_k}] \subseteq (-1,1)$. So each $\phi_k \in \mathcal{D}(-1,1)$. The ϕ_k do not converge in $\mathcal{D}(-1,1)$ for otherwise there would be a compact subset $K \subseteq (-1,1)$ such that supp $\phi_k \subseteq K$ for all k which is impossible. But the $\phi_k \in \mathcal{D}_{K'}(\mathbb{R})$ for all k where $K' = [-2,2]$ and

$$|D^m\phi_k - D^m\phi|_{K'} = \max_{x \in K'} |D^m\phi_k(x) - D^m\phi(x)| = \max |a_k^m \phi^{(m)}(a_k x) - \phi^{(m)}(x)| \to 0$$

as $k \to \infty$ for

$$\begin{aligned}|a_k^m \phi^{(m)}(a_k x) - \phi^{(m)}(x)| &\leqslant |a_k^m \phi^{(m)}(a_k x) - a_k^m \phi^{(m)}(x)| + |a_k^m \phi^{(m)}(x) - \phi^{(m)}(x)| \\ &\leqslant a_k^m |\phi^{(m)}(a_k x) - \phi^{(m)}(x)| + (a_k^m - 1)|\phi^m(x)| \\ &\leqslant \epsilon\end{aligned}$$

for k large by uniform continuity of $\phi^{(m)}$ and boundedness of $\phi^{(m)}$. □

9. Let Ω be a nonempty open subset of $\mathbb{R}^n$. Give $C_c^\infty(\Omega)$ the relative topology of $\mathcal{E}(\Omega)$. Show this topology on $C_c^\infty(\Omega)$ is strictly weaker than the relative topology of $\mathcal{S}(\mathbb{R}^n)$ on $C_c^\infty(\Omega)$.

Solution: We show it is weaker. Let V be a neighborhood of 0 in the relative topology of $\mathcal{E}(\Omega)$. Thus there is a finite set $\alpha_1, \ldots, \alpha_N$ in $\mathbb{N}_0$ and compact subsets $K_1, K_2, \ldots, K_N$ of Ω and an $\epsilon > 0$ such that if $f \in C_c(\Omega)$ and $|f|_{K_i,\alpha_i} < \epsilon$ for $i = 1, 2, \ldots, N$, then $f \in V$. Thus if $W = \{f \mid |f|_{0,\alpha_i} < \epsilon\}$, then W is open in the relative topology of $\mathcal{S}_n$ on $C_c(\Omega)$ and $W \subseteq V$.

To see it is strictly weaker, let $W = \{f \in C_c^\infty(\Omega) \mid |f|_{0,0} < 1\}$. Then W is open in the relative topology of $\mathcal{S}(\mathbb{R}^n)$. If W is a neighborhood of 0 in the relative topology of $\mathcal{E}(\Omega)$, there would be an $\epsilon > 0$ and pairs (K_i, α_i) for $i = 1, 2, \ldots, N$ such that if $|f|_{K_i,\alpha_i} < \epsilon$ for $i = 1, 2, \ldots, N$, then $f \in W$. Set $K = \cup K_i$. Then $K \subseteq \Omega$. Choose a $\phi \in C_c^\infty(\Omega)$ such that $|\phi|_\infty > 1$ and supp $\phi \subseteq \Omega - K$. Then $0 = |\phi|_{K_i,\alpha_i} < \epsilon$ for each i while $|\phi|_{0,0} > 1$. This is a contradiction. □

10. Show there is no countable family of seminorms on $\mathcal{D}_n$ which defines the Schwartz topology on $\mathcal{D}_n$, and as a consequence show the topology on $\mathcal{D}_n$ is not first countable and hence cannot be metrizable.

Solution:

Assume there is a countable collection $|\cdot|_j$ of seminorms which define the topology on $\mathcal{D}_n$. We show this is impossible. To do this, let ϕ be a nonzero C^∞ function with compact support inside $|x| < 1$. For $k = 1, 2, \ldots$, define $\phi_k(x) = \phi(\frac{1}{k}x)$. We use the ϕ_k to define a sequence of sequences. Start with $\phi_{1,k} = c_{1,k}\phi_k$ where $1 \geqslant c_{1,k} > 0$ is chosen so that $|\phi_{1,k}|_1 \leqslant 1$ for all k. Now choose $1 \geqslant c_{2,k} > 0$ so that $\phi_{2,k} = c_{2,k}\phi_{1,k}$ satisfies $|\phi_{2,k}|_2 \leqslant 1$ for all k. Repeating we have $\phi_{j+1,k} = c_{j+1,k}\phi_{j,k}$ where $1 \geqslant c_{j+1,k} > 0$ and $|\phi_{j+1,k}|_{j+1} \leqslant 1$ for all k.

Note $\phi_{j,k} = c_{j,k}c_{j-1,k}\cdots c_{1,k}\phi_k$ and $|\phi_{j,k}|_l = |c_{j,k}c_{j-1,k}\cdots c_{j-l+1,k}\phi_{j-l,j}|_l \leqslant c_{j-l+1,j}|\phi_{j-l,k}|_l \leqslant 1$ for $l = 1, 2, \ldots, j-1$ and also $|\phi_{j,j}|_j \leqslant 1$. Thus one has $|\phi_{k,k}|_j \leqslant 1$ for all $j \leqslant k$ for $k \geqslant 1$. This implies the sequence $\{\phi_{k,k}\}$ is bounded, for if $|\cdot|$ is any continuous seminorm on $\mathcal{D}_n$, there is a N and a $\delta > 0$ such that $|\psi| \leqslant 1$ if $|\psi|_j \leqslant \delta$ for $j = 1, 2, \ldots, N$. In particular, since $|\delta\phi_{k,k}|_j \leqslant \delta$ for all $j \leqslant k$, one has $|\delta\phi_{k,k}|_j \leqslant \delta$ for $j \leqslant N$ if $k \geqslant N$. So $|\delta\phi_{k,k}| \leqslant 1$ for all $k \geqslant N$ and we see $|\phi_{k,k}| \leqslant \frac{1}{\delta}$ for $k \geqslant N$. So by Exercise 2.1.13, the sequence $\{\phi_{k,k}\}$ is bounded and by Proposition 2.70, we see there is a compact subset K of $\mathbb{R}^n$ such that supp $\phi_{k,k} \subseteq K$ for all k. But if $\phi(x) \neq 0$ where $x \neq 0$, then $\phi_{k,k}(kx) = c_{j,k}c_{j-1,k}\cdots c_{1,k}\phi(x) \neq 0$ and clearly $k|x| \to \infty$ and so $kx \notin K$ for large k. But this is impossible.

□

11. A seminorm $||\cdot||$ on $\mathcal{D}(\Omega)$ is said to be admissible if its restriction to every $\mathcal{D}_K(\Omega)$ where $K \subseteq \Omega$ and K is compact is continuous on $\mathcal{D}_K(\Omega)$.

(a) Show the mapping that assigns to each admissible seminorm the open set $\{f \in \mathcal{D}(\Omega) \mid ||f|| < 1\}$ is a one-to-correspondence between all admissible seminorms $||\cdot||$ on $\mathcal{D}(\Omega)$ and all balanced convex sets U contained in $\mathcal{D}(\Omega)$ such that $U \cap \mathcal{D}_K(\Omega)$ is open in $\mathcal{D}_K(\Omega)$ for all compact subsets $K \subseteq \Omega$.

(b) Show a seminorm on $\mathcal{D}(\Omega)$ is continuous if and only if it is admissible.

Solution:

We do (b) first. Let $||\cdot||$ be admissible. We show it is continuous in the inductive limit topology. Namely consider the topology defined by this seminorm. It is locally convex and since $||\cdot||$ is admissible $\{f \in \mathcal{D}_K(\Omega) \mid ||f|| < \epsilon\} = \{f \mid f \in \mathcal{D}(\Omega), ||f|| < \epsilon\} \cap \mathcal{D}_K(\Omega)$ for each $\epsilon > 0$, the relative topology on $\mathcal{D}_K(\Omega)$ is the topology given by the restriction of $||\cdot||$ to $\mathcal{D}_K(\Omega)$. Since this is true for all K, the topology defined by $||\cdot||$ is weaker than the inductive limit topology and therefore $||\cdot||$ is continuous on $\mathcal{D}(\Omega)$.

Conversely, if it is continuous on $\mathcal{D}(\Omega)$, the relative topology from $||\cdot||$ on each $\mathcal{D}_K(\Omega)$ is weaker than the topology of $\mathcal{D}_K(\Omega)$. Therefore $||\cdot||$ restricted to each $\mathcal{D}_K(\Omega)$ is continuous.

For (a), we note by Exercise 2.4.4 that the topology on $\mathcal{D}(\Omega)$ is a limit of a directed collection of locally convex topological spaces $\mathcal{D}_{K_i}(\Omega)$. Since by (b),

$||\cdot||$ is admissible if and only if it continuous, Proposition 2.16 and Lemma 2.6 give the result.

□

12. Let Ω be an open subset of $\mathbb{R}^n$. For each compact subset K of Ω, let $C_K(\Omega)$ be the subset of $C_c(\Omega)$ consisting of those $\phi \in C_c(\Omega)$ with supp $\phi \subseteq K$. Define a norm on this space by $|\phi|_K = \max_{x\in K}|\phi(x)|$. Give $C_c(\Omega)$ the inductive limit topology obtained from the subspaces $C_K(\Omega)$. Show $C_c(\Omega)$ is complete.

Solution:

We first show if B is a bounded subset of $C_c(\Omega)$, there is a compact K with $B \subseteq C_K(\Omega)$. By Exercise 2.4.4, there is a sequence of compact sets K_i with $K_i \subseteq K_{i+1}^\circ$ with $\cup K_i = \Omega$. Assume $B \nsubseteq C_{K_i}(\Omega)$ for all i. Choose $\phi_i \in B$ with $\phi_i \notin C_{K_i}(\Omega)$. Choose $x_i \notin K_i$ with $\phi_i(x_i) \neq 0$. Set $U = \{\phi \mid |\phi(x_j)| < \frac{1}{j}|\phi_j(x_j)| \text{ for all } j\}$. U is a balanced convex subset of $C_c(\Omega)$.

Since the relative topology of $C_{K_i}(\Omega)$ on $C_K(\Omega)$ is the topology of $C_K(\Omega)$ and each compact subset K of Ω is contained in $\cup K_j^\circ$ and hence in K_i for some i, we see the inductive limit topology on $C_c(\Omega)$ is the inductive limit of the $C_{K_i}(\Omega)$. Moreover, the number of x_j in K is finite for $x_i \notin K$ if $K \subseteq K_i$. Thus $U \cap C_K(\Omega)$ is open in $C_K(\Omega)$ for all compact K. By Corollary 2.17, U is open in $C_c(\Omega)$ and since B is bounded, $B \subseteq \lambda U$ for some $\lambda > 0$. This implies $\frac{1}{\lambda}f \in U$ for all $f \in B$. Now $\phi_i \in B$. So $\frac{1}{\lambda}|\phi_i(x_j)| < \frac{1}{j}|\phi_j(x_j)|$ for all j. But this is impossible if $j = i > \lambda$.

Next note the relative topology of $C_c(\Omega)$ on $C_K(\Omega)$ is the topology of $C_K(\Omega)$. Indeed, by definition, we know it is weaker. Now consider the seminorm, $|\cdot|$ on all of $C_c(\Omega)$ given by $|f| = \max_{x\in\Omega}|f(x)|$. It restricted to any $C_{K'}(\Omega)$ is continuous for if $f \in C_{K'}(\Omega)$, $|f| = |f|_{K'}$. Thus by Proposition 2.16, $|\cdot|$ is continuous on $C_c(\Omega)$. Since $|\cdot|_K$ is the restriction of $|\cdot|$ to $C_K(\Omega)$, we see every set open in $C_K(\Omega)$ is open in the relative topology of $C_c(\Omega)$ on $C_K(\Omega)$.

Now let ϕ_k be Cauchy in $C_c(\Omega)$. By Exercise 2.1.14, $\{\phi_k \mid k \in \mathbb{N}\}$ is bounded. Thus there is a compact K with $\phi_k \in C_K(\Omega)$ for all k. Since $C_K(\Omega)$ has the relative topology, ϕ_k is Cauchy in the norm $|\cdot|_K$. But this is the norm of uniform convergence. So $\phi_k \to \phi$ in $C_K(\Omega)$. But then $\phi_k \to \phi$ in $C_c(\Omega)$, because any continuous seminorm on $C_c(\Omega)$ has restriction to $C_K(\Omega)$ which is continuous; i.e. if $|\cdot|$ is a continuous seminorm on $C_c(\Omega)$, its restriction being continuous implies $|\phi_k - \phi| \to 0$ as $k \to \infty$.

□

13. Let Ω be an open subset of $\mathbb{R}^n$. Suppose $\{f_k\}_{k=1}^\infty$ is a sequence of C^∞ functions on Ω such that for each multiindex α, the sequence $\{D^\alpha f_k\}_{k=1}^\infty$ is uniformly Cauchy on Ω. Show there is a C^∞ function f on Ω such that $D^\alpha f_k \to D^\alpha f$ uniformly for all multiindices α.

Solution:

Let Ω be a nonempty subset of $\mathbb{R}^n$ and suppose f_k is a sequence of C^∞ functions such that $D^\alpha f_k$ is uniformly Cauchy for all α. Then each sequence $D^\alpha f_k$ is a uniformly convergent sequence of continuous functions and therefore the limit function g_α is a continuous function. Set $f = g_0$. We show $f \in C^\infty$ by showing $D^\alpha f = g_\alpha$ for all α.

This can be done by induction on $|\alpha|$. Assume true for $|\alpha| \leqslant k$. If $|\beta| = k+1$, then $\beta = \alpha + \epsilon_i$ for some i between 1 and n. Let $x \in \Omega$. Choose $r > 0$ such that the ball of radius r with center x is a subset of Ω. Now by the fundamental theorem of calculus, for $|t| < r$,

$$D^\alpha f_k(x + te_i) = D^\alpha f_k(x) + \int_0^t D^{\epsilon_i}(D^\alpha f_k(x + se_i))\, ds.$$

Thus since $D^{\alpha+\epsilon_i} f_k \to g_{\alpha+\epsilon_i}$ uniformly on Ω, $\int_0^t D^{\alpha+\epsilon_i} f_k(x + se_i)\, ds$ converges to $\int_0^t g_{\alpha+\epsilon_i}(x + se_i)\, ds$ as $k \to \infty$. Taking the limit then gives

$$g_\alpha(x + te_i) = g_\alpha(x) + \int_0^t g_{\alpha+\epsilon_i}(x + se_i)\, ds.$$

Again, by the fundamental theorem,

$$\frac{d}{dt} g_\alpha(x + te_i) = g_{\alpha+\epsilon_i}(x + te_i).$$

So taking $t = 0$, we see $D_i g_\alpha(x) = g_{\alpha+\epsilon_i}(x)$. Since $\beta = \alpha + \epsilon_i$ and $g_\alpha = D^\alpha f$ by the induction hypothesis, we obtain $D^\beta f = g_\beta$.

□

14. Show a subset of $\mathcal{D}(\Omega)$ is compact if and only if it is closed and bounded. (**Hint:** Suppose F is a closed and bounded subset of $\mathcal{D}_(\Omega)$. Then $F \subseteq \mathcal{D}_K(\Omega)$ for some compact subset K, and since this is in a metric space (the relative topology of $\mathcal{D}(\Omega)$ is the topology on $\mathcal{D}_K(\Omega)$), you need only show sequential compactness; use the boundedness of the seminorms defining the topology on $\mathcal{D}_K(\Omega)$ to show one can apply Ascoli's Theorem.)

Solution:

Let F be a closed bounded subset of $\mathcal{D}(\Omega)$. By Proposition 2.70, there is a compact subset K of Ω with $F \subseteq \mathcal{D}_K(\Omega)$. Note by Exercise 2.4.11, the seminorms $|\cdot|_{K,\alpha}$ defined on $\mathcal{D}(\Omega)$ by $|f|_{K,\alpha} = \max_{x \in K} |D^\alpha f(x)|$ are continuous on $\mathcal{D}(\Omega)$ for their restrictions to $\mathcal{D}_{K'}(\Omega)$ are the restrictions of $|\cdot|_{K \cap K',\alpha}$ to this space and on this space $|\cdot|_{K \cap K',\alpha} \leqslant |\cdot|_{K,\alpha}$ (we are using the topology on $\mathcal{D}_K(\Omega)$ is the topology defined by the seminorms $|\cdot|_{K,\alpha}$ on this space). Thus the seminorms $|\cdot|_{K,\alpha}$ are bounded on F. Moreover by Lemma 2.66, $\mathcal{D}_K(\Omega)$ has the relative topology of $\mathcal{D}(\Omega)$. Hence F is closed in $\mathcal{D}_K(\Omega)$. As seen in Exercise 2.4.6, every sequence in F has a subsequence which converges in $\mathcal{D}_K(\Omega)$. The limit is in F since F is closed.

Thus F is a sequentially compact subset of $\mathcal{D}_K(\Omega)$. Since $\mathcal{D}_K(\Omega)$ is metrizable, F is compact in $\mathcal{D}_K(\Omega)$. Now let U_γ be an open cover of F in $\mathcal{D}(\Omega)$. Since the relative topology of $\mathcal{D}(\Omega)$ on $\mathcal{D}_K(\Omega)$ is the topology of $\mathcal{D}_K(\Omega)$, we see $U_\gamma \cap \mathcal{D}_K(\Omega)$ is an open cover of F in $\mathcal{D}_K(\Omega)$. So there is a finite subcover $U_{\gamma_j} \cap \mathcal{D}_K(\Omega)$, $j = 1, 2, \ldots, m$. Consequently U_{γ_j} cover F in $\mathcal{D}(\Omega)$ and so F is compact in $\mathcal{D}(\Omega)$.

Conversely, since every compact subset of a Hausdorff space is closed and every continuous seminorm is bounded on a compact set, we see every compact subset of $\mathcal{D}(\Omega)$ is closed and bounded.

□

15. In Lemma 2.59, we showed $x \mapsto \lambda(x)\phi$ is a continuous mapping from $\mathbb{R}$ into $\mathcal{S}(\mathbb{R}^n)$ for $\phi \in \mathcal{S}(\mathbb{R}^n)$. Suppose $f \in \mathcal{D}(\mathbb{R}^n)$. Show $x \mapsto \lambda(x)f$ is a continuous mapping of $\mathbb{R}$ into $\mathcal{D}(\mathbb{R}^n)$.
Solution: Let G be an open subset of $\mathcal{D}(\mathbb{R}^n)$ and suppose $\lambda(x_0)f \in G$. Then there is an open balanced convex subset U of $\mathcal{D}(\mathbb{R}^n)$ such that $\lambda(x_0)f + U \subseteq G$. We show there is an open neighborhood B of x_0 with $\lambda(x)f \in \lambda(x_0)f + U$ for $x \in B$.

Pick a compact neighborhood N of x_0 and let $K = \operatorname{supp} f + N$. Then K is compact. Since U is open, convex, and balanced in the inductive limit topology, $U_K = U \cap \mathcal{D}_K(\mathbb{R}^n)$ is an open subset of $\mathcal{D}_K(\mathbb{R}^n)$. Thus there is a $\delta > 0$ and a finite subset F of $\mathbb{N}_0^n$ such that if $\phi \in \mathcal{D}_K(\Omega)$ and $|\phi|_{K,\alpha} < \delta$ for $\alpha \in F$, then $\phi \in U$. Now note if $x \in N$, then supp $\lambda(x)f \subseteq K$. Indeed, $f(y-x) \neq 0$ implies $y \in x + \operatorname{supp} f \subseteq N + \operatorname{supp} f = K$. So $\lambda(x)f \in \mathcal{D}_K(\Omega)$ for $x \in N$. Now by uniform continuity of all $D^\alpha f$ for $\alpha \in F$, there is a $\delta' > 0$ such that if $|x - x_0| < \delta'$, then $x \in N$ and $|D^\alpha f(y-x) - D^\alpha f(y - x_0)| < \delta$ for all y. So $|\lambda(x)f - \lambda(x_0)f|_{K,\alpha} < \delta$ for all $x \in B$ where $B = B_{\delta'}(x_0) \subseteq N$. Thus $\lambda(x)f - \lambda(x_0)f \in U$ for $x \in B$. □

16. Let Ω_1 and Ω_2 be nonempty subsets of $\mathbb{R}^m$ and $\mathbb{R}^n$, respectively. Let $\Phi \in \mathcal{D}(\Omega_1 \times \Omega_2)$. For $x \in \Omega_1$, define Φ_x by $\Phi_x(y) = \Phi(x, y)$. Show $x \mapsto \Phi_x$ is continuous mapping from Ω_1 into $\mathcal{D}(\Omega_2)$.
Solution: We need to show $x \mapsto \Phi_x$ is continuous into $\mathcal{D}(\Omega_2)$. Fix $x \in \Omega_1$. Let G be an open subset of $\mathcal{D}(\Omega_2)$ and suppose $\Phi_x \in G$. Then there is an open convex balanced subset U of $\mathcal{D}(\Omega_2)$ with $\Phi_x + U \subseteq G$. Let K be the compact support of Φ. Take a compact product subset $K_1 \times K_2$ of $\Omega_1 \times \Omega_2$ containing K and such that x is interior to K_1. Note $U \cap \mathcal{D}_{K_2}(\Omega_2)$ is an open subset of $\mathcal{D}_{K_2}(\Omega_2)$. Note $\Phi_y \in \mathcal{D}_{K_2}(\Omega_2)$ if $y \in K_1$. Since $U \cap \mathcal{D}_{K_2}(\Omega_2)$ is open in $\mathcal{D}_{K_2}(\Omega_2)$, there is a $\delta > 0$ and finite set F of α of n-tuples such that $\phi \in U$ if $\phi \in \mathcal{D}_{K_2}(\Omega_2)$ and $|\phi|_{K_{2,\alpha}} < \delta$ for $\alpha \in F$. If $\alpha \in F$, set $\alpha' = \overbrace{(0, 0, \ldots, 0)}^{m} \times \alpha$.

Now for $y \in K_1$, $|\Phi_y - \Phi_x|_{K_2,\alpha} = \max_{z \in K_2} |D^{\alpha'}\Phi(y, z) - D^{\alpha'}\Phi(x, z)|$. Since the $D^{\alpha'}\Phi$ are uniformly continuous on $K_1 \times K_2$ and K_1 is a neighborhood of x in Ω_1, there is a $\delta' > 0$ such that $y \in K_1$ and $|\Phi_y - \Phi_x|_{K_2,\alpha} < \delta$ if $|y - x| < \delta'$ and $\alpha \in F$. So $\Phi_y - \Phi_x \in U \cap \mathcal{D}_{K_2}(\Omega_2)$ if $|y - x| < \delta'$. So $\Phi_y \in G$ if $|y - x| < \delta'$ and we see since x is arbitrary, $x \mapsto \Phi_x$ is continuous.

□

17. Let $\Omega \subset \mathbb{R}^n$ be open and non-empty. For $\alpha \in \mathbb{N}_0^n$, let $\mathcal{C}_c^\alpha(\Omega)$ be the space of compactly supported functions such that $D^\beta f \in \mathcal{C}(\Omega)$ for all $\beta \leqslant \alpha$. For each compact set $K \subset \Omega$, let $\mathcal{C}_K^\alpha(\Omega) = \{f \in \mathcal{C}^\alpha(\Omega) \mid \operatorname{supp}(f) \subset K\}$. The seminorms $|\cdot|_{K,\beta}$, $\beta \leqslant \alpha$ define a topology on $\mathcal{C}_K^\alpha(\Omega)$. Denote the corresponding topological vector space by $\mathcal{D}_K^\alpha(\Omega)$. Let $\mathcal{D}^\alpha(\Omega)$ have the inductive limit topology of the subspaces $\mathcal{D}_K^\alpha(\Omega)$ where K is a compact subset of Ω. Show the following:

(a) $\mathcal{D}_K^\alpha(\Omega)$ and $\mathcal{D}^\alpha(\Omega)$ are complete.

(b) The inclusion maps $\mathcal{D}_K(\Omega) \hookrightarrow \mathcal{D}_K^\alpha(\Omega)$ and $\mathcal{D}(\Omega) \hookrightarrow \mathcal{D}^\alpha(\Omega)$ are continuous.

Solution:

The seminorms $|\cdot|_{K,\beta}$ for $\beta \leqslant \alpha$ define a locally convex metrizable topology on $\mathcal{C}_K^\alpha(\Omega)$. This topology is complete for Cauchy sequences ϕ_k satisfy all $D^\beta\phi_k$ converge uniformly to a $D^\beta\phi$ as $k \to \infty$. For details, see the solution to Exercise 2.4.13. $\mathcal{D}^\alpha(\Omega)$ is $\mathcal{C}_c^\alpha(\Omega)$ with the inductive limit topology of the $\mathcal{D}_K^\alpha(\Omega)$. The same argument as in the proof of Lemma 2.66 shows the relative topology of $\mathcal{D}^\alpha(\Omega)$ on $\mathcal{D}_K^\alpha(\Omega)$ is the topology of $\mathcal{D}_K^\alpha(\Omega)$. Moreover, the topology on $\mathcal{D}^\alpha(\Omega)$ is an inductive limit of a directed sequence $\mathcal{D}_{K_i}^\alpha(\Omega)$ where $K_i \subseteq K_{i+1}^\circ$ and $\cup K_i = \Omega$. Thus the argument in the proof of Proposition 2.70 shows every bounded set in $\mathcal{D}^\alpha(\Omega)$ is a subset of $\mathcal{D}_K^\alpha(\Omega)$ for some compact subset K of Ω. The argument in Corollary 2.72 then shows $\mathcal{D}^\alpha(\Omega)$ is complete and this finishes (a).

For (b), note if $\beta \leqslant \alpha$, then the seminorm $|\cdot|_{K,\beta}$ on $\mathcal{D}_K^\alpha(\Omega)$ restricted to $\mathcal{D}_K(\Omega)$ is a continuous seminorm (actually a defining seminorm) on $\mathcal{D}_K(\Omega)$. Thus every neighborhood $\{\phi \in \mathcal{D}_K^\alpha(\Omega) \mid |\phi|_{K,\alpha} < \epsilon\}$, $\epsilon > 0$ of 0 in $\mathcal{D}_K^\alpha(\Omega)$ has intersection with $\mathcal{D}_K(\Omega)$ that is a neighborhood of 0 in $\mathcal{D}_K(\Omega)$. This implies the inclusion mapping from $\mathcal{D}_K(\Omega)$ into $\mathcal{D}_K^\alpha(\Omega)$ is continuous.

To see the inclusion mapping from $\mathcal{D}(\Omega)$ into $\mathcal{D}^\alpha(\Omega)$ is continuous, we can use Proposition 2.15. Indeed note the mapping $\phi \mapsto \phi$ is continuous from $\mathcal{D}_K(\Omega)$ into $\mathcal{D}_K^\alpha(\Omega)$ and $\mathcal{D}_K^\alpha(\Omega)$ has the relative topology from $\mathcal{D}^\alpha(\Omega)$. This implies the mapping $\phi \mapsto \phi$ is continuous from $\mathcal{D}_K(\Omega)$ into $\mathcal{D}^\alpha(\Omega)$.

□

18. Show that $\mathcal{D}_K(\Omega)$ is closed subspace of $\mathcal{D}(\Omega)$ for all compact subsets $K \subset \Omega$. (**Hint:** Let $f \notin \mathcal{D}_K(\Omega)$ and choose $x_0 \notin K$ with $f(x_0) \neq 0$. Let $U = \{\phi \in \mathcal{D}(\Omega) \mid |\phi(x_0)| < |f(x_0)|\}$; show U is open in $\mathcal{D}(\Omega)$ and $f + U$ misses $\mathcal{D}_K(\Omega)$.)

Solution:

Let $f \in \mathcal{D}(\Omega)$ and suppose $f \notin \mathcal{D}_K(\Omega)$ where K is a compact subset of Ω. Let $x \notin K$ with $f(x) \neq 0$. Consider the set $U = \{\phi \in \mathcal{D}(\Omega) \mid |\phi(x)| < |f(x)|\}$. Note U is a balanced convex set. Moreover if K' is a compact subset of Ω, then $U \cap \mathcal{D}_{K'}(\Omega)$ is the set of all ϕ in $\mathcal{D}_{K'}(\Omega)$ with $|\phi(x)| < |f(x)|$. Note this is open in $\mathcal{D}_{K'}(\Omega)$ for if $\phi \in U \cap \mathcal{D}_{K'}(\Omega)$ and $\epsilon = |f(x)| - |\phi(x)|$ and $|\psi - \phi|_{K',0} < \frac{\epsilon}{2}$ where $\psi \in \mathcal{D}_{K'}(\Omega)$, then $|\psi(x)| \leqslant |\psi(x) - \phi(x)| + |\phi(x)| < \frac{\epsilon}{2} + |\phi(x)| - |f(x)| + |f(x)| = -\frac{\epsilon}{2} + |f(x)| \leqslant |f(x)|$. So $\psi \in U \cap \mathcal{D}_{K'}(\Omega)$. This implies $U \cap \mathcal{D}_{K'}(\Omega)$ is open in $\mathcal{D}_{K'}(\Omega)$. Since the collection of all subspaces $\mathcal{D}_{K'}(\Omega)$ is directed, Corollary 2.17 gives that U is an open subset of $\mathcal{D}(\Omega)$. Moreover $f + U$ misses $\mathcal{D}_K(\Omega)$ for $f(x) + \phi(x) \neq 0$ for all $\phi \in U$. This implies the complement of $\mathcal{D}_K(\Omega)$ in $\mathcal{D}(\Omega)$ is an open set. So $\mathcal{D}_K(\Omega)$ is a closed linear subspace of $\mathcal{D}(\Omega)$.

□

19. Let K be a compact subset of an open subset Ω of $\mathbb{R}^n$. Show if $\mathcal{F}$ is a closed subset of $\mathcal{D}_K(\Omega)$ and $\alpha \in \mathbb{N}_0^n$, then $D^\alpha \mathcal{F}$ is a closed subset of $\mathcal{D}_K(\Omega)$.

Solution:

Let $\mathcal{F}$ be a closed subset of $\mathcal{D}_K(\Omega)$ and suppose $\alpha \in \mathbb{N}_0^n$. We prove by induction on $|\alpha|$ that $D^\alpha \mathcal{F}$ is a closed subset of $\mathcal{D}_K(\Omega)$. Indeed, this is true if $|\alpha| = 0$. So assume it is true if $|\alpha| \leqslant k$. If $|\beta| = k+1$, then $\beta = \alpha + \epsilon_i$ for some i where $|\alpha| = k$ and then $D^\beta \mathcal{F} = D_i D^\alpha \mathcal{F} = D_i(D^\alpha \mathcal{F})$. By induction we know $\mathcal{D}^\alpha \mathcal{F}$ is closed. So to show it is true for $k+1$, it suffices to show $D_i \mathcal{F}$ is closed in $\mathcal{D}_K(\Omega)$ if $\mathcal{F}$ is closed in $\mathcal{D}_K(\Omega)$. Next note we may assume we are in $\mathcal{D}_K = \mathcal{D}_K(\mathbb{R}^n)$ by extending each $f \in \mathcal{F}$ to all of $\mathbb{R}^n$ by taking $\tilde{f}(y) = 0$ for $y \notin K$.

Hence suppose $D_i f_k$, $k = 1, 2, \ldots$ is Cauchy in $\mathcal{D}_K$. Hence $|D_i f_k - D_i f_l|_{K,\alpha} \to 0$ as $k, l \to \infty$ for all α. This implies $|f_k - f_l|_{K,\alpha+\epsilon_i} \to 0$ as $k, l \to \infty$ for any α. Thus $|f_k - f_l|_{K,\beta} \to 0$ if $\beta_i \in \mathbb{N}$. Hence assume $\beta_i = 0$. We claim $D^\beta f_k$ and $D_j D^\beta f_k$ are uniformly bounded. First fix a and b so that if x is a point in K, a is smaller than any coordinate of x and b is larger than any coordinate of x.

Since $D_i D^\beta f_k = D^{\beta+\epsilon_i} f_k$ converges uniformly, the functions $|D_i D^\beta f_k|$ have a uniform bound M; i.e. $|D_i D^\beta f_k(x)| \leqslant M$ for all x.

Then

$$
\begin{aligned}
|D^\beta f_k(x)| &= \left| \int_a^{x_i} D_i D^\beta f_k(x_1, x_2, \ldots, x_{i-1}, t, x_{i+1}, \ldots x_n)\, dt \right| \\
&\leqslant \int_a^{x_i} |D^{\beta+\epsilon_i} f_k(x_1, x_2, \ldots, x_{i-1}, t, x_{i+1}, \ldots x_n)|\, dt \\
&\leqslant \int_a^{x_i} M\, dt \leqslant M(x_i - a) \leqslant Mb.
\end{aligned}
$$

So the sequence $\{D^\beta f_k\}$ is uniformly bounded. Replacing β by $\beta + \epsilon_j$ where $j = 1, 2, \ldots, n$ and the same argument show $|D_i D^\beta f_k|$ is uniformly bounded.

We now show each subsequence of the sequence $D^\beta f_k$ has a subsequence which converges uniformly on K. To do this we apply the Ascoli Theorem. We have just showed the $D^\beta f_k$ are uniformly bounded. They are pointwise equicontinuous since all $D_j D^\beta f_k$ are uniformly bounded.

To finish we enumerate the β with $\beta_i = 0$ into a sequence $\alpha_1, \alpha_2, \ldots$ where $\alpha_1 = (0, 0, 0, \ldots, 0)$ and do Cantor diagonalization. Pick a subsequence $\{g_{1,j}\}$ of the $\{f_k\}$ such that the sequence $g_{1,j} = D^{\alpha_1} g_{1,j}$ converges uniformly. Then pick a subsequence $\{g_{2,j}\}$ of $\{g_{1,j}\}$ such that $D^{\alpha_2} g_{2,j}$ converges uniformly. Repeat this process and take the diagonal sequence $\{g_{j,j}\}$. We then have $D^\beta g_{j,j}$ converges uniformly for all β with $\beta_i = 0$. Thus the sequence $\{g_{j,j}\}$ is a subsequence of the sequence $\{f_k\}$ and $D^\alpha g_{j,j}$ converges uniformly for all α. This implies $g_{j,j}$ converges in $\mathcal{D}_K(\mathbb{R}^n)$ to its uniform limit g. Since the $g_{j,j} \in \mathcal{F}$ and $\mathcal{F}$ is closed, $g \in \mathcal{F}$. Moreover, $\lim_k D_i f_k = \lim_k D_i g_{j,j} = Dg$. Thus $D_i \mathcal{F}$ is closed.

□

20. Let $1 \leqslant p < \infty$, and $K \subset L^p(\mathbb{R}^n)$ be a closed and bounded subset. Show K is compact in $L^p(\mathbb{R}^n)$ if and only if for every $\epsilon > 0$ there exists a $\delta = \delta(\epsilon) > 0$

and a compact subset $L \subset \mathbb{R}^n$, $L^o \neq \varnothing$, such that for each $f \in K$ and $y \in \mathbb{R}^n$ with $|y| < \delta$ we have

$$\left(\int |f(x+y) - f(x)|^p \, d\mu(x)\right)^{1/p} < \epsilon \quad \text{and} \quad \left(\int_{L^c} |f(x)|^p \, d\mu\right)^{1/p} < \epsilon\,.$$

Solution:

Let K be a closed bounded subset of $L^P(\mathbb{R}^n)$ and suppose K is compact. For each $\epsilon > 0$, let $U(\delta) = \{f \in L^p \mid |\lambda(y)f - f|_p < \epsilon \text{ for } |y| < \delta\}$. Use Lemma 2.74 or its proof to see the sets $U(\delta)$ for $\delta > 0$ cover $L^p(\mathbb{R}^n)$. We next note each $U(\delta)$ is open. In fact if $f_0 \in U(\delta)$, then $\epsilon_1 = \epsilon - |\lambda(y)f_0 - f_0|_p > 0$ and if $|f - f_0|_p < \frac{\epsilon_1}{2}$, then

$$\begin{aligned}|\lambda(y)f - f|_p &\leqslant |\lambda(y)f - \lambda(y)f_0|_p + |\lambda(y)f_0 - f_0|_p + |f - f_0|_p \\ &= 2|f - f_0|_p + |\lambda(y)f_0 - f_0|_p \\ &< \epsilon_1 + |\lambda(y)f_0 - f_0|_p = \epsilon.\end{aligned}$$

Hence the $U(\delta)$ for $\delta > 0$ cover K. This implies there is a finite subcover, $U(\delta_k)$, $k = 1, 2, \ldots, m$. Set $\delta = \min\{\delta_1, \delta_2, \ldots, \delta_m\}$. Then $|\lambda(y)f - f|_p < \epsilon$ if $|y| < \delta$.

Next for any compact subset L of $\mathbb{R}^n$ with nonempty interior, let $U(L)$ be the subset of $L^p(\mathbb{R}^n)$ consisting of those f with $\left(\int_{L^c} |f(x)|^p \, dx\right)^{1/p} < \epsilon$ where here ϵ is a fixed positive number.

Since $|f|^p$ is integrable, the collection $U(L)$ for L compact covers $L^p(\mathbb{R}^n)$. Moreover, each $U(L)$ is open in $L^p(\mathbb{R}^n)$ for if $f_0 \in U(L)$, and $|f - f_0|_p < \epsilon - \left(\int_{L^c} |f_0(x)|^p \, dx\right)^{1/p}$, then

$$\begin{aligned}\left(\int_{L^c} |f(x)|^p \, dx\right)^{1/p} &\leqslant \left(\int_{L^c} |f(x) - f_0(x)|^p \, dx\right)^{1/p} + \left(\int_{L^c} |f_0(x)|^p \, dx\right)^{1/p} \\ &\leqslant \left(\int_{\mathbb{R}^n} |f(x) - f_0(x)|^p \, dx\right)^{1/p} + \left(\int_{L^c} |f_0(x)|^p \, dx\right)^{1/p} \\ &< \epsilon - \left(\int_{L^c} |f_0(x)|^p \, dx\right)^{1/p} + \left(\int_{L^c} |f_0(x)|^p \, dx\right)^{1/p} = \epsilon.\end{aligned}$$

Since K is compact the $U(L)$ cover K and hence there exist $L_1, L_2, \ldots, L_m$ which are compact sets with interior so that $K \subseteq \bigcup U(L_k)$. Take $L = \bigcup_{k=1}^m L_k$. Then $\left(\int_{L^c} |f(x)|^p \, dx\right)^{1/p} < \epsilon$ if $f \in K$.

For the converse, let K be a closed bounded set and suppose one has for each $\epsilon > 0$, a $\delta > 0$ with $|\lambda(-y)f - f|_p < \epsilon$ if $|y| < \delta$ and a compact subset L of $\mathbb{R}^n$ with interior so that $\left(\int_{L^c} |f(x)|^p \, dx\right)^{1/p} < \epsilon$ for all $f \in K$. Here we are using $\lambda(y)f(x) = f(x - y)$. Note we have $y \mapsto \lambda(y)f$ is continuous from $\mathbb{R}^n$ into $L^p(\mathbb{R}^n)$, for this can be checked for continuous f with compact support and then extend this by applying the density of these functions in L^P for $1 \leqslant p < \infty$. This also follows from Lemma 2.74 in the next section.

Recall that a subset of a metric space is compact if and only if it is complete and totally bounded. Since K is closed in a complete metric space, it is complete. So our goal will be to show it is totally bounded.

Fix $\epsilon > 0$. For $r > 0$, let $B_r(0)$ be the closed ball of radius $r > 0$ having center 0. We can choose a compact subset L of $\mathbb{R}^n$ which contains $B_1(0)$ such that $\left(\int_{L^c} |f(x)|^p \, dx\right)^{1/p} < \frac{\epsilon}{6}$ for $f \in K$.

For $1 > \delta > 0$, we define $A_\delta f(x) = \frac{1}{m(B_\delta(0))} \int_{B_\delta(0)} \lambda(-y) f(x) \, dy$ where here we are using m for Lebesgue measure. We establish some interesting facts about $A_\delta f$. Note if $\frac{1}{p} + \frac{1}{q} = 1$, then

$$\begin{aligned} A_\delta f(x) &= \frac{1}{m(B_\delta(0))} \int_{B_\delta(0)} f(x+y) \, dy \\ &= \int \lambda(-x) f(y) \, \frac{1}{m(B_\delta(0))} \chi_{B_\delta(0)}(y) \, dy \\ &= (\lambda(-x) f, g_\delta) \end{aligned}$$

where g_δ is the $L^q(\mathbb{R}^n)$ function given by $g_\delta(x) = \frac{1}{m(B_\delta(0))} \chi_{B_\delta(0)}(x)$. In fact, note $|g_\delta|_q^q = \frac{m(B_\delta(0))}{m(B_\delta(0))^q} = \frac{1}{m(B_\delta(0))^{q-1}}$ and thus $|g_\delta|_q = \frac{1}{m(B_\delta(0))^{1/p}}$; this working also for the $p = 1$, $q = \infty$ pair. Since the pairing $(f, g) = \int fg \, dm$ is bicontinuous on $L^p \times L^q$ by Hölder's inequality, we see $A_\delta f$ is continuous for all f in L^p. The collection $\{A_\delta f \mid f \in K\}$ is also an equicontinuous family. Indeed,

$$\begin{aligned} |A_\delta(f)(x+z) - A_\delta(f)(x)| &= |(\lambda(-x-z) f, g_\delta) - (\lambda(-x) f, g_\delta)| \\ &= |(\lambda(-x-z) f - \lambda(-x) f, g_\delta)| \\ &\leqslant |\lambda(-x-z) f - \lambda(-x) f|_p |g_\delta|_q \\ &= \left(\int |f(x+z+y) - f(x+y)|^p \, dy\right)^{1/p} |g_\delta|_q \\ &= \left(\int |f(z+y) - f(y)|^p \, dy\right)^{1/p} |g_\delta|_q \end{aligned}$$

which is uniformly small for all x and $f \in K$ if z is near 0.

The $\{A_\delta(f) \mid f \in K\}$ is also uniformly bounded on $\mathbb{R}^n$. Indeed we have seen $A_\delta f(x) = (\lambda(-x) f, g_\delta)$ and since K is bounded in L^p, we see $|A_\delta(f)(x)| \leqslant |\lambda(-x) f|_p |g_\delta|_q = |f|_p |g_\delta|_q$ is uniformly bounded in x and $f \in K$.

Next we note $A_\delta f$ is an L^p function and $|A_\delta f|_p \leqslant |f|_p$. To do this we use

Lemma 2.29

$$\left(\int_{\mathbb{R}^n}\left|\frac{1}{m(B_\delta(0))}\int_{B_r(0)} f(x+y)\,dy - f(x)\right|^p dx\right)^{1/p}$$

$$= \left(\int_{\mathbb{R}^n}\left|\frac{1}{m(B_\delta(0))}\int_{B_r(0)} (f(x+y) - f(x))dy\right|^p dx\right)^{1/p}$$

$$\leqslant \left(\int_{\mathbb{R}^n}\left(\frac{1}{m(B_\delta(0))}\int_{B_r(0)} |f(x+y) - f(x)|dy\right)^p dx\right)^{1/p}$$

$$\leqslant \frac{1}{m(B_\delta(0))}\left(\int_{\mathbb{R}^n}\left(\int_{B_r(0)} |f(x+y) - f(x)|\,dy\right)^p dx\right)^{1/p}$$

$$\leqslant \frac{1}{m(B_\delta(0))}\int_{B_r(0)}\left(\int_{\mathbb{R}^n} |f(x+y) - f(x)|^p dx\right)^{1/p} dy \leqslant \max_{y\in B_\delta(0)} |\lambda(y)f - f|_p.$$

So $A_\delta f - f \in L^p$ and thus $A_\delta f \in L^p$ for $f \in L^p$.

Pick $1 > \delta > 0$ with $|\lambda(y)f - f|_p < \frac{\epsilon}{6}$ if $|y| \leqslant \delta$ and $f \in K$. Then $|A_\delta(f) - f|_p < \frac{\epsilon}{6}$ for $f \in K$. Fix this δ. Consider the family $\mathcal{F} = \{A_\delta(f)|_L \mid f \in K\}$. This is a pointwise uniformly bounded equicontinuous family of functions in $C(L)$. By Ascoli's Theorem its closure is sequentially compact. Thus it is totally bounded. So there exist $g_1, g_2, \ldots, g_m \in C(L)$ such that $\mathcal{F} \subseteq \bigcup_{k=1}^m \{g \mid |g - g_k|_\infty < \frac{\epsilon}{3m(L)^{1/p}}\}$. Thus if $f \in K$, there is a k with $|A_\delta(f)(x) - g_k(x)| < \frac{\epsilon}{3m(L)^{1/p}}$ if $x \in L$. Define G_k on $\mathbb{R}^n$ by $G_k(x) = 0$ off L and $G_k(x) = g_k(x)$ for $x \in L$. Then if $f \in K$,

$$|f - G_k|_p \leqslant |f - A_\delta(f)|_p + |A_\delta(f) - A_\delta(f)\chi_L|_p + |A_\delta(f)\chi_L - G_k|_p$$

$$< \frac{\epsilon}{6} + \left(\int_{L^c} |A_\delta(f)(x)|^p\,dx\right)^{1/p} + \left(\int_L \left(\frac{\epsilon}{3m(L)^{1/p}}\right)^p dx\right)^{1/p}$$

$$= \frac{\epsilon}{6} + \left(\int_{L^c} |A_\delta(f)(x)|^p\,dx\right)^{1/p} + \frac{\epsilon}{3} < \epsilon$$

for again by Lemma 2.29

$$
\begin{aligned}
\left(\int_{L^c} |A_\delta(f)(x)|^p\,dx\right)^{1/p} &= \left(\int_{L^c} \left|\frac{1}{m(B_\delta(0))}\int_{B_\delta(0)} \lambda(-y)f(x)\,dy\right|^p dx\right)^{1/p} \\
&\leqslant \frac{1}{m(B_\delta(0))}\left(\int_{L^c}\left(\int_{B_\delta(0)} |\lambda(-y)f(x)|\,dy\right)^p dx\right)^{1/p} \\
&\leqslant \frac{1}{m(B_\delta(0))}\int_{B_\delta(0)}\left(\int_{L^c} |\lambda(-y)f(x)|^p dx\right)^{1/p} dy \\
&\leqslant \frac{1}{m(B_\delta(0))}\int_{B_\delta(0)} \frac{\epsilon}{3}dy = \frac{\epsilon}{3}
\end{aligned}
$$

since

$$
\begin{aligned}
\left(\int_{L^c} |\lambda(-y)f(x)|^p\,dx\right)^{1/p} &= |(\lambda(-y)f - f)\chi_{L^c} + f\chi_{L^c}|_p \\
&\leqslant |(\lambda(-y)f - f)\chi_{L^c}|_p + |f\chi_{L^c}|_p \\
&\leqslant |\lambda(-y)f - f|_p + |f\chi_{L^c}|_p \\
&\leqslant \frac{\epsilon}{6} + \frac{\epsilon}{6}
\end{aligned}
$$

if $|y| \leqslant \delta$ and $f \in K$. Thus the neighborhoods $\{F \in L^p \mid |F - G_k|_p < \epsilon\}$ for $k = 1, 2, \ldots, m$ cover K. Since $\epsilon > 0$ was arbitrary, K is totally bounded.

□

Exercise Set 2.5

1. Suppose f and g are in $L^1(\mathbb{R}^n)$. Use Fubini's Theorem to show $f * g$ is in $L^1(\mathbb{R}^n)$ and then

$$|f * g|_1 \leqslant |f|_1|g|_1.$$

Solution: Note

$$
\begin{aligned}
\int\int |f(y)g(x-y)|\,dy\,dx &= \int\int |f(y)||g(x)|\,dx\,dy \\
&= |f|_1|g|_1
\end{aligned}
$$

and thus $y \mapsto f(y)g(x-y)$ is integrable for a.e. x. Furthermore we then have $f * g$ is L^1 and $|f * g|_1 \leqslant |f|_1|g|_1$. □

2. Suppose f is in $L^1(\mathbb{R}^n)$ and $g \in L^p(\mathbb{R}^n)$ where $1 < p \leqslant \infty$. Choose q with $\frac{1}{p} + \frac{1}{q} = 1$. Use the fact that the dual of $L^q(\mathbb{R}^n)$ is $L^p(\mathbb{R}^n)$ to show $f * g(x)$ exists a.e. and $|f * g|_p \leqslant |f|_1|g|_p$.

Solution: Define T on L^q by

$$T(h) = \int\int f(y)g(x-y)h(x)\,dy\,dx.$$

Note

$$\begin{aligned}\int\int |f(y)g(x-y)h(x)|\,dy\,dx &= \int |f(y)| \int |\lambda(y)g(x)h(x)|\,dx\,dy \\ &\leqslant \int |f(y)|\,|\lambda(y)g|_p|h|_q\,dy \\ &\leqslant |f|_1|g|_p|h|_q.\end{aligned}$$

By Fubini's Theorem, $y \mapsto f(y)g(x-y)$ is integrable for a.e. x (take h in L^q which is positive everywhere) and T is a bounded linear functional on L^q with $||T|| \leqslant |f|_1|g|_p$.

Thus there is a function F in L^p such that

$$T(h) = \int F(x)h(x)\,dx.$$

So $F = f * g$ and $|F|_p = ||T|| \leqslant |f|_1|g|_p$. □

3. Let

$$f(x) = \begin{cases} 1 & 0 \leqslant x < 1 \\ 0 & \text{otherwise.} \end{cases}$$

For $N \in \mathbb{N}$, define f_N by $f_1 = f$ and $f_{N+1} = f_N * f$.

(a) Evaluate the functions f_2 and f_3.

(b) Show that $\operatorname{supp}(f_N) = [0, N]$ and $f_N(x) > 0$ for $x \in (0, N)$.

(c) For $N \geqslant 2$ show that $f_N \in C^{N-2}(\mathbb{R})$.

(d) If $g \in C(\mathbb{R})$, show

$$\int_{-\infty}^{\infty} g(x)f_N(x)\,dx = \int_0^1 \ldots \int_0^1 g(x_1 + \ldots + x_N)\,dx_1 \ldots dx_N$$

(e) Show $\sum_{k=-\infty}^{\infty} f_N(x-k) = 1$ for all $x \in \mathbb{R}$.

(f) Verify each function f_N is symmetric with respect to the center of $[0, N]$, i.e.,

$$f_N(\frac{N}{2} + x) = f_N(\frac{N}{2} - x)\,.$$

(g) Show the following recursion formula holds:

$$f_{N+1}(x) = \frac{x}{N} f_N(x) + \frac{N+1-x}{N} f_N(x-1)\,.$$

Solution:

First note if g is integrable, then

$$\begin{aligned} f * g(x) &= \int f(y)g(x-y)\,dy \\ &= \int_0^1 g(x-y)\,dy \\ &= \int_{x-1}^{x} g(y)\,dy. \end{aligned}$$

For (a), note

$$\begin{aligned} f_2(x) &= \int f(x-y)f(y)\,dy \\ &= \int_0^1 f(x-y)\,dy \\ &= \int 1_{[0,1]}(y)1_{[0,1]}(x-y)\,dy \\ &= \int 1_{[0,1]\cap[x-1,x]}(y)dy \\ &= \lambda([0,1]\cap[x-1,x]) \\ &= \begin{cases} 0 & \text{if } x<0 \\ x & \text{if } 0\leqslant x\leqslant 1 \\ 2-x & \text{if } 1<x\leqslant 2 \\ 0 & \text{if } 2<x\,. \end{cases} \end{aligned}$$

Next,

$$\begin{aligned} f_3(x) &= \int \chi_{[0,1]}(x-y)f_2(y)\,dy \\ &= \int \chi_{[x-1,x]}(y)f_2(y)\,dy \\ &= \begin{cases} 0 & \text{if } x<0 \\ \int_0^x y\,dy & \text{if } 0\leqslant x\leqslant 1 \\ \int_{x-1}^1 y\,dy + \int_1^x (2-y)\,dy & \text{if } 1<x\leqslant 2 \\ \int_{x-1}^2 (2-y)\,dy & \text{if } 2<x\leqslant 3 \\ 0 & \text{if } 3<x \end{cases} \\ &= \begin{cases} 0 & \text{if } x<0 \\ \frac{x^2}{2} & \text{if } 0\leqslant x\leqslant 1 \\ -\frac{3}{2}+3x-x^2 & \text{if } 1<x\leqslant 2 \\ \frac{(3-x)^2}{2} & \text{if } 2<x\leqslant 3 \\ 0 & \text{if } 3<x\,. \end{cases} \end{aligned}$$

We note $f_3(x)$ is C^1 and vanishes off $[0,3]$.

For (b), we note it is true for $N = 1, 2$. Suppose true for N. Then

$$\begin{aligned} f_{N+1}(x) &= \int 1_{[0,1]}(x-y)F_N(y)\,dy \\ &= \int 1_{[x-1,x]}(y)F_N(y)\,dy \\ &= \begin{cases} 0 & \text{if } x \leqslant 0 \text{ or } x \geqslant N+1 \\ \int_{x-1}^{x} F_N(y)\,dy > 0 & \text{otherwise} \end{cases} \end{aligned}$$

for $F_N(y)$ is a positive continuous function on $(x-1, x) \cap (0, N)$ which is an open interval when $0 < x < N$.

To see (c), first note f_1 is continuous and f_2 has continuous derivative. So assume f_N is C^{N-2}. Then since $f_{N+1}(x) = \int_{x-1}^{x} f_N(y)\,dy$, we have by the fundamental theorem of calculus that

$$f'_{N+1}(x) = f_N(x) - f_N(x-1).$$

This implies f'_{N+1} has $N-2$ continuous derivatives and so f_N has $N-1$ continuous derivatives. Induction gives the result.

To see (d), we again argue by induction. Take $g \in C(\mathbb{R})$. Then

$$\int g(x)f_1(x)\,dx = \int_0^1 g(x)\,dx.$$

Assume

$$\int_{\mathbb{R}} g(x)f_N(x)\,dx = \int_0^1 \cdots \int_0^1 g(x_1 + x_2 + \cdots + x_N)\,dx_1dx_2\cdots dx_N\,.$$

Then by Fubini and the induction hypothesis,

$$\begin{aligned} \int g(x)f_{N+1}(x)\,dx &= \int g(x)\int 1_{[x-1,x]}(y)f_N(y)\,dy\,dx \\ &= \iint g(x)1_{[x-1,x]}(y)f_N(y)\,dy\,dx \\ &= \iint g(x)1_{[y,y+1]}(x)f_N(y)\,dx\,dy \\ &= \int \left(\int_y^{y+1} g(x)\,dx\right) f_N(y)\,dy \\ &= \int G(y)f_N(y)\,dy \\ &= \int_0^1 \cdots \int_0^1 G(x_1 + x_2 + \cdots + x_N)\,dx_1dx_2\cdots dx_N \end{aligned}$$

where $G(y) = \int_y^{y+1} g(x_{N+1})dx_{N+1}$ is a continuous function. Thus using translation invariance and Fubini,

$$\begin{aligned}
\int g(x) f_{N+1}(x)\,dx &= \int_0^1 \cdots \int_0^1 \int_{x_1+x_2+\cdots+x_N}^{x_1+x_2+\cdots+x_N+1} g(x_{N+1})\,dx_{N+1}dx_1dx_2\cdots dx_N \\
&= \int_0^1 \cdots \int_0^1 \int_0^1 g(x_1+x_2+\cdots+x_N+x_{N+1})\,dx_{N+1}dx_1dx_2\cdots dx_N \\
&= \int_0^1 \cdots \int_0^1 g(x_1+x_2+\cdots+x_{N+1})\,dx_1dx_2\cdots dx_{N+1}\,.
\end{aligned}$$

For (e), we note this is true for $N = 0$ for the sum

$$\sum_{k=-\infty}^{\infty} 1_{[0,1)}(x-k) = \sum_k 1_{[k,k+1)}(x) = 1.$$

Assume true for N. Then since the following is a finite sum,

$$\begin{aligned}
\sum_k f_{N+1}(x-k) &= \sum_k \int_{x-k-1}^{x-k} f_N(y)\,dy \\
&= \sum_k \int \chi_{[x-k-1,x-k)}(y) f_N(y)\,dy \\
&= \int \sum_k \chi_{[x-k-1,x-k)}(y) f_N(y)\,dy \\
&= \int \sum_k \chi_{[0,1)}(y) f_N(x-k-1+y)\,dy \\
&= \int_0^1 1\,dy = 1\,.
\end{aligned}$$

To see (f), we again argue by induction. It is true for $N = 0$. So assume true for N. Then

$$\begin{aligned}
f_{N+1}(\frac{N+1}{2}+x) &= \int_{\frac{N+1}{2}+x-1}^{\frac{N+1}{2}+x} f_N(y)\,dy \\
&= \int_{\frac{N}{2}+x-\frac{1}{2}}^{\frac{N}{2}+x+\frac{1}{2}} f_N(y)\,dy \\
&= \int_{-1/2}^{1/2} f_N(\frac{N}{2}+x+y)\,dy \\
&= \int_{-1/2}^{1/2} f_N(\frac{N}{2}-x-y)\,dy \\
&= \int_{-1/2}^{1/2} f_N(\frac{N}{2}-x+y)\,dy \\
&= \int_{\frac{N}{2}-x-\frac{1}{2}}^{\frac{N}{2}-x+\frac{1}{2}} f_N(y)\,dy \\
&= \int_{\frac{N+1}{2}-x-1}^{\frac{N+1}{2}-x} f_N(y)\,dy \\
&= f_{N+1}(\frac{N+1}{2}-x)\,.
\end{aligned}$$

For (g), we check it first for $N = 1$. Note

$$f_2(x) = \begin{cases} 0 & \text{if } x < 0 \\ x & \text{if } 0 \leqslant x \leqslant 1 \\ 2-x & \text{if } 1 < x \leqslant 2 \\ 0 & \text{if } 2 < x \end{cases}$$

and

$$f_3(x) = \begin{cases} 0 & \text{if } x < 0 \\ \frac{x^2}{2} & \text{if } 0 \leqslant x \leqslant 1 \\ -\frac{3}{2}+3x-x^2 & \text{if } 1 < x \leqslant 2 \\ \frac{(3-x)^2}{2} & \text{if } 2 < x \leqslant 3 \\ 0 & \text{if } 3 < x\,. \end{cases}$$

So

$$f_3(x) = \frac{x}{2}f_2(x) + \frac{3-x}{2}f_2(x-1)$$

clearly holds for $x < 0$. Also

$$\frac{x^2}{2} = \frac{x}{2}(x) + \frac{3-x}{2}(0)$$

for $0 \leqslant x < 1$,

$$-\frac{3}{2} + 3x - x^2 = \frac{x}{2}(2-x) + \frac{3-x}{2}(x-1)$$

for $1 \leqslant x < 2$, and

$$\frac{(3-x)^2}{2} = \frac{x}{2}(0) + \frac{3-x}{2}(3-x)$$

if $2 \leqslant x < 3$. Also both sides are 0 for $x \geqslant 3$. Hence assume true for N. Then

$$\begin{aligned}
f_{N+2}(x) &= \int_{x-1}^{x} f_{N+1}(y) \\
&= \int_{x-1}^{x} \left(\frac{y}{N} f_N(y) + \frac{N+1-y}{N} f_N(y-1) \right) dy \\
&= \int_{x-1}^{x} \left(\frac{y}{N} (f_N(y) - f_N(y-1) \right) dy + \frac{N+1}{N} \int_{x-1}^{x} f_N(y-1)\, dy \\
&= \int_{x-1}^{x} \frac{y}{N} f'_{N+1}(y)\, dy + \frac{N+1}{N} \int_{x-2}^{x-1} f_N(y)\, dy \\
&= \left(\frac{y}{N} f_{N+1}(y) \right) \Big|_{y=x-1}^{y=x} - \frac{1}{N} \int_{x-1}^{x} f_{N+1}(y)\, dy + \frac{N+1}{N} f_{N+1}(x-1) \\
&= \frac{x}{N} f_{N+1}(x) - \frac{x-1}{N} f_{N+1}(x-1) - \frac{1}{N} f_{N+2}(x) + \frac{N+1}{N} f_{N+1}(x-1)
\end{aligned}$$

and thus

$$\frac{N+1}{N} f_{N+2}(x) = \frac{x}{N} f_{N+1}(x) - \frac{N+2-x}{N} f_{N+1}(x-1)$$

and so

$$f_{N+2}(x) = \frac{x}{N+1} f_{N+1}(x) - \frac{(N+1)+1-x}{N+1} f_{N+1}(x-1)$$

and the inductive step is established.

□

4. Let f and g be even functions. Show $f * g$ is an even function. That is show if x is in the domain of $f * g$, then $-x$ is in the domain of $f * g$ and then

$$f * g(x) = f * g(-x).$$

Solution:

Note if f and g are even functions and x is in the domain of $f * g$; i.e. $y \mapsto f(y)g(x-y)$ is integrable, then since $f(-y)g(-x+y) = f(y)g(x-y)$, we have $y \mapsto f(-y)g(-x+y)$ is integrable. Since $y \mapsto -y$ preserves Lebesgue measure, $y \mapsto f(y)g(-x-y)$ is integrable and

$$\int f(y)g(-x-y)\, dy = \int f(-y)g(-x+y)\, dy = \int f(y)g(x-y)\, dy\,.$$

Thus x is in the domain of $f * g$ implies $-x$ is in the domain of $f * g$ and $f * g(-x) = f * g(x)$.

□

5. Let $f(x) = e^{-|x|}$ for $x \in \mathbb{R}$. Show $f * f(x) = (1 + |x|)\, e^{-|x|}$.

Solution:

Note f is an even function and thus by Exercise 2.5.4, we need only calculate $f * f(x)$ for $x \geqslant 0$. Note f is integrable, so all x are in the domain. Now if $x \geqslant 0$,

$$\begin{aligned}
f * f(x) &= \int e^{-|y|} e^{-|x-y|}\, dy \\
&= \int_{-\infty}^{0} e^{y} e^{-x+y}\, dy + \int_{0}^{x} e^{-y} e^{-x+y}\, dy + \int_{x}^{\infty} e^{-y} e^{x-y}\, dy \\
&= \int_{-\infty}^{0} e^{-x+2y}\, dy + \int_{0}^{x} e^{-x}\, dy + \int_{x}^{\infty} e^{x-2y}\, dy \\
&= \frac{1}{2} e^{-x+2y}\Big|_{y=-\infty}^{y=0} + xe^{-x} - \frac{1}{2} e^{x-2y}\Big|_{y=x}^{y=\infty} \\
&= \frac{1}{2} e^{-x} + xe^{-x} + \frac{1}{2} e^{-x} \\
&= (1+x)e^{-x} = (1+|x|)e^{-|x|}.
\end{aligned}$$

□

6. Let $f(x) = e^{-x}\chi_{[0,\infty)}(x)$. Set $f_1 = f$ and $f_{N+1} = f * f_N$ for $N > 1$. Evaluate the function f_N for all N.

Solution:

We have $f(x) = e^{-x}\chi_{[0,\infty)}(x)$. So if $f_{N+1} = f * f_N$, we see

$$\begin{aligned}
f_2(x) &= \int_{0}^{\infty} e^{-y} e^{-(x-y)} \chi_{[0,\infty)}(x-y)\, dy \\
&= \begin{cases} 0 & \text{if } x < 0 \\ \int_0^x e^{-x}\, dy = xe^{-x} & \text{if } x \geqslant 0 \end{cases} \\
&= \chi_{[0,\infty)}(x) x e^{-x}
\end{aligned}$$

and

$$\begin{aligned}
f_3(x) &= \int_{0}^{\infty} e^{-y} (x-y) e^{-(x-y)} \chi_{[0,\infty)}(x-y)\, dy \\
&= \chi_{[0,\infty)}(x) \int_{0}^{x} (x-y) e^{-x}\, dy \\
&= \chi_{[0,\infty)}(x) \frac{x^2}{2} e^{-x}.
\end{aligned}$$

We note induction then implies $f_N(x) = \chi_{[0,\infty)}(x) \frac{x^{N-1} e^{-x}}{(N-1)!}$ for all N. In fact, if this holds for N, then

$$\begin{aligned} f_{N+1}(x) &= \int_0^\infty e^{-y}\frac{(x-y)^{N-1}}{(N-1)!}e^{-(x-y)}\chi_{[0,\infty)}(x-y)\,dy \\ &= \chi_{[0,\infty)}(x)e^{-x}\int_0^x \frac{(x-y)^{N-1}}{(N-1)!}e^{-x}\,dy \\ &= \chi_{[0,\infty)}(x)\frac{x^N e^{-x}}{N!}\,. \end{aligned}$$

□

7. Let $f \in L^1_{\text{loc}}(\mathbb{R}^n)$ and $g \in C_c^\infty(\mathbb{R}^n)$. Show $f * g(x)$ exists everywhere and $f * g \in C^\infty(\mathbb{R}^n)$.

Solution:

First we note if f is integrable on every compact subset K of $\mathbb{R}^n$, then if $K = \text{supp}\, g$, one has $f(y)g(x-y) = \chi_{x-K}(y)f(y)g(x-y)$ for all y and thus is integrable in y. So all x are in the domain of $f * g$.

We next show $D^\alpha(f * g)$ exists and equals $f * D^\alpha g$ for all α. By induction it suffices to show $D_i(f * g) = f * D_i g$ for each i.

Fix x and $r > 0$. Set $L = [-r, r]e_i + x - K$. L is compact and for $0 < |h| < r$, one has

$$\frac{f * g(x + he_i) - f * g(x)}{h} = \int \chi_L(y)f(y)\frac{g(x-y+he_i) - g(x-y)}{h}\,dy\,.$$

Now $\frac{g(x-y+he_i)-g(x-y)}{h} \to D_i g(x-y)$ pointwise and uniformly bounded on L for by the mean value theorem, $|\frac{g(x-y+he_i)-g(x-y)}{h}| = |D_i g(x - y + h^* e_i)| \leqslant M$ where M is the maximum of $|D_i g(x - y + he_i)|$ on the compact set $\{(y, h) \mid y \in L, |h| \leqslant r\}$. The Lebesgue dominated convergence theorem then implies

$$\lim_{h\to 0}\frac{f * g(x + he_i) - f * g(x)}{h} = \int f(y)D_i g(x-y)\,dy = f * D_i g(x)\,.$$

So $D_i(f * g) = f * D_i g$.

□

8. Suppose $f(x) = 0$ for $x \notin E$ and $g(x) = 0$ for $x \notin F$. Show $f * g(x) = 0$ for any $x \notin E + F$. In particular, if f and g have compact supports supp f and supp g and $f * g$ is defined everywhere, then $f * g$ has compact support contained in supp f + supp g.

Solution:

Suppose $f(x) = 0$ off E and $g(x) = 0$ off F. Then in the product $f(y)g(x-y)$ we have if $x \notin E+F$, then for $y \in E$, $x-y \notin F$. So the product is 0 if $x \notin E+F$ for all y. Thus $y \mapsto f(y)g(x-y)$ is the zero function and thus $f * g(x) = 0$.

□

9. Suppose $f * g(x)$ is defined a.e. x. Show $\lambda(a)f * \lambda(b)g(x)$ is defined a.e. and

$$\lambda(a)f * \lambda(b)g(x) = \lambda(a+b)(f * g)(x)$$

for a.e. x.

Solution:

Suppose $f * g(x)$ is defined a.e. x. Thus $y \mapsto f(y)g(x-y)$ is integrable for for $x \notin E$ where E is a set of Lebesgue measure 0. Thus $y \mapsto f(y)g(x-y-b-a)$ is integrable for $x \notin E+a+b$. By translation invariance of Lebesgue measure, $y \mapsto f(y-a)g(x-y-b)$ is integrable for $x \notin E+a+b$ and

$$\begin{aligned} f * g(x-b-a) &= \int f(y)g(x-b-y-a)\,dy \\ &= \int f(y-a)g(x-y-b)\,dy \\ &= (\lambda(a)f) * (\lambda(b)g)(x)\,. \end{aligned}$$

So $\lambda(a+b)(f*g)(x) = (\lambda(a)f) * (\lambda(b)g)(x)$ for $x \notin E+a+b$.

□

10. Recall $\delta(a)f(x) = a^{-n/2}f(a^{-1}x)$ for $a > 0$ when f is a function on $\mathbb{R}^n$. Show

$$\delta(a)f * \delta(a)g(x) = (f*g)(a^{-1}x) = a^{n/2}\delta(a)(f*g)(x).$$

Solution: Since Lebesgue measure satisfies $d(ay) = a^n\,dy$ for $a > 0$, we have

$$\begin{aligned} \delta(a)f * \delta(a)g(x) &= \int \delta(a)f(y)\delta(a)g(x-y)\,dy \\ &= \int a^{-n/2}f(a^{-1}y)a^{-n/2}g(a^{-1}x - a^{-1}y)\,dy \\ &= \int f(y)g(a^{-1}x - y)\,dy \\ &= f*g(a^{-1}x) \\ &= a^{n/2}\delta(a)(f*g)(x). \end{aligned}$$

□

11. Show $\chi_{[0,1]} * \chi_{[a,b]}$ is the function defined by:

$$\chi_{[0,1]} * \chi_{[a,b]}(x) = \begin{cases} 0 & \text{if } x \leqslant a \\ x-a & \text{if } a \leqslant x \leqslant \min\{a+1,b\} \\ \min\{1, b-a\} & \text{if } \min\{a+1,b\} \leqslant x \leqslant \max\{a+1,b\} \\ b+1-x & \text{if } \max\{a+1,b\} \leqslant x \leqslant b+1 \\ 0 & \text{if } b+1 \leqslant x. \end{cases}$$

Solution:

$$\begin{aligned} \chi_{[0,1]} * \chi_{[a,b]}(x) &= \int_a^b \chi_{[0,1]}(x-t)\,dt \\ &= \int_a^b \chi_{[x-1,x]}(t)\,dt \\ &= \lambda([a,b] \cap [x-1,x]). \end{aligned}$$

Thus $T_{a,b}(x) = 0$ if $x \leqslant a$ and $T_{[a,b]}(x) = 0$ if $x \geqslant b+1$.

Also $T_{a,b}(x) = x - a$ while $x - 1 \leqslant a \leqslant x \leqslant b$ or

$T_{a,b}(x) = 1$ while $a \leqslant x - 1 \leqslant x \leqslant b$.

$T_{a,b}(x) = b - a$ while $x - 1 \leqslant a \leqslant b \leqslant x$

$T_{a,b}(x) = b + 1 - x$ while $a \leqslant x - 1 \leqslant b \leqslant x$.

This covers all cases. □

12. Let $c > 0$. Show $\chi_{[0,c]} * \chi_{[a,b]}$ is the function defined by

$$\chi_{[0,c]} * \chi_{[a,b]}(x) = \begin{cases} 0 & \text{if } x \leqslant a \\ x - a & \text{if } a \leqslant x \leqslant \min\{a+c, b\} \\ \min\{c, b-a\} & \text{if } \min\{a+c,b\} \leqslant x \leqslant \max\{a+c,b\} \\ b + c - x & \text{if } \max\{a+c,b\} \leqslant x \leqslant b+c \\ 0 & \text{if } b + c \leqslant x. \end{cases}$$

Solution:

One can proceed directly as in the proof of Exercise 2.5.11 or use Exercises 2.5.11 and 2.5.10. Note

$$\delta(c)\chi_{[0,1]}(x) = c^{-1/2}\chi_{[0,1]}(c^{-1}x) = c^{-1/2}\chi_{[0,c]}(x)$$

and

$$\delta(c)\chi_{[\frac{a}{c},\frac{b}{c}]} = c^{-1/2}\chi_{[a,b]}.$$

But

$$c^{-1}\chi_{[0,c]} * \chi_{[a,b]}(x) = c^{1/2}\delta(c)(\chi_{[0,1]} * \chi_{[a,b]})(x) = \chi_{[0,1]} * \chi_{[\frac{a}{c},\frac{b}{c}]}(c^{-1}x)$$

$$= \begin{cases} 0 & \text{if } c^{-1}x \leqslant a \\ c^{-1}x - c^{-1}a & \text{if } \frac{a}{c} \leqslant c^{-1}x \leqslant \min\{\frac{a}{c}+1, \frac{b}{c}\} \\ \min\{1, \frac{b-a}{c}\} & \text{if } \min\{\frac{a}{c}+1, \frac{b}{c}\} \leqslant c^{-1}x \leqslant \max\{\frac{a}{c}+1, \frac{b}{c}\} \\ \frac{b}{c} + 1 - c^{-1}x & \text{if } \max\{\frac{a}{c}+1, \frac{b}{c}\} \leqslant c^{-1}x \leqslant \frac{b}{c}+1 \\ 0 & \text{if } \frac{b}{c} + 1 \leqslant c^{-1}x. \end{cases}$$

So

$$\chi_{[0,c]} * \chi_{[a,b]}(x) = c\begin{cases} 0 & \text{if } x \leqslant ca \\ c^{-1}x - c^{-1}a & \text{if } a \leqslant x \leqslant \min\{a+c, b\} \\ \min\{1, \frac{b-a}{c}\} & \text{if } \min\{a+c,b\} \leqslant x \leqslant \max\{a+c,b\} \\ c^{-1}b + 1 - c^{-1}x & \text{if } \max\{a+c,b\} \leqslant x \leqslant b+c \\ 0 & \text{if } b + c \leqslant x \end{cases}$$

which gives the result.

□

13. Let $l \leqslant m$. Show the convolution of $\chi_{[2^l,2^{l+1}]}$ with $\chi_{[2^m,2^{m+1}]}$ is given by:

$$\chi_{[2^l,2^{l+1}]}*\chi_{[2^m,2^{m+1}]}(x) = \begin{cases} 0 & \text{if } x \leqslant 2^m + 2^l \\ x - 2^m - 2^l & \text{if } 2^m + 2^l \leqslant x \leqslant 2^m + 2^{l+1} \\ 2^l & \text{if } 2^m + 2^{l+1} \leqslant x \leqslant 2^{m+1} + 2^l \\ 2^{m+1} + 2^{l+1} - x & \text{if } 2^{m+1} + 2^l \leqslant x \leqslant 2^{m+1} + 2^{l+1} \\ 0 & \text{if } 2^{m+1} + 2^{l+1} \leqslant x. \end{cases}$$

Solution: One can proceed directly from the definition of convolution or use Exercise 2.5.12 and Exercise 2.5.9.

Remember $l \leqslant m$. Note $\lambda(2^l)\chi_{[0,2^l]}(x) = \chi_{[0,2^l]}(x - 2^l) = \chi_{[2^l,2^l+2^l]}(x) = \chi_{[2^l,2^{l+1}]}(x)$ and similarly $\lambda(2^m)\chi_{[0,2^m]} = \chi_{[2^m,2^{m+1}]}$. Thus by Exercise 2.5.9,

$$\chi_{[2^l,2^{l+1}]} * \chi_{[2^m,2^{m+1}]}(x) = \chi_{[0,2^l]} * \chi_{[0,2^m\}}(x - 2^m - 2^l).$$

Hence

$$\chi_{[2^l,2^{l+1}]} * \chi_{[2^m,2^{m+1}]}(x) =$$

$$= \begin{cases} 0 & \text{if } x - 2^m - 2^l \leqslant 0 \\ x - 2^m - 2^l & \text{if } 0 \leqslant x - 2^m - 2^l \leqslant \min\{2^l, 2^m\} \\ \min\{2^l, 2^m\} & \text{if } \min\{2^l, 2^m\} \leqslant x - 2^m - 2^l \leqslant \max\{2^l, 2^m\} \\ 2^m + 2^l - (x - 2^m - 2^l) & \text{if } \max\{2^l, 2^m\} \leqslant x - 2^m - 2^l \leqslant 2^m + 2^l \\ 0 & \text{if } 2^m + 2^l \leqslant x - 2^m - 2^l \end{cases}$$

$$= \begin{cases} 0 & \text{if } x \leqslant 2^l + 2^m \\ x - 2^m - 2^l & \text{if } 0 \leqslant x - 2^m - 2^l \leqslant 2^l \\ 2^l & \text{if } 2^l \leqslant x - 2^m - 2^l \leqslant 2^m \\ 2^m + 2^l - (x - 2^m - 2^l) & \text{if } 2^m \leqslant x - 2^m - 2^l \leqslant 2^m + 2^l \\ 0 & \text{if } 2^m + 2^l \leqslant x - 2^m - 2^l \end{cases}$$

$$= \begin{cases} 0 & \text{if } x \leqslant 2^m + 2^l \\ x - 2^m - 2^l & \text{if } 2^m + 2^l \leqslant x \leqslant 2^{l+1} + 2^m \\ 2^l & \text{if } 2^{l+1} + 2^m \leqslant x \leqslant 2^{m+1} + 2^l \\ 2^{m+1} + 2^{l+1} - x & \text{if } 2^{m+1} + 2^l \leqslant x \leqslant 2^{m+1} + 2^{l+1} \\ 0 & \text{if } 2^{m+1} + 2^{l+1} \leqslant x\,. \end{cases}$$

□

14. Let $f = \sum_{k=1}^{\infty} 2^{k/2}\chi_{[2^{-k},2^{-k+1}]}$.

(a) Show $f \in L^1(\mathbb{R})$ but $f \notin L^2(\mathbb{R})$.

(b) Show $f * f$ is not continuous.

Solution:

We have $f = \sum_{k=1}^{\infty} 2^{k/2}\chi_{[2^{-k},2^{-k+1}]}$. This converges in $L^1(\mathbb{R})$ because

$$\begin{aligned}\int_0^{\infty} \sum_{k=N+1}^{\infty} 2^{k/2}\chi_{[2^{-k},2^{-k+1}]} &= \sum_{k=N+1}^{\infty} 2^{k/2}2^{-k} \\ &= 2^{-(N+1)/2}\sum_{k=0}^{\infty} \frac{1}{\sqrt{2^k}} \\ &= 2^{-(N+1)/2}\frac{1}{1-\frac{1}{\sqrt{2}}} = \frac{2^{-N/2}}{\sqrt{2}-1} \to 0\end{aligned}$$

as $N \to \infty$. Note f is not in L^2 for $f^2 = \sum_{k=1}^{\infty} 2^k\chi_{[2^{-k},2^{-k+1}]}$ and thus

$$\int f^2\,dx = \sum_k 2^k(2^{-k}) = \infty\,.$$

We claim $f * f$ is not continuous, that is it is not equal a.e. to a continuous function. We note Exercise 2.5.13 holds for any integers k and l. Suppose $f * f = g$ a.e. where g is continuous. Note by Exercise 2.5.13 that $f * f(x) = 0$ for a.e. $x < 0$. This implies $g(x) = 0$ for $x = 0$. Also $f * f \geqslant 2^{l/2}\chi_{[2^{-l},2^{-l+1}]} * 2^{l/2}\chi_{[2^{-l},2^{-l+1}]}$ a.e. for each $l > 0$. But again by this exercise,

$$\chi_{[2^{-l},2^{-l+1}]} * \chi_{[2^{-l},2^{-l+1}]}(x) = x - 2^{-l} - 2^{-l}$$

for $2^{-l} + 2^{-l} \leqslant x \leqslant 2^{-l} + 2^{-l} + 2^{-l}$. Thus for $2^{-l} + 2^{-l} + 2^{-l-1} \leqslant x \leqslant 3(2^{-l})$, we have

$$\begin{aligned}\chi_{[2^{-l},2^{-l+1}]} * \chi_{[2^{-l},2^{-l+1}]}(x) &= x - 2^{-l} - 2^{-l} \\ &\geqslant 2^{-l-1}.\end{aligned}$$

Thus $f*f(x) \geqslant \frac{1}{2}$ a.e. for $5(2^{-l-1}) \leqslant x \leqslant 6(2^{-l-1})$. Thus $g(x) \geqslant \frac{1}{2}$ for some x in this interval. By continuity, $\frac{1}{2} \leqslant \lim_{x\to 0+} g(x) = g(0)$. This is a contradiction.

□

15. Give an example of $f, h \in L^2(\mathbb{R})$ where $f * h(x) = \int f(y)h(x-y)\,d_1y$ is not an L^2 function.

Solution: Let $f(x) = \frac{1}{x^\alpha}\chi_{[1,\infty)}$ where $\frac{3}{4} \geqslant \alpha > \frac{1}{2}$. Then $f \in L^2$ and

$$\begin{aligned}
f * f(x) &= \int_1^\infty \frac{1}{t^\alpha}\frac{1}{(x-t)^\alpha}\chi_{[1,\infty)}(x-t)\,dt \\
&= \begin{cases} \int_1^{x-1}\left(\frac{1}{t(x-t)}\right)^\alpha dt & \text{if } x > 2 \\ 0 & \text{otherwise} \end{cases} \\
&= \chi_{(2,\infty)}(x)\int_1^{x-1}\left(\frac{\frac{1}{x}}{t} + \frac{\frac{1}{x}}{x-t}\right)^\alpha dt \\
&= \frac{1}{x^\alpha}\chi_{(2,\infty)}(x)\int_1^{x-1}\left(\frac{1}{t} + \frac{1}{x-t}\right)^\alpha dt \\
&\geqslant \frac{1}{x^\alpha}\chi_{(2,\infty)}(x)\int_1^{x-1} t^{-\alpha}\,dt \\
&= \frac{1}{x^\alpha}\chi_{(2,\infty)}(x)\frac{1}{1-\alpha}t^{1-\alpha}|_1^{x-1}\,dt \\
&= \frac{1}{x^\alpha}\chi_{(2,\infty)}(x)\frac{1}{1-\alpha}\left((x-1)^{1-\alpha} - 1\right) \\
&= \frac{1}{1-\alpha}\chi_{(2,\infty)}(x)\frac{(x-1)^{1-\alpha}}{x^\alpha} - \frac{1}{1-\alpha}f(x)\chi_{(2,\infty)}(x).
\end{aligned}$$

Thus the convolution will not be L^2 if $\frac{(x-1)^{1-\alpha}}{x^\alpha}$ is not L^2 on $(2,\infty)$. But

$$\begin{aligned}
\frac{(x-1)^{2-2\alpha}}{x^{2\alpha}} &= \frac{(x-1)^2}{(x-1)^{2\alpha}x^{2\alpha}} \\
&\geqslant \frac{(x-1)^2}{x^{2\alpha}x^{2\alpha}} \\
&= \frac{x^2 - 2x + 1}{x^{4\alpha}} \\
&= \frac{1}{x^{4\alpha-2}} - \frac{2}{x^{4\alpha-1}} + \frac{1}{x^{4\alpha}}.
\end{aligned}$$

Since $\frac{3}{4} \geqslant \alpha > \frac{1}{2}$, $4\alpha - 1 > 2 - 1 = 1$, so $\frac{1}{x^{4\alpha-1}}$ and $\frac{1}{x^{4\alpha}}$ are integrable on $(2,\infty)$. But $4\alpha - 2 \leqslant 3 - 2 = 1$. Thus $\frac{1}{x^{4\alpha-2}}$ is not integrable on $(2,\infty)$.

One could also take f where $\hat{f}(y) = \frac{1}{|y|^\beta}\chi_{(-1,1)}$ where $\frac{1}{4} \leqslant \beta < \frac{1}{2}$. Then $\hat{f}$ is L^2, $\hat{f}^2$ is not. Then you would expect $f * f$ not to be L^2. □

16. Show the Hilbert space of Hilbert-Schmidt operators on a Hilbert space $\mathcal{H}$ is an example of a Hilbert algebra. That is show the space $\mathcal{B}_2(\mathcal{H},\mathcal{H})$ is a Banach $*$ algebra with multiplication given by composition and the adjoint given by operator adjoint and then show one has the following additional properties:

(a) For each $A \in \mathcal{B}_2(\mathcal{H},\mathcal{H})$, the linear transformation L_A of $\mathcal{B}_2(\mathcal{H},\mathcal{H})$ given by $L_A(B) = AB$ is bounded and has norm $||A||$.

(b) $L_A^* = L_{A*}$.

(c) The linear span of the set $\{AB \mid A, B \in \mathcal{B}_2(\mathcal{H}, \mathcal{H})\}$ is dense in $\mathcal{B}_2(\mathcal{H}, \mathcal{H})$.

Solution: We have $||AB||_2 \leqslant ||A||\,||B||_2$ so $||L_A|| \leqslant ||A||$. Consider $B = v \otimes \bar{w}$. We know $||B||_2 = ||v||\,||w||$. But $L_A(B) = A \circ (v \otimes \bar{w}) = Av \otimes w$. Thus $||L_A(B)||_2 = ||Av||\,||w||$. Taking v and w with length one and taking the supremum we see $||L_A|| \geqslant ||A||$. So $||L_A|| = ||A||$.

Now if $\{e_\alpha\}$ is an orthonormal basis of $\mathcal{H}$, one has

$$\begin{aligned}(L_A B, C)_2 = (AB, C)_2 &= \sum (ABe_\alpha, Ce_\alpha)\\ &= \sum (Be_\alpha, A^*Ce_\alpha) = (B, A^*C)_2\,.\end{aligned}$$

So $L_A^* = L_{A^*}$.

Finally if one takes $A = v \otimes \bar{w}'$ and $B = v' \otimes \bar{w}$, we see

$$AB = (v', w')v \otimes \bar{w}.$$

Thus the linear span of the AB contains all finite rank operators and hence is a dense subspace of $\mathcal{B}_2(\mathcal{H}, \mathcal{H})$. □

Exercise Set 2.6

1. Using Lemma 2.79, Exercise 2.5.8, and Theorem 2.82, show $C_c^\infty(\mathbb{R}^n)$ is dense in $L^p(\mathbb{R}^n)$ for $1 \leqslant p < \infty$. This gives an alternate proof to the second part of Proposition 2.55.

Solution:

Choose a nonnegative C^∞ function h on $\mathbb{R}^n$ with compact support with $\int h(x)\,dx = 1$. Set $h_t = t^{-n}h(t^{-1}x)$. Theorem 2.82 shows if $f \in L^p$, then $h_t * f \to f$ in L^p as $t \to 0+$. By Lemma 2.79, $h_t * f$ is C^∞. Moreover, if g is a continuous function with compact support, Exercise 2.5.8 shows $h_t * g$ has compact support. Moreover, $h_t * g \in C^\infty$ and for $1 \leqslant p < \infty$, $h_t * g \to g$ in L^p as $t \to 0+$. Now let $\epsilon > 0$ and $f \in L^p$. Choose $g \in C_c(\mathbb{R}^n)$ with $|f - g|_p < \frac{\epsilon}{2}$. Choose $t > 0$ with $|h_t * g - g|_p < \frac{\epsilon}{2}$. Then $|h_t * g - f|_p < \epsilon$.

□

2. Let $f \in L^p(\mathbb{R}^n)$ where $1 \leqslant p < \infty$. For $t > 0$ define $u(t, x)$ by

$$u(t, x) = f * h_t(x)$$

where $h(x) = e^{-\pi|x|^2}$ and $h_t(x) = t^{-n/2}h(x/\sqrt{t})$. Show that $u : \mathbb{R}^+ \times \mathbb{R}^n \to \mathbb{C}$ is smooth, satisfies the heat equation

$$\partial_t u = \frac{1}{4\pi}\Delta u,$$

and

$$\lim_{t \to 0+} u(t, \cdot) = f \text{ in } L^p(\mathbb{R}^n).$$

Solution:

By Theorem 2.82, we know $\lim_{t\to 0+} u(t,\cdot) = f(\cdot)$ in L^p. By Lemma 2.79, one has $\Delta u(x,t) = f * \Delta h_t(x)$. Also the Lebesgue Dominated Convergence Theorem and the Mean Value Theorem can be used to show if $h(t,x) = h_t(x)$, then $\partial_t u(t,\cdot) = f * \partial_t h(t,\cdot)$. Indeed, just follow the argument in the proof of Lemma 2.79. Furthermore, since $h(t,x) = t^{-n/2} e^{-\frac{\pi}{t}|x|^2}$, one has

$$\begin{aligned}\partial_t h(t,x) &= -\frac{n}{2t} h(t,x) + \frac{\pi |x|^2}{t^2} h(t,x) \\ &= \left(\frac{-n}{2t} + \frac{\pi |x|^2}{t^2} \right) h(t,x)\end{aligned}$$

and

$$\begin{aligned}\Delta h_t(x) &= t^{-n/2} \sum_{i=1}^{n} \partial_i^2 (e^{-\frac{\pi}{t}|x|^2}) \\ &= t^{-n/2} \sum_{i=1}^{n} \partial_i \left(-\frac{2\pi x_i}{t} e^{-\frac{\pi}{t}|x|^2}\right) \\ &= \sum_{i=1}^{n} \left(-\frac{2\pi}{t} + \frac{4\pi^2 x_i^2}{t^2} \right) t^{-n/2} e^{-\frac{\pi}{t}|x|^2} \\ &= \left(-\frac{2n\pi}{t} + \frac{4\pi^2 |x|^2}{t^2} \right) h_t(x).\end{aligned}$$

Thus if $g_t(x) = \left(\frac{-n}{2t} + \frac{\pi |x|^2}{t^2} \right) h_t(x)$,

$$\partial_t u(t,x) = f * g_t(x)$$

and

$$\Delta u(t,x) = 4\pi f * g_t(x).$$

So $\partial_t u(t,x) = \frac{1}{4\pi} \Delta u(t,x)$.

□

3. Let Ω be an open subset of $\mathbb{R}^n$. Since $L^p(\Omega)$ is a subspace of $L^p(\mathbb{R}^n)$, $h * f$ is defined for $h \in L^1(\mathbb{R}^n)$. Suppose $h \in C_c^\infty(\mathbb{R}^n)$ is nonnegative and has integral one. Let $f \in \mathcal{D}(\Omega)$. Show $h_t * f \to f$ in $\mathcal{D}(\Omega)$ as $t \to 0+$.

Solution:

Let $f \in \mathcal{D}(\Omega)$ have compact support $K \subseteq \Omega$. By Exercise 2.4.4, there is an open subset U of Ω with $K \subseteq U \subseteq \bar{U} \subseteq \Omega$ with $\bar{U}$ compact. Let $r > 0$ and consider the sets $F_r = B_r(0) + K$ where $B_r(0)$ is the closed ball of radius r and center 0. These sets are compact and

$$\bigcap_{r>0} (F_r \cap U^c) = \varnothing$$

for if $x \in U^c$, there is a r with $x + B_r(0)$ missing K. So $x \notin K + B_r(0) = F_r$. Since this intersection is an intersection of a decreasing sequence of compact sets, the finite intersection property implies there is an $r > 0$ with $F_r \cap U^c = \varnothing$. Thus $F_r \subseteq U$. Now choose $t_0 > 0$ so that supp $h_t \subseteq B_r(0)$ for $0 < t \leqslant t_0$. Thus by Exercise 2.5.8, supp $(h_t * f) \subseteq B_r(0) + K \subseteq U \subseteq \bar{U} \subseteq \Omega$. Thus by Lemma 2.79, $h_t * f \in \mathcal{D}^{\infty}_{F_r}(\Omega)$ for $0 < t \leqslant t_0$ and by Lemma 2.83, $D^\alpha(h_t * f) = h_t * D^\alpha f$ converges uniformly to $D^\alpha f$ on F_r.

This implies $h_t * f \to f$ in $\mathcal{D}_{F_r}(\Omega)$ for $t_0 \geqslant t > 0$ and $t \to 0+$. Now by Lemma 2.66, $\mathcal{D}_{F_r}(\Omega)$ has the relative topology of $\mathcal{D}(\Omega)$. Consequently, $h_t * f \to f$ in $\mathcal{D}(\Omega)$.

□

4. By following the proof of Theorem 2.85, check the validity of Theorem 2.86.
Solution:

We follow and use the proof of Theorem 2.85. The mapping $(f, g) \mapsto f \otimes g$ is clearly bilinear. We claim it is separately continuous from $\mathcal{D}(\Omega_1) \times \mathcal{D}(\Omega_2)$ into $\mathcal{D}(\Omega_1 \times \Omega_2)$. Fix $g \in \mathcal{D}_{K_2}(\Omega_2)$ and consider the mapping $T : \mathcal{D}(\Omega_1) \to \mathcal{D}(\Omega_1 \times \Omega_2)$ given by $T(f) = f \otimes g$. This map is linear. Using Proposition 2.73, T is continuous if when $f_k \mapsto f$ in $\mathcal{D}(\Omega_1)$, then $T(f_k) \mapsto T(f)$ in $\mathcal{D}(\Omega_1 \times \Omega_2)$. But by Corollary 2.72, if $f_k \to f$, there is a compact subset K_1 of Ω_1 such that $f_k \to f$ in $\mathcal{D}_{K_1}(\Omega_1)$. Now

$$\begin{aligned}
|T(f_k) - T(f)|_{K_1 \times K_2, (\alpha,\beta)} &= |T(f_k - f)|_{K_1 \times K_2, (\alpha,\beta)} \\
&= \max_{(x,y) \in K_1 \times K_2} |D^\alpha (f_k - f)(x)\, D^\beta g(y)| \\
&= |f_k - f|_{K_1, \alpha} |g|_{K_2, \beta}
\end{aligned}$$

and so $T : \mathcal{D}_{K_1}(\Omega_1) \to \mathcal{D}_{K_1 \times K_2}(\Omega_1 \times \Omega_2)$ is continuous. By Lemma 2.66, the relative topology of $\mathcal{D}(\Omega_1 \times \Omega_2)$ on $\mathcal{D}_{K_1 \times K_2}(\Omega_1 \times \Omega_2)$ is the topology of $\mathcal{D}_{K_1 \times K_2}(\Omega_1 \times \Omega_2)$, this implies $T : \mathcal{D}(\Omega_1) \to \mathcal{D}(\Omega_1 \times \Omega_2)$ is continuous. Similarly, $g \mapsto f \otimes g$ is continuous.

Let U be an open subset of $\mathcal{D}(\Omega_1 \times \Omega_2)$ containing a function F. As seen in the proof of Theorem 2.85, if $F \in \mathcal{D}(\Omega_1 \times \Omega_2)$, there are C^∞ functions ϕ_1 and ϕ_2 of compact supports K_1 and K_2 contained in Ω_1 and Ω_2, respectively such that for any polynomial $P(x, y)$ and any $p \geqslant 0$, one has an $\phi_1(x)\phi_2(y)F(x, y) = F(x, y)$ and an $M_p > 0$ with

$$|D^{\alpha,\beta}(\phi_1(x)P(x, y)\phi_2(y)) - D^{\alpha,\beta}F(x, y)| \leqslant M_p |P - F|_{p, K_1 \times K_2}$$

for all (α, β) with $|\alpha| + |\beta| \leqslant p$. Clearly the functions $\phi_1(x)P(x, y)\phi_2(y)$ are in the linear span of the $f \otimes g$. Moreover, by Lemma 2.66, $U \cap \mathcal{D}_{K_1 \times K_2}(\Omega_1 \times \Omega_2)$ is open in $\mathcal{D}_{K_1 \times K_2}(\Omega_1 \times \Omega_2)$. Hence there is an $\epsilon > 0$ and a $p \geqslant 0$ so that if $H \in \mathcal{D}_{K_1 \times K_2}(\Omega_1 \times \Omega_2)$ and $|H - F|_{K_1 \times K_2, (\alpha,\beta)} < \epsilon$ for $|\alpha| + |\beta| \leqslant p$, then $H \in U$. By Lemma 2.84, we can choose a $P(x, y)$ with $|P - F|_{p, K_1 \times K_2} < \frac{\epsilon}{M_p}$. Thus the linear span of the range is dense.

□

5. Let M_n be the vector space of all complex Borel measures on $\mathbb{R}^n$. For each measure $\mu \in M_n$, let $|\mu|$ be the variation measure of μ. Then M_n with norm $||\mu|| = |\mu|(\mathbb{R}^n)$ is a Banach space. By the Riesz representation theorem, M_n is the dual space of the Banach space $C_0(\mathbb{R}^n)$ consisting of continuous functions vanishing at ∞. For μ and $\nu \in M_n$ define $\mu^* \in M_n$ and $\mu * \nu \in M_n$ by $\mu^*(f) = \int \overline{f(-x)}\, d\mu(x)$ and $\mu * \nu(f) = \iint f(x+y)\, d\mu(x) d\nu(y)$.

(a) Show M_n is a commutative Banach $*$ algebra.

(b) For $g \in L^1(\mathbb{R}^n)$, define $\lambda_g \in M_n$ by $\lambda_g(f) = \int f(x) g(x)\, dx$. Show

$$\lambda_g * \lambda_h = \lambda_{g*h} \text{ for } g, h \in L^1(\mathbb{R}^n).$$

(c) For $f \in L^1(\mathbb{R}^n)$, show $(\lambda_f)^* = \lambda_{f^*}$.

(d) For $f \in L^1(\mathbb{R}^n)$ and $\mu \in M_n$, define $f * \mu(x) = \int f(x-y)\, d\mu(y)$. Show $f * \mu \in L^1(\mathbb{R}^n)$ and

$$\lambda_f * \mu = \lambda_{f*\mu}.$$

(e) Show $f \mapsto \lambda_f$ is an isometric $*$ homomorphism of $L^1(\mathbb{R}^n)$ under convolution onto a closed ideal in M_n.

(f) Let $f \in \mathcal{S}(\mathbb{R}^n)$ and $\mu \in M_n$. Show $f * \mu$ is a smooth function.

Solution: (a) follows immediately by Fubini. For (b) note for $f \in C_0(\mathbb{R}^n)$ and $g, h \in L^1(\mathbb{R}^n)$, one has using Fubini and translation invariance,

$$\begin{aligned}
\lambda_g * \lambda_h(f) &= \iint f(x+y)\, d\lambda_g(x)\, d\lambda_h(y) \\
&= \int \int g(x) f(x+y)\, dx\, d\lambda_h(y) \\
&= \int \int g(x) h(y) f(x+y)\, dx\, dy \\
&= \int \int g(x-y) h(y) f(x)\, dx\, dy \\
&= \int f(x) \int g(x-y) h(y)\, dy\, dx \\
&= \int f(x) g * h(x)\, dx \\
&= \lambda_{g*h}(f).
\end{aligned}$$

Next, using invariance of Lebesgue measure under inversion $x \mapsto -x$, one has

$$\begin{aligned}(\lambda_f)^*(g) &= \overline{\int \overline{g(-x)}\, d\lambda_f(x)} \\ &= \overline{\int \overline{g(-x)}\, f(x)\, dx} \\ &= \int g(-x)\, \overline{f(x)}\, dx \\ &= \int g(x)\overline{f(-x)}\, dx \\ &= \lambda_{f*}(g)\end{aligned}$$

for $f \in L^1$ and $g \in C_0(\mathbb{R}^n)$. Thus (c) holds.

For (d), note using Fubini and translation invariance,

$$\begin{aligned}\lambda_f * \mu(g) &= \iint g(x+y)\, d\lambda_f(x)\, d\mu(y) \\ &= \int\int g(x+y) f(x)\, dx\, d\mu(y) \\ &= \int\int g(x) f(x-y)\, dx\, d\mu(y) \\ &= \int g(x) \int f(x-y)\, d\mu(y)\, dx \\ &= \int g(x)(f * \mu)(x)\, dx \\ &= \lambda_{f*\mu}(g)\,.\end{aligned}$$

It is clear that $f \mapsto \mu_f$ is linear from $L^1(\mathbb{R}^n)$ into M_n and by (b), (c), and (d) is a $*$ homomorphism onto an ideal in M_n. It is an isometry. Indeed, take a sequence of simple functions s_k with $|s_k(x)| \leqslant 1$ for all x which converges to $\chi_{\{f\neq 0\}} \frac{\bar{f}}{|f|}$. Then $\lambda_f(s_k) \to \int |f(x)|\, dx$ by the Lebesgue Dominated Convergence Theorem. But $s_k = \sum_{p=1}^{m_k} a_{k,p}\chi_{E_{k,p}}$ where the sets $E_{k,p}, p = 1, 2, \ldots, m_k$ partition $\mathbb{R}^n$. Note each $|a_{k,p}| \leqslant 1$. Thus $|\lambda_f(s_k)| \leqslant \sum_{p=1}^{m_k} |\lambda_f(E_{k,p})| \leqslant |\lambda_f|(\mathbb{R}^n)$. Taking limits we have $|f|_1 = \int |f(x)|\, dx \leqslant ||\lambda_f||$. Since $|\lambda_f|(E)$ is the supremum over all measurable partitions $E_1, E_2, \ldots, E_m$ of E and $\sum |\lambda_f(E_i)| \leqslant \sum \int_{E_i} |f(x)|\, dx \leqslant \int_E |f(x)|\, dx$, we have $||\lambda_f|| \leqslant |f|_1$. So $f \mapsto \lambda_f$ is norm preserving.

Finally, for (e), it suffices by induction to show $f * \mu$ is continuous and $D_i(f * \mu) = (D_i f) * \mu$ for each i. Note $f * \mu$ is continuous for if $x_k \mapsto x_0$, then $f(x_k + y) \to f(x_0 + y)$ for all y and these functions are bounded by the $L^1(\mu)$ function $|f|_\infty$. Thus the Lebesgue Dominated Convergence Theorem implies continuity of $\mu * f$. To see $D_i(f * \mu) = D_i f * \mu$, consider

$$\frac{f * \mu(x + he_i) - f * \mu(x)}{h} = \int \frac{f(x + he_i - y) - f(x - y)}{h} \, d\mu(y)$$
$$= \int D_i f(x + h^*(y) e_i - y) d\mu(y)$$

which converges to $\int D_i f(x - y)\, d\mu(y)$ as $h \to 0$, where here we have used the Mean Value Theorem, noting that $h^*(y) \mapsto 0$ as $h \to 0$ and the Lebesgue Dominated Convergence Theorem where the integrands are dominated by $|f|_{0,e_i} = \max |D_i f|$.

□

Chapter 3

The Fourier Transform on $\mathbf{R}^n$

Exercise Set 3.1

1. Find the Fourier transform of the following functions:

(a) $\chi_{[-1/2,1/2]}$;

(b) $e^{-|x|}$;

(c) $e^{-|3x-2|}$;

(d) $e^{-\left(ax^2+bx+c\right)}$ where $a>0$;

(e) $\chi_{[0,\infty]}e^{-x}$;

(f) xe^{-x^2};

(g) $\chi_{[0,1]}(x)\sin(2\pi x)$;

(h) $\frac{1}{(x+i)^n}$, $n \geqslant 2$;

Solution:

For (a) one has if $f=\chi_{[-\frac{1}{2},\frac{1}{2}]}$, then

$$\begin{aligned}
\hat{f}(x) &= \int_{-\frac{1}{2}}^{\frac{1}{2}} e^{-2\pi ixy}\,dy \\
&= \frac{-e^{-2\pi ixy}}{2\pi ix}\Bigg|_{y=-1/2}^{y=1/2} \\
&= \frac{e^{\pi ix}-e^{-\pi ix}}{2\pi ix} = \frac{\sin \pi x}{\pi x}\,.
\end{aligned}$$

For (b), we have if $f(x) = e^{-|x|}$, then

$$\begin{aligned}
\hat{f}(x) &= \int e^{-|y|} e^{-2\pi i x y}\, dy \\
&= \int e^{-|y|} \cos(2\pi x y)\, dy \\
&= 2\int_0^\infty e^{-y} \cos(2\pi x y)\, dy \\
&= \left.\frac{e^{-y}\sin(2\pi x y)}{\pi x}\right|_{y=0}^{y=\infty} + \int_0^\infty \frac{\sin(2\pi x y)}{\pi x} e^{-y}\, dy \\
&= \left.\frac{-e^{-y}\cos(2\pi x y)}{2\pi^2 x^2}\right|_{y=0}^{y=\infty} - \int_0^\infty \frac{\cos(2\pi x y)}{2\pi^2 x^2}\, e^{-y}\, dy \\
&= \frac{1}{2\pi^2 x^2} - \frac{1}{4\pi^2 x^2}\hat{f}(x)\,.
\end{aligned}$$

So $(4\pi^2 x^2 + 1)\hat{f}(x) = 2$. This gives $\hat{f}(x) = \frac{2}{1+4\pi^2 x^2}$.

For (c), note if $g(x) = e^{-|3x-2|}$ and $f(x) = e^{-|x|}$, then $\lambda(2)f(x) = e^{-|x-2|}$ and $\delta(\frac{1}{3})\lambda(2)f(x) = \sqrt{3}(\lambda(2)f)(3x) = \sqrt{3}e^{|3x-2|}$. So $g(x) = \frac{1}{\sqrt{3}}\delta(\frac{1}{3})\lambda(2)f(x)$. Thus using f in part (b) , Lemma 3.3 implies

$$\begin{aligned}
\hat{g}(x) &= \frac{1}{\sqrt{3}}\mathcal{F}(\delta(1/3)\lambda(2)f)(x) \\
&= \frac{1}{\sqrt{3}}\delta(3)\widehat{\lambda(2)f}(x) \\
&= \frac{1}{3}\widehat{\lambda(2)f}(x/3) \\
&= \frac{1}{3}(\tau(-2)\hat{f})(x/3) \\
&= \frac{1}{3}e^{-4\pi i x/3}\hat{f}(x/3) \\
&= \frac{1}{3}\frac{2e^{-4\pi i x/3}}{1+4\pi^2 x^2/9} = \frac{6e^{-4\pi i x/3}}{9+4\pi^2 x^2}\,.
\end{aligned}$$

For (d) we know from Example 3.2 that the Fourier transform of $g(x) = e^{-\pi x^2}$ is $\hat{g}(x) = e^{-\pi x^2}$. Now $\delta(\sqrt{\frac{\pi}{a}})g(x) = \left(\frac{a}{\pi}\right)^{1/4} g(\sqrt{\frac{a}{\pi}}x) = \left(\frac{a}{\pi}\right)^{1/4} e^{-ax^2}$. Thus $\lambda(-\frac{b}{2a})\delta(\sqrt{\frac{\pi}{a}})g(x) = \left(\frac{a}{\pi}\right)^{1/4} e^{-a(x+\frac{b}{2a})^2} = \left(\frac{a}{\pi}\right)^{1/4} e^{-a(x^2+\frac{b}{a}x+\frac{b^2}{4a^2})}$. Thus $\left(\frac{\pi}{a}\right)^{1/4} e^{-c+\frac{b^2}{4a}}\lambda(\frac{-b}{2a})\delta(\sqrt{\frac{\pi}{a}})g(x) = e^{-(ax^2+bx+c)}$.

Hence if $f(x) = e^{-(ax^2+bx+c)}$ and $h(x) = \delta(\sqrt{\frac{\pi}{a}})g(x)$, we see by Lemma 3.3, that

$$\begin{aligned}
\hat{f}(x) &= \left(\frac{\pi}{a}\right)^{1/4} e^{-c+\frac{b^2}{4a}} \tau(\frac{b}{2a})\hat{h}(x) \\
&= \left(\frac{\pi}{a}\right)^{1/4} e^{-c+\frac{b^2}{4a}} e^{\pi i b x/a}\hat{h}(x) \\
&= \left(\frac{\pi}{a}\right)^{1/4} e^{-c+\frac{b^2}{4a}} e^{\pi i b x/a}\delta\left(\sqrt{\frac{a}{\pi}}\right)\hat{g}(x) \\
&= \left(\frac{\pi}{a}\right)^{1/4} e^{-c+\frac{b^2}{4a}} e^{\pi i b x/a}\left(\frac{\pi}{a}\right)^{1/4}\hat{g}\left(\sqrt{\frac{\pi}{a}}x\right) \\
&= \left(\frac{\pi}{a}\right)^{1/2} e^{-c+\frac{b^2}{4a}} e^{\pi i b x/a} e^{-\pi^2 x^2/a}.
\end{aligned}$$

If $f(x) = \chi_{[0,\infty)}(x)e^{-x}$, then

$$\begin{aligned}
\hat{f}(x) &= \int_0^\infty e^{-y}e^{-2\pi i y x}\,dy \\
&= \int_0^\infty e^{-(2\pi i x+1)y}\,dy \\
&= -\frac{1}{2\pi i x+1}\,e^{-(2\pi i x+1)y}\Big|_{y=0}^{y=\infty} \\
&= \frac{1}{2\pi i x+1}.
\end{aligned}$$

So we have (e).

To do (f), note by (d) if $h(x) = e^{-x^2}$, then $\hat{h}(x) = \sqrt{\pi}e^{-\pi^2x^2}$. Now note $h'(x) = e^{-x^2}(-2x)$, so $xe^{-x^2} = -\frac{1}{2}h'(x)$. Thus

$$\begin{aligned}
\hat{f}(x) &= \int ye^{-y^2}e^{-2\pi i y x}\,dy \\
&= -\frac{1}{2}\int h'(y)e^{-2\pi i y x}\,dy \\
&= -\frac{1}{2}\,h(y)e^{-2\pi i y x}\Big|_{y=-\infty}^{y=\infty} + \frac{1}{2}\int h(y)(-2\pi i x)e^{-2\pi i y x}\,dy \\
&= -\pi i x\hat{h}(x) \\
&= -\pi^{3/2} i x e^{-\pi^2 x^2}.
\end{aligned}$$

For (g), we have

$$\begin{aligned}
\hat{f}(x) &= \int_0^1 \sin(2\pi y) e^{-2\pi ixy}\, dy \\
&= \left.\frac{\sin(2\pi y)\exp(-2\pi ixy)}{-2\pi ix}\right|_{y=0}^{y=1} + \frac{2\pi}{2\pi ix}\int_0^1 \cos(2\pi y)\exp(-2\pi ixy)\, dy \\
&= \frac{1}{ix}\left(\left.\frac{\cos(2\pi y)\exp(-2\pi ixy)}{-2\pi ix}\right|_{y=0}^{y=1} - \frac{1}{ix}\int_0^1 \sin(2\pi y)\exp(-2\pi ixy)\, dy\right) \\
&= \frac{1}{ix}\left(\frac{\exp(-2\pi ix)-1}{-2\pi ix} - \frac{1}{ix}\hat{f}(x)\right) \\
&= \frac{-1+\exp(-2\pi ix)}{2\pi x^2} + \frac{1}{x^2}\hat{f}(x)\,.
\end{aligned}$$

This gives $(1-x^2)\hat{f}(x) = \frac{1-\exp(-2\pi ix)}{2\pi}$. Thus

$$\hat{f}(x) = \frac{1-\exp(-2\pi ix)}{2\pi(1-x^2)}.$$

Finally for (h), we do induction. First note if $f_n(x) = \frac{1}{(x+i)^n}$ where $n \geqslant 2$, then $f_n'(x) = -nf_{n+1}(x)$. Now

$$\begin{aligned}
\int f_n'(y)e^{-2\pi ixy}\, dy &= f_n(y)e^{-2\pi ixy}\Big|_{y=-\infty}^{y=\infty} + 2\pi ix\int f_n(y)e^{-2\pi ixy}\, dy \\
&= 2\pi ix\hat{f}_n(x).
\end{aligned}$$

Thus $-n\hat{f}_{n+1}(x) = 2\pi ix\hat{f}_n(x)$. So we have the recursion formula

$$\hat{f}_{n+1}(x) = \frac{-2\pi ix}{n}\hat{f}_n(x).$$

To find $\hat{f}_2$ we use residues from complex function theory. Namely, let $H(z) = \frac{\exp(2\pi izx)}{(z+i)^2}$. The only pole is at $z=-i$. Consider γ_R, the piecewise path $\gamma_1(t) = t-2i$, $-R \leqslant t \leqslant R$, $\gamma_2(t) = R-2i+ti$, $0 \leqslant t \leqslant 2$, $\gamma_3(t) = -t$ for $-R \leqslant t \leqslant R$, $\gamma_4(t) = -R-ti$, $0 \leqslant t \leqslant 2$.

We have

$$\begin{aligned}
\int_{\gamma_R} \frac{1}{(z+i)^2}e^{2\pi izx}\, dz &= 2\pi i \operatorname{Res}_{z=-i} \frac{e^{2\pi izx}}{(z+i)^2} \\
&= 2\pi i(e^{2\pi x}\operatorname{Res}_{z=-i} \frac{e^{2\pi i(z+i)x}}{(z+i)^2}) \\
&= 2\pi i(e^{2\pi x}(2\pi ix)) \\
&= -4\pi^2 xe^{2\pi x}.
\end{aligned}$$

Letting $R \to \infty$ gives

$$\int_{-\infty}^{\infty} \frac{1}{(t-i)^2} e^{2\pi i(t-2i)x} dt + \int_{-\infty}^{\infty} \frac{1}{(-t+i)^2} e^{-2\pi itx}(-dt) = -4\pi^2 x e^{2\pi x}.$$

The change of variables $u = -t$ in both integrals gives

$$e^{4\pi x} \int_{-\infty}^{\infty} \frac{1}{(-u-i)^2} e^{-2\pi iux}\, du - \int_{-\infty}^{\infty} \frac{1}{(u+i)^2} e^{2\pi iux} du = -4\pi^2 x e^{2\pi x}$$

or $e^{4\pi x} \hat{f}_2(x) - \hat{f}_2(-x) = -4\pi^2 x e^{2\pi x}$.

We next show $\hat{f}_2(x) = 0$ for $x \leqslant 0$. Consider

$$G_x(b) = \int_{-\infty}^{\infty} \frac{1}{(t+bi+i)^2} e^{-2\pi i(t+bi)x}\, dt$$

where $b > 0$. We note $G_x(b) = \hat{f}_2(x)$. Indeed, if γ_R is the piecewise curve, $\gamma_1(t) = t$ for $-R \leqslant t \leqslant R$, $\gamma_2(t) = R + ti$ for $0 \leqslant t \leqslant b$, $\gamma_3(t) = bi - t$ for $-R \leqslant t \leqslant R$, and $\gamma_4(t) = -R + (b-t)i$ for $0 \leqslant t \leqslant b$, then $\int_\gamma \frac{1}{(z+i)^2} e^{-2\pi izx}\, dz = 0$. Again, doing the integrals over the curves and letting $R \to \infty$ gives

$$\int_{-\infty}^{\infty} \frac{1}{(t+i)^2} e^{-2\pi itx}\, dt + \int_{-\infty}^{\infty} \frac{1}{(bi-t+i)^2} e^{-2\pi i(bi-t)x}\, (-dt) = 0.$$

Invariance of Lebesgue integration under $t \mapsto -t$ yields

$$\hat{f}_2(x) = G_b(x).$$

Now $G_b(x) = e^{2\pi bx} \int_{-\infty}^{\infty} \frac{1}{(t+bi+i)^2} e^{-2\pi itx}\, dt$. But $G_b(x) \to 0$ as $b \to \infty$ for $x \leqslant 0$. This gives $\hat{f}_2(x) = 0$ for $x \leqslant 0$. For $x > 0$, we then have $e^{4\pi x} \hat{f}_2(x) = e^{4\pi x} \hat{f}_2(x) - \hat{f}_2(-x) = -4\pi^2 x e^{2\pi x}$. So $\hat{f}_2(x) = -4\pi^2 x e^{-2\pi x}$. Concluding we see

$$\begin{aligned}
\hat{f}_2(x) &= \left(-4\pi^2 x\right) e^{-2\pi x} \chi_{[0,\infty)}(x) \\
\hat{f}_3(x) &= \frac{8\pi^3 i x^2}{2} e^{-2\pi x} \chi_{[0,\infty)}(x) \\
\hat{f}_4(x) &= \frac{16\pi^4 x^3}{6} e^{-2\pi x} \chi_{[0,\infty)}(x) \\
\hat{f}_5(x) &= -\frac{32\pi^5 i x^4}{24} e^{-2\pi x} \chi_{[0,\infty)}(x)
\end{aligned}$$

and in general

$$\hat{f}_n(x) = \frac{(-2\pi i)^n x^{n-1} e^{-2\pi x}}{(n-1)!} \chi_{[0,\infty)}(x).$$

One could also use residues on the function $H_n(z) = \frac{1}{(z+i)^n}$ for $n \geqslant 2$.

□

2. Let $\operatorname{sinc}(\omega) = \frac{\sin\omega}{\omega}$ for $\omega \neq 0$. Show sinc is not in $L^1(\mathbb{R})$.
Solution: Let $\omega_k = (2k+1)\frac{\pi}{2}$ and $\delta = \frac{\pi}{4}$. Note $|\sin\omega| \geqslant \frac{1}{\sqrt{2}}$ for $\omega \in [\omega_k - \delta, \omega_k + \delta]$. Hence

$$\begin{aligned}
\int_0^\infty \left|\frac{\sin\omega}{\omega}\right| d\omega &\geqslant \frac{1}{\sqrt{2}} \sum_{k=0}^\infty \int_{\omega_k-\delta}^{\omega_k+\delta} \frac{1}{\omega}\, d\omega \\
&\geqslant \frac{1}{\sqrt{2}} \sum_{k=0}^\infty \int_{\omega_k-\delta}^{\omega_k+\delta} \frac{1}{\omega_k+\delta}\, d\omega \\
&= \frac{1}{\sqrt{2}} \sum_{k=0}^\infty \frac{2\delta}{\omega_k+\delta} \\
&= \frac{1}{\sqrt{2}} \sum_{k=0}^\infty \frac{\pi/2}{(2k+1)\pi/2 + \pi/4} \\
&\geqslant \frac{1}{\sqrt{2}} \sum_{k=0}^\infty \frac{1}{(2k+1)+1} \\
&= \frac{1}{2\sqrt{2}} \sum_{k=0}^\infty \frac{1}{k+1} \\
&= \infty.
\end{aligned}$$

□

3. By Exercise 2.6.5, the space M_n of complex Borel measures on $\mathbb{R}^n$ is a Banach $*$ algebra under convolution. Moreover if $f \in L^1(\mathbb{R}^n)$, then λ_f defined by $\lambda_f(E) = \int_E f(x)\,dx$ is in M_n. For μ in M_n, define

$$\hat{\mu}(\omega) := \int_{\mathbb{R}^n} e^{-2\pi i x\cdot\omega}\, d\mu(x)\,.$$

(a) Show that $\hat{\mu}$ is continuous.

(b) Show if f is in $L^1(\mathbb{R}^n)$, then $\widehat{\lambda_f} = \hat{f}$.

(c) Express the function $\widehat{\mu^*}$ in terms of $\hat{\mu}$.

(d) Show that $\widehat{\mu * \nu} = \hat{\mu}\hat{\nu}$ for $\mu, \nu \in M_n$.

(e) Show $\mu = 0$ if $\hat{\mu} = 0$.

Solution:

(a) follows from the Lebesgue Dominated Convergence Theorem. Indeed, if $\omega_k \to \omega$, then $e^{-2\pi i x\cdot w_k} \to e^{-2\pi i x\cdot w}$ for all x. Moreover, $|e^{-2\pi i x\cdot\omega_k}| \leqslant 1 \in L^1(|\mu|)$. So $\hat{\mu}(\omega_k) \to \hat{\mu}(\omega)$ as $k \to \infty$.

For (b), note $\hat{\lambda}_f(\omega) = \int e^{-2\pi i\omega\cdot x}\, d\lambda_f(x) = \int e^{-2\pi i\omega\cdot x} f(x)\, dx = \hat{f}(\omega)$ while since $\mu^*(E) = \overline{\mu(-E)}$ one has

$$\begin{aligned}\widehat{\mu^*}(\omega) &= \int e^{-2\pi i\omega\cdot x}\, d\mu^*(x) \\ &= \overline{\int e^{-2\pi i\omega\cdot x}\, d\mu(x)} \\ &= \overline{\hat{\mu}(\omega)}\,.\end{aligned}$$

and thus $\widehat{\mu^*} = \overline{\hat{\mu}}$. So we also have (c).

To see (d), we use Fubini.

$$\begin{aligned}\widehat{\mu * \nu}(\omega) &= \int e^{-2\pi i\omega\cdot t}\, d(\mu * \nu)(t) \\ &= \iint e^{-2\pi i\omega\cdot(x+y)}\, d\mu(x)\, d\nu(y) \\ &= \int e^{-2\pi i\omega\cdot y}\left(\int e^{-2\pi i\omega\cdot x}\, d\mu(x)\right) d\nu(y) \\ &= \hat{\mu}(\omega)\int e^{-2\pi i\omega\cdot y}\, d\nu(y) \\ &= \hat{\mu}(\omega)\,\hat{\nu}(\omega)\,.\end{aligned}$$

Finally for (e), we have $\int e^{-2\pi i x\cdot y}\, d\mu(y) = 0$ for all x. Thus

$$\int f(x)\int e^{-2\pi i x\cdot y}\, d\mu(y)\, dx = 0$$

for all $f \in \mathcal{S}(\mathbb{R}^n)$. By Fubini's Theorem, $\iint f(x)e^{-2\pi i x\cdot y}\, dx\, d\mu(y) = 0$ for all such f. Hence $\int \hat{f}(y)\, d\mu(y) = 0$ for all $f \in \mathcal{S}(\mathbb{R}^n)$. So by (2.10), $\int_Q d\mu(y) = 0$ for all rectangles Q. By regularity, we see $\int_U d\mu(y) = 0$ for all open sets U. So $\mu = 0$.

□

4. Find the Fourier transform of χ_B where B is the unit ball in $\mathbb{R}^n$. (**Hint:** Use rotational invariance and the formula $J_\nu(x) = \frac{(x/2)^\nu}{\sqrt{\pi}\Gamma(\nu+\frac{1}{2})}\int_0^\pi \cos(x\cos\theta)\sin^{2\nu}\theta\, d\theta$ when $\nu > -\frac{1}{2}$ for Bessel functions of the first kind.)

Solution:

We first note if R is a rotation on $\mathbb{R}^n$, then

$$\begin{aligned}\mathcal{F}(f)(Ry) &= \int f(x)e^{-2\pi i x\cdot Ry}\, d_n x \\ &= \int f(Rx)e^{-2\pi i Rx\cdot Ry}\, d_n x \\ &= \int f(Rx)e^{-2\pi i x\cdot y}\, d_n x \\ &= \mathcal{F}(f\circ R)(y)\end{aligned}$$

for $f \in L^1(\mathbb{R}^n)$. Hence if $f \circ R = f$ for all R, then $\mathcal{F}(f)$ is rotationally invariant. Taking $f = \chi_B$, we then see

$$\mathcal{F}(f)(x_1, x_2, \ldots, x_n) = \mathcal{F}(f)(0, 0, \ldots, 0, |x|).$$

Thus to determine this Fourier transform it suffices to find $\mathcal{F}(0, 0, \ldots, 0, x_n)$. We note $f^* = f$ and thus $\mathcal{F}(f)$ is real valued. Hence if we write $x = (\xi, x_n)$ where $\xi \in \mathbb{R}^{n-1}$

$$\begin{aligned}
\mathcal{F}(f)(0, 0, \ldots, 0, y) &= \int_{\mathbb{R}^n} f(x) e^{-2\pi i x \cdot (0,0,\ldots,0,y)} \, d_n x \\
&= \int_{-1}^{1} e^{-2\pi i x_n y} \left(\int_{|\xi| \leqslant \sqrt{1 - x_n^2}} d_{n-1}\xi \right) dx_n \\
&= \int_{-1}^{1} e^{-2\pi i x_n y} \operatorname{Vol}_{\mathbb{R}^{n-1}}(\{\xi \mid |\xi| \leqslant \sqrt{1 - x_n^2}) \, dx_n \, .
\end{aligned}$$

Now Exercise 2.2.8 implies the volume of a ball of radius r in $\mathbb{R}^n$ is $\frac{\pi^{n/2} r^n}{\Gamma(n/2+1)}$. Hence one sees

$$\begin{aligned}
\mathcal{F}(f)(0, 0, \ldots, 0, y) &= \int_{-1}^{1} e^{-2\pi i x_n y} \frac{\pi^{(n-1)/2}}{\Gamma((n+1)/2)} (1 - x_n^2)^{(n-1)/2} \, dx_n \\
&= \int_{-1}^{1} \cos(2\pi x_n y) \frac{\pi^{(n-1)/2}}{\Gamma((n+1)/2)} (1 - x_n^2)^{(n-1)/2} \, dx_n \\
&= \frac{\pi^{(n-1)/2}}{\Gamma((n+1)/2)} \int_{\pi}^{0} \cos(2\pi y \cos\theta)(1 - \cos^2\theta)^{(n-1)/2} \, d(\cos\theta) \\
&= \frac{\pi^{(n-1)/2}}{\Gamma((n+1)/2)} \int_{0}^{\pi} \cos(2\pi y \cos\theta) \sin^n\theta \, d\theta \, .
\end{aligned}$$

Now Bessel's function J_ν for order $\nu > -\frac{1}{2}$ is given by

$$J_\nu(x) = \frac{(x/2)^\nu}{\sqrt{\pi}\Gamma(\nu + \frac{1}{2})} \int_0^\pi \cos(x \cos\theta) \sin^{2\nu}\theta \, d\theta.$$

This yields

$$\begin{aligned}
\mathcal{F}(\chi_B)(y_1, y_2, \ldots, y_n) &= \mathcal{F}(\chi_\beta)(0, 0, \ldots, 0, |y|) \\
&= \frac{1}{|y|^{n/2}} \left(\frac{(\pi|y|)^{n/2}}{\sqrt{\pi}\Gamma((n+1)/2)} \right) \int_0^{2\pi} \cos(2\pi|y| \cos\theta) \sin^n\theta \, d\theta \\
&= |y|^{-n/2} J_{n/2}(2\pi|y|) \, .
\end{aligned}$$

□

Exercise Set 3.2

1. For $f \in \mathcal{S}(\mathbb{R}^m \times \mathbb{R}^n)$, show the partial Fourier transform $\mathcal{F}_1 f$ given by

$$\mathcal{F}_1 f(\omega, y) = \int f(x,y) e^{-2\pi i x\cdot\omega}\, dx$$

satisfies

$$\begin{aligned} p(D)\mathcal{F}_1 f(\omega,y) &= \mathcal{F}_1(p(-2\pi i\,\cdot\,)f(\,\cdot\,,\cdot\,))(\omega,y) \text{ and} \\ F_1(p(D)f)(\omega,y) &= p(2\pi i\omega)\mathcal{F}_1 f(\omega,y) \end{aligned}$$

for any polynomial p in x.

Solution:

Start by noting $\mathcal{F}_1 f(\omega,y) = \mathcal{F} f_y(\omega)$ where $\mathcal{F}$ is the Fourier transform on $\mathbb{R}^m$, $f_y(x) = f(x,y)$ and y is fixed. Now Exercise 3.2.1 implies f_y is in $\mathcal{S}(\mathbb{R}^m)$ and so we map apply Theorem 3.4. This gives $p(D)\mathcal{F}(f_y)(\omega) = \mathcal{F}(p(2pi\,\cdot\,)f_y(\,\cdot\,))(\omega)$ and $\mathcal{F}(p(D)f_y)(\omega) = p(2\pi i\omega)\mathcal{F}(f_y)(\omega)$ which is the result. □

2. Show the partial Fourier transform $\mathcal{F}_1$ satisfies properties corresponding to properties (a) to (i) in Lemma 3.3.

Solution:

We use (a) to (i) of Lemma 3.3 with the Fourier transform $\mathcal{F}$ on $\mathbb{R}^m$. Let $f_y(x) = f(x,y)$ and define

$$\begin{aligned} \lambda_1(x_0)f(x,y) &= f(x-x_0,y), \\ \tau_1(x_0)f(x,y) &= e^{2\pi i x_0\cdot x} f(x,y), \\ \delta_1(a)f(x,y) &= a^{-m/2} f(a^{-1}x,y) \text{ for } a>0, \text{ and} \\ f *_1 g(x,y) &= \int f(z,y)g(x-z,y)\, d_m z \end{aligned}$$

when this integral exists. Fubini implies $f_y \in L^1(\mathbb{R}^m)$ for a.e. y for each $f \in L^1(\mathbb{R}^m \times \mathbb{R}^n)$. For (a) one would have $|\mathcal{F}_1 f(\,\cdot\,,y)|_\infty = |\mathcal{F}(f_y)|_\infty \leqslant |f_y|_1 = \int |f(x,y)|\, dx$. So (a) would become

$$|\mathcal{F}_1 f(\,\cdot\,,y)|_\infty \leqslant \int |f(x,y)|\, d_m x = |f(\,\cdot\,,y)|_1$$

for each y. (b) becomes $\mathcal{F}_1(f+g)(\omega,y) = \mathcal{F}_1(f)(\omega,y) + \mathcal{F}_1(g)(\omega,y)$ for all ω for a.e. y; this following from $(f+g)_y = f_y + g_y$. For (c) one has

$$\begin{aligned} \mathcal{F}_1(\lambda_1(x)f)(\omega,y) &= \mathcal{F}(\lambda(x)f_y)(\omega) \\ &= \tau(-x)\mathcal{F} f_y(\omega) \\ &= e^{-2\pi i x\cdot\omega}\mathcal{F}_1 f(\omega,y) \\ &= \tau_1(-x)\mathcal{F}_1 f(w,y) \end{aligned}$$

for all ω for a.e. y. Similarly,

$$\begin{aligned}\mathcal{F}_1(\tau_1(x_0)f)(\omega,y) &= \mathcal{F}_1 f(\omega - x_0, y) \text{ and} \\ \mathcal{F}_1(\delta_1(a)f)(\omega,y) &= \delta_1(1/a)\mathcal{F}_1 f(\omega,y)\end{aligned}$$

for all ω for a.e. y. These are the analogues of (c), (d), (e). (f) becomes $\mathcal{F}_1(\bar{f})(\omega,y) = \overline{\mathcal{F}_1 f(-\omega,y)}$ and (g) and (h) become $\mathcal{F}_1(h)(\omega,y) = \mathcal{F}_1 f(-\omega,y)$ if $h(x,y) = f(-x,y)$ and $\mathcal{F}_1(h)(\omega,y) = \overline{\mathcal{F}_1 f(-\omega,y)}$ if $h(x,y) = \overline{f(-x,y)}$. Finally (i) becomes

$$\mathcal{F}_1(f *_1 g)(\omega,y) = \mathcal{F}_1(f)(\omega,y)\mathcal{F}_1(g)(\omega,y)$$

for all ω for a.e. y. Indeed, for a.e. y, f_y and g_y are simultaneously integrable on $\mathbb{R}^m$ and in this case

$$\mathcal{F}_1(f *_1 g)(\omega,y) = \mathcal{F}(f_y * g_y)(\omega) = \mathcal{F}(f_y)(\omega)\mathcal{F}(g_y)(\omega) = \mathcal{F}_1(f)(\omega,y)\mathcal{F}_1 g(\omega,y)$$

for all ω.

□

3. Using

$$\int \left| \int f(x) e^{-2\pi i x\cdot\omega}\, dx \right|^2 d\omega = \int |f(x)|^2\, dx$$

for $f \in \mathcal{S}(\mathbb{R}^n)$, show

$$\int \left| \int f(x) e^{-aix\cdot\omega} \sqrt{\left(\frac{a}{2\pi}\right)^n}\, dx \right|^2 \sqrt{\left(\frac{a}{2\pi}\right)^n}\, d\omega = \int |f(x)|^2 \sqrt{\left(\frac{a}{2\pi}\right)^n}\, dx$$

for $f \in \mathcal{S}(\mathbb{R}^n)$ and $a > 0$. Now define

$$\mathcal{F}_a(f)(\omega) = \int f(x) e^{-iax\cdot\omega}\, d_a x$$

where $d_a x$ is the measure $\sqrt{\left(\frac{a}{2\pi}\right)^n}\, dx$. Show $\mathcal{F}_a$ is a linear homeomorphism of $\mathcal{S}(\mathbb{R}^n)$ that satisfies:

(a) Show $\mathcal{F}_a^2 f = \check{f}$ for $f \in \mathcal{S}(\mathbb{R}^n)$.

(b) Show $\mathcal{F}_a$ extends to a unitary transformation of the space $L^2(\mathbb{R}^n, d_a x)$.

(c) Show $\mathcal{F}_a(\lambda(y)f)(\omega) = e^{-iay\cdot\omega}\mathcal{F}_a(f)(\omega)$ for $f \in \mathcal{S}(\mathbb{R}^n)$.

(d) Show $\mathcal{F}_a(\delta(b)f) = \delta(\frac{1}{b})\mathcal{F}_a(f)$ for $f \in \mathcal{S}(\mathbb{R}^n)$.

(e) Show $\mathcal{F}_a(f^*) = \overline{\mathcal{F}_a(f)}$ for $f \in \mathcal{S}(\mathbb{R}^n)$.

(f) Show $\mathcal{F}_a(p(D)f)(\omega) = p(ai\omega)\mathcal{F}_a f(\omega)$ for $f \in \mathcal{S}(\mathbb{R}^n)$.

(g) Show $\mathcal{F}_a(pf)(\omega) = p(\frac{iD}{a})\mathcal{F}_a(f)(\omega)$ for polynomials p and $f \in \mathcal{S}(\mathbb{R}^n)$.

The important cases are $a = 1$ and $a = 2\pi$.

Solution:

From

$$\int \left| \int f(x) e^{-2\pi i x\cdot\omega}\, dx \right|^2 d\omega = \int |f(x)|^2\, dx,$$

we make the change in variables from ω to $\frac{a\omega}{2\pi}$ on the left side to give

$$\left(\frac{a}{2\pi}\right)^n \int \left| \int f(x) e^{-iax\cdot\omega}\, dx \right|^2 d\omega = \int |f(x)|^2\, dx.$$

This implies

$$\int \left| \int f(x) e^{-iax\cdot\omega} \sqrt{\left(\frac{a}{2\pi}\right)^n}\, dx \right|^2 \sqrt{\left(\frac{a}{2\pi}\right)^n} d\omega = \sqrt{\left(\frac{a}{2\pi}\right)^n} \int |f(x)|^2\, dx.$$

Thus if $d_a x = \sqrt{\left(\frac{a}{2\pi}\right)^n}\, dx$, we see

$$\int \left| \int f(x) e^{-iax\cdot\omega}\, d_a x \right|^2 d_a\omega = \int |f(x)|^2\, d_a x \tag{A}$$

Now note $\mathcal{F}_a(f)(\omega) = \left(\frac{a}{2\pi}\right)^{n/2} \hat{f}(\frac{a}{2\pi}\omega) = \delta(\frac{2\pi}{a})\mathcal{F}(f)(\omega)$. Thus since $\delta(\frac{a}{2\pi})$ and $\mathcal{F}$ are linear homeomorphisms of $\mathcal{S}(\mathbb{R}^n)$, $\mathcal{F}_a$ is a linear homeomorphism of $\mathcal{S}(\mathbb{R}^n)$ and by (A), it is an isometry in the L^2 norm for the measure $d_a x$. Since $|f|_{2,a}^2 = \left(\frac{a}{2\pi}\right)^n |f|_2$, $\mathcal{S}(\mathbb{R}^n)$ is dense in $L^2(\mathbb{R}^n, d_a x)$ and $\mathcal{F}_a$ extends to a unitary mapping of $L^2(\mathbb{R}^n, d_a x)$, and we have (b).

For (a), note by Lemma 3.3,

$$\begin{aligned}
\mathcal{F}_a^2 f &= \delta\left(\frac{2\pi}{a}\right) \widehat{\mathcal{F}_a f} \\
&= \delta(\frac{2\pi}{a}) \widehat{\delta(\frac{2\pi}{a}) \hat{f}} \\
&= \delta(\frac{2\pi}{a}) \delta\left(\frac{a}{2\pi}\right) \hat{\hat{f}} \\
&= \check{f}\,.
\end{aligned}$$

Again by (c) of this Lemma,

$$\begin{aligned}
\mathcal{F}_a(\lambda(y)f)(\omega) &= \delta\left(\frac{2\pi}{a}\right)\mathcal{F}(\lambda(y)f)(\omega) \\
&= \delta\left(\frac{2\pi}{a}\right)(\tau(-y)\hat{f})(\omega) \\
&= \left(\frac{a}{2\pi}\right)^{n/2}(\tau(-y)\hat{f})(\frac{a}{2\pi}\omega) \\
&= \left(\frac{a}{2\pi}\right)^{n/2} e^{-ia\omega\cdot y}\hat{f}(\frac{a}{2\pi}\omega) \\
&= e^{-ia\omega\cdot y}\delta\left(\frac{2\pi}{a}\right)\hat{f}(\omega) \\
&= e^{-ia\omega\cdot y}\mathcal{F}_a f(\omega)\,.
\end{aligned}$$

To see (d) note

$$\begin{aligned}
\mathcal{F}_a(\delta(b)f) &= \delta\left(\frac{2\pi}{a}\right)\widehat{\delta(b)f} \\
&= \delta\left(\frac{2\pi}{a}\right)\delta\left(\frac{1}{b}\right)\hat{f} \\
&= \delta\left(\frac{1}{b}\right)\left(\delta\left(\frac{2\pi}{a}\right)\hat{f}\right) \\
&= \delta\left(\frac{1}{b}\right)\mathcal{F}_a f.
\end{aligned}$$

Since $\delta(\frac{2\pi}{a})f^* = (\delta(\frac{2\pi}{a})f)^*$, one easily obtains (e). From (a) of Theorem 3.4

$$\begin{aligned}
\delta\left(\frac{2\pi}{a}\right)\widehat{p(D)f}(\omega) &= \left(\frac{a}{2\pi}\right)^{n/2}\widehat{p(D)f}(\frac{a}{2\pi}\omega) \\
&= \left(\frac{a}{2\pi}\right)^{n/2} p(2\pi i(\frac{a}{2\pi}\omega))\hat{f}(\frac{a}{2\pi}\omega) \\
&= p(ai\omega)\left(\delta\left(\frac{2\pi}{a}\right)\hat{f}\right)(\omega) \\
&= p(aiw)\mathcal{F}_a f(\omega)\,.
\end{aligned}$$

Hence $\mathcal{F}_a(p(D)f)(\omega) = p(ai\omega)\mathcal{F}_a f(\omega)$ which is (f). Now letting $f = \mathcal{F}_a\check{h}$, from this we see $\mathcal{F}_a(\check{p}(\frac{iD}{a})\mathcal{F}_a\check{h})(\omega) = \check{p}(-\omega)\mathcal{F}_a^2\check{h}(\omega) = p(\omega)h(\omega)$. Thus $\mathcal{F}_a(ph)(\omega) = \check{p}(\frac{iD}{a})(\mathcal{F}_a\check{h})(-\omega)$. But it easy to note $(\mathcal{F}_a\check{h}) = \widecheck{\mathcal{F}_a h}$ and since $D^\alpha(\check{H})(x_1,\ldots,x_n) = (-1)^{|\alpha|}(D^\alpha H)(-x_1,-x_2,\ldots,-x_n)$, one has $p(D)\check{H} = \widecheck{(\check{p}(D)H)}$. Thus

$$\mathcal{F}_a(ph)(\omega) = p(\frac{iD}{a})\mathcal{F}_a h(\omega)$$

for any $h \in \mathcal{S}(\mathbb{R}^n)$. This is (g). One could also do a direct computation for (g) as in part (b) of Theorem 3.4.

As a different argument (g) follows with some manipulation from (b) of Theorem 3.4 for if one has $q(x) = p(\frac{-x}{2\pi i}) = p(\frac{ix}{2\pi})$, then

$$q(D)\hat{f}(\omega) = \mathcal{F}(q(-2\pi i\cdot\,)f(\,\cdot\,))(\omega) = \widehat{p(\,\cdot\,)f(\,\cdot\,)}(\omega)\,,$$

and thus

$$\begin{aligned}
\mathcal{F}_a(pf)(\omega) &= \delta\left(\frac{2\pi}{a}\right)\widehat{pf}(\omega)\\
&= \mathcal{F}\left(\delta\left(\frac{a}{2\pi}\right)(pf)\right)(\omega)\\
&= \left(\frac{a}{2\pi}\right)^{n/2}\mathcal{F}\left((\delta(\frac{a}{2\pi})p)(\delta(\frac{a}{2\pi})f)\right)(\omega)\\
&= \left(\frac{a}{2\pi}\right)^{n/2}\mathcal{F}\left((\delta(\frac{a}{2\pi})q)(-2\pi i\cdot)\delta(\frac{a}{2\pi})f)\right)(\omega)\\
&= \left(\frac{a}{2\pi}\right)^{n/2}(\delta(\frac{a}{2\pi})q)(D)\widehat{\delta(\frac{a}{2\pi})f}(\omega)\\
&= \left(\frac{a}{2\pi}\right)^{n/2}((\delta(\frac{a}{2\pi})q)(D)\delta(\frac{2\pi}{a})\hat{f}))(\omega)\\
&= \left(\frac{a}{2\pi}\right)^{n/2}(\delta(\frac{a}{2\pi})p)(\frac{iD}{2\pi})(\mathcal{F}_a f)(\omega)\\
&= p(\frac{iD}{a})\mathcal{F}_a f(\omega)
\end{aligned}$$

for

$$\begin{aligned}
\left(\frac{a}{2\pi}\right)^{n/2}(\delta(\frac{a}{2\pi})x^\alpha)(\frac{iD}{2\pi})h(\omega) &= \left(\left(\frac{2\pi x_1}{a}\right)^{\alpha_1}\left(\frac{2\pi x_2}{a}\right)^{\alpha_2}\cdots\left(\frac{2\pi x_n}{a}\right)^{\alpha_n}\right)\left(\frac{iD}{2\pi}\right)h(\omega)\\
&= \frac{(iD_1)^{\alpha_1}\cdots(iD_n)^{\alpha_n}}{a^{|\alpha|}}h(\omega)\\
&= \left(\frac{iD}{a}\right)^\alpha h(\omega)
\end{aligned}$$

implies $\left(\frac{a}{2\pi}\right)^{n/2}(\delta(\frac{a}{2\pi})p)(\frac{iD}{2\pi}) = p(\frac{iD}{a})$ for any polynomial.

□

4. Let f be Schwartz on $\mathbb{R}^m \times \mathbb{R}^n$. Show $y \mapsto f(x,y) = f_x(y)$ is Schwartz on $\mathbb{R}^n$ for each $x \in \mathbb{R}^m$.
Solution: Clearly f_x is C^∞. Now

$$\begin{aligned}
(1+|y|^2)^N|D^{0,\beta}f(x,y)| &\leqslant (1+|x|^2+|y|^2)^N|D^{0,\beta}f(x,y)|\\
&\leqslant |f|_{N,(0,\beta)}.
\end{aligned}$$

□

5. Show $x \mapsto f_x$ is continuous from $\mathbb{R}^m$ into $\mathcal{S}(\mathbb{R}^n)$ for each $f \in \mathcal{S}(\mathbb{R}^m \times \mathbb{R}^n)$.
Solution: We consider $|f_x - f_{x_0}|_{p,\beta}$. We have

$$|p(y)D^\beta f(x,y) - p(y)D^\beta f(x_0,y)| = |p(y)g(x,y) - p(y)g(x_0,y)|$$

where $g(x, y) = D^{\beta} f(x, y)$. So using the Mean Value Theorem, there is an $s \in (0, 1)$ so that

$$\begin{aligned}
|p(y)(g(x,y) - g(x_0, y))| &= |p(y)\nabla g(sx_0 + (1-s)x, y) \cdot (x - x_0)| \\
&= \left| p(y) \left(\sum_{j=1}^{m} D^{e_j} g(sx_0 + (1-s)x, y)(x - x_0)_j \right) \right| \\
&\leqslant \sum_{j=1}^{m} |(x - x_0)^{e_j} p(y) D^{e_j} g(sx_0 + (1-s)x, y)| \\
&\leqslant \sum_{j=1}^{m} \frac{|(x - x_0)^{e_j}|}{1 + |(sx_0 + (1-s)x, y)|^2} |g|_{q,(e_j,0)}
\end{aligned}$$

where $q(x, y) = (1 + |(x, y)|^2)p(y)$.

This clearly goes to 0 as $x \to x_0$ for any $p(y)$. □

6. Let $f \in L^1(\mathbb{R}^n)$ be such that

$$\int_{\mathbb{R}^n} |x|^k \; |f(x)| \; dx < \infty.$$

Show $D^{\alpha} \widehat{f}(\omega)$ exists for all α with $|\alpha| \leqslant k$ and

$$D^{\alpha} \hat{f}(\omega) = (-2\pi i)^{|\alpha|} \int x^{\alpha} f(x) \, e^{-2\pi i x \cdot \omega} \, dx.$$

Solution:

We argue by induction on k. It is clearly true for $k = 0$. Assume true for k and assume $\int |x|^{k+1} |f(x)| \, dx < \infty$. Since $|x|^k |f(x)| \leqslant |x|^{k+1} |f(x)| + |f(x)|$ and f and $|x|^{k+1} |f(x)|$ are integrable, $|x|^k |f(x)|$ is integrable. So we know $D^{\alpha} f(\omega) = (-2\pi i)^{|\alpha|} \int x^{\alpha} f(x) e^{-2\pi i x \cdot \omega} \, dx$ for all $|\alpha| \leqslant k$.

Now if $|\beta| = k + 1$, then $\beta = \alpha + \epsilon_j$ for some j. Moreover, by the induction hypothesis,

$$\begin{aligned}
\frac{D^{\alpha} f(\omega + he_j) - D^{\alpha} f(\omega)}{h} &= (-2\pi i)^{|\alpha|} \int x^{\alpha} f(x) \frac{e^{-2\pi i x \cdot (w + he_j)} - e^{-2\pi i x \cdot \omega}}{h} \, dx \\
&= (-2\pi i)^{|\alpha|+1} \int x^{\alpha + \epsilon_j} f(x) e^{-2\pi i x \cdot \omega} \frac{e^{-2\pi i h x_j} - 1}{-2\pi i h x_j} \, dx.
\end{aligned}$$

But the integrand is dominated by

$$e|x|^{k+1} |f(x)|$$

when $|2\pi i h x_j| < 1$ and converges pointwise to $x^{\beta} f(x) e^{-2\pi i x \cdot \omega}$. Indeed $\frac{e^z - 1}{z} \to 1$ as $z \to 0$ and $\left|\frac{e^z - 1}{z}\right| \leqslant \sum_{k=1}^{\infty} \frac{|z|^{k-1}}{k!} \leqslant \sum_{k=1}^{\infty} \frac{|z|^{k-1}}{(k-1)!} = e^{|z|}$. Let $h \to 0$. By the Lebesgue Dominated Convergence Theorem,

$$D^{\alpha + \epsilon_j} f(\omega) = (-2\pi i)^{|\alpha + \epsilon_j|} \int x^{\alpha + \epsilon_j} f(x) e^{-2\pi i x \cdot \omega} \, dx.$$

□

7. Suppose f is C^∞ and has support inside $\{x \mid |x| \leqslant R\}$. Define $\hat{f}(z) = \int f(x)e^{-2\pi i z\cdot x}\,dx$. Show $\hat{f}(z)$ is a holomorphic function and show for each $N \in \mathbb{N}$, there is a constant C_N such that

$$|\hat{f}(z)| \leqslant C_N(1+|z|^2)^{-N}e^{2\pi R|\operatorname{Im}(z)|}$$

for all $z \in \mathbb{C}^n$.

Solution: Holomorphicity follows from Lemma 4.16. This however is in the first part of Chapter 4. Indeed, $K = \{x \mid |x| \leqslant R\}$ is compact and $U = \mathbb{C}^n$ is open in $\mathbb{C}^n$ and Lebesgue measure dx on K is finite. Moreover, $z \mapsto h(x,z) = f(x)e^{-2\pi i z\cdot x}$ is holomorphic for each $x \in K$. So using the lemma, one has $\hat{f}(z) = \int_K h(x,z)\,dx = \int f(x)e^{-2\pi i z\cdot x}\,dx$ is holomorphic on $\mathbb{C}^n$.

Now recall $\mathcal{F}(p(D_x)f)(z) = p(2\pi i z)\hat{f}(z)$. Take $p(x) = (1 - \frac{x\cdot x}{4\pi^2})^N$. Then $p(2\pi i z) = (1+|z|^2)^N$. Take $h(x) = p(D)f(x)$. Then if $z = a+ib$,

$$\begin{aligned}
(1+|z|^2)^N|\hat{f}(z)| &= \left|\int h(x)e^{-2\pi i z\cdot x}\,dx\right| \\
&= \left|\int_{B_R} h(x)e^{-2\pi i a\cdot x}e^{2\pi b\cdot x}\,dx\right| \\
&\leqslant e^{2\pi|b|R}\int_{B_R} |h(x)|e^{2\pi b\cdot x}e^{-2\pi|b|R}\,dx \\
&\leqslant e^{2\pi|b|R}\int_{B_R} |h(x)|\,dx \\
&= e^{2\pi|b|R}|p(D)f|_1.
\end{aligned}$$

□

8. Let $f \in L^1(\mathbb{R})$ and let $h_t(y) = \frac{1}{t}e^{-\pi y^2/t^2}$ be the heat kernel. Show if

$$\lim_{r\to 0+}\frac{1}{2r}\int_{-r}^{r}(f(x-y)-f(x))\,dy = 0,$$

then

$$\lim_{t\to 0+} h_t * f(x) = f(x).$$

(**Hint:** Consider $\int_{-T}^{T} h_t'(y)F(y)\,dy$ where $F(y) = \int_{-y}^{y}(f(x-z)-f(x))\,dz$.)

Solution: Define $F(y) = \int_{-y}^{y}(f(x-z)-f(x))\,dz$. Note F is absolutely continuous and

$$F'(y) = f(x-y) - f(x) + f(x+y) - f(x).$$

One can also integrate by parts. Thus

$$\int_{-T}^{T} h_t'(y)F(y)\,dy = h_t(y)F(y)|_{-T}^{T} - \int_{-T}^{T} h_t(y)\left(f(x-y)-f(x)+f(x+y)-f(x)\right)dy.$$

Thus

$$\int_{-T}^{T} h_t'(y)F(y)\,dy = h_t(y)F(y)|_{-T}^{T} - 2\int_{-T}^{T} h_t(y)\left(f(x-y) - f(x)\right)dy.$$

Let $T \to \infty$. First note $h_t(T)F(T) \to 0$. Indeed, $|F(T)| \leqslant |f|_1 + 2|Tf(x)|$ and $h_t(T)$ decays exponentially. Thus

$$\int_{-\infty}^{\infty} h_t'(y)F(y)\,dy = -2\int_{-\infty}^{\infty} h_t(y)\left(f(x-y) - f(x)\right)dy.$$

Since $\int_{-\infty}^{\infty} h_t(y)f(x)\,dy = f(x)\int_{-\infty}^{\infty} h_t(y)\,dy = f(x)$, the result will follow if $\lim_{t\to 0+}\int_{-\infty}^{\infty} h_t'(y)F(y)\,dy = 0$. But

$$\begin{aligned}
\int_{-\infty}^{\infty} h_t'(y)F(y)\,dy &= \int_{-\infty}^{\infty} \frac{-2\pi y}{t^3}e^{-\pi y^2/t^2}F(y)\,dy \\
&= \int_{-\infty}^{\infty} -2\pi\frac{y}{t}e^{-\pi y^2}F(ty)\,dy \\
&= -4\pi\int_{-\infty}^{\infty} y^2e^{-\pi y^2}\frac{F(ty)}{2ty}\,dy.
\end{aligned}$$

Now by assumption, $\frac{F(ty)}{2ty} \to 0$ as $t \to 0+$ and since $\left|\frac{F(ty)}{2ty}\right|$ is continuous and satisfies $\left|\frac{F(ty)}{2ty}\right| \leqslant \frac{|f|_1}{2ty} + |f(x)|$, we see there is an $M > 0$ with $\left|\frac{F(ty)}{2ty}\right| \leqslant M$ for all ty. Thus using the Lebesgue Dominated Convergence Theorem, one obtains $\lim_{t\to 0+}\int_{-\infty}^{\infty} h_t'(y)F(y)\,dy = 0$. □

9. Let $f \in L^1(\mathbb{R}^n)$. Show that

$$\lim_{t\to 0} H_t(f)(x) = f(x)$$

for all $x \in \mathbb{R}^n$ such that

$$\lim_{r\to 0}\frac{1}{r^n}\int_{|x-y|\leqslant r}\left(f(y) - f(x)\right)dy = 0.$$

Solution:

We follow the argument in the previous problem and by translating f may assume $x = 0$; i.e. argue $h_t * f(x) = \lambda(-x)(h_t * f)(0) = h_t * \lambda(-x)f(0) \to \lambda(-x)f(0) = f(x)$ as $t \to 0+$. Now recall by Corollary 2.26 that Lebesgue measure in polar coordinates is given by $dx = r^{n-1}\,d\sigma\,dr$ where σ is surface measure on the unit ball S^{n-1} in $\mathbb{R}^n$. We shall use u to signify an element in S^{n-1} and use du for $d\sigma(u)$. Define $F(r) = \int_{B_r(0)}(f(x) - f(0))\,dx$. We know $F(r) = \int_0^r s^{n-1}\int_{S^{n-1}}(f(su) - f(0))\,du\,ds$. Thus F is absolutely continuous and

$$F'(r) = r^{n-1}\int_{S^{n-1}}(f(ru) - f(0))\,du.$$

Now taking $H_t(r) = \frac{1}{t^n}\exp(-\pi r^2/t^2)$, we have

$$h_t(y) = \frac{1}{t^n}\exp(-\pi|y|^2/t^2) = \frac{1}{t^n}\exp(-\pi r^2/t^2) = H_t(r)$$

for $y = ru$. Thus using integration by parts,

$$\begin{aligned}\int_{B_T(0)} h_t(x)(f(x) - f(0))\,dx &= \int_0^T H_t(r)r^{n-1}\int_{S^{n-1}}(f(ru) - f(0))\,du\,dr\\ &= \int_0^T H_t(r)F'(r)\,dr\\ &= H_t(r)F(r)|_{r=0}^{r=T} - \int_0^T F(r)H_t'(r)\,dr\,.\end{aligned}$$

Clearly $H_t(0)F(0) = 0$ and since

$$\begin{aligned}|F(r)| &= \left|\int_{B_r(0)}(f(x) - f(0))\,dx\right|\\ &\leqslant |f|_1 + |f(0)|\int_0^r s^{n-1}\int_{S^{n-1}} du\,ds\\ &= |f|_1 + \frac{r^n\sigma(S^{n-1})|f(0)|}{n},\end{aligned}$$

we have $H_t(T)F(T) \to 0$ as $T \to \infty$. This gives

$$\begin{aligned}h_t * f(0) - f(0) &= \int_{\mathbb{R}^n} h_t(x)(f(0 - x) - f(0))\,dx\\ &= \int_{\mathbb{R}^n} h_t(-x)(f(x) - f(0))\,dx\\ &= \lim_{T\to\infty}\int_{B_T(0)} h_t(x)(f(x) - f(0))\,dx\\ &= -\lim_{T\to 0}\int_0^T H_t'(r)F(r)\,dr\\ &= -\int_0^\infty H_t'(r)F(r)\,dr\,.\end{aligned}$$

Now $H_t'(r) = \frac{-2\pi r}{t^{n+2}}\exp(-\pi r^2/t^2)$. So

$$\begin{aligned}-\int_0^\infty H_t'(r)F(r)\,dr &= \int_0^\infty \frac{2\pi r}{t^{n+2}}\exp(-\pi r^2/t^2)F(r)\,dr\\ &= \int_0^\infty \frac{2\pi(tr)}{t^{n+2}}\exp(-\pi(tr)^2/t^2)F(tr)\,tdr\\ &= \int_0^\infty 2\pi r^{n+1}\exp(-\pi r^2)\frac{F(tr)}{(tr)^n}\,dr\,.\end{aligned}$$

As we have already seen, $|F(r)| \leqslant |f|_1 + Ar^n$ where $A = \frac{\sigma(S^{n-1})|f(0)|}{n}$. So $\frac{|F(tr)|}{(tr)^n}$ is bounded for tr large and by assumption is continuous from the right at 0. So $\frac{|F(tr)|}{(tr)^n}$ is bounded. Thus the Lebesgue Dominated Convergence Theorem implies since $\frac{F(tr)}{(tr)^n} \to 0$ as $t \to 0+$ for each $r > 0$ that

$$\lim_{t \to 0+} h_t * f(0) = f(0).$$

□

10. Let $f \in L^1(\mathbb{R}^n)$ and assume that $\hat{f} \in L^1(\mathbb{R}^n)$. Use H_t to show that

$$f(x) = \int \hat{f}(\omega) e^{2\pi i x \cdot \omega} \, d\omega$$

for almost all $x \in \mathbb{R}^n$. In particular, f is continuous almost everywhere. Note this gives another proof of Theorem 3.13.

Solution:

We remark the h_t in these problems are different than the h_t of Example 3.11. Since almost all points x are Lebesgue points (see Rudin, [41, Chapter 8]), we know almost all x satisfy $\lim_{r \to 0+} \frac{1}{r^n} \int_{B_r(x)} (f(y) - f(x)) \, dy = 0$. Take such an x. By Problem 3.2.9, $h_t * f(x) \to f(x)$ as $t \to 0+$. Now again by Example 3.11 we have $\hat{h}_t(y) = e^{-\pi t^2 |y|^2} = t^{-n} t^n e^{-\pi t^2 |y|^2} = t^{-n} h_{1/t}(y)$. Thus $\hat{\hat{h}}_t(y) = t^{-n} \hat{h}_{1/t}(y) = t^{-n} t^n h_t(y) = h_t(y)$.

Using the Lebesgue Dominated Convergence Theorem, this implies since $\hat{h}_t$ is even, that

$$\begin{aligned}
h_t * f(x) &= \int h_t(y) f(x-y) \, dy \\
&= \int \left(\int \hat{h}_t(z) e^{-2\pi i z \cdot y} \, dz \right) f(x-y) \, dy \\
&= \int \left(\int \hat{h}_t(z) e^{2\pi i z \cdot y} \, dz \right) f(x-y) \, dy \\
&= \int \int \hat{h}_t(z) e^{-2\pi i z \cdot y} \, dz f(x+y) \, dy \\
&= \int \int \hat{h}_t(z) e^{-2\pi i z \cdot (y-x)} f(y) \, dz \, dy \\
&= \int \hat{h}_t(z) e^{2\pi i z \cdot x} \int f(y) e^{-2\pi i y \cdot z} \, dy \, dz \\
&= \int \hat{h}_t(z) \hat{f}(z) e^{2\pi i z \cdot x} \, dz \\
&= \int \exp(-\pi t^2 |z|^2) \hat{f}(z) e^{2\pi i z \cdot x} \, dz \to \int \hat{f}(z) \, e^{2\pi i z \cdot x} \, dz
\end{aligned}$$

as $t \to 0+$. Consequently, $\int \hat{f}(z)e^{2\pi i z\cdot x}\, dz = f(x)$.

$\square$

11. Recall from (3.7) that the Dirichlet kernel is given by

$$D_N(u) = \int_{-N}^{N} e^{2\pi i u x}\, dx = \frac{\sin 2\pi N x}{\pi x} \text{ for } N > 0.$$

Let $f \in L^2(\mathbb{R})$.

(a) Show $D_N * f(x)$ is defined for all x and gives a continuous function.

(b) Show $D_N * f(x) = \int_0^\infty D_N(t)(f(x+t) + f(x-t))\, dt$.

(c) Show $D_N * f$ is in $L^2(\mathbb{R})$ and $D_N * f \to f$ in $L^2(\mathbb{R})$ as $N \to \infty$.

(d) Show $D_N(t) = \mathcal{F}(\chi_{[-N,N]})$, this function is sometimes called the sinc_N function.

Solution: Note (d) follows directly from the definition of D_N giving in (3.7) and can be seen in the argument for (c) below. Also D_N is in L^2 and f is in L^2. So by Lemma 2.76, $D_N * f$ is a continuous function and we have (a). Next note

$$\begin{aligned} D_N * f(x) &= \int_{-\infty}^{\infty} f(x-u)\frac{\sin(2\pi N u)}{\pi u}\, du \\ &= \int_{-\infty}^{0} f(x-u)\,\frac{\sin(2\pi N u)}{\pi u}\, du + \int_0^\infty f(x-u)\frac{\sin(2\pi N u)}{\pi u}\, du \\ &= \int_0^\infty [f(x+u) + f(x-u)]\,\frac{\sin(2\pi N u)}{\pi u}\, du\,. \end{aligned}$$

This is (b).

The argument for part (c) we give will depend on some results from the next section. We state them here. Namely, the Fourier transform extends from $\mathcal{S}(\mathbb{R})$ to a unitary isomorphism of $L^2(\mathbb{R})$. Since $\mathcal{F}(f * h) = \hat{f}\hat{h}$ for Schwarz functions f and h, this implies $\mathcal{F}(f * h) = \hat{f}\hat{h}$ if $f \in L^1$ and $h \in L^2$. Indeed, $f_k * h_k \to f * h$ in L^2 if f_k and h_k are Schwartz functions and $f_k \to f$ in L^1 and $h_k \to h$ in L^2. So $\hat{f}_k\hat{h}_k \to \mathcal{F}(f * h)$ in L^2. But $|\hat{f}_k - \hat{f}|_\infty \leqslant |f_k - f|_1 \to 0$ and $\hat{h}_k \to \hat{h}$ in L^2 implies $\mathcal{F}(f * h) = \hat{f}\hat{h}$. We also use $\mathcal{F}^2 f = \check{f}$ for $f \in L^2$. See in part Theorem 3.18. Now with this information, we know since $D_N(x) = \int \chi_{[-N,N]}(t)e^{2\pi i t x}\, dt = \int \chi_{[-N,N]}(t)e^{-2\pi i t x}\, dt = \mathcal{F}(\chi_{[-N,N]})(x)$ that $D_N \in L^2$ and $\mathcal{F}(D_N * f) = \mathcal{F}(D_N)\mathcal{F}(f) = \mathcal{F}^2(\chi_{[-N,N]})\hat{f} = \check{\chi}_{[-N,N]}\hat{f} = \chi_{[-N,N]}\hat{f}$ for Schwartz functions f. Thus if f_k are Schwartz and $f_k \to f$ in L^2, then since $\chi_{[-N,N]}\hat{f}_k \to \chi_{[-N,N]}\mathcal{F}(f)$ in L^2, we have $D_N * f_k$ is Cauchy in L^2 and $D_N * f_k$ converges in L^2 to some L^2 function h. But $D_N * f_k(x) \to D_N * f(x)$ pointwise. In fact, by Lemma 2.76, $|D_N * f_k(x) - D_N * f(x)| = |D_N * (f_k - f)(x)| \leqslant |D_N|_2|f_k - f|_2 \to 0$ as $k \to \infty$. This implies $D_N * f = h$ and thus $D_N * f$ is an L^2 function satisfying $\mathcal{F}(D_N * f) = \chi_{[-N,N]}\mathcal{F}(f)$. Since $\chi_{[-N,N]}\mathcal{F}(f) \to \mathcal{F}(f)$ in L^2 as $N \to \infty$, $D_N * f \to f$ in L^2. Note (d) was shown above in (c). $\square$

12. Suppose f is a real valued integrable function on $\mathbb{R}$ and suppose $f(x+)$, $f(x-)$, $f'(x+)$, and $f'(x-)$ exist. Define

$$A(p) = \frac{1}{\pi}\int_{-\infty}^{\infty} f(x)\cos px\,dx \text{ and } B(p) = \frac{1}{\pi}\int_{-\infty}^{\infty} f(x)\sin px\,dx.$$

Show

$$\frac{1}{2}(f(x+) + f(x-)) = \lim_{N\to\infty}\int_0^N A(p)\cos px + B(p)\sin px\,dp.$$

Solution: Let $\tilde{f}(x) = \frac{1}{2}(f(x+) + f(x-))$. We have by Corollary 3.16 that

$$\begin{aligned}
\tilde{f}(x) &= \lim_{N\to\infty}\int_{-N}^{N} \operatorname{Re}(\hat{f}(y)e^{2\pi ixy})\,dy \\
&= \lim_{N\to\infty}\frac{1}{2\pi}\int_{-N}^{N} \operatorname{Re}(\hat{f}(y/2\pi)e^{ixy})\,dy \\
&= \lim_{N\to\infty}\frac{1}{2\pi}\int_{-N}^{N} \operatorname{Re}(\hat{f}(y/2\pi))\cos(xy) - \operatorname{Im}(\hat{f}(y/2\pi))\sin(xy)\,dy.
\end{aligned}$$

Now

$$\begin{aligned}
\operatorname{Re}(\hat{f}(y/2\pi)) &= \int_{-\infty}^{\infty} f(x)\cos(xy)\,dx = \pi A(y) \text{ is even in } y \text{ and} \\
\operatorname{Im}(\hat{f}(y/2\pi)) &= -\int_{-\infty}^{\infty} f(x)\sin(xy)\,dx = -\pi B(y) \text{ is odd in } y.
\end{aligned}$$

We thus have

$$\begin{aligned}
\tilde{f}(x) &= \lim_{N\to\infty}\frac{1}{2}\int_{-N}^{N} A(y)\cos(xy) + B(y)\sin(xy)\,dy \\
&= \lim_{N\to\infty}\int_0^N A(p)\cos(px) + B(p)\sin(px)\,dp
\end{aligned}$$

for $y \mapsto A(y)\cos(yx) + B(y)\sin(yx)$ is an even function. □

Exercise Set 3.3

1. Let $\mathcal{H}$ and $\mathcal{H}'$ be Hilbert spaces. Let U be a linear transformation from a dense linear subspace D of $\mathcal{H}$ onto a dense linear subspace D' of $\mathcal{H}'$. Suppose

$$(Uv, Uw)' = (u, w)$$

for all $u, w \in D$.

(a) Show if $\{u_n\}_{n=1}^{\infty}$ is a Cauchy sequence in D, then $\{Uu_n\}_{n=1}^{\infty}$ is a Cauchy sequence in D'.

(b) Show $\tilde{U}$ defined on $\mathcal{H}$ by $\tilde{U}v = \lim Uv_n$ if $v = \lim_n v_n$ where $v_n \in D$ is a unitary isomorphism of $\mathcal{H}$ onto $\mathcal{H}'$.

Solution:

Since $(Uv, Uw)' = (v, w)$, $||Uv||' = \sqrt{(Uv, Uv)'} = \sqrt{(v, v)} = ||v||$. So U is an isometry. Hence $||Uv_m - Uv_n||' = ||U(v_m - v_n)||' = ||v_m - v_n||$. Hence if $\{v_n\}$ is Cauchy in $\mathcal{H}$, $\{Uv_n\}$ is Cauchy in $\mathcal{H}'$. Now since D is dense in $\mathcal{H}$, we see if $v \in \mathcal{H}$, there is a sequence $\{v_n\}$ in D such that $v_n \to v$ as $n \to \infty$. Thus $\{v_n\}$ is Cauchy. So $\{Uv_n\}$ is Cauchy in $\mathcal{H}'$. Since $\mathcal{H}'$ is complete, there is a v' in $\mathcal{H}'$ such that $Uv_n \to v'$. Define $\bar{U}v = v'$. This is well defined for if $w_n \to v$ where $w_n \in D$, then $||w_n - v_n|| \to 0$ and so $||Uw_n - Uv_n||' = ||U(w_n - v_n)||' \to 0$. Thus $\lim Uw_n = \lim Uv_n$. Now if $w_n \to w$ and $v_n \to v$ where $w_n \in D$ and $v_n \in D$, then $\lambda Uw_n + Uv_n = U(\lambda w_n + v_n) \to \bar{U}(\lambda w + v)$. But $\lambda Uw_n + Uv_n \to \lambda \bar{U}w + \bar{U}v$. So $\bar{U}(\lambda w + v) = \lambda \bar{U}w + \bar{U}v$ for $\lambda \in \mathbb{C}$ and $v, w \in \mathcal{H}$. Thus $\bar{U}$ is linear. It also preserves inner products. Indeed if $w_n \to w$ and $v_n \to v$ where $w_n, v_n \in D$, then $(\bar{U}w, \bar{U}v)' = \lim(Uw_n, Uv_n)' = \lim(w_n, v_n) = (w, v)$. So to finish, it suffices to show $\bar{U}$ is onto. Let $v' \in \mathcal{H}'$. Since UD is dense in $\mathcal{H}'$, there is a sequence $v_n \in D$ with $Uv_n \to v'$. Since $||v_m - v_n|| = ||U(v_m - v_n)|| = ||Uv_m - Uv_n||$, the sequence $\{v_n\}$ is Cauchy in $\mathcal{H}$. Let $v = \lim v_n$. Then $\bar{U}v = \lim Uv_n = v'$.

□

2. Let $h \in L^1(\mathbb{R}^n)$ and $f \in L^2(\mathbb{R}^n)$. Show $\mathcal{F}(f * h) = \hat{h}\hat{f}$.

Solution: By (i) of Lemma 3.3 of basic props, $\widehat{f * g} = \hat{f}\hat{g}$ if g is in $\mathcal{S}(\mathbb{R}^n)$. Now $|f * g|_2 \leqslant |f|_1 |g|_2$ by Lemma 2.77 and since $\mathcal{S}(\mathbb{R}^n)$ is dense in $L^2(\mathbb{R}^n)$, we see if $g \in L^2(\mathbb{R}^n)$ and $g_n \in \mathcal{S}(\mathbb{R}^n)$ is chosen so that $g_n \to g$ in L^2 as $n \to \infty$, then since $\mathcal{F}$ is an L^2 isometry, $\widehat{f * g_n} \to \widehat{f * g}$ in L^2. Thus $\hat{f}\hat{g}_n \to \widehat{f * g}$. But since by (a) of Lemma 3.3, $\hat{f}$ is bounded, and $\hat{g}_n \to \hat{g}$ in L^2, we see $\hat{f}\hat{g}_n \to \hat{f}\hat{g}$ in L^2. So $\widehat{f * g} = \hat{f}\hat{g}$. □

3. Let T be a bounded linear operator on a Hilbert space $\mathcal{H}$ satisfying $T^4 = I$. Show $\mathcal{H}$ is the direct sum of four orthogonal eigenspaces for T.

Solution:

If T is a bounded linear operator on $\mathcal{H}$ with $T^4 = I$ and $v \in \mathcal{H}$, then

$$\begin{aligned} v &= \frac{T^2 + I}{2} v + \frac{-T^2 + I}{2} v \\ &= \frac{(T + I)(T^2 + I)}{4} v + \frac{(-T + I)(T^2 + I)}{4} v \quad + \\ &\quad \frac{(T + iI)(-T^2 + I)}{4i} v + \frac{(-T + iI)(-T^2 + I)}{4i} v \\ &= v_1 + v_2 + v_3 + v_4 \end{aligned}$$

where the v_i are the respective summands in the last line. Now

$$\begin{aligned} Tv_1 &= T\left(\frac{(T+I)(T^2+I)}{4}v\right) \\ &= \frac{(T+I)(T^3+T)}{4}v \\ &= \frac{T^3+T+T^4+T^2}{4}v \\ &= \frac{T(T^2+I)+T^2+I}{4}v \\ &= \frac{(T+I)(T^2+I)}{4}v \\ &= v_1\,. \end{aligned}$$

Also note

$$\begin{aligned} T(-T+I)(T^2+I) &= T^3+T-T^4-T^2 \\ &= T(T^2+I)-(T^2+I) \\ &= -(-T+I)(T^2+I)\,, \end{aligned}$$

$$\begin{aligned} \frac{1}{i}T(T+iI)(-T^2+I) &= -T^3+T-iT^2+iT^4 \\ &= T(-T^2+I)+i(-T^2+I) \\ &= (T+iI)(-T^2+I) \\ &= i\left(\frac{1}{i}(T+iI)(-T^2+I)\right)\,, \end{aligned}$$

and

$$\begin{aligned} \frac{1}{i}T(-T+iI)(-T^2+I) &= -T^3+T+iT^2-iT^4 \\ &= T(-T^2+I)-i(-T^2+I) \\ &= (T-iI)(-T^2+I) \\ &= -i\left(\frac{1}{i}(-T+iI)(-T^2+I)\right)\,. \end{aligned}$$

These imply $Tv_2 = -v_2$, $Tv_3 = iv_3$, and $Tv_4 = -iv_4$. Thus each v is the sum of four eigenvectors v_1, v_2, v_3, v_4 with corresponding eigenvalues $1, -1, i, -i$.

This sum is direct for if this sum is 0, one has $T^2v + v = 2(v_1 + v_2)$; so $v_1+v_2 = 0$ and $v_3+v_4 = 0$. But then $T(v_1+v_2) = v_1 - v_2 = 0$ and $T(v_3+v_4) = iv_3 - iv_4 = 0$. These equations together imply $v_1 = v_2 = v_3 = v_4 = 0$.

□

4. The support of a measurable complex valued function f on $\mathbb{R}^n$ is the complement of the union of all open subsets U of $\mathbb{R}^n$ on which $f = 0$ a.e. For $T > 0$, let

$$L^2_T(\mathbb{R}^n) = \left\{ f \in L^2(\mathbb{R}^n) \mid \operatorname{supp}(\hat{f}) \subset [-T, T]^n \right\},$$

the space of T-band limited functions.

(a) Show that the operator defined on the dense subspace $\mathcal{S}(\mathbb{R}^n)$ of $L^2(\mathbb{R}^n)$ by

$$h \mapsto \operatorname{sinc}_T * h$$

where

$$\operatorname{sinc}_T(x) = \prod_{j=1}^{n} \frac{\sin(2\pi T x_j)}{\pi x_j}$$

extends to the orthogonal projection P_T of $L^2(\mathbb{R}^n)$ onto $L^2_T(\mathbb{R}^n)$.

(b) Show that

$$f = \lim_{T \mapsto \infty} P_T(f)$$

for $f \in L^2(\mathbb{R}^n)$.

Solution:

We have the Fourier transform $\mathcal{F}$ is a unitary isomorphism of $L^2(\mathbb{R}^n)$ mapping $\mathcal{S}(\mathbb{R}^n)$ onto $\mathcal{S}(\mathbb{R}^n)$. Now let $Q_T = \prod_{j=1}^n [-T, T] = [-T, T]^n$ and let x_T be the characteristic function of Q_T. Note χ_T is an even function and thus $\mathcal{F}(\chi_T)(x) = \mathcal{F}(\chi_T)(-x)$. So $\mathcal{F}^{-1}(\chi_T) = \mathcal{F}(\chi_T)$. Using Equation 3.7, one can see the Fourier transform of χ_T is sinc_T. So $\mathcal{F}^{-1}(\chi_T) = \operatorname{sinc}_T$. This function is L^2 but not L^1. However, if h is Schwartz, Exercise 3.3.2 implies $\mathcal{F}^{-1}(\chi_T h) = \mathcal{F}^{-1}\chi_T * \mathcal{F}^{-1}h = \operatorname{sinc}_T * \mathcal{F}^{-1}h$. Now $\hat{P}_T$ defined by $\hat{P}_T f = \chi_T f$ is an orthogonal projection of $L^2(\mathbb{R}^n)$ whose range is is $\mathcal{F}(L^2_T(\mathbb{R}^n))$, the space of L^2 functions whose essential support is a subset of Q_T. Note $\hat{P}_T f \to f$ in L^2 as $T \to \infty$. Thus $P_T = \mathcal{F}^{-1}\hat{P}_T\mathcal{F}$ is an orthogonal projection of $L^2(\mathbb{R}^n)$ onto $L^2_T(\mathbb{R}^n)$. On the dense subspace $\mathcal{S}(\mathbb{R}^n)$, P_T is given by

$$\begin{aligned} P_T(h) &= \mathcal{F}^{-1}\hat{P}_T(\mathcal{F}h) \\ &= \mathcal{F}^{-1}(\chi_T \mathcal{F}h) \\ &= \operatorname{sinc}_T * h\,. \end{aligned}$$

Since P_T is a bounded linear operator, we see $h \mapsto \operatorname{sinc}_T * h$ on $\mathcal{S}(\mathbb{R}^n)$ is bounded on L^2 and its extension to L^2 is the orthogonal projection P_T. Moreover, since $\hat{P}_T f \to f$ for each $f \in L^2$, $P_T f = \mathcal{F}^{-1}\hat{P}_T\mathcal{F}f \to f$ in $L^2(\mathbb{R}^n)$ as $T \to \infty$.

□

5. Consider the differential operators $D_t^\alpha = \left(\frac{\partial}{\partial t_1}\right)^{\alpha_1}\left(\frac{\partial}{\partial t_2}\right)^{\alpha_2}\cdots\left(\frac{\partial}{\partial t_n}\right)^{\alpha_n}$ and $D_x^\alpha = \left(\frac{\partial}{\partial x_1}\right)^{\alpha_1}\left(\frac{\partial}{\partial x_2}\right)^{\alpha_2}\cdots\left(\frac{\partial}{\partial x_n}\right)^{\alpha_n}$. For f in $C^\infty(\mathbb{R}^n)$, show

$$D_t^\alpha f(x - t)|_{t=0} = (-1)^{|\alpha|} D_x^\alpha f(x).$$

Solution:

It suffices to show $D_t^\alpha f(x-t) = (-1)^{|\alpha|} D_x^\alpha f(x-t)$ for all α and x and t. We show this by induction: first note by the chain rule,

$$\begin{aligned}
\frac{\partial}{\partial t_i} f(x-t) &= \sum_{j=1}^{n} (\partial_j f)(x-t) \frac{\partial}{\partial t_i}(x-t)_j \\
&= \sum_{j=1}^{n} (\partial_j f)(x-t) \frac{\partial}{\partial t_i}(x_j - t_j) \\
&= -\partial_i f(x-t) \frac{\partial}{\partial x_i}(x_i - t_i) \\
&= -\sum_{j=1}^{n} \partial_j f(x-t) \frac{\partial}{\partial x_i}(x-t)_j \\
&= -\frac{\partial}{\partial x_j} f(x-t) .
\end{aligned}$$

Thus the result holds if $|\alpha| = 1$. Assume true if $|\alpha| = k$. If $|\beta| = k+1$, then $\beta = \alpha + \epsilon_i$ for some i where $|\alpha| = k$. Thus,

$$\begin{aligned}
D_t^\beta f(x-t) &= \frac{\partial}{\partial t_i} D_t^\alpha f(x-t) \\
&= \frac{\partial}{\partial t_i} (-1)^{|\alpha|} D_x^\alpha f(x-t) \\
&= (-1)(-1)^{|\alpha|} \frac{\partial}{\partial x_i} (D_x^\alpha f)(x-t) \\
&= (-1)^{|\beta|} D_x^\beta (f(x-t)) .
\end{aligned}$$

□

6. In many books the Hermite polynomials are defined by using the generating function $e^{t\cdot x - t\cdot t/2}$. In this case, the Hermite polynomials $\tilde{H}_\alpha(x)$ satisfy:

$$e^{t\cdot x - t\cdot t/2} = \sum_\alpha \tilde{H}_\alpha(x) \frac{t^\alpha}{\alpha!} .$$

Show that

$$(4\pi)^{|\alpha|/2} \tilde{H}_\alpha(\sqrt{4\pi} x) = H_\alpha(x) .$$

Solution:

We have

$$e^{4\pi t\cdot x - 2\pi t\cdot t} = \sum_\alpha H_\alpha(x) \frac{t^\alpha}{\alpha!} .$$

Thus

$$\begin{aligned}
e^{t\cdot x - t\cdot t/2} &= e^{4\pi(t/\sqrt{4\pi})\cdot x/\sqrt{4\pi} - 2\pi(t/\sqrt{4\pi}\cdot t/\sqrt{4\pi})} \\
&= \sum_\alpha H_\alpha(x/\sqrt{4\pi})\frac{(t/\sqrt{4\pi})^\alpha}{\alpha!} \\
&= \sum_\alpha (4\pi)^{-|\alpha|/2} H_\alpha(x/\sqrt{4\pi})\frac{t^\alpha}{\alpha!} \\
&= \sum_\alpha \tilde{H}_\alpha(x)\frac{t^\alpha}{\alpha!}\,.
\end{aligned}$$

Thus

$$\tilde{H}_\alpha(x) = (4\pi)^{-|\alpha|/2} H_\alpha(x/\sqrt{4\pi})$$

or

$$H_\alpha(x) = (4\pi)^{|\alpha|/2}\tilde{H}_\alpha(\sqrt{4\pi}x)\,.$$

□

7. Let $0 \leqslant \psi(t) \leqslant 1$ be a smooth function with support in $(1/2, \infty)$ such that $\psi(t) = 1$ for $t \geqslant 1$. Let $p = \left[\frac{n}{2}\right] + 1$. Let

$$f(x) := \frac{1}{|x|^p}\psi(|x|)\,.$$

Show the following:

(a) $f \in L^2(\mathbb{R}^n)$.

(b) $\tilde{H}f(x) = \frac{1}{|x|^{p-2}}\psi(|x|) - \Delta f(x)$ is not in $L^2(\mathbb{R}^n)$.

Solution:

We show $f(x) = |x|^{-p}\psi(|x|)$ is in L^2 if $p = \left[\frac{n}{2}\right] + 1$. Indeed,

$$\begin{aligned}
\int |f(x)|^2\,dx &= \int_0^\infty \int_{S^{n-1}} |f(ru)|^2 r^{n-1}\,d\sigma(u)\,dr \\
&= \sigma(S^{n-1})\int_{1/2}^\infty r^{n-1-2p}\psi(r)^2\,dr \\
&\leqslant \sigma(S^{n-1})\int_{1/2}^\infty r^{n-1-2p}\,dr \\
&= \sigma(S^{n-1})\left.\frac{r^{n-2p}}{n-2p}\right|_{r=1/2}^{r=\infty} \\
&< \infty
\end{aligned}$$

if and only if $n - 2p < 0$ or if and only if $p > \frac{n}{2}$. This occurs when $p = \left[\frac{n}{2}\right] + 1$.

Note f is C^∞ and

$$\begin{aligned}\tilde{H}f(x) &= (|x|^2 - \Delta)f(x)\\ &= |x|^{2-p}\psi(|x|) - \sum \partial_i^2(|x|^{-p}\psi(|x|)).\end{aligned}$$

Now $\partial_i^2(|x|^{-p}\psi(|x|) = 0$ for $|x| < \frac{1}{2}$ and since $\psi(|x|) = 1$ for $|x| > 1$, we have $\partial_i^2(|x|^{-p}\psi(|x|)) = \partial_i^2(|x|^{-p})$ for $|x| > 1$. Now

$$\begin{aligned}\partial_i(|x|^{-p}) &= \partial_i(x_1^2 + x_2^2 + \cdots + x_n^2)^{-p/2}\\ &= -px_i(x_1^2 + x_2^2 + \cdots + x_n^2)^{-p/2-1}\end{aligned}$$

and

$$\begin{aligned}\partial_i^2(|x|^{-p}) &= \partial_i\left(-px_i(x_1^2 + x_2^2 + \cdots + x_n^2)^{-p/2-1}\right)\\ &= -p(x_1^2 + x_2^2 + \cdots + x_n^2)^{-p/2-1} + p(p+2)x_i^2(x_1^2 + x_2^2 + \cdots + x_n^2)^{-p/2-2}.\end{aligned}$$

This implies

$$\begin{aligned}\Delta(|x|^{-p}) &= \sum_i \partial_i^2(|x|^{-p})\\ &= -np(x_1^2 + x_2^2 + \cdots + x_n^2)^{-p/2-1} + p(p+2)(x_1^2 + x_2^2 + \cdots + x_n^2)^{-p/2-1}\\ &= (p^2 + (2-n)p)(x_1^2 + x_2^2 + \cdots + x_n^2)^{-p/2-1}\\ &= (p^2 + (2-n)p)|x|^{(-p-2)/2}\end{aligned}$$

As seen above

$$\int_{|x|\geqslant 1} |\Delta(|x|^{-p})|^2\,dx = \sigma(S^{n-1})(p^2 + (2-n)p)^2 \int_1^\infty r^{-p-2}r^{n-1}\,dr < \infty$$

when $p = \left[\frac{n}{2}\right] + 1$. Thus

$$\int |\Delta f(x)|^2\,dx = \int_{1/2\leqslant|x|\leqslant 1} |\Delta(|x|^{-p}\psi(|x|))|^2\,dx + \int_{|x|\geqslant 1} |\Delta(|x|^{-p})|^2\,dx < \infty.$$

Thus $\tilde{H}f$ is L^2 if and only if $|x|^{2-p}\psi(|x|)$ is in L^2. We saw above this occurs if and only if $p-2$ satisfies $p - 2 > \frac{n}{2}$. But when $p = \left[\frac{n}{2}\right] + 1$, $p < \frac{n}{2} + 1$ and thus $p - 2 < \frac{n}{2}$. Thus $\tilde{H}f \notin L^2(\mathbb{R}^n)$.

□

8. Show that the linear span of the functions

$$(x_1, \ldots, x_n) \mapsto f_1(x_1)\cdot \ldots \cdot f_n(x_n)$$

where $f_1, f, \ldots, f_n$ are in $L^2(\mathbb{R})$ is a dense linear subspace of $L^2(\mathbb{R}^n)$.
Solution:

We do this by induction. It is trivial when $n = 1$. Suppose true for n. Let $f \in L^2(\mathbb{R} \times R^n)$ be perpendicular to all these functions.

By the induction hypothesis, if $\phi \in L^2(\mathbb{R})$ and $F \in L^2(\mathbb{R}^n)$, we can approximate F in $L^2(\mathbb{R}^n)$ by a linear combination of functions of form $y \mapsto f_1(y_1) \cdots f_n(y_n)$. Thus $(x, y) \mapsto \phi(x)F(y)$ is a limit in $L^2(\mathbb{R} \times \mathbb{R}^n)$ of linear combinations of functions of form $(x, y) \mapsto \phi(x)f_1(y_1) \cdots f_n(y_n)$. This implies

$$\int_{\mathbb{R}} \int_{\mathbb{R}^n} f(x, y)\bar{\phi}(x)\bar{F}(y)\, dy\, dx = 0$$

for all $F \in L^2(\mathbb{R}^n)$ and $\phi \in L^2(\mathbb{R})$. So $\int f(x, y)\bar{\phi}(x)\, dx = 0$ a.e. y for each $\phi \in L^2(\mathbb{R})$.

Now set $f_y(x) = f(x, y)$. By Fubini, $f_y \in L^2(\mathbb{R})$ for a.e. $y \in \mathbb{R}^n$ and for each ϕ, $\int f_y(x)\bar{\phi}(x)\, dx = 0$ a.e. y. Choose a countable dense subset $\{\phi_k\}$ of $L^2(\mathbb{R})$. One then has $f_y \perp \phi_k$ for all k a.e. y. Thus $f_y = 0$ in $L^2(\mathbb{R})$ for *a.e.* y. This gives $f(x, y) = 0$ a.e. $(x, y) \in \mathbb{R} \times \mathbb{R}^n$.

□

Chapter 4

Further Topics and Applications

Exercise Set 4.1

1. Show, using the definition, that the function $z \mapsto \bar{z}$, i.e., $x + iy \mapsto x - iy$, is not holomorphic.

Solution:

By Equation 4.4, we know if $f = u + iv$ is holomorphic, then it must satisfy the Cauchy-Riemann equation $\frac{\partial}{\partial \bar{z}} f = \frac{1}{2}(\frac{\partial}{\partial x} - \frac{1}{i}\frac{\partial}{\partial y})f = 0$. Note $\bar{z} = x - iy$, thus $\frac{\partial}{\partial \bar{z}}\bar{z} = \frac{1}{2}(1 - \frac{1}{i}(-i)) = 1$. So it is not holomorphic.

□

2. Let $T : \mathbb{R}^2 \to \mathbb{R}^2$ be a linear map corresponding to the matrix $\left(\begin{smallmatrix} a & b \\ c & d \end{smallmatrix}\right)$ with respect to the standard basis e_1, e_2. Show T is complex linear if and only if $a = d$ and $b = -c$. Use that to show that if $f : \mathbb{C} \to \mathbb{C}$ is differentiable, then f is complex differentiable if and only if $Df(z)$ is complex linear. Derive from this that f is complex differentiable if and only if f satisfies the Cauchy-Riemann differential equation.

Solution: Let $T = \left(\begin{smallmatrix} a & b \\ c & d \end{smallmatrix}\right)$ be a linear mapping of $\mathbb{R}^2$ given by $(x, y) \mapsto (x, y)T$. Identifying $\mathbb{R}^2$ with $\mathbb{C}$ by $(x, y) = x + iy$, then T is complex linear if and only if $T(i(x + iy)) = T(-y + ix) = (-y, x)T = (-ay + cx, -by + dx) = iT(x, y)$. But $iT(x, y) = i(ax + cy + i(bx + dy)) = (-bx - dy, ax + cy)$. So T is complex linear if $-ay + cx = -bx - dy$ and $-by + dx = ax + cy$ for all (x, y). This occurs when $c = -b$ and $a = d$.

Now assume $f = u + iv : \mathbb{C} \to \mathbb{C}$ is differentiable as a mapping of $\mathbb{R}^2$ at the point $z_0 = (x_0, y_0)$. We know the derivative D is the linear mapping $D(a_1, a_2) = (a_1, a_2)f'(x_0, y_0)$ where $f'(x_0, y_0) = \begin{bmatrix} u_x(x_0, y_0) & v_x(x_0, y_0) \\ u_y(x_0, y_0) & v_y(x_0, y_0) \end{bmatrix}$. Note we have the Cauchy-Riemann equations $u_x(x_0, y_0) = v_y(x_0, y_0)$, $u_y(x_0, y_0) = -v_x(x_0, y_0)$ if and only if $f'(x_0, y_0)$ is complex linear. Moreover, by differentiability, $f(z_0 + (h, k)) - f(z_0) = (h, k)f'(z_0) + o(h, k)$ or equivalently, $f(z_0 + z) - f(z_0) =$

$z(u_x(z_0) + iv_x(z_0)) + o(|z|)$ where $z = (h,k) = h + ki$ and $f'(z_0) = u_x(z_0) + iv_x(z_0)$. This gives

$$\lim_{z\to 0} \frac{f(z_0+z) - f(z_0)}{z} = f'(z_0).$$

□

3. Find the points $z = x + iy$ at which the following functions are complex differentiable

(a) $f(z) = x^2 + 2ixy$;

(b) $f(z) = 2xy + i(x + \frac{2}{3}y^3)$;

(c) $f(z) = x^2 + y^2 = z\bar{z}$.

Solution: In order to be complex differentiable we need $\frac{\partial}{\partial \bar{z}} f(x,y) = 0$. Applying this to (a) we need $\frac{\partial}{\partial x} f = \frac{1}{i}\frac{\partial}{\partial y} f = -i\frac{\partial}{\partial y} f$ or $2x + 2iy = -i(2ix)$. Thus $2x + 2iy = 2x$. So $(x,y) = (x,0)$. Since f is differentiable and satisfies the Cauchy-Riemann equations, $x^2 + 2ixy$ is complex differentiable at only the points $(x,0)$.

For (b), we need $2y + i = -i(2x + 2y^2 i) = -2ix + 2y^2$. Thus we have the Cauchy-Riemann equations only if $y = 0$ or $y = 1$ and $x = -\frac{1}{2}$. Since f is real differentiable everywhere, we see the only points where this f is complex differentiable are $(-\frac{1}{2}, 0)$ and $(-\frac{1}{2}, 1)$.

Finally for (c), note $\frac{\partial}{\partial x} f = 2x$ and $-i\frac{\partial}{\partial y} f = -2iy$. Here we obtain $(0,0)$.

□

4. Define $f : \mathbb{C} \to \mathbb{C}$ by

$$f(x+iy) = \begin{cases} 0 & \text{if} \quad x + iy = 0 \\ \frac{(1+i)x^3 - (1-i)y^3}{x^2+y^2} & \text{if} \quad x + iy \neq 0\,. \end{cases}$$

Show f satisfies the Cauchy-Riemann equation at $z = 0$ and f is not differentiable at zero. Why does this not contradict Theorem 4.2?

Solution: In order for $f(z) = f(x+iy) = F(x,y)$ to be complex differentiable at a point, one must have the Cauchy-Riemann equations and F has to be real differentiable. The Cauchy-Riemann equation is given by $\frac{\partial}{\partial x} f = \frac{1}{i}\frac{\partial}{\partial y} f = -i\frac{\partial}{\partial y} f$. Now

$$\begin{aligned} \frac{\partial}{\partial x} f(0,0) &= \lim_{h\to 0} \frac{1}{h}\left(\frac{(1+i)h^3}{h^2} - 0\right) \\ &= 1 + i \end{aligned}$$

and

$$\begin{aligned} i\frac{\partial}{\partial y} f(0,0) &= i \lim_{k\to 0} \frac{1}{k}\left(\frac{-(1-i)k^3}{k^2} - 0\right) \\ &= -1 - i\,. \end{aligned}$$

But $F(x,y) = f(x+iy) = (\frac{x^3-y^3}{x^2+y^2}, \frac{x^3+y^3}{x^2+y^2})$ is not real differentiable at $(0,0)$. Indeed,

$$F'(0,0) = \begin{bmatrix} 1 & 1 \\ -1 & 1 \end{bmatrix}$$

and thus $R(x,y)$ given by

$$\begin{aligned} R(x,y) &= F(x,y) - F(0,0) - (x,y)\begin{bmatrix} 1 & 1 \\ -1 & 1 \end{bmatrix} \\ &= \left(\frac{x^3-y^3}{x^2+y^2}, \frac{x^3+y^3}{x^2+y^2}\right) - (x-y, x+y) \\ &= \left(\frac{x^3-y^3-x^3-xy^2+x^2y+y^3}{x^2+y^2}, \frac{x^3+y^3-x^3-xy^2-yx^2-y^3}{x^2+y^2}\right) \\ &= \left(\frac{-xy^2+x^2y}{x^2+y^2}, \frac{-xy^2-yx^2}{x^2+y^2}\right) \\ &= \left(\frac{xy(x-y)}{x^2+y^2}, \frac{-xy(x+y)}{x^2+y^2}\right) \end{aligned}$$

satisfies $|R(x,y)|^2 = \frac{2x^2y^2}{x^2+y^2}$. So $R(x,y) = \frac{2|xy|}{|(x,y)|}$ is not $o(x,y)$ for $\frac{|R(x,y)|}{|(x,y)|} = \frac{2|xy|}{x^2+y^2} \not\to 0$ as $(x,y) \to (0,0)$.

□

5. Let f be a holomorphic function on an open subset $U \subset \mathbb{C}$. Write $f = u + iv$ as usual. Show $\Delta u = \Delta v = 0$.

Solution:

We have f is holomorphic on an open subset U of $\mathbb{C}$. So f is infinitely differentiable. Now $f = u + iv$ and $\frac{\partial}{\partial z} f = u_x + iv_x = \frac{1}{i}u_y + v_y$. Since u and v are C^∞ on U, and $u_x = v_y$ and $u_y = -v_x$, then $u_{xx} = v_{xy}$, $u_{yy} = -v_{xy}$, $v_{xx} = -u_{xy}$ and $v_{yy} = u_{xy}$. Thus $u_{xx} + u_{yy} = v_{xx} + v_{yy} = 0$.

□

6. Let f be a holomorphic function on an open subset U of $\mathbb{C}$. Show the function $z \mapsto \overline{f(\bar{z})}$ is holomorphic on $\bar{U}$.

Solution: Assume $\bar{z} \in U$. We show $F(w) = \overline{f(\bar{w})}$ is differentiable at z. First note since f is holomorphic on U, the real and imaginary parts u and v of f have continuous partial derivatives and thus so do the real and imaginary parts of F. Thus F is real differentiable. Moreover,

$$\begin{aligned} \frac{F(w) - F(z)}{w - z} &= \frac{\overline{f(\bar{w})} - \overline{f(\bar{z})}}{w - z} \\ &= \overline{\left(\frac{f(\bar{w}) - f(\bar{z})}{\bar{w} - \bar{z}}\right)} \to \overline{f'(\bar{z})} \end{aligned}$$

as $w \to z$. Thus F' is complex differentiable at z and $F'(z)$ equals the conjugate of $f'(\bar{z})$.

□

7. Find the radii of convergence for the following series:

(a) $f(z) = \sum_{n=0}^{\infty} nz^n$;

(b) $f(z) = \sum_{n=0}^{\infty} n!z^n$;

(c) $f(z) = \sum_{n=0}^{\infty} e^n z^n$;

(d) $f(z) = \sum_{n=0}^{\infty} e^{n!} z^n$.

Solution:

For $\sum nz^n$, we have $\rho = \lim_n \left|\frac{a_n}{a_{n+1}}\right| = \lim \frac{n}{n+1} = 1$.

For $\sum n!z^n$ we have $\frac{a_n}{a_{n+1}} = \frac{1}{n+1} \to 0$ and so $\rho = 0$. For $\sum e^n z^n$ we have $\frac{a_n}{a_{n+1}} = \frac{1}{e}$ and so $\rho = 1/e$. Finally for $\sum e^{n!} z^n$, we have $\frac{a_n}{a_{n+1}} = e^{n!-(n+1)n!} = e^{-n(n!)} \to 0$ as $n \mapsto \infty$ and so $\rho = 0$.

□

8. Let $\langle \cdot, \cdot \rangle$ be a positive definite Hermitian form on $\mathbb{C}^n$. Thus $z, w \mapsto \langle z, w \rangle$ is complex linear in the first variable, $\langle z, w \rangle = \overline{\langle w, z \rangle}$ and $\langle z, z \rangle > 0$ for all $z \neq 0$. Show $(x, y) := \mathrm{Re}\langle x, y \rangle$ defines a inner product on $\mathbb{R}^{2n} = \mathbb{C}^n$ as a real vector space. Let $e_1, \ldots, e_n$ be an orthonormal basis of $\mathbb{C}^n$. Show $e_1, \ldots, e_n, ie_1, \ldots, ie_n$ is an orthonormal basis of the real vector space $\mathbb{C}^n$ with inner product $(\cdot, \cdot)$.

Solution:

It is clear $(\cdot, \cdot)$ is real bilinear and positive definite and thus is an real inner product on $\mathbb{C}^n$. Now let $e_1, e_2, \ldots, e_n$ be an orthonormal basis of $\mathbb{C}^n$ for the inner product $\langle \cdot, \cdot \rangle$. Then $\langle e_j, ie_k \rangle = -i \langle e_j, e_k \rangle$ is pure imaginary. So $(e_j, ie_k) = 0$ for all j, k. Also $\langle ie_j, ie_k \rangle = \langle e_j, e_k \rangle = \delta_{j,k}$. These together imply the vectors $e_1, e_2, \ldots, e_n, ie_1, ie_2, \ldots, ie_n$ is an orthonormal basis for the real inner product space $\mathbb{C}^n$.

□

9. Define $J : \mathbb{C}^n \to \mathbb{C}^n$ by $x \mapsto ix$. Show the matrix expression of J with the basis in problem 4.1.8 is given by

$$J = \begin{bmatrix} 0 & -I_n \\ I_n & 0 \end{bmatrix}$$

where I_n is the $n \times n$ identity matrix.

Solution:

Now we have $\mathbb{C}^n = \{(z_1, z_2, \ldots, z_n) \mid z_j \in \mathbb{C}\}$ with $\langle z, w \rangle = \sum z_k \bar{w}_k$. Let $e_j = (\delta_{1,j}, \delta_{2,j}, \ldots, \delta_{n,j})$ be the usual orthonormal basis. Let $Tv = iv$ on $\mathbb{C}^n$. Set $e_{r,k} = e_k$ for $k = 1, 2, \ldots, n$ and $e_{r,n+j} = ie_j$ for $j = 1, 2, \ldots, n$. We know this is a basis for the real inner product space $\mathbb{C}^n$ with inner product $(z, w) = \mathrm{Re}\langle z, w \rangle$. Now $Te_{r,k} = e_{r,n+k}$ and $Te_{r,n+k} = ie_{r,n+k} = i^2 e_{r,k} = -e_{r,k}$.

Thus

$$T(\sum_{j=1}^{2n} a_j e_{r,j}) = \sum_{j=1}^{n} a_j e_{r,n+j} - \sum_{j=1}^{n} a_{j+n} e_{r,n}$$
$$= \sum_{j=1}^{2n} b_j e_{r,j}$$

where

$$(b_1, b_2, \ldots, b_{2n})^T = \begin{bmatrix} 0 & -I_n \\ I_n & 0 \end{bmatrix} (a_1, a_2, \ldots, a_{2n})^T.$$

□

10. Let $T : \mathbb{C}^n \to \mathbb{C}^n$ be $\mathbb{R}$-linear. Show the following three conditions are equivalent:

(a) T is complex linear;

(b) $TJ = JT$;

(c) The matrix representation of T with respect to the basis in problem 4.1.8 is given by

$$T = \begin{bmatrix} A & B \\ -B & A \end{bmatrix}$$

where $A, B \in M_n(\mathbb{R})$.

Solution:

(a) and (b) are clearly equivalent for a real linear transformation T is complex linear if and only if $T(iv) = i(Tv)$ or $TJv = JTv$.

We show (b) and (c) are equivalent. Note $TJ = JT$ if and only if the matrices for T and J commute. Relative to the basis $e_{r,j} = e_j$, $e_{r,n+j} = ie_j$, T has matrix $\left[\begin{smallmatrix} A & B \\ C & D \end{smallmatrix}\right]$ and the matrix for J is $J_m = \left[\begin{smallmatrix} 0 & -I_n \\ I_n & 0 \end{smallmatrix}\right]$. Now

$$\begin{bmatrix} A & B \\ C & D \end{bmatrix} \begin{bmatrix} 0 & -I_n \\ I_n & 0 \end{bmatrix} = \begin{bmatrix} B & -A \\ D & -C \end{bmatrix}$$

and

$$\begin{bmatrix} 0 & -I_n \\ I_n & 0 \end{bmatrix} \begin{bmatrix} A & B \\ C & D \end{bmatrix} = \begin{bmatrix} -C & -D \\ A & B \end{bmatrix}.$$

These are equal when $C = -B$ and $D = A$. Thus the matrix of T has form

$$\begin{bmatrix} A & B \\ -B & A \end{bmatrix}.$$

□

11. Prove Lemma 4.17.

Solution:

This lemma was proved in the main text and thus this problem should be removed.

□

12. Prove the second part of the Paley-Wiener theorem.
Solution:

The proof is now part of the main text and thus this problem should be removed.

□

13. For $a \in \mathbb{C}^n$, define $\tau(a)f(z) = f(z-a)$ and $e_a(z) = e^{ia\cdot z}$. Furthermore, let $p(z)$ be a complex polynomial.

(a) Show $\tau(a)$ is a linear homeomorphism of $\mathrm{PW}_R(\mathbb{C}^n)$.

(b) Assuming $a \in \mathbb{R}^n$, show $f \mapsto e_a f$ is a linear homeomorphism of $\mathrm{PW}(\mathbb{C}^n)$.

(c) Show $f \mapsto pf$ is a continuous linear transformation of each space $\mathrm{PW}_R(\mathbb{C}^n)$.

Solution:

Let $a \in \mathbb{C}^n$. Since $\tau(a)^{-1} = \tau(-a)$, it suffices to show $\tau(a)$ is continuous. Let $f \in \mathrm{PW}_R(\mathbb{C}^n)$. Thus for each N, $\pi_{R,N}(f) < \infty$. Now

$$\begin{aligned}(1+|z|^2)^N|f(z-a)|e^{-R|\operatorname{Im} z|-R|\operatorname{Im} a|} &\leqslant (1+(|z-a|+|a|)^2)^N e^{-R|\operatorname{Im}(z-a)|}|f(z-a)|\\ &\leqslant (2^N(1+|z-a|^2)^N + 2^N(1+|a|^2)^N)e^{-R|\operatorname{Im}(z-a)|}|f(z-a)|\\ &\leqslant 2^N\pi_{R,N}(f) + 2^N(1+|a|^2)^N\pi_{R,0}(f)\end{aligned}$$

for $\operatorname{Im}(z-a) = \operatorname{Im} z - \operatorname{Im} a$ and so $|\operatorname{Im}(z-a)| \leqslant |\operatorname{Im} z| + |\operatorname{Im} a|$. Thus

$$\pi_{R,N}(\tau(a)f) \leqslant 2^N e^{R|\operatorname{Im} a|}(\pi_{R,N}(f) + (1+|a|^2)^N\pi_{R,0}(f)).$$

This implies $f \mapsto \tau(a)f$ is continuous from $\mathrm{PW}_R(\mathbb{C}^n)$ into $\mathrm{PW}_R(\mathbb{C}^n)$.

Let $a \in \mathbb{R}^n$. Now $e^{ia\cdot z} = e^{-\operatorname{Im}(a\cdot z)}e^{i\operatorname{Re}(a\cdot z)}$ and $\operatorname{Im}(a\cdot z) = a\cdot\operatorname{Im}(z)$. Thus $|\operatorname{Im}(a\cdot z)| \leqslant |a|\,|\operatorname{Im} z|$. So

$$\begin{aligned}(1+|z|^2)^N e^{-R|\operatorname{Im} z|}|e_a(z)f(z)| &\leqslant (1+|z|^2)^N e^{-R|\operatorname{Im} z|}e^{-a\cdot\operatorname{Im} z}|f(z)|\\ &\leqslant (1+|z|^2)^N e^{-R|\operatorname{Im} z|}e^{|a|\,|\operatorname{Im} z|}|f(z)|\\ &\leqslant (1+|z|^2)^N e^{-(R-|a|)|\operatorname{Im} z|}|f(z)|\\ &\leqslant \pi_{R-|a|,N}(f)\end{aligned}$$

for $R > |a|$. So $\pi_{R,N}(e_a f) \leqslant \pi_{R-|a|,N}(f)$ for $R > |a|$. Thus $f \mapsto e_a f$ is a continuous linear mapping of $\mathrm{PW}_{R-|a|}(\mathbb{C}^n)$ into $\mathrm{PW}_R(C^n)$ for $R > |a|$. But since $\mathrm{PW}(\mathbb{C}^n)$ has the inductive limit topology of the $\mathrm{PW}_R(\mathbb{C}^n)$, the relative topology of the inductive limit topology is weaker than the topology of $\mathrm{PW}_R(\mathbb{C}^n)$. This implies the mapping $f \mapsto e_a f$ is continuous from $\mathrm{PW}_{R-|a|}(\mathbb{C}^n)$ into $\mathrm{PW}(C^n)$ for $R > |a|$. But by Proposition 2.15, $f \mapsto e_a f$ is continuous on $\mathrm{PW}(\mathbb{C}^n)$ into $\mathrm{PW}(\mathbb{C}^n)$.

For (c), it suffices to show $f \mapsto pf$ is continuous on $\mathrm{PW}_R(\mathbb{C}^n)$ when $p(z) = z^\alpha$. Now note

$$\begin{aligned}(1+|z|^2)^N e^{-R|\operatorname{Im} z|}|p(z)f(z)| &\leqslant (1+|z|^2)^N|z|^{|\alpha|}e^{-R|\operatorname{Im} z|}|f(z)|\\ &\leqslant (1+|z|^2)^{N+|\alpha|/2}e^{-R|\operatorname{Im} z|}|f(z)|\\ &\leqslant (1+|z|^2)^K e^{-R|\operatorname{Im} z|}|f(z)|\end{aligned}$$

where K is the smallest integer greater than $N+|\alpha|/2$. So $\pi_{R,N}(pf) \leqslant \pi_{R,K}(f)$. This implies the linear function $f \mapsto pf$ is continuous at $f = 0$ and thus is continuous.

□

14. Show $\mathrm{PW}_R(\mathbb{C}^n)$ is a Fréchet space.
Solution:

We show $\mathrm{PW}_R(\mathbb{C}^n)$ is a Frechét space. Since it is a locally convex Hausdorff topological vector space with topology defined by countably many seminorms, $\pi_{R,N}$, we know by Exercise 2.1.10 that the topology is metrizable. We now need only show it is complete. Let f_k be a Cauchy sequence. Then $\pi_{R,N}(f_k - f_l) \to 0$ as $k, l \to \infty$. Thus $\pi_{R,0}(f_k - f_l) = \sup_z e^{-R|\operatorname{Im} z|}|f_k(z) - f_l(z)| \to 0$. Thus the f_k converge uniformly on $-a < \operatorname{Im} z < a$ for $a > 0$. Since the uniform limit of holomorphic functions on an open set is holomorphic, the f_k converge uniformly on these bands to a holomorphic function f. Moreover, if $\epsilon > 0$ and $\pi_{R,N}(f_k - f_l) \leqslant \epsilon$ for $k, l \geqslant K$, then $(1 + |z|^2)^N e^{-R|\operatorname{Im} z|}|f_k(z) - f_l(z)| \leqslant \epsilon$ for each z for $k, l \geqslant K$. Letting $l \to \infty$, we see $(1+|z|^2)^N e^{-R|\operatorname{Im} z|}|f_k(z) - f(z)| \leqslant \epsilon$ for $k \geqslant K$. Thus $\pi_{R,N}(f_k - f) \leqslant \epsilon$ for $k \geqslant K$. This implies $f_k \to f$ as $k \to \infty$ in $\mathrm{PW}_R(\mathbb{C}^n)$.

□

15. Show if B is a bounded subset of $\mathrm{PW}(\mathbb{C}^n)$, then there is an $R > 0$ such that $B \subseteq \mathrm{PW}_R(\mathbb{C}^n)$.
Solution: A direct argument would be interesting. We however see this is a consequence of the Paley Wiener Theorem 4.21 and Proposition 2.70.

□

16. Show for each $R > 0$, the space $\mathrm{PW}_R(\mathbb{C}^n)$ has the Heine-Borel property.
Solution:

We could use Exercise 2.4.6. By Theorem 4.21, we know the Fourier transform $\mathcal{F}$ maps $\mathcal{D}_{R/2\pi}(\mathbb{R}^n)$ linearly topologically isomorphicly onto $\mathrm{PW}_R(\mathbb{C}^n)$. So by the a forstated exercise, $\mathcal{D}_{R/2\pi}(\mathbb{R}^n)$ has the Heine-Borel property and thus so does $\mathrm{PW}_R(\mathbb{C}^n)$.

One can also give a proof using holomorphicity and some results from complex holomorphic functions of several variables. Namely, let F be a closed and bounded set in $\mathrm{PW}_R(\mathbb{C}^n)$. Since this space is metrizable (its topology is defined by countably many seminorms), it suffices to show it is sequentially compact and thus any sequence in F has a limit point in F. Since F is bounded, the seminorm $\pi_{R,0}$ is bounded on F. Thus $M = \sup_{f \in F} \sup_{z \in \mathbb{C}^n} e^{-R|\operatorname{Im} z|}|f(z)| < \infty$. This implies if $f \in F$, $|f(z)| \leqslant M e^{R|\operatorname{Im} z|}$ for $z \in \mathbb{C}^n$. Hence the functions in F are uniformly bounded on bounded subsets of $\mathbb{C}^n$. But this implies the family F is equicontinuous on any bounded subset of $\mathbb{C}^n$. (See the proof of the Generalized Vitali Theorem in Gunning and Rossi, [20], page 11.) So any sequence from F on any closed bounded set has a subsequence which converges uniformly on this set. Using a Cantor diagonalization, for any sequence f_k in F, there is a subsequence $f_{1,k}$ converging uniformly on $|z| \leqslant 1$. This subsequence has a subsequence $f_{2,k}$ which converges uniformly on $|z| \leqslant 2$. Repeating one has a doubly indexed family $f_{j,k}$ of functions from F such that $f_{j+1,k}$ for $k = 1, 2, \ldots$ is a subsequence of $f_{j,k}$, $k = 1, 2, \ldots$ and $f_{j,k}$ as $k \to \infty$ converges uniformly on $|z| \leqslant j$. Thus $f_{j,j}$ is a subsequence of f_k converging uniformly on each bounded subset of $\mathbb{C}^n$. This implies the limit f_0 also satisfies Theorem 4.15 and thus is holomorphic on $\mathbb{C}^n$.

We claim $f_{j,j} \to f_0$ in $\mathrm{PW}_R(\mathbb{C}^n)$. Let $N \in \mathbb{N} \cup \{0\}$. Then since the norm $\pi_{R,N+1}$ is bounded on F there is an $M > 0$ with $\pi_{N+1,R}(f) < M$ for $f \in F$. Thus if $\epsilon > 0$, there is a $b > 1$ such that $\frac{M}{b} < \frac{\epsilon}{2}$. Now $(1 + |z|^2)^{N+1} e^{-R|\operatorname{Im} z|} |f(z)| < M$ for all z and all $f \in F$. Thus if $1 + |z|^2 > b$, one has

$$\begin{aligned}(1 + |z|^2)^N e^{-R|\operatorname{Im} z|} |f(z)| &= \frac{1}{1 + |z|^2} (1 + |z|^2)^{N+1} e^{-R|\operatorname{Im} z|} |f(z)| \\ &< \frac{M}{b} < \frac{\epsilon}{2}.\end{aligned}$$

Since f_k converges uniformly to f_0 on $1 + |z|^2 \leqslant b$, there is a k_0 such that

$$(1 + |z|^2)^N e^{-R|\operatorname{Im} z|} |f_k(z) - f_0(z)| \leqslant \epsilon$$

when $k \geqslant k_0$ and $1 + |z|^2 \leqslant b$. Thus $\pi_{R,N}(f_k - f_0) \leqslant \epsilon$ for $k \geqslant k_0$ since

$$\begin{aligned}(1 + |z|^2)^N e^{-R|\operatorname{Im} z|} |f_k(z) - f_0(z)| &\leqslant (1 + |z|^2)^N e^{-R|\operatorname{Im} z|} (|f_k(z)| + |f_0(z)|) \\ &< \frac{\epsilon}{2} + \frac{\epsilon}{2} = \epsilon \text{ for } 1 + |z|^2 > b.\end{aligned}$$

This happening for all N implies $f_k \to f_0$ in $\mathrm{PW}_R(\mathbb{C}^n)$.

□

17. Show the space $\mathrm{PW}(\mathbb{C}^n)$ has the Heine-Borel property.
Solution:

This follows immediately from the Paley-Wiener Theorem 4.21 and Exercise 2.4.14.

□

18. Show the space $\mathrm{PW}(\mathbb{C}^n)$ is complete.
Solution:

This follows directly from Theorem 4.21 and Corollary 2.72

□

19. Let $f : \mathbb{R} \to \mathbb{C}$ be measurable and such that on $\mathbb{R}^+$ we have $|f(t)| \leqslant Ce^{t\beta}$ for some $\beta \in \mathbb{R}$ and $C > 0$. Let

$$\mathbb{C}_\beta := \{z \in \mathbb{C} \mid \operatorname{Re}(z) > \beta\} .$$

Show the **Laplace transform** of f:

$$\mathcal{L}(f)(z) := \int_0^\infty e^{-zt} f(t)\, dt$$

is defined for $z \in \mathbb{C}_\beta$ and that $\mathcal{L}(f) : \mathbb{C}_\beta \to \mathbb{C}$ is holomorphic. Furthermore show that for all $x > \beta$ the following inversion formula holds for $t > 0$:

$$f(t) = \frac{1}{2\pi} \int_{-\infty}^{\infty} \mathcal{L}(f)(x + iy)e^{(x+iy)t}\, dy\,.$$

(Hint: Relate the Laplace transform to the Fourier transform.)

Solution:

$\mathcal{L}(f)(z)$ is defined if the function $t \mapsto e^{-zt} f(t)$ is integrable on $(0, \infty)$. For $z = a + ib$, this occurs if $|e^{-zt} f(t)| = e^{-at}|f(t)|$ is integrable on $\mathbb{R}^+$. But if $|f(t)| \leqslant Ce^{\beta t}$, one has $e^{-at}|f(t)| \leqslant Ce^{(\beta-a)t}$. But $Ce^{(\beta-a)t}$ is integrable on $(0, \infty)$ if $a > \beta$. So $\mathcal{L}(f)(z)$ is defined for $\operatorname{Re} z > \beta$ and hence for $z \in C_\beta$.

To see $\mathcal{L}(f)$ is holomorphic on C_β, it suffices by Theorem 4.12 to show if $z \in C_\beta$, $\mathcal{L}(f)'(z)$ exists. So if $F(z) = \mathcal{L}(f)(z)$, we have

$$\begin{aligned}\frac{F(w) - F(z)}{w - z} &= \int_0^{\infty} f(t)\frac{e^{-wt} - e^{-zt}}{w - z}\, dt \\ &\to -\int_0^{\infty} tf(t)e^{-zt}\, dt\end{aligned}$$

by the Lebesgue Dominated Convergence Theorem. Indeed, pointwise the limit of $\frac{e^{-wt} - e^{-zt}}{w}$ as $w \to 0$ is $-te^{-zt}$. Moreover, if $|w - z| < \delta$, then

$$\begin{aligned}\left|\frac{e^{-wt} - e^{-zt}}{w - z}\right| &= \left|e^{-zt}\frac{e^{-(w-z)t} - 1}{w - z}\right| \\ &= |e^{-zt}| \left|\sum_{k=1}^{\infty} \frac{(-t)^k (w - z)^{k-1}}{k!}\right| \\ &\leqslant |e^{-zt}| \sum_{k=1}^{\infty} \frac{t^k \delta^{k-1}}{k!} \\ &\leqslant t|e^{-zt}| \sum_{k=1}^{\infty} \frac{t^{k-1}\delta^{k-1}}{(k-1)!} \\ &= te^{-t(\operatorname{Re} z - \delta)}\,.\end{aligned}$$

Thus $f(t)\frac{e^{-wt} - e^{-zt}}{w-z}$ for $t > 0$ is dominated by $Cte^{t\beta - t(\operatorname{Re} z - \delta)}$. This is integrable on $(0, \infty)$ if $\beta - \operatorname{Re} z + \delta < 0$. So choose $0 < \delta < \operatorname{Re} z - \beta$.

We now show the inversion formula. Suppose $z = x + iy$ with $x > \beta$. Then $F_{x,y}(t) = \chi_{[0,\infty)}(t) f(t)e^{-t(x+iy)}$ is integrable and square-integrable. Now

$$\begin{aligned}\int F_{x,y}(t)e^{-2\pi i t\omega}\, dt &= \int_0^{\infty} f(t)e^{-t(x+iy)}e^{-2\pi i t\omega}\, dt \\ &= \mathcal{L}(f)(x + iy + 2\pi i\omega)\,.\end{aligned}$$

So

$$\mathcal{F}(F_{x,y})(\omega) = \mathcal{L}(f)(x + iy + 2\pi i\omega).$$

Thus the L^2 Fourier inversion formula holds; i.e.,

$$F_{x,y}(t) = \int \mathcal{L}(f)(x + iy + 2\pi i\omega)e^{2\pi i\omega t}\, d\omega$$

or

$$\begin{aligned} f(t)e^{-t(x+iy)} &= \int \mathcal{L}(f)(x + iy + 2\pi i(\omega - y/2\pi))e^{2\pi i(\omega - y/2\pi)t}\, d\omega \\ &= \int \mathcal{L}(f)(x + 2\pi i\omega)e^{2\pi i\omega t}e^{-iyt}\, d\omega \\ &= \frac{1}{2\pi}e^{-iyt}\int \mathcal{L}(f)(x + iw)e^{iwt}\, dw \end{aligned}$$

and so

$$f(t) = \frac{1}{2\pi}\int_{-\infty}^{\infty} \mathcal{L}(f)(x + iw)e^{(x+iw)t}\, dw.$$

Note this in reality means as a L^2 function $e^{-tx}f(t)$ is the limit in $L^2[0,\infty)$ of the L^2 functions $\frac{1}{2\pi}\int_{-N}^{N} \mathcal{L}(f)(x + iy)e^{iyt}dy$ as $N \to \infty$. Changing measures appropriately, one can make a similar statement about just $f(t)$.

□

20. (Hardy spaces) Let

$$L^2_+(\mathbb{R}) = \left\{f \in L^2(\mathbb{R}) \mid \text{ess-supp}\,(\hat{f}) \subset [0,\infty)\right\}$$

and

$$L^2_-(\mathbb{R}^n) = \left\{f \in L^2(\mathbb{R}) \mid \text{ess-supp}\,(\hat{f}) \subset (-\infty, 0]\right\}.$$

Set $\mathbb{C}_+ = \{z \mid \text{Im}(z) > 0\}$ and $\mathbb{C}_- = \{z \mid \text{Im}(z) < 0\}$. Let $\mathcal{H}(\mathbb{C}_+)$ be the space of all holomorphic functions on $\mathbb{C}_+$ such that $f(\cdot + iy) \in L^2(\mathbb{R})$ for all $y > 0$, i.e.,

$$\int_{-\infty}^{\infty} |f(x + iy)|^2\, dx < \infty \text{ for all } y > 0$$

and such that the L^2-limit

$$\beta_+(f) := \lim_{y \to 0+} f(\cdot + iy)$$

exists. Similarly we define $\mathcal{H}(\mathbb{C}_-)$, to be the space of all holomorphic functions on $\mathbb{C}_-$ such that $f(\cdot - iy) \in L^2(\mathbb{R})$ for all $y > 0$, i.e.,

$$\int_{-\infty}^{\infty} |f(x - iy)|^2\, dx < \infty \text{ for all } y > 0$$

and such that the L^2-limit

$$\beta_-(f) := \lim_{y \to 0+} f(\cdot - iy)$$

exists. Show the following:

(a) Let $f \in L^2_+(\mathbb{R})$. Then the f 'extends' to a holomorphic function $F_+(z)$ on $\mathbb{C}_+$. It is given by

$$F_+(z) = \int \hat{f}(x) e^{2\pi i z x}\, dx\,.$$

Furthermore, $F_+ \in \mathcal{H}(\mathbb{C}_+)$ and $\beta_+(F_+) = f$.

(b) Let $f \in L^2_-(\mathbb{R})$. Then f extends to a function $F_- \in \mathcal{H}(\mathbb{C}_-)$ such that $\beta_-(F_-) = f$.

(c) If $f \in L^2(\mathbb{R})$, then there exists $F \in \mathcal{H}(\mathbb{C}_+)$ and $G \in \mathcal{H}(\mathbb{C}_-)$ such that

$$f = \beta_+(F) - \beta_-(G)\,.$$

Thus any L^2-function on $\mathbb{R}$ is the difference of a boundary value of a holomorphic function on the upper half plane and a holomorphic function on the lower half plane.

Solution:

For (a), we note if $f \in L^2_+(\mathbb{R})$, then if $z = a + ib$ with $b > 0$,

$$\begin{aligned} F_+(z) &= \int \hat{f}(x) e^{2\pi i z x}\, dx \\ &= \int_0^\infty \hat{f}(x) e^{2\pi i a x} e^{-2\pi b x}\, dx \end{aligned}$$

is integrable for $x \mapsto \chi_{[0.\infty)}(x) e^{-2bx}$ is L^2. So $F_+(z)$ is defined on the upper half plane $\mathbb{C}_+$. Note

$$\begin{aligned} \left| \frac{e^{2\pi i w x} - e^{2\pi i z x}}{w - z} \right| &= |e^{2\pi i z x}| \left| \frac{e^{2\pi i (w-z) x} - 1}{w - z} \right| \\ &\leqslant |e^{2\pi i z x}| \sum_{k=1}^{\infty} \left| \frac{(2\pi)^k (w-z)^{k-1} x^k}{k!} \right| \\ &\leqslant 2\pi x |e^{2\pi i z x}| \sum_{k=1}^{\infty} \frac{(2\pi)^{k-1} x^{k-1} |w - z|^{k-1}}{(k-1)!} \\ &\leqslant 2\pi x |e^{2\pi i z x}| e^{2\pi x |w - z|} \\ &= 2\pi x e^{2\pi x (|w - z| - \operatorname{Im} z)} \end{aligned}$$

for $x > 0$. For $z \in \mathbb{C}_+$, restrict w so that $|w - z| < \operatorname{Im} z/2$. Then for $0 \leqslant x < \infty$, we see $\hat{f}(x) \frac{e^{2\pi i w x} - e^{2\pi i z x}}{w - z}$ is dominated by $2\pi x |\hat{f}(x)| e^{-\pi x \operatorname{Im} z}$ which is the product of the two L^2 functions $|\hat{f}(x)|$ and $2\pi x e^{-\pi x \operatorname{Im} z}$. Thus the Lebesgue Dominated

Convergence Theorem implies

$$\begin{aligned} F_+'(z) &= \lim_{w\to z}\int_0^\infty \hat f(x)\frac{e^{2\pi iwx}-e^{2\pi izx}}{w-z}\,dx \\ &= \int_0^\infty \hat f(x)\lim_{w\to z}\frac{e^{2\pi iwx}-e^{2\pi izx}}{w-z}\,dx \\ &= \int_0^\infty 2\pi ix\hat f(x)e^{2\pi izx}dx\,. \end{aligned}$$

So F_+ is holomorphic on $\mathbb{C}_+$.

Next note

$$F_+(\omega+bi)=\mathcal{F}^{-1}(\hat f e_b)(\omega)$$

where $e_b(x)=e^{-2\pi bx}$. Thus

$$\int |F_+(x+bi)|^2\,dx = \int_0^\infty |\hat f(x)|^2 e^{-4\pi bx}\,dx \leqslant \int |\hat f(x)|^2 dx$$

for $b>0$.

Finally we note $F_+(\cdot+bi)\to f(\cdot)$ in L^2 as $b\searrow 0$. Since $\mathcal{F}^{-1}$ is L^2 continuous, it suffices to note $\hat f e_b\to\hat f$ in L^2. Indeed, $|\hat f(x)|^2(1-e_b(x))^2\leqslant|\hat f(x)|^2$ and so by the Lebesgue Dominated Convergence Theorem

$$|\hat f e_b-\hat f|_2^2=\int|\hat f(x)|^2(1-e_b(x))^2\,dx\to 0\,.$$

For (b), we use $F_-(z)=\int_{-\infty}^0\hat f(x)e^{2\pi izx}\,dx$ and mirror the argument for (a). For (c), take $F_+(z)=\int_0^\infty\hat f(x)e^{2\pi izx}\,dx$ and $F_-(z)=\int_{-\infty}^0\hat f(x)e^{2\pi izx}\,dx$. Set $G(z)=-F_-(z)$ for $\operatorname{Im} z<0$. Then $F_+(\cdot+bi)\to f\chi_{[0,\infty)}$ in $L^2(\mathbb{R})$ and $G(\cdot-bi)\to -f\chi_{(-\infty,0]}$ in $L^2(\mathbb{R})$ as $b\to 0+$. This gives

$$f=\beta_+(F)-\beta_-(G).$$

□

21. Let X be a Hausdorff topological space, and let $\mathcal{H}\subset\mathcal{C}(X)$ be a Hilbert space. Assume that the maps

$$\mathrm{ev}_x:\mathcal{H}\to\mathbb{C}\,,\quad f\mapsto f(x)$$

are continuous for all $x\in X$. Show there exists a function $K:X\times X\to\mathbb{C}$ such that the following hold:

(a) If $K_y:X\to\mathbb{C}$ is the function $K_y(x)=K(x,y)$, then $K_y\in\mathcal{H}$ for all y. In particular K_y is continuous.

(b) We have $f(y)=(f\mid K_y)$ for all $y\in X$ and all $f\in\mathcal{H}$.

(c) $K(x,y)=\overline{K(y,x)}$ for all $x,y\in X$. In particular $y\mapsto K(x,y)$ is continuous for all $x\in X$.

(d) $(K_y \mid K_x) = K(x,y)$ for $x, y \in X$. In particular $||K_x||^2 = K(x,x)$ for all $x \in X$.

(e) The linear span

$$\mathcal{H}_0 := \left\{ \sum_{\text{finite}} c_j K_{x_j} \in \mathcal{H} \mid c_j \in \mathbb{C},\ x_j \in X \right\}$$

is dense in $\mathcal{H}$.

(f) The function K is positive definite in the sense that for all $k \in \mathbb{N}$, $x_1, \ldots, x_k \in X$, and $c_1, \ldots, c_k \in \mathbb{C}$ one has

$$\sum_{i,j=1}^{k} c_i \bar{c}_j K(x_j, x_i) \geqslant 0\,.$$

The function K is called a *reproducing kernel* for the Hilbert space $\mathcal{H}$.

Solution:

Since $f \mapsto f(y)$ is a continuous linear functional, there is a $K_y \in \mathcal{H} \subseteq C(X)$ such that $f(y) = (f, K_y)$ for $f \in \mathcal{H}$ and $y \in X$. Define $K(x,y) = K_y(x)$. So (a) and (b) hold. Next note $(K_x, K_y) = K(y,x) = K_x(y) = \overline{(K_y, K_x)} = \overline{K_y(x)} = \overline{K(x,y)}$. So $K(y,x) = \overline{K(x,y)}$. This gives (c) and (d).

Now suppose $f \in \mathcal{H}$ is perpendicular to all K_y. Thus $f(x) = (f, K_x) = 0$ for all x. So $f = 0$. This implies (e).

Finally for (f), note

$$\begin{aligned} 0 &\leqslant \Big(\sum_{i=1}^{k} c_i K_{x_i}, \sum_{j=1}^{k} c_j K_{x_j}\Big) \\ &= \sum_{i,j} c_i \bar{c}_j (K_{x_i}, K_{x_j}) \\ &= \sum_{i,j} c_i \bar{c}_j K(x_j, x_i)\,. \end{aligned}$$

□

22. (The Fock space and the Segal-Bargmann transform) This exercise is a set of problems dealing with the Fock space $\mathcal{F}(\mathbb{C}^n)$ of holomorphic functions on $\mathbb{C}^n$ and the Segal-Bargmann transform $B : L^2(\mathbb{R}^n) \to \mathcal{F}(\mathbb{C}^n)$.

(a) Let $\mathcal{F}(\mathbb{C}^n)$ be the space of holomorphic functions $F : \mathbb{C}^n \to \mathbb{C}$ such that

$$|F|^2 := \pi^{-n} \int |F(z)|^2 e^{-|z|^2}\, dz < \infty$$

where $z = x + iy$ and $dz = dx_1 \cdots dx_n dy_1 \cdots dy_n$. For $F, G \in \mathcal{F}(\mathbb{C}^n)$, let

$$(F \mid G) = \pi^{-n} \int F(z)\, \overline{G(z)}\, e^{-|z|^2}\, dz.$$

Show $\mathcal{F}(\mathbb{C}^n)$ is a Hilbert space.

(b) Show every polynomial p is in $\mathcal{F}(\mathbb{C}^n)$ and the vector space $\mathbb{C}[z_1, \ldots, z_n]$ of polynomials is dense in $\mathcal{F}(\mathbb{C}^n)$.

(c) Evaluate the norm of the monomials $p_\alpha(z) = z^\alpha, \alpha \in \mathbb{N}_0^n$.

(d) Show $\{|p_\alpha|^{-1} p_\alpha \mid \alpha \in \mathbb{N}_0^n\}$ is an orthonormal basis for $\mathcal{F}(\mathbb{C}^n)$.

(e) Define $R : \mathcal{F}(\mathbb{C}^n) \to C^\infty(\mathbb{R}^n)$ by

$$RF(x) = e^{-|x|^2/2} F(x) .$$

Show R is injective and that $Rp \in L^2(\mathbb{R}^n)$ for every polynomial p.

(f) Show $\mathrm{Im}(R)$ is dense in $L^2(\mathbb{R}^n)$. Hence $R : \mathcal{F}(\mathbb{C}^n) \to L^2(\mathbb{R}^n)$ is densely defined. (Hint: Use the Hermite polynomials.)

(g) Show $\mathcal{F}(\mathbb{C}^n) \to \mathbb{C}$, $F \mapsto F(z)$, is continuous for each $z \in \mathbb{C}^n$.

(h) Show the reproducing kernel for $\mathcal{F}(\mathbb{C}^n)$ is given by

$$K(z, w) = e^{z \cdot \overline{w}}$$

where $u \cdot v = \sum_{j=1}^n u_j v_j$.

(i) Show R is a closed operator. (Actually it is bounded.)

(j) As R is closed and densely defined the map $R^* : L^2(\mathbb{R}^n) \to \mathcal{F}(\mathbb{C}^n)$ is well defined. Show that

$$R^* f(z) = e^{z \cdot z/2} f * H(z)$$

where $H(z) = e^{-\frac{1}{2} z \cdot z}$. In particularly $RR^* f(x) = f * H(x)$. (Hint: Use that $R^* f(z) = (R^* f \mid K_z)$ where as usual $K_z(w) = K(w, z)$.)

(k) Using the heat semigroup, show $\sqrt{RR^*}(f)(x) = \left(\frac{2}{\pi}\right)^{n/4} f * L(x)$ where $L(z) = H(\sqrt{2} z)$.

(l) Polarizing R we can write $R^* = B\sqrt{RR^*}$ where $B : L^2(\mathbb{R}^n) \to \mathcal{F}(\mathbb{C}^n)$ is an unitary isomorphism. Show B is given by

$$Bf(z) = \left(\frac{2}{\pi}\right)^{n/4} \int f(x) e^{-x \cdot x + 2z \cdot x - \frac{1}{2} z \cdot z} \, dx .$$

The map B is the well known **Segal-Bargmann transform**.

Solution:

For (a) , all is clear except completeness. Let F_k be a Cauchy sequence. Fix $z \in \mathbb{C}^n$ and consider $A_{1,2} = \prod_{j=1}^n \{w_j \mid 1 \leqslant |w_j| \leqslant 2\}$ and measure $dw = dw_1 \, dw_2 \ldots dw_n = dx_1 \, dy_1 \, dx_2 \, dy_2 \cdots dx_n \, dy_n$ where $w = (w_1, w_2, \ldots, w_n) = (x_1 + iy_1, x_2 + iy_2, \ldots, x_n + iy_n) = (r_1 e^{i\theta_1}, r_2 e^{i\theta_2}, \ldots, r_n e^{i\theta_n})$.

We will be considering the $\xi \in \mathbb{C}^n$ with $|\xi_j - z_j| < \frac{1}{2}$ for $j = 1, 2, \ldots, n$. First note by using Lemma 4.14 and the Cauchy integral formula that

$$F_k(\xi) = \frac{1}{(2\pi i)^n} \int_{I_r} \frac{F_k(w_1, w_2, \ldots, w_n)}{(w_1 - \xi_1)(w_2 - \xi_2) \cdots (w_n - \xi_n)} \, dw_1 \, dw_2 \cdots dw_n$$

where $r = (r_1, r_2, \ldots, r_n)$ where $1 \leqslant r_j \leqslant 2$ for $j = 1, 2, \ldots, n$ and I_r is the range of the mapping

$$\gamma_r(\theta_1, \theta_2, \ldots, \theta_n) = (z_1 + r_1 e^{i\theta_1}, z_2 + r_2 e^{i\theta_2}, \ldots, z_n + r_n e^{i\theta_n}).$$

Thus if $w = (w_1, w_2, \ldots w_n)$, $r = (r_1, r_2, \ldots, r_n)$, $\theta = (\theta_1, \theta_2, \ldots, \theta_n)$ and $re^{i\theta} = (r_1 e^{i\theta_1}, r_2 e^{i\theta_2}, \ldots, r_n e^{i\theta_n})$, then

$$\begin{aligned} F_k(\xi) &= \frac{1}{(2\pi i)^n} \int_{[1,2]^n} \int_{I_r} \frac{F_k(w)}{\prod_{j=1}^n (w_j - \xi_j)} \, dw \, dr \\ &= \frac{1}{(2\pi i)^n} \int_{[1,2]^n} \int_{[0,2\pi]^n} \frac{F_k(z + re^{i\theta})}{\prod_{j=1}^n (z_j + r_j e^{i\theta_j} - \xi_j)} \prod_{j=1}^n (r_j i e^{i\theta_j}) \, d\theta \, dr \\ &= \frac{1}{(2\pi)^n} \int_{[1,2]^n} \int_{[0,2\pi]^n} F_k(z + re^{i\theta}) \prod_{j=1}^n \frac{r_j e^{i\theta_j}}{z_j + re^{i\theta_j} - \xi_j} d\theta \, dr \end{aligned}$$

where $d\theta = d\theta_1 \, d\theta_2 \cdots d\theta_n$ and $dr = dr_1 \, dr_2 \cdots dr_n$. Now consider

$$T(r_1, r_2, \ldots, r_n, \theta_1, \theta_2, \ldots, \theta_n) = (r_1 \cos\theta_1, r_1 \sin\theta_1, \ldots, r_n \cos\theta_n, r_n \sin\theta_n)$$

which maps $[1, 2]^n \times [0, 2\pi)^n$ onto $A_{1,2}$ where pairs $(r_j \cos\theta_j, r_j \sin\theta_j)$ are identified with their corresponding complex numbers $r_j \cos\theta_j + i \sin\theta_j$.

Note the Jacobian of this mapping is given by $r_1 r_2 \cdots r_n$. So we obtain by Theorem 2.22 the formula

$$F_k(\xi) = \frac{1}{(2\pi)^n} \int_{A_{1,2}} \frac{F_k(z + w)}{\prod(z_j + w_j - \xi_j)} \prod_{j=1}^n \frac{w_j}{|w_j|} dw \, .$$

Now F_k is Cauchy in $L^2(\mathbb{C}^n, \frac{1}{\pi^n} e^{-|z|^2} dz)$. Thus it has an L^2 limit H. Thus there is a measurable function such that

$$\lim_{k \to \infty} \int_{\mathbb{C}^n} |F_k(w) - H(w)| e^{-|w|^2} \, dw = 0 \, .$$

Since

$$\begin{aligned} \int_{z + A_{1,2}} |F_k(w) - H(w)|^2 \, dw &\leqslant e^{n(|z|+2)^2} \int_{z + A_{1,2}} |F_k(w) - H(w)|^2 e^{-|w|^2} \, dw \\ &\leqslant e^{n(|z|+2)^2} \int_{\mathbb{C}^n} |F_k(w) - H(w)|^2 e^{-|w|^2} \, dw \, , \end{aligned}$$

we have $F_k\chi_{z+A_{1,2}}$ converges in $L^2(\mathbb{C}^n)$ and hence in $L^1(\mathbb{C}^n)$ to $H\chi_{z+A_{1,2}}$. We hence define for ξ with $|\xi - z| < \frac{1}{2}$, $F(\xi)$ by

$$F(\xi) = \frac{1}{(2\pi)^n}\int_{A_{1,2}} \frac{H(z+w)}{\prod_{j=1}^n (z_j + w_j - \xi_j)} \prod_{j=1}^n \frac{w_j}{|w_j|} dw.$$

The convergence of $F_k\chi_{z+A_{1,2}}$ in $L^1(\mathbb{C}^n)$ to $H\chi_{z+A_{1,2}}$ also implies

$$F_k(\xi) = \frac{1}{(2\pi)^n}\int_{A_{1,2}} \frac{F_k(z+w)}{\prod (z_j + w_j - \xi_j)} \prod \frac{w_j}{|w_j|} dw \to F(\xi)$$

for all ξ with $|\xi - z| < \frac{1}{2}$. This implies we may redefine the function H so that it equals F a.e. on $|\xi - z| < \frac{1}{2}$. Next by Lemma 4.16, H is holomorphic in ξ on $|\xi - z| < \frac{1}{2}$. Doing this for all z, we see the L^2 limit H in $L^2(\mathbb{C}^n, e^{-|w|^2}\,dw)$ can be taken to be a holomorphic function on $\mathbb{C}^n$. This implies $\mathcal{F}(\mathbb{C}^n)$ is a Hilbert space.

Next we show z^α is in $L^2(e^{-|z|^2}\,dz)$ and find its norm. We have

$$\begin{aligned}
\pi^n(z^\alpha \mid z^\alpha) &= \int |z_1|^{2\alpha_1}|z_2|^{2\alpha_2}\cdots|z_n|^{2\alpha_n} e^{-|z|^2}\,dz \\
&= \prod_{j=1}^n \int_{\mathbb{C}} |z_j|^{2\alpha_j} e^{-|z_j|^2}\,dz_j.
\end{aligned}$$

But

$$\begin{aligned}
\int_C |z|^m e^{-|z|^2}\,dz &= \int_0^\infty \int_0^{2\pi} r^m e^{-r^2}(r\,d\theta)\,dr \\
&= 2\pi \int_0^\infty r^{m+1} e^{-r^2}\,dr \\
&= 2\pi \int_0^\infty s^{(m+1)/2} e^{-s} \frac{1}{2} s^{-1/2}\,ds \\
&= \pi \int_0^\infty s^{m/2} e^{-s}\,ds \\
&= \pi\Gamma\left(\frac{m+2}{2}\right).
\end{aligned}$$

So $\int |z|^{2m} e^{-|z|^2}\,dz = \pi\Gamma(m+1) = \pi m!$. This gives

$$(z^\alpha \mid z^\alpha) = \alpha!.$$

In particular, $|z^\alpha| = \sqrt{\alpha!}$ and so answers (c) and the first part of (b). To see the z^α are orthogonal, note

$$\begin{aligned}
\pi^n(z^\alpha \mid z^\beta) &= \int z_1^{\alpha_1}\bar{z}_1^{\beta_1} z_2^{\alpha_1}\bar{z}_2^{\beta_2}\cdots z_n^{\alpha_n}\bar{z}_n^{\beta_n} e^{-|z|^2}\,dz \\
&= \prod_{j=1}^n \int_{\mathbb{C}} z_j^{\alpha_j}\bar{z}_j^{\beta_j} e^{-|z_j|^2}\,dz_j.
\end{aligned}$$

Now suppose $k \neq l$, then

$$\begin{aligned}\int_{\mathbb{C}} z^k \bar{z}^l e^{-|z|^2}\, dz &= \int_0^\infty r^{k+l} e^{-r^2} \int_0^{2\pi} e^{i(k-l)\theta} r\, d\theta\, dr \\ &= \int_0^\infty r^{k+l+1} e^{-r^2}\, dr \int_0^{2\pi} e^{i(k-l)\theta}\, d\theta \\ &= 0\end{aligned}$$

if $k \neq l$. So $(z^\alpha \mid z^\beta) = 0$ if $\alpha \neq \beta$. Putting these together, we see $\frac{1}{\sqrt{\alpha!}} z^\alpha$ form an orthonormal set.

To see the z^α span a dense subspace, we first show if $(F \mid 1) = 0$, then $F(0) = 0$. Indeed, if $\theta = (\theta_1, \theta_2, \ldots, \theta_n)$, $r = (r_1.r_2, \ldots, r_n)$, $d\theta = d\theta_1 \cdots d\theta_n$, $dr = dr_1 dr_2 \cdots dr_n$ and $re^{i\theta} = (r_1 e^{i\theta_1}, r_2 e^{i\theta_2}, \ldots, r_n e^{i\theta_n})$, then

$$\begin{aligned}\pi^n (F \mid 1) &= \int_{\mathbb{C}^n} F(z) e^{-|z|^2}\, dz \\ &= \int_{[0,\infty]^n} \int_{[0,2\pi]^n} F(re^{i\theta}) \prod_j e^{-r_j^2} \prod_j r_j\, d\theta\, dr \\ &= i^{-n} \int_{[0,\infty]^n} \int_{[0,2\pi]^n} \frac{F(re^{i\theta})}{\prod_j (r_j e^{i\theta_j})} (\prod_j i r_j e^{i\theta_j}) d\theta \prod_j (r_j e^{-r_j^2}) dr \\ &= i^{-n} \int_{[0,\infty]^n} (2\pi i)^n F(0) \prod_j (r_j e^{-r_j^2}) dr \\ &= \pi^n F(0).\end{aligned}$$

So we have established:

$$(F \mid 1) = F(0)\,.$$

We next use the above argument with $H(z) = e^{-(z,w)} F(z+w)$ where w is fixed. We first note $H \in \mathcal{F}(\mathbb{C}^n)$. Indeed, it is holomorphic. For the integral,

$$\begin{aligned}\int |e^{-(z,w)} F(z+w)|^2 e^{-|z|^2}\, dz &= \int |e^{-(z-w,w)} F(z)|^2 e^{-|z-w|^2}\, dz \\ &= \int |e^{-(z-w,w)} F(z)|^2 e^{-|z|^2} e^{-|w|^2} |e^{(z,w)} e^{(w,z)}|\, dz \\ &= \int |F(z)|^2 e^{-|z|^2} |e^{-2(z,w)} e^{(z,w)} e^{(w,z)}|\, dz \\ &= \int |F(z)|^2 e^{-|z|^2}\, dz < \infty\,.\end{aligned}$$

Thus

$$(H \mid 1) = \pi^n H(0) = F(w).$$

Thus $F(w) = 0$ if $H \perp 1$. In order to do this we show $e^{(z,w)}$ is in Fock space and the series for it converges to it. It is holomorphic and

$$
\begin{aligned}
e^{(z,w)} &= \sum_{k=0}^{\infty} \frac{\left(\sum_{j=1}^{n} z_j \bar{w}_j\right)^k}{k!} \\
&= \sum_{k=0}^{\infty} \frac{1}{k!} \sum_{k_1+k_2+\cdots+k_n=k} \frac{k!}{k_1!k_2!\cdots k_n!}(z_1\bar{w}_1)^{k_1}(z_2\bar{w}_2)^{k_2}\cdots(z_n\bar{w}_n)^{k_n} \\
&= \sum_{k=0}^{\infty} \sum_{|\alpha|=k} \frac{1}{\alpha!} z^\alpha \bar{w}^\alpha .
\end{aligned}
$$

This implies

$$
e^{(z,w)} - \sum_{k=0}^{N} \sum_{|\alpha|=k} \frac{1}{\alpha!} z^\alpha \bar{w}^\alpha = \sum_{k=N+1}^{\infty} \sum_{|\alpha|=k} \frac{1}{\alpha!} \bar{w}^\alpha z^\alpha
$$

and thus because the z^α are orthogonal with norm $\sqrt{\alpha!}$,

$$
\begin{aligned}
\left| e^{(z,w)} - \sum_{k=0}^{N} \sum_{|\alpha|=k} \frac{1}{\alpha!} z^\alpha \bar{w}^\alpha \right|^2 &= \sum_{k=N+1}^{\infty} \sum_{|\alpha|=k} \frac{|w_1|^{2\alpha_1}|w_2|^{2\alpha_2}\cdots|w_n|^{2\alpha_n}}{(\alpha!)^2} |z^\alpha|^2 \\
&= \sum_{k=N+1}^{\infty} \sum_{|\alpha|=k} \frac{|w_1|^{2\alpha_1}|w_2|^{2\alpha_2}\cdots|w_n|^{2\alpha_n}}{\alpha!} \\
&= \sum_{k=N+1}^{\infty} \frac{1}{k!} \sum_{|\alpha|=k} \frac{k!}{\alpha!} |w_1|^{2\alpha_1}|w_2|^{2\alpha_2}\cdots|w_n|^{2\alpha_n} \\
&= \sum_{k=N+1}^{\infty} \frac{1}{k!} (|w_1|^2 + \cdots + |w_n|^2)^k \to 0
\end{aligned}
$$

as $N \to \infty$. Next if F is perpendicular to all polynomials, then

$$
\begin{aligned}
F(w) &= H(0) \\
&= (H \mid 1) \\
&= (e^{-(z,w)} F(z+w) \mid 1) \\
&= \int e^{-(z,w)} F(z+w) e^{-|z|^2} \, dz \\
&= \int e^{-(z-w,w)} F(z) e^{-(z-w,z-w)} \, dz \\
&= \int e^{-(z,w)} F(z) e^{(z,w)+(w,z)} e^{-|z|^2} \, dz \\
&= \int F(z) e^{(w,z)} e^{-|z|^2} \, dz \\
&= (F \mid e^{(z,w)})
\end{aligned}
$$

$$= \lim_{N\to\infty} (F \mid \sum_{k=0}^{N} \sum_{|\alpha|=k} \frac{1}{\alpha!} z^\alpha \bar{w}^\alpha)$$

$$= 0\,.$$

Since w is arbitrary, $F = 0$ and thus the polynomial functions form a dense linear subspace of $\mathcal{F}(\mathbb{C}^n)$. Thus we have finished (a), (b), (c), and (d).

Note formally a reproducing kernel on a Hilbert space is just the operator $\sum e_\alpha \otimes \bar{e}_\alpha$ which in the strong operator topology is the identity operator. We just need to have convergence in another appropriate topology.

We take $e_\alpha = \frac{1}{\sqrt{\alpha!}} z^\alpha$. Then $e_\alpha \otimes \bar{e}_\alpha$ would formally be the function $(z, w) \mapsto \frac{1}{\alpha!} z^\alpha \bar{w}^\alpha$. We have already shown the sum of the functions $z \mapsto \frac{1}{\alpha!} z^\alpha \bar{w}^\alpha$ converge in $\mathcal{F}(\mathbb{C}^n)$ to $K_w(z) = e^{(z,w)}$. Moreover by continuity of the inner product,

$$(z^\beta \mid K_w) = (z^\beta \mid \sum_\alpha \frac{1}{\alpha!} \bar{w}^\alpha z^\alpha) = \frac{1}{\beta!} w^\beta (z^\beta \mid z^\beta) = w^\beta$$

and since the linear span of the z^β is dense in $\mathcal{F}(C^n)$, one has

$$(F \mid K_w) = F(w)$$

for all $w \in \mathbb{C}^n$ and $F \in \mathcal{F}(\mathbb{C}^n)$.

We note $K_w = \sum \frac{\bar{w}^\alpha}{\sqrt{\alpha!}} e_\alpha$. Thus

$$\begin{aligned}
|K_w|^2 &= \sum \frac{|\bar{w}^\alpha|^2}{\alpha!} \\
&= \sum \frac{\prod |w_j|^{2\alpha_j}}{\alpha!} \\
&= \prod_{j=1}^{n} \sum_{\alpha_j} \frac{|w_j|^{2\alpha_j}}{\alpha_j!} \\
&= \prod_{j=1}^{n} e^{|w_j|^2} = e^{|w|^2}\,.
\end{aligned}$$

This gives (g) and (h).

For (e) we let $D = \{F \in \mathcal{F}(\mathbb{C}^n) \mid x \mapsto F(x)e^{-|x|^2/2} \in L^2(\mathbb{R}^n)\}$ and define $R : D \to L^2(\mathbb{R}^n)$ by $RF(x) = F(x)e^{-|x|^2/2}$. Clearly R is linear and each $RF \in C^\infty(\mathbb{R}^n)$. It is injective for $RF = 0$ implies $F|\mathbb{R}^n = 0$ and since F is holomorphic, $F = 0$. Claim R is densely defined. Indeed note $Rz^\alpha = x^\alpha e^{-|x|^2/2}$ and

$$\begin{aligned}
\int |x^\alpha|^2 e^{-|x|^2}\, dx &= \prod_{j=1}^{n} \int_{-\infty}^{\infty} |t|^{2\alpha_j} e^{-t^2}\, dt \\
&= 2 \prod_{j=1}^{n} \int_0^\infty s^{\alpha_j} e^{-s} \frac{ds}{2\sqrt{s}}
\end{aligned}$$

$$= \prod_{j=1}^{n} \Gamma(\alpha_j + \frac{1}{2}).$$

This implies if $e_\alpha = \frac{1}{\sqrt{\alpha!}} z^\alpha$, then $|Re_\alpha|_2^2 = \prod \frac{\Gamma(\alpha_j+\frac{1}{2})}{\alpha_j!}$. Thus R is defined on the linear span of the e_α and thus is densely defined.

We also consider K_w for $w \in \mathbb{C}^n$. Note $RK_w(x) = e^{(x,w)}e^{-|x|^2/2}$and thus

$$|RK_w|_2^2 = \int e^{2(x,\mathrm{Re}\, w)} e^{-|x|^2}\, dx$$
$$= \prod_{j=1}^{n} \int e^{2x_j \,\mathrm{Re}\, w_j} e^{-x_j^2}\, dx_j\,.$$

Thus we need the value of integrals of form $I(a) = \int_{-\infty}^{\infty} e^{at}e^{-t^2}\, dt$.

Note $I(a) = I(-a)$ and if $a > 0$, then

$$\begin{aligned} I'(a) &= \int_{-\infty}^{\infty} te^{at}e^{-t^2}\, dt \\ &= \int_{-\infty}^{\infty} \frac{d}{dt}\left(-\frac{e^{-t^2}}{2}\right) e^{at}\, dt \\ &= -\frac{e^{-t^2}}{2} e^{at}|_{-\infty}^{\infty} + \int_{-\infty}^{\infty} \frac{e^{-t^2}}{2} ae^{at}\, dt \\ &= \frac{a}{2} I(a)\,. \end{aligned}$$

This implies $\ln I(a) = \frac{a^2}{4} + \ln B$ for some constant B. So $I(a) = Be^{a^2/4}$. Note $I(2) = e\int e^{-(1-2t+t^2)}\, dt = e\int e^{-(t-1)^2}\, dt = e\int e^{-t^2}\, dt = e\sqrt{\pi}$. Thus $e\sqrt{\pi} = Be$. So $B = \sqrt{\pi}$. Thus $I(a) = \sqrt{\pi}e^{a^2/4}$.

We therefore see

$$\begin{aligned} |RK_w|_2^2 &= \prod_{j=1}^{n} I(2\,\mathrm{Re}\, w_j) \\ &= \pi^{n/2} e^{\sum(\mathrm{Re}\, w_j)^2} \\ &= \pi^{n/2} e^{|\,\mathrm{Re}\, w|^2}\,. \end{aligned}$$

Since $|K_w| = e^{|w|^2/2}$ and $|RK_w|_2 = \pi^{n/4}e^{|\,\mathrm{Re}\, w|^2/2}$, we see if $w \in \mathbb{R}^n$, then $\frac{|RK_w|_2}{|K_w|} = \pi^{n/4}$.

For (f), note the proof of Lemma 3.33 can be modified to show the space of functions of form $p(x)e^{-a|x|^2}$ where $a > 0$ is fixed and $p(x)$ is a polynomial is dense in $L^2(\mathbb{R}^n)$. Since the range of R contains all $x^\alpha e^{-|x|^2/2}$ we see the range of R is dense.

We look for some other functions in Fock space. For simplicity assume $n = 1$ and take $F_a(z) = e^{az^2}$. We determine the $a > 0$ for which this is in Fock space.

Indeed,

$$\begin{aligned}
\infty > |F_a|^2 &= \int_{\mathbb{C}} |e^{2az^2}|e^{-|z|^2}\,dz \\
&= \int_{\mathbb{R}^2} e^{2a(x^2-y^2)}e^{-x^2}e^{-y^2}\,dx\,dy \\
&= \int_{\mathbb{R}} e^{-(1-2a)x^2}\,dx \int_{\mathbb{R}} e^{-(1+2a)y^2}dy \\
&= \frac{\pi}{\sqrt{1-2a}\sqrt{1+2a}} \\
&= \frac{\pi}{(1-4a^2)^{1/2}}
\end{aligned}$$

when $a < \frac{1}{2}$ and then $|F_a| = \frac{\sqrt{\pi}}{(1-4a^2)^{1/4}}$.

Now note $RF_a(x) = e^{(a-\frac{1}{2})x^2}$. So then

$$\begin{aligned}
|RF_a|_2^2 &= \int_{\mathbb{R}} e^{-(1-2a)x^2}\,dx \\
&= \frac{\pi}{\sqrt{1-2a}}.
\end{aligned}$$

This gives $|RF_a|_2 = \frac{\sqrt{\pi}}{(1-2a)^{1/4}}$. So

$$\frac{|RF_a|_2}{|F_a|} = \frac{(1-4a^2)^{1/4}}{(1-2a)^{1/4}} = (1+2a)^{1/4}.$$

For (i), let $F_k \to F$ in Fock space and $RF_k \to f$ in $L^2(\mathbb{R}^n)$. Thus $F_k(z) \to F(z)$ for all z and since $RF_k \to f$ in L^2 we have $F_k(x)e^{-|x|^2/2} \to F(x)e^{-|x|^2/2}$ for all x. But a subsequence of RF_k converges pointwise a.e. Thus $f(x) = F(x)e^{-|x|^2/2}$ a.e. x. Consequently, $RF = f$ and R is a closed operator.

For (j), (k), and (l) we shall use some operator theory from functional analysis. Since R is a closed operator, R^* is defined. Its domain is all $f \in L^2(\mathbb{R}^n)$ such that $F \mapsto (RF, f)_2$ is continuous and then R^*f is the unique function in $\mathcal{F}(\mathbb{C}^n)$ with $(RF, f)_2 = (F \mid R^*f)$. But if $F = K_z$, $(F \mid R^*f) = (K_z \mid R^*f) = \overline{R^*f(z)}$. This implies $R^*f(z) = (f, RK_z)_2 = \int f(x)\overline{K_z(x)}e^{-|x|^2/2}\,dx$ for $f \in \mathrm{Dom}(R^*)$. So

$$\begin{aligned}
R^*f(z) &= \int f(x)e^{(z,x)}e^{-|x|^2/2}\,dx \\
&= e^{(z,\bar{z})/2}\int f(x)e^{-(x,x)/2+(z,x)-(z,\bar{z})/2}\,dx \\
&= e^{(z,\bar{z})/2}\int f(y)e^{-(z-x,\bar{z}-x)/2}\,dx \\
&= e^{z\cdot z/2}f * H(z)
\end{aligned}$$

where $H(z) = e^{-\frac{1}{2}z\cdot z}$. Thus $RR^*f(x) = (2\pi)^{n/2} f * h_{2\pi}(x)$ where

$$h_t(y) = t^{-n/2} e^{-\pi|y|^2/t}$$

is the heat kernel described in Example 3.11. From functional analysis, if R is a closed operator, R^* is densely defined and $R^{**} = R$. We note the range of R^* is dense for if $(F \mid R^*f) = (RF, f)_2$ for all f in the domain of R^*, then $RF = 0$ and since R is injective, $F = 0$.

Moreover, $R^* = B\sqrt{RR^*}$ where B is a linear isometry of the closure of the range of R onto the closure of the range of R^*. Since the range of R is dense in $L^2(\mathbb{R}^n)$ and the range of R^* is dense in $\mathcal{F}(\mathbb{C}^n)$, we see B gives a unitary mapping of $L^2(\mathbb{R}^n)$ onto $\mathcal{F}(\mathbb{C}^n)$.

Since $RR^*f = (2\pi)^{n/2} h_{2\pi} * f$ and by Example 3.11, $h_\pi * h_\pi = h_{2\pi}$, we see if $Tf = (2\pi)^{n/4} h_\pi * f$, then

$$\begin{aligned} T^2 f &= (2\pi)^{n/2} h_\pi * h_\pi * f \\ &= (2\pi)^{n/2} h_{2\pi} * f \\ &= RR^* f. \end{aligned}$$

Also note T is a positive operator for one can check

$$(h_\pi * f, f)_2 = (h_{\pi/2} * h_{\pi/2} * f, f)_2 = (h_{\pi/2} * f, h_{\pi/2} * f)_2 \geqslant 0.$$

There is another consequence of this. Note T is a bounded operator on L^2 and thus RR^* is bounded. This implies R^* is bounded and so R is actually bounded. A direct proof of this seems to be difficult. Anyway, this gives (k). Indeed

$$\begin{aligned} \sqrt{RR^*} f(x) &= Tf(x) \\ &= (2\pi)^{n/4} h_\pi * f(x) \\ &= (2\pi)^{n/4} \pi^{-n/2} \int f(y) e^{-|x-y|^2}\, dy \\ &= \left(\frac{2}{\pi}\right)^{n/4} f * L(x) \end{aligned}$$

where $L(y) = e^{-|y|^2} = H(\sqrt{2}y)$.

Finally for (l), we have from $BT = R^*$ that $RBT = RR^* = T^2$. This implies $RB = T$ on the range of T. So $RB = T$ on this dense subspace. Now for f in the range of T, one has

$$\begin{aligned} e^{-|x|^2/2} Bf(x) &= Tf(x) \\ &= \left(\frac{2}{\pi}\right)^{n/4} \int f(y) e^{-|x-y|^2}\, dy \\ &= \left(\frac{2}{\pi}\right)^{n/4} \int f(y) e^{-(x-y)\cdot(x-y)}\, dy\,. \end{aligned}$$

So

$$Bf(x) = \left(\frac{2}{\pi}\right)^{n/4} e^{x\cdot x/2} \int f(y)e^{-(x-y)\cdot(x-y)}\,dy.$$

But Bf is holomorphic and this function on $\mathbb{R}^n$ has one holomorphic extension. Namely

$$\begin{aligned} Bf(z) &= \left(\frac{2}{\pi}\right)^{n/4} e^{z\cdot z/2} \int f(y)e^{-(z-y)\cdot(z-y)}\,dy \\ &= \left(\frac{2}{\pi}\right)^{n/4} \int f(y)e^{-z\cdot z/2}e^{2z\cdot y}e^{-y\cdot y}\,dy \\ &= \left(\frac{2}{\pi}\right)^{n/4} \int f(y)e^{-y\cdot y+2z\cdot y-\frac{1}{2}z\cdot z}\,dy. \end{aligned}$$

We leave the remaining details; namely this works for all $f \in L^2$.

□

Exercise Set 4.2

1. Use the Fourier transform to solve the Cauchy problem

$$\frac{\partial u}{\partial x}(x,t) = \frac{\partial u}{\partial t}(x,t), \quad u(x,0) = f(x)$$

for some $f \in \mathcal{S}(\mathbb{R}^n)$.

Solution:

We have $u_x(x,t) = u_t(x,t)$ with initial condition $u(x,0) = f(x)$ where f is Schwartz. We use the Fourier transform $\hat{u}(x,t) = \mathcal{F}_1 u(\omega,t) = \int u(x,t)e^{-2\pi i x\omega}\,dx$ We obtain $(2\pi i\omega)\hat{u}(\omega,t) = \hat{u}_t(\omega,t)$ with $\hat{u}(\omega,0) = \hat{f}(\omega)$. This implies $\hat{u}(\omega,t) = \hat{u}(\omega,0)e^{2\pi i\omega t}$. So $\hat{u}(\omega,t) = \hat{f}(\omega)e^{2\pi i\omega t}$. Taking the inverse Fourier transform gives

$$u(x,t) = f(x+t)\,.$$

□

Exercise Set 4.3

1. Show if Ω is a nonempty open subset of $\mathbb{R}^n$ and K is a compact subset of Ω, then there is an open set V with compact closure with $K \subseteq V \subseteq \bar{V} \subseteq \Omega$.

Solution: For each point $p \in K$, choose $r_p > 0$ so that $B_{2r_p}(p) \subseteq \Omega$. The set $B_{r_p}(p)$ for $p \in K$ cover K. Choose a finite set $F \subseteq K$ so that $B_{r_p}(p)$, $p \in F$, cover K. Set $V = \cup_{p\in F} B_{r_p}(p)$. Then V is open and $\bar{V} \subseteq \cup_{p\in F} B_{2r_p}(p) \subseteq \Omega$. □

2. Let Ω be a nonempty open subset of $\mathbb{R}^n$. Show the inclusion mapping from $\mathcal{D}(\Omega)$ into $\mathcal{D}(\mathbb{R}^n)$ is continuous and thus the restriction of every distribution to $\mathcal{D}(\Omega)$ is a distribution on Ω.

Solution: This was already shown in Exercise 2.4.8. □

3. Let Ω be a nonempty subset of $\mathbb{R}^n$ and suppose $f : \Omega \to \mathbb{C}$ is a continuous function. Show the support of f as a measurable function is $\overline{\{x \in \Omega \mid f(x) \neq 0\}} \cap \Omega$.

Solution:

Note the support of f as a measurable function or equivalently as a distribution is the complement in Ω of the union of all open subsets V of Ω such that $f = 0$ a.e. on V. But since f is continuous we know if $f(x) \neq 0$ at one point x in V, then f is not 0 on a subset of V of positive measure. So $f = 0$ a.e. on an open subset V of Ω if and only if $f = 0$ everywhere on V. Thus the support of f in Ω is the complement of the union of all open subsets V of Ω on which f vanishes. Thus $y \in \Omega$ is in the support if and only if for every neighborhood V of y, $f|_{V \cap \Omega}$ is not identically 0. That is V meets $\{x \in \Omega \mid f(x) \neq 0\}$. This is equivalent to $y \in \overline{\{x \in \Omega \mid f(x) \neq 0\}} \cap \Omega$.

□

4. Let V be an open subset of $\mathbb{R}^n$ and suppose $f \in L_{\text{loc}}(V)$. Show $\int \phi f \, dx = 0$ for all $\phi \in C_c^\infty(V)$ if and only if $f = 0$ a.e. on V.

Solution: The inverse direction is clear. Assume $\int_V \phi(x) f(x) \, dx = 0$ for all such ϕ. By Lemma 2.53 for any closed rectangle $R = \prod_{i=1}^n [a_i, b_i] \subseteq V$, one can construct a $\phi \in C_c^\infty(\Omega)$ with support inside V such that $0 \leqslant \phi \leqslant 1$ and $\phi(x) = 1$ if and only if $x \in R$. Thus $\phi^k \to \chi_R$ and $|\phi^k f| \leqslant f\chi_{\operatorname{supp}\phi} \in L^1(V)$. Hence the Lebesgue Dominated Convergence Theorem implies $\int_R f \, dx = 0$ for all closed rectangles $R \subseteq V$. This implies $\int_{R'} \chi_R f \, dx = 0$ for all rectangles R'. But then this implies $\chi_R f = 0$ a.e. on V. Now V is a countable union of such rectangles. So $f = 0$ a.e. on V. □

5. Let μ be a Borel measure on Ω, a nonempty open subset of $\mathbb{R}^n$. Then the support of the measure μ (denoted by $\operatorname{supp}\mu$) is the complement of the union of all open subsets of Ω which have measure 0. Show if $\mu(K) < \infty$ for all compact subsets K of Ω, then the distribution T_μ on Ω has compact support if and only if $\operatorname{supp}\mu$ is a compact subset of Ω.

Solution:

Note the support of T_μ is the complement in Ω of all the open subsets U of Ω such that T_μ vanishes on U. But T_μ vanishes on U if and only if for every $f \in C^\infty$ with compact support a subset of U, $T_\mu f = \int f \, d\mu = 0$. Thus by Proposition 4.30 T_μ has compact support a subset of Ω if and only if there is a compact subset K of Ω such that T_μ vanishes on $V = \Omega - K = \Omega \cap K^c$. This occurs if and only if there is an open subset V of Ω whose complement in Ω is compact such that T_μ on V is 0. But T_μ vanishes on V if and only if $\mu(V) = 0$ for if T_μ on V is 0 and $\mu(V) \neq 0$, then there is a compact subset C of V with $0 < \mu(C) < \infty$. By Corollary 2.54, there is a C^∞ function f with $0 \leqslant f \leqslant 1$ with compact support in V and such that $f = 1$ on C. Consequently, $T_\mu f \geqslant \int_C f \, d\mu = \mu(C) > 0$. Thus T_μ has compact support inside Ω if and only if there is an open V of Ω with measure 0 whose complement in V is compact; that is if and only if the union of all open subsets of Ω with μ measure 0 has a compact complement in Ω.

□

6. Let μ be a measure on an open subset Ω of $\mathbb{R}^n$. Show $\operatorname{supp}\mu = \{x \in \Omega \mid \mu(N) > 0 \text{ for every neighborhood } N \text{ of } x\}$.

Solution:

Let W be the support of μ. Then W is the complement in Ω of all open subsets of V of Ω with $\mu(V) = 0$. Thus $x \in W$ if and only if no open subset of Ω containing x has μ measure equal to 0 if and only if $\mu(N) > 0$ for any open neighborhood of x contained in Ω.

□

7. Let μ be a complex Borel measure on a nonempty open set Ω of $\mathbb{R}^n$. Define T_μ by

$$T_\mu(\phi) = \int \phi \, d\mu$$

for $\phi \in \mathcal{D}(\Omega)$. The support of the measure μ is the complement of the union of all open subsets U of Ω such that $\mu(E) = 0$ for all Borel subsets E of U. Show the support of the distribution T_μ is the support of the measure μ.

Solution:

We need to show T_μ vanishes on an open subset U of Ω if and only if $\mu(E) = 0$ for all measurable subsets E of U. This is equivalent to T_μ is zero on U if and only if $|\mu|(U) = 0$. But since $|\int_U f \, d\mu| \leqslant \int_U |f| \, d|\mu|$, one direction is easy. So assume that $\int_U f \, d\mu = 0$ for all f which are C^∞ and have compact support inside U. Since $|\mu|$ and the measures in μ's decomposition $\mu = \mu_1 + i\mu_2 = \mu_1^+ - \mu_1^- + i(\mu_2^+ - \mu_2^-)$ where μ_1 and μ_2 and their Jordan decomposition are all finite measures, we see $\int f \, d\mu_1 = \int f \, d\mu_2 = 0$ for all real valued C^∞ functions with compact support in U. Consequently, by finiteness of the measures and regularity, $\int_U f \, d\mu_j = 0$ for all positive bounded measurable functions f when $j = 1$ or $j = 2$. Since the measures μ_j^+ and μ_j^- are disjoint, we then obtain $\int_U f \, d\mu_j^+ = \int_U f d\mu_j^- = 0$ for all such f and $j = 1, 2$. This implies $\mu_j^+(U) = \mu_j^-(U) = 0$ for $j = 1, 2$. So $|\mu|(U) = 0$.

□

8. Prove Remark 4.34.

Solution:

To finish one needs to show if $f \in L_{\text{loc}}(\Omega)$ and $\int_V f\phi \, dx = 0$ for all $\phi \in C^\infty$ with compact support in V, then $f = 0$ a.e. on V. Indeed, if $K \subseteq V$ is compact, one can find a sequence V_k of open subsets of V with $K = \cap V_k$. For example take $V_k = \{y \in V \mid |y - x| < \frac{1}{k} \text{ for some } x \in K\}$. By Corollary 2.54, we can find a $\phi_k \in C^\infty$ with $0 \leqslant \phi_k \leqslant 1$, $\phi_k(x) = 1$ for $x \in K$ and ϕ_k has compact support in V_k. Set $\Phi_k = \prod_{j=1}^k \phi_k \in C^\infty$. Then $\Phi_k \to \chi_K$ pointwise, $0 \leqslant \Phi_k \leqslant 1$ and $\operatorname{supp}\Phi_k \subseteq V_k \subseteq V$. The Lebesgue Dominated Convergence Theorem implies $\int_K f(x) \, dx = 0$ for all compact subsets K ov V. By inner regularity, this then implies $f(x) = 0$ a.e. on V.

□

9. Show $\operatorname{supp}(\delta_x) = \{x\}$.

Solution:

We need to show δ_x vanishes on an open set V if and only if $x \notin V$. It is clear if $x \notin V$ and ϕ is C^∞ with compact support a subset of V, then $\delta_x(\phi) = 0$. So δ_x vanishes on V if $x \notin V$. Suppose $x \in V$. By Corollary 2.54, there is a $\phi \in C_c^\infty(V)$ with $\phi(x) = 1$. Therefore $\delta_x(\phi) = \phi(x) = 1 \neq 0$. So δ_x does not vanish on V.

□

10. Let p be a real valued nonnegative polynomial on $\mathbb{R}^n$. For $z \in \mathbb{C}$ with $\mathrm{Re}(z) \geqslant 0$ define $P^z : \mathcal{D}(\mathbb{R}^n) \to \mathbb{C}$ by

$$P^z(\varphi) = \int_{\mathbb{R}^n} \varphi(x)p(x)^z \, dx.$$

Show P^z is a distribution, and that

$$\{z \in \mathbb{C} \mid \mathrm{Re}(z) > 0\} \ni z \to P^z(\varphi) \in \mathbb{C}$$

is a holomorphic function in z for all $\varphi \in \mathcal{D}(\mathbb{R}^n)$.
Solution: It easy to see it is a distribution. Now note for $r \geqslant 0$, $z \mapsto r^z$ is holomorphic for $\mathrm{Re} z > 0$. Thus if γ is a simple smooth closed curve lying in the half plane $\mathrm{Re}(z) > 0$, using Fubini one has

$$\begin{aligned}
\int_\gamma P^z(\phi)\, dz &= \int_\gamma \int_{\mathbb{R}^n} \phi(x)p(x)^z \, dx\, dz \\
&= \int_{\mathbb{R}^n} \phi(x) \int_\gamma p(x)^z \, dz\, dx \\
&= 0\,.
\end{aligned}$$

Consequently $z \mapsto P^z(\phi)$ is holomorphic on the half plane $\mathrm{Re}\, z > 0$.

□

Solution: Remark : Bernstein's Result
Let $p(x) > 0$ on $\mathbb{R}^n$. Then there is a differential operator D such that $D(x)p(x)^z = \beta(z)p(x)^{z-1}$ where $\beta(z)$ is a polynomial. This allows one to analytically continue the distribution P^z where $P^z(\phi) = \int p(x)^z \phi(x)\, dx$. In fact, $P^{z-1}(\phi) = \int p(x)^{z-1}\phi(x)\, dx = \int \frac{1}{\beta(z)} Dp(x)^z\, \phi(x)\, dx = \frac{1}{\beta(z)} \int p(x)^z D^*\phi(x)\, dx$. So

$$P^{z-1}(\phi) = \frac{1}{\beta(z)} P^z(\phi)$$

and the zeros of $\beta(z)$ introduce the poles in the analytic continuation. □

11. Suppose that $n = 1$ and let $p(x) = x^2$. Show the following:

(a) $\left(\frac{d}{dx}\right)^2 p(x)^z = 2z(2z-1)p(x)^{z-1}$.

(b) If $\mathrm{Re}(z) > 0$, then

$$P^z(\phi) = \frac{2^{-2}}{(z+1)(z+1/2)} \int \phi''(x)p^{z+1}(x)\, dx\,.$$

(c) The distribution P^z can be extended to $\{z \in \mathbb{C}\backslash\{1/2\} \mid \mathrm{Re}(z) > -1\}$ such that $P^z(f)$ is meromorphic by defining

$$P^z(\varphi) = \frac{2^{-2}}{(z+1)(z+1/2)} \int \varphi''(x) p^{z+1}(x)\, dx\,.$$

(d) P^z can be extended to all $z \in \mathbb{C}$ where $z \notin -\frac{1}{2}\mathbb{N}$.

(e) If $\varphi \in \mathcal{D}(\mathbb{R})$, then

$$\mathbb{C} \ni z \mapsto \frac{1}{\Gamma(2z+1)} P^z(\varphi) \in \mathbb{C}$$

is holomorphic.

Solution:

We note $\frac{d}{dx}(x^2)^z = \frac{d}{dx}\left(e^{z\ln x^2}\right) = e^{z\ln x^2}\frac{z}{x^2}(2x) = \frac{2z}{x}x^{2z} = 2zx^{2z-1} = 2zx^{2z-1}$. Repeating we see $\frac{d^2}{dx^2}x^{2z} = 2z(2z-1)x^{2z-2}$. This gives (a).

We have $p(x) = x^2$. We note

$$\begin{aligned}
P^z(\phi) &= \int_{\mathbb{R}} p(x)^z \phi(x)\, dx \\
&= \int x^{2z}\phi(x)\, dx \\
&= \frac{x^{2z+1}}{2z+1}\phi(x)|_{-\infty}^{\infty} - \int \frac{x^{2z+1}}{2z+1}\phi'(x)\, dx \\
&= -\frac{x^{2z+2}}{(2z+2)(2z+1)}\phi'(x)|_{-\infty}^{\infty} + \int \frac{x^{2z+2}}{(2z+2)(2z+1)}\phi''(x)\, dx \\
&= \frac{1}{(2z+2)(2z+1)} P^{z+1}(\phi'') \\
&= 2^{-2}\frac{1}{(z+1)(z+\frac{1}{2})} P^{z+1}(\phi'')
\end{aligned}$$

if $\phi \in \mathcal{D}(\mathbb{R})$ and $\mathrm{Re}\, z > 0$. This is (b).

For (c), define P^{z-1} for $0 < \mathrm{Re}\, z \leqslant 1$, $z \neq \frac{1}{2}$ by

$$P^{z-1}(\phi) = 2^{-2}\frac{1}{z(z-\frac{1}{2})} P^z(\phi'').$$

Repeat inductively to extend P^z for all z where $z \notin -\frac{1}{2}\mathbb{N}$, namely if P^{z-k} is defined for $\mathrm{Re}\, z > 0$, define P^{z-k-1} for $0 < \mathrm{Re}\, z \leqslant 1$ by

$$P^{z-k-1}(\phi) = \frac{2^{-2}}{(z-k)(z-k-\frac{1}{2})} P^{z-k}(\phi'')\,.$$

Note

$$P^{z-k}(\phi) = \frac{2^{-2}}{(z+1-k)(z+\frac{1}{2}-k)} P^{z+1-k}(\phi'')$$
$$= \frac{2^{-4}}{(z+1-k)(z+1-\frac{1}{2}-k)(z+2-k)(z+2-\frac{1}{2}-k)} P^{z+2-k}(\phi^{(4)})$$
$$= \frac{2^{-6}}{(z+1-k)(z+1-\frac{1}{2}-k)(z+2-k)(z+2-\frac{1}{2}-k)(z+3-k)(z+3-\frac{1}{2}-k)} \times P^{z+3-k}(\phi^{(6)})$$
$$= \cdots = \frac{2^{-2k}}{\prod_{j=1}^{k}(z+j-k)(z+j-\frac{1}{2}-k)} P^{z}(\phi^{(2k)})$$

or

$$P^{z}(\phi) = \frac{2^{-2k}}{\prod_{j=1}^{k}(z+j)(z+j-\frac{1}{2})} P^{z+k}(\phi^{(2k)}).$$

Note we already know $z \mapsto P^z(\phi)$ is holomorphic on $\operatorname{Re} z > 0$. So note if $-k < \operatorname{Re} z < -k+2$ where $k = 1, 2, 3 \ldots$, then

$$P^{z}(\phi) = \frac{2^{-2k-2}}{\prod_{j=1}^{k+1}(z+j)(z+j-\frac{1}{2})} P^{z+k+1}(\phi^{(2k+2)})$$

and $z+k+1$ has real part positive. Thus $P^z(\phi)$ is meromorphic on this band can have simple poles only where the denominator is zero; that is if $z = -k+\frac{1}{2}$, $z = -k+1$, $z = -k+3/2$. Moreover, $\Gamma(2z+1)$ has simple poles at $z = -\frac{1}{2}, -1, -3/2, \cdots$ and so $z \mapsto \frac{P^z(\phi)}{\Gamma(2z+1)}$ has no poles and thus is holomorphic on $-k < \operatorname{Re} z < -k+2$. Thus $z \mapsto \frac{1}{\Gamma(2z+1)} P^z(\phi)$ is holomorphic on $\mathbb{C}$. This completes (d) and (e).

□

12. Let $p(x) = |x|^2$. Find a differential operator D on $\mathbb{R}^n$ and a polynomial $\beta(z)$ such that

$$Dp(x)^z = \beta(z)p(x)^{z-1}.$$

Use this to extend the distribution P^z to almost all $z \in \mathbb{C}$.

Solution:

First note if $p(x) = |x|^2$, then

$$\frac{\partial^2}{\partial x_j \partial x_k}(\sum x_l^2)^z = \frac{\partial}{\partial x_j}((z(|x|^2)^{z-1}(2x_k))$$
$$= z(z-1)(|x|^2)^{z-2}(2x_j)(2x_k) + \delta_{j,k} 2z(|x|^2)^{z-1}$$
$$= 4z(z-1)x_j x_k p(x)^{z-2} + 2z\delta_{j,k} p(x)^{z-1}$$

Thus

$$\partial_j^2 p(x)^z = 4z(z-1)x_j^2 p(x)^{z-2} + 2zp(x)^{z-1}$$

and if $j \neq k$ then

$$\partial_j \partial_k p(x)^z = 4z(z-1)x_j x_k p(x)^{z-2}.$$

Let

$$D = \sum_{j=1}^{n} \partial_j^2 = \Delta.$$

Then

$$Dp(x)^z = 4z(z-1)p(x)^{z-1} + 2nzp(x)^{z-1} = (4z^2 + 2nz - 4z)p(x)^{z-1}.$$

We then have

$$Dp(x)^z = 2z(2z + (n-2))p(x)^{z-1}.$$

Thus the Bernstein polynomial is $\beta(z) = 2z(2z + (n-2))$. Now note that the transpose $D^{\#}$ of D by Equation 4.15 (see the next section) is given by

$$D^{\#} = D.$$

Thus

$$\begin{aligned} P^z(\phi) &= \int \frac{2^{-2}}{(z+1)(z+n/2)} D(p(x)^{z+1})\phi(x)\,dx \\ &= \frac{2^{-2}}{(z+1)(z+n/2)} \int p(x)^{z+1} D\phi(x)\,dx \\ &= \frac{2^{-2}}{(z+1)(z+n/2)} P^{z+1}(D\phi) \end{aligned}$$

Repeating this inductively and using it to define P^z for Re $z > -k$ for $k = -1, -2, \ldots$ we see as in Exercise 4.3.11 that for $-k < \operatorname{Re} z$ one has

$$P^z(\phi) = \frac{2^{-2k}}{\prod_{j=1}^{k}(z+j)(z+j-1+n/2)} P^{z+k}(D^k\phi).$$

Note for $\operatorname{Re} z > -k$, one has poles at $z = -1, -2, \ldots, -k+2, -k+1$ due to the factor $(z+j)$ while for the factor $(z+j-1+n/2)$, one has poles $z = -\frac{n}{2}, \frac{-n}{2} - 1, \ldots, -m+1-\frac{n}{2}$ for the largest m with $-m+1-\frac{n}{2} > -k$. Doing this for all k gives P^z is meromorphic on $\mathbb{C}$ with poles at $-k$ and at $-\frac{n}{2} - k + 1$ for $k = 1, 2, 3, \ldots$. We note if n is odd, all poles are simple while if n is even, one has simple poles at $-1, -2, \ldots, -n/2+1$ and double poles at $-n/2, -n/2-1, -n/2-2, \ldots$

□

13. Define a function $x_+ : \mathbb{R} \to [0, \infty]$ by $x_+ = x$ if $x \geqslant 0$ and $x_+ = 0$ if $x \leqslant 0$. Thus $x_+ = xH(x)$ where $H(x)$ is the Heaviside function. The function x_+^z is locally integrable for $\operatorname{Re}(z) > -1$ and hence defines a distribution which we denote by the same letter:

$$\langle x_+^z, \varphi \rangle = \int_0^\infty x^z \varphi(x)\,dx.$$

Show $z \mapsto x_+^z$ extends to $\{z \in \mathbb{C} \mid z \neq -1, -2, -3, \ldots\} = \mathbb{C} \backslash -\mathbb{N}$ such that

$$z \mapsto \langle x_+^z, \varphi \rangle$$

is holomorphic on the above set and

$$z \mapsto \frac{1}{\Gamma(z+1)} \langle x^z, \varphi \rangle$$

is holomorphic on $\mathbb{C}$.

Solution:

We do this problem in the same way as the previous two exercises. Note the distribution x_+^z is defined for $\operatorname{Re} z > -1$ and is holomorphic there. We shall start with $\operatorname{Re} z > 0$. Since $\frac{d}{dx} x^z = z x^{z-1}$ for $x > 0$ one has

$$\begin{aligned} \int_0^\infty z x^{z-1} \phi(x)\, dx &= x^z \phi(x)|_0^\infty - \int_0^\infty x^z \phi'(x)\, dx \\ &= 0 - x_+^z(\phi') . \end{aligned}$$

So $x_+^{z-1}(\phi) = -\frac{1}{z} x^z(\phi')$ or $x_+^z(\phi) = -\frac{1}{z+1} x_+^{z+1}(\phi')$ for $\operatorname{Re} z > -1$. Assume x_+^z is defined for $\operatorname{Re} z > -m$. Then for $\operatorname{Re} z > -m-1$ define

$$\begin{aligned} x_+^z(\phi) &= -\frac{1}{z+1} x_+^{z+1}(\phi') \\ &= \frac{1}{(z+1)(z+2)} x_+^{z+2}(\phi'') \\ &= \cdots \\ &= \frac{(-1)^{m+1}}{\prod_{k=1}^{m+1} (z+k)} x_+^{z+m+1}(\phi^{(m+1)}) . \end{aligned}$$

One obtains a meromorphic function with simple poles at $-1, -2, \ldots$ and since $\Gamma(z+1)$ has simple poles at $z = -1, -2, \ldots$ and no zeros, one sees $z \mapsto \frac{1}{\Gamma(z+1)} x_+^z(\phi)$ is holomorphic on $\mathbb{C}$.

□

14. Let $F : \mathbb{R}^n \to \mathbb{R}^n$ be linear with $\det F \neq 0$. If S is a distribution define $S \circ F$ by

$$S \circ F(\varphi) := |\det F|^{-1} \int \varphi(F^{-1}(x))\, dS(x) .$$

Show $S \circ F$ is a distribution. If $S = f$ is a locally integrable function, verify $S \circ F = f \circ F$.

Solution:

We first show the mapping $\phi \mapsto \phi \circ F^{-1}$ is a linear homeomorphism of $\mathcal{D}(\mathbb{R}^n)$. To do this it suffices to show $\phi \mapsto \phi \circ A$ is a continuous linear transformation of $\mathcal{D}(\mathbb{R}^n)$ for each invertible linear transformation A. First note since A is a homeomorphism of $\mathbb{R}^n$, $\phi \circ A$ has compact support. Also by Exercise 2.3.7, $\phi \circ A$ is C^∞ and

$$D^\alpha(\phi \circ A) = \sum_{|\beta| \leqslant |\alpha|} c_\beta (D^\beta \phi) \circ A$$

for some real constants c_β. Thus if K is a compact subset of $\mathbb{R}^n$ and $\phi \in \mathcal{D}_K(\mathbb{R}^n)$, then $\phi \circ A \in \mathcal{D}_{A^{-1}K}(\Omega)$ and

$$\begin{aligned}|\phi \circ A|_{A^{-1}K,\alpha} &= \max_{x \in A^{-1}K} |D^\alpha(\phi \circ A)(x)| \\ &\leqslant \sum_{|\beta| \leqslant |\alpha|} |c_\beta| \max_{x \in A^{-1}K} |(D^\beta \phi)(Ax)| \\ &= \sum_{|\beta| \leqslant |\alpha|} |c_\beta|\, |\phi|_{K,\beta}\,.\end{aligned}$$

This implies the mapping $\phi \mapsto \phi \circ A$ is a continuous linear mapping of $\mathcal{D}_K(\mathbb{R}^n)$ into $\mathcal{D}_{A^{-1}K}(\mathbb{R}^n)$. Since by Lemma 2.66, $\mathcal{D}_{A^{-1}K}(\mathbb{R}^n)$ has the relative topology of $\mathcal{D}(\mathbb{R}^n)$, the mapping $\phi \mapsto \phi \circ A$ is continuous from $\mathcal{D}_K(\mathbb{R}^n)$ into $\mathcal{D}(\mathbb{R}^n)$. Finally by Proposition 2.15, $\phi \mapsto \phi \circ A$ is continuous from $\mathcal{D}(\mathbb{R}^n)$ into $\mathcal{D}(\mathbb{R}^n)$. This implies $S \circ F$ where $S \circ F(\phi) = |\det F|^{-1} S(\phi \circ F^{-1})$ is a distribution. Moreover if $S = f$ where f is locally integrable, then by Corollary 2.23,

$$\begin{aligned}f \circ F(\phi) &= |\det F|^{-1} f(\phi \circ F^{-1}) \\ &= |\det F|^{-1} \int f(y)\phi(F^{-1}y)\, dy \\ &= \int f(Fx)\phi(x)\, dx \\ &= f \circ F(\phi)\,.\end{aligned}$$

This completes the result.

□

15. (Definition of the Principal Value) A function $f : \mathbb{R}^n \setminus \{0\} \to \mathbb{C}$ is called **homogeneous of degree** μ if for all $\lambda > 0$

$$f(\lambda x) = \lambda^\mu f(x)\,.$$

Similarly a distribution T is called **homogeneous of degree** μ if for all test functions φ we have

$$\int \lambda^{-n} \varphi(\lambda^{-1}x)\, dS(x) = \lambda^\mu \int \varphi(x)\, dS(x).$$

(a) Show the δ-distribution is homogeneous and find its degree.

(b) Assume $f \in \mathcal{C}(\mathbb{R}^n \setminus \{0\})$ is homogeneous of degree $-n$ and $\int_{S^{n-1}} f(\omega)\, d\sigma(\omega) = 0$. Define $\mathrm{PV}(f) : \mathcal{D}(\mathbb{R}^n) \to \mathbb{C}$ by

$$\langle \mathrm{PV}(f), \varphi \rangle := \lim_{\epsilon \to 0+} \int_{\mathbb{R}^n \setminus B_\epsilon(0)} f(x)\varphi(x)\, dx\,.$$

Show $\mathrm{PV}(f)$ is a distribution. This distribution is called the **principal value** of f.

(c) Show for any $\epsilon > 0$

$$\langle \mathrm{PV}(f), \varphi \rangle = \int_{|x|\leqslant\epsilon} f(x)(\varphi(x) - \varphi(0))\, dx + \int_{|x|\geqslant\epsilon} f(x)\varphi(x)\, dx \,.$$

(d) Show $\mathrm{PV}(f)$ is homogeneous of degree $-n$.

Solution: We show δ is homogeneous of degree $-n$; indeed

$$\int \lambda^{-n}\phi(\lambda^{-1}x)\, d\delta(x) = \lambda^{-n}\phi(0) = \lambda^{-n}\int \phi(x)\, d\delta(x)\,.$$

This gives (a).

For (b), let $f : \mathbb{R}^n - \{0\} \to \mathbb{C}$ be continuous and satisfy $f(\lambda x) = \lambda^{-n} f(x)$ for $\lambda > 0$ and $\int_{S^{n-1}} f(u)\, d\sigma(u) = 0$.

Then

$$\mathrm{PV}(f)(\phi) = \lim_{\epsilon\to 0+}\int_{|x|\geqslant\epsilon} f(x)\phi(x)\, dx$$

defines a distribution on $\mathbb{R}^n$. It is the principal value of f.

To see it is distribution, we need only show it exists. (See Theorem 4.27.) So it suffices to show $\int_{|x|\geqslant\epsilon} f(x)\phi(x)\, dx - \int_{|x|\geqslant\delta} f(x)\phi(x)\, dx \to 0$ as $\epsilon > \delta$ and $\epsilon \to 0+$ and $\delta \to 0+$. So we show $\lim_{\epsilon\to 0+}\int_{\delta\leqslant|x|\leqslant\epsilon} f(x)\phi(x)\, dx = 0$.

We have

$$\begin{aligned}
\int_{\delta\leqslant|x|\leqslant\epsilon} f(x)\phi(x)\, dx &= \int_\delta^\epsilon \int_{S^{n-1}} f(ru)r^{n-1}\phi(ru)\, d\sigma(u)\, dr \\
&= \int_\delta^\epsilon \int_{S^{n-1}} r^{-n} f(u) r^{n-1}\, \phi(ru) d\sigma(u)\, dr \\
&= \int_\delta^\epsilon \int_{S^{n-1}} r^{-1} f(u)\, \phi(ru) d\sigma(u)\, dr \\
&= \int_\delta^\epsilon \int_{S^{n-1}} r^{-1} f(u)\, (\phi(ru) - \phi(0)) d\sigma(u)\, dr \\
&\quad + \int_\delta^\epsilon \int_{S^{n-1}} r^{-1} f(u)\, \phi(0) d\sigma(u)\, dr \\
&= \int_\delta^\epsilon \int_{S^{n-1}} r^{-1} f(u)\, (\phi(ru) - \phi(0)) d\sigma(u)\, dr + 0.
\end{aligned}$$

Now $\phi(ru) - \phi(0) = ru \cdot \nabla\phi(\xi)$ where ξ depends or r and u and lies on the interior of the line segment from 0 to ru.

Thus

$$\begin{aligned}
\left|\int_{\delta\leqslant|x|\leqslant\epsilon} f(x)\phi(x)\, dx\right| &= \left|\int_\delta^\epsilon \int_{S^{n-1}} r^{-1} f(u)\, (ru \cdot \nabla\phi(\xi)) d\sigma(u)\, dr\right| \\
&\leqslant \int_\delta^\epsilon \int_{S^{n-1}} |f(u)\, (u \cdot \nabla\phi(\xi(r,u)))|\, d\sigma(u)\, dr \\
&\leqslant M \int_0^\epsilon \int_{S^{n-1}} d\sigma(u)\, dr \to 0
\end{aligned}$$

as $\epsilon \to 0+$ and $0 < \delta < \epsilon$ where here M is the maximum of $|f(u)|\,|\nabla\phi(y)|$ where $|u| = 1$ and $|y| \leqslant 1$. Thus the principal value exists. We next determine a formula given in (c) for this principal value. The above implies

$$\begin{aligned}
\int_{\delta\leqslant|x|\leqslant\epsilon} f(x)\phi(x)\,dx &= \int_\delta^\epsilon \int_{S^{n-1}} r^{-1} f(u)\,(\phi(ru) - \phi(0))d\sigma(u)\,dr \\
&= \int_\delta^\epsilon \int_{S^{n-1}} r^{n-1} f(ru)\,(\phi(ru) - \phi(0))d\sigma(u)\,dr \\
&= \int_{\delta\leqslant|x|\leqslant\epsilon} f(x)(\phi(x) - \phi(0))\,dx.
\end{aligned}$$

Thus

$$\begin{aligned}
\mathrm{PV}(f)(\phi) &= \lim_{\delta\to 0+} \int_{|x|\geqslant\delta} f(x)\phi(x)\,dx \\
&= \lim_{\delta\to 0+} \int_{\epsilon\geqslant|x|\geqslant\delta} f(x)\phi(x)\,dx + \int_{\epsilon\leqslant|x|} f(x)\phi(x)\,dx \\
&= \int_{|x|\leqslant\epsilon} f(x)(\phi(x) - \phi(0))\,dx + \int_{|x|\geqslant\epsilon} f(x)\phi(x)\,dx.
\end{aligned}$$

For (d), it suffices to note

$$\begin{aligned}
\int_{|x|\geqslant\epsilon} \lambda^{-n} f(x)\phi(\lambda^{-1}x)\,dx &= \int_{|\lambda^{-1}x|\geqslant\epsilon} \lambda^{-n} f(\lambda x)\phi(x)\,d(\lambda x) \\
&= \lambda^{-n} \int_{|x|\geqslant\lambda\epsilon} f(x)\phi(x)\,dx
\end{aligned}$$

and thus

$$\mathrm{PV}(f)(x \mapsto \lambda^{-n}\phi(\lambda^{-1}x)) = \lim_{\epsilon\to 0+} \lambda^{-n} \int_{|x|\geqslant\lambda\epsilon} f(x)\phi(x)\,dx = \lambda^{-n}\,\mathrm{PV}(f)(\phi).$$

□

16. Define distributions

$$\frac{1}{x \pm i0} = \lim_{y\to 0\pm} \frac{1}{x + iy}$$

where the limit is taken in $\mathcal{D}'(\mathbb{R})$. Show

$$\delta_0 = \frac{1}{2\pi i}\left(\frac{-1}{x + i0} + \frac{1}{x - i0}\right).$$

Thus δ_0 is a *difference of boundary values* of a holomorphic function in the upper half plane and a holomorphic function in the lower half plane.

Solution: Let $\phi \in \mathcal{D}(\mathbb{R})$. Let $\delta > 0$. Then

$$\begin{aligned}\left(\frac{1}{x+i0}\right)(\phi) &= \lim_{y\to 0+}\int \frac{1}{x+iy}\phi(x)\,dx \\ &= \lim_{y\to 0+}\int_{|x|\leqslant\delta}\frac{1}{x+iy}\phi(x)\,dx + \int_{|x|\geqslant\delta}\frac{1}{x}\phi(x)\,dx \\ &= \lim_{y\to 0+}\left(\int_{|x|\leqslant\delta}\frac{\phi(x)-\phi(0)}{x+iy}\,dx + \int_{-\delta}^{\delta}\frac{\phi(0)}{x+iy}\,dx\right) + \int_{|x|\geqslant\delta}\frac{1}{x}\phi(x)\,dx.\end{aligned}$$

Now by Taylor's Formula

$$\phi(x) - \phi(0) = xh(x)$$

where $h \in C^\infty$. Thus

$$\left(\frac{1}{x+i0}\right)(\phi) = \int_{|x|\leqslant\delta}\frac{\phi(x)-\phi(0)}{x}\,dx + \int_{|x|\geqslant\delta}\frac{1}{x}\phi(x)\,dx + \lim_{y\to 0+}\int_{-\delta}^{\delta}\frac{\phi(0)}{x+iy}\,dx.$$

Now

$$\begin{aligned}\lim_{y\to 0+}\int_{-\delta}^{\delta}\frac{\phi(0)}{x+iy}\,dx &= \phi(0)\lim_{y\to 0+}\ln(x+iy)|_{-\delta}^{\delta} \\ &= \phi(0)\lim_{y\to 0+}\{\ln(\delta+iy)-\ln(-\delta+iy)\} \\ &= \lim_{y\to 0+}(\phi(0)\left((\arctan(\frac{y}{\delta})i - (\pi - \arctan(\frac{y}{\delta}))i\right) \\ &= \phi(0)\,(-\pi i)\end{aligned}$$

This gives

$$\begin{aligned}\left(\frac{1}{x+i0}\right)(\phi) &= \int_{|x|\leqslant\delta}\frac{\phi(x)-\phi(0)}{x}\,dx + \int_{|x|\geqslant\delta}\frac{1}{x}\phi(x)\,dx + \lim_{y\to 0+}\int_{-\delta}^{\delta}\frac{\phi(0)}{x+iy}\,dx \\ &= \int_{|x|\leqslant\delta}\frac{\phi(x)-\phi(0)}{x}\,dx + \int_{|x|\geqslant\delta}\frac{1}{x}\phi(x)\,dx + \phi(0)(-\pi i)\end{aligned}$$

Similarly,

$$\begin{aligned}\left(\frac{1}{x-i0}\right)(\phi) &= \int_{|x|\leqslant\delta}\frac{\phi(x)-\phi(0)}{x}\,dx + \int_{|x|\geqslant\delta}\frac{1}{x}\phi(x)\,dx + \lim_{y\to 0+}\int_{-\delta}^{\delta}\frac{\phi(0)}{x-iy}\,dx \\ &= \int_{|x|\leqslant\delta}\frac{\phi(x)-\phi(0)}{x}\,dx + \int_{|x|\geqslant\delta}\frac{1}{x}\phi(x)\,dx + \phi(0)(\pi i)\end{aligned}$$

Thus we see we have distributions and

$$\left(\frac{-1}{x+i0} + \frac{1}{x-i0}\right)\phi = 2\pi i\phi(0).$$

□

Exercise Set 4.4

1. Let $y \in \mathbb{R}^n$ and $a > 0$. Show $\lambda(y)$ and $\delta(a)$ are continuous linear homeomorphisms of $\mathcal{D}(\mathbb{R}^n)$. Also show if Ω is an open subset of $\mathbb{R}^n$ and $g \in \mathcal{E}(\Omega)$, then the map from $\mathcal{D}(\Omega)$ to $\mathcal{D}(\Omega)$ given by $\phi \mapsto g\phi$ is a continuous linear transformation.

Solution: Two of these have already been established. Namely, in Exercise 2.4.3 we have proven Proposition 2.69 and so $\lambda(y)$ and $\delta(a)$ are continuous.

Now let $g \in \mathcal{E}(\Omega)$. Set $T(\phi) = g\phi$ for $\phi \in \mathcal{D}(\Omega)$. T is clearly a linear mapping of $\mathcal{D}(\Omega)$ into $\mathcal{D}(\Omega)$.

To see it is continuous it suffices by Proposition 2.15 to show $T|\mathcal{D}_K(\Omega)$ is continuous. But T on this space maps into $\mathcal{D}_K(\Omega)$. Since by Lemma 2.66, the relative topology of $\mathcal{D}(\Omega)$ on $\mathcal{D}_K(\Omega)$ is the topology defined by the seminorms $|\cdot|_{K,\alpha}$, we need only show $T : \mathcal{D}_K(\Omega) \to \mathcal{D}_K(\Omega)$ is continuous. But

$$|T\phi|_{K,\alpha} = \max_{x\in K} |D^\alpha(g\phi)(x)|.$$

Now by Lemma 2.21,

$$\begin{aligned}|T\phi|_{K,\alpha} &= \max_{x\in K}\left|\sum_{\beta\leqslant\alpha}\binom{\alpha}{\beta}D^{\alpha-\beta}g(x)D^\beta\phi(x)\right| \\ &\leqslant \sum_{\beta\leqslant\alpha}\binom{\alpha}{\beta}|g|_{K,\alpha-\beta}\,|\phi|_{K,\beta}\,.\end{aligned}$$

Thus for $\epsilon > 0$ we can choose a $\delta > 0$ so that if $|\phi|_{K,\beta} < \delta$ for $\beta \leqslant \alpha$, then $|T\phi|_{K,\alpha} < \epsilon$. Consequently T is continuous.

□

2. Show if $y \in \mathbb{R}^n$ and $a > 0$, then $\lambda(y)T_f = T_{\lambda(y)f}$ and $\delta(a)T_f = T_{\delta(a)f}$ for $f \in L_{\text{loc}}(\mathbb{R}^n)$. Also show $gT_f = T_{gf}$ whenever $g \in \mathcal{E}(\Omega)$ and $f \in L_{\text{loc}}(\Omega)$.

Solution:

Let f be locally integrable. Then

$$\begin{aligned}\lambda(y)T_f(\phi) &= T_f(\lambda(-y)\phi) \\ &= \int f(x)\phi(x+y)\,dx \\ &= \int f(x-y)\phi(x)\,dx \\ &= \int \lambda(y)f(x)\phi(x)\,dx \\ &= T_{\lambda(y)f}(\phi)\end{aligned}$$

and

$$\begin{aligned}
\delta(a)T_f(\phi) &= T_f(\delta(a^{-1})\phi) \\
&= \int f(x)a^{n/2}\phi(ax)\,dx \\
&= \int f(x/a)a^{n/2}\phi(x)d(x/a) \\
&= \int f(x/a)a^{n/2}\phi(x)\,a^{-n}\,dx \\
&= \int a^{-n/2}f(a^{-1}x)\phi(x)\,dx \\
&= \int \delta(a)f(x)\phi(x)\,dx \\
&= T_{\delta(a)f}(\phi)\,.
\end{aligned}$$

Also if $g \in \mathcal{E}(\Omega)$ and $\phi \in \mathcal{D}(\Omega)$, then

$$\begin{aligned}
gT_f(\phi) &= T_f(g\phi) \\
&= \int f(x)g(x)\phi(x)\,dx \\
&= \int g(x)f(x)\phi(x)\,dx \\
&= T_{gf}(\phi)\,.
\end{aligned}$$

□

3. Let $f : \Omega \to \mathbb{C}$ be m-times continuously differentiable. Show $D^\alpha f$ is the weak derivative of f for all α with $|\alpha| \leqslant m$.

Solution: Let f be in $C^m(\Omega)$; i.e. $D^\alpha f$ exist and are continuous for all α with $|\alpha| \leqslant m$. We need to show $D^\alpha T_f = T_{D^\alpha f}$. We do this by noting it is true if $|\alpha| = 0$. Suppose true for $|\alpha| < m$. We show it is true for $\beta = \alpha + \epsilon_j$. Namely, if $\phi \in \mathcal{D}(\Omega)$, then by integration by parts,

$$\begin{aligned}
D^\beta T_f(\phi) &= (-1)^{|\alpha|+1}T_f(D^\alpha D_j\phi) \\
&= -D^\alpha T_f(D_j\phi) \\
&= -T_{D^\alpha f}(D_j\phi) \\
&= -\int_\Omega D^\alpha f(x)\,D_j\phi(x)\,dx \\
&= \int_\Omega D_j D^\alpha f(x)\phi(x)\,dx \\
&= \int D^\beta f(x)\phi(x)\,dx \\
&= T_{D^\beta f}(\phi)\,.
\end{aligned}$$

□

4. Suppose that the sequence of distributions $\{T_k\}$ converges to $T \in \mathcal{D}'(\mathbb{R}^n)$. Show $D^\alpha T_k \to D^\alpha T$ for all $\alpha \in \mathbb{N}_0^n$.

Solution:

Suppose $T_k \to T$ in the weak topology; i.e., $T_k(\phi) \to T(\phi)$ for each $\phi \in \mathcal{D}(\mathbb{R}^n)$. Then $D^\alpha T_k(\phi) = (-1)^{|\alpha|}T_k(D^\alpha \phi) \to (-1)^{|\alpha|}T(D^\alpha \phi) = D^\alpha T(\phi)$ as $k \to \infty$. So $D^\alpha T_k \to D^\alpha T$.

□

5. Suppose that $n = 1$. Evaluate the distributional derivative of the following functions and distributions:

(a) $f(x) = \text{sign}(x)$

(b) $f(x) = x\,|x|$

(c) $f(x) = xH(x)$

(d) $f(x) = H(x+1)H(1-x)$

(e) δ_a, $a \in \mathbb{R}$

(f) $x\delta_a$, $a \in \mathbb{R}$

(g) $f(x) = x\log(x)H(x)$.

Solution: Since the motivation for the definition uses integration by parts, this is used repeatedly in the following calculations. If $f(x) = \text{sign}(x)$, then

$$\begin{aligned}
\frac{d}{dx}T_f(\phi) &= -T_f(\phi') \\
&= -\int f(x)\phi'(x)\,dx \\
&= \int_{-\infty}^{0} \phi'(x)\,dx - \int_0^\infty \phi'(x)\,dx \\
&= 2\phi(0) \\
&= 2\delta_0(\phi)\,.
\end{aligned}$$

So $f' = 2\delta_0$.

For $f(x) = x|x|$, we see

$$\begin{aligned}
-T_f(\phi') &= \int_{-\infty}^{0} x^2\phi'(x)\,dx - \int_0^\infty x^2\phi'(x)\,dx \\
&= -\int_{-\infty}^{0} 2x\phi(x)\,dx + \int_0^\infty 2x\phi(x)\,dx \\
&= \int_{-\infty}^{\infty} 2|x|\phi(x)\,dx
\end{aligned}$$

and thus $\frac{d}{dx}T_f = g$ where $g(x) = 2|x|$. Next for $f(x) = xH(x)$, we see

$$\begin{aligned}
\frac{d}{dx}T_f(\phi) &= -\int xH(x)\,\phi'(x)\,dx \\
&= -\int_0^\infty x\phi'(x)\,dx \\
&= \int_0^\infty \phi(x)\,dx \\
&= \int \chi_{[0,\infty)}(x)\phi(x)\,dx
\end{aligned}$$

and thus $f' = \chi_{[0,\infty)}$.

Next if $f(x) = H(x+1)H(1-x) = \chi_{[-1,\infty)}(x)\chi_{(-\infty,1]}(x) = \chi_{[-1,1]}(x)$, we have

$$\begin{aligned}
\frac{d}{dx}T_f(\phi) &= -T_f(\phi') \\
&= -\int_{-1}^{1} \phi'(x)\,dx \\
&= \phi(-1) - \phi(1) \\
&= \delta_{-1}(\phi) - \delta_1(\phi)
\end{aligned}$$

and so $f' = \delta_{-1} - \delta_1$.

For δ_a one has

$$\begin{aligned}
\delta_a'(\phi) &= -\delta_a(\phi') \\
&= -\phi'(a)
\end{aligned}$$

while for $x\delta_a$ one has

$$\begin{aligned}
(x\delta_a)'(\phi) &= -(x\delta_a)(\phi') \\
&= -\delta_a(x\phi') \\
&= -a\phi'(a)\,.
\end{aligned}$$

We finish with $f(x) = x\log(x)H(x)$. First note $x\log x$ and $\log x$ are locally integrable on $[0,\infty)$ since $x\log x \to 0$ as $x \to 0+$ and an antiderivative of $\log x$ is $x\log x - x$. Thus

$$\begin{aligned}
f'(\phi) &= -f(\phi') \\
&= -\int_0^\infty x\log x\,\phi'(x)\,dx \\
&= \int_0^\infty (\log x + 1)\phi(x)\,dx \\
&= (H(x) + H(x)\log x)(\phi)\,.
\end{aligned}$$

So as distributions $(x\log xH(x))' = H(x)(1+\log x)$.

□

6. Show there is no locally integrable function f such that $T_f = \delta$.

Solution:

Assume f is a locally integrable function with $T_f = \delta$. Then $\int f(x)\phi(x)\,dx = \phi(0)$ for all $\phi \in C^\infty$ with compact support. As seen in Equation 2.10, there is a sequence ϕ_k of such functions with $0 \leqslant \phi_k \leqslant 1$ and $\phi_k \to \chi_{[a,b]}$ where $a < b$. So $\int_a^b f(x)\,dx = \chi_{[a,b]}(0)$ for all a, b. But this is impossible for a locally integrable function f. □

7. Show every measurable function whose absolute value is majorized by a polynomial is a tempered distribution. If f is such a function, the corresponding distribution is defined by

$$T_f(\phi) = \int f(x)\,\phi(x)\,dx.$$

Solution:

Suppose $|f(x)| \leqslant |p(x)|$ for all $x \in \mathbb{R}^n$. Since $p(x) = \sum_{|\beta| \leqslant k} c_\beta x^\beta$, we see $|p(x)| \leqslant \sum_{|\beta| \leqslant k} |c_\beta|(1 + |x|^2)^{|\beta|} \leqslant K(1 + |x|^2)^k$ where $K = \sum_{|\beta| \leqslant k} |c_\beta|$. By Example 4.37 with $r = -k - 1 - n/2$ and $p = 1$, we have T_f is a tempered distribution if

$$\int |f(x)|\,(1 + |x|^2)^r\,dx \leqslant K \int (1 + |x|^2)^{r+k}\,dx < \infty.$$

But by Remark 2.27, this integral is finite for $r + k = -1 - n/2 < -n/2$. □

8. Let $L(x) = \log(x)\chi_{(0,\infty)}(x)$.

(a) Show L is locally integrable on $\mathbb{R}$, and hence that L defines a distribution on $\mathbb{R}$.

(b) Show

$$L'(\varphi) = \int_0^\infty \frac{1}{x}\varphi(x)\,dx$$

if $\operatorname{supp}(\varphi) \subset (0, \infty)$.

(c) Show $\varphi \mapsto \int_0^\infty \frac{1}{x}\varphi(x)\,dx$ does not define a distribution on $\mathbb{R}$.

(d) Determine the distribution L'.

Solution: Note L is locally integrable for $\int_0^b \log x\,dx = x \log x|_0^b - \int_0^b 1\,dx = b \log b - b$ and thus $\chi_{[a,b]}(x)\chi_{(0,\infty)}(x) \log x$ is integrable for

$-\infty < a < b < \infty$. So L defines a distribution and we have (a). Also

$$\begin{aligned}
L'(\phi) &= -L(\phi') \\
&= -\int_0^\infty \phi'(x)\ln(x)\,dx \\
&= -\int_0^1 \phi'(x)\ln(x)\,dx - \int_1^\infty \phi'(x)\ln(x)\,dx \\
&= \lim_{a\to 0+}\left(\phi(a)\ln(a) + \int_a^1 \frac{\phi(x)}{x}\,dx\right) - \phi(x)\ln(x)|_1^\infty + \int_1^\infty \frac{\phi(x)}{x}\,dx \\
&= \lim_{a\to 0+}\int_a^1 \frac{\phi(x)-\phi(a)}{x}\,dx + \int_1^\infty \frac{\phi(x)}{x}\,dx\,.
\end{aligned}$$

Thus we know L'. Moreover, if $\phi(x) = 0$ for $x < a$ where $a > 0$, one has $L'(\phi) = \int_0^\infty \frac{1}{x}\phi(x)\,dx$ and so (b) holds. Since $\int_0^1 \frac{1}{x}\,dx = \infty$, the formula in (c) cannot be a distribution. □

9. Suppose that the distribution S is homogeneous of degree μ. Show $D^\alpha S$ is homogeneous of degree $\mu - |\alpha|$. (See Exercise 4.3.15 for the definition of homogeneous.)

Solution:

Recall a distribution T is homogeneous of degree μ if

$$\int \lambda^{-n}\phi(\lambda^{-1}x)\,dT(x) = \lambda^\mu T(\phi)\,.$$

Now

$$\begin{aligned}
\int \lambda^{-n}\phi(\lambda^{-1}x)\,d(D^\alpha T)(x) &= (-1)^{|\alpha|}\int \lambda^{-n} D^\alpha(\phi(\lambda^{-1}x))\,dT(x) \\
&= (-1)^{|\alpha|}\int \lambda^{-n}\lambda^{-|\alpha|}(D^\alpha\phi)(\lambda^{-1}x)\,dT(x) \\
&= (-1)^{|\alpha|}\lambda^{\mu-|\alpha|}\int D^\alpha\phi(x)\,dT(x) \\
&= \lambda^{\mu-|\alpha|}D^\alpha T(\phi)\,.
\end{aligned}$$

So $D^\alpha T$ is homogeneous of degree $\mu - |\alpha|$.

□

10. Antiderivatives of distributions: The following steps show how to find an antiderivative of a distribution T.

(a) Show if $h \in \mathcal{D}(\mathbb{R})$ and $\int h(x)\,dx = 1$, then the mapping $\phi \mapsto \Phi$ defined by

$$\Phi(x) = \int_{-\infty}^x (\phi(t) - \left(\int \phi\right) h(t))\,dt$$

is a continuous linear mapping of $\mathcal{D}(\mathbb{R})$ into $\mathcal{D}(\mathbb{R})$; show this mapping is also continuous from $\mathcal{S}(\mathbb{R})$ into $\mathcal{S}(\mathbb{R})$.

(b) Let T be a distribution on $\mathbb{R}$. Define S_h by

$$\langle \phi, S_h \rangle = -\langle \Phi, T \rangle.$$

Show

$$S_h' = T.$$

(c) Show if S_1 and S_2 are distributions on $\mathbb{R}$ with $S_1' = S_2'$, then $S_1 = S_2 + c$ where c is a constant, i.e., show there is a constant c such that

$$S_1(\phi) = S_2(\phi) + T_c(\phi) = S_2(\phi) + c \int \phi(x)\, dx.$$

(d) Show if T is tempered, then the distribution S_h is tempered.

(e) Show if F is absolutely continuous with derivative f and $T = T_f$, then $S_h = F + c$ for some constant c.

Solution: Let $h \in \mathcal{D}(\mathbb{R})$ with $\int h(x)\, dx = 1$.

Then if $\phi \in \mathcal{D}(\mathbb{R})$, $\psi = \phi - \left(\int \phi(t)\, dt\right) h$ has integral 0. Thus $\Phi(x) = \int_{-\infty}^{x} \psi(t)\, dt \in \mathcal{D}(\mathbb{R})$. To see $\phi \mapsto \Phi$ is continuous, it suffices to show $\phi \mapsto \int \phi$ is continuous. We note this is straight forward for $\mathcal{D}(\mathbb{R})$. Indeed, it suffices to show continuity on $\mathcal{D}_K(\mathbb{R})$ and this follows from $|\int_K \phi| \leqslant |\phi|_{K,0} m(K)$.

In the case of $\mathcal{S}(\mathbb{R})$, note if ϕ is Schwartz, then $\Phi'(x) = \phi(x) - (\int \phi) h(x)$ and this function is Schwartz. Thus $(1 + x^2)^n \Phi^{(k)}(x)$ is bounded for $k \geqslant 1$. Thus Φ is Schwartz if $(1 + x^2)^n \Phi(x)$ is bounded for each n. To see this we need only show $x^n \Phi(x)$ is bounded for each n. Note $\Phi(x) \to 0$ if $x \to \infty$ or $x \to -\infty$. Using l'Hôpital's rule we see

$$\begin{aligned} \lim_{x \to \pm\infty} \frac{\Phi(x)}{x^{-n}} &= \lim_{x \to \pm\infty} \frac{\Phi'(x)}{-n x^{-n-1}} \\ &= -\frac{1}{n} \lim_{x \to \pm\infty} x^{n+1} \Phi'(x) = 0\,. \end{aligned}$$

We next show the mapping $\phi \mapsto \Phi$ is continuous on $\mathcal{S}(\mathbb{R})$. First note the mapping $\phi \mapsto \phi - (\int \phi) h$ is continuous. To see this note $\phi \mapsto \int \phi$ is continuous since $|\int \phi| \leqslant \int \frac{1}{1+x^2}(1 + x^2)|\phi(x)|\, dx \leqslant |\phi|_{1,0} \int \frac{1}{1+x^2}\, dx = \pi |\phi|_{1,0}$. Since $\mathcal{S}(\mathbb{R})$ is a topological vector space, one can conclude $\phi \mapsto \phi - (\int \phi) h$ is a continuous linear mapping of $\mathcal{S}(\mathbb{R})$. Thus it suffices to show $\psi \to \Psi$ where $\Psi(x) = \int_{-\infty}^{x} \psi(t)\, dt$ is a continuous mapping from the subspace $\mathcal{S}_0(\mathbb{R})$ consisting of those $f \in \mathcal{S}(\mathbb{R})$ with $\int f = 0$ into the space $\mathcal{S}(\mathbb{R})$.

By Exercise 2.3.5, the seminorms $|f|'_{n,k} = \sup |x^n D^k f(x)|$ define the topology on $\mathcal{S}(\mathbb{R})$. We note $|\Psi|'_{n,k} = |\psi|'_{n,k-1}$ for $k \geqslant 1$ and hence $|\Psi|'_{n,k} < \epsilon$ if $|\psi|'_{n,k-1} < \epsilon$. Thus we need only estimate the size of $x^n \Psi(x)$.

For $n = 0$, we have

$$
\begin{aligned}
|\Psi(x)| &= |\int_{-\infty}^{x} \psi(t)\, dt| \\
&\leqslant \int_{-\infty}^{x} \frac{1}{1+t^2}(1+t^2)|\psi(t)|\, dt \\
&\leqslant |\psi|_{1,0} \int_{-\infty}^{\infty} \frac{1}{1+t^2}\, dt \\
&= \pi|\psi|_{1,0}
\end{aligned}
$$

and so $|\Psi|'_{0,0} \leqslant \pi|\psi|_{1,0}$.

For $n \geqslant 1$, define $\rho(t) = \Psi(\frac{1}{t})$ and $\sigma(t) = t^n$ and set $\rho(0) = 0$. Note ρ is continuous for $t \neq 0$ and the path $\Gamma(t) = (\sigma(t), \rho(t))$ is differentiable for $0 < t < \frac{1}{x}$ when $x > 0$ and for $\frac{1}{x} < t < 0$ when $x < 0$ and continuous on the closures of these intervals. By the Mean Value Theorem, there is a point ξ_x in the interior of these intervals with

$$
(\rho(\frac{1}{x}) - \rho(0))\sigma'(\xi_x) = (\sigma(\frac{1}{x}) - \sigma(0))\rho'(\xi_x)\,.
$$

Thus $\frac{\rho'(\xi_x)}{\sigma'(\xi_x)} = \frac{\rho(\frac{1}{x})}{\sigma(\frac{1}{x})} = x^n\Psi(x)$ and we have

$$
x^n\Psi(x) = \frac{\psi(\frac{1}{\xi_x})\left(\frac{-1}{\xi_x^2}\right)}{n\xi_x^{n-1}} = -\frac{1}{n}\frac{1}{\xi_x^{n+1}}\psi(\frac{1}{\xi_x})\,.
$$

Consequently for $x \neq 0$, there is a θ with θ strictly between 0 and x with $x^n\Psi(x) = -\frac{1}{n}\theta^{n+1}\psi(\theta)$. This yields $|\Psi|_{n,0} \leqslant \frac{1}{n}|\psi|_{n+1,0}$ for $n > 0$ and consequently we have the continuity of $\psi \mapsto \Psi$. This finishes (a).

So S_h defined by $S_h(\phi) = -T(\Phi)$ is a distribution. Now if ϕ is C^∞ with compact support or Schwartz, then ϕ' is C^∞ with compact support or Schwartz and then Φ for ϕ' is given by $\Phi = \phi$. So

$$
S'_h(\phi) = -S_h(\phi') = T(\phi)
$$

and we see $S'_h = T$. Hence (b).

To see (c), note if $S'_1 = S'_2$ and $S = S_1 - S_2$, then $S' = 0$. Now if $S' = 0$, then $S(\phi') = 0$ for all ϕ. Hence $S(\Phi') = 0$. So $S(\phi - (\int\phi)h) = 0$. This gives

$$
S(\phi) = S(h)\int \phi(x)\, dx = c\int \phi(x)\, dx
$$

where $c = S(h)$. So $S_1 - S_2 = S = c$ and we have (c).

(d) follows by the same arguments.

For (e), note if F is absolute continuous, then $f = F'$ is locally integrable and if $T = T_f$, then using integration by parts for absolutely continuous functions, we have

$$\begin{aligned}
S_h(\phi) &= -T_f(\Phi) \\
&= -\int_{-\infty}^{\infty} f(x)\Phi(x)\,dx \\
&= -\int_{-\infty}^{\infty} F'(x)\Phi(x)\,dx \\
&= -F(x)\Phi(x)|_{-\infty}^{\infty} + \int_{-\infty}^{\infty} F(x)\Phi'(x)\,dx \\
&= \int_{-\infty}^{\infty} F(x)\left(\phi(x) - (\int \phi)h(x)\right)\,dx \\
&= \int_{-\infty}^{\infty} F(x)\phi(x)\,dx - \int_{-\infty}^{\infty} F(x)h(x)\,dx\left(\int_{-\infty}^{\infty} \phi(x)\,dx\right) \\
&= \int_{-\infty}^{\infty} (F(x)+c)\phi(x)\,dx
\end{aligned}$$

where $c = -\int F(x)h(x)\,dx$. □

11. Find the antiderivatives of the following distributions:

(a) $T = 1$

(b) δ_a

(c) $\text{sign}(x)$.

Solution:

We find an antiderivative of $T = 1$. One can expect it to be the locally integrable function $f(x) = x$ and you can by direct calculation see that it works. Let h be in $\mathcal{D}(\mathbb{R})$ with $\int h = 1$ and define $\Phi(x) = \int_{-\infty}^{x} \left(\phi(t) - h(t)\int\phi)\right)\,dt$. Note $\Phi'(x) = \phi(x) - h(x)\int\phi$. Now using integration by parts, an antiderivative T_h is given by

$$\begin{aligned}
T_h(\phi) &= -T(\Phi) \\
&= -\int_{-\infty}^{\infty} \Phi(x)\,dx \\
&= -x\Phi(x)|_{-\infty}^{\infty} + \int_{-\infty}^{\infty} x\Phi'(x)\,dx \\
&= \int_{-\infty}^{\infty} x\left(\phi(x) - h(x)\int\phi\right)\,dx \\
&= \int_{-\infty}^{\infty} x\phi(x)\,dx - \int_{-\infty}^{\infty} th(t)\left(\int_{-\infty}^{\infty}\phi(x)\,dx\right)\,dt \\
&= \int_{-\infty}^{\infty}\left(x - \int_{-\infty}^{\infty} th(t)\,dt\right)\phi(x)\,dx\,.
\end{aligned}$$

Thus an antiderivative is $x + c$ where $c = -\int_{-\infty}^{\infty} th(t)\,dt$.

For δ_a, we obtain a translate of a Heaviside function. In fact if T_h is an antiderivative, we see if $c = \int h$, then

$$\begin{aligned} T_h(\phi) &= -\delta_a(\Phi) \\ &= -\Phi(a) \\ &= -\int_{-\infty}^{a} \left(\phi(t) - h(t)\int_{-\infty}^{\infty}\phi(x)\,dx\right)dt \\ &= -\int \chi_{(-\infty,a)}(t)\,\phi(t)dt + \int_{-\infty}^{\infty} h(t)\,dt \int_{-\infty}^{\infty}\phi(x)\,dx \\ &= \int (-\chi_{(-\infty,a)}(x) + c)\phi(x)\,dx \end{aligned}$$

and thus antiderivatives are $-\chi_{(-\infty,a)} + c$. Taking $c = 1$ gives $H(x - a)$ as an antiderivative.

For $\operatorname{sign}(x)$ we expect $|x|$ to be an antiderivative. We can check this works or argue

$$\begin{aligned} T_h(\phi) &= -\int \Phi(x)\operatorname{sign}(x)\,dx \\ &= \int_{-\infty}^{0}\Phi(x)\,dx - \int_{0}^{\infty}\Phi(x)\,dx \\ &= x\Phi(x)|_{-\infty}^{0} - \int_{-\infty}^{0} x\Phi'(x)\,dx - x\Phi(x)|_0^{\infty} + \int_0^{\infty} x\Phi'(x)\,dx \\ &= -\int_{-\infty}^{0} x\left(\phi(x) - h(x)\int\phi\right)dx + \int_0^{\infty} x\left(\phi(x) - h(x)\int\phi\right)dx \\ &= \int |x|\phi(x)\,dx + c\int\phi(x)\,dx \end{aligned}$$

where $c = -\int_{-\infty}^{\infty}|x|h(x)\,dx$. So the antiderivatives are $|x| + c$.

□

12. Let Ω be a connected open subset of $\mathbb{R}^n$. Show if T is a distribution on Ω and $D^\alpha T = 0$ for all α with $|\alpha| = 1$, then $T = c = T_c$ for some constant c. (**Hint:** Show if T is a distribution on the interval $(a, b) \subseteq \mathbb{R}$ with $\frac{d}{dx}T = 0$, then $T = c$. Then use Theorem 2.86).

Solution: We start by considering products $\prod_{i=1}^{n}(a_i, b_i)$ of open intervals and argue by induction. For $n = 1$ and interval (a, b) where $a < b$ and a distribution T on (a, b) with $DT = 0$, we take an $h \in \mathcal{D}(a, b)$ with $\int h = 0$. Set $S(\phi) = T(\Phi)$ where $\Phi(x) = \int_a^x \left(\phi(t) - h(t)\int_a^b \phi(x)\,dx\right)dt$. Then $\Phi'(x) = \phi(x) - h(x)\int_a^b \phi(y)\,dy$. So one has $0 = -D(T)(\Phi) = T(\Phi') = T(\phi) - T(h)\int_a^b \phi(y)\,dy$. Hence

$$T(\phi) = c\int_a^b \phi(y)\,dy \text{ where } c = T(h).$$

Suppose now the result holds for Ω of form $\prod_{i=1}^n (a_i, b_i)$ and T is a distribution on $\prod_{i=1}^n (a_i, b_i) \times (a, b)$ with $D^\alpha T = 0$ whenever $|\alpha| = 1$. Fix $\psi \in \mathcal{D}(\prod_{i=1}^n (a_i, b_i))$ and define $S_\psi(\phi) = T(\psi \otimes \phi)$. Note $DS_\psi(\phi) = D^{\epsilon_{n+1}} T(\psi \otimes \phi) = 0$ and thus there is a constant c_ψ with $S_\psi(\phi) = c_\psi \int \phi(x)\,dx$ for $\phi \in \mathcal{D}(a,b)$. Since by Theorem 2.86, $\psi \mapsto \psi \otimes \phi$ is continuous, the mapping $\psi \mapsto c_\psi$ is a distribution R on $\prod_{i=1}^n (a_i, b_i)$ and one has $D^\alpha R = 0$ if $|\alpha| = 1$. By induction, $c_\psi = c \int_\Omega \psi(x)\,dx$ for some constant c. Thus one has $T(\psi \otimes \phi) = c \int_{\prod(a_i,b_i)\times(a,b)} \psi(x)\phi(y)\,dx\,dy$. Using the density of the linear span referred to in Theorem 2.86, we have the result for open rectangles Ω. Now $\Omega = \cup_j \Omega_j$ where Ω_j are nonempty open rectangles. Thus $T_j = T|_{\Omega_j} = c_j$. Also if $U = \Omega_j \cap \Omega_{j'} \neq \emptyset$, we have $c_j = c_{j'}$ for if $\operatorname{supp} h \subseteq U$ and $\int h\,dx = 1$, then $T_j(h) = T_{j'}(h)$ and so $c_j = c_j \int h = c_{j'} \int h = c_{j'}$. Since Ω is connected, the general result follows. □

13. Let f be a locally integrable function on $\mathbb{R}^n$. Show the support of the distribution T_f is the complement of the union of all open subsets U of $\mathbb{R}^n$ such that $f|_U$ is zero almost everywhere.

Solution: This was already done in Exercise 4.3.8 and thus should be removed.

□

14. Let $f : \Omega \to \mathbb{C}$ be m-times continuously differentiable. Show $D^\alpha f$ is the weak derivative of f for all α with $|\alpha| \leqslant m$. **Remark:** This is a repeat of Exercise 4.4.3 and should be removed.

15. Let D be a differential operator of order N on an open subset Ω of $\mathbb{R}^n$. Show the expression $D = \sum_{|\alpha| \leqslant N} a_\alpha \partial^\alpha$ is unique. (Hint: Apply D to the polynomials $x \mapsto x^\gamma$.)

Solution:

To show the expression $D = \sum_{|\alpha| \leqslant N} a_\alpha \partial^\alpha$ is unique, it suffices to show if $D = 0$, then each $a_\alpha \in \mathcal{E}(\Omega)$ is zero. But if $D = 0$ and $\{x \mid |x-y| < r\} \subseteq \{x \mid |x-y| < R\} \subseteq \Omega$ where y is an interior point of Ω and $0 < r < R$, one has by Lemma 2.53 a C^∞ function ϕ with $0 \leqslant \phi \leqslant 1$ and $\phi(x) = 1$ for $|x-y| \leqslant r$ and $\phi(x) = 0$ for $|x-y| \geqslant R$. Now $\phi_\gamma(x) = x^\gamma \phi(x)$ is in $\mathcal{D}(\Omega)$ and for $|x-y| < r$, one has

$$D\phi_\gamma(x) = \sum_{|\alpha| \leqslant N} a_\alpha(x) D^\alpha(x^\gamma) = \sum_{|\alpha| \leqslant N,\, \alpha \leqslant \gamma} \frac{\gamma!}{(\gamma-\alpha)!} a_\alpha(x) x^{\gamma-\alpha}.$$

First note if $\gamma = 0$, we obtain $\alpha_0(x) = 0$ on $B = \{x \mid |x-y| < r\}$. Assume $a_\alpha(x) = 0$ for $x \in B$ when $|\alpha| \leqslant k < N$, Then if β has length $k+1$, $\beta = \alpha' + \epsilon_j$ for some j and α' with $|\alpha'| = k$. Since $D\phi_\gamma(x) = \sum_{k < |\alpha| \leqslant N, \alpha \leqslant \gamma} \frac{\gamma!}{(\gamma-\alpha)!} a_\alpha(x) x^{\gamma-\alpha} = 0$, we obtain by taking $\gamma = \alpha' + \epsilon_j$ that $a_{\alpha'+\epsilon_j}(x) = 0$ for $x \in B$. So following this finitely many times implies all a_α vanish on B. Since Ω is a union of such sets, we see all a_α vanish on Ω.

□

16. Let Ω be an open subset of $\mathbb{R}^n$ and suppose $p(x) = \sum_{|\alpha| \leqslant N} a_\alpha x^\alpha$ is a polynomial. Then $L = p(D) = \sum a_\alpha D^\alpha$ is a differential operator with constant

coefficients. Show if T is a distribution on Ω, then Definition (4.16) agrees with the natural definition of $p(D)T$ as $\sum a_\alpha D^\alpha T$.
Solution: We have from (4.15) that $L^\sharp = \sum_{|\alpha|\leqslant N}(-1)^{|\alpha|}a_\alpha D^\alpha$. Thus

$$\begin{aligned} LT(\phi) &= T(L^\sharp\phi) \\ &= T(\sum(-1)^{|\alpha|}a_\alpha D^\alpha\phi) \\ &= \sum a_\alpha(-1)^{|\alpha|}T(D^\alpha\phi) \\ &= \sum a_\alpha D^\alpha T(\phi). \end{aligned}$$

□

17. Let T be a distribution on a nonempty open subset Ω of $\mathbb{R}^n$. The distribution T is said to have finite order if there is a $N \in \mathbb{N}_0$ such that for each compact subset K of Ω and each $\epsilon > 0$, there is a $\delta_K > 0$ such that if $\phi \in \mathcal{D}_K(\Omega)$ and $\max_{x\in K}|D^\alpha\phi(x)| < \delta_K$ for $|\alpha| \leqslant N$, then $|T(\phi)| < \epsilon$. The smallest N for which this is true is called the **order** of T. If T has order 0, then T is said to be a **Radon** distribution on Ω.

(a) Show if the order of T is 0 and $T(\phi) \geqslant 0$ if $\phi \geqslant 0$, then there is a Radon measure μ on Ω such that

$$T(\phi) = \int \phi(x)\,d\mu(x)$$

for $\phi \in \mathcal{D}(\Omega)$.

(b) Show if T is a Radon distribution on Ω, then there are Radon measures μ_1, μ_2, μ_3, and μ_4 such that

$$T(\phi) = \int \phi\,d\mu_1 - \int \phi\,d\mu_2 + i\int \phi\,d\mu_3 - i\int \phi\,d\mu_4.$$

(Hint: See Rudin, [41], Real and Complex Analysis on complex measures and the dual of $C_c(\Omega)$.)

Solution:

For (a), it is known that if T is a linear functional on $C_c(\Omega)$ with $T(\phi) \geqslant 0$ for $\phi \geqslant 0$, then there is a Radon measure μ on Ω with $T(\phi) = \int \phi\,d\mu$.

Now let T be a distribution of order 0 on Ω. We wish to extend it to $C_c(\Omega)$. To do this let $f \in C_c(\Omega)$ and choose a compact subset K of Ω with $\operatorname{supp} f \subseteq K$. Then since T has order 0, there is δ_K such that $|T(\phi)| \leqslant \delta_K|\phi|_{0,K}$ for $\phi \in \mathcal{D}_K(\Omega)$. Choose a sequence $\phi_k \in \mathcal{D}_K(\Omega)$ with $\phi_k \to f$ uniformly. Since $|T(\phi_k)-T(\phi_j)| \leqslant \delta_K|\phi_k-\phi_j|_{0,K} \to 0$ as $j,k \to \infty$, $\lim T(\phi_k)$ exist. It is straight forward to see it is independent of the choice of $\{\phi_k\}$ and the K containing the support of f. Define $T(f) = \lim_k T(\phi_k)$. If $f \geqslant 0$, the ϕ_k can be taken to be nonnegative and so $T(f) \geqslant 0$. So a Radon measure μ with $T(\phi) = \int\phi\,d\mu$ exists.

For (b), it is known that if T is a continuous linear functional on $C_c(\Omega)$ with the norm $|\cdot|_{\max}$, then there is a complex measure μ on Ω with $T(\phi) = \int\phi\,d\mu$.

The same argument as in part (a) shows if T is a distribution on Ω with order 0, then T has a continuous extension to $C_c(\Omega)$. Hence a complex measure μ exists. One then has $\mu = \nu_1 + i\nu_2$ where ν_1 and ν_2 are finite signed measures. Then one has the Hahn decompositions, $\nu_1 = \mu_1 - \mu_2$ and $\nu_2 = \mu_3 - \mu_4$ where each μ_j is a positive measure and $\mu_1 \perp \mu_2$ and $\mu_3 \perp \mu_4$. This gives

$$T(\phi) = \int \phi \, d\mu_1 - \int \phi \, d\mu_2 + i \int \phi \, d\mu_3 - i \int \phi \, d\mu_4 \, .$$

□

18. Show if T is a distribution on Ω with compact support, then T has finite order.

Solution:

Suppose T is a distribution on Ω with compact support K. As seen in the argument for Exercise 2.4.4, there are open sets U and V with $K \subseteq U \subseteq \bar{U} \subseteq V \subseteq \bar{V} \subseteq \Omega$ where $\bar{U}$ and $\bar{V}$ are compact. Using Lemma 2.66, one has T restricted to $\mathcal{D}_{\bar{V}}(\Omega)$ is continuous. Since $|\cdot|_{\bar{V},\alpha}$ are the seminorms defining the topology of $\mathcal{D}_{\bar{V}}(\Omega)$, we can find $\alpha_1, \alpha_2, \ldots, \alpha_N$ such that if $\epsilon > 0$, there is a $\delta > 0$ so that if $f \in \mathcal{D}_{\bar{V}}(\Omega)$ and $\max_{x \in \bar{V}} |D^{\alpha_j} f(x)| < \delta$ for $j = 1, 2, \ldots, N$, then $|T(f)| < \epsilon$. By Corollary 2.54, we can find a $\psi \in \mathcal{D}(\Omega)$ such that $\psi = 1$ on $\bar{U}$ and ψ has support inside V. We fix this ψ. Now let $\phi \in \mathcal{D}(\Omega)$. We have $\phi = \psi\phi + (1-\psi)\phi$. Note $(1-\psi)\phi \in \mathcal{D}(\Omega)$ and if $x \in U$, then $(1-\psi(x))\phi(x) = 0$. So $\{x \mid (1-\psi(x))\phi(x) \neq 0\} \subseteq \Omega - U$. Since this is closed in Ω, the support of $(1-\psi)\phi$ is a subset of $\Omega - U$ and thus misses K. Since T vanishes on the open set $\Omega - K$, we see $T((1-\psi)\phi) = 0$. Thus

$$T(\phi) = T(\phi\psi)$$

for all $\phi \in \mathcal{D}(\Omega)$. Clearly $\psi\phi$ has support inside $\bar{V}$. Now let $\epsilon > 0$. Since by Lemma 2.21,

$$D^\alpha(\phi\psi) = \sum_{\beta \leqslant \alpha} \binom{\alpha}{\beta} D^\alpha \phi D^{\alpha-\beta}\psi$$

we can find a $\delta' > 0$ so that if $\max_x |D^{\alpha_j}\phi(x)| < \delta'$ for $j = 1, 2, \ldots, N$, then

$$\sum_{\beta \leqslant \alpha_j} \binom{\alpha}{\beta} \delta' \max_{x \in \Omega} |D^{\alpha_j - \beta}\psi(x)| < \delta \text{ for } j = 1, 2, \ldots, N.$$

So we have $\max_{x \in \Omega} |D^{\alpha_j}(\phi\psi)(x)| < \delta$ for $j = 1, 2, \ldots, N$ if $\max_x |D^{\alpha_j}\phi(x)| < \delta'$ for $j = 1, 2, \ldots, N$. Thus

$$|T(\phi)| < \epsilon \text{ if } \max_x |D^{\alpha_j}(\phi)(x)| < \delta' \text{ for } j = 1, 2, \ldots, N.$$

So T has finite order.

□

19. Show if T is a Radon distribution on Ω with compact support, then there is a complex measure μ on Ω with $T(\phi) = \int \phi \, d\mu$ for $\phi \in \mathcal{D}(\Omega)$.
Solution:

This is just part (b) of Exercise 4.4.17. This problem should be removed.

□

20. Show if T is a distribution on Ω with compact support, then there exists an N and complex Borel measures μ_α on Ω such that

$$T = \sum_{|\alpha| \leqslant N} D^\alpha T_{\mu_\alpha}.$$

(**Hint:** Let N be the order of T and U be an open subset of Ω with $\operatorname{supp} T \subseteq U \subseteq \bar{U} \subseteq \Omega$ and $\bar{U}$ compact. Show the mapping $\{D^\alpha \phi|_U\}_{|\alpha| \leqslant N} \mapsto T(\phi)$ is continuous and linear on the relative topology of the product topology of $\prod_{|\alpha| \leqslant N} C_c(U)$. Then use the Hahn-Banach Theorem.)
Solution:

Suppose T is a distribution on Ω with compact support. By Exercise 4.4.18, T has finite order N for some N. Consider the linear subspace $\mathcal{L} = \prod_{|\alpha| \leqslant N} \{D^\alpha \phi \mid \phi \in \mathcal{D}(\Omega)\}$ of $\prod_{|\alpha| \leqslant N} C_c(\Omega)$ with norm $|\{(\phi_\alpha) \mid |\alpha| \leqslant N\}| = \sum_{|\alpha| \leqslant N} \max_{x \in \Omega} |\phi_\alpha(x)|$. Define $F : \mathcal{L} \to \mathbb{C}$ by

$$F((D^\alpha \phi \mid |\alpha| \leqslant N) = T(\phi).$$

This is well defined for if $D^\alpha \phi = 0$ for $|\alpha| \leqslant N$, then $T(\phi) = 0$. It is also bounded for if $\epsilon = 1$, there is a $\delta > 0$ so that if $\max_x |D^\alpha \phi(x)| < \delta$ for $|\alpha| \leqslant N$, then $|T(\phi)| < 1$. Thus if $|\{(D_\alpha \phi) \mid |\alpha| \leqslant N\}| < \delta$, $|F(\{D^\alpha \phi \mid |\alpha| \leqslant N\})| < 1$. By the Hahn-Banach Theorem, F has a continuous extension which we also call F to $\prod_{|\alpha| \leqslant N} C_c(\Omega)$. Since the continuous linear functionals on $C_c(\Omega)$ are given by $\phi \mapsto \int \phi \, d\mu$ where μ is a complex measure on Ω and the injections I_α which send ϕ to $\{(\delta_{\alpha,\beta}\phi) \mid |\beta| \leqslant N\}$ from $C_c(\Omega)$ into $\prod_{|\beta| \leqslant N} C_c(\Omega)$ are continuous, we see there are complex measures μ_α with

$$F(\{(\phi_\alpha) \mid |\alpha| \leqslant N\} = \sum_{|\alpha| \leqslant N} \int \phi_\alpha \, d\mu_\alpha.$$

Consequently,

$$\begin{aligned}
T(\phi) &= F(\{(D^\alpha \phi) \mid |\alpha| \leqslant N\}) \\
&= \sum_{|\alpha| \leqslant N} \int D^\alpha \phi(x) \, d\mu_\alpha(x) \\
&= \sum_{|\alpha| \leqslant N} (-1)^{|\alpha|} T_{(-1)^{|\alpha|}\mu_\alpha}(D^\alpha \phi) \\
&= \sum_{|\alpha| \leqslant N} D^\alpha T_{(-1)^{|\alpha|}\mu_\alpha}(\phi)
\end{aligned}$$

for $\phi \in \mathcal{D}(\Omega)$.

□

21. Show the following version of Taylor's formula. Suppose f is C^∞. Then

$$f(x) = \sum_{|\alpha| \leqslant N} \frac{x^\alpha}{\alpha!} D^\alpha(f)(0) + (N+1) \sum_{|\alpha|=N+1} \frac{x^\alpha}{\alpha!} \int_0^1 (1-t)^N D^\alpha f(tx)\,dt.$$

Solution: We first do the proof for $\mathbb{R}$.

$$\begin{aligned} f(x) &= f(0) + \int_0^x f'(t)\,dt \\ &= f(0) + (t-x)f'(t)|_0^x + \int_0^x (x-t)f''(t)\,dt \\ &= f(0) + (t-x)f'(t)|_0^x - \frac{(x-t)^2}{2} f''(t)|_0^x + \int_0^x \frac{(x-t)^2}{2} f'''(t)\,dt. \end{aligned}$$

Repeating we see

$$f(x) = \sum_{k=0}^n \frac{f^k(0)}{k!} x^k + \frac{1}{n!} \int_0^x (x-t)^n f^{(n+1)}(t)\,dt.$$

Making the substitution

$$t = xs,$$

we get

$$f(x) = \sum_{k=0}^n \frac{f^k(0)}{k!} x^k + \frac{1}{n!} \int_0^1 (x - sx)^n f^{(n)}(sx)\, xds.$$

Thus

$$f(x) = \sum_{k=0}^n \frac{f^k(0)}{k!} x^k + \frac{x^{n+1}}{n!} \int_0^1 (1-t)^n f^{(n+1)}(tx)\,dt.$$

For $\mathbb{R}^n$, consider the function $g(s) = f(sx)$. Then

$$f(x) = g(1) = \sum_{k=0}^N \frac{g^{(k)}(0)}{k!} + \frac{1}{N!} \int_0^1 (1-t)^N g^{(N+1)}(t)\,dt.$$

Now $g^{(k)}(t) = ((x_1 D_1 + \cdots x_n D_n)^k f)(tx) = \sum_{|\alpha|=k} \frac{k!}{\alpha!} x^\alpha D^\alpha f(tx)$.
Thus

$$f(x) = \sum_{k=0}^N \sum_{|\alpha|=k} \frac{x^\alpha}{\alpha!} D^\alpha f(0) + \frac{1}{N!} \int_0^1 \sum_{|\alpha|=N+1} \frac{(N+1)!}{\alpha!} (1-t)^N x^\alpha D^\alpha f(tx)\,dt$$

□

22. Show if T is a distribution on Ω and T has support $\{x_0\}$ where $x_0 \in \Omega$, then T has form

$$T(\phi) = \sum_{|\alpha| \leqslant N} c_\alpha D^\alpha \phi(x_0)$$

for some finite N. (**Hint:** Exercise 4.4.21 might be useful.)

Solution: We can assume by extending and translating T that $\Omega = \mathbb{R}^n$ and $x_0 = 0$. By Exercise 4.4.18 the distribution T has finite order. Since T has finite order, we know there is an $N \in \mathbb{N}_0$ so that for $\epsilon = 1$, there is a $\delta > 0$ such that $|T(\phi)| < 1$ for $\phi \in \mathcal{D}_n$ if $|\phi|_\alpha \leqslant \delta$ for $|\alpha| \leqslant N$. Hence $|T(\phi)| \to 0$ if $|\phi|_\alpha \to 0$ for $|\alpha| \leqslant N$.

We claim $T(\phi) = 0$ if $\phi \in \mathcal{D}_n$ and $|D^\alpha \phi(0)| = 0$ for $|\alpha| \leqslant N$. Choose a function ψ with $\psi \in \mathcal{D}_n$ such that $0 \leqslant \psi \leqslant 1$ and $\psi = 1$ for $|x| \leqslant \frac{1}{2}$ and $\psi(x) = 0$ if $|x| > 1$. Set $\psi_r(x) = \psi(rx)$ for $1 < r$. Note $\psi_r(x) = 0$ for $|x| \geqslant \frac{1}{r}$. Since $\phi - \psi_r\phi$ has its support missing x_0, we have $T(\phi) = T(\psi_r\phi)$. So to show $T(\phi) = 0$ need only show $T(\psi_r\phi) \to 0$ as $r \to \infty$. It thus suffices to show $|D^\alpha(\psi_r\phi)|_\infty \to 0$ for $|\alpha| \leqslant N$.

But then by the Leibniz rule it suffices to show

$$|D^{\alpha-\beta}\psi_r\, D^\beta\phi|_\infty \to 0$$

whenever $\beta \leqslant \alpha$ and $|\alpha| \leqslant N$. But

$$|D^{\alpha-\beta}\psi_r(x)| = r^{|\alpha-\beta|}|(D^{\alpha-\beta}\psi)(rx)| \leqslant r^{|\alpha-\beta|}|\psi|_{\alpha-\beta}\,.$$

Hence

$$|D^{\alpha-\beta}\psi_r\, D^\beta\phi|_\infty \leqslant r^{|\alpha-\beta|} \max_{|x|\leqslant 1/r} |D^\beta\phi(x)|\,|\psi|_{\alpha-\beta}\,.$$

By Taylor's Formula

$$\phi(x_1, \ldots, x_n) = \sum_{|\gamma|=N+1} \frac{x^\gamma}{\gamma!} g_\gamma(x)$$

where

$$g_\gamma(x) = (N+1)\int_0^1 (1-t)^N D^\gamma\phi(tx)\,dt.$$

Thus is suffices to

$$r^{|\alpha-\beta|} \max_{|x|\leqslant 1/r} |D^\beta(x^\gamma g_\gamma(x))| \to 0$$

as $r \to \infty$ if $\beta \leqslant \alpha$, $|\gamma| = N+1$, and $|\alpha| = N$. But

$$D^\beta(x^\gamma g_\gamma) = \sum_{\rho\leqslant\beta} \binom{\beta}{\rho} D^\rho(x^\gamma) D^{\beta-\rho} g_\gamma.$$

So it suffices to show

$$r^{|\alpha-\beta|} \max_{|x|\leqslant 1/r} |D^\rho x^\gamma| \to 0$$

if $\rho \leqslant \beta \leqslant \alpha$ and $|\alpha| \leqslant N$ and $|\gamma| = N+1$. Note $D^\rho x^\gamma = 0$ if ρ is not less or equal to γ and otherwise $D^\rho x^\gamma = \frac{\gamma!}{(\gamma-\rho)!}x^{\gamma-\rho}$. In either instance we have

$$\begin{aligned} r^{|\alpha-\beta|}\max_{|x|\leqslant 1/r}|D^\rho x^\gamma| &\leqslant r^{|\alpha-\beta|}\frac{\gamma!}{(\gamma-\rho)!}r^{-|\gamma-\rho|} \\ &= \frac{\gamma!}{(\gamma-\rho)!}r^{|\alpha|-|\beta|-|\gamma|+|\rho|} \\ &= \frac{\gamma!}{(\gamma-\rho)!}r^{|\alpha|-N-1}r^{-|\beta|+|\rho|} \to 0 \end{aligned}$$

as $r \to \infty$. □

23. Let Ω be an nonempty open subset of $\mathbb{R}^n$ and suppose $D : \mathcal{D}(\Omega) \to \mathcal{D}(\Omega)$ is a continuous linear mapping with the property that $D(\operatorname{supp}\phi) \subseteq \operatorname{supp}\phi$. Show there are $a_\alpha \in \mathcal{E}(\Omega)$ such that for each compact subset K of Ω there is an N such that

$$D(\phi) = \sum_{|\alpha|\leqslant N} a_\alpha D^\alpha \phi$$

for $\phi \in \mathcal{D}_K(\Omega)$.

Solution: For each $p \in \Omega$, define a distribution T_p on $\mathcal{D}(\Omega)$ by $T_p(\phi) = D(\phi)(p)$. Note T_p has support $\{p\}$, for if ϕ has support missing $\{p\}$, then the support of $D\phi$ misses $\{p\}$. As seen in Exercise 4.4.22, there is an N_p and functions a_α for $|\alpha| \leqslant N_p$ such that $T_p(\phi) = \sum_{|\alpha|\leqslant N_p} a_\alpha(p)D^\alpha\phi(p)$. We claim each $a_\alpha \in \mathcal{E}(\Omega)$. Indeed, given $\epsilon > 0$ small enough, there is a ϕ such that if $|x-p| < \epsilon$, then $\phi(x) = 1$. This gives a_0 is C^∞ on $B_\epsilon(p)$. Suppose a_α is C^∞ for all $|\alpha| < N$ and $|\beta| = N$. Then consider $\phi(x)x^\beta$ near p. We have $D^\alpha(\phi(x)x^\beta) = 0$ near p if $|\alpha| > N$, $D^\alpha(\phi(x)x^\beta) = 0$ near p if $|\alpha| = N$ and $\beta \neq \alpha$ and $D^\beta(\phi(x)x^\beta) = \beta!$ near p. So $D(\phi(x)x^\beta) = \sum_{|\alpha|<N} a_\alpha(x)D^\alpha(\phi(x)x^\beta) + a_\beta(x)\beta!$ for x near p. Since a_α are C^∞ for $|\alpha| < N$, we see a_β is C^∞ near p.

By replacing D be a nonzero multiple of D we may suppose if $K \subseteq \Omega$ is compact, there is an N such that $|\phi|_{K,\alpha} \leqslant 1$ for $|\alpha| \leqslant N$ implies $|D\phi|_{K,0} \leqslant 1$. By subtracting from D all terms $a_\alpha D^\alpha$ for $|\alpha| \leqslant N$, we may suppose $D = \sum_{|\alpha|>N} a_\alpha D^\alpha$. Pick a point $p \in K$ where $D(\phi)(p) \neq 0$. Thus we can make $\sum_{N<|\alpha|\leqslant N_p} a_\alpha(p)D^\alpha(\phi)(p)$ small by making $|\phi|_{K,\beta}$ small for $|\beta| \leqslant N$. But this can be shown to be impossible. Thus $D(\phi) = \sum_{|\alpha|\leqslant N} a_\alpha D^\alpha\phi$ for $\phi \in \mathcal{D}_K(\Omega)$. □

24. Let Ω be a nonempty open subset of $\mathbb{R}^n$. Show there is a nondifferential operator on $\mathcal{D}(\Omega)$ whose restriction to each $\mathcal{D}_K(\Omega)$ is a differential operator.

Solution: We show in the case of $\Omega = (0,1)$. Let $p_n = \frac{2^n-1}{2^n}$ and let a_n be a C^∞ function on $(0,1)$ with $\operatorname{supp} a_n \subseteq (p_n - \frac{1}{4^n}, p_n + \frac{1}{4^n})$ and $a_n(p_n) = 1$. Set $D = \sum_{n=1}^\infty a_n D^n$. Then D has the properties needed. □

Exercise Set 4.5

1. Suppose $f \in L^1(\mathbb{R}^n)$ or $f \in L^2(\mathbb{R}^n)$. Show

$$\mathcal{F}(T_f) = T_{\hat{f}}.$$

Solution: We remark that this also works for $f \in L^2$. Indeed, using Theorem 3.18 one sees

$$\begin{aligned}
\hat{T}_f(\phi) &= \int f(x)\hat{\phi}(x)\,dx \\
&= \int f(x)\overline{\mathcal{F}(\phi^*)}(x)\,dx \\
&= \int \mathcal{F}^{-1}f(x)\overline{\phi^*(x)}\,dx \\
&= \int \mathcal{F}f(-x)\phi(-x)\,dx \\
&= \int \hat{f}(x)\phi(x)\,dx
\end{aligned}$$

for $\hat{\phi}(x) = \int \phi(y)e^{-2\pi i y\cdot x}\,dy = \overline{\int \bar{\phi}(y)e^{2\pi i y\cdot x}\,dy} = \overline{\int \bar{\phi}(-y)e^{-2\pi i y\cdot x}\,dy}$ and so $\hat{\phi} = \overline{\mathcal{F}(\phi^*)}$.

The case for $f \in L^1$ is a simple use of Fubini's Theorem. Indeed,

$$\begin{aligned}
\hat{T}_f(\phi) &= T_f(\hat{\phi}) \\
&= \int f(x)\hat{\phi}(x)\,dx \\
&= \int f(x)\int \phi(y)e^{-2\pi i y\cdot x}\,dy\,dx \\
&= \int \phi(y)\int f(x)e^{-2\pi i y\cdot x}\,dx\,dy \\
&= \int \phi(y)\hat{f}(y)\,dy \\
&= T_{\hat{f}}(\phi)\,.
\end{aligned}$$

□

2. Let $f \in \mathcal{S}(\mathbb{R}^n)$. Show $\lambda(x)(T_f) = T_{\lambda(x)f}$, and $gT_f = T_{gf}$ if g is a tempered C^∞ function.

Solution:

Let $f \in \mathcal{S}(\mathbb{R}^n)$. Then for $\phi \in \mathcal{S}(\mathbb{R}^n)$, one has

$$\begin{aligned} T_{\lambda(x)f}(\phi) &= \int \lambda(x)f(y)\,\phi(y)\,dy \\ &= \int f(y-x)\phi(y)\,dy \\ &= \int f(y)\phi(y+x)\,dy \\ &= \int f(y)\lambda(-x)\phi(y)\,dy \\ &= T_f(\lambda(-x)\phi) \\ &= (\lambda(x)T_f)(\phi) \end{aligned}$$

and

$$\begin{aligned} T_{gf}(\phi) &= \int g(y)f(y)\phi(y)\,dy \\ &= \int f(y)g(y)\phi(y)\,dy \\ &= T_f(g\phi) \\ &= gT_f(\phi) \end{aligned}$$

when g is a tempered C^∞ function.

□

3. Let $x \in \mathbb{R}^n$. Determine the following Fourier transforms:

(a) $\mathcal{F}(e_x)$

(b) $\mathcal{F}(\epsilon_x)$ where ϵ_x is the point mass measure at x.

Explain why one writes $\mathcal{F}(\epsilon_x) = e_{-x}$ and $\mathcal{F}(e_x) = \epsilon_x$.

Solution:

Now $e_x(y) = e^{2\pi i x\cdot y}$. Now let S be the tempered distribution defined by e_x. Then by Proposition 3.8,

$$\begin{aligned} \hat{S}(\phi) &= S(\hat{\phi}) \\ &= \int e^{2\pi i x\cdot y}\hat{\phi}(y)\,dy \\ &= \phi(x) \\ &= \epsilon_x(\phi)\,. \end{aligned}$$

So $\mathcal{F}T_{e_x} = \epsilon_x$. When $S = \epsilon_x$, then

$$\begin{aligned}\hat{S}(\phi) &= S(\hat{\phi}) \\ &= \epsilon_x(\hat{\phi}) \\ &= \hat{\phi}(x) \\ &= \int e^{-2\pi i x\cdot y}\phi(y)\,dy \\ &= T_{e_{-x}}(\phi).\end{aligned}$$

Using e_{-x} to denote the tempered distribution $T_{e_{-x}}$, we would then have $\mathcal{F}(\epsilon_x) = e_{-x}$ and $\mathcal{F}(e_x) = \epsilon_x$.

□

4. Let $p(x)$ be a polynomial on $\mathbb{R}^n$ and for $x \in \mathbb{R}^n$, let ϵ_x be the point mass measure at x. Determine the following:

(a) $\mathcal{F}(p\epsilon_x)$

(b) $\mathcal{F}(p(D)\epsilon_x)$

Solution:

For (a), we have

$$\begin{aligned}\mathcal{F}(p\epsilon_x)(\phi) &= p\epsilon_x(\hat{\phi}) \\ &= \epsilon_x(p\hat{\phi}) \\ &= p(x)\hat{\phi}(x) \\ &= p(x)\int \phi(y)e^{-2\pi i y\cdot x}\,dy \\ &= p(x)T_{e_{-x}}(\phi)\end{aligned}$$

and thus $\mathcal{F}(p\epsilon_x) = p(x)e_{-x}$. One could also use (i) of Theorem 4.48 and Exercise 4.5.3. Namely $\mathcal{F}(p\epsilon_x) = p(-\frac{D}{2\pi i})\mathcal{F}(\epsilon_x) = p(-\frac{D_y}{2\pi i})e^{-2\pi i x\cdot y}$ and then use $D_y^\alpha e^{-2\pi i x\cdot y} = (-2\pi i x)^\alpha e^{-2\pi i x\cdot y}$ to see $\left(-\frac{D_y}{2\pi i}\right)^\alpha e_{-x} = x^\alpha e_{-x}$ which implies the result.

Similarly for (b), we have by (h) of Theorem 4.48 and the previous exercise,

$$\begin{aligned}\mathcal{F}(p(D)\epsilon_x)(\phi) &= \mathcal{F}(p(\frac{D}{2\pi i}2\pi i\cdot)\epsilon_x)(\phi) \\ &= p(2\pi i\cdot)\mathcal{F}(\epsilon_x)(\phi) \\ &= p(2\pi i\cdot)e_{-x}(\phi) \\ &= e_{-x}(p(2\pi i\cdot)\phi) \\ &= \int p(2\pi i y)\phi(y)e^{-2\pi i x\cdot y}\,dy \\ &= T_{p(2\pi i\cdot)e_{-x}}(\phi)\end{aligned}$$

and we write $\mathcal{F}(p(D)\epsilon_x) = p(2\pi i\cdot)e_{-x}(\cdot)$. We could also do this directly by calculating $\mathcal{F}(D^\alpha \epsilon_x)(\phi)$; i.e.,

$$\begin{aligned}
\mathcal{F}(D^\alpha \epsilon_x)(\phi) &= (D^\alpha \epsilon_x)(\hat{\phi}) \\
&= (-1)^{|\alpha|}\epsilon_x(D^\alpha \hat{\phi}) \\
&= (-1)^{|\alpha|}\epsilon_x(y \mapsto D_y^\alpha(\int e^{-2\pi i y\cdot z}\phi(z)\,dz)) \\
&= (-1)^{|\alpha|}\epsilon_x(y \mapsto \int(-2\pi i z)^\alpha e^{-2\pi i y\cdot z}\phi(z)\,dz) \\
&= \int (2\pi i z)^\alpha e^{-2\pi i x\cdot z}\phi(z)\,dz \\
&= T_{(2\pi i\cdot)^\alpha e_{-x}(\cdot)}(\phi)\,.
\end{aligned}$$

□

5. Assume that $n = 1$. Evaluate the Fourier transform of the following distributions:

(a) $f(x) = |x|$

(b) $f(x) = x\,|x|$

(c) $f(x) = xH(x)$

(d) $f(x) = \cos(\lambda x)$ for some $\lambda \in \mathbb{R}$

(e) $f(x) = x\sin(\lambda x)$ for some $\lambda \in \mathbb{R}$

Solution:

We first note that the Fourier transform of 1 is δ for $1(\hat{\phi}) = \int \hat{\phi}(y)\,dy = \phi(0)$. Also if H is the Heaviside function,

$$\begin{aligned}
\hat{H}(\phi) &= \lim_{N\to\infty}\int_0^N \hat{\phi}(x)\,dx \\
&= \lim_{N\to\infty}\int_0^N \int \phi(y)e^{-2\pi i x y}\,dy\,dx \\
&= \lim_{N\to\infty}\int \phi(y)\int_0^N e^{-2\pi i x y}\,dx\,dy \\
&= \lim_{N\to\infty}\int \phi(y)\left.\frac{e^{-2\pi i x y}}{-2\pi i y}\right|_{x=0}^{x=N} dy \\
&= \lim_{N\to\infty}\int \phi(y)\frac{1-e^{-2\pi i N y}}{2\pi i y}\,dy.
\end{aligned}$$

Now by (i) of Theorem 4.48, the Fourier transform of $xH(x)$ as a distribution is given by

$$\mathcal{F}(xH) = -\frac{1}{2\pi i}\frac{d}{dx}\hat{H}\,.$$

So if $f(x) = xH(x)$, then

$$\begin{aligned}\mathcal{F}(f)(\phi) &= -\frac{1}{2\pi i}\hat{H}(-\phi') \\ &= \frac{1}{2\pi i}\lim_{N\to\infty}\int \phi'(y)\frac{1-e^{-2\pi i N y}}{2\pi i y}\,dy\,.\end{aligned}$$

This gives (c). There is a question of whether this can be written in a better form.

For $f(x) = |x| = 2xH(x) - x$ we have again by (i) of 4.48 that

$$\begin{aligned}\hat{f}(\phi) &= \mathcal{F}(2xH(x))(\phi) - \mathcal{F}(x)(\phi) \\ &= \frac{1}{\pi i}\lim_{N\to\infty}\int \frac{1-e^{-2\pi i N}}{2\pi i y}\phi'(y)\,dy + \frac{1}{2\pi i}\left(\frac{d}{dx}\mathcal{F}(1)\right)(\phi) \\ &= \frac{1}{\pi i}\lim_{N\to\infty}\int \frac{1-e^{-2\pi i N}}{2\pi i y}\phi'(y)\,dy + \frac{1}{2\pi i}\left(\frac{d}{dx}\delta\right)(\phi) \\ &= \frac{1}{\pi i}\lim_{N\to\infty}\int \frac{1-e^{-2\pi i N}}{2\pi i y}\phi'(y)\,dy - \frac{1}{2\pi i}\delta(\phi') \\ &= \frac{1}{\pi i}\lim_{N\to\infty}\int \frac{1-e^{-2\pi i N}}{2\pi i y}\phi'(y)\,dy - \frac{1}{2\pi i}\phi'(0)\,.\end{aligned}$$

This answers (a).

Now if $g(x) = xf(x) = x|x|$, we have

$$\begin{aligned}\hat{g}(\phi) &= -\frac{1}{2\pi i}\frac{d}{dx}(\hat{f})(\phi) \\ &= \frac{1}{2\pi i}\hat{f}(\phi') \\ &= \frac{1}{2\pi i}\left(\frac{1}{\pi i}\lim_{N\to\infty}\int \frac{1-e^{-2\pi i N}}{2\pi i y}\phi''(y)\,dy - \frac{1}{2\pi i}\phi''(0)\right).\end{aligned}$$

and we have (b).

For (d) and (e), we first consider $f_\lambda(x) = e^{\lambda i x}$. Note

$$\begin{aligned}
\hat{f}_{2\pi\lambda}(\phi) &= \int e^{2\pi i \lambda x}\hat{\phi}(x)\,dx = \hat{\hat{\phi}}(-\lambda) \\
&= \phi(\lambda) \\
&= \int\int e^{2\pi i\lambda x}\phi(y)e^{-2\pi i x y}\,dy\,dx \\
&= \lim_{N\to\infty}\int \phi(y)\int_{-N}^{N} e^{-2\pi i x(y-\lambda)}\,dx\,dy \\
&= \lim_{N\to\infty}\int \phi(y)\frac{e^{-2\pi i x(y-\lambda)}|_{x=-N}^{x=N}}{-2\pi i(y-\lambda)}\,dy \\
&= \lim_{N\to\infty}\int \phi(y)\frac{e^{-2\pi i N(y-\lambda)} - e^{2\pi i N(y-\lambda)}}{-2\pi i(y-\lambda)}\,dy \\
&= \lim_{N\to\infty} N\int \phi(y)\frac{\sin(2\pi N(y-\lambda))}{\pi N(y-\lambda)}\,dy \\
&= \lim_{N\to\infty}\int \phi(y)(2N\,\mathrm{sinc}(2N(y-\lambda))\,dy.
\end{aligned}$$

So $\hat{f}_{2\pi\lambda} = \delta_\lambda$. So $\hat{f}_\lambda = \delta_{\lambda/2\pi}$. Now Set $c_\lambda(x) = \cos(\lambda x)$ and $s_\lambda(x) = \sin(\lambda x)$. So $f_\lambda = c_\lambda + i s_\lambda$ and $f_{-\lambda} = c_\lambda - i s_\lambda$ Thus $c_\lambda = \frac{1}{2}(f_\lambda + f_{-\lambda})$ and $s_\lambda = \frac{1}{2i}(f_\lambda - f_{-\lambda})$. Thus $\hat{c}_\lambda = \frac{1}{2}(\delta_{\lambda/2\pi} + \delta_{-\lambda/2\pi})$ and $\hat{s}_\lambda = \frac{1}{2i}(\delta_{\lambda/2\pi} - \delta_{-\lambda/2\pi})$. So if $f(x) = \cos\lambda x$, $\hat{f} = \frac{1}{2}(\delta_{\lambda/2\pi} + \delta_{-\lambda/2\pi})$ and this gives (d).

For (e), we have $f(x) = x s_\lambda(x)$. So again using (i) of 4.48,

$$\begin{aligned}
\hat{f}(\phi) &= \left(\frac{-D}{2\pi i}\right)\hat{s}_\lambda(\phi) \\
&= \left(\frac{-D}{2\pi i}\right)(\frac{1}{2i}(\delta_{\lambda/2\pi} - \delta_{-\lambda/2\pi}))(\phi) \\
&= \frac{1}{4\pi i^2}(\delta_{\lambda/2\pi} - \delta_{-\lambda/2\pi})(\phi') \\
&= \frac{1}{4\pi}(\phi'(-\lambda/2\pi) - \phi'(\lambda/2\pi))\,.
\end{aligned}$$

We could write

$$(x\sin\lambda x)' = \frac{1}{4\pi}(\delta'_{\lambda/2\pi} - \delta'_{-\lambda/2\pi})$$

where $\delta'(\phi) = -\delta(\phi')$.

□

6. Let f and g be Schwartz functions. Show using the Fourier transform that if $p(x)$ is a polynomial function, then $p(D)(f * g) = (p(D)f) * g = f * p(D)g$.

Solution:

We need only show $p(D)(f * g) = p(D)f * g$ by the commutativity of convolution. Since the Fourier transform is one-to-one on $\mathcal{S}(\mathbb{R}^n)$, it suffices to show

$$\mathcal{F}(p(D)(f*g)) = \mathcal{F}(p(D)f*g).$$

But by Theorem 3.4 and Lemma 3.3,

$$\mathcal{F}(p(D)(f*g))(\omega) = p(2\pi i\omega)\widehat{f*g}(\omega) = p(2\pi i\omega)\hat{f}(\omega)\hat{g}(\omega)\,.$$

and

$$\begin{aligned}\mathcal{F}(p(D)f*g)(\omega) &= \mathcal{F}(p(D)f)(\omega)\hat{g}(\omega)\\ &= p(2\pi i\omega)\hat{f}(\omega)\hat{g}(\omega)\,.\end{aligned}$$

Thus the result holds.

□

7. This exercise belongs after the next section.

Let T be a distribution on $\mathbb{R}^n$ and let ϕ be a test function. Show if $p(x)$ is a polynomial function, then $p(D)(T*\phi) = (p(D)T)*\phi = T*p(D)\phi$.

Solution:

We need to show $p(D)(T*\phi) = p(D)T*\phi = T*p(D)\phi$. This follows from Proposition 4.57. Indeed by this proposition, it holds for D^α. Now extend linearly.

□

8. One can use the Fourier transform to show each tempered distribution is the difference of a boundary value of a holomorphic function in the upper half plane and one in the lower half plane. This exercise establishes this for a special class of distributions. Let T be a tempered distribution on the real line such that $\hat{T}$ is a continuous function bounded by a polynomial $p(\omega)$. Define $F_\pm$ by

$$F_+(z) := \int_0^\infty e^{2\pi i z\omega}\hat{T}(\omega)\,d\omega\,, \qquad \mathrm{Im}(z) > 0$$

and

$$F_-(z) := -\int_{-\infty}^0 e^{2\pi i z\omega}\hat{T}(\omega)\,d\omega\,, \qquad \mathrm{Im}(z) < 0\,.$$

Show the following:

(a) The function F_+ is holomorphic in $\mathbb{C}_+ = \{z \in \mathbb{C} \mid \mathrm{Im}(z) > 0\}$ and F_- is holomorphic in $\mathbb{C}_- = \{z \in \mathbb{C} \mid \mathrm{Im}(z) < 0\}$.

(b) Define

$$\langle T_\pm, \varphi\rangle = \lim_{y\to 0\pm}\int_{-\infty}^\infty \varphi(x)F_\pm(x+iy)\,dx\,.$$

Show T_+ and T_- are distributions and

$$T = T_+ - T_-\,.$$

(c) Show if $T = \delta_0$ then $T_+ = \frac{-1}{2\pi i}\frac{1}{x+i0}$ and $T_- = \frac{-1}{2\pi i}\frac{1}{x-i0}$.

(d) Assume that $\hat{T}(\omega) = \omega^n$. Find $T_{\pm}$.

Solution: For (a), note we know $\hat{T}(\omega)$ is a continuous function bounded by a polynomial. Thus there is a $B > 0$ and an N such that $|\hat{T}(\omega)| \leqslant B\omega^N$ for $\omega \geqslant 1$. So if $z = a + bi$ with $b > \delta$ where $\delta > 0$ and $\omega \geqslant 1$, one has

$$\begin{aligned}
|e^{2\pi i z\omega}\hat{T}(\omega)| &= |e^{-2\pi b\omega}\hat{T}(\omega)| \\
&\leqslant Be^{-2\pi b\,\omega}\omega^N \\
&\leqslant B(N+2)!\frac{\omega^N}{(2\pi b\omega)^{N+2}} \\
&\leqslant B(N+2)!\frac{\omega^{-2}}{(2\pi\delta)^{N+2}}
\end{aligned}$$

which is integrable in ω on $[1, \infty)$. This implies $(t, \omega) \mapsto e^{2\pi i\gamma(t)\omega}\gamma'(t)\hat{T}(\omega)$ on $[0, 1] \times (0, \infty)$ is integrable for any smooth simple closed curve γ whose range is in the open domain $\operatorname{Im} z > \delta$. Consequently, by Fubini's Theorem,

$$\begin{aligned}
\oint_\gamma F_+(z)\,dz &= \int_0^1 \int_0^\infty e^{2\pi i\gamma(t)\omega}\hat{T}(\omega)\,d\omega\,\gamma'(t)\,dt \\
&= \int_0^\infty \hat{T}(\omega) \int_0^1 e^{2\pi i\gamma(t)\omega}\gamma'(t)\,dt\,d\omega \\
&= \int_0^\infty \hat{T}(\omega) \oint_\gamma e^{2\pi i z\omega}\,dz\,d\omega = 0\,.
\end{aligned}$$

Being true for all $\delta > 0$, we see by Morera's Theorem that $F_+(z)$ is holomorphic on $\operatorname{Im} z > 0$. Similarly $F_-(z)$ is holomorphic on $\operatorname{Im} z < 0$.

For (b) we start by calculating $\int F_+(x + iy)\phi(x)\,dx$. Note by the argument above, $(\omega, x) \mapsto \hat{T}(\omega)e^{2\pi i(x+iy)}\phi(x)$ is integrable for $y > 0$ on $(0, \infty) \times \mathbb{R}$. Again using Fubini,

$$\begin{aligned}
\int F_+(x + iy)\phi(x)\,dx &= \int\int_0^\infty \hat{T}(\omega)e^{2\pi i(x+iy)\omega}\,d\omega\,\phi(x)\,dx \\
&= \int_0^\infty \hat{T}(\omega) \int \phi(x)e^{2\pi i\omega x}e^{-2\pi y\omega}\,dx\,d\omega \\
&= \int_0^\infty \hat{T}(\omega)\hat{\phi}(-\omega)e^{-2\pi y\omega}\,d\omega.
\end{aligned}$$

Thus the limit as $y \to 0+$ is the distribution T_+ where

$$T_+(\phi) = \int_0^\infty \hat{T}(\omega)\hat{\phi}(-\omega)\,d\omega.$$

Similarly

$$\begin{aligned}
T_-(\phi) &= \lim_{y\to 0-} \int F_-(x+iy)\phi(x)\,dx \\
&= -\lim_{y\to 0-} \int\int_{-\infty}^{0} \hat{T}(\omega)e^{2\pi i(x+iy)\omega}\,d\omega\,\phi(x)\,dx \\
&= -\lim_{y\to 0-} \int_{-\infty}^{0} \hat{T}(\omega)e^{-2\pi y\omega} \int e^{2\pi ix\omega}\,\phi(x)\,dx\,d\omega \\
&= -\lim_{y\to 0-} \int_{-\infty}^{0} \hat{T}(\omega)e^{-2\pi y\omega}\hat{\phi}(-\omega)\,d\omega \\
&= -\int_{-\infty}^{0} \hat{T}(\omega)\hat{\phi}(-\omega)\,d\omega
\end{aligned}$$

Thus

$$(T_+ - T_-)(\phi) = \int \hat{T}(\omega)\hat{\phi}(-\omega)\,d\omega = \mathcal{F}^2(T)(\check{\phi}) = \check{T}(\check{\phi}) = T(\phi).$$

Now for (c), we take $T = \delta_0$. One has $\int \hat{T}(\omega)\phi(\omega)\,d\omega = T(\hat{\phi}) = \hat{\phi}(0) = \int \phi(\omega)\,d\omega$. So $\hat{T}(\omega) = 1$. Thus

$$\begin{aligned}
F_+(x+iy) &= \int_0^{\infty} \hat{T}(\omega)e^{2\pi(ix-y)\omega}\,d\omega \\
&= \int_0^{\infty} e^{2\pi(ix-y)\omega}\,d\omega \\
&= \frac{1}{2\pi(ix-y)} e^{2\pi(ix-y)\omega}\Big|_{\omega=0}^{\omega=\infty} \\
&= \frac{-1}{2\pi i(x+iy)}.
\end{aligned}$$

This shows $T_+ = \frac{-1}{2\pi i(x+i0)}$. Similarly,

$$\begin{aligned}
F_-(x+iy) &= -\int_{-\infty}^{0} e^{2\pi i(x+iy)\omega}\hat{T}(\omega)\,d\omega \\
&= -\int_{-\infty}^{0} e^{2\pi i(x+iy)\omega}\,d\omega \\
&= -\frac{1}{2\pi i(x+iy)} e^{2\pi i(x+iy)\omega}\Big|_{\omega=-\infty}^{\omega=0} \\
&= -\frac{1}{2\pi i(x+iy)}.
\end{aligned}$$

So

$$T_- = -\frac{1}{2\pi i(x-i0)}.$$

For a direct argument, note if $\epsilon > 0$, then

$$
\begin{aligned}
(T_+ - T_-)(\phi) &= \lim_{y\to 0+} \int_{-\infty}^{\infty} \left(\frac{-1}{2\pi i(x+iy)} + \frac{1}{2\pi i(x-iy)} \right) \phi(x)\, dx \\
&= \lim_{y\to 0+} \int_{|x|\leqslant\epsilon} \left(\frac{-1}{2\pi i(x+iy)} + \frac{1}{2\pi i(x-iy)} \right) \phi(x)\, dx \\
&+ \lim_{y\to 0+} \int_{|x|\geqslant\epsilon} \left(\frac{-1}{2\pi i(x+iy)} + \frac{1}{2\pi i(x-iy)} \right) \phi(x)\, dx \\
&= \lim_{y\to 0+} \int_{-\epsilon}^{\epsilon} \left(\frac{-1}{2\pi i(x+iy)} + \frac{1}{2\pi i(x-iy)} \right) \phi(x)\, dx \\
&= \frac{1}{2\pi i} \lim_{y\to 0+} \int_{-\epsilon}^{\epsilon} \frac{2iy}{x^2+y^2} \phi(x)\, dx \\
&= \lim_{y\to 0+} \frac{1}{\pi}\phi(0) \int_{-\epsilon}^{\epsilon} \frac{y}{x^2+y^2}\, dx + \lim_{y\to 0+} \frac{1}{\pi} \int_{-\epsilon}^{\epsilon} \frac{(\phi(x)-\phi(0))y}{x^2+y^2}\, dx \\
&= \frac{1}{\pi}\phi(0) \lim_{y\to 0+} \arctan\left(\frac{x}{y}\right)\Big|_{x=-\epsilon}^{x=\epsilon} + 0 \\
&= \phi(0)
\end{aligned}
$$

for $\left| \frac{(\phi(x)-\phi(0))y}{x^2+y^2} \right| \leqslant |\phi'|_\infty \frac{|xy|}{x^2+y^2}$ and thus

$$
\begin{aligned}
\left| \int_{-\epsilon}^{\epsilon} \frac{(\phi(x)-\phi(0))y}{x^2+y^2}\, dx \right| &\leqslant 2y\, |\phi'|_\infty \int_0^{\epsilon} \frac{x}{x^2+y^2}\, dx \\
&\leqslant y\, |\phi'|_\infty \ln(x^2+y^2)|_{x=0}^{x=\epsilon} \\
&= y|\phi'|_\infty \left(\ln(\epsilon^2+y^2) - \ln(y^2) \right) \to 0 \text{ as } y \to 0+\ .
\end{aligned}
$$

We remark part (c) was also shown in Exercise 4.3.16.

Finally for (d), we find

$$
\begin{aligned}
F_+(x+iy) &= \int_0^{\infty} \hat{T}(\omega) e^{2\pi i(x+iy)\omega}\, d\omega \\
&= \int_0^{\infty} \omega^n e^{2\pi i(x+iy)\omega}\, d\omega.
\end{aligned}
$$

This gives

$$
\begin{aligned}
F_+(x+iy) &= \omega^n \frac{e^{2\pi i(x+iy)\omega}}{2\pi i(x+iy)}\Big|_{\omega=0}^{\omega=\infty} - \int_0^{\infty} n\omega^{n-1} \frac{e^{2\pi i(x+iy)\omega}}{2\pi i(x+iy)}\, d\omega \\
&= + \int_0^{\infty} n(n-1)\omega^{n-2} \frac{e^{2\pi i(x+iy)\omega}}{[2\pi i(x+iy)]^2}\, d\omega \\
&\ \ \vdots \\
&= (-1)^n n! \int_0^{\infty} \frac{e^{2\pi i(x+iy)\omega}}{[2\pi i(x+iy)]^n}\, d\omega
\end{aligned}
$$

$$= \frac{(-1)^{n+1} n!}{2^{n+1}\pi^{n+1} i^{n+1}(x+iy)^{n+1}}.$$

In a similar vein, one sees the same formula works for $F_-(x+iy)$. Thus

$$T = \left(\frac{-1}{2\pi i}\right)^{n+1} n! \left(\frac{1}{(x+i0)^{n+1}} - \frac{1}{(x-i0)^{n+1}}\right).$$

□

Exercise Set 4.6

1. Let T be a tempered distribution and let $\varphi \in \mathcal{S}(\mathbb{R}^n)$. Show $\widehat{T * \varphi} = \hat{\varphi}\hat{T}$.
Solution: We start by proving if $H \in C_c^\infty(\mathbb{R}^{2n})$, then

$$\int dT(y) \left(\int H(x,y)\,dx\right) = \int \left(\int dT(y) H(x,y)\right) dx.$$

We first note $\int H(x,y)\,dx$ is C^∞ with compact support in $\mathbb{R}^n$ and

$$D_y^\alpha \left(\int H(x,y)\,dx\right) = \int D_y^\alpha H(x,y)\,dx.$$

Moreover, these integrals can be taken in the Riemann sense. That is

$$\int H(x,y)\,dx = \lim_P \sum H(x_i,y) m(E_i)$$

where E_i are rectangles which form a partition of a fixed rectangle R covering the compact set $\{x : (x,y) \in \text{supp } H \text{ for some } y\}$. We note this limit is in the topology of $\mathcal{D}(\mathbb{R}^n)$. Indeed

$$\left|\int H(x,y)\,dx - \sum H(x_i,y) m(E_i)\right|_\alpha = \max_y \left|\int D_y^\alpha H(x,y)\,dx - \sum m(E_i) D_y^\alpha H(x_i,y)\right|.$$

Replacing $H(x,y)$ by $D_y^\alpha H(x,y)$, we see it suffices to show

$$\max_y \left|\int H(x,y)\,dx - \sum H(x_i,y) m(E_i)\right| \to 0$$

as the partition's norm goes to 0.

But $\int H(x,y)\,dx = \sum \int_{E_i} H(x,y)\,dx$. Thus

$$\max_y \left|\int H(x,y)\,dx - \sum H(x_i,y) m(E_i)\right| = \max_y \left|\sum_i \int_{E_i} H(x,y)\,dx - H(x_i,y)\int_{E_i} dx\right|$$
$$\leqslant \max_y \sum_i \max_{x \in E_i} |H(x,y) - H(x_i,y)| m(E_i) \to 0$$

as the maximum diameter of the E_i goes to 0 for H is uniformly continuous. Thus

$$\int dT(y)\left(\int H(x,y)\,dx\right) = \lim \sum m(E_i)\int dT(y)H(x_i,y)$$
$$= \int\left(\int dT(y)H(x,y)\right)\,dx.$$

We now note if ϕ and f both are C^∞ functions with compact support, then

$$\begin{aligned}(T*\phi)(f) &= \int\left(\int dT(y)\phi(x-y)\right)f(x)\,dx\\ &= \int\left(\int dT(y)(\phi(x-y)f(x)\right)\,dx\\ &= \int dT(y)\left(\int \phi(x-y)f(x)\,dx\right)\\ &= \int dT(y)(\check{\phi}*f)\,.\end{aligned}$$

Hence $(T*\phi)(f) = T(\check{\phi}*f)$. Now suppose $\phi \in \mathcal{S}_n$. Choose $\phi_k \in \mathcal{D}(\mathbb{R}^n)$ with $\phi_k \to \phi$ in $\mathcal{S}_n$. We begin by noting if ϕ_k is close to ϕ in $\mathcal{S}_n$, then $\lambda(x)\check{\phi}_k$ is uniformly close to $\lambda(x)\check{\phi}$ on compact sets K of x. Namely given α and N and a compact subset K of $\mathbb{R}^n$, there is for each $\epsilon > 0$ a $k_{N,\alpha}(\epsilon)$ such that for $k \geqslant k_{N,\alpha}(\epsilon)$ one has $|\lambda(x)\check{\phi}_k - \lambda(x)\check{\phi}|_{N,\alpha} \leqslant \epsilon$ for all $x \in K$. To see this note using Exercise 2.3.4 that

$$\begin{aligned}&(1+|y|^2)^N|D^\alpha(\lambda(x)\check{\phi}_k)(y) - D^\alpha(\lambda(x)\check{\phi})(y)|\\ &\qquad = (1+|y|^2)^N|\lambda(x)\widecheck{D^\alpha\phi_k}(y) - \lambda(x)\widecheck{D^\alpha\phi}(y)|\\ &\qquad = (1+|y|^2)^N|D^\alpha\phi_k(x-y) - D^\alpha\phi(x-y)|\\ &= \left(\frac{1+|y|^2}{1+|x-y|^2}\right)^N (1+|x-y|^2)^N|D^\alpha\phi_k(x-y) - D^\alpha\phi(x-y)|\\ &\leqslant \max_{x\in K}(1+|x|^2)^N\left(\frac{1+|y|^2}{(1+|y-x|^2)(1+|x|^2)}\right)^N|\phi_k-\phi|_{N,\alpha}\\ &\qquad\leqslant 2^N\max_{x\in K}(1+|x|^2)^N|\phi_k-\phi|_{N,\alpha} < \epsilon\end{aligned}$$

if k is large. Since T is tempered and thus continuous in the Schwartz topology, we see if K is a compact subset of $\mathbb{R}^n$, $T(\lambda(x)\check{\phi}_k)$ converges uniformly to $T(\lambda(x)\check{\phi})$ for $x \in K$. This along with continuity of convolution yields

$$T*\phi(f) = \lim_k \int T*\phi_k(x)f(x)\,dx = \lim_k (T*\phi_k)(f) = \lim_k T(\check{\phi}_k * f) = T(\check{\phi}*f)$$

for $f \in \mathcal{D}(\mathbb{R}^n)$.

By Proposition 4.58, $T*\phi \in \mathcal{S}'_\infty(\mathbb{R}^n)$ and thus the function $T*\phi$ is a tempered distribution. Consequently using the density of $\mathcal{D}(\mathbb{R}^n)$ and the continuity of convolution, we obtain $(T*\phi)(f) = T(\check{\phi}*f)$ for all $f, \phi \in \mathcal{S}_n$.

Finally $\mathcal{F}(T*\phi)(f) = T*\phi(\hat{f}) = T(\check{\phi}*\hat{f}) = T(\mathcal{F}(\mathcal{F}^{-1}(\check{\phi})f)) = \hat{T}(\mathcal{F}(\phi)f) = \hat{\phi}\hat{T}(f)$. □

2. Show $\delta_0 * \phi = \phi$ for all test functions ϕ.

Solution:

Now $\delta_0 * \phi$ is given by

$$\delta_0 * \phi(x) = \delta_0((\lambda(x)\check{\phi})) = \lambda(x)\check{\phi}(0) = \check{\phi}(-x) = \phi(x) = \delta_x(\phi)\,.$$

So $\delta_0 * \phi = \phi$. □

3. Let H be the Heaviside function. Show $H * \varphi(x) = \int_{-\infty}^{x} \varphi(y)\,dy$ for $\varphi \in \mathcal{S}(\mathbb{R})$.

Solution:

We have

$$\begin{aligned} H * \phi(x) &= H(\lambda(x)\check{\phi}) \\ &= \int H(y)\lambda(x)\check{\phi}(y)\,dy \\ &= \int \chi_{[0,\infty)}(y)\check{\phi}(y-x)\,dy \\ &= \int \chi_{[0,\infty)}(y)\phi(x-y)\,dy \\ &= \int \chi_{[0,\infty)}(y+x)\phi(-y)\,dy \\ &= \int \chi_{[0,\infty)}(x-y)\phi(y)\,dy \\ &= \int \chi_{(-\infty,x]}(y)\phi(y)\,dy \\ &= \int_{-\infty}^{x} \phi(y)\,dy\,. \end{aligned}$$

One can write this as $H * \phi(x) = T_{\chi_{(-\infty,x]}}(\phi) = ((1-\lambda(x))H)(\phi)$.

□

4. Suppose that S is a compactly supported distribution. Show $\varphi \mapsto S * \varphi$ is a continuous linear transformation of $\mathcal{E}(\mathbb{R}^n)$ into $\mathcal{E}(\mathbb{R}^n)$.

Solution: Let S be a distribution with compact support. Since S is a continuous linear functional on $\mathcal{E}(\mathbb{R}^n)$ and $S*\phi(x) = S(\lambda(x)\check{\phi})$, we see if $x \mapsto \lambda(x)\check{\phi}$ is a continuous mapping of $\mathbb{R}^n$ into $\mathcal{E}(\Omega)$, then $S*\phi$ is continuous.. This, however, is a consequence of

$$\begin{aligned}|\lambda(y)\check{\phi} - \lambda(x)\check{\phi}|_{K,\alpha} &= \max_{z\in K} |D^\alpha(\lambda(y)\check{\phi})(z) - D^\alpha(\lambda(x)\check{\phi})(z)| \\ &= \max_{z\in K} |(\lambda(y)D^\alpha\check{\phi})(z) - (\lambda(x)D^\alpha\check{\phi})(z)| \\ &= \max_{z\in K} |(-1)^{|\alpha|}(\lambda(y)\widetilde{D^\alpha\phi})(z) - (-1)^{|\alpha|}(\lambda(x)\widetilde{D^\alpha\phi})(z)| \\ &= |\lambda(y)(D^\alpha\phi) - \lambda(x)(D^\alpha\phi)|_{-K,0} \to 0\end{aligned}$$

as $y \to x$ by the uniform continuity of $D^\alpha\phi$ on $-K$. So $S * \phi$ is continuous on $\mathbb{R}^n$ for any distribution S of compact support.

Now the same argument as in the proof of Lemma 4.51 shows if $\phi \in \mathcal{E}(\mathbb{R}^n)$, then $\phi_h \to \partial_j\phi$ in $\mathcal{E}(\Omega)$ as $h \to 0$ if $\phi_h(x) = \frac{\phi(x+he_j)-\phi(x)}{h}$. Thus

$$\begin{aligned}\partial_j(S*\phi)(x) &= \lim_{h\to 0}\frac{S*\phi(x+he_j) - S*\phi(x)}{h} \\ &= \lim_{h\to 0}\frac{S(\lambda(x+he_j)\check{\phi}) - S(\lambda(x)\check{\phi})}{h} \\ &= \lim_{h\to 0} S\left(\frac{\lambda(x+he_j)\check{\phi} - \lambda(x)\check{\phi}}{h}\right) \\ &= -\lim_{h\to 0} S\left(\frac{\lambda(he_j)\lambda(x)\check{\phi} - \lambda(x)\check{\phi}}{-h}\right) \\ &= -S(\partial_j(\lambda(x)\check{\phi})) \\ &= -S(\lambda(x)\partial_j\check{\phi}) \\ &= S(\lambda(x)\widetilde{\partial_j\phi}) \\ &= S*\partial_j(\phi)\,.\end{aligned}$$

So by repeating we see $D^\alpha(S*\phi) = S*D^\alpha\phi$ exist and are continuous functions for all α. Thus $S*\phi \in \mathcal{E}(\mathbb{R}^n)$. Moreover, this also implies $D^\alpha(S*\phi) = (D^\alpha S)*\phi$.

To see $\phi \mapsto S*\phi$ is continuous, consider a compact subset K' of $\mathbb{R}^n$ and an $\alpha \in \mathbb{N}_0^n$. Let K be a compact set whose interior contains the support of S. One can check $D^\alpha S$ has compact support in the interior of K and since $D^\alpha S \in \mathcal{E}(\mathbb{R}^n)'$, there is a finite set F of β and a $\delta > 0$ such that

$$|(D^\alpha S)(\phi)| \leqslant 1 \text{ if } |\phi|_{K,\beta} < \delta \ \text{ for } \beta \in F\,.$$

Note

$$\begin{aligned}|D^\beta(\lambda(x)\check{\phi})(y)| &= |\lambda(x)D^\beta\check{\phi}(y)| \\ &= |\lambda(x)\widetilde{D^\beta\phi}(y)| \\ &= |D^\beta\phi(x-y)| \\ &\leqslant |\phi|_{K'-K,\beta}\end{aligned}$$

if $y \in K$ and $x \in K'$. Thus $|\lambda(x)\check{\phi}|_{K,\beta} \leqslant |\phi|_{K'-K,\beta}$ if $x \in K'$. Hence if $|\phi|_{K'-K,\beta} < \delta$ for $\beta \in F$, $|D^\alpha(S * \phi)(x)| = |(D^\alpha S)(\lambda(x)\check{\phi})| \leqslant 1$ for $x \in K'$. This gives continuity.

□

5. Suppose T and S are distributions on $\mathbb{R}^n$ with compact support. Show $\operatorname{supp}(S * T) \subseteq \operatorname{supp} S + \operatorname{supp} T$.

Solution:

We break this down into some simpler steps. Assume T has compact support and ϕ is C^∞ with compact support. Note if $T(\phi) \neq 0$, then $\operatorname{supp} T \cap \operatorname{supp} \phi \neq \varnothing$. This implies $\operatorname{supp}(T * \phi) \subseteq \operatorname{supp} T + \operatorname{supp} \phi$. Indeed, $T * \phi(x) = T(\lambda(x)\check{\phi}) \neq 0$ implies $\operatorname{supp} T \cap \operatorname{supp}(\lambda(x)\check{\phi}) \neq \varnothing$. But $\operatorname{supp}(\lambda(x)\check{\phi}) = x - \operatorname{supp} \phi$. So if $T * \phi(x) \neq 0$, there is a $y \in \operatorname{supp} T \cap (x - \operatorname{supp} \phi)$. Thus $x \in \operatorname{supp} T + \operatorname{supp} \phi$. This implies $\check{T} * \phi$ has support in $\operatorname{supp} \phi - \operatorname{supp} T$.

To finish let T have compact support K and S have compact support K'. Suppose ϕ is a C^∞ with compact support a subset of the complement of $K + K'$. Thus $\check{T} * \phi$ has support in $\operatorname{supp} \phi - K$. This is a compact set missing K'. Since S has compact support K', one has $S(\check{T} * \phi) = 0$. So $(S * T)(\phi) = 0$. Thus the support of $S * T$ is a subset of $K + K'$.

□

6. Let T be a tempered distribution on $\mathbb{R}^n$. Suppose $\lambda(x)T = T$ for all x. Show $T = c\,dx_n$; i.e. T is a multiple of Lebesgue measure.

Solution: Assume $T \neq 0$. Let $\epsilon(x)$ be a C^∞ function of x with $0 \leqslant \epsilon \leqslant 1$ and having support in $|x| \leqslant 1$ and satisfying $\int \epsilon(x)\, d_n x = 1$. Set $\epsilon_k(x) = k^n \epsilon(kx)$. The ϵ_k is a C^∞ function with $0 \leqslant \epsilon_k \leqslant k^n$, $\int \epsilon_k\, d_n x = 1$ and ϵ_k has support in $|x| \leqslant \frac{1}{k}$. By the argument in Exercise 4.6.1, we know

$$T * \epsilon_k(f) = T(\check{\epsilon}_k * f)$$

for $f \in \mathcal{S}_n$. Note if $f \in \mathcal{D}(\mathbb{R}^n)$, then $\check{\epsilon}_k * f \to f$ in $\mathcal{D}(\mathbb{R}^n)$. Indeed, since by Proposition 4.57, $D^\alpha(\check{\epsilon}_k * f) = \check{\epsilon}_k * D^\alpha f$, it suffices to show $|\check{\epsilon}_k * f - f|_0 \to 0$. But

$$\begin{aligned} \left|\int \epsilon_k(y-x) f(y)\, d_n y - f(x)\right| &= \left|\int \epsilon_k(y)(f(y+x) - f(x))\, d_n y\right| \\ &< \epsilon \end{aligned}$$

if $k \geqslant K$ where K is chosen such that $|f(y) - f(z)| < \epsilon$ if $|y - z| < \frac{1}{K}$.

Suppose $f \in \mathcal{D}(\mathbb{R}^n)$. We note $T * \epsilon_k(x) = T(\lambda(x)\check{\epsilon}_k) = (\lambda(-x)T)(\check{\epsilon}_k) = T(\check{\epsilon}_k) = \lambda(0)T(\check{\epsilon}_k) = T * \epsilon_k(0)$. Hence $T * \epsilon_k$ is a constant function c_k. Now $T * \epsilon_k(f) = c_k \int f(x)\, d_n x = T(\check{\epsilon}_k * f) \to T(f)$ since $\check{\epsilon}_k * f \to f$ in $\mathcal{D}(\mathbb{R}^n)$. Thus $T(f) = \lim_k c_k \int f(x)\, d_n x$. In particular, $c = \lim c_k$ exists and $T(f) = c \int f(x)\, d_n x$ for $f \in \mathcal{D}(\mathbb{R}^n)$. Since $\mathcal{D}(\mathbb{R}^n)$ is dense in $\mathcal{S}_n$, the we have $T(f) = c \int f(x)\, d_n x$ for all $f \in \mathcal{S}_n$. □

EXERCISE SET 4.7

1. Let $f \geqslant 0$ be a continuous function such that $\int_{\mathbb{R}^n} f\, dx = 1$. Define $f_m(x) := m^n f(mx)$. Show

$$\lim_{m \to \infty} T_{f_m} = \delta_0.$$

Solution:

We have $f \geqslant 0$ and $\int_{\mathbb{R}^n} f(x)\, dx = 1$, and $f_m(x) = m^n f(mx)$. Let $\phi \in \mathcal{D}(\mathbb{R}^n)$. Then

$$\begin{aligned} \int f_m(x)\phi(x)\, dx &= \int m^n f(mx)\phi(x)\, dx \\ &= \int f(x)\phi(\frac{x}{m})\, dx. \end{aligned}$$

Now $|f(x)\phi(\frac{x}{m})| \leqslant |\phi|_\infty |f(x)|$. So $g_m(x) = f(x)\phi(\frac{x}{m})$ is dominated by an L^1 function and pointwise $g_m(x) \to f(x)\phi(0)$ as $m \to \infty$. Thus by the Lebesgue Dominated Convergence Theorem, $T_{f_m}(\phi) \to \int f(x)\phi(0)\, dx = \phi(0) = \delta_0(\phi)$ as $m \to \infty$.

□

2. Let $s \in (-\infty, \infty)$. Define $\mathcal{H}_s$ to be the Hilbert space of tempered distributions obtained by Fourier transforming the Hilbert space $L^2(\mathbb{R}^n, (1+|x|^2)^s\, dx)$; i.e.,

$$\mathcal{H}_s = \{\mathcal{F}f \mid \int |f(x)|^2 (1+|x|^2)^s\, dx < \infty\}$$

with

$$(\mathcal{F}f, \mathcal{F}g) = \int f(x)\bar{g}(x)(1+|x|^2)^s\, dx.$$

$\mathcal{H}_s$ is the Sobolev space of order s on $\mathbb{R}^n$; it is sometimes denoted by $W^{s,2}(\mathbb{R}^n)$.

(a) Show $\mathcal{H}_s \supseteq \mathcal{H}_{s'}$ if $s < s'$ and $\mathcal{H}_s \subseteq L^2(\mathbb{R}^n)$ if and only if $s \geqslant 0$.

(b) Given an integer $p \geqslant 0$, find the smallest index s such that $\mathcal{H}_{s'} \subseteq C^p(\mathbb{R}^n)$ if $s' > s$.

(c) Show the differential operator D^α is a bounded linear operator from $\mathcal{H}_s$ into $\mathcal{H}_{s-|\alpha|}$.

(d) Let $\alpha \in \mathbb{N}_0^n$. Determine which $\mathcal{H}_s$ contain the distribution $D^\alpha \epsilon_0$ where ϵ_0 is point mass at 0.

(e) Show $(1 - \frac{1}{4\pi^2}\Delta)$ is a unitary mapping of $\mathcal{H}_s$ onto $\mathcal{H}_{s-4}$.

(f) Show $\mathcal{H}_s(+) := \cup_{s'>s}\mathcal{H}_{s'}$ and $\mathcal{H}_s(-) := \cap_{s'<s}\mathcal{H}_{s'}$ are proper subspaces of $\mathcal{H}_s$.

(g) Show $\mathcal{H}_\infty = \cap_s \mathcal{H}_s$ is the subspace of $C^\infty(\mathbb{R}^n)$ consisting of all f with $D^\alpha f \in L^2(\mathbb{R}^n)$ for all $\alpha \in \mathbb{N}_0^n$.

Solution:

For (a), we note if $s < s'$, then $\int |f(x)|^2(1+|x|^2)^s\,dx < \int |f(x)|^2(1+|x|^2)^{s'}\,dx$. Thus $\mathcal{H}_{s'} \subseteq \mathcal{H}_s$. Moreover, if $s \geqslant 0$ and $\mathcal{F}f \in \mathcal{H}_s$, then $f \in L^2$ and so $\mathcal{F}f \in L^2$. Also if $s < 0$, then one can choose $f \notin L^2$ with $x \mapsto f(x)(1+|x|^2)^{s/2}$ in L^2; e.g. take $f(x) = (1+|x|^2)^{r/4}$ where $r + 2s < -n < r$.

To start (b), first we note if $f \in L^2$, then $\mathcal{F}f$ need not be a continuous function. So we need only consider the cases where $s > 0$ and all elements in $\mathcal{H}_s$ are given by L^2 functions. Next by the Sobolev Lemma, we know since for $s > 0$, $\mathcal{F}f$ is given by an L^2 function, that the distribution $\mathcal{F}f$ is given by a C^p function if $D^\alpha T_{\mathcal{F}f}$ is locally an L^2 function for any α with $|\alpha| \leqslant s$ where $s > p + \frac{n}{2}$. But by Theorem 4.48, $D^\alpha T_{\mathcal{F}f} = D^\alpha \mathcal{F}(T_f) = \mathcal{F}((-2\pi i x)^\alpha T_f) = \mathcal{F}(T_{(-2\pi i x)^\alpha f})$ and thus $D^\alpha T_{\mathcal{F}f}$ is given by an L^2 function for all $|\alpha| \leqslant s$ where $s > p + \frac{n}{2}$ if $x^\alpha f \in L^2$ for all such α. This occurs if and only if $(1+|x|^2)^{|\alpha|/2} f \in L^2$ for all such α. Thus $\mathcal{F}f$ is a C^p function whenever $(1+|x|^2)^{s/2} f \in L^2$ for all $s > p + \frac{n}{2}$. This gives $\mathcal{H}_s \subseteq C^p(\mathbb{R}^n)$ for all $s > p + \frac{n}{2}$. So $\cap_s \mathcal{H}_s \subseteq C^\infty$ and all $D^\alpha f \in L^2$ for all α. So we also have (g).

To finish (b), we need to show $p + \frac{n}{2}$ is the smallest such number or equivalently for each $s < p + \frac{n}{2}$, there is an $\mathcal{F}f \in \mathcal{H}_s$ with $\mathcal{F}f \notin C^p$. We start with the case $n = 1$. Consider the sinc function defined in Exercise 3.1.2. Using Example 3.1 one sees if $g = \chi_{[-1/2,1/2]}$, then $\hat{g}(\omega) = \operatorname{sinc}(\pi\omega)$. Thus by Lemma 3.3, $\widehat{\delta(\pi^{-1})g}(\omega) = \delta(\pi)\hat{g}(\omega) = \pi^{-1/2}\hat{g}(\pi^{-1}\omega) = \pi^{-1/2}\operatorname{sinc}(\omega)$ and since $\hat{\hat{g}} = \check{g}$, we have

$$\pi^{-1/2}\mathcal{F}(\operatorname{sinc})(x) = \delta(\pi^{-1})g(-x) = \pi^{1/2}\chi_{[-\frac{1}{2},\frac{1}{2}]}(-\pi x).$$

Thus $\mathcal{F}(\operatorname{sinc}) = \pi\chi_{[-\frac{1}{2\pi},\frac{1}{2\pi}]}$ which is not a continuous function. But $(1 - \chi_{[-1,1]})\operatorname{sinc}(\cdot)$ has Fourier transform $\pi\chi_{[-\frac{1}{2\pi},\frac{1}{2\pi}]} - h = \mathcal{F}(\operatorname{sinc}) - \mathcal{F}(\chi_{[-1,1]}\operatorname{sinc})$ and by (g), h is C^∞ for $\mathcal{F}(\chi_{[-1,1]}\operatorname{sinc}) \in \mathcal{H}_s$ for all s. Set $f(x) = (1 - \chi_{[-1,1]}(x))\frac{\operatorname{sinc}(x)}{x^p}$. Note the above shows the Fourier transform of $x^p f(x)$ is $\pi\chi_{[-\frac{1}{2\pi},\frac{1}{2\pi}]}(\omega) - h(\omega)$. Thus $\mathcal{F}(x^p f(x)) = \frac{D^p}{(-2\pi i)^p}\mathcal{F}(f)$ is discontinuous. We next show $\mathcal{F}f \in \mathcal{H}_s$ for $s < p + \frac{1}{2}$. Note f belongs to $L^2(\mathbb{R}, (1+x^2)^s\,dx)$ if

$$2\int_1^\infty \left|\frac{\operatorname{sinc}(x)}{x^p}\right|^2 (1+x^2)^s\,dx \leqslant 2\int_1^\infty \left|\frac{\sin x}{x^{p+1}}\right|^2 (1+x^2)^s\,dx < \infty.$$

But this occurs if $2s - 2p - 2 < -1$ or $s < p + \frac{1}{2}$. We have been unable to extend this argument to $n > 1$. So for now the case $n > 1$ is unresolved and unanswered though it is we think still true.

For (c), note it suffices to show $x^\alpha f \in \mathcal{F}^{-1}(\mathcal{H}_{s-|\alpha|})$ if $f \in \mathcal{F}^{-1}(\mathcal{H}_s)$. But

$$\begin{aligned}\int |x^\alpha f(x)|^2(1+|x|^2)^{s-|\alpha|}\,dx &\leqslant \int |f(x)|^2|x|^{2|\alpha|}(1+|x|^2)^{s-|\alpha|}\,dx \\ &\leqslant \int |f(x)|^2(1+|x|^2)^{|\alpha|}(1+|x|^2)^{s-|\alpha|}\,dx \\ &\leqslant |\mathcal{F}f|_s^2.\end{aligned}$$

Pre Fourier transforming $D^\alpha \epsilon_0$ essentially gives us the function $f(x) = x^\alpha$. So we need to determine when $\int_{|x|\geqslant 1} |x^\alpha|^2 |x|^{2s} < \infty$. This occurs if $\int \prod |x_j|^{2\alpha_j}(1+|x|^2)^s\,dx$. Using polar coordinates, (see Corollary 2.26 and Exercise 2.2.7), this will be finite if and only if $\int_1^\infty r^{2|\alpha|} r^{2s} r^{n-1}\,dr < \infty$. This occurs if and only if $2|\alpha| + 2s + n - 1 < -1$ or $s < -|\alpha| - n/2$. So we have (d).

For (f), we first consider the function $f_t(x) = |x|^t \frac{1-\chi_{B_e}(x)}{\ln|x|}$ where $B_e = \{x \mid |x| \leqslant e\}$. We determine the t for which $f_t \in L^2(\mathbb{R}^n, (1+|x|^2)^s dx)$. We us polar coordinates. Thus this occurs if

$$\begin{aligned}\infty &> \int |f_t(x)|^2 (1+|x|^2)^s\,dx \\ &= \sigma(S^{n-1}) \int_e^\infty \frac{r^{2t}}{(\ln r)^2}(1+r^2)^s r^{n-1}\,dr\,.\end{aligned}$$

But this happens when $\int_e^\infty \frac{r^{2t+2s+n-1}}{(\ln r)^2}\,dr < \infty$. We claim this is finite if and only if $2t + 2s + n - 1 \leqslant -1$. We note when this is -1 or less, then

$$\begin{aligned}\int_e^\infty \frac{r^{2t+2s+n-1}}{(\ln r)^2}\,dr &\leqslant \int_e^\infty (\ln r)^{-2}\frac{1}{r}\,dr \\ &= \left.\frac{-1}{\ln r}\right|_{r=e}^{r=\infty} = 1\end{aligned}$$

while if this $u-1$ which is greater than -1, one has $u > 0$ and thus $\frac{r^u}{\ln r} \to \infty$ as $r \to \infty$ and so there is an $R > e$ with $\frac{r^u}{\ln r} \geqslant 1$ for $r \geqslant R$. Thus

$$\begin{aligned}\int_e^\infty \frac{r^{u-1}}{(\ln r)^2}\,dr &= \int_e^\infty \frac{r^u}{\ln r}(\ln r)^{-1}\frac{1}{r}\,dr \\ &\geqslant \int_R^\infty (\ln r)^{-1}\frac{1}{r}\,dr \\ &= \ln(\ln r)|_{r=R}^{r=\infty} = \infty\,.\end{aligned}$$

Thus $\mathcal{F}f_t \in \mathcal{H}_s$ if and only if $2t + 2s + n - 1 \leqslant -1$ or $s \leqslant -t - \frac{n}{2}$. So $\mathcal{F}f_t \notin \mathcal{H}_s$ if $s > -t - \frac{n}{2}$ but $f_t \in \mathcal{H}_{-t-\frac{n}{2}}$. This implies $\mathcal{H}_{-t-\frac{n}{2}}(+)$ is a strictly proper subset of $\mathcal{H}_{-t-\frac{n}{2}}$.

For the second part of (f) consider the function $f_t(x) = |x|^t(1 - \chi_{B_e}(x))$. We determine the s for which $\int |f_t(x)|^2(1+|x|^2)^s\,dx < \infty$. Note again using polar coordinates,

$$\begin{aligned}\int |f_t(x)|^2(1+|x|^2)^s\,dx &= \sigma(S^{n-1}) \int_e^\infty r^{2t}(1+r^2)^s\, r^{n-1} dr \\ &= \sigma(S^{n-1}) \int_e^\infty r^{2t+n-1}(1+r^2)^s\,dr\end{aligned}$$

which is finite if and only if $\int_e^\infty r^{2t+2s+n-1}\,dr < \infty$. But

$$\int_e^\infty r^{2t+2s+n-1}\,dr = \begin{cases} \ln r|_e^\infty & \text{if } 2t+2s+n=0 \\ \frac{r^{2t+2s+n}}{2t+2s+n}\Big|_e^\infty & \text{otherwise.} \end{cases}$$

So the integral is finite if and only if $2t+2s+n<0$ or $s<-t-\frac{n}{2}$. Therefore $\mathcal{F}f_t \in \mathcal{H}_s$ for all $s<-t-\frac{n}{2}$ and $\mathcal{F}f_t \notin \mathcal{H}_{-t-\frac{n}{2}}$. So $\mathcal{H}_{-t-\frac{n}{2}}(-) = \cap_{s<-t-\frac{n}{2}}\mathcal{H}_s$ contains $\mathcal{F}f_t$ but $\mathcal{F}f_t \notin \mathcal{H}_{-t-\frac{n}{2}}$. So $\mathcal{H}_{-t-\frac{n}{2}}(-)$ is a proper subspace of $\mathcal{H}_{-t-\frac{n}{2}}$.

We finish with (e). Note $T: L^2(\mathbb{R}^n,(1+|x|^2)^s\,dx) \to L^2(\mathbb{R}^n,(1+|x|^2)^{s-2}\,dx)$ defined by $Tf(x)=(1+|x|^2)f(x)$ is a unitary mapping between Hilbert spaces. Thus $\mathcal{F}T\mathcal{F}^{-1}$ is a unitary mapping of $\mathcal{H}_s$ onto $\mathcal{H}_{s-2}$. Moreover since by Theorem 4.48, $\mathcal{F}(x_j^2 f) = \frac{D_j^2\hat{f}}{(-2\pi i)^2} = -\frac{1}{4\pi^2}D_j^2\hat{f}$, we see $\mathcal{F}(|x|^2 f) = \frac{-\Delta\hat{f}}{4\pi^2}$. Hence $\mathcal{F}T\mathcal{F}^{-1}\hat{f} = \mathcal{F}(Tf) = \mathcal{F}((1+|x|^2)f) = \hat{f} - \frac{1}{4\pi^2}\Delta\hat{f}$. This implies $\hat{f} \mapsto (I - \frac{1}{4\pi^2}\Delta)\hat{f}$ is a unitary mapping of $\mathcal{H}_s$ onto $\mathcal{H}_{s-2}$. □

Exercise Set 4.8

1. Show the space $\mathcal{R}$ of rapidly decreasing n-multi indexed sequences with the Schwartz topology is a Fréchet space.

Solution: We show $\mathcal{R}$ is a Frechét space. One can show the seminorms $|\lambda|_N$ where $|\lambda|_N = \max_{\alpha\in A}(1+|\alpha|^2)^N|\lambda_\alpha|$ define the topology on $\mathcal{R}$ and since these are countable in number, we know by Exercise 2.1.10 that $\mathcal{R}$ is metrizable. Moreover, a function $\alpha \mapsto \lambda_\alpha$ is in $\mathcal{R}$ if and only if $|\lambda|_N < \infty$ for all $N \in \mathbb{N}_0$.

We need to show $\mathcal{R}$ is complete. Let λ^k be a sequence in $\mathcal{R}$ and suppose $|\lambda^k - \lambda^l|_N \to 0$ as $k,l\to\infty$ for all N. Thus $|\lambda_\alpha^k - \lambda_\alpha^l| \leqslant |\lambda^k-\lambda^l|_0 \to 0$ for each $\alpha \in A$ as $k,l\to\infty$. Consequently, for each $\alpha\in A$, the sequence of complex numbers λ_α^k converges to a complex number λ_α as $k\to\infty$. We claim $\lambda^k \to \lambda$ in $\mathcal{R}$. Indeed, it suffices to note if $\epsilon>0$, there is a $K\in\mathbb{N}$ such that if $k,l\geqslant K$, then

$$|\lambda^k-\lambda^l|_N \leqslant \epsilon \text{ for } k,l \geqslant K\,.$$

This implies

$$(1+|\alpha|^2)^N|\lambda_\alpha^k - \lambda_\alpha^l| \leqslant \epsilon \text{ for each } \alpha \text{ for } k,l\geqslant K.$$

Fixing α and letting $l\to\infty$, we obtain $(1+|\alpha|^2)^N|\lambda_\alpha^k-\lambda_\alpha| \leqslant \epsilon$ for all $\alpha\in A$ if $k\geqslant K$. Thus $\lambda^k-\lambda\in\mathcal{R}$ and so $\lambda\in\mathcal{R}$, and furthermore, one has $|\lambda^k-\lambda|_N\leqslant\epsilon$ for $k\geqslant K$. Thus $\lambda^k\to\lambda$ and $\mathcal{R}$ is complete.

□

2. Let e_n, $n=0,1,2,\ldots$ be a Schwartz basis of $L^2(\mathbb{R})$. Show there is a positive polynomial p with $\max_{x\in\mathbb{R}}|e_n(x)| \leqslant p(n)$ for all n.

Solution: Define a seminorm $|\cdot|$ on $\mathcal{S}(\mathbb{R})$ by $|f| = \max_x|f(x)|$. This seminorm is continuous. By Corollary 4.76, the sequence $|e_n|$ is tempered. Thus there is positive polynomial $p(x)$ such that $|e_n|\leqslant p(n)$ for all n. □

3. Show the dual of $\mathcal{R}$ is the vector space of all tempered multi-indexed sequences $\{c_\alpha\}_{\alpha \in A}$.

Solution:

Let $c : A \mapsto \mathbb{C}$ be tempered. Then there is a polynomial p such that $|c_\alpha| \leqslant p(\alpha)$ for all α. Note we can take $p(x) = M(1+|x|^2)^N$ for some $M > 0$ and nonnegative integer N. Define a linear functional f on $\mathcal{R}$ by $f(\lambda) = \sum_\alpha c_\alpha \lambda_\alpha$. We show f is a continuous linear functional on $\mathcal{R}$. By Exercise 4.8.8 which is in this set, we have

$$K = \sum_{\alpha \neq 0} \frac{1}{|\alpha|^{2n}} < \infty.$$

Thus

$$\begin{aligned}
|f(\lambda)| &= |\sum_\alpha c_\alpha \lambda_\alpha| \\
&\leqslant |c_0 \lambda_0| + M \sum_{\alpha \neq 0} (1 + |\alpha|^2)^N |\lambda_\alpha| \\
&\leqslant |c_0 \lambda_0| + M \sum_{\alpha \neq 0} \frac{(1 + |\alpha|^2)^N}{|\alpha|^{2n+2N}} \left((1 + |\alpha|^2)^{N+n} |\lambda_\alpha|\right) \\
&\leqslant |c_0 \lambda_0| + 2^N M \sum_{\alpha \neq 0} \frac{1}{|\alpha|^{2n}} |\lambda|_{N+n} \\
&\leqslant |c_0| \, |\lambda|_0 + 2^N MK \, |\lambda|_{N+n}
\end{aligned}$$

where $|\lambda|_L = \max_\alpha |(1 + |\alpha|^2)^L \lambda_\alpha|$. Thus f is defined and continuous on $\mathcal{R}$.

Finally let f be a continuous linear functional. Let $e_\alpha \in \mathcal{R}$ be defined by $e_\alpha(\beta) = \delta_{\alpha,\beta}$ for $\beta \in A$. Set $c_\alpha = f(e_\alpha)$. We claim c is tempered. Since f is continuous there an N and a $\delta > 0$ such that $|f(\lambda)| \leqslant 1$ if $|\lambda|_N \leqslant \delta$. Note $|e_\alpha|_N = \max_\beta (1 + |\beta|^2)^N |e_\alpha(\beta)| = (1 + |\alpha|^2)^N$. Thus $|(1 + |\alpha|^2)^{-N} \delta e_\alpha|_N = \delta$ and so $|f((1 + |\alpha|^2)^{-N} \delta e_\alpha| \leqslant 1$. This implies $|c_\alpha| \leqslant \delta^{-1}(1 + |\alpha|^2)^N$ and so $\alpha \mapsto c_\alpha$ is tempered. In particular, if f_c is the continuous linear functional on $\mathcal{R}$ defined by $f_c(\lambda) = \sum c_\alpha \lambda_\alpha$, f_c and f agree on the linear span of the e_β for $\beta \in A$. But this linear span is dense. Indeed, if $\lambda \in \mathcal{R}$ and N and $\epsilon > 0$ are given, there is a K such that $(1 + |\alpha|^2)^N |\lambda_\alpha| < \epsilon$ if $|\alpha| > K$ for λ is rapidly decreasing. Thus $\lambda_0 = \sum_{|\alpha| \leqslant K} \lambda_\alpha e_\alpha$ is the linear span and $|\lambda_0 - \lambda|_N = \max_\alpha (1 + |\alpha|^2)^N |\lambda_{0,\alpha} - \lambda_\alpha| = \max_{|\alpha| > K} (1 + |\alpha|^2)^N |\lambda_\alpha| < \epsilon$. This implies the linear span is dense and consequently $f = f_c$ on $\mathcal{R}$ for both f and f_c are continuous.

□

4. Show a rearrangement of a rapidly decreasing sequence need not be rapidly decreasing.

Solution: Let $a_k = 2^{-k}$. This defines a rapidly decreasing sequence. Consider $L = \{4^k \mid k \geqslant 1\}$. Define $b_{4^k} = a_{2k}$ and let $\phi : \{1, 3, 5, \ldots\} \to \mathbb{N} - L$ be a bijection. Define $b_{\phi(j)} = a_j$. Then b is a rearrangement of a. We consider $p(n) b_n$ where $p(n) = n$. Note $4^k b_{4^k} = 4^k a_{2k} = 4^k 2^{-2k} = 1$. Thus $p(n) b_n$ does not go to zero. □

5. Let e_n be the orthonormal sequence of normalized Hermite functions on $\mathbb{R}$. Show $n \mapsto \sup_x |p(x)D^k e_n(x)|$ is tempered for all polynomials p and nonnegative integers k.

Solution:

We use Corollary 4.76. By Corollary 2.61, or by Exercise 2.3.5, we know $f \mapsto \max_x |p(x)D^k f(x)|$ is continuous seminorm on $\mathcal{S}(\mathbb{R})$. Thus by 4.76, $n \mapsto \max_x |p(x)D^k f(x)|$ is tempered.

Note one could also argue using equations (3.11) and (3.12).

□

6. Show the Weyl algebra $\mathcal{W}$ is an algebra; i.e., show it is closed under addition, multiplication by scalars, and composition.

Solution:

It is clear $\mathcal{W}$ is a linear space of differential operators and since they are a linear subspace of the algebra of linear transformations of $C^\infty(\mathbb{R}^n)$, it suffices to show $\mathcal{W}$ is closed under composition. Thus we need only consider $x^\alpha D^\beta x^\gamma D^\delta$. But by Leibniz's Rule (Lemma 2.21),

$$\begin{aligned} D^\beta(x^\gamma D^\delta f) &= \sum_{\mu \leqslant \beta} \binom{\beta}{\mu} D^\mu x^\gamma D^{\beta-\mu} D^\delta f \\ &= \sum_{\mu \leqslant \beta,\, \mu \leqslant \gamma} \binom{\beta}{\mu} \frac{\gamma!}{(\gamma-\mu)!} x^{\gamma-\mu} D^{\beta-\mu+\delta} f\,. \end{aligned}$$

So

$$(x^\alpha D^\beta)(x^\gamma D^\delta) = \sum_{\mu \leqslant \beta,\, \mu \leqslant \gamma} \binom{\beta}{\mu} \frac{\gamma!}{(\gamma-\mu)!} x^{\alpha+\gamma-\mu} D^{\beta-\mu+\delta}\,.$$

But this then implies the linear subspace $\mathcal{W}$ is closed under multiplication and thus is a subalgebra.

□

7. Show $\mathcal{S}(\mathbb{R}^n)$ is the collection of all $f \in L^2(\mathbb{R}^n)$ such that as distributions, $Df \in L^2(\mathbb{R}^n)$ for all D in the Weyl algebra $\mathcal{W}$.

Solution: First assume $f \in \mathcal{S}(\mathbb{R}^n)$. Clearly $x^\alpha D^\beta f \in L^2$ for all α and β for by Propositions 2.57 and 2.58 $x \mapsto x^\alpha D^\beta f(x)$ is Schwartz. Thus $Df \in L^2(\mathbb{R}^n)$ for all $D \in \mathcal{W}$.

Now consider a distribution f where Df as a distribution is in $L^2(\mathbb{R}^n)$ for all $D \in \mathcal{W}$. Since $D^\alpha f$ is in L^2 for all α, the Sobolev Lemma implies f can be redefined on a set of measure 0 so that f and all its derivatives $D^\beta f$ are C^∞. Thus we need only show f is Schwartz.

We start first with the case $n = 1$. We already know $p(x)D^\beta f$ is C^∞ for any p and β. Thus it suffices to show each $pD^\beta f$ is bounded. Letting $\phi = D^\beta f$ and using $D^\alpha(p\phi)$ is L^2 for all polynomials p, we see

$$p(x)\phi(x) = \int_0^x (1+s^2)^{-1}(1+s^2)(p(s)\phi(s))'\,ds + p(0)\phi(0).$$

So

$$
\begin{aligned}
|p(x)\phi(x)| &\leqslant \int_{-\infty}^{\infty} (1+s^2)^{-1}(1+s^2)|D(p(s)\phi(s))|\,ds + |p(0)\phi(0)| \\
&\leqslant \left(\int_{-\infty}^{\infty} (1+s^2)^{-2}\,ds\right)^{\frac{1}{2}} \left(\int_{-\infty}^{\infty} (1+s^2)^2|D(p(s)\phi(s))|^2\,ds\right)^{\frac{1}{2}} + |p(0)\phi(0)|
\end{aligned}
$$

which is finite.

Now that this has been done we show we can extend this to $n > 1$. For simplicity, we do the case $n = 2$ as the general case is a simple extension of the argument we provide.

Since $(1 + x_1^2 + x_2^2)^m D^\alpha f(x_1, x_2)$ is L^2 for all m and α, one has $(1 + x_1^2 + x_2^2)^{-k+m} D^\alpha f(x_1, x_2)$ is in L^1 when $\int_{\mathbb{R}^2} (1 + x_1^2 + x_2^2)^{-2k}\, d_2x < \infty$. This using polar coordinates occurs when $k > \frac{1}{2}$. Thus $(1 + x_1^2 + x_2^2)^m D^\alpha f(x_1, x_2)$ is L^1 for all m. Moreover, by Fubini

$$
\begin{aligned}
\int_{\mathbb{R}}\int_{\mathbb{R}} (1+x_1^2)^{2m}|D^\alpha f(x_1,x_2)|^2\,dx_1\,dx_2 &\leqslant \int_{\mathbb{R}}\int_{\mathbb{R}} (1+x_1^2+x_2^2)^{2m}|D^\alpha f(x_1,x_2)|^2\,dx_1\,dx_2 \\
&\leqslant \int_{\mathbb{R}^2} (1+|x|^2)^{2m}|D^\alpha f(x)|^2\,d_2x \\
&< \infty
\end{aligned}
$$

for each α and m implies there is a set E of Lebesgue measure 0 such that

$$
\int (1+x_1^2)^{2m}|D^\alpha f(x_1,x_2)|^2\,dx_1 < \infty
$$

for any α and m if $x_2 \notin E$. Thus by the $n = 1$ case, $x_1 \mapsto f(x_1, x_2)$ is in $\mathcal{S}(\mathbb{R})$ for $x_2 \notin E$. This allows us to Fourier transform $D_1^k f$. In fact, using Fubini and integration by parts k times and the fact that $D_1^j f(x_1, x_2)$ vanishes at $x_1 = \pm\infty$ for all j when $x_2 \notin E$, one sees

$$
\begin{aligned}
\mathcal{F}(D_1^k f)(y_1,y_2) &= \int_{\mathbb{R}} e^{-2\pi i x_2 y_2} \int_{\mathbb{R}} D_1^k f(x_1,x_2)e^{-2\pi i x_1 y_1}\,dx_1\,dx_2 \\
&= \int_{\mathbb{R}} e^{-2\pi i x_2 y_2}(2\pi i y_1)^k \int_{\mathbb{R}} f(x_1,x_2)e^{-2\pi i x_1 y_1}\,dx_1\,dx_2 \\
&= (2\pi i y_1)^k \int_{\mathbb{R}^2} f(x_1,x_2)e^{-2\pi i(x_1y_1+x_2y_2)}\,d_2x \\
&= (2\pi i y_1)^k \mathcal{F}f(y_1,y_2)\,.
\end{aligned}
$$

Since $D_1^k f$ is in $L^1(\mathbb{R}^2)$, we see $y_1^k \mathcal{F}f(y)$ is bounded on $\mathbb{R}^2$ for any $k \geqslant 0$. Similarly, $y_2^k \mathcal{F}f(y)$ is bounded for each k. This implies $(1 + |y|^2)^k \mathcal{F}f(y)$ is bounded for each k for

$$
\begin{aligned}
(1+y_1^2+y_2^2)^k|\mathcal{F}f(y_1,y_2)| &\leqslant \left((1+2y_1^2)^k + (1+2y_2^2)^k\right)|\mathcal{F}f(y_1,y_2)| \\
&\leqslant \sum_{r=0}^{k} \binom{k}{r}(2^r y_1^{2r} + 2^r y_2^{2r})|\mathcal{F}f(y_1,y_2)|\,.
\end{aligned}
$$

We note since we can apply this argument to $x^\beta f(x)$ for any β for f can be replaced by $x^\beta f(x)$ in the above and since by interchanging limits and integrals one has

$$\mathcal{F}((-2\pi i x)^\beta f) = D^\beta \mathcal{F} f$$

then $y \mapsto (1+|y|^2)^k D^\beta \mathcal{F} f(y)$ is bounded for any k and $\beta \in \mathbb{N}_0 \times \mathbb{N}_0$. Consequently, $\mathcal{F} f$ is Schwartz and thus $f = \widetilde{\mathcal{F}^2 f}$ is Schwartz.

□

8. Let A be the collection of all n-tuples of nonnegative integers. Recall $|\alpha| = \sum a_i$ if $\alpha = (a_1, a_2, \ldots, a_n) \in A$. Show

$$\sum_{|\alpha|>0} \frac{1}{|\alpha|^k} < \infty \text{ if and only if } k > n.$$

Solution: We may assume $k > 0$. Consider the integral $\int_{[1,\infty)^n} f(x)\, d_n x$ where $f(x) = \frac{1}{(x_1+x_2+\cdots+x_n)^k}$. Let $A_+ = \{\alpha \in A \mid |\alpha| > 0\}$.

Then

$$\int_{[1,\infty)^n} f(x)\, dx = \int_1^\infty \int_1^\infty \cdots \int_1^\infty \frac{1}{(x_1+x_2+\cdots+x_n)^k}\, dx_1\, dx_2\, \ldots\, dx_n.$$

Note $[1,\infty]^n = \bigcup_{\alpha \in A_+} ([0,1)^n + \alpha)$. Thus one has:

$$\int_{[1,\infty)^n} f(x)\, dx \leqslant \sum_{\alpha \in A_+} \int_{[0,1)^n+\alpha} \frac{1}{|\alpha|^k}\, dx = \sum_{\alpha \in A_+} \frac{1}{|\alpha|^k}.$$

Set $1_n = (1,1,1,\ldots,1)$. Since $f(x) \geqslant \frac{1}{|\alpha+1_n|}$ on $[0,1)^n + \alpha$, we see

$$\int_{[1,\infty)^n} f(x)\, dx \geqslant \sum_{\alpha \in A_+} \int_{[0,1)^n+\alpha} \frac{1}{|\alpha+1_n|^k}\, dx = \sum_{\alpha \in A_+} \frac{1}{|\alpha+1_n|^k} = \sum_{\alpha - 1_n \in A_+} \frac{1}{|\alpha|^k}.$$

Thus $\sum_{\alpha \in A_+} \frac{1}{|\alpha|^k} < \infty$ if and only if $\int_{[1,\infty)^n} f(x)\, d_n x < \infty$.

We can calculate this integral by Fubini:

$$\int_1^\infty \int_1^\infty \int_1^\infty \cdots \int_1^\infty \frac{1}{(x_1 + x_2 + \cdots + x_n)^k} dx_1\, dx_2\, dx_3 \, \ldots \, dx_n =$$

$$\int_1^\infty \int_1^\infty \cdots \int_1^\infty \frac{1}{(1-k)(x_1 + x_2 + \cdots + x_n)^{k-1}} \Bigg|_{x_1=1}^{x_1=\infty} dx_2\, dx_3 \, \ldots \, dx_n$$

$$= \frac{1}{k-1} \int_1^\infty \int_1^\infty \cdots \int_1^\infty \frac{1}{(1 + x_2 + \cdots + x_n)^{k-1}} dx_2\, dx_3 \, \ldots \, dx_n$$

$$= \cdots$$

$$= \frac{1}{k-1} \cdot \frac{1}{k-2} \cdots \frac{1}{k-n-1} \int_1^\infty \frac{1}{(n-1+x_n)^{k-n+1}} dx_n$$

$$= \frac{1}{k-1} \cdot \frac{1}{k-2} \cdots \frac{1}{k-n-1} \cdot \frac{1}{(n-k)} \cdot \frac{1}{(n-1+x_n)^{k-n}} \Bigg|_1^\infty$$

$$= \frac{1}{(k-1)(k-2)\cdots(k-n-1)(k-n)}$$

provided the integrals at each step are finite. This occurs if and only if $k > 1$, $k > 2$, $\ldots$, $k > n$. □

9. Let A be the collection of all n-tuples consisting of nonnegative integers. Show an orthonormal set $\{e_\alpha\}_{\alpha \in A}$ is a Schwartz basis if and only if the mapping $\Lambda : \mathcal{R} \to \mathcal{S}(\mathbb{R}^n)$ defined by $\Lambda(\lambda) = \sum_{\alpha \in A} \lambda_\alpha e_\alpha$ is a topological isomorphism.

Solution:

Assume Λ is a linear homeomorphism. Then each e_α is a Schwartz function for they are in the range of Λ. Since Λ is onto and e_α are orthonormal one has $(\Lambda(\lambda), e_\alpha) = (\sum_\beta \lambda_\beta e_\beta, e_\alpha) = \lambda_\alpha$ and thus $(\Lambda(\lambda), e_\alpha)$ is rapidly decreasing for $\lambda \in \mathcal{R}$. Since Λ is onto, each $f \in \mathcal{S}(\mathbb{R}^n)$ satisfies (f, e_α) is rapidly decreasing in α and for these f, one has $f = \sum (f, e_\alpha) e_\alpha$. So the linear span of the e_α is dense in $L^2(\mathbb{R}^n)$. Thus e_α is an orthonormal basis. Using Definition 4.68, we see $\{e_\alpha\}$ is a Schwartz basis.

To see the converse, suppose $\{e_\alpha\}$ is a Schwartz basis. Then $\Lambda : \mathcal{R} \to \mathcal{S}(\mathbb{R}^n)$ is a linear mapping of $\mathcal{R}$ onto $\mathcal{S}(\mathbb{R}^n)$. It is clearly one to one for $\Lambda(\lambda) = 0$ if and only if $\sum \lambda_\alpha e_\alpha = 0$ if and only if $\lambda_\alpha = 0$ for all α. Thus we need only show it is a homeomorphism. This, however, was done in the proof of Theorem 4.69. There the Principle of Uniform Boundedness and the Open Mapping Theorem was used to see Λ is bicontinuous.

□

Exercise Set 4.9

1. Let $f \in \mathcal{S}(\mathbb{R})$ with $|f|_2 = 1$. Let $a, b \in \mathbb{R}$. Define g by $g(x) = e^{-2\pi iax} f(x+b)$. Show $m(g) = m(f) - b$, $m(\hat{g}) = m(\hat{f}) - a$, $\nu(g) = \nu(f)$ and $\nu(\hat{g}) = \nu(\hat{f})$.

Solution:

$$\begin{aligned} m(g) &= \int x|g(x)|^2\,dx \\ &= \int x|f(x+b)|^2\,dx \\ &= \int (x-b)|f(x)|^2\,dx \\ &= m(f) - b. \end{aligned}$$

Now $\hat{g}(\omega) = \mathcal{F}(\lambda(-b)f)(\omega + a) = e^{2\pi ib(\omega+a)}\hat{f}(\omega + a)$ and so as above $m(\hat{g}) = m(\hat{f}) - a$.

For the variance note

$$\begin{aligned} \nu(g) &= \int (x + b - m(f))^2|g(x)|^2\,dx \\ &= \int (x + b - m(f))^2|f(x+b)|^2\,dx \\ &= \int (x - m(f))^2|f(x)|^2\,dx \\ &= \nu(f). \end{aligned}$$

Similarly $\nu(\hat{g}) = \nu(\hat{f})$. □

2. Let $f \in L^2(\mathbb{R})$. Suppose $\omega \mapsto \omega\hat{f}(\omega)$ is an L^2 function. Show the distribution Df is an L^2 function f' and

$$\int |f'(x)|^2\,dx = 4\pi^2 \int \omega^2|\hat{f}(\omega)|^2\,d\omega.$$

Solution: Define $F(\omega) = 2\pi i\omega\hat{f}(\omega)$. Note F is an L^2 function and thus $f' = \mathcal{F}^{-1}(F)$ is an L^2 function with the same norm.

We show $DT_f = T_{f'}$. Let ϕ be a test function. Then

$$\begin{aligned}
DT_f(\phi) &= -T_f(\phi') \\
&= -\int f(x)\phi'(x)\,dx \\
&= -\int \mathcal{F}^{-1}(\hat{f})(x)\phi'(x)\,dx \\
&= -\int \hat{f}(\omega)\mathcal{F}^{-1}(\phi')(\omega)\,d\omega \\
&= -\int \hat{f}(\omega)\int \phi'(x)e^{2\pi i x\omega}\,dx\,d\omega \\
&= -\int \hat{f}(\omega)\left(-\int \phi(x)(2\pi i\omega)e^{2\pi i x\omega}\,dx\right)\,d\omega \\
&= \int 2\pi i\omega \hat{f}(\omega)\hat{\phi}(-\omega)\,d\omega \\
&= \int \mathcal{F}(f')(\omega)\hat{\phi}(-\omega)\,d\omega \\
&= \int f'(x)\mathcal{F}^2\check{\phi}(x)\,dx \\
&= \int f'(x)\phi(x)\,dx \\
&= T_{f'}(\phi).
\end{aligned}$$

Since $\mathcal{F}$ is an isometry and $\mathcal{F}(f') = F$, we have the result. □

3. Suppose f and xf are in $L^2(\mathbb{R})$ and the distribution Df is an L^2 function f'. Show there is a sequence f_k in $\mathcal{D}(\mathbb{R})$ such that $f_k \to f$, $f_k' \to f'$, and $xf_k \to xf$ in $L^2(\mathbb{R})$. (**Hint:** Consider $h_q * (k_p f)$ where h_q is an approximate identity and $k_p(x) = k(\frac{x}{p})$ where k is a C^∞ function of compact support in $[-1,1]$, $0 \leqslant k(x) \leqslant 1$, and $k(x) = 1$ for $-1/2 \leqslant x \leqslant 1/2$.)

Solution: For $q \in \mathbb{N}$, set $h_q(x) = qh(qx)$ where $0 \leqslant h$, h is C^∞ with compact support in $[-1,1]$ and $\int h(x)\,dx = 1$. Note the support of each $h_q \subseteq [-\frac{1}{q}, \frac{1}{q}] \subseteq [-1,1]$.

Let $\epsilon > 0$. We claim there is a p and q such that

$$|h_q * (k_p f) - f|_2 < \epsilon,\ |(h_q * (k_p f))' - f'|_2 < \epsilon,\ |x(h_q * k_p f) - xf|_2 < \epsilon.$$

For the first we know $k_p f$ is near f in L^2 if p is large and since $h_q * (k_p f) \to k_p f$ in L^2 as $q \to \infty$, we see if we make p large and then q large, we obtain the first inequality.

For the second note by Lemma 2.79, $h_q * (k_p f)$ is C^∞. Since $h_q * (k_p f) = (k_p f) * h_q$, Proposition 4.57 implies that $(h_q * (k_p f))' = (k_p f)' * h_q$. Now the product rule holds for the product $k_p f$ and thus $(h_q * (k_p f))' = (k_p' f + k_p f') * h_q = h_q * (k_p' f + k_p f')$ and $h_q * (k_p f)$ is C^∞ with support in $[-p-1, p+1]$. Also since $k_p'(x) = \frac{1}{p}k'(\frac{x}{p})$, we see $k_p' f$ is near 0 in L^2 for p large. Since

$|h_q * (k_p' f)|_2 \leqslant |h_q|_1 |k_p' f|_2$, we see $h_q * (k_p' f)$ is uniformly small relative to q, if $k_p' f$ is L^2 small. But again $k_p f'$ is near f' for p large and thus $h_q * (k_p f')$ is near f' in L^2 for p large and then q large.

Finally for p large and fixed, $h_q * (k_p f) \to k_p f$ in L^2 for $q \to \infty$. Since $|x| < p+1$ on the support of $h_q * (k_p f)$, we see $x h_q * (k_p f) \to x k_p f$ as $q \to \infty$. Thus $x h_q * (k_p f)$ is near xf in L^2 for p and q large. □

4. Assume $f \in L^2(\mathbb{R})$ with $|f|_2 = 1$. Show using the previous exercise that

$$\left(\int x^2 |f(x)|^2 \, dx \right) \left(\int \omega^2 |\hat{f}(\omega)|^2 \, d\omega \right) \geqslant \frac{1}{16\pi^2}.$$

Also show one has equality if and only if $f = Ae^{bx^2}$ where $\operatorname{Re} b < 0$. (**Hint:** Show $|f(x)|^2$ is absolutely continuous and use integration by parts.)

Solution: We may assume $xf(x)$ and $y\hat{f}(y)$ are L^2. We also know f' as a distribution is given by an L^2 function and $\mathcal{F}(f')(\omega) = 2\pi i \omega \hat{f}(\omega)$. Now $\hat{f}(\omega)$ is integrable on $[-1,1]$ for it is square integrable and since $\hat{f}(\omega) = \frac{1}{\omega}\omega\hat{f}(\omega)$ for $|\omega| > 1$, we see $\hat{f}$ is integrable on $|\omega| > 1$. Therefore $\hat{f}$ is integrable. Similarly, f is integrable. Since $f = \mathcal{F}^{-1}(\hat{f})$, we have f is continuous. Now since $(f(x)\bar{f}(x))' = f(x)\bar{f}'(x) + f'(x)\bar{f}(x) = 2\operatorname{Re}(f(x)\bar{f}'(x))$ which is an L^1 function. We claim $f\bar{f}$ is absolutely continuous. To see this we first use Exercise 4.9.3 to find a sequence $\{f_k\}$ in $\mathcal{D}(\mathbb{R})$ such that $f_k \to f$, $f_k' \to f'$, and $xf_k \to xf$ in L^2 as $k \to \infty$. But then there exists a subsequence k_j such that f_{k_j} and f_{k_j}' converge pointwise a.e. to f and f', respectively. Using this and the continuity of f one has

$$\begin{aligned} f\bar{f}(x) &= \lim_j f_{k_j}(x)\bar{f}_{k_j}(x) \\ &= \lim_j \int_0^x (f_{k_j}(t)\bar{f}_{k_j}(t))' \, dt + |f_{k_j}(0)|^2 \\ &= \lim_j \int_0^x (f_{k_j}'(t)\bar{f}_{k_j}(t) + f_{k_j}(t)\bar{f}_{k_j}'(t)) \, dt + |f_{k_j}(0)|^2 \\ &= \int_0^x (f'(t)\bar{f}(t) + f(t)\bar{f}'(t)) \, dt + |f(0)|^2. \end{aligned}$$

Thus $|f|^2$ is absolutely continuous and $x|f(x)|^2$ is continuous and integrable. Thus there exist $\alpha_k < \beta_k$ with $\alpha_k \to -\infty$ and $\beta_k \to \infty$ with $\beta_k |f(\beta_k)|^2 \to 0$

and $\alpha_k|f(\alpha_k)|^2 \to 0$. In particular,

$$\begin{aligned}
1 &= \lim \int_{\alpha_k}^{\beta_k} |f(t)|^2\,dt \\
&= \lim \int_{\alpha_k}^{\beta_k} \frac{d}{dt}(t)|f(t)|^2\,dt \\
&= \lim_k \left(t|f(t)|^2\Big|_{\alpha_k}^{\beta_k} - \int_{\alpha_k}^{\beta_k} t\frac{d}{dt}|f(t)|^2\,dt \right) \\
&= \lim_k \left(-\int_{\alpha_k}^{\beta_k} 2t\,\mathrm{Re}(f'(t)\bar{f}(t))\,dt \right) \\
&= -\int_{-\infty}^{\infty} 2t\,\mathrm{Re}(f'(t)\bar{f}(t))\,dt \\
&\leqslant \int 2|f'(t)(t\bar{f}(t))|\,dt \\
&\leqslant 2|f'|_2|tf|_2 \\
&= 2(2\pi)|\omega\hat{f}|_2|xf|_2.
\end{aligned}$$

So

$$\left(\int x^2|f(x)|^2\,dx\right)\left(\int \omega^2|\hat{f}(\omega)|^2\,d\omega\right) \geqslant \frac{1}{16\pi^2}.$$

Moreover as seen in the proof of the uncertainty principle, we have equality if and only if

$$|\int xf(x)\overline{f'(x)}\,dx| = |xf|_2|f'|_2\,.$$

Since in an inner product space, there is equality in Cauchy-Schwarz if and only if one has linear dependence, one sees $f'(x) = 2bxf(x)$ for some b. Again

$$f(x) = \lim_k f_{k_j}(0) + \int_0^x f_k'(t)\,dt = f(0) + \int_0^x f'(t)\,dt.$$

So f is absolutely continuous. Then $f(x)e^{-bx^2}$ has derivative 0 and so is a constant. Thus $f(x) = ae^{bx^2}$ which is in L^2 if and only if $\mathrm{Re}\,b < 0$. □

5. Let $f \in L^2(\mathbb{R})$. Let $a, b \in \mathbb{R}$. Show

$$\left(\int (x-a)^2|f(x)|^2\,dx\right)\left(\int (\omega-b)^2|\hat{f}(\omega)|^2\,d\omega\right) \geqslant \frac{|f|_2^4}{16\pi^2}.$$

Solution:

Note an easy consequence of the previous exercise is

$$\left(\int x^2|h(x)|^2\,dx\right)\left(\int \omega^2|\hat{h}(\omega)|^2\,d\omega\right) \geqslant \frac{|h|_2^4}{16\pi^2}$$

for $h \in L^2(\mathbb{R})$. Now suppose $f \in L^2(\mathbb{R})$. Take

$$h(x) = e^{-2\pi i b x} f(x+a) = \tau(-b)\lambda(-a)f(x).$$

Note using Lemma 3.3 and its natural extension to L^2, one has

$$\hat{h}(\omega) = \lambda(-b)(\widehat{\lambda(-a)f})(\omega) = \tau(a)\hat{f}(\omega+b) = e^{2\pi i a(\omega+b)}\hat{f}(\omega+b).$$

Thus

$$\begin{aligned}\left(\int (x-a)^2|f(x)|^2\,dx\right)\left(\int (\omega-b)^2|\hat{f}(\omega)|^2\,d\omega\right) &= \\ \left(\int x^2|f(x+a)|^2\,dx\right)\left(\int \omega^2|\hat{f}(\omega+b)|^2\,d\omega\right) & \\ = \int x^2|h(x)|^2\,dx \int \omega^2|\hat{h}(\omega)|^2\,d\omega \geqslant \frac{|h|_2^4}{16\pi^2} = \frac{|f|_2^4}{16\pi^2}.\end{aligned}$$

□

Exercise Set 4.10

1. Let f and g be in $L^2(\mathbb{R}^n)$. Show $\int |f(x)\lambda(u)g(x)|\,dx \to 0$ as $u \to \infty$.
Solution: Let $\epsilon > 0$. Choose $f_0, g_0 \in C_c(\mathbb{R}^n)$ with $|f-f_0|_2 < \frac{\epsilon}{2|g|_2+1}$ and $|g-g_0|_2 < \frac{\epsilon}{2|f_0|_2+1}$. Then by Cauchy-Schwarz,

$$\begin{aligned}\int |f(x)\lambda(u)g(x)|\,dx &\leqslant \int |(f(x)-f_0(x))\lambda(u)g(x)|\,dx \\ &\quad + \int |f_0(x)\lambda(u)g(x) - f_0(x)\lambda(u)g_0(x)|\,dx \\ &\quad + \int |f_0(x)\lambda(u)g_0(x)|\,dx \\ &< \frac{\epsilon}{2} + \frac{\epsilon}{2} + \int |f_0(x)\lambda(u)g_0(x)|\,dx \\ &< \epsilon\end{aligned}$$

if u is large for then f_0 and $\lambda(u)g_0$ have disjoint supports. □

2. Show if $f, \psi \in L^2(\mathbb{R}^n)$, then the windowed Fourier transform $\mathcal{S}_\psi f$ is a bounded uniformly continuous function on $\mathbb{R}^n \times \mathbb{R}^n$.
Solution: (First proof). Assume $f, \psi \in C_c(\mathbb{R}^n)$. If K is the compact support of f, then there is a $\delta_1 > 0$ such that $|e^{-2\pi i \omega\cdot x} - e^{-2\pi i \omega_0\cdot x}| < \frac{\epsilon}{2|\psi|_\infty \int |f|+1}$ if $x \in K$ and $|\omega - \omega_0| < \delta_1$. Also since ψ is uniformly continuous, there is a $\delta_2 > 0$ such that $|\psi(x-u) - \psi(x-u_0)| < \frac{\epsilon}{2\int |f|+1}$ if $|u-u_0| < \delta_2$. Hence if $|\omega - \omega_0| < \delta_1$

and $|u-u_0|<\delta_2$, one has

$$\begin{aligned}
&|\mathcal{S}_\psi(f)(u,\omega)-\mathcal{S}_\psi(f)(u_0,\omega_0)|\leqslant\\
&\quad\int|e^{-2\pi i\omega\cdot x}f(x)\overline{\psi(x-u)}-e^{-2\pi i\omega_0\cdot x}f(x)\psi(x-u_0)|\,dx\leqslant\\
&\quad\int|f(x)\left(e^{-2\pi i\omega\cdot x}-e^{-2\pi i\omega_0\cdot x}\right)\overline{\psi(x-u)}|\,dx+\int|f(x)\overline{\psi(x-u)-\psi(x-u_0)}|\,dx\leqslant\\
&\quad|\psi|_\infty\frac{\epsilon}{2|\psi|_\infty\int|f(x)|\,dx+1}\int|f(x)|\,dx+\frac{\epsilon}{2\int|f(x)|\,dx+1}\int|f(x)|\,dx\leqslant\epsilon
\end{aligned}$$

In general, $f,\psi\in L^2$ and $\epsilon>0$. Choose $f_0,\psi_0\in C_c(\mathbb{R}^n)$ with $|f-f_0|_2<\epsilon$ and $|\psi-\psi_0|_2<\epsilon$. Choose δ_1 and δ_2 for f_0 and ψ_0. Then if $|\omega-\omega_0|<\delta_1$ and $|u-u_0|<\delta_2$, one has

$$\begin{aligned}
|\mathcal{S}_\psi(f)(u,\omega)-\mathcal{S}_\psi(f)(u_0,\omega_0)|&\leqslant|\mathcal{S}_\psi(f)(u,\omega)-\mathcal{S}_\psi(f_0)(u,\omega)|\\
&\quad+|\mathcal{S}_\psi(f_0)(u,\omega)-\mathcal{S}_{\psi_0}(f_0)(u,\omega)|\\
&\quad+|\mathcal{S}_{\psi_0}(f_0)(u,\omega)-\mathcal{S}_{\psi_0}(f_0)(u_0,\omega_0)|\\
&\quad+|\mathcal{S}_{\psi_0}(f_0)(u_0,\omega_0)-\mathcal{S}_\psi(f_0)(u_0,\omega_0)|\\
&\quad+|\mathcal{S}_\psi(f_0)(u_0,\omega_0)-\mathcal{S}_\psi(f)(u_0,\omega_0)|\\
&\leqslant|\psi|_2|f-f_0|_2+|\psi-\psi_0|_2|f_0|_2+\epsilon\\
&\quad+|\psi_0-\psi|_2|f_0|_2+|\psi|_2|f_0-f|_2\\
&\leqslant|\psi|_2\epsilon+\epsilon(|f|_2+\epsilon)+\epsilon\\
&\quad+\epsilon(|f|_2+\epsilon)+|\psi|_2\epsilon.
\end{aligned}$$

So $\mathcal{S}_\psi(f)$ is uniformly continuous.

(Alternate proof) It is bounded since $|\mathcal{S}_\psi(f)(u,\omega)|=|(f,e_\omega\lambda(u)\psi)|\leqslant|f|_2|\psi|_2$.

To see uniform continuity, let $\epsilon>0$. We claim there is a $\delta_1>0$ so that if $|u-u_0|<\delta_1$, then $|\lambda(u)\psi-\lambda(u_0)\psi|_2=|\lambda(u-u_0)\psi-\psi|_2<\frac{\epsilon}{2|f|_2+1}$. Note this follows from continuity of $u\to\lambda(u)\psi\in L^2$ at $u=0$. So by the Cauchy-Schwarz inequality, $|f\overline{\lambda(u)\psi}-f\overline{\lambda(u_0)\psi}|_1<\frac{\epsilon}{2}$ when $|u-u_0|<\delta_1$.

By Exercise 4.10.1, there is a compact subset $K_1\subseteq\mathbb{R}^n$ such that

$$\int|\bar{f}(x)\lambda(u_0)\psi(x)|\,dx<\frac{\epsilon}{4}$$

if $u_0\notin K_1$. Now assume $u_0\in K_1$. We can find a compact subset K_2 such that $|\psi\,\chi_{K_2^c}|_2<\frac{\epsilon}{8|f|_2+1}$. Since most of ψ lives on K_2, $\lambda(u_0)\psi$ essentially lives on $u_0+K_2\subseteq K_1+K_2$; i.e. if $K=K_1+K_2$, then $|(\lambda(u_0)\psi)\,\chi_{K^c}|_2<\frac{\epsilon}{8|f|_2+1}$ if $u_0\in K_1$.

Now choose $\delta_2>0$ so that $|e^{2\pi i\omega\cdot x}-e^{2\pi i\omega_0\cdot x}|<\frac{\epsilon}{4|f|_2|\psi|_2+1}$ if $x\in K$ and

$|\omega - \omega_0| < \delta_2$. Then if $|u - u_0| < \delta_1$ and $|\omega - \omega_0| < \delta_2$, one has

$$\begin{aligned}
&|\mathcal{S}_\psi f(u,\omega) - \mathcal{S}_\psi f(u_0,\omega_0)| \leqslant \\
&\quad \int |f(x)\overline{\lambda(u)\psi(x)}e^{-2\pi i\omega\cdot x} - f(x)\overline{\lambda(u_0)\psi(x)}e^{-2\pi i\omega_0\cdot x}|\,dx \leqslant \\
&\qquad \int |f(x)(\overline{\lambda(u)\psi(x) - \lambda(u_0)\psi(x)})|\,dx \quad + \\
&\qquad\quad \int |f(x)\overline{\lambda(u_0)\psi(x)}(e^{-2\pi i\omega\cdot x} - e^{-2\pi i\omega_0\cdot x})|\,dx \leqslant \\
&\qquad |f|_2|\lambda(u)\psi - \lambda(u_0)\psi|_2 + \int |\bar{f}(x)\lambda(u_0)\psi(x)(e^{2\pi i\omega\cdot x} - e^{2\pi i\omega_0\cdot x})|\,dx \\
&\qquad\quad < \frac{\epsilon}{2} + \int |\bar{f}(x)\lambda(u_0)\psi(x)(e^{2\pi i\omega\cdot x} - e^{2\pi i\omega_0\cdot x})\,|dx < \epsilon
\end{aligned}$$

if we know

$$\int |\bar{f}(x)\lambda(u_0)\psi(x)(e^{2\pi i\omega\cdot x} - e^{2\pi i\omega_0\cdot x})|\,dx \leqslant \frac{\epsilon}{2}.$$

First if $u_0 \notin K_1$, we have

$$\begin{aligned}
&\int |\bar{f}(x)\lambda(u_0)\psi(x)(e^{2\pi i\omega\cdot x} - e^{2\pi i\omega_0\cdot x})\,dx| \leqslant \\
&\qquad\qquad 2\int |\bar{f}(x)\lambda(u_0)\psi(x)|\,dx < 2\left(\frac{\epsilon}{4}\right) = \frac{\epsilon}{2}.
\end{aligned}$$

Hence we may assume $u_0 \in K_1$. Then

$$\begin{aligned}
\int |\bar{f}(x)\lambda(u_0)\psi(x)(e^{2\pi i\omega\cdot x} - e^{2\pi i\omega_0\cdot x})|\,dx &\leqslant \int_K |\bar{f}(x)\lambda(u_0)\psi(x)(e^{2\pi i\omega\cdot x} - e^{2\pi i\omega_0\cdot x})|\,dx \\
&\quad + 2\int_{K^c} |\bar{f}(x)\lambda(u_0)\psi(x)|\,dx \\
&\leqslant \frac{\epsilon}{4|f|_2|\psi|_2 + 1}\int_K |\bar{f}(x)\lambda(u_0)\psi(x)|\,dx + 2|f|_2|(\lambda(u_0)\psi)\chi_{K^c}|_2 \\
&\leqslant \frac{\epsilon}{4} + 2|f|_2\frac{\epsilon}{8|f|_2 + 1} \\
&\leqslant \frac{\epsilon}{2}.
\end{aligned}$$

□

3. Let $\psi \in L^2(\mathbb{R}^n)$. Show the range $\mathcal{S}_\psi(L^2(\mathbb{R}^n))$ is a closed subspace of $L^2(\mathbb{R}^n \times \mathbb{R}^n)$.

Solution: By the Plancherel Formula 4.80, we have $|\mathcal{S}_\psi(f)|_2^2 \leqslant |\psi|_2^2|f|_2^2$. So $\mathcal{S}_\psi$ is continuous. Note we may assume $\psi \neq 0$ for otherwise the range is $\{0\}$. Again by 4.80, $\mathcal{S}_\psi^*\mathcal{S}_\psi = |\psi|_2^2 I$. So $\frac{1}{|\psi|_2}\mathcal{S}_\psi$ is an isometry. Thus the range is closed. □

4. Let f and ψ be L^2 functions on $\mathbb{R}^n$. Show

$$\mathcal{S}_\psi(f)(u,\omega) = e^{-\pi i u\cdot\omega}\int_{\mathbb{R}^n} f(x + \frac{u}{2})\,\overline{\psi(x - \frac{u}{2})}\,e^{-2\pi i x\cdot\omega}\,dx\,.$$

Solution:

Let $f, \psi \in L^2$. Then using the integrability of the product of two L^2 functions and translation invariance, one has

$$\begin{aligned}\mathcal{S}_\psi(f)(u,\omega) &= \int f(x)\overline{\psi(x-u)}e^{-2\pi i\omega\cdot x}\,dx \\ &= \int f(x+\frac{u}{2})\overline{\psi(x+\frac{u}{2}-u)}e^{-2\pi i\omega\cdot(x+u/2)}\,dx \\ &= e^{-\pi i\omega\cdot u}\int f(x+\frac{u}{2})\overline{\psi(x-\frac{u}{2})}e^{-2\pi i x\cdot\omega}\,dx\,.\end{aligned}$$

□

5. Let $\psi \in \mathcal{S}(\mathbb{R}^n)$ be nonzero. Show the mapping $\mathcal{S}_\psi$ is a continuous linear linear homeomorphism of $\mathcal{S}(\mathbb{R}^n)$ onto its range which is a closed subspace of $\mathcal{S}(\mathbb{R}^{2n})$. (**Hint:** Use Propositions 2.62 and 4.73.)

Solution: By Proposition 4.73, the mapping $B(f,g) = f \otimes g$ is a continuous bilinear mapping on $\mathcal{S}(\mathbb{R}^n) \times \mathcal{S}(\mathbb{R}^n)$ into $\mathcal{S}(\mathbb{R}^{2n})$. Thus $f \mapsto \Big((u,x) \mapsto f(x)\overline{\psi(u)}\Big)$ is continuous into $\mathcal{S}(\mathbb{R}^{2n})$. One notes the range is closed for if $B(f_n, \bar{\psi})$ converges in $\mathcal{S}(\mathbb{R}^{2n})$, then $f_n \otimes \bar{\psi}$ converges uniformly to a function of form $f \otimes \bar{\psi}$. Since ψ is nonzero this mapping is one-to-one. By the Open Mapping Theorem 2.19, the map is a homeomorphism. Now $(u,x) \mapsto (x-u,x)$ is a continuous invertible linear transformation of $\mathbb{R}^{2n}$. It follows by Proposition 2.62 that

$$f \mapsto \Big(G : (u,x) \mapsto f(x)\overline{\psi(x-u)}\Big)$$

is a homeomorphism onto a closed subspace of $\mathcal{S}(\mathbb{R}^{2n})$. Since $\mathcal{S}_\psi(f)(u,\omega) = \mathcal{F}_2(G)$ and $\mathcal{F}_2$ is a homeomorphism of $\mathcal{S}(\mathbb{R}^{2n})$, we then see $f \mapsto \mathcal{S}_\psi(f)$ is a continuous linear homeomorphism of $\mathcal{S}(\mathbb{R}^n)$ onto a closed linear subspace of $\mathcal{S}(\mathbb{R}^{2n})$. □

6. Let $\phi \in \mathcal{S}(\mathbb{R}^n)$. Show $\mathcal{S}^*_\phi$ is a continuous linear mapping from $\mathcal{S}(\mathbb{R}^{2n})$ into $\mathcal{S}(\mathbb{R}^n)$. Conclude using the previous exercise that if ϕ and ψ are in $\mathcal{S}(\mathbb{R}^n)$ with $(\phi,\psi)_2 = 1$, then $\mathcal{S}_\psi$ is a homeomorphism of $\mathcal{S}(\mathbb{R}^n)$ onto the closed subspace $\mathcal{F} = \mathcal{S}_\psi(\mathcal{S}(\mathbb{R}^n))$ of $\mathcal{S}(\mathbb{R}^{2n})$ which has inverse $\mathcal{S}^*_\phi|_{\mathcal{F}}$. Moreover,

$$f(x) = \iint \mathcal{S}_\psi f(u,\omega)e^{2\pi i\omega\cdot x}\phi(x-u)\,d(u,\omega) \text{ for all } x.$$

Solution: Note by Theorem 4.82, $\mathcal{S}^*_\phi(F)(x) = \int_{\mathbb{R}^n} \mathcal{F}_2F(u,-x)\phi(x-u)\,du$ for all x when $F \in \mathcal{S}(\mathbb{R}^n)$. Since the mapping $F \mapsto ((u,x) \mapsto \mathcal{F}_2F(u,-x))$ is continuous from $\mathcal{S}(\mathbb{R}^{2n})$ into $\mathcal{S}(\mathbb{R}^{2n})$, it suffices to show $F \mapsto \big(x \mapsto \int F(u,x)\phi(x-u)\,du\big)$ is continuous from $\mathcal{S}(\mathbb{R}^{2n})$ into $\mathcal{S}(\mathbb{R}^n)$.

Consider the mapping $F \mapsto G$ where $G(u,x) = F(u,x)\phi(x-u)$. Note $\phi(x-u)$ is a bounded C^∞ function on $\mathbb{R}^{2n}$. By Proposition 2.57, the mapping $F \mapsto G$ is continuous from $\mathcal{S}(\mathbb{R}^{2n})$ into $\mathcal{S}(\mathbb{R}^{2n})$. Hence we need only note that

$$G \mapsto (x \mapsto \int G(u,x)\,du)$$

is continuous from $\mathcal{S}(\mathbb{R}^{2n})$ into $\mathcal{S}(\mathbb{R}^n)$. But this follows from Exercise 2.3.12.

The final equality comes from Theorem 4.84 for the weak L^2 integral $\iint \mathcal{S}_\psi f(u,\omega)[e_\omega \lambda(u)\phi]\, d(u,\omega)$ is given by the Schwartz function

$$x \mapsto \iint \mathcal{S}_\psi f(u,\omega) e^{2\pi i \omega\cdot x} \phi(x-u)\, d(u,\omega)\,.$$

□

7. Let $\psi \in L^2(\mathbb{R}^n)$. Show $\mathrm{Im}(\mathcal{S}_\psi)$ is a reproducing Hilbert space (see Exercise 4.1.21) with reproducing kernel

$$K((u_1,\omega_1),(u,\omega)) = \frac{1}{(\psi,\psi)_2}\int \psi(x-u)\overline{\psi(x-u_1)} e^{2\pi i x\cdot(\omega-\omega_1)} dx\,.$$

Solution: We may assume $\psi \neq 0$. By the Plancherel Theorem 4.80, $f \mapsto \mathcal{S}_\psi(f)$ is a bicontinuous linear mapping of $L^2(\mathbb{R}^n)$ onto $\mathrm{Im}(\mathcal{S}_\psi)$ satisfying $|\mathcal{S}_\psi(f)|_2 = |\psi|_2|f|_2$. Exercises 4.10.2 and 4.10.3 show $\mathcal{S}_\psi(L^2(\mathbb{R}^n))$ is a closed subspace of $L^2(\mathbb{R}^{2n})$ consisting of continuous functions. Since

$$\mathcal{S}_\psi f(u,\omega) = \int f(x)\overline{\psi(x-u)} e^{-2\pi i x\cdot\omega}\, dx$$

we see

$$|\mathcal{S}_\psi f(u,\omega)| \leqslant |f|_2 |\psi|_2 = |\mathcal{S}_\psi(f)|_2$$

and thus $\mathcal{S}_\psi(f) \mapsto \mathcal{S}_\psi f(u,\omega)$ is continuous on $\mathrm{Im}(\mathcal{S}_\psi)$.

We now use Exercise 4.1.21. We need a function K such that

$$\mathcal{S}_\psi f(u,\omega) = (\mathcal{S}_\psi f, K_{u,\omega})_2$$

for $f \in L^2$ where $K_{u,\omega}(u_1,\omega_1) = \mathrm{ev}_{(u_1,\omega_1)}\,\mathrm{ev}^*_{(u,\omega)} : \mathbb{C} \to \mathbb{C}$ and thus is a scalar.

Now

$$\begin{aligned}
(\mathcal{S}_\psi f, \mathrm{ev}^*_{u,\omega}\, c) &= (\mathcal{S}_\psi f(u,\omega), c) \\
&= \bar{c}\mathcal{S}_\psi f(u,\omega) \\
&= \bar{c}(f, \tau(\omega)\lambda(u)\psi)_2 \\
&= \frac{\bar{c}}{(\psi,\psi)_2}(\mathcal{S}_\psi f, \mathcal{S}_\psi[\tau(\omega)\lambda(u)\psi])_2.
\end{aligned}$$

So $\mathrm{ev}^*_{u,\omega}\, c = \frac{c}{(\psi,\psi)_2}\mathcal{S}_\psi[\tau(\omega)\lambda(u)\psi]$. This gives

$$\begin{aligned}
K((u_1,\omega_1),(u,\omega)) &= \frac{1}{(\psi,\psi)_2}\mathcal{S}_\psi[\tau(\omega)\lambda(u)\psi](u_1,\omega_1) \\
&= \frac{1}{(\psi,\psi)_2}\int \tau(\omega)\lambda(u)\psi(x)\overline{\psi(x-u_1)} e^{-2\pi i \omega_1\cdot x}\, dx \\
&= \frac{1}{(\psi,\psi)_2}\int e^{2\pi i\omega\cdot x}\psi(x-u)\overline{\psi(x-u_1)} e^{-2\pi i\omega_1\cdot x}\, dx \\
&= \frac{1}{(\psi,\psi)_2}\int e^{2\pi i(\omega-\omega_1)\cdot x}\psi(x-u)\overline{\psi(x-u_1)}\, dx.
\end{aligned}$$

□

8. Assume $\mathcal{H}$ is a separable Hilbert space, Ω is open in $\mathbb{R}^d$, and $F : \Omega \to \mathcal{H}$ is weakly integrable; i.e. $x \mapsto (v, F(x))$ is integrable for each $v \in \mathcal{H}$. Assume $\int_\Omega ||F(x)||\, dx < \infty$. Show if $\{e_j\}$ is an orthonormal basis for $\mathcal{H}$, then $F(x) = \sum_j (F(x),\, e_j)e_j$ for a.e. x and the weak Hilbert space integral satisfies

$$\int F(x)dx = \sum_j \left(\int_\Omega (F(x),\, e_j)\, dx \right) e_j\, .$$

Solution:

The fact that $F(x) = \sum (F(x), e_j)e_j$ for a.e. x follows since $F(x) \in \mathcal{H}$ a.e. and the e_j form an orthonormal basis.

Now the weak integral $\int F(x)\, dx$ is the unique vector v (if it exists) in $\mathcal{H}$ satisfying

$$(w, v) = \int (w, F(x))\, dx$$

for all $w \in \mathcal{H}$. Note $w \mapsto \int_\Omega (w, F(x))\, dx$ is a well defined linear functional and

$$\begin{aligned} |\int (w, F(x))\, dx| &\leqslant \int |(w, F(x))|\, dx \\ &\leqslant \int ||w||\ ||F(x)||\, dx \\ &= \left(\int ||F(x)||\, dx \right) ||w||. \end{aligned}$$

By the Riesz Theorem, the weak integral v exists. Note $(e_j, v) = \int (e_j, F(x))\, dx$. So $(v, e_j) = \int (F(x), e_j)\, dx$. Hence

$$\begin{aligned} \int_\Omega F(x)\, dx &= v \\ &= \sum_j (v, e_j)e_j \\ &= \sum_j \left(\int_\Omega (F(x), e_j)\, dx \right) e_j\, . \end{aligned}$$

□

Exercise Set 4.11

1. In many applications one would prefer to use positive numbers when doing dilations. In this exercise we show what changes are necessary. For $\epsilon = (\epsilon_1, \ldots, \epsilon_n) \in \{-1, 1\}^n$ let

$$\mathbb{R}^n_\epsilon = \{x \in (\mathbb{R}^*)^n \mid \forall j \in \{1, \ldots, n\} : \text{sign}(x_j) = \epsilon_j\}\, .$$

Define a wavelet function to be a function $\psi \in L^2(\mathbb{R}^n)$ such that there exists a constant $0 < C_\psi < \infty$ such that for all $\epsilon \in \{-1, 1\}^n$ we have

$$\int_{\mathbb{R}^n_\epsilon} \frac{|\hat{\psi}(\omega)|^2}{|\det \omega|}\, d\omega = C_\psi\,.$$

Show the Plancherel and Inversion formula still hold.

Solution:

For $\mathbb{R}^n_\epsilon$ one can follow the arguments through each statement from 4.30 on and see the arguments are precisely the same in this case. Namely one still has Lemmas 4.88 and 4.89 for the Fourier transform is the same and the change of variables gives the same result. The proof of the Plancherel Theorem again is the same for it is completely based on Fubini's Theorem and Lemma 4.89.

Moreover, the inversion formula follows directly from the Plancherel Theorem as the argument in Theorem 4.91 shows.

□

2. Let ψ be a wavelet function such that $C_{\psi,\psi} = 1$. Prove the following:

(a) $H_\psi := \operatorname{Im}(W_\psi)$ is a closed subspace of $L^2(X, d\mu)$ and hence is a Hilbert space.

(b) $H_\psi \subset \mathcal{C}(X)$ and the point evaluation maps $\mathrm{ev}_x : H_\psi \to \mathbb{C},\ F \mapsto F(x)$ are continuous for each $x \in X$.

(c) The reproducing kernel for H_ψ is given by

$$K(x, y) = W_\psi \psi_y(x)\,.$$

Solution:

By Theorem 4.80, W_ψ is a unitary map of $L^2(\mathbb{R}^n)$ into $L^2((\mathbb{R}^*)^n \times \mathbb{R}^n, \frac{da\,db}{|\det a|^2})$ when

$$C_\psi = \int_{(\mathbb{R}^*)^n} |\hat{\psi}(\omega)|^2 \frac{d\omega}{|\det \omega|} = 1.$$

Its image H_ψ is a Hilbert space.

For (b), we have

$$W_\psi f(a, b) = (f, \psi_{a,b})$$

and this is clearly continuous if f is continuous with compact support. Next recall $\psi_{a,b}(x) = |\det a|^{-1/2}\psi(M(a)^{-1}(x-b))$ and thus $|\psi_{a,b}|_2 = |\psi|_2$ for all (a, b). Thus for general f, let $\epsilon > 0$. Choose $f_0 \in C_c(\mathbb{R}^n)$ such that $|f - f_0|_2 < \frac{\epsilon}{3|\psi|_2}$. Pick $\delta > 0$ so that $|W_\psi f_0(a', b') - W_\psi f_0(a, b)| < \frac{\epsilon}{3}$ if $|a' - a| + |b' - b| < \delta$. Then

$$\begin{aligned}
&|W_\psi f(a', b') - W_\psi f(a, b)| = \\
&|(f, \psi_{a',b'}) - (f_0, \psi_{a',b'}) + (f_0, \psi_{a',b'}) - (f_0, \psi_{a,b}) + (f_0, \psi_{a,b}) - (f, \psi_{a,b})| \\
&\leqslant |(f - f_0, \psi_{a',b'})| + |W_\psi f_0(a', b') - W_\psi f_0(a, b)| + |(f_0 - f, \psi_{a,b})| \\
&< 2|f - f_0|_2|\psi|_2 + \frac{\epsilon}{3} < \epsilon\,.
\end{aligned}$$

Moreover by (c) of Theorem 4.80, $|W_\psi f| = |f|_2$ for $f \in L^2(\mathbb{R}^n)$ and thus

$$\begin{aligned}
|\operatorname{ev}_{(a,b)}(W_\psi f)| &= |W_\psi f(a,b)| \\
&= |(f, \psi_{a,b})_2| \\
&\leqslant |f|_2 \, |\psi_{a,b}|_2 \\
&= |f|_2 \, |\psi|_2 \\
&= |W_\psi f|_2 \, |\psi|_2
\end{aligned}$$

and so $W_\psi f \mapsto \operatorname{ev}_{(a,b)} W_\psi f$ is a continuous linear functional on H_ψ. Exercise 4.1.21 implies H_ψ is a reproducing Hilbert space and the function $K(a', b', a, b) = K_{a,b}(a', b')$ which satisfies

$$(W_\psi f, K_{a,b}) = \operatorname{ev}_{a,b}(W_\psi f)$$

is the reproducing kernel for this Hilbert space.

But since W_ψ is unitary,

$$\begin{aligned}
W_\psi f(a,b) &= (f, \psi_{a,b})_2 \\
&= (W_\psi f, W_\psi(\psi_{a,b}))\,.
\end{aligned}$$

This implies

$$K(a', b', a, b) = W_\psi(\psi_{a,b})(a', b') = (\psi_{a,b}, \psi_{a',b'}).$$

□

Exercise Set 4.12

1. Use the Fourier transforms given in Exercise 3.2.3 to show if $f \in L^2(\mathbb{R}^n)$, then $\mathcal{F}f = \mathcal{F}_{2\pi} f$ has essential support in Q_T where $T > 0$ if and only if $\mathcal{F}_a f$ has essential support in $Q_{\frac{a}{2\pi}T}$.

Solution: Note $\mathcal{F}f(\omega) = 0$ for a.e. $\omega \notin Q_T$ if and only if $\int f(x) e^{-2\pi i x \cdot \omega}\, dx = 0$ for a.e. $\omega \notin Q_T$ if and only if $\int f(x) e^{-i(ax \cdot \frac{2\pi}{a}\omega)}\, dx = 0$ for a.e. $\omega \notin Q_T$ if and only if $\mathcal{F}_a f(\frac{2\pi}{a}\omega) = 0$ for a.e. $\omega \notin Q_T$ if and only if $\mathcal{F}_a f(\omega') = 0$ for a.e. $\omega' \notin \frac{a}{2\pi} Q_T = Q_{\frac{a}{2\pi}T}$. □

2. Suppose f is a C^∞ square integrable function on $\mathbb{R}$ with $\operatorname{supp} \hat{f} \subseteq [a, a+2b]$. Show

$$f(x) = \sum_k e^{-\pi i k(\frac{a}{b}+1)} f(\frac{k}{2b}) e^{2\pi i (a+b) x} \operatorname{sinc}(2\pi b x - k\pi)$$

pointwise uniformly and in $L^2(\mathbb{R})$.

Solution: We consider the function $g(x) = e^{-2\pi i (a+b) x} f(x)$. We have $\hat{g}(\omega) = \int e^{-2\pi i (a+b) x} f(x) e^{-2\pi i \omega x}\, dx = \hat{f}(\omega + a + b)$. So $\hat{g}(\omega) = 0$ a.e. where $\omega + a + b \notin$

$[a, a+2b]$; i.e. where $\omega \notin [-b, b]$. Hence by the Shannon Sampling Theorem 4.96,

$$\begin{aligned} g(x) &= \sum_{\gamma \in \frac{\mathbb{Z}}{2b}} g(\gamma) \operatorname{sinc}_b(x-\gamma) \\ &= \sum_{k \in Z} g(\frac{k}{2b}) \operatorname{sinc}_b(x - \frac{k}{2b}) \\ &= \sum_k e^{-2\pi i \frac{k(a+b)}{2b}} f(\frac{k}{2b}) \operatorname{sinc}_b(x - \frac{k}{2b}) \\ &= \sum_k e^{-\pi i k(\frac{a}{b}+1)} f(\frac{k}{2b}) \operatorname{sinc}(2\pi b x - k\pi) \end{aligned}$$

Thus

$$f(x) = e^{2\pi i(a+b)x} g(x) = \sum_k e^{-\pi i k(\frac{a}{b}+1)} f(\frac{k}{2b}) e^{2\pi i(a+b)x} \operatorname{sinc}(2\pi b x - k\pi).$$

□

3. Show

$$\frac{\sin \pi x}{2x} + \frac{\cos \pi x - 1}{\pi x^2} = \frac{4x \sin \pi x}{\pi^2} \sum_{n=1}^{\infty} \frac{1}{(2n+1)^2} \left(\frac{1}{(2n+1)^2 - x^2} \right)$$

pointwise uniformly in x. (**Hint:** Apply Shannon's sampling to the function f where $\hat{f}(\omega) = |2\pi\omega| \chi_{[-\frac{1}{2}, \frac{1}{2}]}(\omega)$.)

Solution: $f(x) = \int_{-.5}^{.5} |2\pi\omega| e^{2\pi i \omega x} \, d\omega = 2 \operatorname{Re} \int_0^{.5} 2\pi\omega e^{2\pi i \omega x} \, d\omega$. So

$$\begin{aligned} f(x) &= 4\pi \operatorname{Re} \left(\omega \frac{e^{2\pi i \omega x}}{2\pi i x} \Bigg|_{\omega=0}^{\omega=.5} - \int_0^{.5} \frac{e^{2\pi i \omega x}}{2\pi i x} d\omega \right) \\ &= 4\pi \operatorname{Re} \left(\frac{e^{\pi i x}}{4\pi i x} \right) - 4\pi \operatorname{Re} \left(\frac{e^{2\pi i \omega x}}{(2\pi i x)^2} \Bigg|_{\omega=0}^{\omega=.5} \right). \end{aligned}$$

Thus

$$f(x) = 4\pi \left(\frac{\sin \pi x}{4\pi x} \right) + 4\pi \operatorname{Re} \left(\frac{e^{\pi i x} - 1}{4\pi^2 x^2} \right) = 4\pi \left(\frac{\sin \pi x}{4\pi x} \right) + 4\pi \frac{\cos \pi x - 1}{4\pi^2 x^2}.$$

But by the Shannon Sampling Theorem 4.96,

$$f(x) = \sum_k f(k) \operatorname{sinc}_{.5}(x-k) = \frac{\pi}{2} \operatorname{sinc}_{.5}(x) + \sum_{k \neq 0} \frac{(-1)^k - 1}{\pi k^2} \operatorname{sinc}_{.5}(x-k).$$

So

$$\frac{\sin \pi x}{x} + \frac{\cos \pi x - 1}{\pi x^2} = \frac{\pi}{2} \frac{\sin(\pi x)}{\pi x} + \sum_{n \in \mathbb{Z}} \frac{-2}{(2n+1)^2 \pi} \operatorname{sinc}_{.5}(x - (2n+1)) \text{ or}$$

$$\frac{\sin \pi x}{2x} + \frac{\cos \pi x - 1}{\pi x^2} = \sum_{n \in \mathbb{Z}} \frac{-2}{(2n+1)^2 \pi} \frac{\sin(\pi(x - (2n+1)))}{\pi(x - (2n+1))}$$
$$= \sum_{n=1}^{\infty} \frac{-2}{(2n+1)^2 \pi^2} \frac{2x \sin \pi x}{x^2 - (2n+1)^2}$$
$$= \frac{-4x \sin(\pi x)}{\pi^2} \sum_{n=1}^{\infty} \frac{1}{(2n+1)^2} \left(\frac{1}{x^2 - (2n+1)^2} \right).$$

□

Exercise Set 4.13

1. Show the Poisson summation formula holds if there exist positive constants $A, B > 0$, and $p > n/2$ such that

$$|f(x)| \leqslant A(1 + |x|^2)^{-p} \text{ and } \left|\hat{f}(\omega)\right| \leqslant B(1 + |\omega|^2)^{-p}.$$

More specifically show $\sum_{k \in \mathbb{Z}^n} f(x+k)$ and $\sum_{k \in \mathbb{Z}^n} \hat{f}(k) e^{2\pi i x \cdot k}$ converge uniformly absolutely and

$$\sum_{k \in \mathbb{Z}^n} f(x+k) = \sum_{k \in \mathbb{Z}^n} \hat{f}(k) e^{2\pi i x \cdot k}.$$

In particular,

$$\sum_{k \in \mathbb{Z}^n} f(k) = \sum_{k \in \mathbb{Z}^n} \hat{f}(k).$$

Solution: We have $|f(x)| \leqslant A(1 + |x|^2)^{-p}$ and $|\hat{f}(\omega)| \leqslant B(1 + |\omega|^2)^{-p}$ where $p > \frac{n}{2}$. This implies f and $\hat{f}$ are integrable and square integrable. Thus $f(x) = \int \hat{f}(\omega) e^{2\pi i \omega \cdot x} \, d\omega$ and $\hat{f}(\omega) = \int f(x) e^{-2\pi i \omega \cdot x} \, dx$ and so we may assume f and $\hat{f}$ are continuous.

We claim $\sum_k f(x+k)$ converges uniformly on Q. For $k \in \mathbb{Z}^n$, let $M(k) = \sup_{x \in [0,1]^n} \frac{A}{(1+|x+k|^2)^p}$. We show $\sum_k M(k) < \infty$. But this is true if

$$\sum_{|k| \geqslant 2\sqrt{n}} M(k) < \infty.$$

Taking $k \geqslant 2\sqrt{n}$, note since $||x| - |k|| \leqslant |x+k|$, $M(k) \leqslant \sup_{x \in Q} \frac{A}{(1+(|k|-|x|)^2)^p} = \frac{A}{(1+(|k|-\sqrt{n})^2)^p} \leqslant \frac{A}{(1+(|k|-1/2|k|)^2)^p}$. So $M(k) \leqslant \frac{4^p A}{(4+|k|^2)^p}$. Now Lemma 4.101 implies

$$\sum_{|k| \geqslant 2\sqrt{n}} \frac{1}{(4 + |k|^2)^p} < \infty.$$

Thus $f_a(x)$ is a $\mathbb{Z}^n$ periodic continuous function. As can be easily seen (see proof of Poisson Summation Theorem), we have

$$\hat{f_a}(k) = \hat{f}(k)$$

for all $k \in \mathbb{Z}^n$. Now $|\hat{f}(k)| \leqslant \frac{B}{(1+|k|^2)^p}$. Now Lemma 4.101 implies $\sum |\hat{f}(k)| < \infty$ and one has the uniform convergence of $\sum \hat{f}(k)e^{2\pi ik\cdot x}$. This must converge to f_a for it is its Fourier series. □

2. Let $f : \mathbb{R}^n \to \mathbb{C}$ be $\mathbb{Z}^n$ periodic and satisfy

$$\sum_{k\in\mathbb{Z}^n} |\hat{f}(k)|\, |k|^m < \infty.$$

Show $f \in C^m(\mathbb{R}^n)$ and

$$\sum_k \hat{f}(k)(2\pi ik)^\alpha e^{2\pi ik\cdot x}$$

converges uniformly to $D^\alpha f$ if $|\alpha| \leqslant m$.

Solution: Assume $|\alpha| \leqslant m$. We know $\sum_k |\hat{f}(k)k^\alpha| < \infty$. Thus

$$\sum_k \hat{f}(k)(2\pi ik)^\alpha e^{2\pi ik\cdot x}$$

converges uniformly for $|\alpha| \leqslant m$.

Since the series $\sum_{k\in\mathbb{Z}^n} \hat{f}(k)e^{2\pi ik\cdot x}$ converges uniformly and is the Fourier series for f, we have $\sum_k \hat{f}(k)(2\pi ik)^\alpha e^{2\pi ik\cdot x}$ converges uniformly to $D^\alpha f(x)$ for $\alpha = 0$. Now the fundamental theorem of calculus and interchange of limits with uniform convergence and induction imply

$$D^{\alpha-e_i} f(x) = D^{\alpha-e_i} f(x - he_i) + \int_{x_i-he_i}^{x_i} \sum_k \hat{f}(k)(2\pi ik)^\alpha e_i^{2\pi ik\cdot(x-he_i+te_i)}\, dt\,.$$

Using $D^\alpha f(x) = \lim_{h\to 0} \frac{D^{\alpha-e_i} f(x)-D^{\alpha-e_i} f(x-he_i)}{h}$, we obtain

$$D^\alpha f(x) = \sum_k \hat{f}(k)(2\pi ik)^\alpha e^{2\pi ik\cdot x}$$

uniformly in x for $|\alpha| \leqslant m$. □

3. Suppose f is an L^2 function on $\mathbb{R}^n$, $m \in \mathbb{N}$, and there are positive constants A, $B > 0$, and $p > \frac{m+n}{2}$ with

$$|f(x)| \leqslant A(1 + |x|^2)^{-p}, \quad |\hat{f}(\omega)| \leqslant B(1 + |\omega|^2)^{-p}.$$

Show f is in $C^m(\mathbb{R}^n)$ and for $|\alpha| \leqslant m$ one has

$$D^\alpha\left(\sum_{k\in\mathbb{Z}^n} f(x+k)\right) = \sum_{k\in\mathbb{Z}^n} D^\alpha f(x+k) = \sum_{k\in\mathbb{Z}^n} \hat{f}(k)(2\pi ik)^\alpha e^{2\pi ik\cdot x}.$$

Solution:

Note we are assuming $p > \frac{m+n}{2}$ and $|\alpha| \leqslant m$. By Exercise 4.13.1, $f_a(x) = \sum_k f(x+k) = \sum \hat{f}(k)e^{2\pi ix\cdot k}$ and this series converges uniformly. Also $\mathcal{F}(f_a)(k) =$

$\hat{f}(k)$ where $\mathcal{F}$ is the Fourier transform on $L^2(Q)$. By Exercise 4.13.2, f_a will be in C^m if $\sum_k |\mathcal{F}(f_a)(k)|\,|k|^m < \infty$ or equivalently $\sum_k |\hat{f}(k)|\,|k|^m < \infty$ and then $\sum_k \mathcal{F}(f_a)(k)(2\pi i k)^\alpha e^{2\pi i k\cdot x}$ converges uniformly to $D^\alpha(f_a)(x)$. But

$$\begin{aligned}\sum_k |\hat{f}(k)|\,|k|^m &\leqslant B\sum_k (1+|k|^2)^{-p}|k|^m\\ &\leqslant B\sum_k (1+|k|^2)^{-p}(1+|k|^2)^{m/2}\\ &= B\sum_k (1+|k|^2)^{-p+m/2} < \infty\end{aligned}$$

if $p - \frac{m}{2} > \frac{n}{2}$ by Lemma 4.101. Thus Exercise 4.13.2 gives $f_a \in C^m(\mathbb{R}^n)$ and

$$D^\alpha f_a(x) = \sum_k \hat{f}(k)(2\pi i k)^\alpha e^{2\pi i k\cdot x}$$

uniformly.

We remark that we do not know if $\sum_k D^\alpha f(x+k)$ converges and think this may not be the case. If one could show $|D^\alpha f(x)| \leqslant A_\alpha(1+|x|^2)^{-p_\alpha}$ where $p_\alpha > \frac{n}{2}$, then this series would converge uniformly and then we would have

$$D^\alpha\left(\sum_{k\in\mathbb{Z}^n} f(x+k)\right) = \sum_{k\in\mathbb{Z}^n} (D^\alpha f)(x+k).$$

□

4. Introduce the *Wiener space* $W(\mathbb{R}^n)$ as the space of functions $f \in L^\infty(\mathbb{R}^n)$ such that

$$||f||_W := \sum_{k\in\mathbb{Z}^d} \operatorname*{ess\,sup}_{x\in Q} |\lambda(k)f(x)| < \infty\,.$$

Show the following:

(a) $W(\mathbb{R}^n) \subset L^1(\mathbb{R}^n)$.

(b) $\mathcal{S}(\mathbb{R}^n) \subset W(\mathbb{R}^n)$.

(c) If f and $\hat{f}$ are in $W(\mathbb{R}^n)$, then $\sum_{k\in\mathbb{Z}^n} f(x-k)$ and $\sum_{k\in\mathbb{Z}^n} \hat{f}(k)e^{2\pi i k\cdot x}$ converge uniformly absolutely.

(d) The Poisson summation formula holds if $f, \hat{f} \in W(\mathbb{R}^n)$; i.e.,

$$\sum_{k\in\mathbb{Z}^n} f(x-k) = \sum_{k\in\mathbb{Z}^n} \hat{f}(k)e^{2\pi i k\cdot x}$$

for $x \in \mathbb{R}^n$.

Solution: We note $W(\mathbb{R}^n) \supseteq \mathcal{S}(\mathbb{R}^n)$. Indeed, this was shown in the proof of Proposition 4.102.

To see $W(\mathbb{R}^n) \subseteq L^1(\mathbb{R}^n)$, we have

$$\begin{aligned}
\int |f(x)|\,dx &= \sum_{k\in\mathbb{Z}^n} \int_{Q+k} |f(x)|\,dx \\
&\leqslant \sum_{k\in\mathbb{Z}^n} \int \operatorname*{ess\,sup}_{x\in Q+k} |f(x)|\,dx \\
&= \sum_{k\in\mathbb{Z}^n} \operatorname*{ess\,sup}_{x\in Q+k} |f(x)| \\
&= \sum_{k\in\mathbb{Z}^n} \operatorname*{ess\,sup}_{y\in Q} |f(y+k)| \\
&= \sum_{k\in\mathbb{Z}^n} \operatorname*{ess\,sup}_{y\in Q} |\lambda(-k)f(y)| \\
&< \infty.
\end{aligned}$$

So $f \in L^1$.

If f and $\hat{f}$ are in $W(\mathbb{R}^n)$, both are in $L^1(\mathbb{R}^n)$ and thus both f and $\hat{f}$ are continuous. So we have $\sum_k f(x+k)$ converges pointwise uniformly absolutely on Q and thus by periodicity pointwise uniformly absolutely on $\mathbb{R}^n$. We claim $\sum_k |\hat{f}(k)|$ is finite. Indeed, we know $\sum_k \max_{x\in Q} |\hat{f}(x+k)|$ converges. Since $|\hat{f}(k)| \leqslant \max_{x\in Q} |\hat{f}(x+k)|$, we see $\sum_k |\hat{f}(k)|$ converges. This implies $\sum_k \hat{f}(k)e^{2\pi i k\cdot x}$ converges absolutely uniformly.

Let $f_a(x) = \sum_k f(x+k)$. We know f_a is continuous and $\mathbb{Z}^n$ periodic. Consequently $\hat{f}_a(x) = \sum_k \hat{f}_a(k)e^{2\pi i k\cdot x}$ in $L^2(Q)$. Since $\hat{f}_a(k) = \hat{f}(k)$, we have

$$\hat{f}_a(x) = \sum_k \hat{f}(k)e^{2\pi i k\cdot x}$$

in $L^2(Q)$. Since $\sum_k \hat{f}(k)e^{2\pi i k\cdot x}$ and $\sum_k f(x+k)$ converge uniformly and are thus continuous and give the same L^2 function, they are equal everywhere. □

Chapter 5

Topological Groups

Exercise Set 5.1

1. Let G be a group with topology τ. Show that G is a topological group if and only if the map

$$G \times G \ni (x, y) \mapsto xy^{-1} \in G$$

is continuous.

Solution:

Assume G is a topological group. Then $y \mapsto y^{-1}$ and $(x, y) \mapsto xy$ are continuous. This implies $(x, y) \mapsto (x, y^{-1})$ is continuous from $G \times G$ into $G \times G$. Composing this function with $(x, y) \mapsto xy$ gives the function $(x, y) \mapsto xy^{-1}$. Since compositions of continuous functions are continuous, we see $(x, y) \mapsto xy^{-1}$ is continuous.

Conversely, suppose $(x, y) \mapsto xy^{-1}$ is continuous. Since $y \mapsto (e, y)$ is continuous, the composition with $(x, y) \mapsto xy^{-1}$ gives the function $y \mapsto y^{-1}$ is continuous. This implies $(x, y) \mapsto (x, y^{-1})$ is continuous and again composition with $(x, y) \mapsto xy^{-1}$ gives $(x, y) \mapsto xy$ is continuous. Thus G is a topological group.

□

2. Let G be a topological group. Show G is T_1 if and only if $\{e\}$ is a closed set in G.

Solution:

Recall a space is T_1 if and only if for any two distinct points x and y, there is a neighborhood U of x not containing y.

Assume $\{e\}$ is closed and $x \neq y$. Then $x^{-1}y \neq e$. Thus there is a neighborhood U of e with $x^{-1}y \notin U$. So $y \notin xU$. But xU is a neighborhood of x. Thus G is T_1.

Assume G is T_1. Let $x \neq e$. Then there is a neighborhood U of x with $e \notin U$. Thus x is not a limit point of $\{e\}$. So the closure of $\{e\}$ is $\{e\}$ and we see $\{e\}$ is a closed set.

□

3. Let G be a topological group. Let $U \subset G$ be open and let $a \in U$. Show there exists an open set $V \ni e$ such that $aV \subset U$.

Solution:

Assume U is open and $a \in U$. Since $(x, y) \mapsto xy$ is continuous at (a, e), there is a neighborhood of form $W \times V$ where W is an open set containing a and V is an open set containing e such that $xy \in U$ if $x \in W$ and $y \in V$. Since $a \in W$, we have $ay \in U$ for $y \in V$. So $aV \subseteq U$.

□

4. Let G and H be topological groups. Show a function $f : G \to H$ is continuous if and only if for each x in G and each neighborhood V of the identity in H, there is a neighborhood U of the identity in G with

$$f(xU) \subseteq f(x)V.$$

Solution:

Let $f : G \to H$ be continuous. Fix x. Then if V is a neighborhood of e in H, there is an open neighborhood V' of e contained in V. Thus $f(x)V'$ is an open set. Since f is continuous, its preimage W under f is an open subset of G containing x. Now $y \mapsto xy$ is continuous on G. Thus the preimage U of W under this mapping is open and contains e. Consequently $xU \subseteq W \subseteq f^{-1}(f(x)V')$ and thus $f(xU) \subseteq f(x)V' \subseteq f(x)V$.

Conversely assume for each $x \in G$ and for each neighborhood V of e in H, there is a neighborhood U of e in G with $f(xU) \subseteq f(x)V$. Let W be an open subset of H and suppose $x \in f^{-1}(W)$. So $f(x) \in W$ and thus there is a neighborhood V of e in H such that $f(x)V \subseteq W$. Consequently, there is a neighborhood U of e in G such that $f(xU) \subseteq f(x)V$. Thus $xU \subseteq f^{-1}(f(x)V) \subseteq f^{-1}(W)$. Thus each x in $f^{-1}(W)$ is an internal point of $f^{-1}(W)$. So $f^{-1}(W)$ is open and we see f is continuous.

□

5. Let G be a topological group. Show that $\rho(a) : G \to G$ is a homeomorphism.

Solution:

We first note $\rho(a)$ is continuous. Indeed, it is the composition of the continuous mappings $x \mapsto (x, a)$ and $(x, y) \mapsto xy$. So $\rho(a) : G \to G$ defined by $\rho(x) = xa$ is continuous. Moreover $\rho(a)\rho(a^{-1})x = (xa^{-1})a = x$ and $\rho(a^{-1})\rho(a)x = (xa)a^{-1} = x$. Thus $\rho(a)$ is one-to-one and onto and has continuous inverse $\rho(a^{-1})$. So $\rho(a)$ is a homeomorphism.

□

6. Let G be a topological group and $H \subset G$ an open subgroup. Show H is also closed.

Solution:

Let H be an open subgroup of G. Then each of the cosets aH are open subsets of G. But $H^c = \cup_{a \notin H} aH$. So H^c is open. Thus H is closed.

□

7. Let H be a subgroup of a topological group G. Show the closure $\bar{H}$ is a group and the interior H° of H is either empty or is all of H.

Solution:

Let H be a subgroup of G. We claim its closure $\bar{H}$ is a subgroup. It suffices to show if $x, y \in \bar{H}$, then $xy^{-1} \in \bar{H}$. Let N be a neighborhood of xy^{-1} in G. Since $(a, b) \mapsto ab^{-1}$ is continuous at (x, y), there is a neighborhood N_x of x and a neighborhood N_y of y such that $ab^{-1} \in N$ if $a \in N_x$ and $b \in N_y$. Since $x \in \bar{H}$ and $y \in \bar{H}$, there are an a and b with $a \in H \cap N_x$ and $b \in H \cap N_y$. So $ab^{-1} \in H \cap N$. This implies $xy^{-1} \in \bar{H}$. So $\bar{H}$ is a subgroup.

Now suppose H° is nonempty. Let $a \in H^\circ$. Thus there is a neighborhood N of a such that $N \subseteq H$. This implies $ba^{-1}N$ is a neighborhood of b contained in H for each $b \in H$. Thus if $H^\circ \neq \varnothing$, then $H^\circ = H$ and is thus a subgroup of G.

□

Exercise Set 5.2

1. Let X be a topological space. Show the following:

(a) $T_4 \Rightarrow T_3$;

(b) $T_3 \Rightarrow T_2$;

(c) Completely regular $\Rightarrow$ regular.

Solution:

For (a), assume X is T_4. Since X is T_4, it is T_1. Thus to see X is T_3, it suffices to show it is regular. Thus let A be a closed subset of X and suppose $x \notin A$. Since X is T_1, $\{x\}$ is a closed set disjoint from A. Thus there are open sets U and V with $A \subseteq U$ and $\{x\} \subseteq V$ with $U \cap V = \varnothing$. This gives X is regular.

For (b), assume X is T_3. Then since it is T_1, singleton sets are closed. So if $x \neq y$, $\{x\}$ is closed and $y \notin \{x\}$. Thus there are disjoint open sets U and V with $\{x\} \subseteq U$ and $y \in V$. Thus X is T_2.

Finally for (c), assume X is completely regular and A is a closed subset and $x \notin A$. Then there is a continuous function f with $f(x) = 1$ and $f(y) = 0$ for $y \in A$. Set $U = \{z \mid f(z) > \frac{1}{2}\}$ and $V = \{z \mid f(z) < \frac{1}{2}\}$. U and V are open, $x \in U$, $A \subseteq V$, and $U \cap V = \varnothing$. So X is regular.

□

2. Show any closed subspace of a Lindelöf space is Lindelöf. Give an example of a Lindelöf space with a subspace which is not Lindelöf.

Solution:

Let X be a Lindelöf space and suppose Y is a closed subspace. Give Y the relative topology and suppose U_α, $\alpha \in A$ are relatively open subsets of Y with $Y \subseteq \cup U_\alpha$. Choose open subsets V_α of X with $V_\alpha \cap Y = U_\alpha$. These sets along with $X - Y$ form an open cover of X. Thus there is a countable subset $I \subseteq A$ with $X = (\cup_{\alpha \in I} V_\alpha) \cup (X - Y)$. This implies $Y \subseteq \cup_{\alpha \in I}(V_\alpha \cap Y) = \cup_{\alpha \in I} U_\alpha$. Thus Y is Lindelöf.

Now let Y be an uncountable space with the discrete topology. Let $X = Y \cup \{\infty\}$ be the one point compactification of Y; i.e., a subset U of X is open if and only if $U \subseteq Y$ or if $\infty \in U$, then the complement of U is a finite set. This is a compact space and thus is Lindelöf. Now the relative topology of X on Y is the discrete topology and since Y is uncountable, it cannot be Lindelöf.

□

3. Suppose that G is a topological group and H a subgroup. Let τ be the collection of all sets $U \subset G/H$ such that $\kappa^{-1}(U) \subset G$ is open in G. Show that τ is a topology on G/H.

Solution:

τ is the collection of all sets U such that $\kappa^{-1}(U)$ is open in G. Note $\kappa^{-1}(\varnothing) = \varnothing$ and $\kappa^{-1}(G/H) = G$. So $\varnothing$ and G/H are in τ. Let U_1 and U_2 be in τ. Then $\kappa^{-1}(U_1 \cap U_2) = \kappa^{-1}(U_1) \cap \kappa^{-1}(U_2)$ for $g \in \kappa^{-1}(U_1 \cap U_2)$ if and only if $\kappa(g) \in U_1 \cap U_2$ if and only if $\kappa(g) \in U_1$ and $\kappa(g) \in U_2$ if and only if $g \in \kappa^{-1}(U_1) \cap \kappa^{-1}(U_2)$. So $U_1 \cap U_2 \in \tau$. Assume $U_i \in \tau$ where $i \in I$. Then $\kappa(g) \in \cup U_i$ if and only if $\kappa(g) \in U_i$ for some i if and only if $g \in \kappa^{-1}(U_i)$ for some i. So $\kappa^{-1}(\cup U_i) = \cup_i \kappa^{-1}(U_i)$ is open in G. So $\cup U_i \in \tau$. Thus τ is a topology.

□

4. Let H be a subgroup of a topological group G, and let X be a topological space.

(a) Show that a function $f : G/H \to X$ is continuous if and only if the map $f \circ \kappa : G \to X$ is continuous.

(b) Suppose a function f maps G/H onto X. Show f is open if and only if the map $f \circ \kappa : G \to X$ is open.

(c) Assume that G acts separately continuously and transitively on X. Show the following are equivalent.

- There exists a $x \in X$ such that κ_x is open.
- There exists a $x \in X$ such that $\pi_x : G/G_x \to X$ is a homeomorphism.
- κ_x is open for all $x \in X$.
- $\pi_x : G/G_x \to X$ is a homeomorphism for all $x \in X$.

Solution:

For (a), if G/H has the quotient topology, a function $f : G/H \to X$ is continuous if and only if $f^{-1}(U)$ is open in G/H for each open subset U of X. This happens if and only if $\kappa^{-1}(f^{-1}(U))$ is open in G for each open set U and since $\kappa^{-1}(f^{-1}(U)) = (f \circ \kappa)^{-1}(U)$, we see f is continuous if and only if $f \circ \kappa$ is continuous.

To see (b), note if $f : G/H \to X$, then f is open if and only if $f(V)$ is open in X for each open subset V of G/H. But κ is an open mapping and the open subsets of G/H are the sets $\kappa(U)$ where U is open in G. Thus f is open if and only if $f \circ \kappa$ is open.

Now assume G acts separately continuously and transitively on X. Assume x is given and $\kappa_x : G \to X$ is open. First note κ_x is onto by transitivity and thus $\pi_x : G/G_x \to X$ is onto. Moreover, this map is continuous by (a) for $g \mapsto g \cdot x$ is continuous. Since this map is one-to-one, (b) implies it is an open mapping. Thus π_x is a homeomorphism.

Suppose π_x is a homeomorphism. Let $y \in X$. Choose $h \in G$ with $h \cdot x = y$. Then $\kappa_y(U) = U \cdot y = U \cdot (h \cdot x) = \kappa_x(Uh)$. So $\kappa_y(U) = \pi_x(UhG_x)$ is open for UhG_x is open in G/G_x. Thus κ_y is an open mapping for all y. The argument given earlier then implies $\pi_y : G/G_y \to X$ is a homeomorphism for all y in G.

Finally if π_y is a homeomorphism for all y, then π_x is an open mapping. Thus $\kappa_x(U) = \pi_x(UG_x)$ is open for any open subset U of G. So κ_x is open. Putting this all together gives (c). □

5. Let $G = \mathbb{R}$, and let τ be an irrational number.

(a) Show $H = \mathbb{Z} + \mathbb{Z}\tau$ is dense in $\mathbb{R}$. (**Hint:** Show H has no smallest positive member.)

(b) Show that G acts continuously on $\mathbb{T}^2$ by

$$x \cdot (e^{iu}, e^{iv}) := (e^{i(u+x)}, e^{i(v+\tau x)}) \text{ for } x, u, v \in \mathbb{R}\,.$$

(c) Let $z \in \mathbb{T}^2$. Show $G_z = \{0\}$ and $G \cdot z$ is dense in $\mathbb{T}^2$.

(d) Show show the orbit $G \cdot z$ contains no open subset of $\mathbb{T}^2$ and thus the mapping $\kappa_z : x \mapsto x \cdot z$ is not an open mapping.

Solution:

$H = \mathbb{Z} + \mathbb{Z}\tau$ is a subgroup of $\mathbb{R}$. We note if H has a least positive element a, then $H = \mathbb{Z}a$ for if $w \in G$, choose $n \in \mathbb{Z}$ such that $na \leqslant w < na + a$. Then if $w \neq na$, $0 < w - na < a$. But $w - na \in H$, a contradiction. So $w = na$. This gives $1 = ma$ and $\tau = na$ for some $m, n \in \mathbb{Z}$. So $\tau = \frac{n}{m}$ is rational, a contradiction. Thus H has no least positive element. This implies H is dense in $\mathbb{R}$. So (a) holds.

For (b), the left action of $\mathbb{R}$ on $\mathbb{T}^2$ by $x \cdot (e^{iu}, e^{iv}) = (e^{i(u+x)}, e^{i(v+x\tau)})$ is continuous for

$$\begin{aligned} |x \cdot (e^{iu}, e^{iv}) - x_0 \cdot (e^{iu_0}, e^{iv_0})| &= |(e^{i(u+x)}, e^{i(v+x\tau)}) - (e^{i(u_0+x_0)}, e^{i(v_0+x_0\tau)})| \\ = |e^{i(u_0+u-u_0+x-x_0+x_0)} - e^{i(u_0+x_0)}, &e^{i(v_0+v-v_0+(x-x_0)\tau+x_0\tau)} - e^{i(v_0+x_0\tau)})| \\ &= |e^{i(u-u_0+x-x_0)} - 1, e^{i(v-v_0+(x-x_0)\tau)} - 1)| \\ &\leqslant |u - u_0 + x - x_0| + |v - v_0 + (x - x_0)\tau| \end{aligned}$$

for $|e^{i\theta} - 1| \leqslant \theta$. Thus $|x \cdot (e^{iu}, e^{iv}) - x_0 \cdot (e^{iu_0}, e^{iv_0})| < \epsilon$ if $|u - u_0| < \frac{\epsilon}{4}$, $|v - v_0| < \frac{\epsilon}{4}$ and $|x - x_0| < \frac{1}{4}\min\{\epsilon, \frac{\epsilon}{\tau}\}$. Thus the action is continuous.

Next note if $z \in \mathbb{T}^2$, then $x \cdot z = (e^{ix}z_1, e^{i\tau x}z_2) = (z_1, z_2)$ if and only if x and τx are multiples of 2π; i.e. $x = m2\pi$ and $\tau x = n2\pi$ where $m, n \in \mathbb{Z}$. So if $x \neq 0$, then $\tau = \frac{2\pi n}{x} = \frac{2\pi n}{2\pi m} = \frac{n}{m}$ which would be rational. So $G_z = \{0\}$.

Now the orbit $G \cdot z$ consists of all the point $(e^{ix}z_1, e^{i\tau x}z_2)$ or all points $(e^{ix}, e^{i\tau x})(z_1, z_2)$ where $x \in \mathbb{R}$. So the orbit is dense if $H = \{(e^{ix}, e^{i\tau x}) \mid x \in \mathbb{R}\}$ is a dense subgroup of $\mathbb{T}^2$. Note $x \mapsto ax$ is an isomorphism of $(\mathbb{R}, +)$ for $a > 0$. By (a), $a\mathbb{Z}+a\mathbb{Z}\tau$ is dense in $\mathbb{R}$. So the image of $2\pi\mathbb{Z}+2\pi\mathbb{Z}\tau$ in $\mathbb{R}/2\pi\mathbb{Z}$ is dense. Since the mapping $\theta + 2\pi\mathbb{Z} \to e^{i\theta}$ is a homeomorphism of $\mathbb{R}/2\pi\mathbb{Z}$ onto $\mathbb{T}$, the points $e^{2\pi in\tau}$ form a dense subset of $\mathbb{T}$. Similarly the points $e^{2\pi in/\tau}$ form a dense subset of $\mathbb{T}$. The image of the set $2\pi\mathbb{Z}$ under $x \mapsto (e^{ix}, e^{i\tau x})$ is $\{(1, e^{2\pi in\tau}) \mid n \in \mathbb{Z}\}$ and so the closure of the range of the mapping $x \mapsto (e^{ix}, e^{i\tau x})$ contains $\{1\} \times \mathbb{T}$. Similarly, the image of the set $2\tau^{-1}\pi\mathbb{Z}$ are the points $(e^{2\pi in/\tau}, 1)$ and thus the closure contains $\mathbb{T}\times\{1\}$. Since the image is a subgroup, the closure is a subgroup and thus contains $(\mathbb{T} \times \{1\})(\{1\} \times \mathbb{T}) = \mathbb{T}^2$. So the image is a dense subgroup of $\mathbb{T}^2$ and (c) holds.

Finally for (d), it suffices to show $\mathbb{R} \cdot z$ contains no nonempty open subset of $\mathbb{T}^2$. If this set contains an open subset in $\mathbb{T}^2$ with $(e^{i\theta}z_1, e^{i\theta\tau}z_2)$ as an element, then $(e^{ia}e^{i\theta}z_1, e^{ia}e^{i\theta\tau}z_2)$ would be in this set if a is near 0. Pick $1 > a > 0$, $a = q\pi$ where q is rational and θ_1 so this is true. Thus $(e^{ia}e^{i\theta_1}z_1, e^{ia}e^{i\tau\theta_1}z_2) = (e^{i\theta_2}z_1, e^{i\theta_2\tau}z_2)$ for some θ_2. This would imply $a + \theta_1 = \theta_2 + 2\pi m$ and $a + \tau\theta_1 = \theta_2\tau + 2\pi n$ for some $m, n \in \mathbb{Z}$. So $\theta_2 - \theta_1 = a - 2\pi m = \pi q_1$ and $(\theta_2 - \theta_1)\tau = a - 2\pi n = \pi q_2$ where q_1 and q_2 are rational. Note $\theta_2 \neq \theta_1$ for $a - 2\pi m \neq 0$. This gives $\tau = \frac{\pi q_2}{\theta_2 - \theta_1} = \frac{\pi q_2}{\pi q_1} = \frac{q_2}{q_1}$ and we would see τ is rational. So the orbit contains no open set.

□

6. Prove the basic topological statements highlighted on page 246.
Solution:

- Let $L \subseteq K$ be a closed subset of a compact subset K of X. If U_α cover L where each U_α is open, then the sets U_α and $(X - L)$ cover K. Since L is closed, this is an open cover of K. Thus finitely many of these cover K and thus cover L. Since $X - L$ is not needed in a cover of L, we see finitely many U_α cover L and so L is compact.

- Let $f : X \to Y$ be continuous and suppose V_α for $\alpha \in A$ are an open cover of the set $f(K)$. Then the open subsets $f^{-1}(V_\alpha)$ of X cover K. So there is a finite set F of α's such that $K \subseteq \cup_{\alpha\in F} f^{-1}(V_\alpha)$. This implies $f(K) \subseteq \cup_{\alpha\in F} V_\alpha$. Hence $f(K)$ is compact.

- Suppose X is Hausdorff and $x \notin K$. For $y \in K$ choose open sets U_y and V_y with $x \in U_y$, $y \in V_y$, and $U_y \cap V_y = \varnothing$. Note $K \subseteq \cup_{y\in K} V_y$. By compactness, there are $y_1, y_2, \ldots, y_m$ in K with $K \subseteq \cup_{i=1}^m V_{y_i}$. Set $U = \cap_{i=1}^m U_{y_i}$. Then U is an open subset of X containing x with $U \cap K = \varnothing$. So x is an interior point of the complement of K. Thus the complement of K is open and we see K is a closed set.

□

7. Suppose G is a topological group and X and Y are G-spaces. Assume that $\varphi : X \to Y$ is an G-isomorphism. Show that $\varphi^{-1} : Y \to X$ is also a G-isomorphism.

Solution:

Let $\phi : X \to Y$ be a G-isomorphism. Thus ϕ is a homeomorphism and $\phi(g \cdot x) = g \cdot \phi(x)$ for all x. Note ϕ^{-1} is a homeomorphism and $\phi^{-1}(g \cdot \phi(x)) = g \cdot x = g \cdot \phi^{-1}(\phi(x))$. Since ϕ is onto, $\phi^{-1}(g \cdot y) = g \cdot \phi^{-1}(y)$ for all $y \in Y$. Thus ϕ^{-1} is a G-isomorphism.

□

8. Let G be a topological group and let X be a topological G-space. Set $G\backslash X = \{Gx \mid x \in X\}$ to be the space of orbits. Let $\pi : X \to G\backslash X$ by $\pi(x) = Gx$. Define a set U in $G\backslash X$ to be open if $\pi^{-1}(U)$ is open in X.

(a) Show the set of open sets is a topology on $G\backslash X$. It is called the *quotient topology.*

(b) Show $\pi : X \to G\backslash X$ is an open mapping.

(c) Give an example where G and X are Hausdorff, but $G\backslash X$ is not Hausdorff.

(d) Suppose G and X are locally compact Hausdorff spaces. Show $G\backslash X$ is compact if and only if there is a compact set $C \subseteq X$ with $GC = X$.

Solution:

Note if U and V are open in $G\backslash X$, then $\pi^{-1}(U \cap V) = \pi^{-1}(U) \cap \pi^{-1}(V)$ is open in X. So $U \cap V$ is open in $G\backslash X$. Clearly $\varnothing$ and $G\backslash X$ are open in $G\backslash X$ for $\pi^{-1}(\varnothing) = \varnothing$ and $\phi^{-1}(G\backslash X) = X$ are open in X. Finally, if U_i are open sets in $G\backslash X$, then $\pi^{-1}(\cup_i U_i) = \cup_i \pi^{-1}(U_i)$ is open in X. So $\cup U_i$ is an open set in $G\backslash X$. These together give (a).

To see (b), let U be an open set in X. Then $\pi^{-1}(\pi(U)) = G{\cdot}U = \cup_{g\in G}(g{\cdot}U)$ is open in X for $x \mapsto g \cdot x$ is a homeomorphism of X for each $g \in G$. Thus π is an open mapping.

For (c), let $X = \mathbb{T}$, $G = \mathbb{Z}$, and $n \cdot z = e^{in}z$. Let U and V be open nonempty subsets in $G\backslash X$. Then $\pi^{-1}(U)$ and $\pi^{-1}(V)$ are open G invariant subsets of $\mathbb{T}$. But each orbit in $\mathbb{T}$ is dense. Thus $\pi^{-1}(U) \cap \pi^{-1}(V) \neq \varnothing$. Hence $U \cap V \neq \varnothing$.

Finally, first suppose there is a compact C with $GC = X$. Then $\pi(C) = X$ and since π is continuous, $G\backslash X$ is compact. Conversely, suppose $G\backslash X$ is compact. For each x, choose a compact neighborhood N_x of x. Then $\cup \pi(N_x)^\circ = G\backslash X$. Hence there is a finite set F with $\cup_{x\in F}\pi(N_x) = G\backslash X$. Let $C = \cup_{x\in F} N_x$. So one has (d). □

9. Let G be a topological group and let H be a subgroup with the relative topology.

(a) Show if H is locally compact and Hausdorff, then H is closed in G. In particular, if H is discrete in the relative topology, then H is closed.

(b) Give an example of a discrete subspace M of some topological space X which is not closed.

Solution: Let x be a limit point of H. Since H is locally compact, there is a compact neighborhood C in H about e. Choose U open in G with $e \in U$

and $U \cap H \subseteq C$. Choose V open and symmetric about e with $V^2 \subseteq U$. Then $Vx \cap H \neq \varnothing$. Then $(Vx \cap H)(Vx \cap H)^{-1} \subseteq V^2 \cap H \subseteq C$. This implies $Vx \cap H \subseteq Cy$ for each $y \in Vx \cap H$. Let W be any neighborhood of x in G. Then $W \cap Vx \cap H \neq 0$. Hence $W \cap Cy \neq \varnothing$. Thus x is a limit point of Cy. Since H is Hausdorff, Cy is closed. Thus $x \in Cy \subseteq H$. Hence H is closed.

Now for an example, take $X = \mathbb{N} \cup \{\infty\}$ to be the one-point compactification of $\mathbb{N}$. Hence a subset is open if it is a subset of $\mathbb{N}$ or if it is the complement of a finite subset of $\mathbb{N}$. Then $\mathbb{N}$ with the relative topology of X is discrete. But $\mathbb{N}$ is not closed for every neighborhood of ∞ meets $\mathbb{N}$.

□

10. Suppose G is a locally compact Hausdorff group and K is a compact subgroup. Show if A is compact in G/K, then $\kappa^{-1}(A)$ is compact in G.

Solution:

Since K is compact, it is closed. Thus by Lemma 5.13, G/K is Hausdorff. Since A is compact, it is closed. Write $G = \cup V_x$ of compact neighborhoods, one for each x in G. Then $\kappa(G) = \cup \kappa(V_x)^\circ$ is an open cover of A. So $A \subseteq \cup_{x\in F} \kappa(V_x)$ for some finite subset F of G. Hence $\kappa^{-1}(A) \subseteq \cup_{x\in F} V_x K$. Thus $\kappa^{-1}(A)$ is a closed subset of the compact set $\cup_{x\in F} V_x K$ and hence is compact. □

11. Let G be a compact Hausdorff group. Show a subgroup D has discrete relative topology if and only if D if finite.

Solution:

Suppose the relative topology on D is discrete. Choose an open subset U in G with $U \cap D = \{e\}$. Take an open symmetric subset V with $e \in V$ and $V^2 \subseteq U$. Cover G by $\cup_{x\in F} Vx$ where F is finite. Consider $(Vx \cap D)(Vx \cap D)^{-1} \subseteq V^2 \cap D \subseteq U \cap D = \{e\}$. Thus $Vx \cap D$ has most one point. So $D \subseteq \cup_{x\in F}(Vx \cap D)$ is finite. □

12. Let D be a subgroup of G where G is locally compact and Hausdorff. Show D is discrete if and only if $D \cap K$ is finite for every compact subset K of G.

Solution:

Suppose D is discrete. Then by (a) of Exercise 5.2.9 above, D is closed. Hence $D \cap K$ is compact and discrete and Hausdorff and thus must be finite.

Suppose $D \cap K$ is finite for all compact subsets K. Pick a compact neighborhood V of e in G. Then $D \cap V$ is a finite neighborhood in D of e. Since D is Hausdorff, this makes e isolated in D. Hence D is discrete. □

13. A dyadic rational is a number of form $\frac{k}{2^n}$ where k and n are integers. Consider a topological space X, and let F be a closed set and G be an open set with $F \subseteq G$. Suppose V_d is an open set for each d in D, the set of dyadic rationals, and $V_0 = \varnothing$, $V_1 = X$, $F \subseteq V_d \subseteq G$ when $0 < d < 1$, and $\bar{V}_d \subseteq V_{d'}$ for $d < d'$ in D. Show the function f defined by $f(x) = \inf\{d \mid x \in V_d\}$ is a continuous function that satisfies $0 \leqslant f \leqslant 1$, $f = 0$ on F, and $f = 1$ off G. Moreover, if $\cap_{d>0} V_d = F$, show $f(x) = 0$ if and only if $x \in F$.

Solution: Clearly $0 \leqslant f \leqslant 1$ and if $x \in F$, then $f(x) = 0$. If $x \notin G$, then $f(x) = 1$.

We show f is continuous. Note $f(x) < \alpha$ if and only if there is a $d \in D$ with $d < \alpha$ and $x \in V_d$. Thus $f^{-1}(-\infty, \alpha) = \cup_{d<\alpha} V_d$.

Now $f(x) \leqslant \alpha$ if and only if $f(x) < \alpha + \epsilon$ for all $\epsilon > 0$ if and only if for all $\epsilon > 0$, $x \in V_d$ for some $d < \alpha + \epsilon$. But this is equivalent to $x \in V_d$ for all $d > \alpha$. Thus $f^{-1}(-\infty, \alpha] = \cap_{d>\alpha} V_d = \cap_{d>\alpha} \bar{V}_d$. Hence f is continuous.

Now suppose $\cap_{d>0} V_d = F$. If $x \notin F$, then there is a $d > 0$ such that $x \notin V_d$. Thus $x \notin V_{d'}$ for $d' < d$. So $f(x) \geqslant d > 0$. Hence $f(x) = 0$ if and only if $x \in F$. □

14. Let X be a locally compact Hausdorff space and let V be an open set containing a compact subset K. Show there is a an open subset G of V with compact closure such that $K \subseteq \bar{G} \subseteq V$.

Solution:

For each $x \in K$, by Lemma 5.14, there is a compact neighborhood N_x of x such that $N_x \subseteq V$. The interiors N_x° of the sets N_x for $x \in K$ cover K. Thus since K is compact, there is a finite set $F \subseteq K$ with $K \subseteq \cup_{x \in F} N_x^\circ$. Set $G = \cup_{x \in F} N_x^\circ$. Note G is open, G contains K and $\bar{G} \subseteq \cup_{x \in F} N_x \subseteq V$. But the set $\cup_{x \in F} N_x$ is compact. So $\bar{G}$ is compact.

□

15. Let X be a σ-compact locally compact Hausdorff space. Show there is a sequence U_n of precompact open subsets which cover X and has the property $\bar{U}_n \subseteq U_{n+1}$.

Solution: Let C_k be a sequence of compact subsets which cover X. By Exercise 5.2.14, we can find an open subset U_1 with $C_1 \subseteq U_1$ and $\bar{U}_1$ is compact. Then consider $\bar{U}_1 \cup C_2$. Again find an open set U_2 with $\bar{U}_1 \cup C_2 \subseteq U_2$ and $\bar{U}_2$ is compact. Repeat this process and we find a sequence of open sets U_n with $\bar{U}_n$ compact and with $\bar{U}_{n-1} \cup C_n \subseteq U_n$. □

16. Let X be a locally compact Hausdorff space. Show a subset K is a compact G_δ set if and only if there is a continuous function $0 \leqslant \phi \leqslant 1$ on X having compact support with $\phi^{-1}\{1\} = K$.

Solution:

One direction was established in Proposition 5.23. Namely if K is a compact G_δ, then there is a ϕ which is continuous, has compact support, and satisfies $0 \leqslant \phi \leqslant 1$ with $\phi^{-1}(1) = K$.

The converse is the easier direction. Namely suppose such a ϕ exists. For $k = 1, 2, 3, \ldots$, set $U_k = \{x \mid \phi(x) > 1 - \frac{1}{k}\}$. Each U_k is open and contains K. Moreover, if $x \in U_k$ for all k, $\phi(x) = 1$ and thus $x \in K$. So $K = \cap_k U_k$ is a G_δ.

□

Exercise Set 5.3

1. Show the topology defined by the norm $||\cdot||_2$ on $M(m \times n, \mathbb{F})$ is the product topology.

Solution: We show the product topology is given by the norm $||\cdot||_2$. It suffices to show a point X in a subset A of $M(m \times n, \mathbb{F})$ is interior in the product topology if and only if it is interior in the norm topology $||\cdot||_2$.

Note $X = [x_{i,j}]$ is interior in the product topology if there is a $\delta > 0$ such that if $Y = [y_{i,j}]$ satisfies $|y_{i,j} - x_{i,j}| < \delta$ for all i, j, then $Y \in A$. But this implies if $||Y - X||_2 = \left(\sum_{i,j} |y_{i,j} - x_{i,j}|^2\right)^{1/2} < \delta$, then $Y \in A$. So X would be interior in A in the norm topology.

Conversely, if X is interior to A in the norm topology, there is a $\delta > 0$ such that $||Y - X||_2 < \delta$ implies $Y \in A$. But then $|y_{i,j} - x_{i,j}| < \frac{\delta}{\sqrt{mn}}$ for all i, j implies $||Y - X||_2 = \left(\sum_{i,j} |y_{i,j} - x_{i,j}|^2\right)^{1/2} < \left(\sum_{i,j} \frac{\delta^2}{mn}\right)^{1/2} = \delta$ and thus $Y \in A$. So X is interior to A in the product topology.

□

2. Show that the standard norm on $M(n \times m, \mathbb{F})$ is given by

$$||X||_2 = \sqrt{\mathrm{Tr}(XX^*)}, \quad X \in M(n \times m, \mathbb{F})$$

where $X^* = (\bar{x}_{ji}) = \bar{X}^t$.

Solution:

Note $X^* = [x^*_{i,j}]$ where $x^*_{i,j} = \bar{x}_{j,i}$. Thus

$$\begin{aligned} ||X||_2^2 &= \sum_{i,j} |x_{i,j}|^2 \\ &= \sum_{i,j} x_{i,j}\bar{x}_{i,j} \\ &= \sum_{i,j} x_{i,j}x^*_{j,i} \\ &= \sum_i (XX^*)_{i,i} \\ &= \mathrm{Tr}(XX^*)\,. \end{aligned}$$

So the result holds.

□

3. Prove Lemma 5.34. This exercise does not refer to an actual lemma and therefore no solution is given.

4. Let $\mathbb{F}$ be the reals or the complexes.

(a) Show $\mathrm{GL}(n, \mathbb{F})$ is locally compact and Lindelöf.

(b) Show all closed subgroups of $\mathrm{GL}(n, \mathbb{F})$ are locally compact and Lindelöf.

Solution:

By Exercise 5.3.1, $M(n \times n, \mathbb{F})$ has the product topology of $\mathbb{F}^{n^2}$ under the identification

$$[x_{i,j}] \mapsto (x_{1,1}x_{1,2}, \ldots, x_{1,n}, x_{2,1}, x_{2,2}, \ldots, x_{2,n}, \cdots, x_{n,1}, x_{n,2}, \ldots, x_{n,n})$$

and GL$(n,\mathbb{F})$ is the subset of all X with $\det(X) \neq 0$. Since det is continuous, GL$(n,\mathbb{F})$ is an open subset of $\mathbb{F}^{n^2}$ and since $\mathbb{F}^{n^2}$ is locally compact, GL$(n,\mathbb{F})$ is locally compact. Moreover, since $\mathbb{F}^{n^2}$ is second countable, GL$(n,\mathbb{F})$ is second countable and thus is Lindelöf. Now a closed subset of a Hausdorff locally compact space is locally compact and a subspace with the relative topology of a second countable space is second countable. This implies closed subgroups of GL$(n,\mathbb{F})$ are locally compact and Lindelöf.

□

5. Suppose β is a sesquilinear form on a Hilbert space V given by

$$\beta(u,v) = (Tu,v)$$

where T is a linear transformation of V. Show $||\beta|| = ||T||$.
Solution:
Suppose β is a sesquilinear form on a Hilbert space V given by $\beta(u,v) = (Tu,v)$ where T is a linear transformation of V. Then $||\beta|| = \sup\{|(Tu,v)| \mid ||u|| = ||v|| = 1\}$. Using the Cauchy-Schwarz inequality, T is bounded if and only if

$$\begin{aligned} \infty > ||T|| &= \sup_{||u||=1} ||Tu|| \\ &= \sup\{|(Tu,v)| \mid ||u|| = ||v|| = 1\} \\ &= ||\beta|| . \end{aligned}$$

So β is bounded if and only if T is bounded and then $||\beta|| = ||T||$.

□

6. Let $\mathbb{F} = \mathbb{C}$ and let V be a complex vector space. Let β be a Hermitian bilinear form on V. Show that $\beta(u,u) \in \mathbb{R}$ for all $u \in V$.
Solution:
Let β be a Hermitian bilinear form on a complex linear space V. Then $\beta(u,v) = \overline{\beta(v,u)}$ for $u, v \in B$. Thus $\beta(u,u) = \overline{\beta(u,u)}$ for $u \in V$. So $\beta(u,u)$ is real for all u.

□

7. Show there is a conjugation on a Hilbert space which is not symmetric; i.e., (u, jv) need not always be (v, ju). In view of Lemma 5.33, there is no orthonormal basis $\{e_m\}_{m \in I}$ with $je_m = e_m$ for all $m \in M$.
Solution:
Let V be $\mathbb{C}^2$ with inner product $((u_1,v_1),(u_2,v_2)) = u_1\bar{u}_2 + v_1\bar{v}_2$. Define j by $j(u,v) = (\bar{u} - 2\bar{v}, -\bar{v})$. Note j is real linear, $j(iu,iv) = -ij(u,v)$, and

$$j^2(u,v) = j(\bar{u} - 2\bar{v}, -\bar{v}) = (u,v).$$

Therefore j is a conjugation. But

$$\begin{aligned} ((u_1,v_1), j(u_2,v_2))_2 &= ((u_1,v_1),(\bar{u}_2 - 2\bar{v}_2, -\bar{v}_2))_2 \\ &= u_1u_2 - 2u_1v_2 - v_1v_2 \end{aligned}$$

while

$$\begin{aligned}((u_2, v_2), j(u_1, v_1))_2 &= ((u_2, v_2), (\bar{u}_1 - 2\bar{v}_1, -\bar{v}_1)_2 \\ &= u_2 u_1 - 2u_2 v_1 - v_1 v_2\end{aligned}$$

which are not equal. Thus j is not symmetric. □

8. Suppose j is a conjugation on a Hilbert space V defined by $j(\sum \lambda_m e_m) = \sum \bar{\lambda}_m e_m$ where $\{e_m\}$ is an orthonormal basis. Let A be a complex bounded linear operator on V. Show the matrix of A^t is the transpose of the matrix for A; i.e., show $(Ae_m, e_k) = (A^t e_k, e_m)$.

Solution:

We have e_k is an orthonormal basis and $j(\sum \lambda_k e_k) = \sum \bar{\lambda}_k e_k$. By Lemma 5.31, the transformation A^t satisfies

$$(Au, jv) = (A^t v, ju).$$

Thus $(Ae_k, je_m) = (A^t e_m, je_k)$. Hence $(Ae_k, e_m) = (A^t e_m, e_k)$ for all m and k. □

9. Let V be an infinite dimensional complex Hilbert space. Show there is a conjugate linear transformation j of V satisfying $j^2 = I$ that is not a conjugation.

Solution:

Take a countable orthonormal set $\{e_k\}_{k=1}^{\infty}$ and extend it to a Hamel basis; $\{e_k\}_k \cup \{e_\alpha\}_A$; i.e. a vector space basis. Thus every vector v has a unique finite expansion of form $v = \sum_{k \in F_1} \lambda_k e_k + \sum_{\alpha \in F_2} \lambda_\alpha e_\alpha$ where F_1 is a finite subset of $\mathbb{N}$ and F_2 is a finite subset of A. Set $F_{1,e} = F_1 \cap 2\mathbb{N}$ and $F_{1,o} = F_1 \cap (2\mathbb{N} - 1)$. Define j on V by

$$jv = \sum_{k \in F_{1,e}} k\bar{\lambda}_k e_{k+1} + \sum_{i \in F_{1,o}} \frac{1}{k-1} \bar{\lambda}_k e_{k-1} + \sum_{\alpha \in A} \bar{\lambda}_\alpha e_\alpha.$$

Note j is the conjugate linear transformation of V with $je_\alpha = e_\alpha$ for $\alpha \in A$ and $j(e_{2k}) = 2ke_{2k+1}$ and $j(e_{2k+1}) = \frac{1}{2k} e_{2k}$ for $k \in \mathbb{N}$. One has $j^2 = I$ for $j^2(e_{2k}) = j(2ke_{2k+1}) = e_{2k}$ and $j^2(e_{2k+1}) = j(\frac{1}{2k} e_{2k}) = e_{2k+1}$. But j is not continuous for $\frac{1}{2k} e_{2k} \to 0$ in V while $j(\frac{1}{2k} e_{2k}) = e_{2k+1}$ does not converge to 0. So j satisfies all the properties to be a conjugation except continuity. □

10. Let j be a conjugation on a complex Hilbert space V. Show $V_{\mathbb{R}}$ and $iV_{\mathbb{R}}$ are orthogonal relative to the real inner product $(v, v')_{\mathbb{R}} = \mathrm{Re}(v, v')$ if and only if j is symmetric.

Solution:

By Lemma 5.33, j is symmetric if and only if there is an orthonormal basis $\{e_k\}_{k \in K}$ such that $je_k = e_k$. But this is equivalent to the existence of a real orthonormal basis $\{e_k\}$ of $V_{\mathbb{R}}$ such that $\mathrm{Im}(e_k, e_l) = 0$ for all $k, l \in K$. But since $V_{\mathbb{R}}$ is the closure of the real linear span of the e_k, such a basis exists if and only if $\mathrm{Re}(V_{\mathbb{R}}, iV_{\mathbb{R}}) = \mathrm{Im}(V_{\mathbb{R}}, V_{\mathbb{R}}) = 0$. Indeed, any real orthonormal basis of $V_{\mathbb{R}}$ is an orthonormal basis of V when $\mathrm{Im}(V_{\mathbb{R}}, V_{\mathbb{R}}) = 0$.

□

11. Show that $\mathrm{O}(\beta, \mathbb{F})$ is a group.
Solution:
Let β be a symmetric bilinear mapping on $\mathbb{F}^n$ where $\mathbb{F} = \mathbb{R}$ or $\mathbb{F} = \mathbb{C}$. $\mathrm{O}(\beta, \mathbb{F})$ is the set of all invertible matrices $a \in \mathrm{GL}(n, \mathbb{F})$ such that $\beta(a^t u, a^t v) = \beta(u, v)$ for all $u, v \in \mathbb{F}^n$. Clearly $I \in \mathrm{O}(\beta, \mathbb{F})$. Also if $a \in \mathrm{O}(\beta, \mathbb{F})$, then $\beta(u, v) = \beta((a^{-1}a)^t u, (a^{-1}a)^t v) = \beta(a^t(a^{-1})^t u, a^t(a^{-1})^t v) = \beta((a^{-1})^t u, (a^{-1})^t v)$ for all u, v. So $a^{-1} \in \mathrm{O}(\beta, \mathbb{F})$. Finally, if $a, b \in \mathrm{O}(\beta, \mathbb{F})$, then $\beta((ab)^t u, (ab)^t v) = \beta(b^t a^t u, b^t a^t v) = \beta(a^t u, a^t v) = \beta(u, v)$ for all $u, v \in \mathbb{F}^n$. Thus $\mathrm{O}(\beta, \mathbb{F})$ is a group.

□

12. Show the action of $\mathrm{GL}(n, \mathbb{F})$ on $\mathrm{Symb}_n(\mathbb{F})$ given by

$$a \cdot \beta(u, v) = \beta(a^t u, a^t v)$$

is continuous.
Solution: Let e_k, $k = 1, \ldots, n$ be the standard basis of $\mathbb{F}^n$. Set $\beta_{j,k}$ to be the unique bilinear form on $\mathbb{F}^n$ satisfying

$$\beta_{j,k}(e_l, e_m) = \delta_{j,l}\delta_{k,m}.$$

The $\beta_{j,k}$ form a basis of the space of bilinear forms on $\mathbb{F}^n$ and if $\beta \in \mathrm{Symb}_n(\mathbb{F})$, then β can be written uniquely as $\beta = \sum \beta(e_j, e_k)\beta_{j,k}$ where $\beta_{j,k} = \beta_{k,j}$ for all j, k. Thus $\mathrm{Symb}_n(\mathbb{F})$ is homeomorphic to the subspace of $M(n \times n)$ consisting of all symmetric matrice $B = [x_{i,j}]$ under the mapping $\beta \mapsto [\beta(e_j, e_k)]$. Under this correspondence we have $a \cdot \beta$ is given by

$$\begin{aligned} a \cdot \beta(e_j, e_k) &= \beta(a^t e_j, a^t e_k) \\ &= \sum_{l,m} a_{j,l} a_{k,m} \beta(e_l, e_m) \end{aligned}$$

or more specifically, $a \cdot B = aBa^t$ where $B = [\beta(e_j, e_k)]$. □

13. Let β be a bilinear symmetric form on $\mathbb{C}^n$. Show there is a unique r such that β is in the $\mathrm{GL}(n, \mathbb{C}^n)$ orbit of the bilinear form β_r defined on $\mathbb{C}^n$ by

$$\beta_r(u, v) = \sum_{j=r+1}^{n} u_j v_j.$$

Solution:
By Lemma 5.36, there is an r and a basis $e_1, e_2, \ldots, e_r, e_{r+1}, \ldots, e_n$ such that $\beta(v, e_j) = 0$ for all v if $j \leqslant r$ and $\beta(e_i, e_j) = \delta_{i,j}$ if $i, j > r$. If we set $N = \{v \mid \beta(v, w) = 0 \text{ for all } w \in V\}$, then N is a subspace of $\mathbb{C}^n$ and it is easy to check $r = \dim N$. Let f_j be the usual basis of $\mathbb{C}^n$; i.e., $f_j = (\delta_{1,j}, \delta_{2,j}, \ldots, \delta_{n,j})^t$ and let a be the matrix $[e_1, e_2, \ldots, e_n]$. Note $af_j = e_j$

Recall vectors in $\mathbb{C}^n$ are column vectors. Note $a^t \cdot \beta(f_i, f_j) = \beta(af_i, af_j) = \beta(e_i, e_j)$ and $\beta_r(f_i, f_j) = \sum_{k>r}(f_i)_k(f_j)_k = \sum_{k>r} \delta_{k,i}\delta_{k,j} = \delta_{i,j}$ if $i, j > r$ and 0 otherwise. Thus on the basis $f_1, f_2, \ldots, f_n$, one has $a^t \cdot \beta(f_i, f_j) = \beta_r(f_i, f_j)$. Thus $a^t \cdot \beta = \beta_r$ and so $\beta = (a^{-1})^t \cdot \beta_r$ and we see β is in the orbit of β_r.

Since r is the dimension of N, r is unique.

□

14. Let β be a symmetric bilinear form on $\mathbb{R}^n$.

(a) Show the rank and signature of β are unique. (**Hint:** Show if V and W are subspaces of $\mathbb{R}^n$ and β is positive definite on V and negative definite on W, then the sum $V + W$ is direct.)

(b) Show if β is a symmetric bilinear form of rank $n - r$ and signature $p - q$, then there is an $a \in \mathrm{GL}(n, \mathbb{R})$ such that

$$a \cdot \beta = \beta_{r,p,q}$$

where $\beta_{r,p,q}$ is the bilinear form defined by

$$\beta_{r,p,q}(u, v) = \sum_{k=r+1}^{r+p} u_k v_k - \sum_{k=r+p+1}^{r+p+q} u_k v_k.$$

Solution:

For (a), we note by Lemma 5.38 that the nullity r is the dimension of R where $R = \{v \mid \beta(v, w) = 0 \text{ for all } w \in \mathbb{R}^n\}$ and $r + p + q = n$. Thus r is unique. Note β is positive definite on the linear span V of $\{e_{r+1}, e_{r+2}, \ldots, e_{r+p}\}$ and negative definite on the linear span W of $\{e_{r+p+1}, e_{r+p+2}, \ldots, e_{r+p+q}\}$. Now if we had another decomposition, $\mathbb{R}^n = R + V' + W'$ given by Lemma 5.38, the sum $R + V + W'$ is direct for if $u + v + w' = 0$, then $0 \leqslant \beta(v, v) = \beta(v+u, v+u) = \beta(-w', -w') \leqslant 0$. So $\beta(v, v) = \beta(w', w') = 0$. By definiteness, $v = w' = 0$ and consequently u in R is also 0. So $r + p + q' \leqslant n = r + p + q$. So $q' \leqslant q$. By symmetry, $q \leqslant q'$. So $q = q'$ and thus $p = p'$. This gives the uniqueness of the rank $p + q$ and signature $p - q$.

For (b), we use the same argument as in Exercise 5.3.13 using Lemma 5.38 instead of Lemma 5.36. Namely, if β has rank $p + q$, nullity r and signature $p - q$ where $r + p + q = n$, then by Lemma 5.38, there is a basis $e_1, e_2, e_3, \ldots, e_n$ of $\mathbb{R}^n$ such that $\beta(e_i, e_j) = 0$ if $i \neq j$, $\beta(e_i, e_i) = 0$ for $i \leqslant r$, $\beta(e_i, e_i) = 1$ for $i = r+1, r+2, \ldots, r+p$, and $\beta(e_i, e_i) = -1$ for $i = r+p+1, r+p+2, \ldots, n$. Now let $f_1, f_2, \ldots, f_n$ be the usual basis of $\mathbb{R}^n$. Thus $f_i = [\delta_{i,1}, \delta_{i,2}, \ldots, \delta_{i,n}]^t$ and let a be the matrix in $\mathrm{GL}(n, \mathbb{R})$ which satisfies $a^t e_i = f_i$ for $i = 1, 2, \ldots, n$; i.e., a is the matrix $(E^{-1})^t$ where E is the matrix $[e_1, e_2, \ldots, e_n]$ for $[f_1, f_2, \ldots, f_n]$ is the identity matrix. Thus $a \cdot \beta_{r,p,q}(e_i, e_j) = \beta_{r,p,q}(a^t e_i, a^t e_j) = \beta_{r,p,q}(f_i, f_j) = \sum_{k=r+1}^{r+p} \delta_{i,k}\delta_{j,k} - \sum_{k=r+p+1}^{n} \delta_{i,k}\delta_{j,k}$. But the difference of these two sums is 0 if $i \neq j$, is 1 if $i = j$ and $r < i \leqslant r + p$, and is -1 if $i = j$ and $r + p < i \leqslant n$. So we see $a \cdot \beta_{r,p,q}$ and β agree on pairs from a basis. Consequently, $a \cdot \beta_r = \beta$.

□

15. Let β be a bilinear form on a finite dimensional vector space V. Show β is nondegenerate if and only if $\beta(v, w) = 0$ for all w implies $v = 0$.

Solution:

Assume V is finite dimensional and β is nondegenerate. Thus for each $v \neq 0$, there is a u and a w with $\beta(u, v) \neq 0$ and $\beta(v, w) \neq 0$. So if $\beta(v, w) = 0$ for all w, then $v = 0$.

Conversely, since V is linearly isomorphic to $\mathbb{F}^n$ where $\mathbb{F} = \mathbb{R}$ or $\mathbb{F} = \mathbb{C}$, we may suppose $V = \mathbb{F}^n$. Then assume $\beta(v, w) = 0$ for all w implies $v = 0$. Thus if $v \neq 0$, then $\beta(v, w) \neq 0$ for some w. Let $e_1, e_2, \ldots, e_n$ be the basis of $\mathbb{F}^n$ given by $e_i = [\delta_{1,i}, \delta_{2,i}, \ldots, \delta_{n,i}]^t$ and set $B_{i,j} = \beta(e_i, e_j)$. Thus

$$\begin{aligned}\beta(u, v) &= \beta(\sum u_i e_i, \sum v_j e_j)\\ &= \sum_{i=1}^{n}\sum_{j=1}^{n} u_i B_{i,j} v_j\\ &= u^t B v\end{aligned}$$

where $B = [B_{i,j}]$. Now since B is an $n \times n$ matrix and $v^t B u = 0$ for all u implies $v = 0$, we see $v^t B = 0$ implies $v = 0$. So B^t is invertible. So if $v \neq 0$, there is a $u \in \mathbb{F}^n$ with $B^t u = v$ and thus $\beta(u, v) = u^t B v = v^t v \neq 0$. This is equivalent to $\beta(u, v) = 0$ for all u implies $v = 0$. So all this together gives β is nondegenerate.

□

16. Let $\beta_A(u, v) = \beta(Au, v)$ where $\beta(u, v) = \sum_{j=1}^{n} u_j v_j$. Show the following:

(a) Assume that A and B are symmetric and $B = aAa^t$ for some $a \in \mathrm{GL}(n, \mathbb{F})$. Then

$$\mathrm{O}(\beta_B, \mathbb{F}) = a\,\mathrm{O}(\beta_A, \mathbb{F})a^{-1}$$

(b) Assume that A is in $\mathrm{GL}(n, \mathbb{F})$. Then $\mathrm{O}(\beta_A, \mathbb{F}) := \{a \in \mathrm{GL}(n, \mathbb{F}) \mid aAa^t = A\}$, where a^t stands for the transposed matrix $a^t_{ij} = a_{ji}$.

Solution:

Let A and B be symmetric matrices with $B = aAa^t$. Then

$$\begin{aligned}
\mathrm{O}(\beta_B, \mathbb{F}) &= \{b \in \mathrm{GL}(n, \mathbb{F}) \mid \beta_B(b^t u, b^t v) = \beta_B(u, v) \text{ for all } u, v\} \\
&= \{b \in \mathrm{GL}(n, \mathbb{F}) \mid (Bb^t u)^t b^t v = (Bu)^t v \text{ for all } u, v\} \\
&= \{b \in \mathrm{GL}(n, \mathbb{F}) \mid u^t bB^t b^t v = u^t B^t v \text{ for all } u, v\} \\
&= \{b \in \mathrm{GL}(n, \mathbb{F}) \mid u^t baA^t a^t b^t v = u^t aA^t a^t v \text{ for all } u, v\} \\
&= \{b \in \mathrm{GL}(n, \mathbb{F}) \mid ((a^{-1})^t u)^t baA^t a^t b^t (a^t)^{-1} v = \\
&\qquad ((a^{-1})^t u)^t aA^t a^t (a^t)^{-1} v \text{ for all } u, v\} \\
&= \{b \in \mathrm{GL}(n, \mathbb{F}) \mid (u^t (a^{-1}) baA^t a^t b^t (a^{-1})^t v = u^t A^t v \text{ for all } u, v\} \\
&= \{aca^{-1} \in \mathrm{GL}(n, \mathbb{F}) \mid u^t cA^t c^t v = u^t A^t v \text{ for all } u, v\} \\
&= \{aca^{-1} \in \mathrm{GL}(n, \mathbb{F}) \mid (Ac^t u)^t c^t v = (Au)^t v \text{ for all } u, v\} \\
&= \{aca^{-1} \in \mathrm{GL}(n, \mathbb{F}) \mid \beta_A(c^t u, c^t v) = \beta_A(u, v) \text{ for all } u, v\} \\
&= a\,\mathrm{O}(\beta_A, \mathbb{F})a^{-1}.
\end{aligned}$$

So we have (a).

Now $\mathrm{O}(\beta_A, \mathbb{F})$ consists of all $a \in \mathrm{GL}(n, \mathbb{F})$ such that $a \cdot \beta_A(u, v) = \beta_A(u, v)$ for all u and v. Thus $\beta_A(a^t u, a^t v) = (Aa^t u)^t (a^t v) = (Au)^t v = \beta_A(u, v)$ for all u and v. So this is equivalent to $u^t aA^t a^t v = u^t A^t v$ for all u and v or equivalently to $aA^t a^t = A^t$ or $A = aAa^t$. This gives (b). Note this can also be used to give (a). Indeed, if $B = aAa^t$, then $\mathrm{O}(\beta_B, \mathbb{F})$ is the collection of all $b \in \mathrm{GL}(n, \mathbb{F})$ such that $bBb^t = B$ or the set of all b with $baAa^t b^t = aAa^t$ or $a^{-1} baAa^t b^t (a^{-1})^t = A$. But these b's are the matrices of form aca^{-1} where $c \in \mathrm{O}(\beta_A, \mathbb{F})$. So $\mathrm{O}(\beta_B, \mathbb{F}) = a\,\mathrm{O}(\beta_A, \mathbb{F})a^{-1}$.

□

17. Let $A \in \mathrm{O}(p, q)$. Show $\det(A) = \pm 1$ and if $A \in \mathrm{Sp}(n, \mathbb{R})$, then $\det(A) = 1$.
Solution: As seen on page 265, $\mathrm{O}(p, q)$ consists of all invertible $n \times n$ matrices A with $AI_{p,q}A^t = I_{p,q}$. So $\det(AI_{p,q}A^t) = \det(I_{p,q})$. This implies $\det(A)^2 = 1$. So $\det A = \pm 1$.

For $\mathrm{Sp}(n, \mathbb{R})$ we deal with forms. Take J to be the matrix $\begin{bmatrix} 0 & I_n \\ -I_n & 0 \end{bmatrix}$ and let ω be the alternating bilinear form on $\mathbb{R}^{2n} \times \mathbb{R}^{2n}$ (as row vectors) defined by $\omega(x, y) = xJy^t$. Set $e_i = (\delta_{1,i}, \delta_{2,i}, \ldots, \delta_{n,i}, 0, 0, \ldots, 0)$ and $f_i = (0, 0, \ldots, 0, \delta_{1,i}, \delta_{2,i}, \ldots, \delta_{n,i})$ where i varies from 1 to n. Then $\omega(e_i, e_j) = \omega(f_i, f_j) = 0$ for all i and j and $\omega(e_i, f_j) = \delta_{i,j}$; thus these vectors form a symplectic basis relative to ω for $\mathbb{R}^{2n}$. Indeed, using wedge products

$$\omega = e^1 \wedge f^1 + e^2 \wedge f^2 + \cdots + e^n \wedge f^n$$

where e^i and f^i are the dual vectors; i.e. $e^i(f_j) = 0$ for all j and $e^i(e_j) = \delta_{i,j}$; similarly for f^i. Here $e^i \wedge f^j$ is the alternating 2 form with $e^i \wedge f^j(e_i, f_j) = 1$ and at all basis pairs v_1, v_2 differing from e_i and f_j evaluate to 0. Now a linear transformation C (equivalently a matrix) is in $\mathrm{Sp}(n, \mathbb{R})$ if C preserves ω. That is $\omega(Cv, Cw) = \omega(v, w)$ for all $v, w \in \mathbb{R}^n$. We form ω^n, that is $\omega \wedge \omega \wedge \cdots \wedge \omega$ and note C preserves ω^n; i.e., $\omega^n(Cv_1, \ldots, Cv_{2n}) = \omega^n(v_1, \ldots, v_{2n})$ for all vectors $v_1, v_2, \ldots, v_{2n}$. The alternating multilinear form ω^n is defined by

$$\omega^n(v_1, \ldots, v_{2n}) = \sum_{\sigma \in S_{2n}} \operatorname{sgn}(\sigma)\omega(v_{\sigma 1}, v_{\sigma 2})\omega(v_{\sigma 3}, v_{\sigma 4}) \cdots \omega(v_{\sigma(2n-1)}, v_{\sigma(2n)})$$

where $\operatorname{sgn}(\sigma)$ is ± 1 depending on whether σ is an even or odd permutation. Recall the sequence $\sigma 1, \sigma 2, \cdots, \sigma(2n)$ can be reordered as $1, 2, \cdots, 2n$ with $\sum_{i<j,\, \sigma i > \sigma j} 1$ transpositions and therefore $\operatorname{sgn}(\sigma) = \prod_{i<j,\, \sigma i > \sigma j}(-1)$.

If S is the set of permutations σ such that

$$\{\{\sigma(2i-1), \sigma(2i)\} \mid i = 1, 2, \ldots n\} = \{\{1, 1+n\}, \{2, 2+n\}, \ldots, \{n, 2n\}\}$$

and S' is the set of permutations σ with

$$\{(\sigma(2i-1), \sigma(2i)) \mid i = 1, 2, \ldots, n\} = \{(1, 1+n), (2, 2+n), \ldots, (n, 2n)\},$$

then since for each $\sigma' \in S'$, there are 2^n permutations σ in S where $\sigma = \delta\sigma'$, $\delta = (\delta_1, \delta_2, \ldots, \delta_n) \in S_2^n$, and

$$(\sigma 1, \sigma 2, \ldots, \sigma(2n)) = (\delta_1\sigma' 1, \delta_1\sigma' 2, \delta_2\sigma' 3, \delta_2\sigma' 4, \ldots, \delta_n\sigma'(2n-1), \delta_n\sigma'(2n))$$

and $\delta_k\sigma'(2k-1) = \sigma'(2k-1)$ and $\delta_k\sigma'(2k) = \sigma'(2k)$ if δ_k is the identity in S_2 and $\delta_k\sigma'(2k-1) = \sigma'(2k)$ and $\delta_k\sigma'(2k) = \sigma'(2k-1)$ if δ_k is the nontrivial permutation in S_2, we see

$$\begin{aligned} \omega^n(e_1, f_1, e_2, f_2, \ldots, e_n, f_n) &= \sum_{\sigma \in S} \operatorname{sgn}(\sigma) \prod_{1 \leqslant i \leqslant n,\, \sigma(2i-1) > \sigma(2i)} (-1)\omega(e_i, f_i) \\ &= 2^n \sum_{\sigma' \in S'} \operatorname{sgn}(\sigma') \prod_{i=1}^{n} \omega(e_i, f_i) \\ &= 2^n(-1)^{\frac{n(n-1)}{2}} |S'|, \end{aligned}$$

for it takes an even number of transpositions to change a $\sigma' \in S'$ into the permutation $\sigma : 2i-1 \mapsto i;\ 2i \mapsto i+n$ for $i = 1, \ldots, n$, and the permutation σ is even if and only if $\frac{n(n-1)}{2}$ is even. Now the only alternating $2n$ form up to a scalar on $(R^{2n})^{2n}$ is the determinant function $\det(\cdot)$. So

$$\omega^n(v_1, \ldots, v_{2n}) = (-1)^{\frac{n(n-1)}{2}} 2^n |S'| \det \begin{bmatrix} v_1 \\ v_2 \\ \vdots \\ v_{2n} \end{bmatrix}.$$

Now

$$\omega^n(Cv_1, \ldots, Cv_{2n}) = \det(C)\omega^n(v_1, \ldots, v_{2n})$$

for any linear transformation C. So if $C \in \mathrm{Sp}(n, \mathbb{R})$, $\omega^n(Cv_1, \ldots, Cv_{2n}) = \omega^n(v_1, \ldots, v_{2n})$. Since $\omega^n \neq 0$, this implies $\det C = 1$. So $\mathrm{Sp}(n, \mathbb{R}) \subseteq \mathrm{SL}(2n, \mathbb{R})$.

There is an alternate related way using Pfaffians. To do this one must be familiar with some of their properties. Namely, the Pfaffian $\mathrm{Pf}(A)$ of a square skew-symmetric matrix A is a polynomial in the entries in A and satisfies $\mathrm{Pf}(A)^2 = \det(A)$ and $\mathrm{Pf}(CAC^T) = \det(C)\,\mathrm{Pf}(A)$. So $\mathrm{Pf}(A) \neq 0$ if and only if A is invertible. Now $C \in \mathrm{Sp}(n, \mathbb{R})$ if and only if $CJC^t = J$ and since $\det(J) = 1$, $\mathrm{Pf}(J) = \pm 1$ and $\det(C)\,\mathrm{Pf}(J) = \mathrm{Pf}(J)$. So $\det(C) = 1$.

□

18. Assume $\mathbf{n} = (n_1, \ldots, n_k) \in \mathbb{N}^k$ and $n_1 < n_2 < \ldots < n_k$. Let G_j be a subgroup of $\mathrm{GL}(n_j - n_{j-1}, \mathbb{R})$. Define

$$S(G_1 \times \ldots \times G_k) := \left\{ \begin{pmatrix} A_1 & 0 & 0 \\ 0 & \ddots & 0 \\ 0 & 0 & A_k \end{pmatrix} \mid A_j \in G_j,\ \det(A_1) \cdot \ldots \cdot \det(A_k) = 1 \right\}.$$

Show that $\mathrm{SO}(n)$ acts transitively on $\mathrm{Flag}(\mathbf{n})$ and that

$$\mathrm{Flag}(\mathbf{n}) \simeq \mathrm{SO}(n)/\,\mathrm{S}(\mathrm{O}(n_1) \times \ldots \times \mathrm{O}(n_k - n_{k-1}))).$$

Solution:

This was essentially shown on page 269 by noting $\mathrm{SO}(n) \cap \mathrm{P}(\mathbf{n}) = \mathrm{S}(G_1 \times G_2 \times \cdots G_k)$ where $G_k = \mathrm{O}(n_k - n_{k-1})$ with $n_0 = 0$. We redo the argument here. Let $\mathbf{n} = (n_1, n_2, \ldots, n_k)$ with $1 \leqslant n_1 < n_2 < \ldots < n_k = n$ and $\mathrm{Flag}(\mathbf{n})$ be the collection of all k-tuples of vector subspace chains $V_1 \subseteq V_2 \subseteq \cdots \subseteq V_k$ with $\dim V_j = n_j$. Define E_k to be the linear span of the usual basis vectors e_k for $k = 1, 2, \ldots, n_k$. Then $\mathbf{E}$ is a flag. If $\mathbf{V}$ is a flag and $f_1, f_2, \ldots, f_n$ is chosen to be an orthonormal basis of $\mathbb{R}^n$ with $V_j = \langle f_1, f_2, \ldots, f_{n_j} \rangle$, then the matrix T which transforms e_j into f_j for each j is an orthogonal transformation satisfying $TE_k = V_k$ for each k and thus $T \cdot \mathbf{E} = \mathbf{V}$, the flag given by the chain $V_1, V_2, \ldots, V_k$. Since T is orthogonal $TT^t = I$ and so $\det T = \pm 1$. If T has determinant -1, then changing T so that $Te_n = -f_n$ makes T have determinant 1. This implies $\mathrm{SO}(n)$ acts transitively on $\mathrm{Flag}(\mathbf{n})$.

To finish we need only find the stabilizer of $\mathbf{E}$. Note if $T \in \mathrm{SO}(n)$ and $T \cdot \mathbf{E} = \mathbf{E}$, then $\det T = 1$ and $TE_j = E_j$ for each j. But since T is orthogonal $TE_1^{\perp} \subseteq E_1^{\perp}$ and thus $T(E_2 \oplus E_3 \oplus \cdots \oplus E_k) \subseteq E_2 \oplus E_3 \oplus \cdots \oplus E_k$. But repeating this we see $T(\langle e_{n_j+1}, e_{n_j+2}, \ldots, e_{n_{j+1}} \rangle) \subseteq \langle e_{n_j+1}, e_{n_j+2}, \ldots, e_{n_{j+1}} \rangle$ for each $j = 0, 1, \ldots, k-1$. So $T = \begin{bmatrix} T_1 & 0 & \cdots & 0 \\ 0 & T_2 & \cdots & 0 \\ \vdots & \vdots & \cdots & \vdots \\ 0 & 0 & \cdots & T_k \end{bmatrix}$ where $T_j \in \mathrm{O}(n_{j+1} - n_j)$. Since $\det T = 1$, $T \in \mathrm{S}(\mathrm{O}(n_1) \times \mathrm{O}(n_2 - n_1) \times \cdots \times \mathrm{O}(n_k - n_{k-1}))$. This gives the result.

□

19. Show that $E(n)$ is the group of distance preserving affine transformations of $\mathbb{R}^n$.

Solution:

We recall an affine transformation of $\mathbb{R}^n$ is a transformation of form $Tx = b + Ax$ where A is $n \times n$ matrix. Now T is distance preserving if and only if $|b + Ax - (b + Ay)| = |x - y|$. This implies the linear transformation A is distance preserving and hence satisfies $|Ax| = |x|$ for all x. So $A \in \mathrm{O}(n)$. Thus the distance preserving affine transformations are the transformations $T_{b,A}$ given by $x \mapsto b + Ax$ where $A \in \mathrm{O}(n)$. Note $T_{b,A} \mapsto (b, A)$ is a group isomorphism between this transformation group and $E(n)$.

□

20. Let $\mathrm{Sym}^+(n, \mathbb{R})$ be the set of positive definite symmetric matrices. Show that $\mathrm{Sym}^+(n, \mathbb{R}) \simeq \mathrm{GL}(n, \mathbb{R})/\mathrm{O}(n)$.

Solution:

By Proposition 5.40 the mapping $S \mapsto \beta_S$ where $\beta_S(v, w) = v^t S w$ is a linear isomorphism of the symmetric matrices S onto the space of bilinear symmetric forms on $\mathbb{R}^n$. Now S is positive definite if and only if β_S is a positive definite symmetric bilinear form. Thus β_S is nondegenerate and so the nullity is 0 and the rank is n. Hence $r = 0$, $p = n$, and $q = 0$, and we see by Lemma 5.38 that there is a basis $f_1, f_2, \ldots, f_n$ of $\mathbb{R}^n$ with $\beta_S(f_i, f_j) = \delta_{i,j}$ for all i, j. Let $A \in \mathrm{GL}(n, \mathbb{R})$ be the matrix with rows f_j^t. Note $A^t e_j = f_j$ where $e_j = (\delta_{1,j}, \delta_{2,j}, \cdots, \delta_{n,j})^t$; recall we are using column vectors for $\mathbb{R}^n$. Thus $A \cdot \beta_S(e_i, e_j) = \beta_S(A^t e_i, A^t e_j) = \beta_S(f_i, f_j) = \delta_{i,j} = \beta_I(e_i, e_j)$. So $A \cdot \beta_S = \beta_I$ which implies $\beta_S = A^{-1} \cdot \beta_I$. Thus the orbit of β_I under the action of $\mathrm{GL}(n, \mathbb{R})$ is all symmetric positive definite bilinear forms. Since $A \cdot \beta_S = \beta_{ASA^t}$, we see the orbit of I under the action $A \cdot I = AIA^t$ is all positive definite matrices. Since the stabilizer is all A with $AA^t = I$, one has

$$\mathrm{Sym}^+(n, \mathbb{R}) = \mathrm{GL}(n, \mathbb{R})/\mathrm{O}(n).$$

□

21. Show that the map $\left(\begin{smallmatrix} a & b \\ c & d \end{smallmatrix}\right) \cdot z = \frac{az+b}{cz+d}$ satisfies $AB \cdot z = A \cdot (B \cdot z)$ wherever it is defined.

Solution:

Let $A = \left[\begin{smallmatrix} a & b \\ c & d \end{smallmatrix}\right]$ and $B = \left[\begin{smallmatrix} e & f \\ g & h \end{smallmatrix}\right]$ and suppose $z \in \mathbb{C}$ with $gz + h \neq 0$. So $B \cdot z = w$ where $w = \frac{ez+f}{gz+h}$. Now suppose $cw + d \neq 0$. Then

$$\begin{aligned} A \cdot w &= \frac{aw + b}{cw + d} \\ &= \left(a\left(\frac{ez + f}{gz + h}\right) + b\right)\left(c\left(\frac{ez + f}{gz + h}\right) + d\right)^{-1} \\ &= \left(\frac{aez + af + bgz + bh}{gz + h}\right)\left(\frac{cez + cf + dgz + dh}{gz + h}\right)^{-1} \\ &= \frac{(ae + bg)z + af + bh}{(ce + dg)z + cf + dh}. \end{aligned}$$

But

$$AB = \begin{bmatrix} ae + bg & af + bh \\ ce + dg & cf + dh \end{bmatrix}$$

and $AB \cdot z$ is defined if $(ce + dg)z + cf + hd \neq 0$ and then

$$AB \cdot z = \frac{(ae + bg)z + af + bh}{(ce + dg)z + cf + dh}.$$

Thus whenever $B \cdot z$, $A \cdot (B \cdot z)$, and $AB \cdot z$ are defined, one has

$$(AB) \cdot z = A \cdot (B \cdot z).$$

$\square$

22. Let $\left(\begin{smallmatrix} a & b \\ c & d \end{smallmatrix}\right)z = \frac{az+b}{cz+d}$ be the action of $\mathrm{SL}(2, \mathbb{R})$ on the upper half plane H_+. Let $A = \{a = \left(\begin{smallmatrix} x & 0 \\ 0 & x^{-1} \end{smallmatrix}\right) \mid x > 0\}$ and $N = \{n = \left(\begin{smallmatrix} 1 & x \\ 0 & 1 \end{smallmatrix}\right) \mid x \in \mathbb{R}\}$.

(a) Show AN is a closed subgroup of $\mathrm{SL}(2, \mathbb{R})$.

(b) Show the mapping $g \mapsto g \cdot i$ is a homeomorphism from AN onto H_+.

Solution:

Let $A = \{\left[\begin{smallmatrix} x & 0 \\ 0 & x^{-1} \end{smallmatrix}\right] \mid x > 0\}$ and $N = \{\left[\begin{smallmatrix} 1 & x \\ 0 & 1 \end{smallmatrix}\right] \mid x \in \mathbb{R}\}$. Note

$$\left[\begin{smallmatrix} a & 0 \\ 0 & a^{-1} \end{smallmatrix}\right]\left[\begin{smallmatrix} 1 & x \\ 0 & 1 \end{smallmatrix}\right] = \left[\begin{smallmatrix} a & ax \\ 0 & a^{-1} \end{smallmatrix}\right]$$

for $a > 0$ and $x \in \mathbb{R}$. So $AN = \{\left[\begin{smallmatrix} a & y \\ 0 & a^{-1} \end{smallmatrix}\right] \mid a > 0,\, y \in \mathbb{R}\}$ and one can easily check this set of matrices is closed under multiplication and inversion. So it is a subgroup of $\mathrm{SL}(n, \mathbb{R})$. To see it is closed, note if $\left[\begin{smallmatrix} a_k & y_k \\ 0 & a_k^{-1} \end{smallmatrix}\right]$ is a sequence converging to a matrix $\left[\begin{smallmatrix} a & b \\ c & d \end{smallmatrix}\right]$ in M_2, then $c = 0$, $a_k \to a$, $a_k^{-1} \to d$ and $y_k \to b$. Since $a_k a_k^{-1} = 1$ for all k, $ad = 1$ and so the limit is $\left[\begin{smallmatrix} a & b \\ 0 & a^{-1} \end{smallmatrix}\right]$ which is in AN. This gives (a).

For (b), note the mapping $g \mapsto g \cdot i$ is continuous by Lemma 5.25 and the image of $\left[\begin{smallmatrix} a & b \\ 0 & a^{-1} \end{smallmatrix}\right]$ is $\frac{ai+b}{a^{-1}} = a^2 i + ab$. So if $x + iy \in H_+$, $y = a^2$ for a unique $a > 0$ and then if

$$g(x, y) = \begin{bmatrix} \sqrt{y} & \frac{1}{\sqrt{y}}x \\ 0 & \frac{1}{\sqrt{y}} \end{bmatrix},$$

$g(x, y) \cdot i = x + iy$. Thus $g \mapsto g \cdot i$ is an onto mapping. The stabilizer of i is all $\left[\begin{smallmatrix} a & b \\ 0 & a^{-1} \end{smallmatrix}\right]$ with $a^2 = 1$ and $b = 0$. Thus $g \mapsto g \cdot i$ is a continuous one-to-one mapping of AN onto H_+. The inverse mapping as seen above is $x + iy \mapsto g(x, y)$ which is continuous. Thus $g \mapsto g \cdot i$ is a homeomorphism.

$\square$

23. Let $\mathcal{V}(l)$ be the collection of all subspaces of the Hilbert space $\mathbb{F}^n$ having dimension l. If $V \in \mathcal{V}(l)$, let P_V be the orthogonal projection onto V. Define ρ_l by $\rho_l(V, V') = ||P_V - P_{V'}||$. Show ρ_l is a metric on $\mathcal{V}(l)$.

Solution: Note $V = V'$ if and only if $P_V = P_{V'}$. So $\rho_l(V, V') = 0$ if and only if $V = V'$. Clearly $\rho_l(V, V') = \rho_l(V', V)$ and

$$\begin{aligned} \rho_l(V, V'') &= ||P_V - P_{V'} + P_{V'} - P_{V''}|| \\ &\leqslant ||P_V - P_{V'}|| + ||P_{V'} - P_{V''}|| \\ &= \rho_l(V, V') + \rho_l(V', V''). \end{aligned}$$

So ρ_l is a metric. $\square$

24. Define d on $\operatorname{Flag}(\mathbf{n})$ by

$$d((V_1, V_2, \ldots, V_k), (V_1', V_2', \ldots, V_k')) = \sum_{i=1}^{k} \rho_{n_i}(V_i, V_i').$$

where ρ_{n_i} is the metric defined in Exercise 5.3.23.

(a) Show d is a metric.

(b) Let $\mathbb{F} = \mathbb{R}$. Show the topology on $\operatorname{Flag}(\mathbf{n})$ given by the metric d is the topology on $\operatorname{Flag}(\mathbf{n})$ given by the quotient topology on $\mathrm{SO}(n)/\mathrm{S}(\mathrm{O}(n_1) \times \mathrm{O}(n_2 - n_1) \times \cdots \times \mathrm{O}(n_k - n_{k-1}))$ under the isomorphism given in Theorem 5.43.

(c) Show the topology on $\operatorname{Flag}(\mathbf{n})$ and the quotient topology of $\mathrm{GL}(n, \mathbb{R})/\mathrm{P}(\mathbf{n})$ are the same. (Similar statements hold in the case $\mathbb{F} = \mathbb{C}$.)

Solution:

We first show d is a metric. Note $d((V_1, V_2, \ldots, V_k), (V_1', V_2', \ldots, V_k')) = 0$ if and only if $\rho_{n_i}(V_i, V_i') = 0$ for $i = 1, 2, \ldots, k$ if and only if $V_i = V_i'$ for $i = 1, 2, \ldots, k$ if and only if $(V_1, V_2, \ldots, V_k) = (V_1', V_2', \ldots, V_k')$. Since $\rho_{n_i}(V_i, V_i') = \rho_{n_i}(V_i', V_i)$, we have

$$d((V_1, V_2, \ldots, V_k), (V_1', V_2', \ldots, V_k')) = d((V_1', V_2', \ldots, V_k'), (V_1, V_2, \ldots, V_k)).$$

The triangle inequality follows from

$$\rho_{n_i}(V_i, V_i'') \leqslant \rho_{n_i}(V_i, V_i') + \rho_{n_i}(V_i', V_i'')$$

for each i.

Now $G = \mathrm{SO}(n)$ is a Lindelöf space and the action of G on $\operatorname{Flag}(\mathbf{n})$ is continuous. To see this note first if $T \in \mathrm{SO}(n)$ and V is a closed subspace of $\mathbb{R}^n$, then $P_{TV} = TP_VT^*$. So

$$\begin{aligned}
\rho_{n_i}(TV_i, SV_i') &= ||P_{TV_i} - P_{SV_i'}|| \\
&= ||TP_{V_i}T^* - SP_{V_i'}S^*|| \\
&\leqslant ||TP_{V_i}T^* - TP_{V_i}S^*|| + ||TP_{V_i}S^* - SP_{V_i}S^*|| + ||SP_{V_i}S^* - SP_{V_i'}S^*|| \\
&\leqslant ||T||\,||P_{V_i}||\,||T^* - S^*|| + ||T - S||\,||P_{V_i}||\,||S^*|| + ||S||\,||P_{V_i} - P_{V_i'}||\,||S^*|| \\
&\leqslant ||T^{-1} - S^{-1}|| + ||T - S|| + ||P_{V_i} - P_{V_i'}||
\end{aligned}$$

which can be made small if T is close to S in $\mathrm{SO}(n)$ and V_i is close to V_i' in the ρ_{n_i} topology. But since $d((TV_1, TV_2, \ldots, TV_k), (SV_1', SV_2', \ldots, SV_k')) = \sum_{i=1}^{k} \rho_{n_i}(TV_i, SV_i')$, we see $(T, (V_1, V_2, \ldots, V_k)) \mapsto (TV_1, TV_2, \ldots, TV_k)$ is continuous from $\mathrm{SO}(n) \times \operatorname{Flag}(\mathbf{n})$ into $\operatorname{Flag}(\mathbf{n})$.

Recall from the material beginning at page 269 that this action is transitive and the stabilizer of $\mathbf{E}$ is $\mathrm{S}(\mathrm{O}(n_1) \times \mathrm{O}(n_2 - n_1) \times \cdots \times \mathrm{O}(n_k - n_{k-1}))$. It follows by Theorem 5.15 that under the isomorphism given in Theorem 5.43 that the topology on Flag($\mathbf{n}$) given by d is the quotient topology on the homogeneous space $\mathrm{SO}(n)/\mathrm{S}(\mathrm{O}(n_1) \times \mathrm{O}(n_2 - n_1) \times \cdots \times \mathrm{O}(n_k - n_{k-1}))$. This gives (b).

We outline two ways to do (c). As seen on page 269, the mapping $AP(\mathbf{n}) \mapsto A \cdot \mathbf{E}$ is an isomorphism of $\mathrm{GL}(n, \mathbb{R})/P(\mathbf{n})$ onto Flag($\mathbf{n}$). Also the mapping $(T, A) \mapsto TA$ is continuous from $\mathrm{SO}(n) \times \mathrm{GL}(n, \mathbb{R})$ into $\mathrm{GL}(n, \mathbb{R})$ and consequently the mapping $(T, AP(\mathbf{n})) \mapsto TAP(\mathbf{n})$ is a continuous action of $\mathrm{SO}(n)$ on $\mathrm{GL}(n, \mathbb{R})/P(\mathbf{n})$. Next note the stabilizer of the coset $P(\mathbf{n})$ in $\mathrm{SO}(n)$ is all orthogonal matrices T of determinant 1 such that $TP(\mathbf{n}) = P(\mathbf{n})$. Thus the stabilizer is $P(\mathbf{n}) \cap \mathrm{SO}(n)$ which is the subgroup $\mathrm{SO}(n)_{IP(\mathbf{n})} = \mathrm{S}(\mathrm{O}(n_1) \times \mathrm{O}(n_2 - n_1) \times \cdots \times \mathrm{O}(n_k - n_{k-1}))$. Also $\mathrm{SO}(n)$ acts transitively on $\mathrm{GL}(n, \mathbb{R})/P(\mathbf{n})$. Indeed, if $A \in \mathrm{GL}(n, \mathbb{R})$, then $\mathbf{F} = A \cdot \mathbf{E}$ is a flag for which there is a $T \in \mathrm{SO}(n)$ so that for each i, the linear span of $Te_1, Te_2, \ldots, Te_{n_i}$ is V_i. This implies $T^{-1} A \cdot \mathbf{E} = \mathbf{E}$ and thus $T^{-1}A \in P(\mathbf{n})$. So $AP(\mathbf{n}) = TP(\mathbf{n})$. Thus the $\mathrm{SO}(n)$ orbit of the coset $IP(\mathbf{n})$ is $\mathrm{GL}(n, \mathbb{R})/P(\mathbf{n})$ and the action is transitive. By Theorem 5.15, the mapping from $\mathrm{SO}(n)/\mathrm{S}(\mathrm{O}(n_1) \times \mathrm{O}(n_2 - n_1) \times \cdots \times \mathrm{O}(n_k - n_{k-1}))$ is an $\mathrm{SO}(n)$ space homeomorphism onto $\mathrm{GL}(n, \mathbb{R})/P(\mathbf{n})$. Using this with Theorem 5.43 shows both spaces $\mathrm{GL}(n, \mathbb{R})/P(\mathbf{n})$ and $\mathrm{SO}(n)/\mathrm{S}(\mathrm{O}(n_1) \times \mathrm{O}(n_2 - n_1) \times \cdots \times \mathrm{O}(n_k - n_{k-1}))$ are G-space homeomorphic to Flag($\mathbf{n}$) with the topology given by d.

For (c) one could also use Theorem 5.15 by showing the action

$$(A, (V_1, V_2, \ldots, V_k)) \mapsto (AV_1, AV_2, \ldots, AV_k))$$

of $\mathrm{GL}(n, \mathbb{R})$ on Flag($\mathbf{n}$) is separately continuous. To see this map is continuous in A note if V is a subspace of $\mathbb{R}^n$ with orthonormal basis $v_1, v_2, \ldots, v_m$, then the orthogonal projection P_V onto V is given by $x \mapsto \sum_{j=1}^{m} (x, v_j) v_j$. The matrix of this linear transformation can easily be determined, the column vectors are the images of the standard basis $e_1, e_2, \ldots, e_n$. Now if A is an invertible matrix, the matrix of the orthogonal projection onto the range of AV is the matrix obtained by doing a Gram-Schmidt orthonormalization of the linearly independent vectors $Av_1, Av_2, \ldots, Av_m$. The orthonormalization process proceeds as follows: $w_1 = \frac{Av_1}{||Av_1||}$, $w_2 = \frac{Av_2 - (Av_2, w_1)w_1}{||Av_2 - (Av_2, w_1)w_1||}, \ldots, w_m = \frac{Av_m - \sum_{j=1}^{m-1} (Av_m, w_m) w_m}{||Av_m - \sum_{j=1}^{m-1} (Av_m, w_m) w_m||}$. In terms of the matrix A, all w_j have entries that depend continuously on the elements $a_{i,j}$ of A. The resulting orthogonal projection $x \mapsto \sum_{j=1}^{m} (x, w_j) w_j$ then has a matrix (the matrix of P_{AV}) whose entries depend continuously on A. Consequently $\rho_m(BV, AV) = ||P_{BV} - P_{AV}|| \to 0$ as $B \to A$ in $\mathrm{GL}(n, \mathbb{R})$. Since $d(B(V_1, V_2, \ldots, V_k), A(V_1, V_2, \ldots, V_k)) = \sum_{j=1}^{k} ||P_{BV_j} - P_{AV_j}||$, we see $B(V_1, V_2, \ldots, V_k) \to A(V_1, V_2, \ldots, V_k)$ in Flag($\mathbf{n}$) if $B \to A$ in $\mathrm{GL}(n, \mathbb{R})$.

We next indicate how one can show $A(V_1', V_2', \ldots, V_k') \to A(V_1, V_2, \ldots, V_k)$ if $(V_1', V_2', \ldots, V_k') \to (V_1, V_2, \ldots, V_k)$ in Flag($\mathbf{n}$). Suppose $P_{V'} \to P_V$ and $v_1, v_2, \ldots, v_m$ is an orthonormal basis of V. Then $w_1', w_2', \ldots, w_m'$ given by $w_i' = P_{V'} v_i$ have inner products and norms closer and closer to those corresponding to the v_i's. Thus the Gram-Schmidt orthonormalizations of the vectors $Aw_1', Aw_2', \ldots, Aw_m'$ become in norm and inner products closer to the corresponding Gram-Schmidt orthonormalization of the vectors $Av_1, Av_2, \ldots, Av_m$. This could then be used to see $P_{AV'}x \to P_{AV}x$ for each vector x and so

$||P_{AV'} - P_{AV}|| \to 0$ as $V' \to V$. This seems to be a messy process. As already seen, the first argument using homogeneous spaces is easier but less direct. □

25. Let H and K be topological Hausdorff groups. Suppose $\alpha : H \to \mathrm{Aut}(K)$ is a homomorphism of H into the automorphism group of K and the mapping $(h, k) \mapsto \alpha(h)(k)$ is continuous. Define $G = K \times_\alpha H$ to be the set $K \times H$ with product topology and multiplication defined by

$$(k, h) \cdot (k', h') = (k\alpha_h(k'), hh').$$

(a) Show G is a topological Hausdorff group.

(b) Show the mapping $(k, h) \cdot k' = k\alpha_h(k')$ defines a continuous G action on K.

(c) Show
$$G/(\{e_K\} \times H) \cong K$$
as G spaces; i.e.; the map $(k, h)(\{e_K\} \times H) \mapsto k$ is a bicontinuous G map. The group $K \times_\alpha H$ is a **semi-direct product group** of K and H.

Solution:

First note G is a group; namely one has the associative law and inverses. For associativity,

$$\begin{aligned}
(k, h) \cdot ((k', h') \cdot (k'', h'')) &= (k, h) \cdot (k'\alpha_{h'}(k''), h'h'') \\
&= (k\alpha_h(k'\alpha_{h'}(k'')), hh'h'') \\
&= (k\alpha_h(k')\alpha_h(\alpha_{h'}(k'')), hh'h'') \\
&= (k\alpha_h(k')\alpha_{hh'}(k''), (hh')h'') \\
&= (k\alpha_h(k'), hh') \cdot (k'', h'') \\
&= ((k, h) \cdot (k', h')) \cdot (k'', h'')\,.
\end{aligned}$$

One also can check (e_K, e_H) is a multiplicative identity. From this one sees:

$$\begin{aligned}
(\alpha_{h^{-1}}(k^{-1}), h^{-1}) \cdot (k, h) &= (\alpha_{h^{-1}}(k^{-1})\alpha_{h^{-1}}(k), h^{-1}h) \\
&= (\alpha_{h^{-1}}(k^{-1}k), e_H) \\
&= (e_K, e_H)
\end{aligned}$$

and

$$\begin{aligned}
(k, h) \cdot (\alpha_{h^{-1}}(k^{-1}), h^{-1}) &= (k\alpha_h(\alpha_{h^{-1}}(k^{-1})), hh^{-1}) \\
&= (k\alpha_{hh^{-1}}(k^{-1}), hh^{-1}) \\
&= (kk^{-1}, hh^{-1}) \\
&= (e_K, e_H)\,.
\end{aligned}$$

So G is a group. Since the product topology of Hausdorff spaces is Hausdorff, G is a Hausdorff space.

Now note a function $(f_1, f_2) : K \times H \to K \times H$ is continuous in the product topology where $f_1 : K \times H \to K$ and $f_2 : K \times H \to H$ if and only if f_1 and f_2 are continuous. Thus $(k, h) \mapsto (\alpha_{h^{-1}}(k), h^{-1})$ is continuous if and only if $(k, h) \mapsto \alpha_{h^{-1}}(k)$ and $(k, h) \mapsto h^{-1}$ are continuous. The second clearly is for the preimage of an open set V in H is $K \times V^{-1}$ which is open in the product topology. The first is for the preimage of an open set U in K consists of the pairs (k, h^{-1}) with $\alpha_h(k) \in U$ and thus is the image of the open set $\{(h, k) \mid \alpha_h(k) \in U\}$ under the open mapping $(k, h) \mapsto (k, h^{-1})$. Thus $(k, h) \mapsto (k, h)^{-1}$ is continuous.

For multiplication the mappings

$$((k, h), (k', h')) \mapsto k\alpha_h(k') \text{ and } ((k, h), (k', h')) \mapsto hh'$$

are continuous. They are continuous for the first is a composition of the continuous mappings $((k, h), (k', h')) \mapsto (k, \alpha_h(k'))$ and $(k_1, k_2) \mapsto k_1k_2$, and the second is the composition of continuous mappings $((k, h), (k', h')) \mapsto (h, h') \mapsto hh'$. So G is a topological Hausdorff group.

For (b) and (c), we note the mapping $(k, h)(\{e_K\} \times H) \mapsto k$ is homeomorphism of $G/(\{e_K\} \times H)$ onto K. It is clearly one-to-one and onto for $(k, e_H)(\{e_K\} \times H) = (k, e_H)\cdot(e_K, h)(\{e_K\} \times H) = (k, h)(\{e_K\} \times H)$. To see it is a homeomorphism, the preimage of an open subset U of K is $(U \times \{e_H\})(\{e_K\} \times H)$. The preimage of this set in $K \times H$ is $U \times H$ for $(k, h) = (u, e_H)(e_K, h')$ if and only if $h = h'$ and $u = k$. So the map $(k, h)(\{e_K\} \times H) \mapsto k$ is continuous. It is an open mapping for if W is an open subset of $G/(\{e_K\} \times H)$, then the set O consisting of all points (k, h) with $(k, h)(\{e_K\} \times H) \in W$ is an open subset of $K \times H$ and its image in K under the mapping $(k, h) \mapsto k$ is open in K. But this set is the image of W under the mapping $(k, h)(\{e_K\} \times H) \mapsto k$. So the mapping $(k, h)(\{e_K\} \times H) \mapsto k$ is a homeomorphism.

By Lemma 5.13, we know the action of G on $G/(\{e_K\} \times H)$ is continuous. But under this homeomorphism $(k, h) \cdot k' \leftrightarrow (k, h) \cdot (k', e_H)(\{e_K\} \times H) = (k\alpha_h(k'), h)(\{e_K\} \times H) \leftrightarrow k\alpha_h(k')$. This shows the action of K on the cosets of $\{e_K\} \times H$ under the homeomorphism is the action of $K \times H$ on K given by $(k, h) \cdot k' = \alpha_h(k)k'$. So the action of his is a continuous action.

□

26. Let H be the Euclidean group (page 269) $E(n) = \mathbb{R}^n \rtimes \mathrm{O}(n)$ with multiplication $(v, A)(v', A') = (v + Av', AA')$ and let K be the group $\mathbb{R}^n \times \mathbb{R}$ under addition. Define $\alpha_{(v,A)}(x, t) = (Ax + tv, t)$ for $(t, A) \in H$.

(a) Show $\alpha : H \to \mathrm{Aut}(K)$ is a homomorphism and determine multiplication in the group $K \times_\alpha H$.

(b) Show $((y, s), (v, A)) \cdot (x, t) = (y + Ax + tv, s + t)$ is a continuous action of the Gallilean group $K \times_\alpha H$ on $\mathbb{R}^n \times \mathbb{R}$ or Gallilean space-time.

Solution:

We use the previous exercise. Now $K = \mathbb{R}^n \times \mathbb{R}$ with multiplication $(x, t) \cdot (x', t') = (x+x', t+t')$ and from page 269, the Euclidean group $E(n) = \mathbb{R}^n \times \mathrm{O}(n)$

has multiplication $(v, A) \cdot (v', A') = (v + Av', AA')$. Using $\alpha_{(v,A)}(y, t) = (Ay + tv, t)$, one sees

$$\begin{aligned}\alpha_{(v,A)}(y + y', t + t') &= (A(y + y') + (t + t')v, t + t') \\ &= (Ay + tv, t) + (Ay' + t'v, t') \\ &= \alpha_{(v,A)}(y, t) + \alpha_{(v,A)}(y', t')\end{aligned}$$

and $\alpha_{(v,A)}^{-1} = \alpha_{(-A^{-1}v, A^{-1})}$ for

$$\begin{aligned}\alpha_{(-A^{-1}v,A^{-1})}(Ay + tv, t) &= (A^{-1}(Ay + tv) + t(-A^{-1}v), t) \\ &= (y, t)\,.\end{aligned}$$

Thus $\alpha_{(v,A)} \in \mathrm{Aut}(K)$. Furthermore,

$$\begin{aligned}\alpha_{(v,A)} \circ \alpha_{(v',A')}(y, t) &= \alpha_{(v,A)}(A'y + tv', t) \\ &= (A(A'y + tv') + tv, t) \\ &= (AA'y + t(Av' + v), t) \\ &= \alpha_{(v+Av',AA')}(y, t) \\ &= \alpha_{(v,A)(v',A')}(y, t)\end{aligned}$$

Clearly $((v, A), (x, t)) \mapsto \alpha_{(v,A)}(x, t) = (Ax + tv, t)$ is continuous.

Now multiplication on $K \times E(n)$ is given by

$$\begin{aligned}((x, t), (y, A)) \cdot ((x', t'), (y', A')) &= ((x, t)\alpha_{(y,A)}(x', t'), (y, A)(y', A')) \\ &= ((x, t)(Ax' + t'y, t'), (y + Ay', AA')) \\ &= ((x + Ax' + t'y, t + t'), (y + Ay', AA'))\,.\end{aligned}$$

For (b), we know the action of $K \times H$ on K is given by

$$\begin{aligned}((y, s), (v, A)) \cdot (x, t) &= (y, s)\alpha_{(v,A)}(x, t) \\ &= (y, s)(Ax + tv, t) \\ &= (y + Ax + tv, t + s)\,.\end{aligned}$$

□

27. Let $\alpha : \mathbb{R} \to \mathrm{Aut}(\mathbb{C}^2, +)$ by

$$\alpha_t(z_1, z_2) = (e^{it}z_1, e^{\pi it}z_2).$$

Show this defines a semi-direct product group $\mathbb{C}^2 \times_\alpha \mathbb{R}$. Write out explicitly the multiplication for this semi-direct product and find $(z_1, z_2, t)^{-1}$.

Solution:

Again we use Exercise 5.3.25 with $K = \mathbb{C}^2$ under addition and $\alpha_t(z_1, z_2) = (e^{it}z_1, e^{\pi it}z_2)$. Indeed, it is easy to check α_t is an automorphism of $\mathbb{C}^2$ and

$$\begin{aligned}\alpha_t \circ \alpha_{t'}(z_1, z_2) &= \alpha_t(e^{it'}z_1, e^{i\pi t'}z_2) \\ &= (e^{i(t+t')}z_1, e^{i\pi(t+t')}z_2) \\ &= \alpha_{t+t'}(z_1, z_2).\end{aligned}$$

Since $(t, z_1, z_2) \mapsto (e^{\pi it}z_1, e^{\pi it}z_2)$ is continuous, one has the topological group $\mathbb{C}^2 \times_\alpha \mathbb{R}$ with multiplication given by

$$\begin{aligned}((z_1, z_2), t) \cdot ((z_1', z_2'), t') &= ((z_1, z_2) + \alpha_t(z_1', z_2'), t + t') \\ &= ((z_1 + e^{it}z_1', z_2 + e^{i\pi t'}z_2'), t + t').\end{aligned}$$

The multiplicative inverse of $((z_1, z_2), t)$ is $((-e^{-t}z_1, -e^{-i\pi t}z_2), -t)$ for

$$\begin{aligned}((-e^{-it}z_1, -e^{-i\pi t}z_2), -t) \cdot ((z_1, z_2), t)) &= ((-e^{-it}z_1 + e^{-it}z_1, -e^{-i\pi t}z_2 + e^{-i\pi t}z_2), 0) \\ &= ((0, 0), 0) \\ &= ((z_1 + e^{it}(-e^{-it}z_1), z_2 + e^{i\pi t}(-e^{-i\pi t}z_2), 0) \\ &= ((z_1, z_2), t) \cdot ((-e^{-it}z_1, -e^{-i\pi t}z_2), -t).\end{aligned}$$

We remark the action of $\mathbb{C}^2 \times_\alpha \mathbb{R}$ on $\mathbb{C}^2$ is given by

$$((z_1, z_2), t) \cdot (w_1, w_2) = \alpha_t(z_1, z_2) + (w_1, w_2) = (w_1 + e^{it}z_1, w_2 + e^{i\pi t}z_2)$$

and thus the orbit of (w_1, w_2) under the restriction to $\mathbb{R}$ of this action is a translate by (w_1, w_2) of a dense winding line on the torus

$$\{z \mid |z| = |z_1|\} \times \{z \mid |z| = |z_2|\}.$$

□

28. Let ϕ be a transformation of $\mathbb{R}^n$ onto $\mathbb{R}^n$ that preserves the inner product $(\cdot \mid \cdot)$. Show there is an matrix $T \in \mathrm{O}(n, \mathbb{R})$ so that $\phi(x) = Tx$ for $x \in \mathbb{R}^n$.
Solution: Note $\phi(0) = 0$. Also $(\phi(x + y) \mid \phi(z)) = (x + y \mid z) = (x \mid z) + (y \mid z) = (\phi(x) \mid \phi(z)) + (\phi(y) \mid \phi(z))$. Since ϕ is onto, this implies

$$\phi(x + y) = \phi(x) + \phi(y)$$

for all x and y. Also let $\alpha \in \mathbb{R}$. Then

$$(\phi(\alpha x) \mid \phi(z)) = (\alpha x \mid z) = \alpha(\phi(x) \mid \phi(z))$$

for all z. Again since ϕ is onto, we see

$$\phi(\alpha x) = \alpha\phi(x).$$

Thus $\phi(x) = Tx$ for some $n \times n$ matrix T. Note

$$(Tx \mid Ty) = (x \mid y)$$

and so $T \in O(n, \mathbb{R})$. □

Chapter 6

Basic Representation Theory

Exercise Set 6.1

In the following exercises, unless otherwise stated, X and Y denote locally compact Hausdorff spaces and G will denote a locally compact Hausdorff topological group.

1. Use Proposition 5.23 to show every every compact subset K of X is contained in a compact G_δ.

Solution:

By Proposition 5.23, if K is compact, there is a $f \in C_c(X)$ with $0 \leqslant f \leqslant 1$ and $f(x) = 1$ for $x \in K$. But $\{x \mid f(x) = 1\} = \cap_{n=1}^{\infty}\{x \mid 1 - \frac{1}{n} < f(x) < 1 + \frac{1}{n}\}$ is a compact G_δ.

□

2. Show if E is a Baire subset of X, then E or $X - E$ is σ-bounded.

Solution:

Consider the σ-algebra of Baire subsets of X. It is the smallest σ-algebra containing all compact G_δ subsets.

Let $\mathcal{A}$ be the collections of all subsets E of X such that E or $X - E$ is σ-bounded. Note $\mathcal{A}$ contains all compact G_δ subsets K of X. We claim $\mathcal{A}$ is a σ-algebra. It clearly contains $\varnothing$ and is closed under complements. We claim it is closed under countable unions. Suppose E_i for $i \in \mathbb{N}$ are in $\mathcal{A}$. If each E_i is contained in a countable union of compact sets, then $\cup_{i=1}^{\infty} E_i$ is a subset of a countable union of compact sets and so $\cup E_i \in \mathcal{A}$. If E_k is not in a countable union of compact subsets for some k, then $X - E_k$ is a subset of a countable union of compact sets. But then $X - \cup E_i \subseteq X - E_k$ is a subset of a countable union of compact sets. Hence $\cup E_i \in \mathcal{A}$. Since the Baire sets form the σ-algebra on X generated by the compact G_δ sets, every Baire set in $\mathcal{A}$. □

3. Let X be a second countable locally compact Hausdorff space.

(a) Show X has a countable dense subset.

(b) Show there is a countable base for the topology of X consisting of open sets whose closures are compact.

(c) Show X is σ-compact.

Solution:

Let $\mathcal{C}$ be a countable base for the topology. For each $C \in \mathcal{C}$, choose $x \in C$ and form the set D of these x's. D is countable. Let U be a nonempty open set. Then $U \supseteq C$ for some C. So $D \cap U \neq \varnothing$. Thus D is dense.

Let $\mathcal{U}$ be the set of $C \in \mathcal{C}$ with $\bar{C}$ compact. Since by Exercise 5.2.14, for each open set V with $x \in V$, there is an open set U with $x \in U \subseteq \bar{U} \subseteq V$ and $\bar{U}$ compact, we see $\mathcal{U}$ satisfies (b). Clearly (c) follows from (b). □

4. Let X be a locally compact Hausdorff space with Radon measure μ. Show if ν is the unique regular extension of μ to the Borel subsets of X, then for each Borel subset E of ν finite measure, there is a Baire subset E' such that $\nu(E - E') + \nu(E' - E) = 0$; i.e., E and E' are the same sets in the measure algebra of ν.

Solution: We first note if E is a Borel subset of finite measure, then there is a Baire set E' such that E and E' are the same sets in the measure algebra of μ. Indeed, for each n we can choose $O_n \supseteq E$ where O_n is open and $\nu(O_n - E) < \frac{1}{2^n}$. We may assume O_n is decreasing. By using inner regularity, we may choose a compact G_δ set $K_n \subseteq O_n$ with $\nu(O_n - K_n) < \frac{1}{2^n}$. Now $\nu(E - K_n) \leqslant \nu(O_n - K_n) < \frac{1}{2^n}$ and $\mu(K_n - E) < \nu(O_n - E) < \frac{1}{2^n}$.

Let $F = \cap_{n=1}^{\infty}(\cup_{k=n}^{\infty} K_n)$. Note F is Baire. Also $F - E \subseteq \cup_{k=n}^{\infty} K_k - E \subseteq \cup_{k=n}^{\infty}(K_k - E)$ for all n. So $\nu(F - E) < \sum_{k=n}^{\infty} \frac{1}{2^k} = \frac{1}{2^{n-1}}$ for all n. So $\nu(F - E) = 0$.

Now if $x \in E - F$, then $x \in E$ and $x \notin F$. So for each n, $x \in E$ and $x \notin K_k$ for some $k \geqslant n$. So for each n, $x \in \cup_{k \geqslant n}(E - K_k)$. Thus $\nu(E - F) \leqslant \sum_{k=n}^{\infty} \nu(E - K_k) < \sum_{k=n}^{\infty} \frac{1}{2^k} = \frac{1}{2^{n-1}}$ for all n. Thus $\nu(E - F) = 0$. □

5. Let μ be a Radon measure on a locally compact Hausdorff space X and let ν be the regular Borel measure measure which extends μ. Show $L^p(\mu) = L^p(\nu)$ for $1 \leqslant p < \infty$ while $L^\infty(\mu)$ might be strictly smaller than $L^\infty(\nu)$.

Solution: We do the proof for real valued functions. We need to show in L^p where p is finite, that for each Borel measurable function f in $L^p(\nu)$, there is a Baire measurable function f' in $L^p(\mu)$ equal to f in $L^p(\nu)$. First note if $f \in L^p(\nu)$, then $W = \{x \mid f(x) \neq 0\}$ is σ-finite. By partitioning f, we may assume W has finite measure. Also be doing the positive and negative parts of f separately, we may assume $f \geqslant 0$. For each positive rational q, set $E_q = \{f > q\}$. E_q is Borel and has finite ν measure. Also $E_r \subseteq E_s$ if $r \geqslant s$. Using Exercise 6.1.4, we can choose Baire sets E'_r with $E_r = E'_r$ in the measure algebra of ν. We would like the sets E'_r to be decreasing. So define $W_r = \cup_{q \geqslant r} E'_q$. The W_r are decreasing. Moreover, $W_r \supseteq E'_r$ and thus $\nu(E_r - W_r) \leqslant \nu(E_r - E'_r) = 0$.

We claim $W_r - E_r$ has ν measure 0. Indeed, if $x \in W_r - E_r$, then $x \in E'_q - E_r$ for some $q \geqslant r$. But $E'_q - E_r \subseteq E'_q - E_q$. So $W_r - E_r \subseteq \cup_{q \geqslant r}(E'_q - E_q)$. Thus $\nu(W_r - E_r) = 0$. Set $W_0 = X$.

Now define f' by $f'(x) = \sup\{r \geqslant 0 \mid x \in W_r\}$. Then f' is Baire for each W_r is a Baire set. In fact, $f'(x) > r$ if and only if there is a $q > r$ with $x \in W_q$ if and only if $x \in \cup_{q>r} W_q$ which is Baire for $r \in \mathbb{Q}$. Finally we show $f' = f$ a.e. ν. Indeed, let $N = \cup_{r>0}((E_r - W_r) \cup (W_r - E_r))$. N has ν measure 0. If $x \notin N$ and $f(x) = 0$, then $x \notin E_r$ for all $r > 0$. So $x \notin W_r$ for all $r > 0$; otherwise $x \in W_r - E_r$ for some $r > 0$ and then $x \in N$. Thus $f'(x) = 0$. Similarly, if $x \notin N$, then $x \in W_r$ if and only if $x \in E_r$. But if $f(x) \neq 0$, then

$$\begin{aligned} f(x) &= \sup\{r \in \mathbb{Q}^+ \mid f(x) > r\} = \sup\{r \in \mathbb{Q}^+ \mid x \in E_r\} \\ &= \sup\{r \in \mathbb{Q}^+ \mid x \in W_r\} = f'(x). \end{aligned}$$

So $f = f'$ in $L^p(\nu)$.

We now consider the problem when $p = \infty$. Take an uncountable discrete space X. The Baire sets for X are the subsets W of X such that W or W^c is countable. The Borel subsets of X are all subsets of X. Let μ be counting measure. Then ν is counting measure. Divide X into two uncountable disjoint subsets X_1 and X_2. Let f be the characteristic function of X_1. It is in $L^\infty(\nu)$. Suppose $f' \in L^\infty(\mu)$ and equals f a.e. ν. Then $\{x \mid f'(x) = 1\}$ and X_1 are equal in the measure algebra of ν. But $\{x \mid f'(x) = 1\}$ is either countable or has countable complement. Since X_1 and X_1^c are both uncountable, this is impossible. □

6. Show m is a left invariant regular Borel measure on G if and only if m^* defined by

$$m^*(E) = m(E^{-1})$$

is a right invariant regular Borel measure on G. Then show the modular function for m^* and m satisfy $\Delta^* = \check{\Delta}$; i.e., show if $m(Eg) = \Delta(g)m(E)$ for all Borel sets E, then $m^*(gE) = \Delta(g^{-1})m^*(E)$ for all Borel sets E.

Solution:

Since $x \mapsto x^{-1}$ is a homeomorphism, a subset E is Borel if and only if E^{-1} is a Borel subset. Thus m^* is a Borel measure if m is a Borel measure. Furthermore m is left invariant if and only if $m(gE) = m(E)$ for all Borel subsets E and all $g \in G$ if and only if $m((Eg)^{-1}) = m(g^{-1}E^{-1}) = m(E^{-1})$ for all E and g. Thus m is left invariant if and only if $m^*(Eg) = m^*(E)$ for all E and g.

Now m^* is regular if and only if m is regular for the mappings $K \mapsto K^{-1}$ and $U \mapsto U^{-1}$ are bijections between the collection of compact sets and itself and the collection of open sets and itself. Moreover $K \subseteq E$ if and only if $K^{-1} \subseteq E^{-1}$ and $E^{-1} \subseteq U^{-1}$ if and only if $E \subseteq U$. So (a), (b), and (c) on page 276 defining regularity hold for m if and only if they hold for m^*.

Now if Δ is the modular function for m, we have $m(Eg) = \Delta(g)m(E)$ for Borel subsets E of G and $g \in G$. So $m^*(gE) = m(E^{-1}g^{-1}) = \Delta(g^{-1})m(E^{-1}) = \check{\Delta}(g)m^*(E)$.

□

7. Show Haar measure on the multiplicative group of positive reals is given by $\frac{dx}{x}$. That is show $\int_{\mathbb{R}+} f(x)\, dm(x) = \int_0^\infty \frac{f(x)}{x}\, dx$.
Solution: Let $m(E) = \int_E \frac{1}{x} dx$ for $E \subseteq \mathbb{R}^+$ and E Borel. Now by Lemma 6.4 and its proof, the Baire sets for $\mathbb{R}^+$ are the Borel sets and every compact subset is Baire and clearly $m(K) < \infty$ if K is compact. So m is a Baire measure. Moreover, if $a > 0$,

$$\begin{aligned} m(aE) &= \int \frac{\chi_{aE}(x)}{x}\, dx = \int \frac{\chi_E(a^{-1}x)}{x}\, dx \\ &= \int \frac{\chi_E(x)}{ax}\, a\, dx = \int_E \frac{1}{x}\, dx \\ &= m(E) \end{aligned}$$

and so m is left invariant. Thus m is a Haar measure. Note the Radon-Nikodym derivative of m relative to dx is $\frac{1}{x}$. This gives

$$\int f(x)\, dm(x) = \int f(x) \frac{1}{x}\, dx$$

for positive Borel functions f or functions $f \in L^1(m)$.

□

8. Let G be the $ax + b$ group; that is the subset $\{(a, b) \mid a > 0,\ b \in \mathbb{R}\}$ of $\mathbb{R}^2$ with the relative topology and with multiplication defined by

$$(a, b)(a', b') = (aa', b + ab').$$

Show that I defined by

$$I(f) = \int_0^\infty \int_{-\infty}^\infty f(a, b)\, \frac{1}{a^2}\, db\, da$$

is a positive left invariant integral on $C_c(G)$. Then show the modular function for G is given by $\Delta(a, b) = \frac{1}{a}$. In particular, the $ax + b$ group is nonunimodular.
Solution: Multiplication on the $ax+b$ group is given by $(a, b)(a', b') = (aa', b+ ab')$. The integral defined by $I(f) = \int_0^\infty \int_{-\infty}^\infty f(a, b) \frac{1}{a^2}\, db\, da$ is left invariant for

$$\begin{aligned} \int_0^\infty \int_{-\infty}^\infty f(a'a, b' + a'b) \frac{1}{a^2}\, db\, da &= \int_0^\infty \int_{-\infty}^\infty f(a, b' + b) \frac{1}{(a/a')^2}\, (\frac{1}{a'}\, db)\, (\frac{1}{a'}\, da) \\ &= \int_0^\infty \int_{-\infty}^\infty f(a, b' + b) \frac{1}{a^2}\, db\, da \\ &= \int_0^\infty \int_{-\infty}^\infty f(a, b) \frac{1}{a^2}\, db\, da \end{aligned}$$

and thus I is left invariant on $C_c(G)$. It is clearly positive and thus is a Haar integral. We use (e) of Proposition 6.3 to find Δ.

$$\begin{aligned}\int_0^\infty \int_{-\infty}^\infty f((a,b)(a',b'))\frac{1}{a^2}\,db\,da &= \int_0^\infty \int_{-\infty}^\infty f(aa', b+ab')\frac{1}{a^2}\,db\,da \\ &= \int_0^\infty \int_{-\infty}^\infty f(aa', b)\frac{1}{a^2}\,db\,da \\ &= \int_0^\infty \int_{-\infty}^\infty f(a,b)\frac{1}{(a/a')^2}\,db\,(\frac{1}{a'}\,da) \\ &= a'\int_0^\infty \int_{-\infty}^\infty f(a,b)\frac{1}{a^2}\,db\,da \\ &= \Delta((a',b')^{-1})\int_0^\infty \int_{-\infty}^\infty f(a,b)\frac{1}{a^2}\,db\,da\end{aligned}$$

implies $\Delta(a',b') = \frac{1}{a'}$.

□

9. Let G be a discrete topological group. Determine the Baire sets for G. Then show counting measure on the Baire sets is both left and right invariant.

Solution:

We first note the compact G_δ sets are the compact sets which are the finite sets. This implies every countable set or its complement is a Baire set. But the family $\mathcal{B}$ of all sets which are countable or have countable complement is a σ-algebra. Thus the σ-algebra of Baire sets is this algebra. Now let m be counting measure on $\mathcal{B}$. It is clearly a Baire measure for it is finite on compact sets, and since $m(Eg) = m(gE)$ for sets $E \in \mathcal{B}$, it is a Haar measure.

□

10. Let $C_0(X)$ be the space of all continuous functions f on X that vanish at ∞; i.e., for each $\epsilon > 0$, $\{x \mid |f(x)| \geqslant \epsilon\}$ is compact. Show $C_0(X)$ is a Banach space when equipped with norm $|\cdot|_\infty$ where $|f|_\infty = \max_{x\in X}|f(x)|$. Then show $C_c(X)$ is dense in $C_0(X)$.

Solution:

We note convergence in $|\cdot|_\infty$ is uniform convergence. Hence the limit of any sequence in $C_0(X)$ is a continuous function. We claim any limit also vanishes at ∞. Indeed, if $|f_n - f|_\infty \to 0$ where each $f_n \in C_0(X)$, then if $\epsilon > 0$, there is an N with $|f_N - f|_\infty < \frac{\epsilon}{2}$. But the set of x with $|f_N(x)| \geqslant \epsilon/2$ is compact. But then $\{x \mid |f(x)| \geqslant \epsilon\} \subseteq \{x \mid |f_N(x)| \geqslant \frac{\epsilon}{2}\}$ for if $|f_N(x)| < \frac{\epsilon}{2}$, then $|f(x)| \leqslant |f(x) - f_N(x)| + |f_N(x)| \leqslant |f - f_N|_\infty + |f_N(x)| < \epsilon$. Since f is continuous, $\{x \mid |f(x)| \geqslant \epsilon\}$ is a closed set contained in the compact set $\{x \mid |f_N(x)| \geqslant \frac{\epsilon}{2}\}$ and thus is compact.

Now let $f \in C_0(X)$ and suppose $\epsilon > 0$. Set $K = \{x \mid |f(x)| \geqslant \epsilon\}$. K is compact and by Proposition 5.23 there is function ϕ with $0 \leqslant \phi \leqslant 1$ and having compact support such that $\phi(x) = 1$ for $x \in K$. Set $f_0 = f\phi$. Then f_0 is continuous and has compact support. Moreover, if $x \in K$, $f_0(x) = f(x)$ and if

$x \notin K$,

$$\begin{aligned} |f(x) - f_0(x)| &= |(1-\phi)(x)|\,|f(x)| \\ &\leqslant |f(x)| < \epsilon. \end{aligned}$$

These together imply $|f - f_0|_\infty \leqslant \epsilon$. So the continuous functions with compact support are dense in $C_0(X)$.

□

11. Let $X_\infty = X \cup \{\infty\}$ be the one-point compactification of X and let $C(X_\infty)$ be the space of continuous functions on X_∞. Show that

$$C_0(X) = \{f|_X \mid f \in C(X_\infty) \text{ and } f(\infty) = 0\}$$

and that $C(X_\infty)$ corresponds to the space of bounded continuous functions on X such that $\lim_{x\to\infty} f(x)$ exists. Finally show that every finite Borel measure on X gives by integration a continuous linear functional on $C(X_\infty)$.

Solution:

Recall a subset U of X_∞ is open if and only if U is an open subset of X or the complement of U in X is compact.

First note if $f \in C(X_\infty)$ and $f(\infty) = 0$, then since the inclusion map from X into X_∞ is continuous, $f|_X$ is continuous. Moreover, $f|_X$ vanishes at ∞, for if $\epsilon > 0$, $\{x \mid |f(x)| \geqslant \epsilon\}$ is a closed set in X_∞. Thus $\{x \mid |f(x)| < \epsilon\}$ is an open subset of X_∞ containing the point ∞. But the open sets in X_∞, the one point compactification, containing ∞ are the complements of compact subsets of X. So $\{x \in X \mid |f(x)| \geqslant \epsilon\}$ is compact. Thus $f|_X$ vanishes at ∞ if $f(\infty) = 0$. But if $h \in C_0(X)$, define f on X_∞ by $f(x) = h(x)$ if $x \in X$ and $f(\infty) = 0$. Note f is continuous for if U is an open subset of $\mathbb{C}$ and $f(p) \in U$ and $p \in X$, then since h is continuous $f^{-1}(U)$ contains $h^{-1}(U)$ which is open in X and thus $f^{-1}(U)$ contains an open subset of X_∞ containing p. If $f(\infty) = 0 \in U$, there is an $\epsilon > 0$ with $c \in U$ if $|c| < \epsilon$. Thus $\{x \mid |h(x)| \geqslant \epsilon\}$ is a compact subset of X and $f^{-1}(U)$ contains the complement in X_∞ of this set which is an open subset of $f^{-1}(U)$ in the topology of X_∞ containing ∞. So $f^{-1}(U)$ is open in X_∞. Thus $f \in C(X_\infty)$ and $f|_X = h$. Also note if $h \in C(X)$ and $\lim_{x\to\infty} h(x)$ exists, then $h - \lim_{x\to\infty} h(x)$ is in $C_0(X)$ and thus is $F|_X$ for some $F \in C(X_\infty)$ with $F(\infty) = 0$. So $h = (F + \lim_{x\to\infty} h(x))|_X$ and $\lim_{x\to\infty} h(x) = F(\infty)$.

Now let μ be a finite Borel measure on X. Let $f \in C(X_\infty)$, then f is bounded and so $f|_X$ is bounded and continuous. Hence $I(f) = \int_X f(x)\,d\mu(x)$ is defined and is a positive integral on $C(X_\infty)$. We note it is continuous for $|I(f)| \leqslant \int_X |f(x)|\,d\mu(x) \leqslant \int_X |f|_\infty\,d\mu = \mu(X)\,|f|_\infty$.

□

12. Let λ be the Lebesgue measure on $\mathbb{R}$.

(a) Show μ defined by $\mu(E) = \int_E \frac{1}{x}\,d\lambda(x)$ is a Radon measure on $\mathbb{R}^+$.

(b) Show the measure μ can not be extended to a Radon measure on $\mathbb{R}$.

Solution:

For (a), since in a second countable space, the G_δ compact subsets are the compact subsets, one has for each compact subset $K \subseteq \mathbb{R}^+$ a minimum a and a maximum b. Thus $0 < a < b < \infty$ and $\mu(K) = \int_K \frac{1}{x}\,dx \leqslant \int_a^b \frac{1}{x}\,dx < \infty$. So μ is a Radon measure.

For any extension $\bar{\mu}$ of μ to $\mathbb{R}$, one has

$$\begin{aligned}\bar{\mu}([0,1]) &\geqslant \sup_{0<a<1} \int_a^1 \frac{1}{x}\,dx \\ &\geqslant \sup\{-\ln a \mid 0 < a < 1\} \\ &= \infty\,.\end{aligned}$$

Thus no extension could be a Radon measure.

□

13. Let G be a locally compact Hausdorff topological group and let m be a left Haar measure on G. Show the following:

(a) G is compact if and only if $m(G) < \infty$. (**Hint:** If G is not compact, show if V is a compact symmetric neighborhood of e, there is a sequence g_k where the sets g_kV are disjoint.)

(b) G is discrete if and only if $m(\{e\}) > 0$. (**Hint:** Show if $m(\{e\}) > 0$ and G is not discrete, then there is a compact subset of infinite measure.)

Solution:

Clearly $m(G) < \infty$ if G is compact for m is a Radon measure. Conversely, assume G is not compact. Take a compact symmetric neighborhood V of e. Set $g_1 = e$. Note V^2 is compact and thus there is a g_2 with $g_2 \notin V^2$. This implies $g_2V \cap V = \varnothing$. Suppose now $g_1, g_2, \ldots, g_k$ have be chosen so that the sets $g_1V, g_2V, \ldots, g_kV$ are disjoint. Set $W = \bigcup_{i=1}^k g_iV$. Note W^2 is compact. So there is a g_{k+1} with $g_{k+1} \notin W^2$. So $g_{k+1}V \cap g_iV = \varnothing$ for $i = 1, 2, \ldots, k$. By induction, we obtain a sequence g_k with the compact sets g_kV disjoint. But since V is a neighborhood of e, we know by (b) of Proposition 6.3 that $m(V) > 0$. So $m(G) \geqslant \sum_i m(g_iV) = \sum_i m(V) = \infty$. Thus we have (a).

For (b), since m is counting measure if G is discrete, one has $m(\{e\}) > 0$. Now suppose G is not discrete. Let V be a compact neighborhood of e. Then V is infinite for if V were finite, using the fact that G is Hausdorff, there would be a neighborhood U of e missing all points in V but e. Now choose g_k in V for $k = 1, 2, \ldots$ with $g_k \neq g_l$ for $k \neq l$. Then $\sum_{k=1}^\infty m(\{e\}) = \sum_{k=1}^\infty m(\{g_k\}) \leqslant m(V) < \infty$. This implies $m(\{e\}) = 0$. □

14. Let μ be a Radon measure on X. Show $C_c(X)$ is dense in $L^p(X)$ for $1 \leqslant p < \infty$.

Solution:

We know the linear span of the characteristic functions of Baire sets E of finite measure is dense in L^p for $1 \leqslant p < \infty$. But since $\mu(E)$ is the supremum of all $\mu(K)$ where K is a compact G_δ subset of E, χ_E can be approximated in

L^p by a χ_K where K is a compact G_δ. It therefore suffices to show χ_K can be approximated in L^p by an $f \in C_c(X)$. Now by Proposition 5.23, there is a $\phi \in C_c(X)$ with $\phi^{-1}(1) = K$ and $0 \leqslant \phi \leqslant 1$. So $\phi^k \to \chi_K$ as $k \to \infty$. Thus

$$|\phi^k - \chi_K|_p^p = \int_X |\phi^k(x) - \chi_K(x)|^p \, d\mu(x) \to 0$$

as $k \to \infty$ for $|\phi^k(x) - \chi_K(x)|^p \leqslant 2^p \phi(x)^p$ and thus one can use the Lebesgue Dominated Convergence Theorem. So if $\epsilon > 0$ and k is sufficiently large, $f = \phi^k$ is in $C_c(X)$ and $|f - \chi_K|_p < \epsilon$.

□

15. Let K be a compact subgroup of G. Set $X = G/K$, and let $\kappa : G \to X$ be the canonical map $a \mapsto aK$. Fix a left Haar measure m for G. Define a linear functional I_K on $C_c(X)$ by

$$I_K(f) = \int f \circ \kappa(a) \, dm(a) .$$

Show that I_K is a positive, G-invariant integral on $C_c(X)$. Then determine an expression for the Radon measure μ on X for this integral in terms of m.

Solution:

Since κ is continuous, $f \circ \kappa$ is continuous. Also $f \circ \kappa(g) \neq 0$ only if $\kappa(g) \in C$, the compact support of f. Thus $g \in \kappa^{-1}(C)$. But by Exercise 5.2.10, $\kappa^{-1}(C)$ is compact. Hence $f \circ \kappa \in C_c(G)$ and $I_K(f) = \int f \circ \kappa(g) \, dm(g)$ is defined. Now let $\lambda(a) f(gH) = f(agH)$ and $\lambda'(a)\phi(g) = \phi(ag)$ for functions f on X and ϕ on G. For $f \in C_c(X)$,

$$\begin{aligned} I_K(\lambda(a)f) &= \int (\lambda(a)f) \circ \kappa(g) \, dm(g) \\ &= \int \lambda'(a)(f \circ \kappa)(g) \, dm(g) \\ &= \int f \circ k(g) \, dm(g) \\ &= I_K(f). \end{aligned}$$

Thus I_K is G-invariant and so is given by a Radon measure μ on X. If C is a compact G_δ subset of X, using Proposition 5.23, there is an $f \in C_c(X)$ with $0 \leqslant f \leqslant 1$ and $f^{-1}(1) = C$. So

$$\begin{aligned} \mu(C) &= \lim_k \int_X f(x)^k \, d\mu(x) \\ &= \lim_k \int_G (f \circ \kappa)^k(g) \, dm(g) \\ &= \int \chi_{\kappa^{-1}(C)}(g) \, dm(g) \\ &= m(\kappa^{-1}(C)). \end{aligned}$$

Also $\kappa^{-1}(C)$ is a G_δ for it is $\cap_m\{x \mid |f \circ \kappa(x)| > 1 - \frac{1}{m}\}$. This implies $\kappa^{-1}(C)$ is a compact G_δ subset of G for any compact G_δ subset of X. But the set of all Baire subsets E of X such that $\kappa^{-1}(E)$ is a Baire subset of G is a σ-algebra. Thus κ is Baire measurable. Note

$$\begin{aligned}\mu(E) &= \sup\{\mu(C) \mid C \subseteq E \text{ and } C \text{ is a compact } G_\delta\}\\ &= \sup\{m(\kappa^{-1}(C)) \mid C \subseteq E \text{ and } C \text{ is a compact } G_\delta\}\\ &\leqslant m(\kappa^{-1}(E))\end{aligned}$$

for all Baire subsets E of X. Suppose $m(\kappa^{-1}(E)) > \mu(E)$. Now $\kappa^{-1}(E)$ is a Baire subset of G and thus there is a compact G_δ subset F of $\kappa^{-1}(E)$ with $m(F) > \mu(E)$. Since κ is an open mapping, $\kappa(F)$ is a compact G_δ in X and thus $\mu(E) \geqslant \mu(\kappa(F)) = m(\kappa^{-1}(\kappa(F))) = m(F) > \mu(E)$ which is impossible. So $\mu(E) = m(\kappa^{-1}(E))$ for all Baire subsets E of X.

□

16. By Lemma 5.25, $\mathrm{SL}(2,\mathbb{R})$ acts on $H_+ = \{z \in \mathbb{C} \mid \mathrm{Im}(z) > 0\}$ by $\left(\begin{smallmatrix} a & b \\ c & d \end{smallmatrix}\right) z = \frac{az+b}{cz+d}$, has stabilizer $\mathrm{SO}(2)$, and $H_+ = \mathrm{SL}(2,\mathbb{R})/\mathrm{SO}(2)$. Define a linear functional I on $C_c(H_+)$ by

$$I(f) = \int_{-\infty}^{\infty}\int_0^{\infty} f(x+iy)\,\frac{dydx}{y^2}\ .$$

Show that I is positive, non-zero, and $\mathrm{SL}(2,\mathbb{R})$-invariant.

Solution:

Let $G = \mathrm{SL}(2,R)$ and $K = \mathrm{SO}(2)$. By Lemma 5.25 and Exercise 5.3.22, if A is the subgroup of G consisting of the matrices $\left(\begin{smallmatrix} a & 0 \\ 0 & a^{-1} \end{smallmatrix}\right)$ where $a > 0$ and N is the subgroup consisting of the matrices $\left(\begin{smallmatrix} 1 & x \\ 0 & 1 \end{smallmatrix}\right)$, then $\left(\begin{smallmatrix} a & b \\ c & d \end{smallmatrix}\right)\cdot z = \frac{az+b}{cz+d}$ defines an action of G on H_+ that is transitive and the mapping $g \mapsto g \cdot i$ is a homeomorphism of the subgroup AN of G onto H_+. Moreover, the stabilizer of i in G is K. We could argue directly that the integral is invariant by showing the integral is invariant under A, N, and K. Instead, we use Corollary 6.18 and the previous exercise. Now AN is the $ax+b$ group; i.e., there is a topological isomorphism of the $ax+b$ group onto AN, namely take $\phi(a,b) = \begin{pmatrix} \sqrt{a} & \frac{b}{\sqrt{a}} \\ 0 & \frac{1}{\sqrt{a}} \end{pmatrix} = \begin{pmatrix} \sqrt{a} & 0 \\ 0 & \frac{1}{\sqrt{a}} \end{pmatrix}\begin{pmatrix} 1 & \frac{b}{a} \\ 0 & 1 \end{pmatrix}$.

Note

$$\begin{aligned}\phi(a,b)\phi(a',b') &= \begin{pmatrix} \sqrt{a} & \frac{b}{\sqrt{a}} \\ 0 & \frac{1}{\sqrt{a}} \end{pmatrix}\begin{pmatrix} \sqrt{a'} & \frac{b'}{\sqrt{a'}} \\ 0 & \frac{1}{\sqrt{a'}} \end{pmatrix}\\ &= \begin{pmatrix} \sqrt{aa'} & \sqrt{\frac{a}{a'}}b' + \frac{1}{\sqrt{aa'}}b \\ 0 & \frac{1}{\sqrt{aa'}} \end{pmatrix}\\ &= \begin{pmatrix} \sqrt{aa'} & \frac{1}{\sqrt{aa'}}(b+ab') \\ 0 & \frac{1}{\sqrt{aa'}} \end{pmatrix}\end{aligned}$$

and so ϕ is a homomorphic homeomorphism.

Now by Exercise 6.1.8 the Haar integral on the $ax+b$ groups is given by

$$I(f) = \int_{-\infty}^{\infty}\int_0^{\infty} f(a,b)\frac{1}{a^2}\,da\,db$$

and by the argument for Exercise 5.3.22, the inverse of the mapping $g \mapsto g \cdot i$ from AN into H_+ is given by $g(x+iy) = \begin{pmatrix} \sqrt{y} & \frac{1}{\sqrt{y}}x \\ 0 & \frac{1}{\sqrt{y}} \end{pmatrix} = \phi(y,x)$. So an AN invariant integral on H_+ would be given by

$$\begin{aligned} I(f) &= \int_{-\infty}^{\infty}\int_0^{\infty} f \circ \phi^{-1} \circ g(x+iy)\,\frac{1}{y^2}\,dy\,dx \\ &= \int_{-\infty}^{\infty}\int_0^{\infty} f \circ \phi^{-1} \circ \phi(y,x)\frac{1}{y^2}\,dy\;dx \\ &= \int_{-\infty}^{\infty}\int_0^{\infty} f(x+iy)\frac{1}{y^2}\,dy\;dx. \end{aligned}$$

Note this integral is also invariant under K. In fact, using the homeomorphism from AN on H_+ and Corollary 6.18, there is only one up to a scalar multiple AN invariant integral on H_+. But the G invariant integral given by $I_K(f) = \int f(g \cdot i)\,dm(g)$ in Exercise 6.1.15 also gives an AN invariant integral and since G/K is also topologically isomorphic to H_+ as G spaces, the integral I_K being AN invariant, must be a multiple of I. Since I_K is K invariant, I must also be K-invariant.

We remark with some argument, the above can be used to show a Haar integral on $\mathrm{SL}(2,\mathbb{R})$ is given by

$$I_G(f) = \frac{1}{2\pi}\int_0^{2\pi}\int_{-\infty}^{\infty}\int_0^{\infty} f\left(\left(\begin{smallmatrix} a & 0 \\ 0 & a^{-1}\end{smallmatrix}\right)\left(\begin{smallmatrix} 1 & x \\ 0 & 1\end{smallmatrix}\right)\left(\begin{smallmatrix} \cos\theta & \sin\theta \\ -\sin\theta & \cos\theta\end{smallmatrix}\right)\right)\frac{1}{a^2}\,da\,dx\,d\theta.$$

One way to start is to consider the function $m_E(F) = m(EF)$ where E and F are appropriate Borel subsets of AN and K, respectively.

□

17. Let H be a locally compact Hausdorff group and suppose the modular function for H is Δ. Define a multiplication on $G = H \times \mathbb{R}$ by

$$(x,s)(y,t) = (xy, \Delta(y^{-1})s + t).$$

Then show if dy is a left invariant Haar measure on H, then

$$I(f) = \int_G \int_{-\infty}^{\infty} f(y,t)\,dt\,dy$$

defines a left and right invariant integral on G. In particular, every locally compact Hausdorff group is a closed subgroup of a unimodular group.

Solution: We show G is a group. We could do a direct argument. However, this is also a semi-direct product group. Let $\alpha : H \to \mathrm{Aut}(\mathbb{R},+)$ be defined by $\alpha_h(t) = \Delta(h)t$. Since Δ is continuous, $(h,t) \mapsto \alpha_h(t)$ is continuous. By Exercise 5.3.25, $(t,x)\cdot(s,y) = (t+\alpha_x(s), xy) = (t+\Delta(x)s, xy)$ defines a Hausdorff topological group $\mathbb{R} \times_\alpha H$. Now ϕ defined by $\phi(x,s) = (s\Delta(x), x)$ is a homeomorphism from G onto $\mathbb{R} \times_\alpha H$ and

$$\begin{aligned}\phi(x,s)\phi(y,t) &= (s\Delta(x),x)\cdot(t\Delta(y),y)\\ &= (s\Delta(x)+\Delta(x)t\Delta(y),xy)\\ &= (s\Delta(x)+t\Delta(xy),xy)\\ &= (s\Delta(xy)\Delta(y^{-1})+t\Delta(xy),xy)\\ &= \phi(xy,\Delta(y^{-1})s+t)\,.\end{aligned}$$

This implies G is a topological group and ϕ is a topological group isomorphism of G onto $\mathbb{R} \times_\alpha H$. In particular G is a group.

We now show I is left and right invariant. Let $f \in C_c(G)$. Then by Fubini's Theorem and left invariance,

$$\begin{aligned}\int_H\int_{-\infty}^{\infty} f((x,s)\cdot(y,t))\,dt\,dy &= \int_H\int_{\mathbb{R}} f(xy,\Delta(y^{-1})s+t)\,dt\,dy\\ &= \int_H\int_{\mathbb{R}} f(xy,t)\,dt\,dy\\ &= \int_{\mathbb{R}}\int_H f(xy,t)\,dy\,dt\\ &= \int_{\mathbb{R}}\int_H f(y,t)\,dy\,dt\\ &= \int_H\int_{\mathbb{R}} f(y,t)\,dt\,dy\end{aligned}$$

and so I is left invariant.

Also by Fubini and (e) of Proposition 6.3,

$$\begin{aligned}\int_H\int_{\mathbb{R}} f((y,t)\cdot(x,s))\,dt\,dy &= \int_H\int_{\mathbb{R}} f(yx,\Delta(x^{-1})t+s)\,dt\,dy\\ &= \int_H \Delta(x)\int_{\mathbb{R}} f(yx,t+s)\,dt\,dy\\ &= \int_H \Delta(x)\int_{\mathbb{R}} f(yx,t)\,dt\,dy\\ &= \int_{\mathbb{R}}\int_H \Delta(x)f(yx,t)\,dy\,dt\\ &= \int_{\mathbb{R}}\int_H f(y,t)\,dy\,dt\\ &= I(f)\,.\end{aligned}$$

Thus I is also right invariant. Concluding, G is unimodular.

□

18. A $2n+1$ dimensional Heisenberg group H is the Euclidean space $\mathbb{R}^{2n}\times\mathbb{R}$ with multiplication defined by

$$(x,t)(y,s) = (x+y,t+s+\frac{1}{2}B(x,y))$$

where B is a nondegenerate alternating bilinear form on $\mathbb{R}^{2n}$. See Example 5.9. Show Lebesgue measure on $\mathbb{R}^{2n} \times \mathbb{R}$ is both a left and right Haar measure for H. In particular, Heisenberg groups are unimodular.

Solution:

Lebesgue measure $\lambda_{2n} \times \lambda$ on the Borel subsets is a Baire measure and one has for $f \in C_c(H)$,

$$\begin{aligned}
\int_{\mathbb{R}^{2n}} \int_{\mathbb{R}} f((x,t)(y,s))\, ds\, dy &= \int_{\mathbb{R}^{2n}} \int_{\mathbb{R}} f(x+y, t+s+\frac{1}{2}B(x,y))\, ds\, dy \\
&= \int_{\mathbb{R}^{2n}} \int_{\mathbb{R}} f(x+y, s)\, ds\, dy \\
&= \int_{\mathbb{R}^{2n}} \int_{\mathbb{R}} f(y, s)\, ds\, dy \\
&= \int_{\mathbb{R}^{2n}} \int_{\mathbb{R}} f(x, t)\, dt\, dx \\
&= \int_{\mathbb{R}} \int_{\mathbb{R}^{2n}} f(x+y, t)\, dx\, dt \\
&= \int_{\mathbb{R}^{2n}} \int_{\mathbb{R}} f(x+y, t+s+\frac{1}{2}B(x,y))\, dt\, dx \\
&= \int_{\mathbb{R}^{2n}} \int_{\mathbb{R}} f((x,t)(y,s))\, dt\, dx\,,
\end{aligned}$$

where we have used Fubini and translation invariance of both the measures λ_{2n} and λ. So the measure $\lambda_{2n} \times \lambda$ is both right and left invariant.

□

19. Show extension groups of closed unimodular groups by compact groups are unimodular. More specifically, let H be a closed normal subgroup of a locally compact Hausdorff group G. Suppose H is unimodular and the group G/H is compact. Show G is unimodular.

Solution: We know that G/H has an invariant measure. By Theorem 6.17, this implies the modular functions for G and H agree on H. Hence the modular function Δ for G is 1 on H. It thus factors to give a continuous homomorphism $\bar{\Delta}$ of G/H into $\mathbb{R}^+$. But since G/H is compact, $\bar{\Delta} = 1$. So $\Delta = 1$ and we see G is unimodular. □

20. Use Example 6.9 to show the group $\mathrm{GL}(n, \mathbb{R})$ is unimodular.

Solution: One can use the same argument as in the example to show the transformation $X \mapsto XC$ has determinant $\det(C)^n$ and thus the measure μ in the example is also right invariant. One could also the invariance of Lebesgue measure under the mapping $X \mapsto X^t$ on $\mathbb{R}^{n^2}$ (its matrix has determinant 1) and thus

$$\begin{aligned}\int f(XC)\,d\mu(X) &= \int f(XC)|\det X|^n\,d\lambda(X)\\ &= \int f(X^tC)|\det X^t|^n\,d\lambda(X)\\ &= \int f((C^tX)^t)|\det X|^n\,d\lambda(X)\\ &= \int f((C^tX)^t)\,d\mu(X)\\ &= \int f(X^t)\,d\mu(X)\\ &= \int f(X^t)|\det X|^n\,d\lambda(X)\\ &= \int f(X)|\det X^t|^n\,d\lambda(X)\\ &= \int f(X)|\det X|^n\,d\lambda(X)\\ &= \int f(X)\,d\mu(X)\end{aligned}$$

for $f \in C_c(\mathrm{GL}(n,\mathbb{R}))$ and thus one has right invariance. □

21. Let G be the group of invertible upper triangular matrices. Thus G is collection of all real $n \times n$ matrices with 0's below the diagonal having nonzero diagonal elements. This is naturally identified with an open subset of $\mathbb{R}^k$ where $k = \frac{1}{2}n(n+1)$. With the relative topology and matrix multiplication, this is a locally compact Hausdorff group. Define a homomorphism H on G into $\mathbb{R}^+$ by

$$H(A) = |A_{11}^n A_{22}^{n-1} \cdots A_{nn}^1|.$$

Show a left Haar integral on G is given by

$$I(f) = \int \frac{f(X)}{H(X)}\,dX$$

where dX is Lebesgue measure on G. Moreover, show the modular function Δ is given by

$$\Delta(X) = \prod_{i<j} \left|\frac{X_{j,j}}{X_{i,i}}\right|.$$

Solution: We have

$$\begin{aligned}\int f(C^{-1}X)\frac{1}{H(X)}\,dX &= \int f(X)\frac{1}{H(CX)}\,d(CX)\\ &= \frac{1}{H(C)}\int f(X)\frac{1}{H(X)}\,d(CX).\end{aligned}$$

Now the nonzero entries in the kth row of CX are $(CX)_{k,l}$ where $l \geqslant k$. But when $k \leqslant l$,

$$(CX)_{k,l} = c_{k,k}x_{k,l} + c_{k,k+1}x_{k+1,l} + \cdots + c_{k,l}x_{l,l}.$$

This implies the Jacobian matrix of the transformation $X \mapsto CX$ in terms of the coordinates $x_{k,l}$ for $k \leqslant l$ is lower triangular and for each coordinate $x_{k,l}$, k fixed, there are $n+1-k$ entries with $c_{k,k}$ on the diagonal. All this together shows the Jacobian of $X \mapsto CX$ is $\prod_{k=1}^{n} c_{kk}^{n+1-k}$. Hence $d(CX) = |H(C)|\, dX$ and we have left invariance.

Now for the modular function. We have

$$\int \frac{f(XC^{-1})}{H(X)}\, dX = \Delta(C) \int \frac{f(X)}{H(X)}\, dX.$$

So

$$\int \frac{f(X)}{H(XC)}\, d(XC) = \Delta(C) \int \frac{f(X)}{H(X)}\, dX.$$

Thus $\frac{d(XC)}{H(C)} = \Delta(C)\, dX$. Now the possible nonzero elements of the lth column of XC are

$$(XC)_{1,l},\ (XC)_{2,l}, \ldots, (XC)_{l,l}.$$

Since

$$(XC)_{k,l} = \sum_{r=k}^{l} x_{k,r}c_{r,l}\,,$$

we see $x_{k,l}$ for $k \leqslant l$ is multiplied by $c_{l,l}$. This implies for each l, there are l entries in the triangular Jacobian matrix for the transformation $X \mapsto XC$ with $c_{l,l}$ on the diagonal. This gives

$$d(XC) = |c_{1,1}c_{2,2}^2 \cdots c_{n,n}^n|\, dX.$$

So

$$\Delta(C) = \frac{|c_{1,1}c_{2,2}^2 \cdots c_{n,n}^n|}{H(C)} = \frac{|c_{1,1}c_{2,2}^2 c_{3,3}^3 \cdots c_{n,n}^n|}{|c_{1,1}^n c_{2,2}^{n-1} c_{3,3}^{n-2} \cdots c_{n,n}|} = |c_{1,1}^{1-n} c_{2,2}^{3-n} \cdots c_{n-1,n-1}^{n-3} c_{n,n}^{n-1}|.$$

□

22. Show an infinite dimensional Hilbert space with the weak topology is an example of a σ-compact Hausdorff space X which is not locally compact.

Solution:

It is well known that every locally compact topological vector space is finite dimensional. For completeness, we give the following argument:

Let X be a Hausdorff topological vector space and suppose V is an open neighborhood of 0 whose closure is compact. Since $x + \frac{1}{2}V$ for $x \in \bar{V}$ cover $\bar{V}$, $\bar{V} \subseteq \cup_{i=1}^{l}(x_i + \frac{1}{2}V)$ for some finite sequence $x_1, x_2, \ldots, x_l$. Let Y be the linear span of the vectors x_i. Y is finite dimensional and thus is a closed subset of

X. Let $x \in X$. Choose $\lambda \neq 0$ and $\frac{1}{\lambda}x \in \bar{V}$. We claim $\frac{1}{\lambda}x \in Y + \frac{1}{2^n}V$ for every n. This is clearly true for $n = 1$. If it is true for n, then $\frac{1}{\lambda}x \in Y + \frac{1}{2^n}V \subseteq Y + \frac{1}{2^n} \cup_{i=1}^{n} (x_i + \frac{1}{2}V) \subseteq Y + \frac{1}{2^{n+1}}V$. By induction, the claim holds. Let U be a neighborhood of 0. Note there is a $\delta > 0$ such that $\mu V \subseteq U$ if $|\mu| < \delta$. Indeed, for each $x \in \bar{V}$, there is a $\delta_x > 0$ and a neighborhood N_x of x with $\mu N_x \subseteq U$ if $|\mu| < \delta_x$. Take a finite subcover N_{x_i} and then set δ equal to the smallest δ_{x_i}. Thus $\frac{1}{2^n}v_n \in U$ for large n. So every neighborhood $\frac{1}{\lambda}x - U$ of $\frac{1}{\lambda}x$ meets Y. This implies $\frac{1}{\lambda}x$ is a limit point of Y and since Y is closed $\frac{1}{\lambda}x \in Y$. Thus x is in Y and we see X is finite dimensional.

Now the weak topology on a Hilbert space is a Hausdorff locally convex topology and thus is locally compact if and only if $\mathcal{H}$ is finite dimensional. However, the balls $\{x \mid ||x|| \leqslant n\}$ for $n = 1, 2, \ldots$ are compact in the weak topology and cover $\mathcal{H}$. So if $\mathcal{H}$ is infinite dimensional, $\mathcal{H}$ in the weak topology is σ-compact but not locally compact.

□

23. Show every σ-compact open subset of a locally compact Hausdorff space X is a countable union of compact G_δ subsets and hence is Baire.

Solution: Let $U = \cup K_n$ where each K_n is a compact subset. By Proposition 5.23, there is for each n a continuous function f_n which satisfies $0 \leqslant f_n \leqslant 1$, f_n has compact support contained in U and $f_n(x) = 1$ for $x \in K_n$. Set $F_n = \{x \mid f_n(x) \geqslant \frac{1}{2}\} = \cap_k \{x \mid f_n(x) > \frac{1}{2} - \frac{1}{k}\}$. We see F_n is a compact G_δ for each n. Now $K_n \subseteq F_n \subseteq U$ and thus $U = \cup F_n$. □

24. Let W be a σ-compact open subset of a locally compact Hausdorff space X. Show every Baire subset of the topological space W is a Baire subset of X.

Solution: Let $\mathcal{A}$ be the collection of subsets E of W that are Baire subsets of X. Since W is a Baire in X (previous exercise), we see $\mathcal{A}$ contains W and is closed under complements in W. Being closed under countable unions shows $\mathcal{A}$ is a σ-algebra. Also every compact G_δ subset of W is a Baire set in X and thus belongs to $\mathcal{A}$. Hence every Baire subset of W is a Baire set in X. □

25. Let X be an uncountable set with the discrete topology. Give $X \times X$ the product topology. Show the σ-algebra of Baire sets for $X \times X$ is strictly smaller than $\mathcal{A} \times \mathcal{A}$ where $\mathcal{A}$ is the σ-algebra of Baire sets on X.

Solution:

X is a uncountable set with the discrete topology. As seen in Exercise 6.1.9, the Baire sets on X are the subsets which are countable or have countable complement. Now the product topology on $X \times X$ is the discrete topology. So the Baire sets for this space are the countable or co-countable subsets. This implies $\{x_0\} \times X$ is not Baire for $X \times X$ for it and its complement are both uncountable. But $\{x_0\} \times X$ is in the product algebra $\mathcal{A} \times \mathcal{A}$.

□

26. Let X and Y be locally compact Hausdorff spaces and consider $X \times Y$ with the product topology. Let $\mathcal{A}$ be the algebra of Borel sets on X and $\mathcal{B}$ be the algebra of Borel sets for Y. Show $\mathcal{A} \times \mathcal{B}$ need not be the σ-algebra of Borel subsets for $X \times Y$. (**Hint:** Let $X = Y$ be a set with cardinality greater than $\mathfrak{c}$ with the discrete topology. Show $D = \{(x,x) \mid x \in X\}$ is not in the product algebra by first establishing if not there are countably many measurable rectangles $A_i \times B_i$ such that D is in the smallest σ-algebra containing these rectangles.)

Solution: Let $\mathcal{W}$ be the union of all σ-algebras $\mathcal{C}$ on $X \times X$ for which there is a countable sequence A_i of Borel subsets of X such that $\mathcal{C}$ is generated by the countably many rectangles $A_i \times A_j$. Clearly $\mathcal{W}$ is closed under complements and contains $X \times X$. Let E_k be sequence in $\mathcal{W}$. Then it by taking the countable union over k of the countable sets of form A_j for which $A_i \times A_j$ generate the σ-algebra in which E_k lies, we see $\cup E_k$ is in a σ-algebra which is generated by countably many sets of form $A_r \times A_s$. Also note $\mathcal{W} \supseteq \mathcal{A} \times \mathcal{A}$.

Assume $D \in \mathcal{A} \times \mathcal{A}$. Then a countable sequence of Borel subsets A_i of X exist for which the sets $A_i \times A_j$ generate a σ-algebra with D as an element. Define $h(x) = \{(i,j) \mid (x,x) \in A_i \times A_j\} = \{(i,j) \mid x \in A_i \cap A_j\}$. Since the collection of all subsets of $\mathbb{N} \times \mathbb{N}$ has cardinality $\mathfrak{c}$, we see h cannot be one-to-one. Thus there are x and y with $h(x) = h(y)$ and $x \neq y$. So $x \in A_i \cap A_j$ if and only if $y \in A_i \cap A_j$. In particular, $x \in A_i$ if and only if $y \in A_i$. This implies $(x,y) \in A_i \times A_j$ if and only if $(y,x) \in A_i \times A_j$.

Let $\mathcal{M}$ be the collection of all subsets U of $X \times X$ such that all or non of the points (x,x), (y,y), (x,y), (y,x) are in U. $\mathcal{M}$ is a σ-algebra and $\mathcal{M}$ contains all the sets $A_i \times A_j$. So $\mathcal{M}$ must contain D, a contradiction for D must then contain both (x,y) and (y,x). □

27. Show if μ is a Radon measure on G/H where G is a locally compact Hausdorff group and H is a closed subgroup, then the measure $x\mu$ defined by (6.1) is a Radon measure on G/H and satisfies (6.2).

Solution:

First since $yH \mapsto x^{-1}yH$ is a homeomorphism, $x^{-1}E$ is a Baire subset of G/H if and only if E is a Baire subset of G/H and $x^{-1}E$ is a compact G_δ subset if and only if E is a compact G_δ subset. Thus $x\mu(E) = \mu(x^{-1}E)$ is finite on the compact G_δ subsets and

$$\begin{aligned}\mu(x^{-1}E) &= \sup\{\mu(K) \mid K \subseteq x^{-1}E,\ K \text{ is a compact } G_\delta\}\\ &= \sup\{\mu(x^{-1}(xK)) \mid xK \subseteq E,\ K \text{ is a compact } G_\delta\}\\ &= \sup\{x\mu(K') \mid K' \subseteq E,\ K' \text{ is a compact } G_\delta\}.\end{aligned}$$

So $x\mu$ is a Radon measure.

To see $\int_{G/H} f(xyH)\,d\mu(yH) = \int_{G/H} f(yH)\,d(x\mu)(yH)$ for all $f \in C_c(G/H)$, it suffices to see $\int_{G/H} \chi_E(xyH)\,d\mu(yH) = \int_{G/H} \chi_E(yH)\,d(x\mu)(yH)$. But

$$\int \chi_E(xyH)\,d\mu(yH) = \int \chi_{x^{-1}E}(yH)\,d\mu(yH) = \mu(x^{-1}E) = (x\mu)(E).$$

□

28. Let X be a locally compact Hausdorff space and suppose U is an open subset of X with the relative topology.

(a) Show every Baire subset of U has form $B \cap U$ where B is a Baire subset of X.

(b) Let $Y = [0,1]$ with the discrete topology have one point compactification $\bar{Y}$. Consider $X = \{0,1\} \times \bar{Y}$ with the product topology. Show $\{0,1\} \times Y$ is an open subset of X, $B = \{1\} \times \bar{Y}$ is a Baire subset of X, and $B \cap (\{0,1\} \times Y)$ is not a Baire subset of $\{0,1\} \times Y$.

Solution:

First we note the compact G_δ subsets of U are the compact G_δ subsets of X which are subsets of U. This is clear for if $K = \cap U_n$ and $K \subseteq U$, then $K = \cap(U_n \cap U)$. Thus $\mathcal{A} = \{B \cap U \mid B \text{ is a Baire subset of } X\}$ is a σ-algebra on U which contains the Baire subsets of U. So $\mathcal{A}_U \subseteq \mathcal{A}$ where $\mathcal{A}_U$ is the σ-algebra of Baire subsets of U. Thus every Baire subset of U has form $B \cap U$ for some Baire subset of X. This gives (a).

We next show in the example stated, there is a Baire subset B of X such that $B \cap U$ is not a Baire subset of U. Namely $Y = [0,1]$ is open in $\bar{Y}$ where Y has the discrete topology. So $\{0,1\} \times Y$ is open in $X = \{0,1\} \times \bar{Y}$. Now we note $B = \{1\} \times \bar{Y}$ is a Baire subset of X for it is compact and open. However, $B \cap (\{0,1\} \times Y) = \{1\} \times Y$ is not a Baire subset of $\{1\} \times Y$ for as seen in the argument for Exercise 6.1.9, a subset of a discrete topological space is Baire if and only if it is countable or co-countable. So we have (b).

□

29. Let H be a closed subgroup of a locally compact Hausdorff group G. Show if f is a Baire measurable function on G and F is a Baire measurable function on G/H, the function $f(F \circ \kappa)$ is a Baire function on G.

Solution: Let $f \in C_c(G)$. Let K be a compact G_δ subset of G/H. By Proposition 5.23, there is a function $F \in C_c(G/H)$ with $0 \leqslant F \leqslant 1$ and $F^{-1}(1) = K$. The sequence $F_n = F^n$ of nonnegative functions in $C_c(G/H)$ decreases pointwise to χ_K. Thus $(F_n \circ \kappa)f \in C_c(G)$ and $(F_n \circ \kappa)f \to (\chi_K \circ \kappa)f$ pointwise. So $(\chi_K \circ \kappa)f$ is Baire measurable. Now consider the collection $\mathcal{A}$ of all Baire subsets E of G/H for which $(\chi_E \circ \kappa)f$ is Baire measurable on G. This collection is closed under complements and countable disjoint unions. Since $(\chi_E \circ \kappa)(\chi_{E'} \circ \kappa) = \chi_{E \cap E'} \circ \kappa$, we see this collection is an algebra and hence consequently a σ-algebra. Since the compact G_δ's of G/H are in this algebra, we have $(\chi_E \circ \kappa)f$ is a Baire function on G for all Baire subsets E of G/H.

Now if K is compact G_δ subset of G, again we know there is a decreasing sequence $\{f_n\}$ in $C_c(G)$ such that $f_n \to \chi_K$ pointwise. Thus if E is a Baire subset of G/H, each $(\chi_E \circ \kappa)f_n$ is a Baire function on G and $(\chi_E \circ \kappa)f_n \to (\chi_E \circ \kappa)\chi_K$ pointwise. Thus $(\chi_E \circ \kappa)\chi_K$ is a Baire function on G. Arguing as before but using the subsets F of G for which $(\chi_E \circ \kappa)\chi_F$ are Baire as the algebra $\mathcal{A}$ shows $(\chi_E \circ \kappa) \circ \chi_F$ is a Baire function on G for each Baire subset E of G/H and any Baire subset F of G. One finishes the argument by taking simple functions which converge pointwise to f and F. □

30. Show every σ-compact locally compact Hausdorff space X is paracompact. (**Hint:** Use Exercise 5.2.15.)

Solution: Exercise 5.2.15 gives a sequence V_i of open subsets with $X = \cup V_i$ where $\bar{V}_i \subseteq V_{i+1}$ and all $\bar{V}_i$ are compact. Thus $X = \cup_{i=1}^{\infty}(V_{i+1} - \bar{V}_{i-1})$ where we are using $V_{-1} = V_0 = \emptyset$. Now let $\mathcal{F}$ be an open cover of X. For each i choose a finite subcover $\mathcal{F}_i$ for $\bar{V}_{i+1}$. Then the sets $W \cap (V_{i+1} - \bar{V}_{i-1})$ for $W \in \mathcal{F}_i$ form an open cover for $V_{i+1} - \bar{V}_{i-1}$ subordinate to the open cover $\{W \cap (V_{i+1} - \bar{V}_{i-1}) \mid W \in \mathcal{F}\}$. Then the family $\{W \cap (V_{i+1} - \bar{V}_{i-1}) \mid W \in \mathcal{F}_i, i = 0, 1, \ldots\}$ is a locally finite open cover of X subordinate to $\mathcal{F}$. □

31. Let G be a locally compact Hausdorff group.

(a) Show the subgroup generated by a compact neighborhood of e is σ-compact and is open and closed.

(b) Show if H is a closed subgroup of G, then G/H is paracompact. (**Hint:** Show G/H is a disjoint union of open and closed σ-compact sets.)

Solution: Let N be a compact neighborhood of e. Set $U = N \cup N^{-1}$ and let $G_0 = \cup_k U^k$. Then G_0 is a subgroup. Also if $y \in U^k$, then U^{k+1} contains an open subset containing y. Thus G_0 is an open and thus closed subgroup of G. Moreover, G_0 is σ-compact. This gives (a).

For (b), we consider the subsets G_0xH of G/H. We note they are open since the map $g \mapsto gH$ is open; they are also σ-compact. Moreover, these sets partition G/H. Indeed, if $G_0xH \cap G_0yH \neq \emptyset$, then $g_1xH = g_2yH$ for some g_1 and g_2 in G_0. Consequently, $G_0xH = G_0g_1xH = G_0g_2yH = G_0yH$. Thus the sets G_0xH are open and closed. Thus if V_α is an open cover of G/H, the open sets $V_\alpha \cap G_0xH$ over α and x form a refinement which cover G/H. For each G_0xH, since G_0xH is σ-compact, there is a locally finite open refinement $W_{G_0xH,i}$ which cover G_0xH. The sets $W_{G_0xH,i}$ over all the i for all the G_0xH then is a locally finite open cover of G/H which refines the cover V_α. □

32. Suppose X is a locally compact Hausdorff space which is a union of disjoint open Baire subsets X_α of X.

(a) Show the Baire sets for X consists of all those sets E such that $E \cap X_\alpha$ is Baire in X_α for all α and either $E \cap X_\alpha \neq \emptyset$ for countably many α or $(X - E) \cap X_\alpha \neq \emptyset$ for countably many α.

(b) Show a measure μ on X is a Radon measure if and only if there is for each α a Radon measure μ_α on X_α such that

$$\mu = \sum_\alpha \mu_\alpha$$

(c) A real valued function h on X is said to be Baire decomposable if $h|_{X_\alpha}$ is Baire measurable for each α. If $h \geqslant 0$ and μ is a Radon measure on X, then $\int h\,d\mu$ is defined to be $\sum_\alpha \int_{X_\alpha} h\,d\mu$. Assume each X_α is σ-compact. Show if ν and μ are Radon measures on X and $\nu \ll \mu$, then there is a Baire decomposable $f \geqslant 0$ on X such that

$$\nu(E) = \int_E f\,d\mu = \int \chi_E f\,d\mu$$

for all Baire subsets E of X.

Solution: To see (a), it is easy to check the family of such E is a σ-algebra. It contains the Baire sets, for if K is a compact G_δ in X, then $K \cap X_\alpha$ is a compact G_δ in X_α for all α and thus is Baire in X_α for all α. Also since the X_α are an open cover of K, $X_\alpha \cap K \neq \varnothing$ for only finitely many α. Thus this σ-algebra contains the Baire sets. Moreover, since each X_α is a closed Baire G_δ subset of X, Lemma 6.6 shows each Baire subset of the space X_α is a Baire set in X. Thus countable unions and complements of countable unions of Baire sets individually from different X_α are Baire in X. But these sets are this σ-algebra.

Now if μ is a Radon measure on X, then since each X_α is a closed and open Baire subset of X, the measure μ_α on X given by $\mu_\alpha(E) = \mu(E \cap X_\alpha)$ is a Radon measure on X_α. Moreover, if E is a Baire subset of X, then by (a), $E \cap X_\alpha$ is nonempty for only countably many α or $(X - E) \cap X_\alpha$ is nonempty for countably many α.

Now assume E is a Baire subset of X with $\mu(E) = \infty$. Then for each $M > 0$, there is a compact G_δ subset K of E with $\mu(K) > M$. Since the X_α form an open cover of X, there is a finite subset F so that $K \subseteq \cup_{\alpha \in F} X_\alpha$. Hence K is the disjoint union of the compact G_δ subsets $K \cap X_\alpha$ and so $\sum_\alpha \mu_\alpha(E) = \sum_\alpha \mu(E \cap X_\alpha) \geqslant \sum_{\alpha \in F} \mu(K \cap X_\alpha) = \mu(K) > M$. This implies $\sum \mu_\alpha(E) = \infty$.

Next assume $\mu(E) < \infty$. We note $\mu(E \cap X_\alpha) > 0$ for only countably many α for otherwise $\sum_\alpha \mu(E \cap X_\alpha) = \infty$ and then again if $M > 0$ we can find a compact G_δ subset K of E with $\mu(K) = \sum_{\alpha \in F} \mu(K \cap X_\alpha) > M$ for some finite set F of α. Consequently, $\mu(E) > M$ for all M and we see $\mu(E) = \infty$. Let $C = \{\alpha \mid \mu(E \cap X_\alpha) > 0\}$. Set $X' = X - \cup_{\alpha \in C} X_\alpha$. X' is a Baire set. If $K \subseteq E \cap X'$, then $\mu(K \cap X_\alpha) = 0$ for all $\alpha \notin C$. Since the X_α for $\alpha \notin C$ is an open cover of K, there exist $\alpha_1, \alpha_2, \ldots, \alpha_m$ such that $K \subseteq \cup_{i=1}^m X_{\alpha_i}$. Thus $\mu(K) = \sum \mu(K \cap X_{\alpha_i}) = 0$. This implies $\mu(E \cap X') = 0$ and so $\mu(E) = \mu(E \cap (\cup_{\alpha \in C} X_\alpha)) + \mu(E \cap X') = \sum_{\alpha \in C} \mu(E \cap X_\alpha) = \sum_\alpha \mu_\alpha(E)$. We hence have (b).

Now (c) follows easily. Note if $\nu \ll \mu$, then $\nu_\alpha \ll \mu_\alpha$ for all α. By the Radon-Nikodym Theorem, there is for each α a nonnegative Baire measurable function f_α with $\nu_\alpha(E) = \int_E d\mu_\alpha$ for all Baire subsets E of X_α. Define f by $f|_{X_\alpha} = f_\alpha$. Now $\nu(E) = \sum_\alpha \nu_\alpha(E) = \sum_\alpha \int_{E \cap X_\alpha} f_\alpha \, d\mu_\alpha = \sum_\alpha \int \chi_{E \cap X_\alpha} f \, d\mu = \sum_\alpha \int_{X_\alpha} \chi_E f \, d\mu = \int_X \chi_E f \, d\mu$. □

33. Show if μ and ν are nonzero left quasi-invariant Radon measures on a locally compact Hausdorff group G, then μ and ν are equivalent. (**Hint:** Use an open and closed σ-compact subgroup of G.)

Solution: Let G_0 be a σ-compact open closed subgroup of G. By Corollary 6.22, μ and ν are equivalent on G_0. Now we know G is a union of disjoint cosets $x_\alpha G_0$ of G. Exercise 6.1.23 implies every coset $x_\alpha G_0$ is Baire. Let E be a Baire subset of G and assume $\mu(E) = 0$. By Exercise 6.1.32, $E \cap x_\alpha G_0 \neq \varnothing$ for countably many α or $(G - E) \cap E_\alpha \neq \varnothing$ for countably many α. We have $\mu(E \cap x_\alpha G_0) = 0$ for all α. Since μ is left quasi-invariant, $\mu(x_\alpha^{-1} E \cap G_0) = 0$ for all α. Consequently, we see $\nu(x_\alpha^{-1} E \cap G_0) = 0$ for all α. Thus $\nu(E \cap x_\alpha G_0) = 0$ for all α. Now if E meets only countably many $x_\alpha G_0$, we see $\nu(E) = 0$. If this is not the case, then there are uncountably many α and the complement of E meets only countably many $x_\alpha G_0$. This implies there exists an α with $E \cap x_\alpha G_0 = x_\alpha G_0$. Consequently, $\mu(x_\alpha G_0) = 0$. Thus $\mu(G_0) = 0$. This implies μ vanishes on all compact G_δ sets K for they are covered by finitely many G_0 cosets. Hence μ would be 0; a contradiction. So $\mu(E) = 0$ implies $\nu(E) = 0$. Conversely, we have $\mu(E) = 0$ if $\nu(E) = 0$. □

34. Show any two nonzero left quasi-invariant Radon measures on G/H where G is a locally compact Hausdorff group and H is a closed subgroup are equivalent. More precisely, show the Baire sets of measure 0 in G/H are the sets whose preimages in G have Haar measure 0.

Solution:

By Exercise 6.1.31, there is a σ-compact open closed subgroup G_0 of G and the sets G_0xH are equal or disjoint open closed subsets of G/H whose union is G/H. Moreover, the sets G_0x for x in G partition G into a collection of open closed σ-compact subsets.

Now let μ be a quasi-invariant Radon measure on G/H and consider μ_x, the restriction of μ to G_0xH. Then μ_x is a G_0 quasi-invariant measure on the transitive G_0 space G_0xH. Note the stabilizer of xH is $G_0 \cap xHx^{-1}$. Now by Exercise 6.1.32, a Baire subset of G/H has μ measure 0 if and only if its intersection with each G_0xH has μ_x measure 0.

But by Corollary 6.25 which was established for σ-compact groups in the proof of Theorem 6.24, a subset E of G_0xH has μ_x measure 0 if and only if the preimage in G_0 has G_0-Haar measure 0. But Haar measure on G_0 is the restriction of Haar measure on G to G_0. Thus a subset E of G/H has μ measure 0 if and only if for each x, the G Haar measure of $\{g \in G_0 \mid gxH \in E\}$ is 0. But this is equivalent to $\{gx \mid gxH \in E,\ g \in G_0\}$ has G Haar measure 0 for all x. So E has μ measure 0 if and only if $G_0x \cap \kappa^{-1}(E)$ has G Haar measure 0 for all x. Again using Exercise 6.1.32, this occurs if and only if $\kappa^{-1}(E)$ has G Haar measure 0.

□

35. Let G and H be locally compact Hausdorff groups and suppose $\mathcal{A}$ is the σ-algebra of Baire subsets of G and $\mathcal{B}$ is the σ-algebra of Baire subsets of H. Use Exercise 6.1.32 to show the σ-algebra of Baire subsets of $G \times H$ is $\mathcal{A} \times \mathcal{B}$.

Solution: Note Exercise 6.1.25 implies this problem is incorrect. We repeat the argument in this context.

Let G be an uncountable group with the discrete topology. It is locally compact and Hausdorff. Moreover, a subset E is Baire if and only if E or E^c is countable. Now $G \times G$ also has the discrete topology. So a subset of $G \times G$ is Baire if it is countable or co-countable. Now $G \times \{e\}$ is in $\mathrm{Ba}(G) \times \mathrm{Ba}(G)$ and is not countable or co-countable in $G \times G$.

□

36. Let F be the free abelian group with $\mathfrak{c}$ generators and with the discrete topology. Show multiplication $(x, y) \mapsto xy$ is not Baire measurable on F.

Solution: The Baire sets for both F and $F \times F$ are the countable or co-

countable sets. Let A be a well ordered index set of cardinality $\mathfrak{c}$. Let E be the Baire subset of F consisting of 1. If $w \in F$, then there is a unique n and unique $a_1 < a_2 < \cdots < a_n$ and $(k_1, k_2, \ldots, k_n) \in \mathbb{Z}^n$ with $w = a_1^{k_1} a_2^{k_2} \cdots a_n^{k_n}$. If $f(x, y) = xy$, then $f^{-1}(E) = \{(w, w^{-1}) \mid w = a_1^{k_1} a_2^{k_2} \cdots a_n^{k_n}\}$. Thus $f^{-1}(E)$ is neither countable nor co-countable. □

Exercise Set 6.2

1. Let π be a representation of G on a locally convex topological vector space V. Let V^* be the dual space of V with the weak $*$ topology, see Exercise 2.1.17. Define $\check{\pi}$ on V^* by

$$\langle v, \check{\pi}(g)f \rangle = \langle \pi(g^{-1})v, f \rangle;$$

i.e., $\check{\pi}(g)f(v) = f(\pi(g^{-1}v))$. Show $\check{\pi}$ is a representation of G. It is the representation *contragredient* to the representation π.

Solution:

First note $v \mapsto \check{\pi}(g)f(v) = f(\pi(g^{-1})v)$ is a continuous linear transformation of V being a composition of continuous linear functions. So $\check{\pi}(g)f \in V^*$ for $f \in V^*$. It is easy to check $\check{\pi}(g)$ is linear. Since

$$\begin{aligned} \langle v, \check{\pi}(g_1)\check{\pi}(g_2)f \rangle &= \langle \pi(g_1^{-1})v, \check{\pi}(g_2)f \rangle \\ &= \langle \pi(g_2^{-1})\pi(g_1^{-1})v, f \rangle \\ &= \langle \pi((g_1 g_2)^{-1})v, f \rangle \\ &= \langle v, \check{\pi}(g_1 g_2)f > \end{aligned}$$

and

$$\langle v, \check{\pi}(e)f \rangle = \langle \pi(e)v, f \rangle = \langle v, f \rangle,$$

we have $\check{\pi}(g_1)\check{\pi}(g_2)f = \check{\pi}(g_1 g_2)f$ and $\check{\pi}(e)f = f$. Thus

$$\check{\pi}(g)\check{\pi}(g^{-1}) = \check{\pi}(g^{-1})\check{\pi}(g) = \check{\pi}(e) = I$$

and we see $\check{\pi}(g) \in \mathrm{GL}(V^*)$. Consequently, $\check{\pi}$ is a homomorphism of G into $\mathrm{GL}(V^*)$.

The weak $*$ topology on V^* is given by the seminorms $|\cdot|_v$ defined by $|f|_v = |\langle v, f \rangle|$. Now $|\check{\pi}(g)f - \check{\pi}(e)f|_v = |\langle \pi(g^{-1})v - v, f \rangle|$. Since $g \mapsto g^{-1}$ and $g \mapsto \pi(g)v$ is continuous and f is continuous on V, one see if $\epsilon > 0$, there is a neighborhood U of e in G such that

$$|\check{\pi}(g)f - f|_v < \epsilon \text{ if } g \in U.$$

Thus $g \mapsto \check{\pi}(g)f$ is continuous at e for any $f \in V^*$. Now $g \mapsto \check{\pi}(g)f = \check{\pi}(gg_0^{-1})\check{\pi}(g_0)f$ is continuous at g_0 for any f, for $g \mapsto gg_0^{-1}$ is continuous at g_0 and $g \mapsto \check{\pi}(g)\check{\pi}(g_0)f$ is continuous at $g = e$. Concluding, $\check{\pi}$ is a representation of G on V^*.

□

2. Let π be a unitary representation of a topological group G on a Hilbert space $\mathcal{H}$. Recall the dual space $\mathcal{H}^*$ of $\mathcal{H}$ is the Hilbert space $\bar{\mathcal{H}}$ under the identification $\bar{v} \to f_v$ where $f_v(w) = (w, v)$. The representation $\bar{\pi}$ of G is defined by $\bar{\pi}(g)\bar{v} = \overline{\pi(g)v}$. Let $\check{\pi}$ be the contragredient representation to π. Show $\check{\pi}$ and $\bar{\pi}$ are equivalent representations of G.
Solution: Define $T : \bar{\mathcal{H}} \to \mathcal{H}^*$ by $T(\bar{v}) = f_v$. We know T is a one-to-one linear isomorphism. For $\epsilon > 0$, one has $|\bar{v}_1 - \bar{v}_2| < \epsilon$ if and only if $|(w, v_1) - (w, v_2)| < \epsilon$ for all $w \in \mathcal{H}$ with $|w| \leqslant 1$ if and only if $|f_{v_1} - f_{v_2}|_w < \epsilon$ for all w with $|w| \leqslant 1$. This implies both T and T^{-1} are continuous. Thus T is a linear topological isomorphism of $\bar{\mathcal{H}}$ onto $\mathcal{H}^*$. Also one has

$$\begin{aligned}
T\bar{\pi}(g)T^{-1}f_v(w) &= T(\bar{\pi}(g)\bar{v})(w) \\
&= T(\overline{\pi(g)v})(w) \\
&= f_{\pi(g)v}(w) \\
&= (w, \pi(g)v) \\
&= (\pi(g^{-1})w, v) \\
&= f_v(\pi(g^{-1})w) \\
&= \check{\pi}(g)f_v(w)\,.
\end{aligned}$$

So $T\bar{\pi}(g)T^{-1} = \check{\pi}(g)$ for all g and we see $\bar{\pi}$ and $\check{\pi}$ are equivalent. □

3. Let V be the space of analytic functions f on $\mathbb{R}$ with seminorms $|\cdot|_K$ where $|f|_K = \max_{x \in K} |f(x)|$ and K is a compact subset. Show the representation m given by $m(x)f(y) = e^{2\pi ixy}f(y)$ has no one dimensional invariant subspaces while the representation λ where $\lambda(x)f(y) = f(y - x)$ has an uncountable number of distinct one dimensional invariant subspaces.
Solution: Suppose f gives a one dimensional invariant subspace under m. Thus for each x, there is a scalar $\Lambda(x)$ so that $m(x)f = \Lambda(x)f$. Hence $e^{2\pi iyx}f(y) = \Lambda(x)f(y)$ for all x and y. Where $f(y) \neq 0$, we see $\Lambda(x) = e^{2\pi iyx}$. Since y where $f(y) \neq 0$ is open, one must have $f(y) = 0$ for all y. Now $\lambda(x)e_z(y) = e_z(y - x) = e^{2\pi iz(y-x)} = e^{-2\pi ixz}e_z(y)$. □

4. Show if $\mathcal{H}$ is a Hilbert space and $\{v_n\}_{n=1}^{\infty}$ is a sequence in $\mathcal{H}$ satisfying $(v_n, w) \to (v, w)$ for all $w \in \mathcal{H}$ and $||v_n|| \to ||v||$, then $v_n \to v$ in $\mathcal{H}$.
Solution:
Note

$$\begin{aligned}
||v_n - v||^2 &= (v_n - v, v_n - v) \\
&= (v_n, v_n) - (v_n, v) - (v, v_n) + (v, v) \\
&= ||v_n||^2 - (v_n, v) - \overline{(v_n, v)} + ||v||^2 \\
&\to ||v||^2 - (v, v) - (v, v) + ||v||^2 = 0
\end{aligned}$$

as $n \to \infty$. So $v_n \to v$ in $\mathcal{H}$.

□

5. Let $\mathcal{H}$ be a Hilbert space. The weak operator topology on $\mathcal{B}(\mathcal{H})$ is defined by the seminorms $|\cdot|_{v,w}$ where $v, w \in \mathcal{H}$ and

$$|T|_{v,w} = |\langle Tv, w\rangle|.$$

Show a sequence (more generally a net) T_n converges to T in the strong operator topology if and only if $T_n \to T$ weakly and $||T_n v|| \to ||Tv||$ for all $v \in \mathcal{H}$.
Solution: Assume $T_n \to T$ weakly and $||T_n v|| \to ||Tv||$ for all v. Then

$$\begin{aligned}||T_n v - Tv||^2 &= (T_n v, T_n v) - (T_n v, Tv) - (Tv, T_n v) + (Tv, Tv)\\ &= ||T_n v||^2 - (T_n v, Tv) - (Tv, T_n v) + ||Tv||^2\\ &\to ||Tv||^2 - (Tv, Tv) - (Tv, Tv) + ||Tv||^2\\ &= 0.\end{aligned}$$

□

6. Show a homomorphism π of G into the unitary group of a Hilbert space is strongly continuous if and only if π is weakly continuous.
Solution:

Let π be strongly continuous. Recall the seminorms $|T|_{v,w} = |(Tv, w)|$ define the weak operator topology on $\mathcal{B}(\mathcal{H})$ and the strong operator topology is defined by the seminorms $|T|_v = ||Tv||$. Since $|T|_{v,w} \leqslant |T|_v\, ||w||$, one sees the weak operator topology is weaker than the strong topology. So π is weakly continuous.

Now suppose π is weakly continuous. To show π is strongly continuous, we need only show if a net g_i converges to g in the topology of G, then $\pi(g_i)$ converges to $\pi(g)$ in the strong topology. But by Exercise 6.2.5, since $||\pi(g_i)v|| = ||v|| = ||\pi(g)v||$, we see $\pi(g_i)$ converges to $\pi(g)$ in the strong operator topology.

□

7. Let V be a locally convex Hausdorff topological vector space. Let $\mathcal{F} = \{|\cdot|_i \mid i \in I\}$ be the collection of all continuous seminorms on V. Let $\mathcal{B}(V)$ be the algebra of continuous linear transformations of V. Define seminorms $|\cdot|_{i,v}$ on $\mathcal{B}(V)$ by

$$|T|_{i,v} = |Tv|_i.$$

The topology defined by these seminorms where $v \in V$ and $i \in I$ is called the *strong operator* topology on $\mathcal{B}(V)$.

(a) Show $\mathcal{B}(V)$ is a Hausdorff.

(b) Show a homomorphism $\pi : G \to \mathrm{GL}(V)$ is continuous in the strong operator topology if and only if $g \mapsto \pi(g)v$ is continuous for each $v \in V$.

(c) Show $(A, B) \mapsto AB$ from $\mathcal{B}(V) \times \mathcal{B}(V)$ into $\mathcal{B}(V)$ is continuous in the strong operator topology.

(d) Let $\mathcal{U}$ be the collection of all sets of form $U \cap U^{-1}$ where U is a open neighborhood of I in the relative strong topology from $\mathcal{B}(V)$. Show $\mathcal{U}$ has the properties of Theorem 5.11 and thus defines a neighborhood base at I for a topological group topology on $\mathrm{GL}(V)$. Then show with this topology, one obtains a topological Hausdorff group.

Solution:

For (a), recall a locally convex topological vector space is Hausdorff if and only if it is separated, and a locally convex topological vector space is separated if and only if for each $v \neq 0$, there is a continuous seminorm $|\cdot|$ with $|v| > 0$. So if $T \in \mathcal{B}(V)$ and $T \neq 0$, there is a $v \in V$ with $Tv \neq 0$. Since V is Hausdorff, there is an i with $|Tv|_i > 0$. So $|T|_{i,v} > 0$ and we see $\mathcal{B}(V)$ is separated.

For (b), recall $\mathrm{GL}(V)$ is the subspace of $\mathcal{B}(V)$ consisting of the T which are continuous and have continuous inverses. Note $g \mapsto \pi(g)v$ is continuous for all v if and only if $g \mapsto \pi(g)$ is strongly continuous for $|\pi(g)v - \pi(g_0)v|_i$ can be made small for each i for g near g_0 if and only if $|\pi(g) - \pi(g_0)|_{i,v}$ can be made small.

Now for (c) one has $|AB - A_0B_0|_{i,v} = |ABv - AB_0v|_i + |(A - A_0)B_0v|_i$. Since A is continuous, there is an i_0 such that $|Av|_i < 1$ when $|v|_{i_0} < 1$. Thus $|ABv - AB_0v|_i = |A(B - B_0)v|_i \leqslant |(B - B_0)v|_{i_0}$. Consequently,

$$|AB - A_0B_0|_{i,v} \leqslant |Bv - B_0v|_{i_0} + |(A - A_0)B_0v|_i$$

and thus $|AB - A_0B_0|_{i,v} \leqslant |B - B_0|_{i_0,v} + |A - A_0|_{i,B_0v}$. This implies multiplication is continuous.

Now take $\mathcal{U}$ to be the collection of all sets of form $U \cap U^{-1}$ where U is a open neighborhood of I in the relative strong topology from $\mathcal{B}(V)$. We show $\mathcal{U}$ has the properties of Theorem 5.11 and thus defines a neighborhood base at I for a topological group topology on $\mathrm{GL}(V)$.

Note (a), (b), and (d) of Theorem 5.11 are immediate. For (c) of Theorem 5.11, we have already shown $(A, B) \mapsto AB$ is continuous from $\mathcal{B}(V) \times \mathcal{B}(V)$ into $\mathcal{B}(V)$. So if W is an open neighborhood of I in $\mathcal{B}(V)$, there are strongly open neighborhoods W_1 and W_2 of I in $\mathcal{B}(V)$ with $W_1W_2 \subseteq W$ and $W_2W_1 \subseteq W$. Thus if $G = \mathrm{GL}(V)$, $(W_1 \cap G)(W_2 \cap G) \subseteq W \cap G$ and $(W_2 \cap G)(W_1 \cap G) \subseteq W$. Taking $U_1 = W_1 \cap G$, $U_2 = W_2 \cap G$, and $U = W \cap G$, we see

$$\begin{aligned}(U_1 \cap U_1^{-1})(U_2 \cap U_2^{-1}) &\subseteq (U_1U_2) \cap (U_1U_2^{-1}) \cap (U_1^{-1}U_2) \cap (U_1^{-1}U_2^{-1}) \\ &\subseteq (U_1U_2) \cap (U_2U_1)^{-1} \\ &\subseteq U \cap U^{-1}.\end{aligned}$$

Let U be a strongly open subset of $\mathcal{B}(V)$ and let $A \in \mathrm{GL}(V)$. We show AU and UA are strongly open. Indeed, let $B_0 \in U$. Since U is open, there are $v_1, v_2, \ldots, v_k$ and $i_1, i_2, \ldots, i_k$ in I so that $B \in U$ if $|Bv_j - B_0v_j|_{i_j} < 1$ for $j = 1, 2, \ldots, k$. Set $w_j = A^{-1}v_j$. Suppose $|B'w_j - B_0Aw_j|_{i_j} < 1$ for $j = 1, 2, \ldots, k$. Then $|B'A^{-1}v_j - B_0v_j|_{i_j} < 1$. So $B'A^{-1} \in U$. Thus $B' \in UA$. So UA is open.

Consider $B_0 \in U$ and $AB_0 \in AU$. Since A^{-1} is continuous, for the seminorms $|\cdot|_{i_j}$ on V, $j = 1, 2, \ldots, k$, there are other seminorms $|\cdot|_{r_s}$ where $s = 1, 2, \ldots, m$ and $r_s \in I$ such that if $|v|_{r_s} < 1$ for $s = 1, 2, \ldots, m$, then $|A^{-1}v|_{i_j} < 1$ for $j = 1, 2, \ldots, k$.

Thus if $|B'v_j - AB_0v_j|_{r_s} < 1$ for $j = 1, 2, \ldots, k$ and $s = 1, 2, \ldots, m$, one has $|A^{-1}B'v_j - B_0v_j|_{i_j} < 1$ for $j = 1, 2, \ldots, k$ and consequently $A^{-1}B' \in U$. Hence $B' \in AU$ under the condition $|B' - AB_0|_{r_s, v_j} < 1$ for $s = 1, 2, \ldots, m$ and $j = 1, 2, \ldots, k$. So B' is interior and AU is open.

Now let W be a strongly open subset of $\mathcal{B}(V)$ containing I. We now know if $A \in \mathrm{GL}(V)$, AWA^{-1} is a strongly open subset of $\mathcal{B}(V)$ containing I. Thus if $U = W \cap \mathrm{GL}(V)$, then

$$\begin{aligned} A(U \cap U^{-1})A^{-1} &= (AUA^{-1}) \cap (AU^{-1}A^{-1}) \\ &= (AUA^{-1}) \cap (AUA^{-1})^{-1} \in \mathcal{U}. \end{aligned}$$

Thus $\mathcal{U}$ satisfies (d) of Theorem 5.11 and the topology consisting of the subsets W of $\mathrm{GL}(V)$ so that if $A \in W$, there is a $E \in \mathcal{U}$ with $AE \subseteq W$ makes $\mathrm{GL}(V)$ a topological group.

To see it is Hausdorff, by Lemma 5.17, it suffices to show $\{I\}$ is a closed set. But if $A \neq I$ is in $\mathrm{GL}(V)$, there is a strongly open subset W of $\mathcal{B}(V)$ containing I which doesn't contain A. So if $U = W \cap \mathrm{GL}(V)$, then $A \notin U \cap U^{-1}$ and thus $A(U \cap U^{-1})$ is a neighborhood of A not containing I. So the complement of $\{I\}$ is closed.

□

Exercise Set 6.3

1. Let G be a compact Hausdorff group. Suppose π is a representation of G on a Banach space X. Using the finiteness of Haar measure and the Uniform Boundedness Principle, show there is an equivalent norm on X such that each $\pi(g)$ is an isometry.

Solution: Take $m(G) = 1$. Define $|v| = \int_G ||\pi(g)v||\, dg$. Since G is compact, $g \mapsto ||\pi(g)v||$ is bounded for each v. By the uniform boundedness principle, there is an M with $||\pi(g)|| \leqslant M$ for all $g \in G$. So $|v| = \int ||\pi(g)v||\, dg \leqslant \int M||v||\, dg \leqslant M||v||$. Now suppose v_n is Cauchy in $|\cdot|$. Then $\int_G ||\pi(g)v_k - \pi(g)v_l||\, dg \to 0$ as $k, l \to \infty$. Choose k_n so that $\int_G ||\pi(g)v_k - \pi(g)v_l||\, dg < \frac{1}{2^{2n}}$ if $k, l \geqslant k_n$. Thus $\{g \mid ||\pi(g)v_k - \pi(g)v_l|| > \frac{1}{2^n}\}$ has measure less than $\frac{1}{2^n}$ for each $k, l \geqslant k_n$. Thus $m\{g \mid ||\pi(g)v_k - \pi(g)v_l|| \leqslant \frac{1}{2^n}\} > 1 - \frac{1}{2^n}$ if $k, l \geqslant k_n$.

Set $S_n = \{g \mid ||\pi(g)v_{k_n} - \pi(g)v_{k_{n+1}}|| \leqslant \frac{1}{2^n}\}$. So $m(S_n) > 1 - \frac{1}{2^n}$. Note Note $G - (S_1 \cap S_2 \cap \cdots \cap S_m \cap \cdots) \subseteq (G - S_1) \cup \cdots \cup (G - S_m) \cup \cdots$ So $m(G - \cap S_j) < \sum(1 - (1 - \frac{1}{2^m})) = 1$. So $m(\cap S_n) > 0$. So choose $x \in \cap S_j$. Have $||\pi(x)v_{k_n} - \pi(x)v_{k_{n+1}}|| \leqslant \frac{1}{2^n}$. This implies $\pi(x)v_{k_n}$ is Cauchy. So $\pi(x)v_{k_n} \to w \in X$. So $v_{k_n} \to v = \pi(x^{-1})w$; *i.e.*, $||v_{k_n} - v|| \to 0$. Consequently $|v - v_{k_n}| = \int |\pi(g)v - \pi(g)v_{k_n}|\, dg \leqslant M||v - v_{k_n}|| \to 0$ as $n \to \infty$. So $(X, |\cdot|)$ is a Banach space. The identity mapping $(X, ||\cdot||) \to (X, |\cdot|)$ is continuous and onto. So by the open mapping theorem, it is open. Thus the inverse map is continuous.

So there is an M' with $||v|| \leqslant M'|v|$. We hence see $\frac{1}{M'}||v|| \leqslant |v| \leqslant M||v||$ for all v. □

2. Let $\mathcal{H} = \oplus \mathcal{H}_i$ be an orthogonal direct sum of Hilbert spaces $\mathcal{H}_i$. For each i, let $A_i \in \mathcal{B}(\mathcal{H}_i)$. Define linear transformation $A = \oplus A_i$ on $\mathcal{H}$ by

$$A(v_i) = (A_i v_i).$$

(a) Show A is bounded if and only if $\sup_i ||A_i|| < \infty$ and then $||A|| = \sup_i ||A_i||$.

(b) Show if $\oplus A_i$ is bounded, then $(\oplus A_i)^* = \oplus(A_i^*)$.

(c) Show A is unitary if and only each A_i is unitary.

Solution:

The norm on $\mathcal{H}$ satisfies $||v||^2 = \sum ||v_i||_i^2$. Thus A is bounded if and only if $||A||^2 = \sup_{||v||=1} \sum ||A_i v_i||_i^2 < \infty$. But $\sum ||A_i v_i||_i^2 \leqslant \sum ||A_i||^2 \, ||v_i||_i^2 \leqslant \sup_i ||A_i||^2 \sum ||v_i||_i^2 \leqslant \sup ||A_i||^2 ||v||^2$. Thus A is bounded if $||A_i||$ are bounded and $||A|| \leqslant \sup ||A_i||$. But conversely, if A is bounded, by taking $v_j = 0$ for $j \neq i$ and $v_i = w$ where $||w||_i = 1$, we see $||A||^2 \geqslant \sup_{w \in \mathcal{H}_i, \, ||w||=1} ||A_i w||_i^2 = ||A_i||^2$. Thus $||A_i||$ are bounded and $||A|| \geqslant \sup_i ||A_i||$. This gives (a).

Assume A is bounded, then since $||A_i^*|| = ||A_i||$, we see $\oplus A_i^*$ is a bounded operator on $\mathcal{H}$ and

$$\begin{aligned}(Av, w)_{\mathcal{H}} &= \sum_i (A_i v_i, w_i)_i \\ &= \sum_i (v_i, A_i^* w_i)_i \\ &= (v, \oplus A_i^*(w))_{\mathcal{H}} .\end{aligned}$$

So $(\oplus A_i)^* = \oplus A_i^*$ and (b) holds.

For (c), note A is unitary if and only if $A^* = A^{-1}$. Now taking $v_j = 0$ for $j \neq i$ and $v_i = w$ and doing this for each i and using $A^* = \oplus A_i^*$, we see $A^* = A^{-1}$ if and only if $A_i^* = A_i^{-1}$ for each i. So A is unitary if and only if each A_i is unitary. □

Exercise Set 6.4

1. Let U be a bounded linear transformation from a Hilbert space $\mathcal{H}$ into a Hilbert space $\mathcal{K}$. Show U is a partial isometry if and only if $E = U^*U$ and $F = UU^*$ are projections. Then show $E\mathcal{H} = (\ker U)^\perp$ and $F\mathcal{K} = U(\mathcal{H})$.

Solution:

By definition U is a partial isometry if U restricted to $(\ker U)^\perp$ is an isometry into $\mathcal{K}$. But $\mathcal{H} = \ker U \oplus (\ker U)^\perp$. For $v \in \mathcal{H}$, take $v = v_1 + v_2$ where $v_1 \in \ker U$

and $v_2 \in (\ker U)^\perp$.

$$\begin{aligned}(v, U^*Uv)_{\mathcal{H}} &= (Uv, Uv)_{\mathcal{K}} \\ &= (Uv_2, Uv_2)_{\mathcal{K}} \\ &= (v_2, v_2)_{\mathcal{H}} \\ &= (v_1 + v_2, v_2)_{\mathcal{H}} \\ &= (v, v_2)_{\mathcal{H}}.\end{aligned}$$

So U^*U is the orthogonal projection onto $(\ker U)^\perp$.

Next consider the decomposition of $\mathcal{K}$ given by $\mathcal{K} = \overline{U(\mathcal{H})} \oplus U(\mathcal{H})^\perp$ and let $w = w_1 + w_2$ where $w_1 \in \overline{U(\mathcal{H})}$ and $w_2 \in (U\mathcal{H})^\perp$. Note $UU^*Uv = Uv_2 = U(v_1 + v_2) = Uv$. By taking limits, we see $UU^*w = w$ if $w \in \overline{U\mathcal{H}}$. Now if $w_2 \in (U\mathcal{H})^\perp$, $(v, U^*w_2)_{\mathcal{H}} = (Uv, w_2)_{\mathcal{K}} = 0$ for all $v \in \mathcal{H}$ and so $U^*w_2 = 0$. Thus $UU^*(w_1 + w_2) = w_1$ and we see UU^* is the orthogonal projection of $\mathcal{K}$ onto $\overline{U\mathcal{H}}$.

Conversely, suppose $E = U^*U$ and $F = UU^*$ are projections. Then $(UEv, UEv)_{\mathcal{K}} = (U^*UEv, Ev)_{\mathcal{H}} = (E^2v, Ev)_{\mathcal{H}} = (Ev, Ev)_{\mathcal{H}}$ and thus U is an isometry of the range $E\mathcal{H}$ into $\mathcal{K}$. Similarly, U^* is an isometry of the range $F\mathcal{K}$ into $\mathcal{H}$. But by Exercise 2.2.13, $(U^*\mathcal{K})^\perp = \ker U$ and $(U\mathcal{H})^\perp = \ker U^*$. Now $\ker U = \ker(U^*U)$ and $\ker U^* = \ker(UU^*)$, e.g. $Uv = 0$ if and only if $(Uv, Uv') = 0$ for all $v' \in \mathcal{H}$ if and only if $(U^*Uv, v') = 0$ for all $v' \in \mathcal{H}$ if and only if $U^*Uv = 0$. So $\ker U = \ker E = (I - E)\mathcal{H}$ and $\ker U^* = \ker F = (I - F)\mathcal{K}$ So $(\ker U)^\perp = (\ker E)^\perp = E\mathcal{H}$ and $F\mathcal{K} = (\ker F)^\perp = (\ker U^*)^\perp = U(\mathcal{H})$.

□

2. Let V and W be finite dimensional vector spaces. Show that every bilinear mapping $B : V \times W \to F$ where F is a vector space can be written uniquely in the form $B(v, w) = T(v \otimes w)$ where T is a linear transformation of $V \otimes W$ into F.

Solution:

Recall $V \otimes W = \mathcal{L}(W^*, V)$. Let B be a bilinear mapping of $V \times W$ into F. Let $e_1, e_2, \ldots, e_m$ be a basis of V and $f_1, f_2, \ldots, f_n$ be a basis of W. Consider the pairing $\langle \cdot, \cdot \rangle$ on $W \times W^*$ given by $\langle w, w^* \rangle = w^*(w)$. Then if $A \in \mathcal{L}(W^*, V)$, $A = \sum_{i,j} a_{i,j} \langle f_j, \cdot \rangle e_i = \sum_{i,j} a_{i,j} e_i \otimes f_j$ for unique constants $a_{i,j}$. Define $T(A) = \sum_{i,j} a_{i,j} B(e_i, f_j)$. This is well defined for the constants $a_{i,j}$ are unique. Note one can easily check T is linear. Now $v \otimes w$ is the mapping $v \otimes w(w^*) = w^*(w)v$ and so if $v = \sum_{i=1}^m a_i e_i$ and $w = \sum_{j=1}^n b_j f_j$, one has $v \otimes w = \sum_{i,j} a_i b_j e_i \otimes f_j$. Thus $T(v \otimes w) = \sum_i \sum_j a_i b_j B(e_i, f_j) = B(v, w)$. Note T is unique since $T(e_i \otimes f_j) = B(e_i, f_j)$ and the decomposition $A = \sum a_{i,j} e_i \otimes f_j$ is unique for each A.

□

3. Show Schur's Lemma given in Proposition 6.49 also holds if π and ρ are representations of $*$ algebras on Hilbert spaces $\mathcal{H}$ and $\mathcal{K}$.

Solution:

We follow the proof of Proposition 6.49. So we have $\theta : D \to \mathcal{K}$ is a closed linear operator from a π invariant linear subspace D of $\mathcal{H}$ and $\theta\pi(x)v = \rho(x)\theta v$

for $v \in D$ and $x \in \mathcal{A}$. Again take $\tilde{D} = \{(v, \theta v) \mid v \in D\}$ with inner product $(\tilde{v}_1, \tilde{v}_2) = (v_1, v_2) + (\theta v_1, \theta v_2)$. The same argument as in the proof of 6.49 shows $\tilde{D}$ is complete.

We then define a representation $\tilde{\pi}$ of $\mathcal{A}$ on $\tilde{D}$ by

$$\tilde{\pi}(x)\tilde{v} = \widetilde{\pi(x)v} = (\pi(x)v, \theta\pi(x)v) = (\pi(x)v, \rho(x)\theta v).$$

$\tilde{\pi}$ is a $*$-representation. To see this note that $\tilde{\pi}(x)$ is a linear transformation and

$$\begin{aligned}||\tilde{\pi}(x)\tilde{v}||_{\tilde{D}}^2 &= ||\pi(x)v||^2 + ||\rho(x)\theta v||^2 \\ &\leqslant ||\pi(x)||^2 \, ||v||^2 + ||\rho(x)||^2 ||\theta v||^2 \\ &\leqslant \max\{||\pi(x)||, \, ||\rho(x)||\}^2 (||v||^2 + ||\theta v||^2) \\ &\leqslant \max\{||\pi(x)||, \, ||\rho(x)||\}^2 ||\tilde{v}||_{\tilde{D}}^2 .\end{aligned}$$

Thus each $\tilde{\pi}(x)$ is a bounded linear operator. Moreover, it is easy to check $x \mapsto \tilde{\pi}(x)$ is linear. Next note

$$\begin{aligned}(\tilde{\pi}(x)\tilde{v}, \tilde{w})_{\tilde{D}} &= (\pi(x)v, w)_{\mathcal{H}} + (\rho(x)\theta v, \theta w)_{\mathcal{K}} \\ &= (v, \pi(x^*)w)_{\mathcal{H}} + (\theta v, \rho(x^*)\theta w)_{\mathcal{K}} \\ &= (\tilde{v}, \tilde{\pi}(x^*)\tilde{w})_{\tilde{D}}\end{aligned}$$

and so $\tilde{\pi}(x)^* = \tilde{\pi}(x^*)$.

Again set $A\tilde{v} = v$. A is clearly linear and since

$$||A\tilde{v}||_{\mathcal{H}} = ||v||_{\mathcal{H}} \leqslant ||\tilde{v}||_{\tilde{D}},$$

A is bounded on $\tilde{D}$. Also $A\tilde{\pi}(x)\tilde{v} = A(\widetilde{\pi(x)v}) = \pi(x)v = \pi(x)A\tilde{v}$ and we obtain $A\tilde{\pi}(x) = \pi(x)A$. Consequently, $\tilde{\pi}(x^*)A^* = A^*\pi(x^*)$ for all x and thus

$$AA^*\pi(x)v = A\tilde{\pi}(x)A^*v = \pi(x)AA^*v$$

and so AA^* is a bounded linear operator on $\mathcal{H}$ commuting with $\pi(x)$ for all x. By Proposition 6.48, $AA^* = C^2I$ for some C where again $C \geqslant 0$. If $C = 0$, then $A^* = 0$ and so $A = 0$. The only way this happens is if $D = \{0\}$. Thus we may suppose $C > 0$ and $D \neq \{0\}$. So $||A^*v||_{\tilde{D}}^2 = (A^*v, A^*v)_{\tilde{D}} = (AA^*v, v)_{\mathcal{H}} = C^2(v, v)_{\mathcal{H}}$. This implies as before the range D of A is a closed linear subspace of $\mathcal{H}$. Indeed, if $A\tilde{v}_k \to w$ in $\mathcal{H}$, then $AA^*A\tilde{v}_k \to C^2w$. But since $A\tilde{v}_k$ is Cauchy in $\mathcal{H}$ and A^* is bounded, $A^*A\tilde{v}_k$ is Cauchy in $\tilde{D}$. But $\tilde{D}$ is complete. So $A^*A\tilde{v}_k \to \tilde{v}$ for some $v \in D$. So $AA^*A\tilde{v}_k \to A\tilde{v} = v$ and $v = C^2w$. Thus $w \in D$. Now $\pi(x)v = \pi(x)A\tilde{v} = A\tilde{\pi}(x)\tilde{v}$ implies D is invariant under π. Since π is irreducible, $D = \mathcal{H}$. As before, the operator θ is bounded and since $\rho(x)\theta = \theta\pi(x)$ for all x, we have $\theta^*\rho(x^*) = \pi(x^*)\theta^*$ and thus $\theta^*\theta \in \mathrm{Hom}(\pi, \pi)$. The rest follows from Proposition 6.48.

□

4. Let π be a finite dimensional representation of a group G on vector space V. Let V_0 be a subspace of V. Define $V_0^\perp$ to be the subspace of V^* consisting of those f for which

$$\langle v_0, f\rangle = 0 \text{ for all } v_0 \in V_0.$$

(a) Using the identification $V^{**} = V$, show $(V_0^\perp)^\perp = V_0$.

(b) Show V_0 is π invariant if and only if $V_0^\perp$ is $\check{\pi}$ invariant.

Solution:

We show (a). $v \in (V_0^\perp)^\perp$ if and only if $f(v) = 0$ for all $f \in V_0^\perp$. This clearly occurs if $v \in V_0$ and if $v \notin V_0$, there is an f with $f(v) \neq 0$ and $f|_{V_0} = 0$. So $v \notin (V_0^\perp)^\perp$. Putting these together we see $(V_0^\perp)^\perp = V_0$.

Using (a), V_0 is π invariant if and only if $\pi(x)V_0 = V_0$ for all $x \in G$ if and only if $(\pi(x)V_0)^\perp = V_0^\perp$ for all x. But $f \in (\pi(x)V_0)^\perp$ if and only if $0 = \langle \pi(x)v, f\rangle = \langle v, \check{\pi}(x^{-1})f\rangle$ for all $v \in V_0$ if and only if $\check{\pi}(x^{-1})f \in V_0^\perp$. Thus V_0 is invariant under $\pi(x)$ for all x if and only if $\check{\pi}(x^{-1})V_0^\perp = V_0^\perp$ for all x.

□

5. Let $\overline{\mathcal{H}}$ be the conjugate Hilbert space to a finite dimensional Hilbert space $\mathcal{H}$ and let A be a linear transformation of $\mathcal{H}$. Show if $v_1, v_2, \ldots, v_n$ is a basis of $\mathcal{H}$, then the matrix of $\bar{A}$ relative to the basis $\bar{v}_1, \bar{v}_2, \ldots, \bar{v}_n$ of $\overline{\mathcal{H}}$ is the conjugate of the matrix of A relative to the basis $v_1, v_2, \ldots, v_n$.

Solution:

Consider the Hilbert space $\bar{\mathcal{H}}$ and a linear transformation $A : \mathcal{H} \to \mathcal{H}$. By definition $\bar{A}\bar{v} = \overline{Av}$. Hence if $v_1, v_2, \ldots, v_n$ is a basis of $\mathcal{H}$, then every vector v can be written uniquely as $a_1v_1 + a_2v_2 + \cdots + a_nv_n$. And so every vector $\bar{v}$ can be written uniquely as $\overline{a_1v_1 + a_2v_2 + \cdots + a_nv_n} = \overline{a_1v_1} + \overline{a_2v_2} + \cdots + \overline{a_nv_n} = \bar{a}_1\bar{v}_1 + \bar{a}_2\bar{v}_2 + \cdots + \bar{a}_n\bar{v}_n$. Thus $\bar{v}_1, \bar{v}_2, \ldots, \bar{v}_n$ is a basis of $\bar{\mathcal{H}}$.

Now the matrix of A is given by $[a_{i,j}]$ where $Av_j = \sum_{i=1}^n a_{i,j}v_j$. Since $\bar{A}\bar{v} = \overline{Av}$, we see $\bar{A}\bar{v}_j = \overline{Av_j} = \overline{\sum_{i=1}^n a_{i,j}v_j} = \sum_{i=1}^n \overline{a_{i,j}v_j} = \sum_{i=1}^n \bar{a}_{i,j}\bar{v}_i$. So the matrix of $\bar{A}$ relative to the basis $\bar{v}_1, \bar{v}_2, \ldots, \bar{v}_n$ is the matrix $[\bar{a}_{i,j}]$.

□

6. Let $A, B, U \in \mathcal{B}(\mathcal{H})$ where $\mathcal{H}$ is a Hilbert space. Show $\bar{A}\bar{B} = \overline{AB}$ and U is unitary if and only if $\overline{U}$ is unitary.

Solution:

Note by definition $\bar{A}\bar{v} = \overline{Av}$. So $\bar{A}\bar{B}\bar{v} = \bar{A}\overline{Bv} = \overline{ABv} = \overline{AB}\bar{v}$. This gives $\bar{A}\bar{B} = \overline{AB}$. Now U is unitary if U is onto and $(Uv, Uw) = (v, w)$ for all $v, w \in \mathcal{H}$. But U is onto if and only if $\overline{U}$ is onto for $v = Uw$ if and only if $\bar{v} = \bar{U}\bar{w}$ and $(Uv, Uw) = (v, w)$ for all $v, w \in \mathcal{H}$ if and only if $\overline{(Uv, Uw)} = \overline{(v, w)}$ for all v, w. This occurs if and only if $(\bar{U}\bar{v}, \bar{U}\bar{w})_{\bar{\mathcal{H}}} = (\overline{Uv}, \overline{Uw})_{\bar{\mathcal{H}}} = (Uw, Uv)_{\mathcal{H}} = (w, v)_{\mathcal{H}} = (\bar{v}, \bar{w})_{\bar{\mathcal{H}}}$ for all $\bar{v}$, $\bar{w} \in \bar{\mathcal{H}}$. Thus U is unitary if and only if $\bar{U}$ is unitary.

□

7. Let π be a finite dimensional unitary representation. Show $\pi \otimes \bar{\pi}$ and $\pi \otimes \check{\pi}$ are equivalent.

Solution:

By Proposition 6.67, the mapping $A : \bar{\mathcal{H}} \to \mathcal{H}^*$ given by $\langle v_1, A(\bar{v}_2)\rangle = (v_1, v_2)_{\mathcal{H}}$ is a linear isomorphism that satisfies $A\bar{\pi}(g) = \check{\pi}(g)A$. Consider the linear transformation $I \otimes A$ which maps $\mathcal{H} \otimes \bar{\mathcal{H}}$ into $\mathcal{H} \otimes \mathcal{H}^*$. It is invertible and onto for $(I \otimes A)^{-1} = I \otimes A^{-1}$. Note

$$\begin{aligned}(I \otimes A)(\pi(g_1) \otimes \bar{\pi}(g_2))(v \otimes \bar{w}) &= (I \otimes A)(\pi(g_1)v \otimes \bar{\pi}(g_2)\bar{w}) \\ &= \pi(g_1)v \otimes A\bar{\pi}(g_2)\bar{w} \\ &= \pi(g_1)v \otimes \check{\pi}(g_2)A\bar{w} \\ &= (\pi(g_1) \otimes \check{\pi}(g_2))(v \otimes A\bar{w}) \\ &= (\pi(g_1) \otimes \check{\pi}(g_2))(I \otimes A)(v \otimes \bar{w}).\end{aligned}$$

Since the linear span of the $v \otimes \bar{w}$ is $\mathcal{H} \otimes \bar{\mathcal{H}}$, we see $I \otimes A$ gives an equivalence between the representations $\pi \times \bar{\pi}$ and $\pi \times \check{\pi}$. Since $\pi \otimes \bar{\pi}$ and $\pi \otimes \check{\pi}$ are restrictions of these representations to the diagonal in $G \times G$, we see $\pi \otimes \bar{\pi}$ and $\pi \otimes \check{\pi}$ are equivalent.

□

8. Let $G = \mathrm{SU}(2)$ be the compact group of 2×2 unitary matrices. Thus $G = \left\{\left[\begin{smallmatrix} a & b \\ -\bar{b} & \bar{a}\end{smallmatrix}\right] \mid |a|^2 + |b|^2 = 1\right\}$. Define π by $\pi(g) = g$ for $g \in G$.

(a) Determine $\bar{\pi}$ as a matrix representation.

(b) Show π and $\bar{\pi}$ are unitarily equivalent.

Solution:

π is the representation of G on $\mathbb{C}^2$ given by $\pi(g)v = gv$ where $g \in \mathrm{SU}(2)$ and $v \in \mathbb{C}^2$ is a column vector. So $\mathcal{H} = \mathbb{C}^2$ and $\bar{\mathcal{H}} = \mathbb{C}^2$ but with $\lambda \cdot \bar{v} = \overline{\bar{\lambda} v}$. Thus if $g = \left[\begin{smallmatrix} a & b \\ -\bar{b} & \bar{a}\end{smallmatrix}\right]$ and $e_1 = \left[\begin{smallmatrix} 1 \\ 0\end{smallmatrix}\right]$ and $e_2 = \left[\begin{smallmatrix} 0 \\ 1\end{smallmatrix}\right]$, we see $\pi(g)e_1 = \left[\begin{smallmatrix} a \\ -\bar{b}\end{smallmatrix}\right] = ae_1 - \bar{b}e_2$ and similarly $\pi(g_2)e_2 = be_1 - \bar{a}e_2$. So the matrix of $\pi(g)$ relative to the basis e_1, e_2 is g. Now $\bar{\pi}(g)\bar{v} = \overline{\pi(g)v}$. But $\bar{e}_1, \bar{e}_2$ is a basis of $\bar{\mathcal{H}}$ and $\bar{\pi}(g)\bar{e}_1 = \overline{\pi(g)e_1} = \overline{ae_1 - \bar{b}e_2} = \bar{a}\bar{e}_1 - b\bar{e}_2$. Similarly $\bar{\pi}(g)\bar{e}_2 = \overline{\pi(g)e_2} = \overline{be_1 + \bar{a}e_2} = \bar{b}\bar{e}_1 + a\bar{e}_2$. This shows the matrix of $\bar{\pi}(g)$ relative to the basis $\bar{e}_1$, $\bar{e}_2$ is $\left[\begin{smallmatrix} \bar{a} & \bar{b} \\ -b & a\end{smallmatrix}\right]$ which is the conjugate of the matrix g.

To see π and $\bar{\pi}$ are unitarily equivalent consider the unitary operator U from $\mathcal{H}$ to $\bar{\mathcal{H}}$ given by $Ue_1 = \bar{e}_2$ and $Ue_2 = -\bar{e}_1$. Then for $g = \left[\begin{smallmatrix} a & b \\ -\bar{b} & \bar{a}\end{smallmatrix}\right]$, one has $U\pi(g)e_1 = U(ae_1 - \bar{b}e_2) = a\bar{e}_2 + \bar{b}\bar{e}_1 = \bar{\pi}(g)\bar{e}_2 = \bar{\pi}(g)Ue_1$ and $U\pi(g)e_2 = U(be_1 + \bar{a}e_2) = b\bar{e}_2 - \bar{a}\bar{e}_1 = \bar{\pi}(g)(-\bar{e}_1) = \bar{\pi}(g)Ue_2$. So $U\pi(g)$ and $\bar{\pi}(g)U$ agree on a basis. Thus $U\pi(g) = \bar{\pi}(g)U$ and we see π and $\bar{\pi}$ are unitarily equivalent.

□

9. Let π be a unitary representation of G on Hilbert space $\mathcal{H}$. Let I be the identity representation of G defined on Hilbert space $\mathcal{H}'$ whose orthonormal basis has dimension cardinality n. Show $\pi \otimes I \cong n\pi$.

Solution:

We have $\mathcal{H} \otimes \mathcal{H}' = \mathcal{B}_2(\bar{\mathcal{H}}', \mathcal{H})$ and by Proposition 2.42, if e_α for $\alpha \in A$ is an orthonormal basis of $\mathcal{H}$ and f_β where $\beta \in B$ and B has cardinality n is an orthonormal basis of $\mathcal{H}'$, then $e_\alpha \otimes f_\beta$ for $(\alpha, \beta) \in A \times B$ is an orthonormal basis of $\mathcal{H}\otimes\mathcal{H}'$. Consider the Hilbert space $\mathcal{K} = n\mathcal{H} = \oplus_{\beta\in B}\mathcal{H}$ and the unitary mapping U defined using bases by $U(e_\alpha \otimes f_\beta) = (\delta_{\beta',\beta}e_\alpha)$, that is $U(e_\alpha \otimes f_\beta)$ is the vector $(v_{\beta'})$ in $\oplus_{\beta\in B}\mathcal{H}$ satisfying $v_{\beta'} = 0$ for $\beta' \neq \beta$ and $v_\beta = e_\alpha$. Since U on these vectors takes an orthonormal basis of $\mathcal{H} \otimes \mathcal{H}'$ to an orthonormal basis of $n\mathcal{H}$, extending U linearly gives a unitary isomorphism. Moreover, it is a unitary equivalence of $\pi \otimes I$ onto $n\pi$. Indeed, by linearity and completeness, $U(v \otimes f_\beta) = (\delta_{\beta,\beta'}v)$ and thus $U(\pi(g) \otimes I(g))(v \otimes f_\beta) = U(\pi(g)v \otimes f_\beta) = (\delta_{\beta',\beta}\pi(g)v) = n\pi(g)(\delta_{\beta',\beta}v) = n\pi(g)U(v \otimes f_\beta)$. By linearity of $n\pi(g)$ and $\pi(g) \otimes I$ and using completeness and the unitarity of U, we then see

$$U(\pi(g) \otimes I(g))w = n\pi(g)Uw$$

for all $w \in \mathcal{H} \otimes \mathcal{H}'$. So $\pi \otimes I$ is unitarily equivalent to $n\pi$.

□

10. Let π be a unitary representation of G. Show $A \mapsto \bar{A}$ is a conjugate linear $*$ algebra isomorphism of $\mathrm{Hom}_G(\pi, \pi)$ onto $\mathrm{Hom}_G(\bar{\pi}, \bar{\pi})$.

Solution:

By Exercise 6.4.6, $\overline{AB} = \bar{A}\bar{B}$. Moreover $\overline{(A + B)}\bar{v} = \overline{Av} + \overline{Bv} = \bar{A}\bar{v} + \bar{B}\bar{v}$ and $\overline{(cA)}\bar{v} = \overline{cAv} = \bar{c}\overline{Av} = \bar{c}\bar{A}\bar{v}$. Also since $(\bar{A}\bar{v}, \bar{w})_{\bar{\mathcal{H}}} = (\overline{Av}, \bar{w})_{\bar{\mathcal{H}}} = (w, Av)_{\mathcal{H}} = (A^*w, v)_{\mathcal{H}} = (\bar{v}, \overline{A^*w})_{\bar{\mathcal{H}}} = (\bar{v}, \overline{A^*}\bar{w})_{\bar{\mathcal{H}}}$, we have $(\bar{A})^* = \overline{A^*}$, and thus $A \mapsto \bar{A}$ is a conjugate linear $*$-isomorphism of the $*$ algebra $\mathcal{B}(\mathcal{H})$ onto $\mathcal{B}(\bar{\mathcal{H}})$. It is onto and one-to-one. Now $A \in \mathrm{Hom}_G(\pi, \pi)$ if and only if $A\pi(g) = \pi(g)A$ for all g if and only if $\overline{A\pi(g)} = \overline{\pi(g)A}$ for all g if and only if $\bar{A}\bar{\pi}(g) = \bar{\pi}(g)\bar{A}$ for all g. Thus $A \in \mathrm{Hom}_G(\pi, \pi)$ if and only if $\bar{A} \in \mathrm{Hom}_G(\bar{\pi}, \bar{\pi})$.

□

11. Let π_1, π_2, π_3, and π_4 be unitary representations of G. Show

$$\mathrm{Hom}_{G\times G}(\pi_1 \times \pi_2, \pi_3 \times \pi_4) = 0$$

if and only if $\mathrm{Hom}_G(\pi_1, \pi_3) = \{0\}$ or $\mathrm{Hom}_G(\pi_2, \pi_4) = \{0\}$.

Solution:

Assume $A \in \mathrm{Hom}_G(\pi_1, \pi_3)$ and $B \in \mathrm{Hom}_G(\pi_2, \pi_4)$ are both nonzero. Then $A \otimes B \in \mathrm{Hom}_{G\times G}(\pi_1 \times \pi_2, \pi_3 \times \pi_4)$ and $A \otimes B \neq 0$.

Suppose $T \in \mathrm{Hom}_{G\times G}(\pi_1 \times \pi_2, \pi_3 \times \pi_4)$ is nonzero. Thus there is a pair v', w' with $T(v' \otimes w') \neq 0$. Since $T(v' \otimes w') \in \mathcal{B}_2(\bar{\mathcal{H}}_4, \mathcal{H}_3)$, $T(v' \otimes w')$ is a nonzero Hilbert-Schmidt operator from $\bar{\mathcal{H}}_4$ into $\mathcal{H}_3$. Since it is nonzero, there is a vector $\bar{w}_4$ with $T(v' \otimes w')(\bar{w}_4) \neq 0$. Define $Av = T(v \otimes w')(\bar{w}_4)$. Note A is linear and nonzero for $Av' \neq 0$. Also A is bounded for by 2.3

$$\begin{aligned} ||Av||_{\mathcal{H}_3} &\leqslant ||T(v \otimes w')||\,||w_4|| \\ &\leqslant ||T(v \otimes w')||_2||w_4|| \\ &\leqslant ||T||\,||v \otimes w'||_2||w_4|| \\ &\leqslant ||T||\,||v||\,||w'||\,||w_4||. \end{aligned}$$

Also $A \in \mathrm{Hom}_G(\pi_1, \pi_3)$ for by Definition 6.69

$$\begin{aligned} A\pi_1(g)v &= T(\pi_1(g)v \otimes I(e)w')(\bar{w}_4) \\ &= T(\pi_1 \times \pi_2)(g,e)(v \otimes w')(\bar{w}_4) \\ &= (\pi_3(g) \otimes \pi_4(e))T(v \otimes w')(\bar{w}_4) \\ &= \pi_3(g)T(v \otimes w')\pi_4(e)^*(\bar{w}_4) \\ &= \pi_3(g)T(v \otimes w')(\bar{w}_4) \\ &= \pi_3(g)Av\,. \end{aligned}$$

Similarly, we can define B by $Bw = \overline{(T(v' \otimes w))^*(v_3)}$ where v_3 can be chosen so that B is nonzero. In fact, since $T(v' \otimes w')$ is a nonzero Hilbert-Schmidt operator in $\mathcal{B}_2(\bar{\mathcal{H}}_4, \mathcal{H}_3)$, the operator $T(v' \otimes w')^*$ is a nonzero operator in $\mathcal{B}_2(\mathcal{H}_3, \bar{\mathcal{H}}_4)$ and so one can pick a vector v_3 in $\mathcal{H}_3$ with $(T(v' \otimes w'))^*(v_3) \neq 0$. As before using $||L^*||_2 = ||L||_2$, we have that B is bounded. Note B is linear since conjugation and adjoint are conjugate linear. Moreover,

$$\begin{aligned} B\pi_2(g)w &= \overline{(T(v' \otimes \pi_2(g)w))^*(v_3)} \\ &= \overline{(T(\pi_1(e)v' \otimes \pi_2(g)w))^*(v_3)} \\ &= \overline{(T(\pi_1 \times \pi_2)(e,g)(v' \otimes w))^*(v_3)} \\ &= \overline{(\pi_3 \times \pi_4)(e,g)(T(v' \otimes w))^*(v_3)} \\ &= \overline{((\pi_3(e) \otimes \pi_4(g))(T(v' \otimes w)))^*(v_3)} \\ &= \overline{\left(I(T(v' \otimes w))\overline{\pi_4(g)}^*\right)^*(v_3)} \\ &= \overline{\left(\overline{\pi_4(g)}(T(v' \otimes w))^*\right)(v_3)} \\ &= \pi_4(g)\overline{(T(v' \otimes w))^*(v_3)} \\ &= \pi_4(g)Bw\,. \end{aligned}$$

So if there is a nonzero T in $\mathrm{Hom}_{G \times G}(\pi_1 \times \pi_2, \pi_3 \times \pi_4)$, then the operators A and B defined above are nonzero and in $\mathrm{Hom}_G(\pi_1, \pi_3)$ and $\mathrm{Hom}_G(\pi_2, \pi_4)$, respectively.

□

12. Let G_1 and G_2 be locally compact Hausdorff groups and let ρ_1 and ρ_2 be discretely decomposable unitary representations of G_1 and G_2, respectively. Show the unitary representation $\rho_1 \times \rho_2$ is discretely decomposable and if π_1 and π_2 are irreducible unitary representations of G_1 and G_2, then the primary projection $P(\pi_1 \times \pi_2)$ is given by $P(\pi_1 \times \pi_2) = P(\pi_1) \otimes P(\pi_2)$.

Solution:

Since ρ_1 and ρ_2 are discretely decomposable, there exist families $\pi_{1,i}$, $i \in I$ and $\pi_{2,j}$, $j \in J$ of irreducible unitary representations of G_1 and G_2 such that $\rho_1 = \oplus_i \pi_{1,i}$ and $\rho_2 = \oplus_j \pi_{2,j}$. Thus the Hilbert space $\mathcal{H}_1$ for ρ_1 is an internal orthogonal direct sum of ρ_1 invariant Hilbert subspaces $\mathcal{H}_{1,i}$ such that ρ_1 on $\mathcal{H}_{1,i}$

is unitarily equivalent to $\pi_{1,i}$. We have a similar decomposition $\mathcal{H}_2 = \oplus_j \mathcal{H}_{2,j}$ for ρ_2 with ρ_2 on $\mathcal{H}_2$ unitarily equivalent to $\pi_{2,j}$. Now one can show with an easy argument that $\mathcal{B}_2(\bar{\mathcal{H}}_2, \mathcal{H}_1)$ can be unitarily mapped to the external Hilbert space direct sum $\oplus_{i,j}\mathcal{B}_2(\bar{\mathcal{H}}_{2,j}, \mathcal{H}_{1,i})$. Indeed if for each i, Q_i is the orthogonal projection of $\mathcal{H}_1$ onto $\mathcal{H}_{1,i}$, one can check that a linear operator T from $\bar{\mathcal{H}}_2$ to $\mathcal{H}_1$ is Hilbert-Schmidt if and only if $Q_iT|_{\bar{\mathcal{H}}_{2,j}}$ is Hilbert-Schmidt from $\bar{\mathcal{H}}_{2,j}$ into $\mathcal{H}_{1,i}$ and $||T||_2^2 = \sum_{i,j}||Q_iT|_{\bar{\mathcal{H}}_{2,j}}||^2$. In particular, under this correspondence, the unitary representation $\rho_1 \times \rho_2$ corresponds to the external direct sum $\oplus_{i,j}(\pi_{1,i} \times \pi_{2,j})$. So $\rho_1 \times \rho_2$ is discretely decomposable.

Now let π_1 and π_2 be irreducible unitary representations. By Proposition 6.75, $\pi_1 \times \pi_2$ is a unitary irreducible representation. Also $\pi_1 \times \pi_2$ is unitarily equivalent to $\pi_{1,i} \times \pi_{2,j}$ if and only if π_1 is unitarily equivalent $\pi_{1,i}$ and $\pi_{2,j}$ is unitarily equivalent to π_2. Now let $I_1 = \{i \mid \pi_{1,i} \text{ is unitarily equivalent to } \pi_1\}$ and $J_2 = \{j \mid \pi_{2,j} \text{ is unitarily equivalent to } \pi_2\}$. Using the argument in the proof of Proposition 6.58 we see $P(\pi_1 \times \pi_2)$ is the orthogonal projection onto the subspace $\oplus_{i,j\in I_1\times J_2}(\mathcal{H}_{1,i} \otimes \mathcal{H}_{2,j})$ of $\oplus_{i,j}(\mathcal{H}_{1,i} \otimes \mathcal{H}_{2,j})$. Now $P(\pi_1) \times I$ is an orthogonal projection and its range is $\oplus_{i\in I_1,j}(\mathcal{H}_i \otimes \mathcal{H}_j)$ for as seen in Exercise 6.4.9, $\pi_{1,i} \times I$ on $\mathcal{H}_{1,i} \otimes \mathcal{H}_{2,j}$ is is unitarily equivalent to $n_j\pi_{1,i}$ where n_j is the cardinality of an orthonormal basis of $\mathcal{H}_{2,j}$. Similarly, $I \otimes P(\pi_2)$ is the orthogonal projection onto $\oplus_{i,j\in J_2}(\mathcal{H}_{1,i} \otimes \mathcal{H}_{2,j})$. The projections $P(\pi_1) \otimes I$ and $I \otimes P(\pi_2)$ commute and thus $P(\pi_1) \otimes P(\pi_2) = (P(\pi_1) \otimes I)(I \otimes P(\pi_2))$ is an orthogonal projection. Its range is $\oplus_{i,j\in I_1\times J_2}(\mathcal{H}_{1,i} \otimes \mathcal{H}_{2,j})$. Thus $P(\pi_1 \times \pi_2) = P(\pi_1) \otimes P(\pi_2)$.

□

13. A finite dimensional representation π of a group G is said to be **completely reducible** if every invariant subspace has a complementary invariant subspace. Show if π is completely reducible, the contragredient representation $\check{\pi}$ is completely reducible.

Solution: Let V_0^* be an invariant subspace of $\check{\pi}$. Set $V_0 = \{v \mid \langle v, v_0^*\rangle = 0 \text{ for all } v_0^* \in V_0^*\}$. We claim V_0 is π invariant. Indeed, let $v \in V_0$. Then $\langle \pi(g)v, v_0^*\rangle = \langle v, \check{\pi}(g^{-1})v_0^*\rangle = 0$. Now π is completely reducible. Thus there is a complementary invariant subspace V_1 to V_0. Now let $V_1^* = \{v^* \mid \langle v_1, v^*\rangle = 0 \text{ for all } v_1 \in V_1\}$. Note $\dim V_1^* = n - \dim V_1 = \dim V_0 = n - \dim V_0^*$ and if $v^* \in V_0^* \cap V_1^*$, then $\langle v, v^*\rangle = 0$ for all $v \in V_0$ and all $v \in V_1$. So $\langle v, v^*\rangle = 0$ for all $v \in V$. Thus $v^* = 0$ and we have $V^* = V_0^* \oplus V_1^*$. Finally, we show V_1^* is invariant under $\check{\pi}$. Indeed, if $v^* \in V_1^*$, then

$$\langle v_1, \check{\pi}(g)v^*\rangle = \langle \pi(g^{-1})v_1, v^*\rangle = 0.$$

□

14. Let π_1 be an irreducible representation of a group G_1 on a finite dimensional complex vector space V_1 and let I be the identity representation of G_2 on a finite dimensional complex vector space V_2. Show

$$\mathrm{Hom}_{G_1\times G_2}(\pi_1 \times I, \pi_1 \times I) = \{I \otimes B \mid B \in \mathcal{L}(V_2)\}.$$

Solution: Let $A \in \mathrm{Hom}(\pi_1 \times I, \pi_1 \times I)$. Thus A is a linear transformation from $L(V_2^*, V_1)$ into $L(V_2^*, V_1)$. Consequently, $A(v \otimes v_2) \in L(V_2^*, V_1)$. So if $v_2^* \in V_2^*$, then $A(v \otimes v_2)(v_2^*) \in V_1$. Define $Tv = A(v \otimes v_2)(v_2^*)$. Then T is a linear transformation of V_1. It commutes with π_1 for

$$\begin{aligned} T\pi_1(g_1)v &= A(\pi_1(g_1)v \otimes v_2)(v_2^*) \\ &= A(\pi_1(g_1)v \otimes \pi_2(e_2)v_2)(v_2^*) \\ &= (\pi(g_1) \otimes I_2)A(v \otimes v_2)(v_2^*) \\ &= \pi(g_1)A(v \otimes v_2)(v_2^*) \\ &= \pi(g_1)Tv. \end{aligned}$$

Since we are over the complexes, it has eigenspaces. Since eigenspaces are invariant, T is scalar. So

$$A(v \otimes v_2)(v_2^*) = c(v_2, v_2^*)v.$$

Now c is bilinear on $V_2 \times V_2^*$. This implies there is a $B : V_2 \to V_2$ such that $c(v_2, v_2^*) = v_2^*(Bv_2)$. Indeed, since V_2 is finite dimensional, the dual of V_2^* is isomorphic to V_2 under the mapping $v_2 \to v_2^{**}$ where $v_2^{**}(v_2^*) = v_2^*(v_2)$. Thus $A(v \otimes v_2)(v_2^*) = v_2^*(Bv_2)v$.

Thus $A = I \otimes B$ on the simple vectors $v \otimes v_2$. In fact,

$$\begin{aligned} (I \otimes B)(v \otimes v_2)(v_2^*) &= I(v \otimes v_2)B^t(v_2^*) \\ &= B^t(v_2^*)(v_2)(v) \\ &= v_2^*(Bv_2)v \\ &= A(v \otimes v_2)(v_2^*). \end{aligned}$$

□

15. Let π_1 and π_2 be finite dimensional irreducible complex representations of groups G_1 and G_2. Show $\mathrm{Hom}_{G_1 \times G_2}(\pi_1 \times \pi_2, \pi_1 \times \pi_2) = \mathbb{C}I$.

Solution: Let $A \in \mathrm{Hom}(\pi_1 \times \pi_2, \pi_1 \times \pi_2)$. Then $A \in \mathrm{Hom}(\pi_1 \times I, \pi_1 \times I)$. So by Exercise 6.4.14, $A = I_1 \otimes B$. We claim B is scalar. Indeed, $(I_1 \otimes B)(I_1 \otimes \pi_2(g_2)) = (I_1 \otimes \pi_2(g_2))(I_1 \otimes B)$. Thus $I_1 \otimes (B\pi_2(g_2)) = I_1 \otimes (\pi_2(g_2)B)$. Now

$$\begin{aligned} I_1 \otimes (B\pi_2(g_2))(v_1 \otimes v_2)(v_2^*) &= (v_1 \otimes v_2)((B\pi_2(g_2))^t v_2^*) \\ &= (B\pi_2(g_2))^t v_2^*(v_2)v_1 \\ &= v_2^*(B\pi_2(g_2)v_2)v_1 \end{aligned}$$

and

$$\begin{aligned} I_1 \otimes (\pi_2(g_2)B)(v_1 \otimes v_2)(v_2^*) &= (v_1 \otimes v_2)((\pi_2(g_2)B)^t v_2^*) \\ &= (\pi_2(g_2)B)^t v_2^*(v_2)v_1 \\ &= v_2^*(\pi_2(g_2)Bv_2)v_1 \text{ for all } v_2^*,\ v_2, \text{ and } v_1, \end{aligned}$$

and hence we see $\pi_2(g_2)B = B\pi_2(g_2)$ for all g_2. By irreducibility, $B = cI_2$. So $A = cI_1 \otimes I_2 = cI$. □

16. In Exercise 6.4.13, we defined a completely reducible representation. A group G is said to be completely reducible if every complex finite dimensional representation of G is completely reducible. Let G be completely reducible.

(a) Show every finite dimensional complex representation of G is a direct sum of irreducible representations.

(b) Show a complex finite dimensional representation π of G is irreducible if and only if $\mathrm{Hom}_G(\pi, \pi) = \mathbb{C}I$.

Solution:

We show finite dimensional representations π of a completely reducible group G are direct sums of irreducible representations. We argue by induction on $\dim \pi$. This is clear if $\dim \pi = 1$. So assume true if the dimension is less than n. Let π be a representation on a vector space V of dimension n. Then by choosing a subspace V_0 of smallest positive dimension invariant under π, we see one has an irreducible invariant subspace. If $V = V_0$ we are done. Otherwise, since π is completely reducible, there is a complementary invariant subspace W. Clearly $\dim W < n$. By induction $W = \bigoplus_{k=1}^{m} V_k$ where each V_k is an irreducible invariant subspace for π. Thus $V = \bigoplus_{i=0}^{m} V_i$ is a direct sum of irreducible invariant subspaces. So (a) holds.

For (b), note if $B \in \mathrm{Hom}_G(\pi, \pi)$ is not a scalar multiple of I, then any eigenspace $V_\lambda = \{w \mid Bw = \lambda w\}$ is invariant and proper. So π is not irreducible. Now suppose $\mathrm{Hom}_G(\pi, \pi) = \mathbb{C}I$. Suppose V_0 is a nonzero invariant subspace of V. Since π is completely reducible, there is a complementary invariant subspace V_1. For $v = v_0 + v_1$ where $v_0 \in V_0$ and $v_1 \in V_1$, define $Pv = v_0$. Note $P\pi(x)v = P(\pi(x)v_0 + \pi(x)v_1) = \pi(x)v_0 = \pi(x)Pv$. Thus $P \in \mathrm{Hom}_G(\pi, \pi)$. Thus $P = \lambda I$ for some $\lambda \neq 0$. $P^2 = P$ then implies $\lambda = 1$ and thus $P = I$ and $V_0 = V$. So V is irreducible.

□

17. Let π be a nonzero representation of a topological group on a locally convex topological vector space V. A projection is a continuous linear operator P on V satisfying $P^2 = I$.

(a) Suppose P is a projection. Show the range of P is a closed subspace of V which is invariant under π if and only if $\pi(x)P = P\pi(x)P$ for all $x \in G$.

(b) Show if P is a projection and $\pi(x)P = P\pi(x)$ for all x, then the range of $I - P$ is invariant under π and $V = PV + (I - P)V$ is a direct sum decomposition of V into closed invariant subspaces.

(c) Show if P is projection satisfying $P\pi(x) = \pi(x)P$ for all x and v is a cyclic vector for π, then Pv is a cyclic vector for the representation π on PV.

(d) Give an example of a representation with cyclic vector v and a projection P such that PV is an invariant subspace for π but Pv is not a cyclic vector for the restriction of π to PV. (**Hint:** Consider the $ax + b$ group.)

Solution: We first note the range of P is a closed subspace of V. Indeed, suppose $Pv \neq v$ and thus $v \notin PV$. Since V is Hausdorff, there is a continuous seminorm $|\cdot|$ with $|Pv - v| = \epsilon > 0$. P being continuous implies there is a second continuous seminorm $||\cdot||$ on V and a $\delta > 0$ so that if $||w - v|| < \delta$, then $|Pw - Pv| < \epsilon/2$. Thus if $||w - v|| < \delta$ and $|w - v| < \frac{\epsilon}{2}$, then $Pw \neq w$ for otherwise $|Pv - v| \leqslant |Pv - Pw| + |Pw - v| = |Pv - Pw| + |w - v| < \frac{\epsilon}{2} + \frac{\epsilon}{2} = \epsilon$. Hence $\{w \mid ||w - v|| < \delta, |w - v| < \frac{\epsilon}{2}\}$ is a neighborhood of v missing the set PV. So PV contains its limit points and thus is closed.

(a) is now immediate for PV is invariant if and only if $P\pi(x)Pw = \pi(x)Pw$ for all $w \in V$.

For (b), note if P is a projection, then $(I - P)^2 = I - 2P + P^2 = I - P$. So $(I - P)V$ is closed linear subspace of V. Now if $P\pi(x) = \pi(x)P$, then $(I - P)\pi(x) = \pi(x)(I - P)$ for all x. So $(I - P)\pi(x)(I - P) = \pi(x)(I - P)^2 = \pi(x)(I - P)$ for all x. By (a), $(I - P)V$ is invariant under π. Finally, the sum $PV + (I - P)V$ is direct for if $Pv + (I - P)w = 0$, then $P^2v + P(I - P)w = 0$. So $Pv = 0$ and $(I - P)w = 0$.

Now consider a cyclic vector v for π and a projection P satisfying $P\pi(x) = \pi(x)P$ for all x. Take V_0 be PV. Then $(I - P)V_0 = 0$. Consider $w = \sum_{i=1}^k a_i\pi(x_i)v = \sum_{i=1}^k (Pa_i\pi(x_i)v + (I - P)a_i\pi(x_i)v)$. The w's of this form are a dense linear subspace of V. Since P is continuous and $PV = V_0$, the Pw's where w has this form is dense in V_0. But $Pw = \sum_{i=1}^k P(a_i\pi(x_i)v) = \sum_{i=1}^k a_iP\pi(x_i)v = \sum_{i=1}^k a_i\pi(x_i)Pv$. Consequently Pv is a cyclic vector for π on V_0. This gives (c).

For (d), let $G = \{(a, b) \mid a > 0, b \in \mathbb{R}\}$ with $(a, b)(a', b') = (aa', b + ab')$. This is the $ax + b$ group (e.g. Exercise 6.1.8). Define π on $\mathbb{C}^2$ by $\pi(a, b)\left(\begin{smallmatrix} x \\ y \end{smallmatrix}\right) = \begin{pmatrix} \sqrt{a} & b \\ 0 & \frac{1}{\sqrt{a}} \end{pmatrix}\left(\begin{smallmatrix} x \\ y \end{smallmatrix}\right) = \begin{pmatrix} \sqrt{a}x + by \\ \frac{y}{\sqrt{a}} \end{pmatrix}$. π is a representation and $V_0 = \{\left(\begin{smallmatrix} x \\ 0 \end{smallmatrix}\right) \mid x \in \mathbb{R}\}$ is a closed invariant subspace. Let $P\left(\begin{smallmatrix} x \\ y \end{smallmatrix}\right) = \left(\begin{smallmatrix} x \\ 0 \end{smallmatrix}\right)$. P is a continuous projection onto V_0 and $P\pi(a, b)P = \pi(a, b)P$ for all $(a, b) \in G$. The vector $\left(\begin{smallmatrix} 0 \\ 1 \end{smallmatrix}\right)$ is a cyclic vector for $\pi(a, b)\left(\begin{smallmatrix} 0 \\ 1 \end{smallmatrix}\right) = \begin{pmatrix} b \\ \frac{1}{\sqrt{a}} \end{pmatrix}$ and the linear span of these vectors is $\mathbb{C}^2$. But $P\left(\begin{smallmatrix} 0 \\ 1 \end{smallmatrix}\right) = \left(\begin{smallmatrix} 0 \\ 0 \end{smallmatrix}\right)$ is not a cyclic vector for V_0. □

18. Let π be a unitary representation of a topological group G. A vector v in $\mathcal{H}_\pi$ is said to be a separating vector for $\mathrm{Hom}_G(\pi, \pi)$ if $A = 0$ whenever $A \in \mathrm{Hom}_G(\pi, \pi)$ and $Av = 0$. Show v is a separating vector for $\mathrm{Hom}_G(\pi, \pi)$ if and only if v is a cyclic vector for π.

Solution: A vector v is a cyclic for π if and only if whenever $w \in \mathcal{H}$ satisfies $(w, \pi(x)v)_{\mathcal{H}} = 0$ for all x, then $w = 0$. Consider $A \in \mathrm{Hom}_G(\pi, \pi)$. One has $(\pi(x^{-1})w, Av) = (w, \pi(x)Av)_{\mathcal{H}} = (w, A\pi(x)v)_{\mathcal{H}} = (A^*w, \pi(x)v)$. Thus if v is a cyclic vector and $Av = 0$, $A^*w = 0$ for all w and we see $A = 0$. So v is a separating vector for $\mathrm{Hom}_G(\pi, \pi)$.

Conversely, if v is a separating vector and $(w, \pi(x)v)_{\mathcal{H}} = 0$ for all $x \in G$, then the orthogonal projection P onto the closure of the linear span of all vectors $\pi(x)w$ for $x \in G$ is in $\mathrm{Hom}_G(\pi, \pi)$ and since $(\pi(x^{-1})w, v)_{\mathcal{H}} = 0$ for all x, $(Pu, v)_{\mathcal{H}} = 0$ for all $u \in \mathcal{H}$. Thus $(u, Pv)_{\mathcal{H}} = 0$ for all $u \in \mathcal{H}$ and so $Pv = 0$. Since v is a separating vector, $P = 0$, and thus $w = Pw = 0$. So v is a cyclic vector for π. $\square$

19. Let π and π' be unitary representations of a topological group G with $\mathrm{Hom}_G(\pi, \pi') = \{0\}$. Show $\mathrm{Hom}_G(\pi \oplus \pi', \pi \oplus \pi') \cong \mathrm{Hom}_G(\pi, \pi) + \mathrm{Hom}_G(\pi', \pi')$ as $*$ algebras.

Solution: Let $\mathcal{H}$ and $\mathcal{H}'$ be the Hilbert spaces for π and π'. Note a bounded linear transformation T from $\mathcal{H} \oplus \mathcal{H}'$ into itself can be written in matrix form as $T = \left[\begin{smallmatrix} A & C \\ D & B \end{smallmatrix}\right]$. Namely $T(v, v') = (Av + Dv', Cv + Bv')$ where $A : \mathcal{H} \to \mathcal{H}$, $C : \mathcal{H} \to \mathcal{H}'$, $D : \mathcal{H}' \to \mathcal{H}$, and $B : \mathcal{H}' \to \mathcal{H}'$ are bounded linear operators. Now $T \in \mathrm{Hom}_G(\pi \oplus \pi', \pi \oplus \pi')$ if and only if for all $x \in G$, one has for all $(v, v') \in \mathcal{H} \oplus \mathcal{H}'$,

$$(A\pi(x)v + D\pi'(x)v', C\pi(x)v + B\pi'(x)v') = \\ (\pi(x)Av + \pi(x)Dv', \pi'(x)Cv + \pi'(x)Bv').$$

This is equivalent to $A\pi(x)v = \pi(x)Av$, $D\pi'(x)v' = \pi(x)Dv'$, $C\pi(x)v = \pi'(x)Cv$, and $B\pi'(x)v' = \pi'(x)Bv'$ for all x, v, and v'. Since $\mathrm{Hom}_G(\pi, \pi') = \{0\}$ one has $\mathrm{Hom}_G(\pi', \pi) = \{0\}$ and so $C = D = 0$. Thus we see $T \in \mathrm{Hom}_G(\pi \oplus \pi', \pi \oplus \pi')$ if and only if $A \in \mathrm{Hom}_G(\pi, \pi)$, $B \in \mathrm{Hom}_G(\pi', \pi')$ and $C = D = 0$. The mapping $T \mapsto (A, B)$ is clearly linear, multiplicative, and adjoint preserving and thus gives a $*$-algebra isomorphism of $\mathrm{Hom}_G(\pi \oplus \pi', \pi \oplus \pi')$ onto $\mathrm{Hom}_G(\pi, \pi) + \mathrm{Hom}_G(\pi', \pi')$.

$\square$

20. Suppose v is a cyclic vector for unitary representation π and v' is a cyclic vector for unitary representation π'. If $\mathrm{Hom}_G(\pi, \pi') = \{0\}$, show $v \oplus v'$ is a cyclic vector for $\pi \oplus \pi'$.

Solution:

By Exercise 6.4.18, we know $v \oplus v' = (v, v')$ is a cyclic vector for $\pi \oplus \pi'$ if and only if it is a separating vector for $\mathrm{Hom}_G(\pi \oplus \pi', \pi \oplus \pi')$. Using Exercise 6.4.19, (v, v') is a separating vector for this algebra if and only if $(A, B)(v, v') = (A \oplus B)(v \oplus v') = (Av, Bv') = (0, 0)$ implies $A = 0$ and $B = 0$ where $A \in \mathrm{Hom}_G(\pi, \pi)$ and $B \in \mathrm{Hom}_G(\pi', \pi')$. So (v, v') is a cyclic vector if and only if v is a separating vector for $\mathrm{Hom}_G(\pi, \pi)$ and v' is a separating vector for $\mathrm{Hom}_G(\pi', \pi')$. But again using Exercise 6.4.18, this occurs if and only if v is a cyclic vector for π and v' is a cyclic vector for π'.

$\square$

21. Let V be finite dimensional complex vector space with basis $e_1, \ldots, e_n$. Let $\mathrm{GL}(n, \mathbb{C})$ be the group of invertible $n \times n$ matrices under multiplication.

(a) Show there is a one-to-one onto correspondence between the representations π of G on V and the continuous homomorphisms $\Pi : G \to \mathrm{GL}(n, \mathbb{C})$, (determined from the basis $e_1, \ldots, e_n$). Such Π are called matrix representations of G.

(b) Show if V is a Hilbert space and $e_1, \ldots, e_n$ is an orthonormal basis, then π is unitary if and only if $\Pi(G) \subseteq U(n)$, the group of $n \times n$ unitary matrices.

Solution:

For (a), note one can easily check the correspondence $\pi \mapsto \Pi$ where π is a representation of G on $\mathbb{C}^n$ and $\Pi : G \mapsto \mathrm{GL}(n, \mathbb{C})$ given by $\Pi(g) = [\pi(g)e_1, \pi(g)e_2, \ldots, \pi(g)e_n]$ where $\mathbb{C}^n$ is made up of column vector is one-to-one and onto. Note the inverse is the mapping $\Pi \mapsto \pi$ where $\pi(g)(v) = \Pi(g)v$. Finally (b) follows from the fact that each linear transformation $B : \mathbb{C}^n \to \mathbb{C}^n$ is given by a matrix $C \in M_{n\times n}(\mathbb{C})$ where $B(v) = Cv$ and B is a unitary transformation if and only if C is an $n \times n$ unitary matrix.

□

22. Let π be an irreducible unitary representation. Let n be a natural number. Show $\mathrm{Hom}_G(n\pi, n\pi) \cong M_{n\times n}(\mathbb{C})$, the algebra of n by n complex matrices.

Solution:

Let π be a unitary irreducible representation. Consider $n\pi$ which acts on $\mathcal{H}^n$ where $\mathcal{H}$ is the Hilbert space of π. Note a bounded linear transformation A of $\mathcal{H}^n$ has form $A(v_1, v_2, \ldots, v_n) = (\sum_{i=1}^n A_{i,1}v_i, \sum_{i=1}^n A_{i,2}v_i, \ldots, \sum_{i=1}^n A_{i,n}v_i)$ where $A_{i,j} : \mathcal{H} \to \mathcal{H}$ is a bounded linear transformation. We claim $A \in \mathrm{Hom}_G(n\pi, n\pi)$ if and only if $A_{i,j} \in \mathrm{Hom}_G(\pi, \pi)$. In fact in this case

$$\begin{aligned} A(n\pi)(g)(v_1, v_2, \ldots, v_n) &= A(\pi(g)v_1, \pi(g)v_2, \ldots, \pi(g)v_n) \\ &= (\sum_{i=1}^n A_{i,1}\pi(g)v_i, \sum_{i=1}^n A_{i,2}\pi(g)v_i, \ldots, \sum_{i=1}^n A_{i,n}\pi(g)v_i) \\ &= (n\pi)(g)(\sum_{i=1}^n A_{i,1}v_i, \ldots, \sum_{i=1}^n A_{i,n}v_n) \\ &= (\sum_{i=1}^n \pi(g)A_{i,1}v_i, \sum_{i=1}^n \pi(g)A_{i,2}v_i, \ldots, \sum_{i=1}^n \pi(g)A_{i,n}v_i) \end{aligned}$$

for all v and g. But this occurs if and only if for each i and any $v_i \in \mathcal{H}$, $A_{i,j}\pi(g)v_i = \pi(g)A_{i,j}v_i$ for all $g \in G$. So $A \in \mathrm{Hom}_G(n\pi, n\pi)$ if and only if $A_{i,j} \in \mathrm{Hom}_G(\pi, \pi)$ for all i, j. Since π is unitary and irreducible, $\mathrm{Hom}_G(\pi, \pi) = \mathbb{C}I$. Thus each $A_{i,j} = c_{i,j}I$ for some $c_{i,j} \in \mathbb{C}$. Consequently any $A \in \mathrm{Hom}_G(n\pi, n\pi)$ has form $A(v_1, v_2, \ldots, v_n) = (\sum_{i=1}^n c_{i,1}v_i, \sum_{i=1}^n c_{i,2}v_i, \ldots, \sum_{i=1}^n c_{i,n}v_i)$ and one can easily check $A \leftrightarrow [c_{i,j}]$ is a $*$-algebra isomorphism of $\mathrm{Hom}_G(n\pi, n\pi)$ onto $M_{n\times n}(\mathbb{C})$.

□

23. Let π be an irreducible unitary representation of finite dimension d. Show $n\pi$ is cyclic if and only if $n \leqslant d$.

Solution: Assume $n \leqslant d$. Let $e_1, e_2, \ldots, e_d$ be an orthonormal basis of $\mathcal{H}_\pi$. One can show $\mathrm{Hom}_G(n\pi, n\pi) = M_n(\mathcal{H}_\pi) = \{[c_{i,j}I] \mid c_{i,j} \in \mathbb{C}, 1 \leqslant i, j \leqslant n\}$. By Exercise 6.4.18, we need to find a separating vector for this algebra. Take $v = e_1 \oplus e_2 \oplus \cdots \oplus e_n$. Suppose $A \in \mathrm{Hom}_G(n\pi, n\pi)$ and $Av = 0$. This gives $\sum_{j=1}^n c_{i,j}e_j = 0$ for $i = 1, 2, \ldots, n$. By linear independence, $c_{i,j} = 0$ for all i and j. So $A = 0$.

Now suppose $n > d$. We show $M_n(\mathcal{H}_\pi)$ cannot have a separating vector. Indeed, suppose $v_1 \oplus v_2 \oplus \cdots \oplus v_n$ is a separating vector. Then for any $n \times n$

matrix, $[c_{i,j}]$, $\sum_{j=1}^{n} c_{i,j}v_j = 0$ for all i from 1 to n implies all $c_{i,j}$ are 0. In particular, the vectors $v_1, v_2, \ldots, v_n$ are linearly independent, a contradiction of $n > d$. □

24. Let $\pi_1, \pi_2, \ldots, \pi_s$ be pairwise inequivalent finite dimensional irreducible unitary representations of topological group G. Show

$$\bigoplus_{j=1}^{s} n_j\pi_j \text{ is cyclic if and only if } n_j \leqslant d(\pi_j) \text{ for } j = 1, 2, \ldots .s.$$

Solution:

Exercise 6.4.20 implies $\bigoplus_{j=1}^{s} n_j\pi_j$ has a cyclic vector if and only if each $n_j\pi_j$ has a cyclic vector for $\text{Hom}_G(n_i\pi_i, n_j\pi_j) = \{0\}$ for $i \neq j$. Now by Exercise 6.4.23, $n_j\pi_j$ has a cyclic vector if and only if $n_j \leqslant d_j$. So $\bigoplus_{j=1}^{s} n_j\pi_j$ has a cyclic vector if and only if $n_j \leqslant d_j$ for $j = 1, 2, \ldots, s$.

□

Exercise Set 6.5

1. Schur's Lemma 6.48 shows that for unitary or $*$ representations π, one has π is irreducible if and only if $\text{Hom}(\pi, \pi) = \mathbb{C}I$. There is a finite dimensional version of Schur's Lemma.

(a) Show if π is an irreducible complex finite dimensional representation of a group or an algebra, then $\text{Hom}(\pi, \pi) = \mathbb{C}I$.

(b) Show the converse to (a) is false. (**Hint:** Consider the group of invertible upper triangular 2×2 matrices in $\text{GL}(2, \mathbb{R})$.)

Solution:

For (a), note if π is finite dimensional and irreducible and $A \in \text{Hom}(\pi, \pi)$, then since this is over the complexes, there is an eigenvalue λ for A; and thus $\{v \in V \mid Av = \lambda v\}$ is a nonzero linear subspace of V and since $Av = \lambda v$ implies $A\pi(x)v = \pi(x)Av = \lambda\pi(x)v$, we see it is invariant. Being closed and nonzero, the irreducibility of π gives $Av = \lambda v$ for all $v \in V$. Thus $A = \lambda I$ and $\text{Hom}(\pi, \pi) = \mathbb{C}I$.

Let G be the group of all matrices $g = \left[\begin{smallmatrix} a & b \\ 0 & c \end{smallmatrix}\right]$ where $ac \neq 0$. Take $V = \mathbb{C}^2$ and $\pi(g)\left[\begin{smallmatrix} c_1 \\ c_2 \end{smallmatrix}\right] = \left[\begin{smallmatrix} a & b \\ 0 & c \end{smallmatrix}\right]\left[\begin{smallmatrix} c_1 \\ c_2 \end{smallmatrix}\right] = \left[\begin{smallmatrix} ac_1+bc_2 \\ cc_2 \end{smallmatrix}\right]$. Note $V_0 = \mathbb{C} \oplus \{0\} = \{\left[\begin{smallmatrix} c_1 \\ 0 \end{smallmatrix}\right] \mid c_1 \in \mathbb{C}\}$ is invariant. However, if $A = \left[\begin{smallmatrix} a_{11} & a_{12} \\ a_{21} & a_{22} \end{smallmatrix}\right] \in \text{Hom}(\pi, \pi)$, then

$$\left[\begin{smallmatrix} a_{11} & a_{12} \\ a_{21} & a_{22} \end{smallmatrix}\right]\left[\begin{smallmatrix} a & b \\ 0 & c \end{smallmatrix}\right] = \left[\begin{smallmatrix} a & b \\ 0 & c \end{smallmatrix}\right]\left[\begin{smallmatrix} a_{11} & a_{12} \\ a_{21} & a_{22} \end{smallmatrix}\right]$$

for all a, b, c where $ac \neq 0$. This gives $a_{11}a = aa_{11} + ba_{2,1}$, $a_{11}b + a_{12}c = aa_{12} + ba_{22}$, $a_{21}a = ca_{21}$, and $ba_{21} + ca_{22} = ca_{22}$ when $ac \neq 0$. So $a_{2,1} = 0$, $a_{1,2} = 0$, and $a_{11} = a_{22}$. Thus $A = cI$ and we see $\text{Hom}(\pi, \pi) = \mathbb{C}I$. So the converse to (a) is false.

□

2. Let π be a unitary or $*$ representation. Let π_0 be an irreducible unitary or $*$ representation and let P_0 be the π_0 primary projection in $\mathrm{Hom}(\pi, \pi)$. Show P_0 commutes with every operator in $\mathrm{Hom}(\pi, \pi)$. Such a projection is said to be *central.*

Solution:

By Corollary 6.54 we know that P_0 is the orthogonal projection onto the linear span of the ranges of all A in $\mathrm{Hom}(\pi_0, \pi)$. Let $B \in \mathrm{Hom}(\pi, \pi)$. Then $BA \in \mathrm{Hom}(\pi_0, \pi)$ and so $P_0BP_0A = BP_0A$. This gives $P_0BP_0 = BP_0$ for all $B \in \mathrm{Hom}(\pi, \pi)$. Since $\mathrm{Hom}(\pi, \pi)$ is closed under adjoints, we see $P_0B^*P_0 = P_0B^*$ for all $B^* \in \mathrm{Hom}(\pi, \pi)$. This gives $P_0BP_0 = BP_0 = BP_0$ and so P_0 commutes with all $B \in \mathrm{Hom}(\pi, \pi)$.

□

3. Let π and ρ be unitary representations of a group G. The representations π and ρ are said to be **disjoint** if $\mathrm{Hom}_G(\pi, \rho) = \{0\}$. Show $\mathrm{Hom}_G(\pi, \rho) = \{0\}$ if and only if π and ρ have no unitarily equivalent subrepresentations.

Solution:

Assume A is a nonzero operator in $\mathrm{Hom}_G(\pi, \rho)$. By Theorem 6.47, π restricted to $(\ker A)^\perp$ is unitarily equivalent to ρ restricted to the closure of the range of A.

Conversely, if $\mathcal{H}_0$ is a closed nonzero π linear invariant subspace of $\mathcal{H}_\pi$ and $\mathcal{K}_0$ is a closed nonzero ρ invariant subspace of $\mathcal{K}_\rho$ and $U : \mathcal{H}_0 \to \mathcal{K}_0$ is a unitary equivalence of $\pi_0 = \pi|_{\mathcal{H}_0}$and $\rho_0 = \rho|_{\mathcal{K}_0}$, then the operator A defined by $A(v_0 + v_0^\perp) = Uv_0$ where we are using the direct sum decomposition $\mathcal{H} = \mathcal{H}_0 \oplus \mathcal{H}_0^\perp$ is a nonzero element in $\mathrm{Hom}_G(\pi, \rho)$. This is a consequence of the invariance of $\mathcal{H}_0^\perp$ under π.

□

4. Let π and ρ be unitary representations of a topological group G. Show if π and ρ_1 and π and ρ_2 are disjoint, then π and $\rho_1 \oplus \rho_2$ are disjoint.

Solution:

Assume A is a nonzero operator in $\mathrm{Hom}_G(\pi, \rho_1 \oplus \rho_2)$. If $\mathcal{H}_1$ is the Hilbert space for ρ_1 and $\mathcal{H}_2$ is the Hilbert space for ρ_2, then we have projection operators $P_1 : \mathcal{H}_1 \oplus \mathcal{H}_2 \to \mathcal{H}_1$ and $P_2 : \mathcal{H}_1 \oplus \mathcal{H}_2 \to \mathcal{H}_2$ given by $P_i(v_1 \oplus v_2) = v_i$. An easy calculation shows $P_i \in \mathrm{Hom}_G(\rho_1 \oplus \rho_2, \rho_i)$. Using Lemma 6.46, we thus have $P_iA \in \mathrm{Hom}_G(\pi, \rho_i)$ for $i = 1, 2$. Now $A = P_1A \oplus P_2A$ implies $P_1A \neq 0$ or $P_2A \neq 0$. Hence $\mathrm{Hom}_G(\pi, \rho_1) \neq \{0\}$ or $\mathrm{Hom}_G(\pi, \rho_2) \neq \{0\}$.

□

5. Let ρ be a discretely decomposable unitary representation. Show $\mathrm{Hom}(\rho, \rho)$ is commutative if and only if the multiplicity of any irreducible unitary representation π in ρ is either 1 or 0.

Solution: Assume $m(\pi, \rho)$ is either 1 or 0 for all $\pi \in \hat{G}_c$. Thus $\rho = \oplus \pi_i$ where π_i are mutually nonequivalent unitary representations. Let $A \in \mathrm{Hom}(\rho, \rho)$ and let $A_{i,j} = E_iAE_j$ where E_j is the orthogonal projection onto $\mathcal{H}_j$. Then $A_{i,j}\pi_j(x)(v) = A_{i,j}\rho(x)v = E_iAE_j\rho(x)v = \rho(x)E_iAE_jv = \pi_i(x)A_{i,j}v$. So $A_{i,j} = 0$ for $i \neq j$ and $A_{i,i} = c_iE_i$ for some c_i. Thus $\mathrm{Hom}(\rho, \rho) = \{\sum c_iE_i \mid c_i \in$

$\mathbb{C}\}$. So this algebra is commutative.

Conversely suppose this algebra is commutative and $m(\pi, \rho) > 1$ for some $\pi \in \hat{G}_c$. Let P be the π primary projection. Then ρ restricted to $P\mathcal{H}$ is $n\pi$ for some $n > 1$. Now look at $n\pi$ on $P\mathcal{H}$. We have an invariant subspace of form $\mathcal{H}_1 \oplus \mathcal{H}_2$ of $P\mathcal{H}$ where $\rho|_{\mathcal{H}_i} \cong \pi$. Let $U_i \in \mathrm{Hom}(\pi, \rho|_{\mathcal{H}_i})$ be unitary and let E_i be the orthogonal projections onto $\mathcal{H}_i$. Consider all operators of form $c_{11}E_1 + c_{12}U_1U_2^{-1}E_2 + c_{21}U_2U_1^{-1}E_1 + c_{22}E_2$. These commute with ρ and are noncommutative. Indeed $U_1U_2^{-1}E_2U_2U_1^{-1}E_1 = E_1$ and $U_2U_1^{-1}E_1U_1U_2^{-1}E_2 = E_2$ □

6. Let ρ be a discretely decomposable unitary representation. Show the center of the algebra $\mathrm{Hom}(\rho, \rho)$ consists of only the operators cI where c is complex if and only if ρ is unitarily equivalent to $n\pi$ for some unique π in $\hat{G}_c$ and some unique cardinal n.

Solution:

Let ρ be discretely decomposable. So $\rho = \oplus n_i\pi_i$ where $\mathrm{Hom}(\pi_i, \pi_j) = 0$ for $i \neq j$ and each π_i is irreducible. By Exercise 6.5.2, if P_i is the primary projection for the representation π_i, then P_i is in the center of $\mathrm{Hom}(\rho, \rho)$ for each i. Thus $P_i = c_iI$ for some c_i for each i. But $P_iP_j = 0$ for $i \neq j$. So there is an i_0 with $P_{i_0} = I$ and using Theorem 6.53, $\rho = n_{i_0}\pi_{i_0}$.

Conversely, suppose $\rho = n\pi$ where π is an irreducible unitary representation. Hence $\mathcal{H}_\rho = n\mathcal{H}_\pi$. Let E_j be the orthogonal projection on the ith entry in the decomposition of $n\mathcal{H}_\pi$. Since π is irreducible, $\mathrm{Hom}(\pi, \pi) = \mathbb{C}I_\pi$. Now let A be in the center of $\mathrm{Hom}(\rho, \rho)$. Consider E_jAE_i. Since $E_i \in \mathrm{Hom}(\rho, \rho)$, A commutes with this operator. So $E_jAE_i = AE_jE_i = 0$ if $j \neq i$ and this implies $A = \sum E_iAE_i$.

Also $\rho(x)E_iAE_i = E_i\rho(x)AE_i = E_iA\rho(x)E_i = E_iAE_i\rho(x)$ and since $\rho(x)$ is irreducible on $\mathcal{H}_i$, Schur's Lemma gives $E_iAE_i = c_iE_i$ for some $c_i \in \mathbb{C}$. Next let $U_{i,j}$ be a nonzero intertwiner between ρ on $\mathcal{H}_j$ and ρ on $\mathcal{H}_i$ and set $T = E_iU_{i,j}E_j$. Then T belongs to $\mathrm{Hom}(\rho, \rho)$ and thus commutes with A. But $AT = (\oplus c_kE_k)E_iU_{i,j}E_j = c_iE_iU_{i,j}E_j$ and $TA = E_iU_{i,j}E_j(\oplus c_kE_k) = E_iU_{i,j}(c_jE_j) = c_jE_iU_{i,j}E_j$. This implies $c_i = c_j$ for $i \neq j$. Consequently, if $c = c_i = c_j$ for all i, j, one has $A = \oplus(c_iE_i) = c(\oplus E_i) = cI$.

□

7. Let $\hat{G}_1$ be the one dimensional unitary representations of G; i.e., the collection of one-dimensional characters of G. For $\chi, \chi' \in \hat{G}_1$, define $(\chi\chi')(g) = \chi(g)\chi'(g)$. Show with this multiplication $\hat{G}_1$ is a group. Next define a topology on $\hat{G}_1$ by defining a nonempty subset U of $\hat{G}_1$ to be open if for each χ_0 in U, there is a compact subset K of G and an $\epsilon > 0$ such that $\{\chi \mid |\chi(g) - \chi_0(g)| < \epsilon \text{ for } g \in K\} \subseteq U$. Show with this topology $\hat{G}_1$ is a Hausdorff topological group.

Solution:

It is easy to see $\hat{G}_1$ is a group. It is T_1 because if $\chi_1 \neq \chi_0$, there is a g with $|\chi_1(g) - \chi_0(g)| = \epsilon > 0$. Take $K = \{g\}$. Then $\{\chi \mid |\chi(g) - \chi_1(g)| < \epsilon\}$ is a neighborhood of χ_1 which does not contain χ_0.

We show $\chi \mapsto \chi^{-1}$ is continuous. Fix χ_0, $\epsilon > 0$ and K a compact subset of G. Set $V = \{\chi \mid |\chi(g) - \chi_0^{-1}(g)| < \epsilon \text{ for } g \in K\}$. This is a neighborhood of χ_0^{-1}.

Now K^{-1} is a compact subset and $U = \{x \mid |\chi(g) - \chi_0(g)| < \epsilon \text{ for } g \in K^{-1}\}$ is a neighborhood of χ_0. Note $U^{-1} \subseteq V$ for if $|\chi(g) - \chi_0(g)| < \epsilon$ for $g \in K^{-1}$, then $|\chi^{-1}(g) - \chi_0^{-1}(g)| = |\chi(g^{-1}) - \chi_0(g^{-1})| < \epsilon$ for $g \in K$.

Let χ_1 and χ_2 be in $\hat{G}_1$. Suppose $\epsilon > 0$ and $K \subseteq G$ is compact. Consider $V = \{x \mid |\chi(g) - \chi_1(g)\chi_2(g)| < \epsilon \text{ for } g \in K\}$. Then if $\epsilon > 0$ and K is a compact subset of G and χ_1, $\chi_2 \in \hat{G}_1$ are fixed, $V = \{\chi \mid |\chi(g) - \chi_1(g)\chi_2(g)| < \epsilon \text{ for } g \in K\}$ is a neighborhood of $\chi_1\chi_2$. Let $U_i = \{\chi \mid |\chi(g) - \chi_i(g)| < \frac{\epsilon}{2} \text{ for } g \in K\}$ for $i = 1, 2$. Then if $\chi \in U_1$ and $\chi' \in U_2$ and $k \in K$,

$$\begin{aligned} |\chi(g)\chi'(g) - \chi_1(g)\chi_2(g)| &\leqslant |\chi(g)\chi'(g) - \chi_1(g)\chi'(g)| + |\chi_1(g)\chi'(g) - \chi_1(g)\chi_2(g)| \\ &= |\chi(g) - \chi_1(g)| + |\chi'(g) - \chi_2(g)| \\ &< \epsilon \end{aligned}$$

for $|\chi| = 1$ for any $\chi \in \hat{G}_1$. So $U_1U_2 \subseteq V$ and multiplication is continuous.

Since $\hat{G}_1$ is a T_1 topological group, Lemma 5.17 shows it is Hausdorff.

□

8. Show the topologies given in Exercise 6.5.7 for the dual groups of $G = \mathbb{R}$, $G = \mathbb{T}$, and $G = \mathbb{Z}$ where these groups have their usual topologies are the usual topologies on $\mathbb{R}$, $\mathbb{Z}$, and $\mathbb{T}$, respectively.

Solution:

For $G = \mathbb{R}$ we have $\hat{G} = \mathbb{R}$ under the identification $\omega \mapsto e_\omega$ where $e_\omega(x) = e^{2\pi i w x}$. Now a neighborhood of e_0 is the set U of all e_ω where $|e_\omega(x) - 1| < \epsilon$ for $x \in K$. So $|e^{2\pi i \omega x} - 1| < \epsilon$ for $x \in K$. Since K is compact, $K \subseteq [-a, a]$ for some $a > 0$ and thus U contains all e_w with $|e^{2\pi i \omega x} - 1| < \epsilon$ for $|x| \leqslant a$. Now $|e^{iy} - 1| \leqslant |y|$ for $y \in \mathbb{R}$. So U contains the e_ω with $|2\pi w a| < \epsilon$ or $\omega \in (\frac{-\epsilon}{2\pi a}, \frac{\epsilon}{2\pi a})$ and thus corresponds to a neighborhood of 0 in $\mathbb{R}$. Conversely, consider $(-a, a)$ for $a > 0$. Take $\epsilon = 1$, $K = [-\frac{1}{4a}, \frac{1}{4a}]$ and suppose $|e_\omega(x) - 1| < \epsilon$ for $x \in K$. We claim this implies $\omega \in (-a, a)$. In fact, if not, $|\omega| \geqslant a$ and if $x = \frac{1}{4\omega}$, then $x \in K$ and so $|e^{2\pi i \omega (\frac{1}{4\omega})} - 1| = |i - 1| = \sqrt{2} > 1$, a contradiction. So $(-a, a)$ contains a neighborhood of 0 in the topology of Exercise 6.5.7 and thus the neighborhoods of 0 which correspond to e_0 are the same.

For $\mathbb{Z}$, we have the identification $\hat{\mathbb{Z}} = \mathbb{T}$ under the mapping $z \leftrightarrow e_z$ where $e_z(k) = z^k$. Fix $\epsilon > 0$. Consider the neighborhood U of 1 in $\mathbb{T}$ consisting of the z with $|z^k - 1| < \epsilon$ for k in a finite subset F of $\mathbb{Z}$. Since $z \mapsto z^k$ is continuous at $z = 1$, there is a $\delta > 0$ such that if $|z - 1| < \delta$, $|z^k - 1| < \epsilon$ for $k \in F$. Thus U contains a neighborhood of 1 in the usual topology of $\mathbb{T}$.

Conversely consider $V = \{z \in \mathbb{T} \mid |z - 1| < \epsilon\}$ where $\epsilon > 0$. Then this set is a neighborhood of 1 in the topology of Exercise 6.5.7, namely take $K = \{1\}$. So the group topologies are the same.

Finally for $\mathbb{T}$, we have $k \leftrightarrow e_k$ where $e_k(z) = z^k$. Note if $|e_k(z) - 1| < 1$ for $z \in \mathbb{T}$, then $|z^k - 1| < 1$ for $|z| = 1$. This implies $k = 0$ for $z \mapsto z^k$ is an onto mapping whenever $k \neq 0$. So this neighborhood in the topology of Exercise 6.5.7 consists only of the point 0. This implies $\hat{\mathbb{T}}$ has the discrete topology which is the topology of $\mathbb{Z}$.

□

9. Let $\mathbb{R}^*$ be the nonzero reals under multiplication with the relative topology. Determine the character group of this abelian group.

Solution: Let χ be a character of $G = \mathbb{R}^*$ and define $\pi(x) = \chi(e^x)$. Then π is a character of $\mathbb{R}$. So there is a ω with $\chi(e^x) = \pi(x) = e^{2\pi\omega x i} = (e^x)^{2\pi\omega i}$. Thus $\chi(a) = a^{2\omega i}$ for $a > 0$. Now $\chi(-1)^2 = 1$. So $\chi(-1) = \pm 1$. Thus each character χ of G is given by either $\chi_{+,\omega}$ or $\chi_{-,\omega}$ where $\chi_{+,\omega}(a) = |a|^{2\pi\omega i}$ and $\chi_{-,\omega}(a) = \frac{a}{|a|}(|a|^{2\pi\omega i})$. □

10. Determine the character group of the nonzero complex numbers under multiplication.

Solution: We know χ restricted to $\mathbb{R}^*$ is given by $\chi(a) = a^{2\pi i\omega}$ for $a > 0$ and χ on $\mathbb{T}$ is given by $\chi(u) = u^m$ for some $m \in \mathbb{Z}$. Thus $\chi(au) = a^{2\pi i\omega}u^n$. It is easy to check this is a character. □

11. Let G be the finite cyclic group $\mathbb{Z}/m\mathbb{Z}$. Find the character group of G.

Solution:

$G = \mathbb{Z}/m\mathbb{Z}$ where we assume $m > 1$. Note if χ is a character of G, then since $m(1 + m\mathbb{Z}) = 0 + mZ$, that if $\chi(1 + m\mathbb{Z}) = z$ where $|z| = 1$, then $\chi(1 + m\mathbb{Z})^m = \chi(m(1 + m\mathbb{Z})) = \chi(0 + m\mathbb{Z}) = 1$. So $\chi(1 + m\mathbb{Z})$ is an mth root of unity. These roots of unity form a multiplicative group which we denote by $\hat{G}$. For each $z \in \hat{G}$, define $\chi_z(k + m\mathbb{Z}) = \chi_z(k(1 + m\mathbb{Z})) = z^k$. Each χ_z is a character and every character has this form.

□

12. Let G be a locally compact Hausdorff group with center $Z(G)$.

(a) Show the regular representation is faithful; i.e., $\lambda(x) = I$ if and only if $x = e$.

(b) Show if the regular representation is discretely decomposable, then the central characters on G separate the points of $Z(G)$; that is show if $g \neq e$ is in the center, there is an irreducible unitary representation π of G such that the central character χ_π for π satisfies $\chi_\pi(g) \neq 1$.

Solution:

Note if $x \neq e$, then by local compactness and Proposition 5.23, there is an $f \in C_c(G)$ with $f(e) = 1$ and $f(x) = 0$. So $f \neq 0$ in $L^2(G)$. Thus $\lambda(x)f \neq f$ for otherwise $\lambda(x)f(y) = f(x^{-1}y) = f(y)$ a.e. But f is continuous. This implies $f(x^{-1}y) = f(y)$ everywhere. Taking $y = x$, we would have $1 = 0$. Thus $\lambda(x) \neq I$ if $x \neq e$. This implies λ is faithful for any locally compact group.

Note if $g \neq e$ is in $Z(G)$, then $\lambda(g) \neq I$. But λ is discretely decomposable; i.e. unitarily equivalent to a direct sum of irreducible unitary representations. This implies there is an irreducible representation π of G with $\pi(g) \neq I$. But by Corollary 6.50, π on the center is given by a central character x; i.e. $\pi(g) = \chi(g)I$. So $\chi(g) \neq 1$.

□

13. Let $\mathcal{H}$ be a Hilbert space. For $A \in \mathcal{B}(\mathcal{H})$, define $\pi(A) = A$. Show π is an irreducible representation.

Solution:

It is easy to check $\pi(A) = A$ is a $*$-representation of the C^* algebra $\mathcal{B}(\mathcal{H})$. Now let $\mathcal{H}_0$ be a closed invariant subspace of $\mathcal{H}$ under π. Thus $A\mathcal{H}_0 \subseteq \mathcal{H}_0$ for all bounded linear operators A on $\mathcal{H}$. This is impossible unless $\mathcal{H}_0 = \{0\}$ or $\mathcal{H}_0 = \mathcal{H}$; otherwise take a continuous linear functional f on $\mathcal{H}$ such that $f|_{\mathcal{H}_0} \neq 0$ and a nonzero vector $v \in \mathcal{H}_0^{\perp}$ and define $Aw = f(w)v$. Then $\{0\} \neq A\mathcal{H}_0 \subseteq \mathcal{H}_0^{\perp}$.

□

14. Show all finite dimensional unitary representations of a group G are discretely decomposable.

Solution:

We note if ρ is a finite dimensional unitary representation of G, then if $\mathcal{H}$ is its Hilbert space and $\mathcal{H}_0$ is an invariant subspace, $\mathcal{H}_0^{\perp}$ is also invariant. So let $\mathcal{H}_1, \mathcal{H}_2, \ldots, \mathcal{H}_k$ be a maximal collection of pairwise orthogonal irreducible invariant subspaces of $\mathcal{H}$. Then $\mathcal{H}_0 = \oplus_{i=1}^{k} \mathcal{H}_i$ is invariant and if $\mathcal{H}_0^{\perp} \neq \{0\}$, then any nonzero invariant subspace $\mathcal{H}_{k+1}$ of minimum dimension would be an irreducible invariant subspace, and we could enlarge the collection $\mathcal{H}_1, \mathcal{H}_2, \ldots, \mathcal{H}_k$. Thus $\mathcal{H}_0 = \mathcal{H}$.

□

15. Recall the $ax + b$ group is the set of all pairs $(a, b) \in \mathbb{R}^2$ with $a > 0$ with the relative topology and has multiplication defined by $(a, b)(a', b') = (aa', b + ab')$. Show every finite dimensional unitary representation of this group is one dimensional. (**Hint:** Use the discrete decomposability of a finite dimensional unitary representation on the subgroup $x + b = \{(1, b) \mid b \in \mathbb{R}\}$.)

Solution: By the previous problem, finite dimensional unitary representations are discretely decomposable. Thus if $H = \{(1, b) \mid b \in \mathbb{R}\}$, then π restricted to H decomposes into $\pi|_H = n(1)e_{\omega(1)} \oplus n(2)e_{\omega(2)} \oplus \cdots \oplus n(k)e_{\omega(k)}$. We claim $k = 1$ and $\omega(1) = 0$. If not, we may assume $\pi|_H$ has a nontrivial one dimensional invariant subspace $\mathcal{H}_0 = \langle v \rangle$. So $\pi(1, b)v = e^{2\pi i\omega b}v$ for all b where $\omega \neq 0$. Thus $\pi(a, 0)v \in \mathcal{H}$ and $\pi(1, b)\pi(a, 0)v = \pi(a, 0)\pi((a^{-1}, 0)(1, b)(a, 0))v = \pi(a, 0)\pi(1, a^{-1}b)v = e^{2\pi i\omega a^{-1}b}\pi(a, 0)v$. Thus $\pi(a, 0)v$ is in the $e_{a^{-1}\omega}$ primary subspace for $\pi|_H$. Since $\omega \neq 0$, one would then have $\mathcal{H}$ is infinite dimensional. □

16. Show every complex irreducible finite dimensional representation of the $ax + b$ group is one dimensional and is given by $\pi(a, x) = a^c$ for some complex number c. In particular, determine the one-dimensional characters for this group.

Solution: For each e_c where c is complex, let $V_c = \{v \mid \pi(1, b)v = e_c(b)v = e^{cb}v \text{ for } b \in \mathbb{R}\}$. Note V_c is a subspace of V. Also the V_c's are linearly independent. Indeed, if $\sum \lambda_j v_{c_j} = 0$ where the c_j are distinct, then there is a b such that e^{bc_j} are distinct. Indeed, just choose $b > 0$ so that $|b(c_i - c_j)| < 2\pi$ for all pairs i, j. Next note if $e^{bc} = e^{bc'}$ and $c \neq c'$, then $0 \neq bc - bc' = 2\pi ik$. So $0 < |b(c - c')| = 2\pi|k| \geqslant 2\pi$. Thus the v_{c_j} are eigenvectors corresponding to the distinct eigenvalues e^{bc_j} for the operator $\pi(1, b)$. So they are linearly independent.

We now claim if $V_c \neq \{0\}$, then $V_{ac} \neq \{0\}$ for $a > 0$. Indeed, $\pi(a^{-1}, 0)V_c \subseteq V_{ac}$ since

$$\begin{aligned}\pi(1,b)\pi(a^{-1},0)v &= \pi(a^{-1},0)\pi(a,0)\pi(1,b)\pi(a^{-1},0)v \\ &= \pi(a^{-1},0)\pi(1,ab)v = \pi(a^{-1},0)e^{cab}v \\ &= e_{ac}(b)\pi(a^{-1},0)v.\end{aligned}$$

Since V is finite dimensional, we have $V_c = \{0\}$ for $c \neq 0$.

Now let b_k be a countable dense subset of $\mathbb{R}$. Let V_1 be a nonzero eigenspace for $\pi(1, b_1)$ with eigenvalue λ_1. Note $\pi(1, b_1)\pi(1, b_2)v = \pi(1, b_2)\pi(1, b_1)v = \lambda_1\pi(1, b_2)v$ for all $v \in V_1$. So $\pi(1, b_2)V_1 \subseteq V_1$. Thus there is a nonzero subspace V_2 of V_1 which is the eigenspace for $\pi(1, b_2)$ on V_1 with corresponding eigenvalue λ_2. Repeating one obtains a decreasing sequence V_k of nonzero subspaces of V with $\pi(1, b_k)v = \lambda_k v$ for $v \in V_k$. By the finite dimensionality of V, there is a k with $V_k = V_j$ for $j > k$. Set $W = V_k$. Then $\pi(1, b_j)w = \lambda_j w$ for all j and $w \in W$. By continuity of π and density of the collection $\{b_k\}$ we then have for each $b \in \mathbb{R}$, there is a $\lambda(b) \in \mathbb{C}$ with $\pi(1, b)w = \lambda(b)w$ for all b and $w \in W$. Since $\pi(1, b_1)\pi(1, b_2) = \pi(1, b_1 + b_2)$, one then has $\lambda(b_1 + b_2) = \lambda(b_1)\lambda(b_2)$ for all b_1 and b_2. Consequently, λ is a continuous homomorphism of $\mathbb{R}$ into $\mathbb{C}^*$. Moreover, if $F(t) = \int_0^t \lambda(s)\,ds$, then

$$\begin{aligned}\lambda(s)F(t) &= \int_0^t \lambda(s)\lambda(u)\,du \\ &= \int_0^t \lambda(s+u)\,du \\ &= \int_s^{s+t} \lambda(u)\,du \\ &= F(s+t) - F(s)\end{aligned}$$

and in particular $\lambda'(s)F(t) = \lambda(s+t) - \lambda(s) = \lambda(s)(\lambda(t) - 1)$. This implies

$$\frac{\lambda'(s)}{\lambda(s)} = \frac{\lambda(t) - 1}{F(t)} = c$$

for some constant c. Setting $H(s) = e^{-cs}\lambda(s)$, we have

$$H'(s) = -ce^{-cs}\lambda(s) + e^{-cs}\lambda'(s) = -ce^{-cs}\lambda(s) + ce^{-cs}\lambda(s) = 0$$

for all s. This yields $\lambda(s) = ce^{cs}$ for all s. Since $\lambda(0) = 1$, $c = 1$ and $\lambda(s) = e^{cs}$. Thus for $w \in W$, $\pi(1, b)w = e^{cb}w$. In particular, $w \in V_c$. Since we saw $V_c = \{0\}$ when $c \neq 0$, we then conclude $\pi(1, b)w = w$ for all b if $w \in W$.

As seen earlier, $\pi(a, 0)V_0 \subseteq V_0$ and thus since π is irreducible, $V = V_0$. Now let $W_c = \{v \in V \mid \pi(a, 0)v = a^c v \text{ for all } a > 0\}$. Again we have $\pi(a', 0)W_c \subseteq W_c$ for $a' > 0$. Arguing as before if a_k is a dense sequence in $\mathbb{R}^+$, there is a decreasing sequence of nonzero linear subspaces W_k of V with $\pi(a_k, 0)w = \lambda_k w$

for $w \in W_k$. By finite dimensionality, there again exists a k with $W_i = W_k$ for $i > k$. The continuity and same argument shows there is a continuous homomorphism $\lambda : \mathbb{R}^+ \to \mathbb{C}^*$ such that $\pi(a, 0)w = \lambda(a)w$ for $w \in W_k$. Thus if w is a nonzero vector in W_k, the linear span of w is a closed invariant subspace. Since π is irreducible, $V = \langle w \rangle$. Now $t \mapsto \lambda(e^t)$ is a continuous homomorphism of $\mathbb{R}$ into $\mathbb{C}^*$. As noted earlier, this must have form $\lambda(e^t) = e^{ct} = (e^t)^c$ for some $c \in \mathbb{C}$. Thus the irreducible finite dimensional complex representations all have form $\pi(a, b) = a^c$ on a one dimensional vector space.

The representation $\pi(a, b) = a^c$ is unitary and hence a character if $|a^c| = 1$ for all $a > 0$. This occurs if $\mathrm{Re}\,(c \ln a) = 0$ for all $a > 0$. So c is pure imaginary and thus $\pi(a, b) = a^{iy}$ for some real number y. □

17. Let D_4 be the dihedral group. Thus D_4 has eight elements and two generators a and b where $a^2 = b^4 = e$ and $aba = b^3$. Find the center of D_4 and then determine the irreducible unitary representations of D_4 which have trivial central character.

Solution:

Note b^2 commutes with all powers of b and since

$$\begin{aligned} b^2 a &= a^2 b^2 a \\ &= a^2 b a^2 b a \\ &= a(aba)(aba) \\ &= a b^3 b^3 \\ &= a b^6 \\ &= a b^2 , \end{aligned}$$

b^2 commutes with a and hence all powers of a. So b^2 is in the center of G. Note the center has either 1, 2, or 4 elements. It can't have four elements for then G would be abelian since G mod the center would have two elements and thus would be commutative. So $\{1, b^2\}$ is the center of G.

To finish, we need to find the irreducible unitary representations π of G with $\pi(b^2) = I$. Such a π would factor to an irreducible unitary representation $\bar{\pi}$ of $G/Z(G) = \{eZ(G), aZ(G), bZ(G), abZ(G)\} \cong \mathbb{Z}_2 \times \mathbb{Z}_2$. Since this group is abelian, by Corollary 6.51 the representation $\bar{\pi}$ is one dimensional. Thus $\bar{\pi}(aZ(G)) = \pm 1$ and $\bar{\pi}(bZ(G)) = \pm 1$. So there are four one dimensional irreducible unitary representations on G which have trivial central character. These are given on the generators a and b by

$$\begin{aligned} &\pi_1(a) = 1,\ \pi_1(b) = 1 \\ &\pi_2(a) = 1,\ \pi_2(b) = -1 \\ &\pi_3(a) = -1,\ \pi_2(b) = 1 \text{ and} \\ &\pi_4(a) = -1,\ \pi_2(b) = -1\,. \end{aligned}$$

□

18. Let X be a Hausdorff topological space with disjoint compact subsets K_1 and K_2. Show there are disjoint open subsets V_1 and V_2 with $K_1 \subseteq V_1$ and $K_2 \subseteq V_2$.

Solution:

Let K_1 and K_2 be disjoint compact sets in a Hausdorff space X. First fix $x \in K_1$. For each $y \in K_2$ pick open neighborhoods U_y of x and V_y of y with $U_y \cap V_y = \emptyset$. The V_y for $y \in K_2$ cover K_2. Thus there is a finite set $y_1, y_2, \ldots, y_m$ with $K_2 \subseteq V_x$ if $V_x = \cup_{i=1}^m V_{y_i}$. Set $N_x = \cap_{i=1}^m U_{y_i}$. Then N_x is an open neighborhood of x and V_x is an open subset of X with $N_x \cap V_x = \emptyset$ and $K_2 \subseteq V$. The N_x for $x \in K_1$ cover K_1. Again we find a finite set $x_1, x_2, \ldots, x_n$ in K_1 with $K_1 \subseteq W_1$ where $W_1 = \cup_{i=1}^n N_{x_i}$. Set $W_2 = \cap_{i=1}^n V_{x_i}$. Note W_2 is open and contains K_2 and $W_1 \cap W_2 \subseteq \cup_{i=1}^n (N_{x_i} \cap V_{x_i}) = \emptyset$.

□

19. Let A be a bounded linear operator on $\mathcal{H}$. Show $||A^*A|| = ||A||^2$.

Solution: Clearly $||A^*A|| \leqslant ||A^*||\,||A|| = ||A||^2$. But

$$\begin{aligned} ||A^*A|| &= \sup\{|(A^*Av, w)|\, ||v|| = ||w|| = 1\} \\ &= \sup\{|(Av, Aw) \mid ||v|| = ||w|| = 1\} \\ &\geqslant \sup\{|(Av, Av) \mid ||v|| = 1\} \\ &= \sup\{||Av||^2 \mid ||v|| = 1\} \\ &= ||A||^2. \end{aligned}$$

□

20. Let $\mathcal{B}$ be a Banach $*$ algebra with no identity.

(a) Define $\mathcal{B}_e$ to be the set of all expressions of form $\lambda e + x$ for $\lambda \in \mathbb{C}$ and $x \in \mathcal{B}$. Define addition, multiplication, involution, and a norm on $\mathcal{B}_e$ by $(\lambda e+x)+(\mu e+y) = (\lambda+\mu)e+(x+y)$, $(\lambda e+x)(\mu e+y) = \lambda\mu e+(\lambda y+\mu x+xy)$, $(\lambda e + x)^* = \bar{\lambda} e + x^*$, and $||\lambda e + x|| = |\lambda| + ||x||$. Show $\mathcal{B}_e$ is a Banach $*$ algebra with an identity.

(b) Let π be a representation of the $*$ algebra $\mathcal{B}$ on a Hilbert space $\mathcal{H}$. Show $\tilde{\pi}$ defined by $\tilde{\pi}(\lambda e + x) = \lambda I + \pi(x)$ is a representation of the $*$ algebra $\mathcal{B}_e$.

Solution:

Showing $\mathcal{B}_e$ is a $*$-algebra is easy and it straight forward to check the norm satisfies the triangle inequality. Also note one has $||(\lambda e + x)^*|| = |\bar{\lambda}| + ||x^*|| = |\lambda| + ||x|| = ||\lambda e + x||$ and

$$\begin{aligned} ||(\lambda e + x)(\mu e + y)|| &= |\lambda\mu| + ||\mu x + \lambda y + xy|| \\ &\leqslant |\lambda\mu| + |\mu|\,||x|| + |\lambda|\,||y|| + ||x||\,||y|| \\ &\leqslant (|\lambda| + ||x||)(|\mu| + ||y||) \\ &= ||\lambda e + x||\,||\mu e + y||\,. \end{aligned}$$

To finish (a) we show completeness. Let $\lambda_n e + x_n$ be Cauchy. Note λ_n is bounded for the sequence of norms $|\lambda_n| + ||x_n||$ is bounded. By taking a subsequence, we may assume λ_n converges to λ. This implies x_n is Cauchy for

$$||x_m - x_n|| \leqslant || - (\lambda_m - \lambda_n)e|| + ||(\lambda_m - \lambda_n)e + (x_m - x_n)|| \to 0\,.$$

So $x_n \to x$ and we see $\lambda_n e + x_n \to \lambda e + x$.

The linearity of $\tilde{\pi}$ is easy. Also

$$\begin{aligned}\tilde{\pi}((\lambda e + x)(\lambda' e + x')) &= \tilde{\pi}(\lambda\lambda' e + \lambda x' + \lambda' x + x x') \\ &= \lambda\lambda' I + \lambda\pi(x') + \lambda'\pi(x) + \pi(x x') \\ &= \lambda\lambda' I + \lambda\pi(x') + \lambda'\pi(x) + \pi(x)\pi(x') \\ &= (\lambda I + \pi(x))(\lambda' I + \pi(x')) \\ &= \tilde{\pi}(\lambda e + x)\tilde{\pi}(\lambda' e + x')\end{aligned}$$

and thus $\tilde{\pi}$ is multiplicative. Finally $\tilde{\pi}(\lambda e + x)^* = (\lambda I + \pi(x))^* = \bar{\lambda} I + \pi(x^*) = \tilde{\pi}(\bar{\lambda} e + x^*) = \tilde{\pi}((\lambda e + x)^*)$ shows $\tilde{\pi}$ is a $*$-representation. So (b) holds.

□

21. Let X be a compact Hausdorff space and let μ be a regular Borel measure on X. Define $\rho(f)(\phi) = f(x)\phi(x)$ for $f \in C(X)$ and $\phi \in L^2(X, \mu)$.

(a) Show ρ is a representation of the $*$ algebra $C(X)$.

(b) Show the algebra $\operatorname{Hom}(\rho, \rho)$ is commutative.

(c) Let $p \in X$ and set $\pi(f) = f(p)$ for $f \in C(X)$. Determine the primary projection $P(\pi)$ for ρ and in particular, determine when $P(\pi) \neq 0$.

Solution:

For (a), note $\rho(f)$ is a bounded linear operator. Linearity is easy and $|\rho(f)\phi|_2^2 = \int |f(x)\phi(x)|^2\, d\mu(x) \leqslant |f|_\infty^2 |\phi|_2^2$. Also $\rho(f_1 + f_2) = \rho(f_1) + \rho(f_2)$, and $\rho(f_1 f_2)\phi = f_1 f_2 \phi = \rho(f_1)(f_2\phi) = \rho(f_1)\rho(f_2)\phi$. Finally,

$$\begin{aligned}(\rho(f)\phi_1, \phi_2)_2 &= \int f(x)\phi_1(x)\overline{\phi_2(x)}\, d\mu(x) \\ &= \int \phi_1(x)\overline{\bar{f}(x)\phi_2(x)}\, d\mu(x) \\ &= (\phi_1, \rho(\bar{f})\phi_2)_2 \\ &= (\phi_1, \rho(f^*)\phi_2)_2\end{aligned}$$

and thus $\rho(f^*) = \rho(f)^*$.

To see $\operatorname{Hom}(\rho, \rho)$ is commutative, note if $A \in \operatorname{Hom}(\rho, \rho)$, then if $A1 = h \in L^2(X, \mu)$, we have $A(\rho(f)1) = \rho(f)A1 = fh$. This implies $Af = fh$ for $f \in C(X)$. Using Exercise 6.1.14, one sees $C(X)$ is dense in L^2. Hence if $\phi \in L^2$, there is a sequence f_k in $C(X)$ converging to ϕ in $L^2(X, \mu)$. Using a subsequence, we may suppose f_k converges pointwise to ϕ a.e. μ. By continuity of A, $Af_k \to A\phi$. Hence $f_k h \to A\phi$ in L^2 and since we have pointwise convergence a.e., $A\phi = \phi h$.

Next note if $\{x \mid |h(x)| > M\}$ has positive measure, regularity implies there is a compact subset K of positive measure with $|h(x)| > M$ on K. Take $\phi = \frac{1}{\sqrt{\mu(K}}\chi_K$. Then $|\phi|_2 = 1$ and $|A\phi|_2^2 \geqslant \frac{1}{\mu(K)}\int M^2\chi_K\,d\mu = M^2$. Thus $||A|| \geqslant M$. Consequently, $|h|_\infty \leqslant ||A||$. One can easily check, $||A|| \leqslant |h|_\infty$. Concluding with some minor argument, we have $\mathrm{Hom}(\rho, \rho) \cong L^\infty(X, \mu)$ under $h \mapsto A$ where $A\phi = M_h\phi = h\phi$ is a $*$-algebra isometric isomorphism. Clearly this algebra is commutative for $M_h M_{h'} = M_{hh'} = M_{h'}M_h$.

Continuing, let $\pi(f) = f(p)$ for $f \in C(X)$. This is easily checked to be a $*$-representation of $C(X)$. Let $T \in \mathrm{Hom}(\pi, \rho)$. Then $T : \mathbb{C} \to L^2(X, \mu)$ and $T\pi(f) = \rho(f)T$ for all $f \in C(X)$. Consequently if $\phi = T1$, $T(f(p)1) = T\pi(f)1 = \rho(f)T1$ and thus $f(p)\phi = f\phi$ for all $f \in C(X)$. So $(f - f(p))\phi(x) = 0$ a.e. x. Since this holds for all $f \in C(X)$, ϕ must vanish a.e. on $X - \{p\}$. Indeed, if not, by regularity, there is a compact subset K of positive measure missing $\{p\}$ on which $\phi(x) \neq 0$. Now by Proposition 5.23, there is an $f \in C(X)$ with $f = 1$ on K and $f(p) = 0$. But then $(f - f(p))\phi \neq 0$ in L^2. Thus the linear span of the ranges of all such T's is the subspace of $L^2(X, \mu)$ consisting of the functions $c\chi_{\{p\}}$ where $c \in \mathbb{C}$. This subspace is nonzero if and only if $\mu\{p\} > 0$ and by Corollary 6.54, the projection $P(\pi)$ is the orthogonal projection given by $P(\pi)\phi = \frac{1}{\mu(\{p\})}(\phi, \chi_{\{p\}})\chi_{\{p\}}$. It is clear $P(\pi)$ is nonzero if and only if $\mu(\{p\}) > 0$.

□

22. Show the regular representation λ of $\mathbb{T}$ is discretely decomposable and determine all the primary projections for λ. (**Hint:** Use Theorem 1.20.)

Solution:

By the Plancherel Theorem, the functions $e_n(z) = z^n$ for $n \in \mathbb{Z}$ form an orthonormal basis for $L^2(\mathbb{T})$ and since $\lambda(w)e_n(z) = e_n(w^{-1}z) = w^{-n}z^n = \omega^{-n}e_n(z)$, we see $\{ce_n \mid c \in \mathbb{C}\}$ is a one dimensional invariant subspace of $L^2(\mathbb{T})$ and λ restricted to this subspace is unitarily equivalent to the representation χ_{-n} where $\chi_{-n}(w) = w^{-n}$. Thus λ is discretely decomposable. Moreover, the subspaces $\langle e_{-n}\rangle$ determine a pairwise orthogonal decomposition of $L^2(\mathbb{T})$ and λ on each of these subspaces is equivalent to the unitary representation χ_n of $\mathbb{T}$. In particular, the primary projection $P(\chi_n)$ is the operator $P(\chi_n)f)(z) = (f, e_{-n})_2 e_{-n}(z) = \frac{1}{2\pi}z^{-n}\int_0^{2\pi} f(e^{i\theta})e^{in\theta}\,d\theta$.

□

23. Show the regular representation λ of $\mathbb{R}$ is totally continuous; i.e. $P(\pi) = 0$ for each irreducible unitary representation of $\mathbb{R}$. (**Hint:** Use the Fourier transform.)

Solution:

Let $\pi = e_\omega$. So π is the 1-dimensional representation given by $\pi(x) = e^{2\pi i\omega x}$. We could find $P(\pi)$ for the regular representation λ, but it is slightly easier to use $\hat{\lambda}$. As seen in Example 6.90, $\hat{\lambda}(x)f(y) = e^{-2\pi ixy}f(y)$ for $f \in L^2(\mathbb{R})$. Now if $A \in \mathrm{Hom}(\pi, \hat{\lambda})$, then $A : \mathbb{C} \to L^2(\mathbb{R})$ is an intertwining operator and $f = A1 \in$

L^2 satisfies $\hat{\lambda}(x)f = \hat{\lambda}(x)A1 = A\pi(x)1 = A(e^{2\pi i\omega x}1) = e^{2\pi i\omega x}A1 = e^{2\pi i\omega x}f$. Thus $e^{-2\pi ixy}f(y) = e^{2\pi i\omega x}f(y)$ for a.e. y for each x. Since $e^{-2\pi ixy} \neq e^{2\pi i\omega x}$ unless $x(\omega + y) \in \mathbb{Z}$, this implies $f(y) = 0$ for a.e. y. So the range of A is $\{0\}$ and the result follows by Corollary 6.54.

□

24. irreducible representations!of $C_0(X)$ Let X be a locally compact Hausdorff space and let $C_0(X)$ be the space of continuous functions f such that $\{x \mid |f(x)| \geqslant \epsilon\}$ is compact for each $\epsilon > 0$. With pointwise multiplication and addition and complex conjugation for the adjoint, show $C_0(X)$ is a commutative C^* algebra when equipped with norm $|f| = \max\{|f(x)| \mid x \in X\}$. Then determine the irreducible representations of this $*$ algebra. Note $C_0(X)$ has no identity when X is not compact.

Solution:

It is easy to check $C_0(X)$ is a $*$-algebra and it clearly satisfies $|ff^*| = |f\bar{f}| = |f|^2$. We show completeness. Let f_k be Cauchy. Thus $|f_k - f_l| = \max|f_k(x) - f_l(x)| \to 0$ as $k, l \to \infty$. So the sequence f_k converges uniformly on X and thus has a continuous limit f. We note f vanishes at ∞ for if $\epsilon > 0$ and f_k is chosen so that $|f_k(x) - f(x)| < \frac{\epsilon}{2}$ for all x, then since $K = \{x \mid |f_k(x)| \geqslant \frac{\epsilon}{2}\}$ is compact, we have if $|f(x)| \geqslant \epsilon$, then $|f_k(x)| = |f_k(x) - f(x) + f(x)| \geqslant |f(x)| - |f_k(x) - f(x)| \geqslant \frac{\epsilon}{2}$. Thus the x with $|f(x)| \geqslant \epsilon$ is a closed subset of K and thus is compact. So $f \in C_0(X)$ and $f_k \to f$.

Let π be an irreducible $*$-representation of $C_0(X)$. Then π is 1-dimensional and since for $f \geqslant 0$, one has $\pi(f) = \pi(\sqrt{f})\overline{\pi(\sqrt{f})} \geqslant 0$, π is a positive integral. By the Riesz Theorem 6.1, since π is nonzero, there is a positive nonzero regular Borel measure μ on X with $\pi(f) = \int f\,d\mu$. Thus $(\int f_1\,d\mu)(\int f_2\,d\mu) = \int f_1 f_2\,d\mu$ for all $f_1, f_2 \in C_0(X)$. Now let K_1 and K_2 be disjoint compact subsets of X. By Exercise 6.5.18, there are disjoint open subsets V_1 and V_2 containing K_1 and K_2, respectively. Taking continuous functions $f_1, f_2 \geqslant 0$ with compact supports in V_1,V_2 such that $f_1 = 1$ on K_1 and $f_2 = 1$ on K_2, we see $\mu(K_1)\mu(K_2) \leqslant \int(f_1f_2)\,d\mu = 0$. So $\mu(K_1)\mu(K_2) = 0$. This implies using property (b) of regularity on page 276 that $\mu(V)\mu(K) = 0$ if V is an open subset and K is a compact subset disjoint from V.

Now consider the collection $\mathcal{F}$ of all compact subsets K with $\mu(K) > 0$. Since $\mu \neq 0$, this collection is nonempty. We claim $\cap_{\mathcal{F}}K$ is a nonempty subset. To see this we show it has the finite intersection property. In fact if $K_1, K_2, \ldots, K_m \in \mathcal{F}$ and $\cap_{i=1}^m K_i = \varnothing$, then $\prod \mu(K_i) = 0$ which is impossible. Thus $\cap_{\mathcal{F}}K = F$ is a nonempty compact subset of X. Pick $p \in F$. From the above, $\mu(X - \{p\})\mu(\{p\}) = 0$. If $\mu(\{p\}) = 0$ and $\mu(X - \{p\}) > 0$, there is a compact subset K_0 of $X - \{p\}$ of positive μ measure. But then $K_0 \in \mathcal{F}$ and we would have $p \notin \cap_{\mathcal{F}}K$. So we can conclude $\mu(\{p\}) > 0$. But then $\mu(X - \{p\}) = 0$. Thus $\mu = c\epsilon_p$ where $c > 0$ and ϵ_p is the point mass measure at p. Hence $\pi(f) = cf(p)$. Now if $f(p) = 1$, $\pi(f^2) = c^2 = c = \pi(f)$. This implies $c = 1$. So $\pi(f) = f(p)$ for some unique $p \in X$.

□

25. Let X be a measure space with a σ-finite measure μ. Show $L^\infty(\mu)$ is a commutative C^* algebra with identity. Next define a representation π of $L^\infty(\mu)$ on $L^2(\mu)$ by $\pi(f)(\phi)(x) = f(x)\phi(x)$ if $f \in L^\infty$ and $\phi \in L^2$. Show $\pi(L^\infty(\mu)) = \mathrm{Hom}(\pi, \pi)$.

Solution:

$L^\infty(X)$ is complete. Clearly one has $|f|_\infty = |\bar{f}|_\infty$ and $|f_1 f_2|_\infty \leqslant |f_1|_\infty |f_2|_\infty$ and $|ff^*|_\infty = |f\bar{f}|_\infty = \operatorname{essup}|f|^2 = \sup_{M \geqslant 0}\{M^2 \mid \mu(\{x \mid |f(x)|^2 \geqslant M^2\} > 0\}$. But if $M > 0$, $\{x \mid |f(x)|^2 \geqslant M^2\}$ has positive measure if and only if $\{x \mid |f(x)| > M\}$ has positive measure. This implies

$$|ff^*|_\infty = \left(\sup_{M \geqslant 0}\{M \mid \mu\{x \mid |f(x)| \geqslant M\} > 0\}\right)^2 = |f|_\infty^2 .$$

All this together and the fact the $L^\infty(X)$ is a complex algebra shows $L^\infty(X)$ is a C^* algebra.

Let $\pi(f)\phi = f\phi$ for $f \in L^\infty$ and $\phi \in L^2$. This can easily be checked to be a representation and since L^∞ is commutative, $\pi(f) \in \mathrm{Hom}(\pi, \pi)$ for all $f \in L^\infty$. Let $A \in \mathrm{Hom}(\pi, \pi)$. Write $X = \cup_{i=1}^\infty E_i$ where the E_i are disjoint and have finite measure. Let $L^2(E_i)$ be the closed subspace of $L^2(X)$ consisting of the L^2 functions equal to zero a.e. off E_i. For each i, set $h_i = A\chi_{E_i}$. Note if $f \in L^\infty$, then $f\chi_{E_i}h_i = \pi(f)\pi(\chi_{E_i})A\chi_{E_i} = A(\pi(f)\pi(\chi_{E_i})x_{E_i}) = A(f\chi_{E_i})$. Taking $f = 1$, we see $A\chi_{E_i} = \chi_{E_i}h_i = h_i$, and thus h_i vanishes a.e. off E_i. Furthermore, the linear span of the functions $f\chi_{E_i}$ where f is in L^∞ is dense in $L^2(E_i)$. Consequently, $A(\phi) = \phi h_i$ if $\phi \in L^2(E_i)$ and thus $L^2(E_i)$ is an invariant subspace under A. If $A_i = A|_{L^2(E_i)}$, then A_i is bounded and maps $L^2(E_i)$ into itself. Now $\phi \mapsto \phi h_i$ is bounded implies $h_i \in L^\infty$. To see this note if $|h_i|_\infty > M$, there is a subset F of E_i with finite positive measure. So $g = \frac{1}{\sqrt{\mu(F)}}\chi_F \in L^2(E_i)$ has norm 1 and

$$|A_i g|_2^2 \geqslant \frac{1}{\mu(F)} \int_F |h_i(x)|^2 \, d\mu(x) \geqslant M^2$$

and thus $||A_i|| \geqslant M$. This cannot be true for all $M > 0$ and so $|h_i|_\infty < \infty$. Moreover, since $A = A_i$ on $L^2(E_i)$, this implies since $||A|| \geqslant ||A_i||$ for all i that $|h_i|_\infty \leqslant ||A||$. Define $h = \sum h_i$. We thus have $|h|_\infty \leqslant ||A||$ and since $A = \bigoplus A_i$ as an internal direct sum, $A\phi = h\phi$ for all $\phi \in L^2(X)$.

□

26. Let μ be a regular finite Borel measure on a locally compact Hausdorff space X. Let $M(X, \mu)$ be the measure algebra given by μ. Thus $M(X, \mu)$ is the collection of equivalence classes of the equivalence relation defined on the Borel subsets of X by $E_1 \sim E_2$ if and only if $\mu(E_1 + E_2) = \mu(E_1 - E_2) + \mu(E_2 - E_1) = 0$.

(a) Show $M(X, \mu)$ is a complete metric space under the metric

$$d([E_1], [E_2]) = \mu(E_1 + E_2).$$

(b) Define a filter $\mathcal{F}$ on $M(X)$ to be a nonempty subfamily of subsets of $M(X)$ with the properties $0 \notin \mathcal{F}$, $E_1 \cap E_2 \in \mathcal{F}$ if E_1 and E_2 are in $\mathcal{F}$, and $E_2 \in \mathcal{F}$ whenever $E_1 \in \mathcal{F}$, $E_2 \in M(X)$, and $E_1 \subseteq E_2$. Show every filter $\mathcal{F}$ on $M(X)$ is contained in a hyperfilter; i.e., every filter is contained in a maximal filter.

(c) Let $\mathcal{F}$ be a hyperfilter. Define ν on $\mathcal{F}$ by $\nu(E) = 0$ if $E \in M(X) - \mathcal{F}$ and $\nu(E) = 1$ if $E \in \mathcal{F}$. Show ν is a finitely additive measure on $M(X)$ with the property $\nu(E_1 \cap E_2) = \nu(E_1)\nu(E_2)$.

(d) Let ν be the measure defined in (c). Show there is a unique one-dimensional representation π of $L^\infty(\mu)$ such that $\pi(\chi_E) = \nu(E)$ for $E \in M(X)$.

(e) Show every representation π in the Gelfand spectrum Δ of $L^\infty(\mu)$ comes from a measure ν defined in terms of a hyperfilter $\mathcal{F}$.

(f) If you know the definition of an ultra filter from point set topology, what is a hyperfilter?

Solution:

For (a) to see it is a metric space, we look only at the triangle inequality. Note

$$\begin{aligned} d([E_1],[E_3]) &= \mu(E_1 - E_3) + \mu(E_3 - E_1) \\ &\leqslant \mu(E_1 - E_2) + \mu(E_2 - E_3) + \mu(E_3 - E_2) + \mu(E_2 - E_1) \\ &= d([E_1],[E_2]) + d([E_2],[E_3]) \end{aligned}$$

for $E_1 - E_3 \subseteq (E_1 - E_2) \cup (E_2 - E_3)$ and $E_3 - E_1 \subseteq (E_3 - E_2) \cup (E_2 - E_1)$.

Now for completeness, let $[E_k]$ be Cauchy. By taking a subsequence we may assume it is rapidly Cauchy, i.e., $d([E_k],[E_l]) \leqslant \frac{1}{2^k}$ if $l \geqslant k$. Set $E = \cap_k \cup_{l \geqslant k} E_l$. E is Borel. Then

$$\begin{aligned} E_m - E &= E_m - (\cap_k \cup_{l \geqslant k} E_l) \\ &\subseteq \cup_{l \geqslant m}(E_l - E_{l+1}) \end{aligned}$$

and thus $\mu(E_m - E) \leqslant \sum_{l \geqslant m} \mu(E_l - E_{l+1}) \leqslant \sum_{l \geqslant m} \frac{1}{2^l} = \frac{1}{2^{m-1}}$. Similarly,

$$\begin{aligned} E - E_m &= (\cap_k \cup_{l \geqslant k} E_l) - E_m \\ &\subseteq \cup_{l \geqslant m}(E_{l+1} - E_l) \end{aligned}$$

and so $\mu(E - E_m) \leqslant \frac{1}{2^{m-1}}$. Putting these together we have $d([E_m],[E]) \leqslant \frac{1}{2^{m-2}}$ and thus $[E_m] \to [E]$ in $M(X,\mu)$.

Note (b) follows from a direct argument using Zorn's Lemma.

For (c), let $\mathcal{F}$ be a hyperfilter. Consider an $F \in M(X)$ with $F \neq 0$. Suppose $X - F \notin \mathcal{F}$. Define $\mathcal{F}'$ to be the collection of all subsets E' which contain $E \cap F$ for some $E \in \mathcal{F}$. Note if $E \in \mathcal{F}$ and $E \cap F = \varnothing$, then $E \subseteq X - F$ and thus $X - F \in \mathcal{F}$. So $\mathcal{F}'$ does not contain 0. Also $E_1', E_2' \in \mathcal{F}'$ implies $E_i \cap F \subseteq E_i'$

for some $E_i \in \mathcal{F}$. So $E_1 \cap E_2 \in \mathcal{F}$ and thus $E_1' \cap E_2' \supseteq (E_1 \cap E_2) \cap F$. So $E_1' \cap E_2' \in \mathcal{F}'$. It is clear that if $E_1' \subseteq E_2'$ and $E_1' \in \mathcal{F}'$, then $E_2' \in \mathcal{F}'$. By maximality, $\mathcal{F}' = \mathcal{F}$. So $F = X \cap F \in \mathcal{F}'$. Thus for each $F \in M(X)$, either $F \in \mathcal{F}$ or $X - F \in \mathcal{F}$.

We note ν defined in (c) is finitely additive. Indeed, if F_1 and F_2 are in $M(X)$ and $F_1 \cap F_2 = 0$, then if neither F_1 nor F_2 is in $\mathcal{F}$, we have both $X - F_1$ and $X - F_2$ are in $\mathcal{F}$ and so $(X - F_1) \cap (X - F_2) \in \mathcal{F}$. But this is $X - (F_1 \cup F_2)$. Consequently $F_1 \cup F_2 \notin \mathcal{F}$ and we see $\nu(F_1 \cup F_2) = 0 = \nu(F_1) + \nu(F_2)$. Note F_1 and F_2 cannot both be in $\mathcal{F}$ for they are disjoint and thus if $F_i \in \mathcal{F}$ and $\{j\} = \{1, 2\} - \{i\}$, we would have $F_i \subseteq X - F_j$ and so $X - F_j \in \mathcal{F}$. Hence only one of F_1 and F_2 could be in $\mathcal{F}$ and then $1 = \nu(F_1 \cup F_2) = \nu(F_i) + \nu(F_j) = 1 + 0$.

Now since the measure ν is finitely additive, one can easily define $\int s\, d\nu$ for simple functions s and since any L^∞ function f can be uniformly a.e. approximated by a measurable simple function, this integral extends to a finitely additive integral on L^∞ and thus $\pi(f) = \int f\, d\nu$ is defined for $f \in L^\infty(\mu) \subseteq L^1(\mu)$. Clearly $\pi(\chi_E) = \nu(E)$ and $\pi(f^*) = \pi(\bar{f}) = \overline{\pi(f)}$. To be a representation, we need to see π is multiplicative. But this will follow if $\pi(s_1)\pi(s_2) = \pi(s_1 s_2)$ for simple functions s_1 and s_2. Now if $s_i = \sum_{j=1}^{m_i} a_{i,j}\chi_{E_{i,j}}$ where the $E_{i,j}$ for distinct j are disjoint, then at most one $E_{i,j}$ is in $\mathcal{F}$ and then $\nu(E_{i,j}) = 1$. So we may assume $s_1 = a_1\chi_{E_1}$ and $s_2 = a_2\chi_{E_2}$ and $s_1 s_2 = a_1 a_2 \chi_{E_1 \cap E_2}$. If either $\nu(E_1) = 0$ or $\nu(E_2) = 0$, we have $\pi(s_1)\pi(s_2) = 0 = \pi(s_1 s_2)$ for $\nu(E_1 \cap E_2) = 0$. If both are 1, then $E_1 \cap E_2 \in \mathcal{F}$ and thus $\nu(E_1 \cap E_2) = 1$ and we would have $\pi(s_1 s_2) = a_1 a_2 = \pi(s_1)\pi(s_2)$.

Next if π is a representation of $L^\infty(\mu)$, then if $E \in M(X, \mu)$, $\chi_E \chi_E = \chi_E$ implies $\pi(\chi_E)^2 = \pi(\chi_E)$. So $\pi(\chi_E) = 0$ or $\pi(\chi_E) = 1$. It's easy to check $\nu(E) = \pi(\chi_E)$ is a measure and the collection of all E with $\pi(\chi_E) = 1$ is a hyperfilter. This essentially gives (e).

Finally, in topology a filter is a collection $\mathcal{F}$ of subsets such that $\varnothing \notin \mathcal{F}$, if $A, B \in \mathcal{F}$, then $A \cap B \in \mathcal{F}$, and $A \in \mathcal{F}$ and $A \subseteq B$ implies $B \in \mathcal{F}$. These filters are used when dealing with convergence. Ultrafilters are maximal under inclusion and compactness can be characterized in terms of these.

□

27. Find three irreducible inequivalent finite dimensional representations of the symmetric group S_3.

Solution:

The elements of S_3 are $e = (1)$, $p_2 = (1, 2)$, $p_3 = (1, 2, 3)$, $p_4 = (1, 3, 2)$, $p_5 = (1, 3)$, and $p_6 = (2, 3)$. $A_3 = \{e, p_3, p_4\}$ is the alternating normal subgroup, and S_3/A_3 is a cyclic group of order 2.

Next note if π is a one-dimensional representation of S_3, then $\pi(\sigma)^2 = 1$ for the transpositions $(1, 2)$, $(2, 3)$, and $(1, 3)$ for their squares are the identity. Thus $\pi(\sigma) = \pm 1$ if σ is a transposition. We also note each permutation in A_3 is a square for $p_3^2 = p_4$ and $p_4^2 = p_3$. So $\pi(\sigma) = 1$ for $\sigma \in A_3$. Since $S_3/A_3 = \{A_3, (1, 2)A_3\}$, we then have two one-dimensional representations which we call π_+ and π_-. They are given by $\pi_\pm(\sigma) = 1$ if $\sigma \in A_3$ and $\pi_\pm(\sigma) = \pm 1$ if $\sigma \in (1, 2)A_3$ or equivalently σ is a transposition. Note these correspond to the two one-dimensional representations of the cyclic group S_3/A_3 of order 2 given in Exercise 6.5.11.

Any other irreducible representation π must not be I on A_3 and must have dimension greater than 1. We hence try to use the fact that S_3 is a permutation group. Taking $\sigma \in S_3$, we let $\rho(\sigma)$ be the linear transformation of $\mathbb{C}^3$ such that on the standard basis e_1, e_2, e_3 one has $\rho(\sigma)e_i = e_{\sigma i}$. Note $\rho(\sigma_1\sigma_2)e_i = e_{\sigma_1\sigma_2 i} = \rho_{\sigma_1}(\rho_{\sigma_2}e_i)$. This implies $\rho(\sigma_1\sigma_2) = \rho(\sigma_1)\rho(\sigma_2)$ and thus ρ is a representation of S_3. This representation is not irreducible for it has a one dimensional invariant subspace. Namely let $W = \mathbb{C}(e_1 + e_2 + e_3)$. This space is invariant under ρ. Hence the quotient vector space $V = \mathbb{C}^3/W$ is two dimensional and the representation ρ factors to a representation π on this space; i.e.,

$$\pi(\sigma)(v + W) = \rho(\sigma)v + W.$$

This representation has no 1-dimensional invariant subspace; for if it did, there would be a vector $v_0 \notin W$ such that $\mathbb{C}v_0 + W$ is invariant under ρ. As seen above, one would have $\rho(\sigma)(v_0 + W) = v_0 + W$ if $\sigma \in A_3$ and $\rho(\sigma)(v_0 + W) = v_0 + W$ or $\rho(\sigma)(v_0 + W) = -v_0 + W$ if $\sigma \in (1, 2)A_3$. Now suppose $v_0 = c_1e_1 + c_2e_2 + c_3e_3$. In the first case, we would have

$$\begin{aligned} \rho(1, 2)v &= c_2e_1 + c_1e_2 + c_3e_3 \text{ and} \\ \rho(2, 3)v &= c_1e_1 + c_3e_2 + c_2e_3 \end{aligned}$$

and thus $\rho(1, 2)v_0 - v_0 = (c_2 - c_1)e_1 + (c_1 - c_2)e_2$ and $\rho(2, 3)v_0 - v_0 = (c_3 - c_2)e_2 + (c_2 - c_3)e_3$ are in W. Consequently, $c_1 = c_2 = c_3$ and we see $v_0 \in W$, a contradiction.

In the second case we would have

$$\begin{aligned} \rho(1, 2)v_0 + v_0 &= (c_2 + c_1)e_1 + (c_1 + c_2)e_2 + 2c_3e_3 \text{ and} \\ \rho(2, 3)v_0 + v_0 &= 2c_1e_1 + (c_3 + c_2)e_2 + (c_2 + c_3)e_3 \end{aligned}$$

are in W. Consequently, $2c_3 = c_1 + c_2$ and $2c_1 = c_2 + c_3$. This implies $c_1 = c_2 = c_3$ and again we would have $v_0 \in W$. Thus π has no one-dimensional invariant subspace and we see π is irreducible.

As noted, this representation is not I on A_3 and one can find determine $\pi(\sigma)$ on the basis vectors $e_1 + W$ and $e_2 + W$. Note first if $\sigma 1 = 1$ or $\sigma 1 = 2$, then $\pi(\sigma)(e_1 + W) = e_{\sigma 1} + W$. Otherwise $\sigma 1 = 3$, and $\pi(\sigma)(e_1 + W) = e_3 + W = e_3 - (e_1 + e_2 + e_3) + W = (-1)(e_1 + W) + (-1)(e_2 + W)$. Working in this way, the matrices $\pi_m(\sigma)$ for $\pi(\sigma)$ can be determined to be

$$\pi_m(e) = \begin{bmatrix} 1 & 0 \\ 0 & 1 \end{bmatrix}, \quad \pi_m(1, 2) = \begin{bmatrix} 0 & 1 \\ 1 & 0 \end{bmatrix}, \quad \pi_m(1, 2, 3) = \begin{bmatrix} 0 & -1 \\ 1 & -1 \end{bmatrix}$$

$$\pi_m(1, 3) = \begin{bmatrix} -1 & 0 \\ -1 & 1 \end{bmatrix}, \quad \pi_m(2, 3) = \begin{bmatrix} 1 & -1 \\ 0 & -1 \end{bmatrix}, \quad \pi_m(1, 3, 2) = \begin{bmatrix} -1 & 1 \\ -1 & 0 \end{bmatrix}$$

and it can be shown $\pi_\pm$ and π are up to equivalence the only irreducible representations of S_3.

□

28. The right regular representation ρ is given by $\rho(a)f(x) = \Delta(a)^{1/2}f(xa)$ for $f \in L^2(G)$. Show that the map $L^2(G) \to L^2(G)$, $f \mapsto \Delta^{-1/2}\check{f}$, is an unitary equivalence between the left regular representation λ and the right regular representation ρ.

Solution:

We have a left Haar measure on G and recall it satisfies $m(Ea) = \Delta(a)m(E)$ for Borel subsets E. By Proposition 6.3, $\int f(xa)\,dm(x) = \Delta(a)^{-1}\int f(x)\,dm(x)$ and $\int f(x^{-1})\Delta(x^{-1})\,dm(x) = \int f(x)\,dm(x)$ for integrable functions f. Thus the mapping $Tf(x) = \Delta^{-1/2}(x)\check{f}(x) = \Delta^{-1/2}(x)f(x^{-1})$ for $f \in L^2$ is unitary for $\|Tf\|_2^2 = \int \Delta(x^{-1})|f(x^{-1})|^2\,dm(x) = \int |f(x)|^2\,dm(x)$ and $T^2 = I$. Furthermore,

$$\begin{aligned}
T\lambda(a)f(x) &= \Delta(x)^{-1/2}(\lambda(a)f)(x^{-1}) \\
&= \Delta(x)^{-1/2}f(a^{-1}x^{-1}) \\
&= \Delta(x)^{-1/2}\check{f}(xa) \\
&= \Delta(a)^{1/2}\Delta(xa)^{-1/2}\check{f}(xa) \\
&= \Delta(a)^{1/2}Tf(xa) \\
&= \rho(a)Tf(x)\,.
\end{aligned}$$

Thus $T\lambda(a)T^{-1} = \rho(a)$ for each $a \in G$ and we see ρ and λ are unitarily equivalent.

□

29. Let G be a locally compact Hausdorff group. Let λ be the left regular representation of G on $L^2(G,m)$ where m is a left Haar measure on G. Show that the representation λ is irreducible if and only if $G = \{e\}$. (**Hint:** Consider the right regular representation.)

Solution: Since $L^2(G) = \mathbb{C}$ if $G = \{e\}$, we see λ is irreducible if $G = \{e\}$. Conversely, assume λ is irreducible. Let ρ be the right regular representation of G given in Exercise 6.5.28. We note $\rho(a)\lambda(b) = \lambda(b)\rho(a)$ for $a, b \in G$. Namely,

$$\begin{aligned}
\rho(a)\lambda(b)f(x) &= \Delta(a)^{1/2}\lambda(b)f(xa) \\
&= \Delta(a)^{1/2}f(b^{-1}xa) \\
&= \rho(a)f(b^{-1}x) \\
&= \lambda(b)\rho(a)f(x)\,.
\end{aligned}$$

Thus if λ is irreducible, we know by Schur's Lemma that $\rho(a) = \chi(a)I$ where $\chi : G \to \mathbb{T}$ is a one-dimensional character. By Exercise 6.5.28, since ρ is unitarily equivalent to λ, ρ is irreducible and thus the Hilbert space for ρ is one dimensional. However, $L^2(G)$ has dimension 1 if and only if $G = \{e\}$. □

30. Let μ be a measure on space X. Show the mapping $f \mapsto \bar{f}$ is a unitary isomorphism of $L^2(X,\mu)$ onto the conjugate Hilbert space $\overline{L^2(X,\mu)}$.

Solution:

This follows directly. In fact if $f \in \mathcal{H}$ where $\mathcal{H} = L^2(X)$ and f_- denotes the same function except as a member of $\bar{\mathcal{H}}$, then we have the mapping T :

$f_- \mapsto \bar{f}$ from $\bar{\mathcal{H}}$ to $\mathcal{H}$. Note clearly, $T(f_- + h_-) = T(f_-) + T(h_-)$. Also $T(cf_-) = \overline{T((\bar{c}f)_-)} = \overline{\bar{c}f} = c\bar{f}$ and thus T is linear. Also $(Tf_-, Th_-)_2 = (\bar{f}, \bar{h})_2 = \overline{(f,h)_2} = (f_-, h_-)_{\bar{\mathcal{H}}}$. So T is unitary.

□

31. Show the left regular representation λ is unitarily equivalent to $\bar{\lambda}$.
Solution:

By Exercise 6.5.30, $U : f \mapsto \bar{f}$ is a unitary isomorphism of $L^2(G)$ onto $\overline{L^2(G)}$. Since $\bar{\lambda}(g)\bar{f} = \overline{\lambda(g)f}$, we see $U(\lambda(g)f) = \bar{\lambda}(g)Uf$ and thus λ and $\bar{\lambda}$ are unitarily equivalent.

□

32. Let π_1 and π_2 be unitary representations of topological groups G_1 and G_2 on the same Hilbert space $\mathcal{H}$. Assume $\pi_1(g_1)\pi_2(g_2) = \pi_2(g_2)\pi_1(g_1)$ for all $g_1, g_2 \in G$. Define $\pi_1 \cdot \pi_2$ on $G_1 \times G_2$ by

$$(\pi_1 \cdot \pi_2)(g_1, g_2) = \pi_1(g_1)\pi_2(g_2).$$

Show $\pi_1 \cdot \pi_2$ is a unitary representation of $G_1 \times G_2$.
Solution:

We have π_1 and π_2 commute. Clearly $\pi_1 \cdot \pi_2(e_1, e_2) = I$ on $\mathcal{H}$ and

$$\begin{aligned}(\pi_1 \cdot \pi_2)(g_1g_1', g_2g_2') &= \pi_1(g_1g_1')\pi_2(g_2g_2') \\ &= \pi_1(g_1)\pi_1(g_1')\pi_2(g_2)\pi_2(g_2') \\ &= \pi_1(g_1)\pi_2(g_2)\pi_1(g_1')\pi_2(g_2') \\ &= (\pi_1 \cdot \pi_2)(g_1, g_2)(\pi_1 \cdot \pi_2)(g_1', g_2').\end{aligned}$$

Also

$$\begin{aligned}\pi_1 \cdot \pi_2((g_1, g_2)^{-1}) &= \pi_1 \cdot \pi_2(g_1^{-1}, g_2^{-1}) \\ &= \pi_1(g_1^{-1})\pi_2(g_2^{-1}) \\ &= \pi_2(g_2)^{-1}\pi_1(g_1)^{-1} \\ &= (\pi_1(g_1)\pi_2(g_2))^{-1} \\ &= (\pi_1 \cdot \pi_2)(g_1, g_2)^{-1}.\end{aligned}$$

Thus $\pi_1 \cdot \pi_2$ is a homomorphism into $\mathcal{U}(\mathcal{H})$.

To see strong continuity, note the mapping $(g_1, g_2) \mapsto (\pi_1(g_1), \pi_2(g_2))$ is continuous from $G_1 \times G_2$ into $\mathcal{U}(\mathcal{H}) \times \mathcal{U}(\mathcal{H})$ where we are taking the product topologies. By Proposition 6.32, $\mathcal{U}(\mathcal{H})$ is a topological group when using the strong operator topology. Thus $(U_1, U_2) \mapsto U_1U_2$ is continuous. Composition of these maps shows $(g_1, g_2) \mapsto (\pi_1 \cdot \pi_2)(g_1, g_2)$ is continuous.

□

33. Use Exercise 6.1.34 and the exercises preceding it to show if ν is a quasi-invariant regular Borel measure on G/H where G is a locally compact Hausdorff space and H is a closed subgroup, then λ' defined in (6.16) is a unitary representation which is unitarily equivalent to the quasi-regular representation λ

defined in (6.15). (**Note:** Part of this exercise is defining the Radon-Nikodym derivative $\frac{d(x\nu)}{d\nu}$. As a hint consider an open closed σ-compact subgroup G_0 of G and a decomposition of G/H into a disjoint union of sets of form $G_0 x H$.)

Solution:

The Radon-Nikodym derivative $\frac{d(x\nu)}{d\nu}$ is defined relative to a decomposition of G/H into a disjoint union of open closed σ-compact sets $G_0 x_\alpha H$ where G_0 is a open-closed σ compact subgroup of G. Namely ν restricted to $G_0 x_\alpha H$ and $x\nu$ on this set are equivalent and being σ-finite, there is a Radon-Nikodym function $\left(\frac{dx\nu}{d\nu}\right)_\alpha$ on $G_0 x_\alpha H$ with $x\nu(E) = \int_E \left(\frac{dx\nu}{d\nu}\right)_\alpha d\nu$ for Borel subsets E of $G_0 x_\alpha H$. Then $\frac{dx\nu}{d\nu}$ is defined on all of G/H by taking $\frac{dx\nu}{d\nu}(yx_\alpha H) = \left(\frac{dx\nu}{d\nu}\right)_\alpha (yx_\alpha H)$ for $y \in G_0$.

We let μ be the quasi-invariant measure given by the continuous rho function ρ. The Radon-Nikodym derivative on all of G/H of $x\mu$ with respect to μ is given by $\frac{d(x\mu)}{d\mu}(yH) = \frac{\rho(xy)}{\rho(y)}$ and thus this function restricted to $G_0 x_\alpha H$ is the Radon-Nikodym derivative for $(x\mu)_\alpha$, the restriction of $x\mu$ to $G_0 x_\alpha H$ relative to μ_α, the restriction of μ to $G_0 x_\alpha H$. By Corollary 6.22 the restrictions of these measures to the Baire sets are equivalent. Exercise 6.1.4 implies as Borel measures, they remain equivalent for each Borel subset of finite measure is equal to a Baire set up to a set of measure 0. These Borel measures being equivalent and σ-finite imply there is a almost everywhere positive Borel function F_α on $G_0 x_\alpha H$ with $(F_\alpha \nu_\alpha) = \mu_\alpha$. Next note

$$L^2(G/H, \mu) = \bigoplus_\alpha L^2(G_0 x_\alpha H, \mu_\alpha) \text{ and } L^2(G/H, \nu) = \bigoplus_\alpha L^2(G_0 x_\alpha H, \nu_\alpha).$$

To define a unitary equivalence, it suffices to do so for each α. Namely, define $T_\alpha : L^2(G_0 x_\alpha H, \mu_\alpha) \to L^2(G_0 x_\alpha H, \nu_\alpha)$ by

$$T_\alpha f_\alpha(yx_\alpha H) = F_\alpha(yx_\alpha H)^{1/2} f_\alpha(yx_\alpha H).$$

Note

$$\begin{aligned} |T_\alpha f_\alpha|_2^2 &= \int F_\alpha(yx_\alpha H)|f_\alpha(yx_\alpha H)|^2 \, d\nu_\alpha(yx_\alpha H) \\ &= \int |f_\alpha(yx_\alpha H)|^2 \, d\mu_\alpha(yx_\alpha H) \\ &= |f_\alpha|_2^2 \end{aligned}$$

and $T_\alpha(F_\alpha^{-1} g_\alpha) = g_\alpha$ for $g_\alpha \in L^2(G_0 x_\alpha H, \nu_\alpha)$ shows T_α is unitary on each summand. By Exercise 6.3.2, $T = \bigoplus T_\alpha$ is unitary from $L^2(G/H, \mu)$ onto $L^2(G/H, \nu)$. To finish it suffices to show $T(\lambda(x)f) = \lambda'(x)Tf$ where $f \in L^2(G/H, \mu)$ and f vanishes off $G_0 x_\alpha H$. First note $xG_0 x_\alpha H$ is a σ-compact open closed subset of G/H and thus there is a countable collection α_i with

$xG_0x_\alpha H \subseteq \dot\cup G_0x_{\alpha_i}H$. Now $\lambda(x)f = \oplus_i(\lambda(x)f)|_{G_0x_{\alpha_i}H}$ and thus

$$\begin{aligned} T(\lambda(x)f) &= \oplus_i T_{\alpha_i}((\chi_{G_0x_{\alpha_i}H})\lambda(x)f) \\ &= \oplus_i\left(\sqrt{F_{\alpha_i}}\chi_{(G_0x_{\alpha_i}H)\cap xG_0x_\alpha H}\lambda(x)f\right) \\ &= \oplus_i\left(\sqrt{F_{\alpha_i}(\,\cdot\,)}\chi_{(G_0x_{\alpha_i}H)\cap xG_0x_\alpha H}(\,\cdot\,)\sqrt{\Gamma_x(\,\cdot\,)}f(x^{-1}\,\cdot\,)\right) \\ &= \sqrt{F_{\alpha_i}(\,\cdot\,)}\sqrt{\Gamma_x(\,\cdot\,)}f(x^{-1}\,\cdot\,) \end{aligned}$$

where $\cdot$ stands for the generic element yH and $\Gamma_x(yH) = \frac{\rho(x^{-1}y)}{\rho(y)}$ gives the Radon-Nikodym derivative $\frac{d(x\mu)}{d\mu}(yH)$. Similarly if $\Gamma'_x(yH) = \left(\frac{dx\nu}{d\nu}\right)_\alpha(yH)$, then $\lambda'(x)Tf$ is given by

$$\begin{aligned} \lambda'(x)Tf(\,\cdot\,) &= \lambda'(x)T_\alpha f(\,\cdot\,) \\ &= \sqrt{\Gamma'_x(\,\cdot\,)}\left(\sqrt{F_\alpha(x^{-1}\,\cdot\,)}f(x^{-1}\,\cdot\,)\right) \\ &= \sqrt{\Gamma'_x(\,\cdot\,)}\bigoplus_i\left(\chi_{(G_0x_{\alpha_i}H)\cap xG_0x_\alpha H}(\,\cdot\,)\sqrt{F_\alpha(x^{-1}\,\cdot\,)}f(x^{-1}\,\cdot\,)\right) \\ &= \sqrt{\Gamma'_x(\,\cdot\,)}\sqrt{F_\alpha(x^{-1}\,\cdot\,)}f(x^{-1}\,\cdot\,) \end{aligned}$$

and so it suffices to show if f vanishes off $G_0x_\alpha H$, then

$$\sqrt{\Gamma'_x(yH)}\sqrt{F_\alpha(x^{-1}yH)}f(x^{-1}yH) = \sqrt{\Gamma_x(yH)}\sqrt{F_{\alpha_i}(yH)}f(x^{-1}yH)$$

μ a.e. yH in $\cup_i G_0x_{\alpha_i}H$. But in this case we are dealing with σ-finite measures and their Radon-Nikodym derivatives. We have $(F_\alpha\nu_\alpha) = \mu_\alpha$, $x\mu_\alpha = \Gamma_x\mu_\alpha$, and $x\nu_\alpha = \Gamma'_x\nu_\alpha$. So if E is a Borel subset of $G_0x_\alpha H$, then

$$\begin{aligned} &\int\chi_E(yH)\Gamma_x(yH)\,d\mu(yH) = (x\mu_\alpha)(E) \\ &= \mu_\alpha(x^{-1}E) \\ &= \int\chi_{x^{-1}E}(yH)\,d\mu_\alpha(yH) \\ &= \int\chi_{x^{-1}E}(yH)F_\alpha(yH)\,d\nu_\alpha(yH) \\ &= \int\chi_{x^{-1}E}(x^{-1}yH)F_\alpha(x^{-1}yH)\,d\nu_\alpha(x^{-1}yH) \\ &= \int\chi_E(yH)F_\alpha(x^{-1}yH)\Gamma'_x(yH)\,d\nu_\alpha(yH) \\ &= \int\chi_E(yH)F_\alpha(x^{-1}yH)\Gamma'_x(yH)F_\alpha(yH)^{-1}(F_\alpha(yH)\,d\nu_\alpha(yH)) \\ &= \int\chi_E(yH)F_\alpha(x^{-1}yH)\Gamma'_x(yH)F_\alpha(yH)^{-1}\,d\mu_\alpha(yH)\,. \end{aligned}$$

By a.e. uniqueness of Radon-Nikodym derivatives, one then obtains

$$\Gamma_x(yH) = F_\alpha(x^{-1}yH)\Gamma'_x(yH)F_\alpha(yH)^{-1} \text{ for } \mu_\alpha \text{ a.e. } yH \text{ in } G_0x_\alpha H.$$

This gives the result.

We remark that historically this has been used in the second countable case and thus these measure theoretical difficulties are not encountered.

□

34. Find the Fourier transform of the representation ρ given in (6.19) of Example 6.95. That is find the representation $(a,b) \mapsto \mathcal{F}L(a,b)\mathcal{F}^{-1}$ of the $ax+b$ group.

Solution: Let f be a Schwartz function. Then

$$\begin{aligned}
\mathcal{F}\rho(a,b)\mathcal{F}^{-1}f(x) &= \int \rho(a,b)\mathcal{F}^{-1}f(y)e^{-2\pi ixy}\,dy \\
&= \int \frac{1}{\sqrt{a}}\mathcal{F}^{-1}f(\frac{y+b}{a})e^{-2\pi ixy}\,dy \\
&= \int \frac{1}{\sqrt{a}}\mathcal{F}^{-1}f(\frac{y}{a})e^{-2\pi ix(y-b)}\,dy \\
&= e^{2\pi ixb}\sqrt{a}\int \mathcal{F}^{-1}f(y)e^{-2\pi ix(ay)}\,dy \\
&= \sqrt{a}e^{2\pi ixb}f(ax).
\end{aligned}$$

□

Exercise Set 6.6

1. Let $\mu = \operatorname{Re}\mu + i\operatorname{Im}\mu$ be a complex Borel measure on a space X where $\operatorname{Re}\mu$ and $\operatorname{Im}\mu$ are signed measures having Hahn decompositions $\operatorname{Re}\mu = \mu_1 - \mu_2$ and $\operatorname{Im}\mu = \mu_3 - \mu_4$. Show $|\mu|$ is regular if and only if each of the measures μ_j are regular.

Solution:

We first recall if μ is a complex measure, $|\mu|$ and μ_j for $j = 1,2,3,4$ are finite measures. Note $|\mu| \leqslant \mu_1+\mu_2+\mu_3+\mu_4$ for if $E_1, E_2, \cdots, E_m$ are disjoint Borel sets of a Borel subset W, then

$$\begin{aligned}
\sum |\mu(E_j)| &= |\mu_1(E_j) - \mu_2(E_j) + i(\mu_3(E_j) - \mu_4(E_j))| \\
&\leqslant \sum_{j=1}^{m}\sum_{k=1}^{4} \mu_k(E_j) \leqslant \mu_1(W)+\mu_2(W)+\mu_3(W)+\mu_4(W)
\end{aligned}$$

and thus $|\mu|(W)| \leqslant \sum_{j=1}^m \mu_j(W)$.

Assume each μ_j is regular, E is a Borel subset, and $\epsilon > 0$. Then we know the existence of an open set U_j containing E and a compact subset K_j of E with $\mu_j(U_j - K_j) < \frac{\epsilon}{4}$. Set $U = \cap U_j$ and $K = \cup K_j$. Clearly $K \subseteq E \subseteq U$ and

$$\begin{aligned}
|\mu|(U-K) &\leqslant \sum \mu_j(U-K) \\
&\leqslant \sum_j \mu_j(U_j - K_j) < \epsilon\,.
\end{aligned}$$

and this gives both inner and outer regularity.

To see the converse, assume $|\mu|$ is regular. It suffices to show μ_1 is regular, (just reapply to $-\mu$, $i\mu$ and $-i\mu$). Now $|\mu(E)| \leqslant |\mu|(E)$ for all Borel subsets E. This implies $|(\operatorname{Re}\mu)(E)| \leqslant |\mu|(E)$. Now $\operatorname{Re}\mu = \mu_1 - \mu_2$ is the Jordan decomposition. Thus there are disjoint Borel subsets A and B with $A \cup B = X$ and $\mu_1(E) = (\operatorname{Re}\mu)(E \cap A)$ and $\mu_2(E) = -(\operatorname{Re}\mu)(E \cap B)$. From this we see $\mu_1(E) \leqslant |\mu|(E \cap A) \leqslant |\mu|(E)$. Let $\epsilon > 0$ and E be given. Since $|\mu|$ is regular there is a compact subset $K \subseteq E$ and an open set $U \supseteq E$ with $|\mu|(E) - |\mu|(K) < \epsilon$ and $|\mu|(U) - |\mu|(E) < \epsilon$. Hence $\mu_1(E) - \mu_1(K) \leqslant |\mu|(E) - |\mu|(K) < \epsilon$ and $\mu_1(U) - \mu_1(E) \leqslant |\mu|(U) - |\mu|(E) < \epsilon$ and so μ_1 is regular.

□

2. Let X be a locally compact Hausdorff space. Exercise 6.1.11 shows $C_0(X)$, the space of complex valued continuous functions on X vanishing at ∞ is a Banach $*$ algebra (actually a C^* algebra). Show $C_0(X)$ is the completion of $C_c(X)$.

Solution:

We must identify Cauchy sequences $\{f_n\}$ in $C_c(X)$ with elements f in $C_0(X)$ such that if $\{f_n\}$ and $\{h_n\}$ are equivalent Cauchy sequences (i.e., $|f_n - h_n|_\infty \to 0$ as $n \to \infty$), then $f = h$ and show if $f \in C_0(X)$, there is a Cauchy sequence $\{f_n\}$ identified with f.

For $\{f_n\}$ we take $f = \lim_n f_n$. Note f is continuous for it is a pointwise uniform limit of continuous functions. Also $f \in C_0(X)$ for if $\epsilon > 0$, there is an N with $|f_N - f|_\infty < \epsilon$. Thus if $|f(x)| \geqslant \epsilon$, $f_N(x) \neq 0$ and so $x \in \operatorname{supp} f_N$, which is a compact subset. So $f \in C_0(X)$. Clearly if $|f_n - h_n|_\infty \to 0$, then $\lim_n f_n = \lim_n h_n$.

Finally assume $f \in C_0(X)$. Set $K_n = \{x \mid |f(x)| \geqslant \frac{1}{n}\}$. Each K_n is compact. Moreover, by Proposition 5.23, there is a function $\phi_n \in C_c(X)$ with $\phi_n = 1$ on K_n and $0 \leqslant \phi_n \leqslant 1$. Take $f_n = \phi_n f$. Note $f_n \in C_c(X)$ and $|f_n(x) - f(x)| = (1 - \phi_n(x))|f(x)| = \chi_{X-K_n}(x)|f(x)| < 1/n$ for all x. Thus $|f_n - f|_\infty \to 0$ as $n \to \infty$.

□

3. Let X be a locally compact Hausdorff space. Show every function $f \in C_0(X)$ is Baire measurable on X.

Solution:

The argument for Exercise 6.6.2 shows every function $f \in C_0(X)$ is the uniform limit of a sequence $\{f_n\}$ of functions in $C_c(X)$ and hence is a pointwise limit of the f_n. But each f_n is a Baire measurable function. So f is Baire measurable.

□

4. Use Lemma 6.4 and Proposition 6.7 to show that every complex Borel measure on a second countable locally compact Hausdorff space is regular.

Solution:

Let X be a second countable locally compact Hausdorff space. Lemma 6.4 shows the Baire sets are the Borel sets. By Exercise 6.1.3, X is σ-compact. Consequently, every Borel subset is σ-bounded. Now consider a complex measure μ on X and the decomposition $\mu = \mu_1 - \mu_2 + i(\mu_3 - \mu_4)$ described in Exercise 6.6.1. This exercise shows $|\mu|$ is regular if and only if each μ_j is regular. Thus it suffices to check the regularity of the finite measure $|\mu|$. Using Proposition 6.7 for the measure $|\mu|$ and the fact every Borel subset is Baire and σ-bounded, we see $|\mu|$ is inner regular on all Borel subsets. Using the finiteness of $|\mu|$, we see if E is a Borel subset of X and $\epsilon > 0$, there is a compact subset K of E with $|\mu|((X - E) - K) < \epsilon$. Thus $|\mu|((X - K) - E) < \epsilon$. So $U = X - K$ is an open subset containing E and $|\mu|(E) > \mu(U) - \epsilon$. So $|\mu|$ is outer regular.

□

5. Let X be a locally compact Hausdorff space. Let I be a positive bounded integral on $C_c(X)$. Show the unique Radon measure μ on X with $I(f) = \int f\,d\mu$ is finite and $||I|| = \mu(X)$.

Solution:

We show μ is finite. By inner regularity of the Radon measure μ there is a sequence K_n of compact G_δ subsets such that $\mu(K_n) \to \mu(X)$. By Proposition 5.23, for each n there is a function $f_n \in C_c(X)$ with $f_n = 1$ on K_n and $0 \leqslant f_n \leqslant 1$. Thus $0 \leqslant \mu(K_n) \leqslant \int f_n\,d\mu = I(f_n) = |I(f_n)| \leqslant ||I||\,|f_n|_\infty = ||I||$. So $\mu(X) = \lim \mu(K_n) \leqslant ||I||$. Since $|I(f)| = |\int f\,d\mu| \leqslant |f|_\infty \mu(X)$, we also have $||I|| \leqslant \mu(X)$. So $||I|| = \mu(X)$.

□

6. Let X be a locally compact Hausdorff space and let $I : C_c(X) \to \mathbb{C}$ be a bounded linear functional which is real valued on real valued functions.

(a) Define I^+ on the nonnegative functions f in $C_c(X)$ by

$$I^+(f) = \sup_{0 \leqslant g \leqslant f} I(g).$$

Show $I^+(cf) = cI^+(f)$ and $I^+(f_1 + f_2) = I^+(f_1) + I^+(f)$ when $c \geqslant 0$ and f, f_1 and f_2 are all nonnegative.

(b) Define $I^+(f) = I^+(f^+) - I^+(f^-)$ for f a real valued function on X. Show I^+ is positive and linear and bounded on the real valued functions.

(c) Let f be real valued. Define $I^-(f) = -(I(f) - I(f^+))$. Then $I = I^+ - I^-$. Show I^- is positive and linear and $||I|| = ||I^+|| + ||I^-||$.

Solution: For (a), first note $I^+(f) \geqslant 0$ for $f \geqslant 0$ since $I(0) = 0$. One can easily verify $I^+(cf) = cI^+(f)$ for $c \geqslant 0$ and $f \geqslant 0$. Consider $f_1, f_2 \geqslant 0$. Note if $0 \leqslant g_1 \leqslant f_1$ and $0 \leqslant g_2 \leqslant f_2$, then $0 \leqslant g_1 + g_2 \leqslant f_1 + f_2$. So $I^+(f_1 + f_2) \geqslant I(g_1) + I(g_2)$. Being true for all g_1 and g_2 gives $I^+(f_1 + f_2) \geqslant I^+(f_1) + I^+(f_2)$. Now let $0 \leqslant g \leqslant f_1 + f_2$. Set $g_1 = \min\{f_1, g\}$. Note $0 \leqslant g_1 \leqslant f_1$. Set $g_2 = g - g_1$. Clearly $0 \leqslant g_2$. Now if $g_1(x) = f_1(x) \leqslant g(x)$, then $g_2(x) = g(x) - f_1(x) \leqslant f_1(x) + f_2(x) - f_1(x) \leqslant f_2(x)$, while if $g_1(x) = g(x)$, then $g_2(x) = 0 \leqslant f_2(x)$. So $0 \leqslant g_2(x) \leqslant f_2(x)$. Hence $I(g) = I(g_1) + I(g_2) \leqslant I^+(f_1) + I^+(f_2)$. Taking the supremum over g gives $I^+(f_1 + f_2) \leqslant I^+(f_1) + I^+(f_2)$.

For (b), it's just a matter of checking. Note $f^+ = f \wedge 0$ and $f^- = (-f) \wedge 0$. Now if $c \geqslant 0$, $(cf)^+ = cf^+$ and $(cf)^- = cf^-$. So $I^+(cf) = I^+(cf^+) - I^+(cf^-) = cI^+(f) - cI^-(f^-) = cI^+(f)$. Now if $c < 0$, $(cf)^+ = -cf^-$ and $(cf)^- = -cf^+$. Thus $I(cf) = I^+(-cf^-) - I^+(-cf^+) = cI^+(f^+) - cI^+(f^-) = cI^+(f)$. Now consider $f_1 + f_2$. We have $(f_1 + f_2)^+ - (f_1 + f_2)^- = f_1^+ + f_2^+ - f_1^- - f_2^-$. So $(f_1 + f_2)^+ + f_1^- + f_2^- = f_1^+ + f_2^+ + (f_1 + f_2)^-$. This gives

$$I^+((f_1 + f_2)^+) + I^+(f_1^-) + I^+(f_2^-) = I^+(f_1^+) + I^+(f_2^+) + I^+((f_1 + f_2)^-).$$

So

$$\begin{aligned} I^+(f_1 + f_2) &= I^+((f_1 + f_2)^+) - I^+((f_1 + f_2)^-) \\ &= I^+(f_1^+) - I^+(f_1^-) + I^+(f_2^+) - I^+(f_2^-) \\ &= I^+(f_1) + I^+(f_2). \end{aligned}$$

Since $0 \leqslant g \leqslant f$ implies $|g|_\infty \leqslant |f|_\infty$ we have $|I(g)| \leqslant ||I||\,|g|_\infty \leqslant ||I||\,|f|_\infty$ for all such g. So $|I^+(f)| \leqslant ||I||\,|f|_\infty$ for $f \geqslant 0$.

Now $|I^+(f^+) - I^+(f^-)| \leqslant I^+(f^+ + f^-) = I^+(|f|) \leqslant ||I||\,|f|_\infty$ for all real valued f. Thus $||I^+|| \leqslant ||I||$.

Now if $I^-(f) = -(I(f) - I^+(f))$, we have I^- is bounded and linear. We claim I^- is positive. Let $f \geqslant 0$. Note $0 \leqslant f \leqslant f$, so $I^+(f) \geqslant I(f)$. So $-(I(f) - I^+(f)) \geqslant 0$. Clearly $||I|| \leqslant ||I^+|| + ||I^-||$. Since

$$\begin{aligned} I^-(f) &= -(I(f) - I^+(f)) \\ &= \sup_{0 \leqslant g \leqslant f} I(g) - I(f) \\ &= \sup_{-f \leqslant h \leqslant 0} I(h) \\ &= \sup\{-I(g) \mid 0 \leqslant g \leqslant f\} \end{aligned}$$

for $f \geqslant 0$, we have $||I^\pm|| = \sup\{\pm I(f) \mid 0 \leqslant f \leqslant 1\}$. Let $\epsilon > 0$. Pick $0 \leqslant g_1 \leqslant 1$ with $I(g_1) \geqslant ||I^+|| - \epsilon$ and $0 \leqslant g_2 \leqslant 1$ with $-I(g_2) \geqslant ||I^-|| - \epsilon$. Consider $g = g_1 - g_2$. We have $|g| \leqslant 1$ and $I(g) = I(g_1) - I(g_2) \geqslant ||I^+|| - \epsilon + ||I^-|| - \epsilon$. This implies $||I|| \geqslant ||I^+|| + ||I^-|| - 2\epsilon$. This gives (c). □

7. Show if μ_1 and μ_2 are finite positive measures on a locally compact Hausdorff space X and I defined on $C_c(X)$ by $I(f) = \int f\,d\mu_1 - \int f\,d\mu_2$ has norm $||I|| = \mu_1(X) + \mu_2(X)$, then the measures μ_1 and μ_2 are disjoint and thus $\mu = \mu_1 - \mu_2$ is the Hahn decomposition of μ.

Solution: Let A and B be disjoint Borel sets such that $\mu = \mu_1 - \mu_2$ is positive on A and negative on B and $A \cup B = X$. We claim $\mu_1(B) = \mu_2(A) = 0$.

Note

$$\begin{aligned} |I(f)| &= |\int_A f\,d\mu + \int_B f\,d\mu| \leqslant |f|\mu(A) - |f|\mu(B) \\ &= |f|\,(\mu_1(A) - \mu_2(A) - \mu_1(B) + \mu_2(B)) \\ &\leqslant |f|\,(\mu_1(A) + \mu_2(A) + \mu_1(B) + \mu_2(B) - 2\mu_2(A) - 2\mu_1(B)) \\ &\leqslant |f|\,(\mu_1(X) + \mu_2(X) - 2\mu_2(A) - 2\mu_1(B)). \end{aligned}$$

Thus $||I|| = \mu_1(X) + \mu_2(X) \leqslant \mu_1(X) + \mu_2(X) - 2\mu_2(A) - 2\mu_1(B)$ and so $\mu_2(A) = \mu_1(B) = 0$. □

8. Let X be a locally compact Hausdorff space. Using the previous exercises, show the dual of $C_c(X)$ is $M(X)$; i.e., every bounded complex linear functional I on $C_c(X)$ is given by a unique complex Radon measure μ by

$$I(f) = \int f\,d\mu.$$

Moreover, show $||I|| = ||\mu|| = |\mu|(X)$.

Solution: Let I be such an integral. We may by taking real and imaginary parts restrict to the case I is real valued on real valued functions. Then by Exercise 6.6.6 we obtain I^+ and I^-. These are bounded. By the Riesz Theorem, they are given by unique positive regular Borel measures μ^+ and μ^-. We claim $\mu^+(X) = ||I^+||$. Clearly $\mu^+(X) \geqslant ||I^+||$. Since μ is regular, there is an increasing sequence of compact subsets K_n with $\mu(K_n) \to \mu(X)$. Pick $0 \leqslant f_n \leqslant 1$ with $f_n = 1$ on K_n. Then $||I^+|| \geqslant |I^+(f_n)| \geqslant \mu(K_n)$. Taking a limit gives $||I^+|| = \mu^+(X)$. So $I(f) = \int f\,d\mu^+ - \int f\,d\mu^-$. Set $\mu = \mu^+ - \mu^-$. Since $||I|| = ||I^+|| + ||I^-|| = \mu^+(X) + \mu^-(X)$, we have by the previous exercise that $\mu = \mu^+ - \mu^-$ is the Hahn decomposition. So since μ^+ and μ^- are regular, so is μ.

Continuing we see each I in $C_c(X)^*$ is given by a complex regular Borel measure μ and $\mu = \mu_1 - \mu_2 + i(\mu_3 - \mu_4)$ where $\mu_1 \perp \mu_2$ and $\mu_3 \perp \mu_4$ and all are finite regular Borel measures. Next, we show $||I|| = |\mu|(X)$. Indeed, note each $\mu_j \prec |\mu|$ and by Radon-Nikodym Theorem, there is an $h \in L^1(|\mu|)$ such that $\mu(E) = \int_E h\,d|\mu|$. Since $|\mu(E)| \leqslant |\mu|(E)$ for each E, we see $\left|\int h\chi_E\,d|\mu|\right| \leqslant \int \chi_E d|\mu|$ for all E. This implies $|h| \leqslant 1$. Suppose $|h| < 1$ on a set of positive measure E. Then $|\mu|(E) \leqslant \int_E |h|\,d|\mu| < |\mu|(E)$. So $|h| = 1$ a.e. Hence $|I(f)| = \left|\int hf\,d|\mu|\right| \leqslant |f| \int |h|\,d|\mu| = |f|\,||\mu||$. So $||I|| \leqslant ||\mu||$. Now let $\epsilon > 0$. Pick disjoint Borel sets $E_1, E_2, \ldots, E_m$ whose union is X and $\sum |\mu(E_j)| \geqslant ||\mu|| - \epsilon$. Now choose compact G_δ sets $K_j \subseteq E_j$ with $|\mu|(E_j - K_j) < \frac{\epsilon}{m}$. Take $0 \leqslant f_j \leqslant 1$ and $f_j = 1$ exactly on K_j. Set $F_k = \sum \frac{\bar{\mu}(E_j)}{|\mu(E_j)|} f_j^k \in C_c(X)$. Note $|F_k(x)| = 1$ if $x \in \cup K_j$ and if $x \notin K = \cup\mathrm{supp}(f_j)$, then $F_k(x) = 0$. Furthermore, if $x \notin \cup K_j$, then $|f_j(x)| < 1$ for each j and thus $|F_k(x)|$ decreases monitonically to 0 as $k \to \infty$. Thus $\cap_k \{x \mid |F_k(x)| \geqslant 1 + \epsilon\} = \varnothing$. By the finite intersection property, we then would have a l with $|F_l(x)| < 1 + \epsilon$ for all x. Thus for $k \geqslant l$, $(1+\epsilon)||I|| \geqslant |I(F_k)|$. Let $k \to \infty$. $F_k \to \sum \frac{\bar{\mu}(E_j)}{|\mu(E_j)|}\chi_{K_j}$. Thus $I(F_k) \to \sum \frac{\bar{\mu}(E_j)}{|\mu(E_j)|}\mu(K_j) = -\sum u_j \mu(E_j - K_j) + \sum |\mu(E_j)|$ where $u_j = \frac{\bar{\mu}(E_j)}{|\mu(E_j)|}$. Thus $||I|| + \epsilon \geqslant \sum_j |\mu(E_j)| - \sum |\mu(E_j - K_j)| \geqslant \sum |\mu(E_j)| - \epsilon$. Letting $\epsilon \to 0$ and taking the supremum over all sets E_j gives $||I|| \geqslant |\mu|(X)$. □

9. Let G be a locally compact Hausdorff group. Show the mappings $\mu \mapsto x\mu$ and $\mu \mapsto \mu x$ are linear isometries of $M(G)$.

Solution:

It is clear the mappings $\mu \mapsto x\mu$ and $\mu \mapsto \mu x$ are linear. To see these are isometries, it suffices by Exercise 6.6.8 to show $f \mapsto \int f\,d(x\mu)$ and $f \mapsto \int f\,d\mu$

have the same norm on $C_c(G)$. But $\int f\,d(x\mu) = \int f(xg)\,d\mu(g) = \int \lambda(x^{-1})f\,d\mu$ where $\lambda(x)f(g) = f(x^{-1}g)$. Note since $|\lambda(x^{-1})f|_\infty = |f|_\infty$ and $\lambda(x)(\lambda(x^{-1})f) = f$, we have $B = \{f \mid |f|_\infty = 1\} = \{\lambda(x^{-1})f \mid |f|_\infty = 1\}$. So $\sup_{f\in B} |\int f\,d\mu| = \sup_{f\in B} |\int \lambda(x^{-1})f\,d\mu| = \sup_{f\in B} |\int f\,d(x\mu)|$. So they have the same norms. A similar argument shows $\mu \mapsto \mu x$ is a linear isometry.

□

10. Show $x \mapsto x\mu$ and $x \mapsto \mu x$ are continuous from G into $M(G)$ where $M(G) = C_c(G)^*$ has the weak $*$ topology.

Solution:

Recall the weak $*$ topology is a locally convex topology and is defined by the seminorms $|\cdot|_f$, $f \in C_c(G)$ where $|\mu|_f = |\int f\,d\mu|$ for $\mu \in M(G)$. Now $|x\mu - y\mu|_f = |\int f(xg)\,d\mu(g) - \int f(yg)\,d\mu(g)| \leqslant \int |f(xg) - f(yg)|\,d|\mu|$. Let $\epsilon > 0$. By the left uniform continuity of f (see Lemma 5.24), there is a neighborhood U of e such that $|f(uxg) - f(xg)| < \frac{\epsilon}{1+|\mu|}$ for all g if $u \in U$ Thus if $y \in Ux$, we have $|x\mu - y\mu|_f \leqslant \frac{|\mu|\epsilon}{1+|\mu|} < \epsilon$. This being true for each $f \in C_c(G)$ implies $x \mapsto x\mu$ is continuous.

Similarly, $x \mapsto \mu x$ is continuous.

□

11. Show if μ and ν are regular complex Borel measures on G, and E is a countable union of compact G_δ sets, then $x \mapsto \nu(x^{-1}E)$ and $y \mapsto \mu(Ey^{-1})$ are Borel measurable and

$$\mu * \nu(E) = \int \nu(x^{-1}E)\,d\mu(x) = \int \mu(Ey^{-1})\,d\nu(y).$$

Solution:

Let K be a compact G_δ. By Proposition 5.23, there is a continuous function ϕ with compact support such that $0 \leqslant \phi \leqslant 1$ and $\phi^{-1}(1) = K$. Thus $\phi^n \to 1_K$ pointwise every where. So $\mu y(K) = \lim_n \int \phi^n(x)\,d\mu y(x) = \lim_n \int \phi^n(xy)\,d\mu(x)$ and each function $y \mapsto \int \phi^n(xy)\,d\mu(y) = \mu y(\phi^n)$ is continuous by Exercise 6.6.10. Thus $y \mapsto \mu y(K)$ is Borel for it is a pointwise limit of continuous functions. Similarly $x \mapsto x\mu(K)$ is Borel. By Lemma 6.5, since a finite union of compact G_δ subsets is a compact G_δ subset, we have if E is a union of compact G_δ subsets, there is an increasing sequence K_n of compact G_δ subsets with $E = \cup K_n$. Thus $\mu y(E) = \lim_n \mu y(K_n)$ and $x\mu(E) = \lim_n x\mu(K_n)$ show $x \mapsto x\mu(E)$ and $y \mapsto \mu y(E)$ are Borel functions. This gives $x \mapsto \mu(x^{-1}E)$ and

$y \mapsto \mu(Ey^{-1})$ are Borel functions for such E. Also for K,

$$\begin{aligned}\mu * \nu(K) &= \lim_n \int \phi^n(z)\, d(\mu * \nu)\\ &= \lim_n \int\int \phi^n(xy)\, d\mu(x)\, d\nu(y)\\ &= \lim_n \int\int \phi^n(x)\, d\mu(xy^{-1})\, d\nu(y)\\ &= \int\int \chi_K(x)\, d\mu y(x)\, d\nu(y)\\ &= \int \mu y(K)\, d\nu(y)\end{aligned}$$

and using Fubini,

$$\begin{aligned}\mu * \nu(K) &= \lim_n \int\int \phi^n(xy)\, d\nu(y)\, d\mu(x)\\ &= \lim_n \int\int \phi^n(y)\, d(x\nu)(y)\, d\mu(x)\\ &= \int x\nu(K)\, d\mu(x)\,.\end{aligned}$$

Using the Monotone Convergence Theorem and the increasing sequence K_n, we see

$$\mu * \nu(E) = \int \mu y(E)\, d\nu(y) = \int x\nu(E)\, d\mu(y).$$

□

12. Show if μ is a regular complex Borel measure on a locally compact Hausdorff group, then the measure $E \mapsto \overline{\mu(E^{-1})}$ is a regular Borel measure.

Solution:

By Exercise 6.6.1, it suffices to show the result for nonnegative finite real valued regular measures μ. Thus it suffices to show $\check{\mu}(E) = \mu(E^{-1})$ is a regular Borel measure. Since $E \mapsto E^{-1}$ takes Borel sets to Borel sets, $\check{\mu}$ is a Borel measure. Also K^{-1} is compact if and only if K is compact. So $\check{\mu}(K) < \infty$ for compact subsets. Now $\check{\mu}(E) = \mu(E^{-1}) = \inf\{\mu(U) \mid E^{-1} \subseteq U, U \text{ is open}\} = \inf\{\mu(U^{-1}) \mid E \subseteq U, U^{-1} \text{ is open}\}$ implies $\check{\mu}$ is outer regular. Inner regularity follows similarly for if E is a Borel subset, then E^{-1} is a Borel subset and $\mu(E^{-1}) < \infty$. By inner regularity, $\check{\mu}(E) = \mu(E^{-1}) = \sup\{\mu(K) \mid K \subseteq E^{-1}, K \text{ is compact}\} = \sup\{\mu(K^{-1}) \mid K^{-1} \subseteq E^{-1}, K \text{ is compact}\}$. This gives $\check{\mu}(E) = \sup_{K \subseteq E} \check{\mu}(K)$.

□

13. Let X be a locally compact Hausdorff space. Let $\mathcal{F}(X)$ be the vector space of all bounded complex Borel functions on X and define $||f|| = \sup\{|f(x)| \mid x \in X\}$ for $f \in \mathcal{F}(X)$.

(a) Show $\mathcal{F}(X)$ is a Banach space.

(b) Show the dual space of $\mathcal{F}(X)$ is the space of all finitely additive complex Borel measures on X; i.e., show each continuous linear functional I on $\mathcal{F}(X)$ has form $I(f) = \int f\,d\mu$ for a unique finitely additive measure. (Here one needs to define $\int f\,d\mu$ appropriately.)

(c) If $I(f) = \int f\,d\mu$ where μ is finitely additive, show $||I|| = \sup\{\sum_{k=1}^{m} |\mu(E_k)| \mid E_1, E_2, \ldots, E_m$ are disjoint and have union $X\}$.

Solution: Clearly $||\cdot||$ is a norm. We show completeness. Let f_k be Cauchy. Then $|f_k(x) - f_m(x)| \leqslant ||f_k - f_m|| \to 0$ as $k, m \to \infty$. So $f_k(x) \to f(x)$ for each x. Thus f is Borel being a pointwise limit of Borel functions. Also if $||f_k - f_m|| \leqslant \epsilon$ where $\epsilon > 0$ for $k, m \geqslant N$ then $|f_k(x) - f(x)| = \lim_m |f_k(x) - f_m(x)| \leqslant \limsup_m ||f_k - f_m|| \leqslant \epsilon$ for $k \geqslant N$. So $\mathcal{F}(X)$ is complete.

Now let I be a continuous linear functional on $\mathcal{F}(X)$. For E a Borel subset of X, define $\mu(E) = I(\chi_E)$. Clearly, $\mu(\phi) = 0$ and $\mu(E) \in \mathbb{C}$ for each E. Moreover, μ is finitely additive. Claim $I(f) = \int f\,d\mu$. For each $\epsilon > 0$, one can choose a simple Borel function s with $||f - s|| \leqslant \epsilon$. Then $|I(f) - I(s)| \leqslant ||I||\,||f - s||$. Thus $I(s) \to I(f)$ as $s \to f$. But $I(s) = \int s\,d\mu$. So if we define $\int f\,d\mu = \lim_{s \to f} \int s\,d\mu$ we see $I(f) = \int f\,d\mu$.

For the norm of I, note $|I(f)| \leqslant |I(f - s)| + |I(s)| \leqslant \epsilon + \sum |a_k|\,|\mu(E_k)| \leqslant \epsilon + ||f|| \sum |\mu(E_k)| \leqslant \epsilon + |\mu|(X)\,||f||$ if s is close to f. This implies $||I|| \leqslant ||\mu||$. Let $E_1, E_2, \ldots, E_m$ be disjoint Borel subsets with union X. Take $f = \sum \frac{\overline{\mu(E_j)}}{|\mu(E_j)|} \chi_{E_j}$. Then $||f|| \leqslant 1$ and so $||I|| \geqslant I(f) = \sum |\mu(E_j)|$. So $||I|| \geqslant ||\mu||$. □

Exercise Set 6.7

In the following exercises G will always denote a σ-compact locally compact Hausdorff group and λ will be a left-invariant Haar measure on the locally compact Hausdorff topological group G.

1. Let $1 < p \leqslant \infty$, $f \in L^p(G)$, and $\mu \in M(G)$. Establish the following:

(a) $\mu * f \in L^p(G)$ and $||\mu * f||_p \leqslant ||f||_p ||\mu||$.

(b) $x \mapsto \int f(xy^{-1}) \Delta(y)^{-1/p}\,d\mu(y)$ is in $L^p(G)$ and has norm less than $||f||_p ||\mu||$. In particular if G is unimodular, $f * \mu \in L^p(G)$ and $||f * \mu||_p \leqslant ||f||_p ||\mu||$.

Solution:

We let $\mathcal{B}$ be the σ-algebra of Baire sets on G. By Proposition 6.12, $\mathcal{B} \times \mathcal{B}$ is the σ-algebra of Baire sets for $G \times G$ and by Lemma 6.13, $(x, y) \mapsto y^{-1}x$ is Baire measurable. Now let $1 < p \leqslant \infty$ and $f \in L^p(G)$ with f Baire measurable. Choose q with $\frac{1}{p} + \frac{1}{q} = 1$. For $\mu * f$, define T on $L^q(G)$ by

$$T(h) = \int \int h(yx)\, f(x) d\mu(y)\, d\lambda(x).$$

To see this makes sense, we can use Fubini's Theorem since μ is finite and λ is σ-finite. Note

$$\begin{aligned}\int\int |h(yx)f(x)|\, d\lambda(x)\, d|\mu|(y) &\leqslant \int ||x \mapsto h(yx)||_q ||f||_p \, d|\mu|(y) \\ &= \int ||h||_q ||f||_p \, d|\mu|(y) \\ &= ||h||_q ||f||_p ||\mu||.\end{aligned}$$

Now applying Fubini again, we see

$$\left|\int h(x) \int f(y^{-1}x)\; d\mu(y)\, d\lambda(x)\right| \leqslant ||h||_q ||f||_p ||\mu||$$

for $h \in L^q(G)$. This implies $x \mapsto \int f(y^{-1}x)\, d\mu(y)$ is finite a.e. and defines a function in L^p with norm less than $||f||_p||\mu||$.

For the second part, define T on $L^q(G)$ by

$$T(h) = \int\int h(xy)f(x)\Delta(y)^{1/q} d\mu(y)\, d\lambda(x).$$

Note

$$\begin{aligned}|T(h)| &\leqslant \int\int |h(xy)|\, |f(x)|\Delta(y)^{1/q} d\lambda(x)\, d|\mu|(y) \\ &\leqslant \int ||f||_p ||x \mapsto h(xy)\Delta(y)^{1/q}||_q \, d|\mu|(y) \\ &= ||f||_p ||h||_q \int d|\mu|(y) \\ &= ||f||_p ||h||_q ||\mu||\end{aligned}$$

for

$$\int |h(xy)|^q \Delta(y)\, d\lambda(x) = \int |h(x)|^q \Delta(y)\Delta(y)^{-1}\, d\lambda(x) = ||h||_q^q.$$

Thus T is bounded and the integrals here exist. But by Fubini,

$$\begin{aligned}T(h) &= \int\int h(x)f(xy^{-1})\Delta(y)^{1/q}\Delta(y)^{-1} d\lambda(x)\, d\mu(y) \\ &= \int h(x) \int f(xy^{-1})\Delta(y)^{-1/p}\, d\mu(y)\, d\lambda(x).\end{aligned}$$

This implies $x \mapsto \int f(xy^{-1})\Delta(y)^{-1/p}\, d\mu(y)$ is in $L^p(G)$ and has norm smaller than $||f||_p||\mu||$. In particular when G is unimodular, $\Delta = 1$ and we have $f * \mu \in L^p(G)$.

□

2. Let f, g, and h be complex Borel functions on G. Show if $f * g$ and $(f * g) * h$ are defined a.e. λ, then $f * (g * h)$ is defined a.e. λ and $(f * g) * h = f * (g * h)$ a.e. λ.

Solution:

We are dealing with σ-finite measure spaces so the Fubini Theorem applies. Now $f * g$ existing a.e. implies $y \mapsto f(y)g(y^{-1}x)$ is integrable for a.e x and defines a Borel function. Similarly $(f*g)*h$ existing a.e. implies $(f*g)(z)h(z^{-1}x)$ is integrable a.e. x. Thus for a.e. x, one has $\int h(z^{-1}x)\int f(y)g(y^{-1}z)\,d\lambda(y)\,d\lambda(z)$ exists; thus integrable. By Fubini, $\int\int h(z^{-1}x)f(y)g(y^{-1}z)\,d\lambda(z)\,d\lambda(y)$ is integrable for a.e. x. This implies $\int f(y)\int g(z)h(z^{-1}y^{-1}x)\,d\lambda(z)\,d\lambda(y)$ is integrable for a.e. x and has value $(f * g) * h(x)$ a.e. x. Thus $z \mapsto g(z)h(z^{-1}y^{-1}x)$ is integrable a.e. x and by left invariance, $z \mapsto g(z)h(z^{-1}x)$ is integrable for a.e. x. So $g * h(x)$ is defined a.e. and $y \mapsto f(y)g * h(y^{-1}x)$ is integrable a.e. x, and

$$\begin{aligned} f * (g * h)(x) &= \int f(y) \int g * h(y^{-1}x)\,d\lambda(y) \\ &= \int f(y) \int g(z)h(z^{-1}y^{-1}x)\,d\lambda(z)\,d\lambda(y) \\ &= (f * g) * h(x) \end{aligned}$$

for a.e. x.

□

3. Suppose $f*g$ and $f*h$ are defined a.e. λ and a and b are complex numbers. Show $f * (ag + bh) = a(f * g) + b(f * h)$.

Solution:

Suppose $f * g(x)$ and $f * h(x)$ are defined for all $x \in E$ where the complement of E has measure 0. Thus $y \mapsto f(y)g(y^{-1}x)$ and $y \mapsto f(y)h(y^{-1}x)$ are integrable for $x \in E$. So $y \mapsto f(y)(ag(y^{-1}x) + bh(y^{-1}x))$ is integrable for $x \in E$ and $\int f(y)(ag(y^{-1}x) + bh(y^{-1}x))\,d\lambda(y) = a\int f(y)g(y^{-1}x)\,d\lambda(y) + b\int f(y)h(y^{-1}x)\,d\lambda(y)$. Thus $f * (ag + bh)(x)$ exists and $f * (ag + bh)(x) = af * g(x) + bf * h(x)$ for $x \in E$.

□

4. Assume G is unimodular. Let $1 \leqslant p \leqslant \infty$ and $1 \leqslant q \leqslant \infty$ with $\frac{1}{p} + \frac{1}{q} = 1$. Show if $f \in L^p(\lambda)$ and $g \in L^q(\lambda)$, then $f*g$ is a continuous function and satisfies $|(f * g)(x)| \leqslant |f|_p|g|_q$ for all x.

Solution: Consider the mapping $L(g)$ on $L^p(G)$ defined by $L(g)f(x) = f(g^{-1}x)$. Note $|L(g)f|_p^p = \int |f(g^{-1}x)|^p\,d\lambda(x) = \int |f(x)|^p\,d\lambda(x) = |f|_p^p$ and thus $L(g)$ is a linear isometry. Note formally, $f * g(x) = \int f(y)g(y^{-1}x)\,d\lambda(y) = \int f(xy)\check{g}(y)\,d\lambda(y) = \int \lambda(x^{-1})f(y)\check{g}(y)\,d\lambda(y)$. By unimodularity, $\check{g} \in L^q$ and $|\check{g}|_q = |g|_q$.

First assume $1 \leqslant p < \infty$. The second part of the argument in the proof of Proposition 6.87 shows $y \mapsto \mathrm{L}(y)f$ is a continuous mapping of G into $L^p(G)$. (This also follows from Exercise 6.7.8 below.) The only change needed to be made is that $|f_0(y^{-1}x) - f_0(x)|^p < \frac{\epsilon^p}{3\,\lambda(N_0\,\mathrm{supp}(f_0))}$ for $y \in N$. (Note this does not work for $p = \infty$). Thus by Hölder's inequality, $y \mapsto f(y)g(y^{-1}x)$ is in $L^1(G)$

and $f * g(x)$ exists and $|f * g(x)| \leqslant |f|_p |g|_q$. Moreover, Hölder's inequality yields

$$\begin{aligned} |f * g(x) - f * g(x_0)| &= \left| \int (L(x^{-1}) f(y) \check{g}(y) - L(x_0^{-1}) f(y) \check{g}(y))\, d\lambda(y) \right| \\ &\leqslant \int |(L(x^{-1}) f - L(x_0^{-1}) f)(y)|\, |\check{g}(y)|\, d\lambda(y) \\ &\leqslant |L(x^{-1} f) - L(x_0^{-1}) f|_p |g|_q. \end{aligned}$$

The continuity of $x \mapsto L(x) f$ then implies the continuity of $f * g$.

Now assume $f \in L^\infty(G)$ and $g \in L^1(G)$. Note

$$\begin{aligned} f * g(x) &= \int f(y) g(y^{-1} x)\, d\lambda(y) \\ &= \int f(y) \check{g}(x^{-1} y)\, d\lambda(y) \\ &= \int f(y) \lambda(x) \check{g}(y)\, d\lambda(y)\,. \end{aligned}$$

Thus $f * g(x)$ exists, $|f * g(x)| \leqslant |f|_\infty |\lambda(x) \check{g}|_1 = |f|_\infty |g|_1$ and since $x \mapsto \lambda(x) \check{g}$ is continuous into $L^1(G)$, one has

$$|f * g(x) - f * g(x_0)| \leqslant |f|_\infty |L(x) \check{g} - L(x_0) \check{g}|_1$$

and so again $x \mapsto f * g(x)$ is continuous.

□

5. Suppose $f \in L^1_{\text{loc}}(\lambda)$ and $g \in C_c(G)$. Show $f * g \in C(G)$.
Solution: Consider $y \mapsto f(y) g(y^{-1} x)$. It is integrable; so $f * g(x)$ exists everywhere. Fix a compact neighborhood N_0 of e. Set $K = x N_0 \operatorname{supp}(g)^{-1}$. Now g is uniformly continuous. So given $\epsilon > 0$, there is a neighborhood N of e with $N \subseteq N_0$ such that $|g(y^{-1} x u) - g(y^{-1} x)| < \frac{\epsilon}{|f \chi_K|_1 + 1}$ if $u \in N$. Now $y^{-1} x u \in \operatorname{supp}(g)$ if and only if $y \in x u \operatorname{supp}(g)^{-1} \subseteq x N \operatorname{supp}(g)^{-1} \subseteq K$. So $|f * g(xu) - f * g(x)| \leqslant \int_K \frac{\epsilon}{|f \chi_K|_1 + 1} |f(y)|\, dy \leqslant \epsilon$. Thus $f * g$ is continuous at x. □

6. Let f and g be complex Borel functions such that $f * g(w)$ exists for λ a.e. w. Let $L_x f(y) = f(x^{-1} y)$. Show $(L_x f) * g(w) = L_x (f * g)(w)$ a.e. w.
Solution:

$$\begin{aligned} L_x f * g(w) &= \int L_x f(u) g(u^{-1} w)\, du \\ &= \int f(x^{-1} u) g(u^{-1} w)\, du \\ &= \int f(u) g(u^{-1} x^{-1} w)\, du \\ &= (f * g)(x^{-1} w) \\ &= L_x (f * g)(w). \end{aligned}$$

□

7. Let H be a closed subgroup of G and let $X = G/H$. Assume that μ is a positive and regular G-invariant measure on X. Show that if $f \in L^1(G)$ and $g \in L^1(X, \mu)$ then

$$G \ni a \mapsto f(a)g(a^{-1}x) \in \mathbb{C}$$

is integrable with respect to λ for μ-almost all $x \in X$. Furthermore, show

(a) $X \ni x \mapsto f * g(x) := \int f(a)g(a^{-1}x)\, d\lambda(a)$ is integrable with respect to μ and

(b) $||f * g||_1 \leqslant ||f||_1 ||g||_1$.

Solution:

We recall G is σ-compact and hence so is G/H. By regularity of the measures, λ and μ are σ-finite. Using Fubini's Theorem and the invariance of μ,

$$\begin{aligned}\int \left| \int f(a)g(a^{-1}x)\, d\lambda(a) \right| d\mu(x) &\leqslant \int \int |f(a)g(a^{-1}x)|\, d\lambda(a)\, d\mu(x) \\ &\leqslant \int \int |f(a)g(a^{-1}x)|\, d\mu(x)\, d\lambda(a) \\ &\leqslant \int \int |f(a)g(x)|\, d\mu(x)\, d\lambda(a) \\ &= |f|_1 |g|_1 < \infty .\end{aligned}$$

Thus $a \mapsto f(a)g(a^{-1}x)$ is integrable for μ a.e. x and if we take $f * g(x)$ to be $\int f(a)g(a^{-1}x)\, d\lambda(a)$ when it exists, then $f * g$ is integrable and $|f * g|_{L^1(X)} \leqslant |f|_{L^1(G)} |g|_{L^1(X)}$.

□

8. Let $1 \leqslant p < \infty$. Show that the map $L : G \to B(L^p(G))$, $L_a f(b) := f(a^{-1}b)$ is strongly continuous, i.e.,

$$\lim_{a \to b} |L_a f - L_b f|_p = 0$$

for all $f \in L^p(G)$.

Solution:

We use the density of $C_c(G)$ in $L^p(G)$ for $1 \leqslant p < \infty$. So assume f is in $L^p(G)$ and $\epsilon > 0$. Choose a nonzero f_0 in $C_c(G)$ with $||f - f_0||_p < \frac{2\epsilon}{3}$. Fix a compact neighborhood N_0 of e. Using Lemma 5.24, the left uniform continuity of f_0 implies there is a neighborhood N of e with $N \subseteq N_0$ and such that

$$|f_0(n^{-1}x) - f_0(x)|^p \leqslant \frac{\epsilon^p}{3^p\, \lambda(bN_0 \text{ supp } f_0)} \quad \text{for } x \in G \text{ and } n \in N .$$

Thus $|f_0(n^{-1}b^{-1}x) - f_0(b^{-1}x)| \leqslant \frac{\epsilon^p}{3\, \lambda(bN_0 \text{ supp } f_0)}$ for all x and $n \in N$. This implies

$$\begin{aligned}&||L_{bn} f - L_b f||_p \leqslant ||L_{bn} f - L_{bn} f_0||_p + ||L_{bn} f_0 - L_b f_0||_p + ||L_b f_0 - L_b f||_p \\ &\leqslant \frac{2\epsilon}{3} + \left(\int_{bN_0 \text{ supp } f_0} |f_0(n^{-1}b^{-1}x) - f_0(b^{-1}x)|^p\, dx \right)^{1/p} < \epsilon\end{aligned}$$

Thus $||L_{b_n}f - L_bf||_p < \epsilon$ if $n \in N$. So $a \mapsto L_af$ is continuous at b for all b.

□

9. Assume G is unimodular. Let $f, g \in L^2(G)$. Show $f * g$ is defined and $f * g \in C_0(G)$.

Solution:

By Exercise 6.7.4, $f * g$ is continuous. In this exercise we showed $f * g(x) = \int \lambda(x^{-1})f(y)\check{g}(y)\,d\lambda(y) = (\lambda(x^{-1})f, g^*)_2$ for $f^* = \overline{\check{f}}$ when G is unimodular. Pick f_0 in $C_c(G)$ with $|f - f_0|_2 < \frac{\epsilon}{2|g|_2}$. Now pick a compact subset K of G such that $\check{g}1_{G-K}$ is in L^2 and has L^2 norm less than $\frac{\epsilon}{2|f_0|_2}$. Set $K_0 = \operatorname{supp} f_0$. Note if $xy \in K_0$ and $y \in K$, then $x \in K_0K^{-1}$. Hence if $x \notin K_0K^{-1}$, $f_0(xy) = 0$ or $y \notin K$. So for $x \notin K_0K^{-1}$, $|(\lambda(x^{-1})f_0, g^*)| = |\int f_0(xy)\check{g}(y)\,d\lambda(y)| = |\int_{G-K} f_0(xy)\check{g}(y)\,d\lambda(y)| \leqslant |\lambda(x^{-1})f_0|_2|\check{g}1_{G-K}|_2 < \frac{\epsilon}{2}$. Thus for $x \notin K_0K^{-1}$,

$$\begin{aligned}|f * g(x)| &\leqslant |(f - f_0) * g(x)| + |f_0 * g(x)| \\ &= |(\lambda(x^{-1})(f - f_0), g^*)_2| + |(\lambda(x^{-1})f_0, g^*)_2| \\ &< |f - f_0|_2|g^*|_2 + \frac{\epsilon}{2} < \epsilon\,.\end{aligned}$$

So $f * g$ vanishes at ∞.

□

10. Assume that there exists a sequence of open set $U_n \in \mathcal{N}(e)$, $n \in \mathbb{N}$, such that $\cap U_n = \{e\}$. Show there exists a sequence of continuous compactly supported functions $\{f_n\}_{n\in\mathbb{N}}$ such that $\operatorname{supp}(f_n) \subset U_n$ and $\int f_n\,d\lambda = 1$ with:

(a) for $f \in C_c(G)$, $f_n * f$ converges uniformly to f.

(b) for $f \in L^1(G)$, $f_n * f$ converges to f in L^1.

Solution:

For each n, by local compactness, there is a compact neighborhood V_n of e with $V_n \subseteq U_n$. By Urysohn, there is a sequence ϕ_n in $C_c(G)$ with $\phi_n = 1$ on V_n, $0 \leqslant \phi_n \leqslant 1$ and $\operatorname{supp}\phi_n \subseteq U_n$. Set $\Phi_n = \prod_{k=1}^n \phi_k$ and $K_n = \operatorname{supp}\Phi_n$. Note $0 \leqslant \Phi_n \leqslant 1$, $K_{n+1} \subseteq K_n$ and $\Phi_n(e) = 1$ for all n. Define $f_n = \frac{\Phi_n}{\int \Phi_n\,d\lambda}$. Then $\operatorname{supp} f_n \subseteq U_n$. Also if $f \in C_c(G)$ and $K = \operatorname{supp} f$, then

$$f_n * f(x) = \int f_n(y)f(y^{-1}x)\,d\lambda(y)$$

is a continuous function with compact support a subset of the compact subset K_nK. Also

$$\begin{aligned}|f_n * f(x) - f(x)| &= \left|\int f_n(y)f(y^{-1}x)\,d\lambda(y) - \int f_n(y)f(x)\,d\lambda(y)\right| \\ &\leqslant \int f_n(y)\,|f(y^{-1}x) - f(x)|\,d\lambda(y)\,.\end{aligned}$$

By the left uniform continuity of f, there is an open neighborhood U of e with $|f(y^{-1}x) - f(x)| < \epsilon$ for all $y \in U$.

Note $X - U$ is closed and $\cap((X - U) \cap K_n) = \varnothing$ for $\cap K_n = \{e\}$. By the finite intersection property and compactness, some finite intersection of these sets is empty . So $(X - U) \cap K_n = \varnothing$ for some n. Hence $K_m \subseteq U$ for some m. This implies if $n \geqslant m$, then

$$|f_n * f(x) - f(x)| \leqslant \int_{K_n} f_n(y)\epsilon \, d\lambda(y) = \epsilon\,.$$

Thus $f_n * f$ converges uniformly to f. Since the support of every $f_n * f$ is contained in K_1K, one also has $|f_n * f - f|_1 \to 0$ as $n \to \infty$.

Now let $f \in L^1(G)$ and let $\epsilon > 0$. Choose $f_0 \in C_c(G)$ with $|f - f_0|_1 < \frac{\epsilon}{3}$ and pick n_0 with $|f_n * f_0 - f_0|_1 < \frac{\epsilon}{3}$ for $n \geqslant n_0$. Then by Equation 6.27,

$$\begin{aligned} |f_n * f - f|_1 &\leqslant |f_n * (f - f_0)|_1 + |f_n * f_0 - f_0|_1 + |f_0 - f|_1 \\ &\leqslant |f_n|_1 \, |f - f_0|_1 + \frac{2\epsilon}{3} \\ &< \epsilon \end{aligned}$$

for $n \geqslant n_0$. So $f_n * f \to f$ in $L^1(G)$.

(**Remark:** Note the argument here shows if G is a locally compact Hausdorff group and if there is a sequence U_n of open neighborhoods of e with $\cap U_n = \{e\}$, then G is first countable.) □

Exercise Set 6.8

In the following exercises G unless otherwise stated will be a σ-compact locally compact Hausdorff group.

1. Show a representation π on a $*$ algebra $\mathcal{A}$ is nondegenerate if and only if there is no nonzero vector v with $\pi(x)v = 0$ for all $x \in \mathcal{A}$.

Solution: Let $\mathcal{H}_0 = \{w \in \mathcal{H} \mid \pi(a)w = 0 \text{ for all } a \in \mathcal{A}\}$. Then $\mathcal{H}_0$ and $\mathcal{H}_0^\perp$ are invariant subspaces and $\pi(x)\mathcal{H} = \pi(x)\mathcal{H}_0 + \pi(x)\mathcal{H}_0^\perp \subseteq \mathcal{H}_0^\perp$. Thus the linear span of all $\pi(a)v$ for $a \in \mathcal{A}$ and $v \in \mathcal{H}$ is a dense subspace of $\mathcal{H}$ if and only if $\mathcal{H}_0 = \{0\}$. □

2. Let π be a nondegenerate representation of a Banach $*$ algebra having an identity e. Show $\pi(e) = I$.

Solution:

We have the linear span of the vectors $\pi(a)v$ for all $a \in \mathcal{A}$ and $v \in \mathcal{H}$ is dense in $\mathcal{H}$. Since $\pi(e)\pi(a)v = \pi(ea)v = \pi(a)v$ and $\pi(e)$ is linear, $\pi(e)w = w$ for all w is this linear span. The operator $\pi(e)$ being continuous then implies $\pi(e)v = v$ for all $v \in \mathcal{H}$.

□

3. Suppose $U(x)$ and $V(x)$ are bounded continuous functions and μ is a Radon measure on a locally compact Hausdorff space X with the property $\mu(O) > 0$ for every nonempty open Baire subset O of X. Show if $\int f(x)U(x)\,d\mu(x) = \int f(x)V(x)\,d\mu(x)$ for every $f \in C_c(X)$, then $U(x) = V(x)$ for all x.

Solution: First by taking real and imaginary parts of U and V, we may assume U and V are real valued. Suppose $U(x_0) = V(x_0) + \delta$ where $\delta > 0$ for some x_0. Then $x_0 \in N \subseteq \{x \mid U(x) - V(x) > \frac{\delta}{2}\}$ for some compact neighborhood N of x_0. By Proposition 5.23 there is an $f \in C_c(X)$ with $0 \leqslant f \leqslant 1$, $\operatorname{supp} f \subseteq \{x \mid U(x) - V(x) > 0\}$ and $f(x) = 1$ for $x \in N$. Thus $f(x)(U(x) - V(x)) \geqslant 0$ and $O = \{x \mid f(x)(U(x) - V(x)) > \frac{\delta}{2}\}$ is a nonempty open Baire subset of X. This implies $\int f(x)(U(x) - V(x))\,d\mu(x) \geqslant \frac{\delta}{2}\mu(O) > 0$, a contradiction. Thus $U(x) \leqslant V(x)$ for all x. Similarly $V(x) \leqslant U(x)$ for all x. □

4. Let $f \in L^1(G)$ and for $a, b \in G$, define ${}_af_b$ by ${}_af_b(x) = f(a^{-1}xb)\Delta(b)$ where Δ is the modular function for G. Show

(a) $f \mapsto {}_af_b$ is a linear isometry of $L^1(G)$ onto $L^1(G)$.

(b) Show $({}_af_b)^* = {}_b(f^*)_a$ and $({}_af_b) * ({}_bg_c) = {}_a(f * g)_c$ for $a, b, c \in G$ and $f, g \in L^1(G)$.

(c) Show if π is a unitary representation of G, then

$$\pi({}_af_b) = \pi(a)\pi(f)\pi(b^{-1})$$

Solution: To see (a) note $f \mapsto {}_af_b$ is linear and

$$\begin{aligned} |{}_af_b|_1 &= \int |f(a^{-1}xb)|\Delta(b)\,dx \\ &= \int |f(x)|\,dx \\ &= |f|_1. \end{aligned}$$

Also ${}_a({}_{a^{-1}}f_{b^{-1}})_b = f$, so the map is onto.

To see (b), note

$$\begin{aligned} ({}_af_b)^*(x) &= \overline{{}_af_b(x^{-1})}\,\Delta(x^{-1}) \\ &= \overline{f(a^{-1}x^{-1}b)}\,\Delta(b)\Delta(x^{-1}) \\ &= \overline{f((b^{-1}xa)^{-1})}\,\Delta(b)\Delta(x^{-1})\Delta(a^{-1})\Delta(a) \\ &= f^*(b^{-1}xa)\Delta(a) \\ &= {}_b(f^*)_a(x). \end{aligned}$$

Now

$$\begin{aligned}({}_af_b)*({}_bg_c)(x) &= \int f(a^{-1}yb)\Delta(b)g(b^{-1}y^{-1}xc)\Delta(c)\,dy\\ &= \int f(yb)\Delta(b)g(b^{-1}y^{-1}a^{-1}xc)\Delta(c)\,dy\\ &= \int f(y)g(y^{-1}a^{-1}xc)\,dy\,\Delta(c)\\ &= (f*g)(a^{-1}xc)\Delta(c)\\ &={}_a(f*g)_c(x).\end{aligned}$$

Finally for (c), one has

$$\begin{aligned}(\pi({}_af_b)v,w) &= \int \Delta(b)f(a^{-1}xb)(\pi(x)v,w)\,dx\\ &= \int \Delta(b)f(x)(\pi(axb^{-1})v,w)\,d(xb^{-1})\\ &= \int \Delta(b)f(x)(\pi(a)\pi(x)\pi(b)^{-1}v,w)\ \Delta(b^{-1})\,dx\\ &= \int f(x)(\pi(x)\pi(b)^{-1}v,\pi(a)^*w)\,dx\\ &= (\pi(f)\pi(b)^{-1}v,\pi(a)^*w)\\ &= (\pi(a)\pi(f)\pi(b)^{-1}v,w)\end{aligned}$$

for $v, w \in \mathcal{H}_\pi$. Thus $\pi({}_af_b) = \pi(a)\pi(f)\pi(b)^{-1}$. □

5. Let G be a nondiscrete locally compact Hausdorff group. Show there is a representation of $M(G)$ which is nonzero and whose restriction to $L^1(G)$ is trivial. (Hint: Show the atom free measures in $M(G)$ form an ideal in $M(G)$.)
Solution: Let $M(G)$ be the regular Borel complex measures on G. We know $M(G)$ is a Banach $*$ algebra. Let $M_c(G)$ be the subset of $M(G)$ consisting of those measures μ satisfying $\mu(\{x\}) = 0$ for all $x \in G$. We claims $M_c(G)$ is a ideal in $M(G)$. Indeed let $\mu \in M_c(G)$ and $\nu \in M(G)$. Then $\mu * \nu$ is the measure on G having the property $\mu * \nu(\{z\}) = \int\int 1_{\{z\}}(xy)\,d\mu(x)\,d\nu(y) = \int \mu\{zy^{-1}\}\,d\nu(y) = 0$. Similarly $\nu * \mu\{z\} = \int\int 1_{\{z\}}(xy)\,d\nu(x)\,d\mu(y) = \int\int 1_{\{z\}}(xy)d\mu(y)\,d\nu(x) = \int \mu\{x^{-1}z\}\,d\nu(x) = 0$. Thus $M_c(G)$ is an ideal. (It is clearly closed under addition and scalar multiplication.) Note $M_c(G)$ is also closed under the involution $*$. In fact if $\mu \in M_c(G)$, then $\mu^*\{x\} = \overline{\mu\{x^{-1}\}} = 0$ for all x. Hence $M(G)/M_c(G)$ is a $*$ algebra. But $M(G)/M_c(G)$ is isomorphic to the subalgebra of $M(G)$ consisting of measures of form $\sum c_i\epsilon_{x_i}$. Hence if π is a unitary representation of G, we can extend π to $M(G)/M_c(G)$ by $\pi(\sum c_i\epsilon_{x_i}) = \sum c_i\pi(x_i)$. This sum converges in $B(\mathbb{H})$ since $\sum |c_i| < \infty$ and one can check it defines a $*$-representation of $M(G)/M_c(G)$. Hence composing π with the natural projection $p : M(G) \to M(G)/M_c(G)$ gives a representation $\pi \circ p$ of $M(G)$ which is nonzero and vanishes on $M_c(G) \supset L^1(G)$. □

6. Show $L^\infty(\mathbb{R}^n)$ is a C^* algebra under pointwise multiplication and addition, adjoint $h \mapsto \bar{h}$, and norm $||h|| = |h|_\infty$. Then show the mapping $h \mapsto M_h$ is an isometric $*$ isomorphism of the algebra $L^\infty(\mathbb{R}^n)$ onto a norm closed $*$ subalgebra of $\mathcal{B}(L^2(\mathbb{R}^n))$.

Solution:

We already know $L^\infty(\mathbb{R}^n)$ is a Banach space. It is closed under multiplication for $|f(x)g(x)| \leqslant |f|_\infty|g|_\infty$ a.e. implies $|fg|_\infty \leqslant |f|_\infty|g|_\infty$. Also $|f^*|_\infty = |\bar{f}|_\infty = |f|_\infty$ and $|ff^*|_\infty = |f|_\infty^2$ for $|ff^*|_\infty \leqslant |f|_\infty|f^*|_\infty = |f|_\infty^2$ and if $M = |f|_\infty > 0$ and $M > \epsilon > 0$, the set E of those x with $|f(x)| > M - \epsilon$ has positive measure. Thus for $x \in E$, $|f(x)f^*(x)| = |f(x)|^2 \geqslant (M-\epsilon)^2$. So $|ff^*|_\infty \geqslant (M-\epsilon)^2$ for all $M > \epsilon > 0$. This gives $|ff^*|_\infty \geqslant |f|_\infty^2$ and so $L^\infty(\mathbb{R}^n)$ is a C^* algebra.

For $h \in L^\infty(\mathbb{R}^n)$, M_h is a bounded linear operator on $L^2(\mathbb{R}^n)$ and $||M_h|| \leqslant |h|_\infty$ for $\int |h(x)f(x)|^2\, d\lambda(x) \leqslant \int |h|_\infty^2 |f(x)|^2\, d\lambda(x) = |h|_\infty^2|f|_2^2$. To see $||M_h|| = |h|_\infty$, we may assume $|h|_\infty > 0$. Let $|h|_\infty > \epsilon > 0$. Consider the subset E of x with $|h(x)| > |h|_\infty - \epsilon$. It is a Borel set and has positive Lebesgue measure. By regularity there is a compact subset K of E with positive measure. Taking $\frac{0}{0} = 0$, we see since $\lambda(K) < \infty$, that $x \mapsto f(x) = \frac{\bar{h}(x)}{|h(x)|\sqrt{\lambda(K)}}\chi_K(x)$ is in L^2 and has L^2 norm 1. Thus

$$\begin{aligned} ||M_h||^2 \geqslant |M_h f|^2 &= \int_K \frac{|h(x)|^4}{|h(x)|^2\lambda(K)}\, d\lambda(x) \\ &\geqslant \frac{1}{\lambda(K)}\int_K (|h|_\infty - \epsilon)^2\, d\lambda(x) = (|h|_\infty - \epsilon)^2 . \end{aligned}$$

Letting $\epsilon \to 0$, we have $||M_h|| = |h|_\infty$. Thus $h \mapsto M_h$ is an isometry. Furthermore, one can check directly $M_{ah_1+h_2} = aM_{h_1} + M_{h_2}$ and $M_{h_1}M_{h_2} = M_{h_1}M_{h_2}$ for $a \in \mathbb{C}$, h_1 and h_2 in $L^\infty(\mathbb{R}^n)$. Finally,

$$\begin{aligned} (M_h f_1, f_2)_2 &= \int h(x)f_1(x)\bar{f_2}(x)\, d\lambda(x) \\ &= \int f_1(x)\overline{\bar{h}(x)\, f_2(x)}\, d\lambda(x) \\ &= \int f_1(x)\overline{h^*(x)f_2(x)}\, d\lambda(x) \\ &= (f_1, M_{h^*}f_2)_2 \end{aligned}$$

and thus $(M_h)^* = M_{h^*}$. So $h \mapsto M_h$ is a $*$-isometric isomorphism of $L^\infty(\mathbb{R}^n)$ onto closed $*$ subalgebra of $\mathcal{B}(L^2(\mathbb{R}^n))$. Its image is closed for its image is complete.

□

In the following 5 exercises μ and ν are regular Borel measures on $\mathbb{R}^n$ and λ_n is Lebesgue measure on $\mathbb{R}^n$.

7. Define for each x, an operator $\pi_\mu(x)$ on $L^2(\mathbb{R}^n, \mu)$ by

$$\pi_\mu(x)f(y) = e^{-2\pi i x\cdot y}f(y) .$$

(a) Show π_μ is a unitary representation of $\mathbb{R}^n$.

(b) Find $\pi_\mu(f)$ for $f \in L^1(\mathbb{R}^n)$.

(c) Show π_μ is unitarily equivalent to π_ν if and only if μ and ν are equivalent measures.

Solution: It is easy to see π_μ is a homomorphism from $\mathbb{R}^n$ into the unitary group. We show it is strongly continuous. Let $f \in L^2(\mathbb{R}^n, \mu)$ and $\epsilon > 0$. Since μ is regular, $C_c(\mathbb{R}^n)$ is dense in $L^2(\mu)$. Let $f_0 \in C_c(\mathbb{R}^n)$ with $|f - f_0|_2 \leqslant \frac{\epsilon}{3}$. Set K be its support. We can pick a neighborhood N_0 of 0 with $|e^{-2\pi i x\cdot y} - 1| < \frac{\epsilon}{3|f_0|_2}$ if $x \in N_0$ and $y \in K$. For $x \in N_0$,

$$|\pi_\mu(x)f_0 - \pi_\mu(0)f_0|_2^2 \leqslant \int_K |e^{-2\pi i x\cdot y} - 1|^2 |f_0(y)|^2 \, d\mu(y) \leqslant \frac{\epsilon^2}{9}.$$

Thus $|\pi_\mu(x)f_0 - \pi_\mu(0)f_0|_2 \leqslant \frac{\epsilon}{3}$ and we see

$$\begin{aligned}|\pi_\mu(x)f - \pi_\mu(0)f|_2 &\leqslant |\pi_\mu(x)f - \pi_\mu(x)f_0|_2 + |\pi_\mu(x)f_0 - \pi_\mu(0)f_0|_2 \\ &\quad + |\pi_\mu(0)f_0 - \pi_\mu(0)f|_2 \leqslant 2|f - f_0|_2 + \frac{\epsilon}{3} = \epsilon.\end{aligned}$$

This gives (a).

We find $\pi_\mu(f)h$ for $f \in L^1(\mathbb{R}^n)$. Note

$$\begin{aligned}(\pi_\mu(f)h, g)_{L^2(\mu)} &= \int f(x)(\pi_\mu(x)h, g)_{L^2(\mu)} \, dx \\ &= \int f(x) \int e^{-2\pi i x\cdot y} h(y)\overline{g(y)} \, d\mu(y) \, dx \\ &= \int \left(\int f(x) e^{-2\pi i x\cdot y} \, dx \right) h(y)\overline{g(y)} \, d\mu(y) \\ &= \int \hat{f}(y)h(y)\,\overline{g(y)} \, d\mu(y) \\ &= (\hat{f}h, g)_{L^2(\mu)}.\end{aligned}$$

So $\pi_\mu(f)(h) = \hat{f}h$ for $h \in L^2(\mu)$.

If μ and ν are equivalent measures, there is a Borel function h with $h(x) > 0$ a.e. μ with $\nu(E) = \int_E h \, d\mu$ for Borel subsets E. Define $T : L^2(\nu) \to L^2(\mu)$ by $Tf = \sqrt{h}\, f$. Note T is an isometry for

$$|Tf|^2 = \int h|f|^2 \, d\mu = \int |f|^2 \, d\nu.$$

Note $T^{-1}f = \frac{1}{\sqrt{h}} f$ is an isometry from $L^2(\mu)$ to $L^2(\nu)$. Indeed,

$$\int \frac{1}{h}|f|^2 \, d\nu = \int \frac{1}{h} h|f|^2 \, d\mu = \int |f|^2 \, d\mu.$$

Now, an easy argument shows $T\pi_\nu(x) = \pi_\mu(x)T$ for all x. So π_μ and π_ν are equivalent.

Conversely, suppose $T : L^2(\nu) \to L^2(\mu)$ is unitary and $T\pi_\nu(x) = \pi_\mu(x)T$ for all x. This implies $T\pi_\nu(f) = \pi_\mu(f)T$ for all $f \in L^1(\mathbb{R}^n)$. Since $\hat{L}_1(\mathbb{R}^n)$ contains the Fourier transform of the Schwartz functions, we see $T(hf) = T(h)f$ for all $f \in C_c^\infty(\mathbb{R}^n)$. Let $f \in C_c(\mathbb{R}^n)$. By Corollary 2.54, there is a $\phi \in C_c^\infty(\mathbb{R}^n)$ with $0 \leqslant \phi(x) \leqslant 1$ and $\phi(x) = 1$ if and only if $x \in \operatorname{supp}(f)$. The sequence $f_k = \phi^k f$ converges uniformly to f and thus converges to f in $L^2(\nu)$. By the continuity of T, one then obtains $T(hf) = T(h)f$ for all $f \in C_c(\mathbb{R}^n)$. Using the density of $C_c(\mathbb{R}^n)$ in $L^2(\nu)$, one has $T(hf) = T(h)f$ for all bounded functions f in $L^2(\nu)$ for all $h \in L^2(\nu)$. Let $Q = \prod_{i=1}^n [0,1)$. Then the sets $Q+k$ for $k \in \mathbb{Z}^n$ are disjoint rectangles with union $\mathbb{R}^n$. So $T(\chi_{Q+k}f) = T(\chi_{Q+k}^2 f) = \chi_{Q+k}T(\chi_{Q+k})f$ for bounded $L^2(\nu)$ functions f. Define $\phi = \sum_{k\in\mathbb{Z}^n} \chi_{Q+k}T(\chi_{Q+k})$. Note $\phi \in L^2_{\text{loc}}(\mu)$. Also if f is a bounded $L^2(\nu)$ function, $\sum_k \chi_{Q+k}f$ converges to f in $L^2(\nu)$. Thus $T(f) = \sum_k T(\chi_{Q+k}f) = \sum_k \chi_{Q+k}T(\chi_{Q+k})f$ in $L^2(\nu)$. So $T(f) = \phi f$ for bounded $L^2(\nu)$ functions f. Since T is bounded, ϕ is essentially μ bounded. Taking limits again gives $T(f) = \phi f$ for all $f \in L^2(\nu)$. T being unitary implies

$$\int |\phi(x)|^2 \chi_E^2(x)\, d\mu(x) = \int \chi_E(x)^2\, d\nu(x)$$

for each Borel subset E of finite ν measure. Thus $\nu(E) = \int_E |\phi(x)|^2\, d\mu(x)$. In particular, $\nu(E) = 0$ if $\mu(E) = 0$. Similarly, using T^{-1} we see $\mu(E) = 0$ if $\nu(E) = 0$. Thus the measures are equivalent.

□

8. Show π_μ is unitarily equivalent to a subrepresentation of the left regular representation of $\mathbb{R}^n$ if and only if $\mu < \lambda_n$.

Solution:

We know the left regular representation of $\mathbb{R}^n$ is unitarily equivalent to the representation on $L^2(\mathbb{R}^n, \lambda_n)$ given by

$$\pi(x)f(y) = e^{-2\pi i x\cdot y} f(y)$$

and the representation π_μ is given by the same formula except that f is in $L^2(\mathbb{R}^n, \mu)$. Now let $\mu < \lambda_n$ and set $h = \frac{d\mu}{d\lambda_n}$ be the Radon-Nikodym derivative. Define $T : L^2(\mu) \to L^2(\lambda_n)$ by

$$Tf(x) = \sqrt{h(x)}f(x)\,.$$

T is a linear isometry for

$$|Tf|_{\lambda_n}^2 = \int h(x)|f(x)|^2\, d\lambda_n(x) = \int |f(x)|^2\, d\mu(x).$$

Also

$$T\pi_\mu(x)f(y) = \sqrt{h(y)}(e^{-2\pi i x\cdot y} f(y)) = e^{-2\pi i x\cdot y}Tf(y) = \pi(x)Tf(y)\,.$$

This implies $T\pi_\mu(x) = \pi(x)T$ for $x \in \mathbb{R}^n$. Thus T is a unitary transformation of $L^2(\mu)$ onto a closed invariant subspace for π.

Conversely, assume $A : L^2(\mu) \to L^2(\lambda_n)$ is a unitary transformation onto a closed invariant subspace of $L^2(\lambda_n)$ and $A\pi_\mu(x) = \pi(x)A$ for all x. This implies $A\pi_\mu(h) = \pi(h)A$ for all $h \in L^1(\mathbb{R}^n, \lambda_n)$. So $A(\hat{h}f) = \hat{h}A(f)$ for $f \in L^2(\mathbb{R}^n, \mu)$ and $h \in L^1(\mathbb{R}^n, \lambda_n)$. Thus $A(hf) = hA(f)$ for all Schwarz functions h. Then as seen in the argument preceding Theorem 6.114, $A(f) = hf$ for some measurable function h. Since A is unitary,

$$\int |f(x)|^2\, d\mu(x) = \int |f(x)|^2 |h(x)|^2\, d\lambda_n(x)$$

for all $f \in L^2(\mu)$. This implies $\int \chi_E(x)\, d\mu(x) = \int \chi_E(x)\, |h(x)|^2\, d\lambda_n(x)$ and thus $h(x) = \frac{d\mu}{d\lambda_n}$ and $\mu < \lambda_n$.

□

9. Show π_μ is always a cyclic representation. (**Hint:** Use Exercise 3.1.3.)
Solution: Let h be in $L^2(\mu)$. Then h is cyclic if and only if there is no nonzero f in $L^2(\mu)$ with $(\pi(x)h, f)_{L^2(\mu)} = 0$ for all x. This implies

$$\int e^{-2\pi i x\cdot y} h(y)\overline{f(y)}\, d\mu(y) = 0$$

for all x. Thus if σ is the complex measure given by $\sigma(E) = \int_E h(y)\overline{f(y)}\, d\mu(y)$, then $\hat{\sigma} = 0$. It follows by Exercise 3.1.3 that $\sigma = 0$. Since $\mathbb{R}^n$ is μ σ-finite, there is a function $h \in L^2(\mu)$ with $h(x) \neq 0$ for all x. Thus $h(y)\overline{f(y)}$ is 0 for μ a.e. y if and only if $f = 0$ in $L^2(\mu)$. □

10. Show $\mathrm{Hom}(\pi_\mu, \pi_\mu) = \{M_f \mid f \in L^\infty(\mu)\}$ where $M_f h = fh$ for $h \in L^2(\mu)$.
Solution: If $A \in \mathrm{Hom}(\pi_\mu, \pi_\mu)$, then $A\hat{f}h = \hat{f}Ah$ for all f in $L^1(\mathbb{R}^n)$ and $h \in L^2(\mu)$. So $A(\phi f) = \phi Af$ for all Schwartz functions ϕ. Taking ϕ with $\phi^{-1}(1) = Q$ and $0 \leqslant \phi(x) \leqslant 1$ for all x, one has $A(\chi_Q\chi_{Q'}) = \lim_k A((\phi)^k\chi_{Q'}) = \lim_k \phi^k A\chi_{Q'} = \chi_Q A\chi_{Q'}$ for all rectangles Q and Q'.

Let $\mathcal{M}$ be the set of all Borel measurable subsets E of S such that $\chi_E Af = A(\chi_E f)$ for all $f \in L^2(\mu)$. Note $\chi_{E^c}Af = (1-\chi_E)Af = Af - A\chi_E f = A(1-\chi_E)f$ and so $\mathcal{M}$ is closed under complements. Since $\sum \chi_{E_i}h \to \chi_{\cup_i E_i}h$ in $L^2(\mu)$ for a countable disjoint union of Borel sets E_i, it is closed under countable disjoint unions. Now $A\chi_{E_1}\chi_{E_2}f = \chi_{E_1}A\chi_{E_2}f = \chi_{E_1}\chi_{E_2}Af$ if $E_1, E_2 \in \mathcal{M}$. So $E_1 \cap E_2 \in \mathcal{M}$. Hence $\mathcal{M}$ contains all Borel subsets. Thus $M_h A = AM_h$ for all simple Borel functions h. But if $h \in L^\infty(\mu)$, there is a simple s with $|s-h|_\infty < \epsilon$. This implies $M_h A = AM_h$ for all $h \in L^\infty(\mu)$. One can assume μ is finite. Thus $Ah1 = hA1$ for all simple functions $h \in L^2(\mu)$. This implies $Af = f(A1)$ a.e. for every f in $L^2(\mu)$. Now let $c > ||A||$ and suppose $E = \{x \mid |A1(x)| \geqslant c\}$ has positive measure. Set $h = \frac{\chi_E(x)}{\sqrt{\mu(E)}}\frac{\overline{A1(x)}}{|A1(x)|}$. Note $|h|_2 = 1$ and $A1(x)h(x) = \frac{\chi_E(x)}{\sqrt{\mu(E)}}|A1(x)|$. Thus we have $||A||^2 \geqslant |Ah|_2^2 = \frac{1}{\mu(E)}\int_E |A1(x)|^2\, d\mu(x) \geqslant c^2 > ||A||^2$, which is impossible. So E has measure 0. This being true for all $c > ||A||$ implies $|A1|_\infty \leqslant ||A||$. □

11. Show $\text{Hom}(\pi_\mu, \pi_\mu)$ is maximal abelian, i.e., show if $A \in \mathcal{B}(L^2(\mu))$ and $AB = BA$ for all $B \in \text{Hom}(\pi_\mu, \pi_\mu)$, then $A \in \text{Hom}(\pi_\mu, \pi_\mu)$.

Solution:

Assume A is a bounded operator which commutes with each operator in $\text{Hom}(\pi_\mu, \pi_\mu)$. By Exercise 6.8.10, $AM_f = M_f A$ for all $f \in L^\infty(\mu)$. Thus $A(\chi_E h) = \chi_E A(h)$ for all $h \in L^2(\mu)$ and all measurable subsets E. Take a countable family E_i of disjoint measurable sets with union $\mathbb{R}^n$ and $\mu(E_i) < \infty$ for all i. Then $A(h) = \sum A(\chi_{E_i} h) = \sum \chi_{E_i} A(\chi_{E_i} h) = \sum \chi_{E_i} h A(\chi_{E_i})$ for any function $h \in L^2(\mu) \cap L^\infty(\mu)$. So $Ah = \phi h$ where $\phi = \sum_i \chi_{E_i} A(\chi_{E_i})$ and h is any such function. If $C = ||A||$, we have $|Ah|_2 \leqslant C|h|_2$. Thus if $\epsilon > 0$ and E is a Borel subset of $\{x \mid |\phi(x)| \geqslant C + \epsilon\}$ with positive finite measure, then $h(x) = \chi_E(x) \frac{\bar{\phi}(x)}{\sqrt{\mu(E)}|\phi(x)|}$ has $L^2(\mu)$ length 1. Thus $|Ah|_2 \leqslant C$. But

$$\begin{aligned} |Ah|_2^2 &= \int_E |\phi(x)h(x)|^2 \, d\mu(x)| \\ &= \int_E \frac{1}{\mu(E)} |\phi(x)|^2 \, d\mu(x) \\ &\geqslant (C + \epsilon)^2 . \end{aligned}$$

So $\mu(E) = 0$ and $\phi \in L^\infty(\mu)$. Since $L^2(\mu) \cap L^\infty(\mu)$ is dense in $L^2(\mu)$, this implies $Ah = M_\phi h$ for all $h \in L^2(\mu)$. Thus $\text{Hom}(\pi_\mu, \pi_\mu)$ is maximal abelian.

□

12. A function $\varphi : G \to \mathbb{C}$ is called **positive definite** if for all $x_1, \ldots, x_n \in G$ the matrix $[\varphi(x_i^{-1} x_j)]$ is positive semi-definite. Show the following:

(a) The function φ is positive definite if and only if $\sum_{i,j=1}^n c_i \bar{c}_j \varphi(x_i^{-1} x_j) \geqslant 0$ for all $x_1, \ldots, x_n \in G$ and all $c_1, \ldots, c_n \in \mathbb{C}$.

(b) One has $\phi(e) \geqslant 0$ and $\phi(x^{-1}) = \overline{\phi(x)}$ for all $x \in G$.

(c) Let E be the space of functions $f : G \to \mathbb{C}$ of the form $f = \sum_{j=1}^n c_j L(x_j)\varphi$ where $L(x)\phi(y) = \phi(x^{-1}y)$. Define a two form $(\cdot\,, \cdot)$ on E by

$$(\sum_{i=1}^n c_i L(x_i)\varphi, \sum_{j=1}^k d_j L(y_j)\varphi) := \sum_{i,j} c_i \bar{d}_j \varphi(x_i^{-1} y_j) .$$

Show this is a well defined sesquilinear form on E and $(f, f) \geqslant 0$ for all $f \in E$.

(d) Use the Cauchy-Schwarz inequality to show

$$|\phi(x) - \phi(y)|^2 \leqslant 2\phi(e)[\phi(e) - \text{Re}\,\phi(x^{-1}y)]$$

for $x, y \in G$.

(e) Let $N = \{f \in E \mid (f, f) = 0\}$. Show $(f + N, g + N) = (f, g)$ defines an inner product on E/N. Let $|| \cdot ||$ be the corresponding norm on E/N.

(f) For $f \in E$ and $x \in G$, let $\pi(x)f(y) = f(x^{-1}y)$. Then $\pi(G)N \subset N$.

(g) Define $\pi_\varphi(x) : E/N \to E/N$ by $\pi_\varphi(x)(f + N) := \pi(x)f + N$. Then $||\pi_\varphi(x)(f + N)|| = ||f + N||$. In particular π_φ extends to a continuous map on V, the completion of E/N in the norm $||\cdot||$ to V. Show π_ϕ is strongly continuous and thus is a unitary representation of G if and only if ϕ is a continuous function on G.

Solution:

For (a) note a square $n \times n$ matrix A is positive definite if and only if for every row vector $c \in \mathbb{C}^n$, $cA\bar{c}^T \geqslant 0$. Thus ϕ is positive definite if and only if $\sum_{i,j} c_i\phi(x_i^{-1}x_j)\bar{c}_j \geqslant 0$ for all $c \in \mathbb{C}^n$ if and only if the matrix $[\phi(x_i^{-1}x_j)]$ is positive definite for any finite sequence $x_1, x_2, \dots x_n$.

To see (b), we know the matrix $\begin{bmatrix} \phi(e) & \phi(x^{-1}) \\ \phi(x) & \phi(e) \end{bmatrix}$ is positive semidefinite. Thus it is skew symmetric and so $\phi(x^{-1}) = \overline{\phi(x)}$. Also $(c, 0)\begin{bmatrix} \phi(e) & \phi(x^{-1}) \\ \phi(x) & \phi(e) \end{bmatrix}\begin{bmatrix} \bar{c} \\ 0 \end{bmatrix} = |c|^2\phi(e) \geqslant 0$ for all $c \in \mathbb{C}$. This gives $\phi(e) \geqslant 0$.

For (c), we need to show $(\cdot, \cdot)$ is well defined, sesquilinear and positive semidefinite. We first show it is well defined. Assume $\sum c_jL(x_j)\phi = 0$. Then $\sum c_j\phi(x_j^{-1}y) = 0$ for all y. So the $\sum c_j\bar{d}_i\phi(x_j^{-1}y_i) = 0$ for any $y_i \in G$ and $d_i \in \mathbb{C}$. Thus if $d_1, d_2, \dots, d_m \in \mathbb{C}$ and $y_1, y_2, \dots, y_m \in G$, we obtain by adding these together over i that $\sum_{i,j} c_j\bar{d}_i\phi(x_j^{-1}y_i) = 0$ and so $(\sum c_jL(x_j)\phi, \sum d_iL(y_i)\phi) = 0$.

Similarly,

$$\begin{aligned}(\textstyle\sum d_iL(y_i)\phi, \sum c_jL(x_j)\phi) &= \sum_i d_i \sum_j \bar{c}_j\phi(y_i^{-1}x_j) \\ &= \sum_i d_i \sum_j \bar{c}_j\overline{\phi(x_j^{-1}y_i)} \\ &= \sum_i d_i\overline{\sum_j c_j\phi(x_j^{-1}y_i)} \\ &= 0.\end{aligned}$$

These together imply $(\cdot, \cdot)$ is well defined. Note $f \mapsto (f, g)$ is clearly linear and if $f = \sum c_jL(x_j)\phi$ and $g = \sum d_iL(y_i)\phi$, then

$$\begin{aligned}(f, g) &= \sum_{i,j} c_j\bar{d}_i\phi(x_j^{-1}y_i) \\ &= \overline{\sum_{i,j} \bar{c}_jd_i\phi(y_i^{-1}x_j)} \\ &= \overline{(g, f)}.\end{aligned}$$

So $g \mapsto (g, f)$ is conjugate linear and thus $(\cdot, \cdot)$ is a sesquilinear form.

Also

$$(\textstyle\sum c_iL(x_i)\phi, \sum c_jL(x_j)\phi) = \sum c_i\bar{c}_j\phi(x_i^{-1}x_j) \geqslant 0$$

for any $x_1, x_2, \ldots, x_n \in G$ and $c_1, c_2, \ldots, c_n \in \mathbb{C}$. Thus $(\cdot\,,\,\cdot)$ is a well defined positive semidefinite sesquilinear form on E.

For (d), we first note since $(\cdot\,,\,\cdot)$ is positive semidefinite and sesquilinear, it satisfies the Cauchy-Schwarz inequality; i.e., $|(f,g)|^2 \leqslant (f,f)(g,g)$ for $f, g \in E$. We also see

$$|\phi(x)-\phi(y)|^2 = |(\phi, L(x)\phi - L(y)\phi)|^2 \leqslant (\phi,\phi)(L(x)\phi - L(y)\phi, L(x)\phi - L(y)\phi)\,.$$

This gives $|\phi(x) - \phi(y)|^2 \leqslant \phi(e)\left[2\phi(e) - \phi(x^{-1}y) - \phi(y^{-1}x)\right]$. So $|\phi(x) - \phi(y)|^2 \leqslant \phi(e)\left[2\phi(e) - \phi(x^{-1}y) - \overline{\phi(x^{-1}y)}\right] = \phi(e)\left[2\phi(e) - 2\operatorname{Re}\phi(x^{-1}y)\right]$. Thus

$$|\phi(x) - \phi(y)|^2 \leqslant 2\phi(e)[\phi(e) - \operatorname{Re}\phi(x^{-1}y)]$$

for all $x, y \in G$.

For (e), note $(g,f) = 0$ if $g \in N$ and $f \in E$ for $|(g,f)|^2 \leqslant (g,g)\,(f,f) = 0$. This implies N is a linear subspace for $(cg_1 + g_2, cg_1 + g_2) = |c|^2(g_1,g_1) + c(g_1,g_2) + \bar{c}(g_2,g_1) + (g_2,g_2) = 0$ for $g_1, g_2 \in N$ and $c \in \mathbb{C}$. Hence taking $(f+N, g+N) = (f,g)$ for $f, g \in E$, one sees one obtains a well defined inner product on E/N.

For (f), let $f \in E$ and $x \in G$. Then $f = \sum c_j L(x_j)\phi$ and $(f,f) = \sum_{i,j} c_j \bar{c}_i \phi(x_j^{-1}x_i)$. Now

$$\pi_\phi(x)L(x_j)\phi(y) = L(x_j)\phi(x^{-1}y) = \phi(x_j^{-1}x^{-1}y) = L(xx_j)\phi$$

gives $\pi_\phi(x)f = \sum_j c_j L(xx_j)\phi$ and so

$$(\pi_\phi(x)f, \pi_\phi(x)f) = \sum_{i,j} c_j\bar{c}_i\phi((xx_j)^{-1}(xx_i)) = \sum_{i,j} c_j\bar{c}_j\phi(x_jx_i^{-1}) = (f,f)\,.$$

In particular, $(\pi_\phi(x)f, \pi_\phi(x)f) = 0$ if and only if $(f,f) = 0$. So $\pi_\phi(x)N \subseteq N$.

(g) now follows easily for $\pi_\phi(x)$ is well defined on E/N and $(\pi_\phi(x)(f+N), \pi_\phi(x)(f+N)) = (\pi_\phi(x)f, \pi_\phi(x)f) = (f,f) = (f+N, f+N)$. So $||\pi_\phi(x)(f+N)|| = ||f+N||$. So $\pi_\phi(x)$ extends to a linear isometry of the Hilbert space V obtained by completing the space E/N. One sees easily that $\pi_\phi(e) = I$ and since $\pi_\phi(x)\pi_\phi(y) = \pi_\phi(xy)$, one has $\pi_\phi(x)^{-1} = \pi_\phi(x^{-1})$. So each $\pi_\phi(x)$ is a unitary operator.

In particular, π_ϕ is a unitary representation if $x \mapsto \pi_\phi(x)(f+N)$ is continuous. This is equivalent to $x \mapsto ||\pi_\phi(x)(f+N) - \pi_\phi(y)(f+N)||^2$ being continuous at y for every $y \in G$. But this mapping is

$$x \mapsto (f,f) - (\pi_\phi(x)f, \pi_\phi(y)f) - (\pi_\phi(y)f, \pi_\phi(x)f) + (f,f).$$

But if $f = \sum c_j L(x_j)\phi$, this mapping is

$$\begin{aligned} x \mapsto\, & 2(f,f) - \Big(\sum_j c_j L(xx_j)\phi, \sum_i c_i L(yx_i)\phi\Big) - \Big(\sum_j c_j L(yx_j)\phi, \sum_i c_i L(xx_i)\phi\Big) \\ &= 2(f,f) - \sum_{i,j} c_j\bar{c}_i\phi(x_j^{-1}x^{-1}yx_i) - \sum_{i,j} c_j\bar{c}_i\phi(x_j^{-1}y^{-1}xx_i). \end{aligned}$$

Hence if ϕ is continuous we see if $x \to y$, this mapping tends to 0 and we have continuity at y and thus π_ϕ is a unitary representation.

Conversely, assume π_ϕ is a unitary representation. Then $(\phi + N, \pi_\phi(x)\phi + N) \to (\phi + N, \pi_\phi(y)\phi + N)$ as $x \to y$. So $\phi(x) = (\phi, L(x)\phi) \to (\phi, L(y)\phi) = \phi(y)$ as $x \to y$ and ϕ is continuous.

□

Exercise Set 6.9

1. Let $\mathcal{H} \otimes \overline{\mathcal{H}}$ be the $*$ algebra of Hilbert-Schmidt operators on a Hilbert space $\mathcal{H}$. Show if Φ is a $*$ algebra onto isomorphism of $\mathcal{H} \otimes \overline{\mathcal{H}}$ which preserves inner products, then there is a unitary operator U on $\mathcal{H}$ such that $\Phi(T) = UTU^*$ for $T \in \mathcal{H} \otimes \bar{\mathcal{H}}$.

Solution: Let Φ be a $*$ algebra isomorphism of $\mathcal{H} \otimes \bar{\mathcal{H}}$ which preserves inner products. Let $\{e_\alpha \mid \alpha \in A\}$ be an orthonormal basis of $\mathcal{H}$. Since $e_\alpha \otimes \bar{e}_\alpha$ is a rank one orthogonal projection, $\Phi(e_\alpha \otimes \bar{e}_\alpha)$ is an orthogonal projection P_α onto some closed subspace $\mathcal{H}_\alpha$ of $\mathcal{H}$. Now $1 = ||\Phi(e_\alpha \otimes \bar{e}_\alpha)||_2^2 = \mathrm{Tr}(P_\alpha^* P_\alpha) = \mathrm{Tr}(P_\alpha)$. This implies the range of each P_α is one dimensional. Fix α_0 and pick f_{α_0} a unit vector with $\Phi(e_{\alpha_0} \otimes \bar{e}_{\alpha_0}) = f_{\alpha_0} \otimes \bar{f}_{\alpha_0}$ and consider e_α for $\alpha \neq \alpha_0$.

Note since

$$\begin{aligned}\Phi(e_{\alpha_0} \otimes \bar{e}_\alpha)\Phi(e_{\alpha_0} \otimes \bar{e}_\alpha)^* &= \Phi((e_{\alpha_0} \otimes \bar{e}_\alpha)(e_\alpha \otimes \bar{e}_{\alpha_0})) \\ &= \Phi(e_{\alpha_0} \otimes \bar{e}_{\alpha_0}) \\ &= f_{\alpha_0} \otimes \bar{f}_{\alpha_0}\end{aligned}$$

and

$$\begin{aligned}\Phi(e_{\alpha_0} \otimes \bar{e}_\alpha)^*\Phi(e_{\alpha_0} \otimes \bar{e}_\alpha) &= \Phi((e_\alpha \otimes \bar{e}_{\alpha_0})(e_{\alpha_0} \otimes \bar{e}_\alpha)) \\ &= \Phi(e_\alpha \otimes \bar{e}_\alpha) \\ &= P_\alpha \, ,\end{aligned}$$

we know by Exercise 6.4.1 that $\Phi(e_{\alpha_0} \otimes \bar{e}_\alpha)$ is a partial isometry of $P_\alpha\mathcal{H}$ onto $\mathbb{C}f_{\alpha_0}$. This implies

$$\Phi(e_{\alpha_0} \otimes \bar{e}_\alpha) : \mathcal{H} \to \mathbb{C}f_{\alpha_0}$$

and thus by the Riesz Theorem, there is a unique vector f_α with $\Phi(e_{\alpha_0} \otimes \bar{e}_\alpha)(v) = (v, f_\alpha)f_{\alpha_0} = f_{\alpha_0} \otimes \bar{f}_\alpha(v)$. Since Φ is an isometry, (a) of Proposition 2.42 implies $||f_{\alpha_0} \otimes \bar{f}_\alpha||_2 = ||f_{\alpha_0}|| \, ||f_\alpha|| = ||f_\alpha|| = ||e_{\alpha_0} \otimes \bar{e}_\alpha||_2 = 1$. We claim $f_\alpha \perp f_\beta$ if $\alpha \neq \beta$. Now again using (a) of Proposition 2.42, $(e_{\alpha_0} \otimes \bar{e}_\alpha, e_{\alpha_0} \otimes \bar{e}_\beta)_2 = (e_{\alpha_0}, e_{\alpha_0})(e_\beta, e_\alpha) = 0$. So $f_{\alpha_0} \otimes \bar{f}_\alpha = \Phi(e_{\alpha_0} \otimes e_\alpha)$ is perpendicular to $f_{\alpha_0} \otimes \bar{f}_\beta = \Phi(e_{\alpha_0} \otimes \bar{e}_\beta)$. Thus $(f_{\alpha_0} \otimes \bar{f}_\alpha, f_{\alpha_0} \otimes \bar{f}_\beta)_2 = (f_{\alpha_0}, f_{\alpha_0})(f_\beta, f_\alpha) = 0$. So $(f_\alpha, f_\beta) = 0$ and the vectors f_α form an orthonormal set. We claim their linear span is dense. First note

$$\begin{aligned}\Phi(e_\alpha \otimes \bar{e}_\beta) &= \Phi((e_\alpha \otimes \bar{e}_{\alpha_0})(e_{\alpha_0} \otimes \bar{e}_\beta)) \\ &= \Phi((e_{\alpha_0} \otimes \bar{e}_\alpha)^*(e_{\alpha_0} \otimes \bar{e}_\beta)) \\ &= (f_{\alpha_0} \otimes \bar{f}_\alpha)^*(f_{\alpha_0} \otimes \bar{f}_\beta) \\ &= (f_\alpha \otimes \bar{f}_{\alpha_0})(f_{\alpha_0} \otimes \bar{f}_\beta) \\ &= (f_{\alpha_0}, f_{\alpha_0}) f_\alpha \otimes \bar{f}_\beta \\ &= f_\alpha \otimes \bar{f}_\beta\,.\end{aligned}$$

Now assume there is a unit vector f orthogonal to all the vectors f_α. Since Φ is onto and Φ^{-1} is a $*$ isometric isomorphism, $\Phi^{-1}(f \otimes \bar{f}) = e \otimes \bar{e}$ for some unit vector e. Note

$$\begin{aligned}|(e, e_\alpha)|^2 &= (e_\alpha, e)(e, e_\alpha) \\ &= (\Phi(e_\alpha \otimes \bar{e}_\alpha), \Phi(e \otimes \bar{e}))_2 \\ &= (f_\alpha \otimes \bar{f}_\alpha, f \otimes \bar{f})_2 \\ &= (f_\alpha, f)(f, f_\alpha) \\ &= 0\end{aligned}$$

for all α. Since the e_α form an orthonormal basis, $e = 0$ and thus $f = 0$, a contradiction.

Hence the f_α form an orthonormal basis and thus there is a unitary mapping U satisfying $Ue_\alpha = f_\alpha$ for all α. Now by Equation 2.6, $\Phi(e_\alpha \otimes \bar{e}_\beta) = f_\alpha \otimes \bar{f}_\beta = Ue_\alpha \otimes \overline{Ue_\beta} = U(e_\alpha \otimes \bar{e}_\beta)U^*$. By Proposition 2.32, $T \mapsto UTU^*$ is continuous on $\mathcal{H} \otimes \bar{\mathcal{H}}$ for $||UTU^*||_2 \leqslant ||T||_2$ and since the $e_\alpha \otimes \bar{e}_\beta$ form an orthonormal basis of $\mathcal{H} \otimes \bar{\mathcal{H}}$, we see

$$\Phi(T) = UTU^* \text{ for all } T \in \mathcal{H} \otimes \bar{\mathcal{H}}\,.$$

□

2. Show $L^1(G)_T$ is a $*$ ideal in $L^1(G)$.
Solution: $L^1(G)_T$ is all $f \in L^1$ where $\pi(f)$ is a trace class operator. But by Corollary 2.41 and Exercises 2.2.24 and 2.2.25, the trace class operators are a $*$ ideal subalgebra of $\mathcal{B}(\mathcal{H})$ where $\mathcal{H}$ is the Hilbert space for π. Since π is a $*$ homomorphism of the Banach $*$ algebra $L^1(G)$ into $\mathcal{B}(\mathcal{H})$, the preimage $L^1(G)_T$ under π of the algebra of trace class operators is a $*$ ideal. □

3. Show if π is finite dimensional, then

$$\Theta_\pi(f) = \int_G f(x)\chi_\pi(x)\,dx$$

where χ_π is the character function for π.

Solution: We have

$$\begin{aligned}\Theta_\pi(f) &= \mathrm{Tr}(\pi(f))\\ &= \sum_{k=1}^{n} (\pi(f)e_i, e_i)\\ &= \sum_{k=1}^{n} \int f(x)(\pi(x)e_i, e_i)\, dx\\ &= \int f(x) \sum_{k=1}^{n} (\pi(x)e_i, e_i)\, dx\\ &= \int f(x)\chi_\pi(x)\, dx.\end{aligned}$$

□

4. Burnside Theorem: Let G be a completely reducible group (see Exercise 6.4.16) with an irreducible representation π on a finite dimensional complex vector space V. Show the linear span of the $\pi(g)$ for $g \in G$ is $\mathcal{L}(V)$. (**Hint:** Use Exercise 6.4.15 and the irreducibility of $\check{\pi}$.)

Solution: By Exercises 6.4.13, 6.4.15, and 6.4.16, we see $\pi \times \check{\pi}$ is irreducible. Let $\mathcal{A}$ be the subspace of $\mathcal{L}(V) = V \otimes V^*$ spanned by the $\pi(g)$. $\mathcal{A}$ is finite dimensional and nonzero. We recall $\pi \times \check{\pi}$ is the representation of $G \times G$ on $V \otimes V^*$ given on tensors $v \otimes v^* \in V \otimes V^* = \mathcal{L}(V, V)$ by $\pi \otimes \check{\pi}(g_1, g_2)(v \otimes v^*) = \pi(g_1)v \otimes \check{\pi}(g_2)v^*$. Now $v \otimes v^*(w) = v^*(w)v$ for $w \in V$. Hence

$$\begin{aligned}(\pi \otimes \check{\pi}(g_1, g_2))(v \otimes v^*)(w) &= (\pi(g_1)v \otimes \check{\pi}(g_2)v^*)(w)\\ &= (\check{\pi}(g_2)v^*)(w)(\pi(g_1)v)\\ &= v^*(\pi(g_2^{-1})w)(\pi(g_1)v)\\ &= \pi(g_1)(v \otimes v^*)\pi(g_2^{-1})(w).\end{aligned}$$

Consequently $(\pi \times \check{\pi})(g_1, g_2)(A) = \pi(g_1)A\pi(g_2^{-1})$ for all $A \in V \otimes V^*$. In particular, the subspace $\mathcal{A}$ is closed, invariant and nonzero and thus must be all of $\mathcal{L}(V)$. □

5. Assume π and ρ are unitary representations such that $\pi(f)$ and $\rho(f)$ are Hilbert-Schmidt for all $f \in C_c(G)$. Show if $\Theta_\pi(f^* * f) = \Theta_\rho(f^* * f)$ for $f \in C_c(G)$, then $\mathrm{Hom}_G(\pi, \rho) \neq \{0\}$. In particular, if π and ρ are irreducible, then they are unitarily equivalent.

Solution: Consider the algebras $\mathcal{A}_1 = \pi(C_c(G))$ and $\mathcal{A}_2 = \rho(C_c(G))$. Let $\mathcal{H}$ be the Hilbert space for π and $\mathcal{K}$ be the Hilbert space for ρ. We have $\mathcal{A}_1$ is a $*$ subalgebra of $\mathcal{H} \otimes \overline{\mathcal{H}}$ and $\mathcal{A}_2$ is a $*$ subalgebra of $\mathcal{K} \otimes \bar{\mathcal{K}}$. Define $\Phi : \mathcal{A}_1 \to \mathcal{A}_2$ by $\Phi(\pi(f)) = \rho(f)$. We claim Φ is well defined. Indeed, if $\pi(f) = 0$, then $\pi(f^* * f) = 0$ and thus $\Theta_\pi(f^* * f) = 0$. Thus $\Theta_\rho(f^* * f) = 0$. So $||\rho(f)||_2^2 = \mathrm{Tr}(\rho(f)^*\rho(f)) = 0$ and we see $\rho(f) = 0$. So Φ is well defined. Now $\mathcal{A}_1$ is invariant under $\pi \times \bar{\pi}$, for $\pi(g_1) \times \bar{\pi}(g_2)(\pi(f)) =$

$\pi(g_1)\pi(f)\pi(g_2)^* = \int \pi(g_1)f(x)\pi(x)\pi(g_2^{-1})\,dx = \Delta(g_2)\int f(g_1^{-1}xg_2)\pi(x)\,dx = \pi(\sqrt{\Delta(g_2)}B(g_1,g_2)f) \in \mathcal{A}_1$. Similarly, $\mathcal{A}_2$ is invariant under $\rho \times \bar{\rho}$. Now Φ is a $*$ algebra isometry from $\mathcal{A}_1$ onto $\mathcal{A}_2$. Indeed,

$$\begin{aligned}
||\Phi(\pi(f))||_2^2 &= \mathrm{Tr}(\rho(f)^*\rho(f)) \\
&= \mathrm{Tr}(\rho(f^* * f)) \\
&= \Theta_\rho(f^* * f) \\
&= \Theta_\pi(f^* * f) \\
&= \mathrm{Tr}(\pi(f^* * f)) \\
&= \mathrm{Tr}(\pi(f)^*\pi(f)) \\
&= ||\pi(f)||_2^2.
\end{aligned}$$

Moreover,

$$\begin{aligned}
\Phi((\pi \times \bar{\pi})(g_1,g_2)\pi(f)) &= \Phi(\pi(\Delta(g_2)^{1/2}B(g_1,g_2)f)) \\
&= \rho(\Delta(g_2)^{1/2}B(g_1,g_2)f) \\
&= (\rho \otimes \bar{\rho})(g_1,g_2)\rho(f) \\
&= (\rho \otimes \bar{\rho})(g_1,g_2)\Phi(\pi(f)).
\end{aligned}$$

Thus the linear mapping Φ extends to a linear isometric intertwiner of the invariant subspace $\bar{\mathcal{A}}_1$ of $\mathcal{H}\otimes\overline{\mathcal{H}}$ onto the $\rho \times \bar{\rho}$ invariant subspace $\bar{\mathcal{A}}_2$ of $\mathcal{K}\otimes\overline{\mathcal{K}}$. Extend Φ to $\mathcal{H}\otimes\bar{\mathcal{H}}$ by defining $\Phi(T+S) = \Phi(T)$ for $T \in \bar{\mathcal{A}}_1$ and $S \in \bar{\mathcal{A}}_1^\perp$. Then $\Phi \in \mathrm{Hom}_{G\times G}(\pi \times \bar{\pi}, \rho \times \bar{\rho})$. Since $\Phi \neq 0$, Lemma 6.77 implies $\mathrm{Hom}_G(\pi,\rho) \neq \{0\}$. □

Exercise Set 6.10

1. Show one need not assume G is σ-compact in the Proposition 6.130. (**Hint:** Use Exercises 6.1.31, 6.1.32, and 6.1.34.)

Solution:

Using Theorem 6.15 there is a unique regular Borel measure μ giving the integral I on $C_c(G/H)$ defined by $I(f_H) = \int \rho(x)f(x)\,dm(x)$ where m is a fixed Haar measure on G and $f_H(xH) = \int f(xh)\,dh$ and dh corresponds to a fixed Haar measure for the subgroup H. One has

$$d\mu(xyH) = \frac{\rho(xy)}{\rho(y)}\,d\mu(yH)$$

or

$$\int F(x^{-1}yH)\,d\mu(yH) = \int \frac{\rho(xy)}{\rho(y)}F(yH)\,d\mu(yH)$$

for integrable functions F on G/H.

As in the solution of Exercise 6.5.33, G/H is a union of disjoint open-closed σ compact subsets $G_0x_\alpha H$ where G_0 is a σ-compact open closed subgroup of

G. Note as a subset of G, the set $G_0x_\alpha H$ need not be σ-compact. Being open, it is a Borel subset.

Now if a function $f : G \to \mathcal{H}$ is Borel and satisfies $f(xh) = \pi(h^{-1})f(x)$ for all h a.e. x, then $f_\alpha = f|_{G_0x_\alpha H}$ is Borel on $G_0x_\alpha H$ for each α and satisfies $f(xh) = \pi(h^{-1})f(x)$ for m a.e. x in the open closed subset $G_0x_\alpha H$ of G. The same argument as in the proof of Proposition 6.130 shows if $f : G \to \mathcal{H}$ and $f' : G \to \mathcal{H}$ are Borel, then $x \mapsto (f(x), f'(x))_{\mathcal{H}}$ is Borel and this depended on the separability of the Hilbert space $\mathcal{H}$. Since $xH \mapsto ||f(x)||^2_{\mathcal{H}} = ||f(xh)||^2_{\mathcal{H}}$ for any $h \in H$ is Borel on G/H and thus is μ integrable if and only if for countably many α, $\int_{G_0x_\alpha H} ||f(xh)||^2 \, d\mu(xH) > 0$ and then $\sum_\alpha \int ||f(xH)||^2 \, d\mu(xH) < \infty$. Thus $L^2_\pi(G, \mathcal{H})$ can be taken to be $\bigoplus_\alpha L^2_\pi(G_0x_\alpha H, \mathcal{H})$ where $L^2_\pi(G_0x_\alpha H, \mathcal{H})$ is the space of L^2 Borel functions $f_\alpha : G_0x_\alpha H \to \mathcal{H}$ satisfying $f_\alpha(xh) = \pi(h^{-1})f(x)$ for all h for a.e. x in $G_0x_\alpha H$.

Now we follow the proof of Proposition 6.130. That $L^2_\pi(G, \mathcal{H})$ is complete is a verbatim copy. Next for a fixed $f \in L^2_\pi(G, \mathcal{H})$, let C to be the countable subset of the α with $f_\alpha \neq 0$; i.e., $\int_{G_0x_\alpha H} ||f(x)||^2_{\mathcal{H}} \, d\mu(xH) > 0$. Thus we may take $f(x) = 0$ for all $x \notin \cup_{\alpha \in C} G_0x_\alpha H$.

Let $\epsilon > 0$. Again there is an m and a compact subset K_n of G/H such that if $S_m = \{x \mid ||f(x)||_{\mathcal{H}} \leqslant m\}$, then $f_0(x) = \chi_{S_m}(x)f(x)\chi_{K_n}(xH)$ is in $L^2_\pi(G, \mathcal{H})$ and $|f - f_0|_{L^2_\pi(G,\mathcal{H})} < \frac{\epsilon}{3}$. Using Lemma 6.129, one has a bounded Borel function $F : G \to \mathcal{H}$ which vanishes off a compact subset $\widetilde{W}$ of G with $F_H = f_0$ where $F_H(x) = \int \pi(h)F(xh) \, dh$. Continuing to follow the argument in the proof, one finds an n, $M > 0$, and and $M' > 0$ so that $F_n(x) = \sum_{k=1}^n (F(x), e_k)_{\mathcal{H}} e_k$ satisfies $|(F_n)_H - F_H|_2 < \frac{\epsilon}{3}$ and $||F_n(x)||_{\mathcal{H}} \leqslant M$ and $||(F_n)_H(xH)||_{\mathcal{H}} \leqslant M'$ for all x. Let $a_k(x) = (F(x), e_k)$. Note $|a_k(x)| \leqslant M$ for all k and x. Fixing a precompact open subset V with $V \supseteq \widetilde{W}$, we can find continuous functions b_k on G with support inside V satisfying $|b_k(x)| \leqslant M$ for all x and

$$\int |a_k(x) - b_k(x)| \, \rho(x) \, dx < \frac{\epsilon}{3n(n+1)Mm(V^{-1}V \cap H)} \text{ for } k = 1, 2, \ldots, n.$$

The argument is completed using the last paragraph in Proposition 6.130's proof.

□

2. Let G be a σ-compact locally compact Hausdorff space with a closed subgroup H. Suppose ρ is a continuous rho function defining a corresponding quasi-invariant measure μ on G/H. Show if π is a unitary representation of H on a separable Hilbert space $\mathcal{H}$ and ν is a regular quasi-invariant measure on G/H, then π^G is unitarily equivalent to the representation L defined by

$$L(x)f(y) = \sqrt{\frac{d(x\nu)}{d\nu}}(yH)f(x^{-1}y)$$

on the Hilbert space of Borel functions f on G satisfying $f(xh) = \pi(h^{-1})f(x)$ for all $h \in H$ for a.e. x and $\int ||f(y)||^2_{\mathcal{H}} \, d\nu(yH) < \infty$.

Solution: By Theorem 6.24, μ and ν are equivalent measures. Thus there is a positive Borel function D such that $\frac{d\nu}{d\mu} = D$.

Define $T : L^2_\pi(G, \mathcal{H}, \mu) \to L^2_\pi(G, \mathcal{H}, \nu)$ by $Tf(y) = D^{-1/2}(yH)f(y)$. Note T is unitary for

$$\begin{aligned}
|Tf|_2^2 &= \int D^{-1}(yH)|f(y)|^2_{\mathcal{H}}\, d\nu(yH) \\
&= \int D^{-1}(yH)|f(y)|^2_{\mathcal{H}}\, D(yH)\, d\mu(yH) \\
&= \int |f(y)|^2\, d\mu(yH) \\
&= |f|_2^2.
\end{aligned}$$

Now

$$\begin{aligned}
T\pi^G(x)f(y) &= D^{-1/2}(yH)\sqrt{\frac{\rho(x^{-1}y)}{\rho(y)}}\, f(x^{-1}y) \\
&= D^{-1/2}(yH)\sqrt{\frac{\rho(x^{-1}y)}{\rho(y)}}\, D^{1/2}(x^{-1}yH)\, D^{-1/2}(x^{-1}yH) f(x^{-1}y) \\
&= \sqrt{F(x, yH)}\, Tf(x^{-1}y)
\end{aligned}$$

where $F(x, yH) = D^{-1}(yH)\frac{\rho(x^{-1}y)}{\rho(y)}D(x^{-1}yH)$. Thus $T\pi^G(x)f = L(x)Tf$ if and only if $F(x, yH) = \frac{d(x\nu)}{d\nu}(yH)$ for ν a.e yH for each $x \in G$. But

$$\begin{aligned}
x\nu(E) &= \nu(x^{-1}E) \\
&= \int D(yH)\chi_{x^{-1}E}(yH)\, d\mu(yH) \\
&= \int D(yH)\chi_E(xyH)\, d\mu(yH) \\
&= \int D(x^{-1}xyH)\chi_E(xyH)\, d\mu(yH) \\
&= \int D(x^{-1}yH)\chi_E(yH)\, d\mu(x^{-1}yH) \\
&= \int D(x^{-1}yH)\chi_E(yH)\frac{\rho(x^{-1}y)}{\rho(y)}\, d\mu(yH) \\
&= \int D(x^{-1}yH)\chi_E(yH)\frac{\rho(x^{-1}y)}{\rho(y)} D^{-1}(yH)\, d\nu(yH)
\end{aligned}$$

for all Borel subsets $E \subseteq G/H$. So $\frac{d(x\nu)}{d\nu}(yH) = D^{-1}(yH)\frac{\rho(x^{-1}y)}{\rho(y)}D(x^{-1}yH)$ for ν a.e. yH for each $x \in G$. □

3. Suppose G is a second countable locally compact Hausdorff group with a closed subgroup H and a continuous rho function ρ. Let μ be the corresponding quasi-invariant measure on G/H. Assume there is a Borel function $\sigma : G/H \to G$ satisfying $\sigma(xH)H = xH$ for all x in G. Show if π is a unitary representation of H on a separable Hilbert space $\mathcal{H}$, then π^G is unitarily equivalent to the representation L defined on $L^2(G/H, \mathcal{H})$ by

$$L(x)f(yH) = \sqrt{\frac{\rho(x^{-1}y)}{\rho(y)}}\pi(\sigma(yH)^{-1}x\sigma(x^{-1}yH))f(x^{-1}yH).$$

Solution: Define $T : L^2_\pi(G, \mathcal{H}) \to L^2(G/H, \mathcal{H})$ by $Tf(yH) = f(\sigma(yH))$. Note Tf is Borel and $|Tf|_2 = |f|_2$. So T is an isometry. Now let $f \in L^2(G/H, \mathcal{H})$. Define $f_0(x) = f_0(\sigma(xH)\sigma(xH)^{-1}x) = \pi(x^{-1}\sigma(xH))f(xH)$. Clearly $f_0(xh) = \pi(h^{-1})f_0(x)$. Note $x \mapsto (x^{-1}, \sigma(xH))$ is a Borel mapping of G into $G \times G/H$. Indeed, since G is second countable, so is G/H, and thus so is $G \times G/H$. In fact there is a countable basis of open sets of form $U_i \times V_j$ where U_i is open in G and V_j is open in G/H. Now the preimage of $U_i \times V_j$ is all x with $x^{-1} \in U_i$ and $\sigma(xH) \in V_j$ and thus is $U_i^{-1} \cap \sigma^{-1}(V_j)$ and thus is Borel. Since every open subset of $G \times G/H$ is a union of a subset of the countable collection $U_i \times V_j$, we see $x \mapsto (x^{-1}, \sigma(xH))$ is Borel. We now claim $x \mapsto \pi(x^{-1}\sigma(xH))f(xH)$ is Borel from G into $\mathcal{H}$. Indeed, if $\{e_k\}_{k=1}^{\infty}$ is an orthonormal basis of $\mathcal{H}$, then

$$\begin{aligned}\pi(x^{-1}\sigma(xH))f(xH) &= \sum_{k=1}^{\infty}(\pi(x^{-1}\sigma(xH))f(xH), e_k)_{\mathcal{H}}e_k \\ &= \sum_{k=1}^{\infty}(f(xH), \pi(\sigma(xH)^{-1}x)e_k)_{\mathcal{H}}e_k\end{aligned}$$

is Borel if $x \mapsto (f(xH), \pi(\sigma(xH)^{-1}x)e_k)_{\mathcal{H}}$ is Borel for each k. But this by strong continuity will follow if $x \mapsto \sigma(xH)^{-1}x$ is Borel into G. Now let U be an open subset of G. Then the set $(x, y) \in G \times G$ with $x^{-1}y \in U$ is open in $G \times G$. But this set has form a countable union of products $U_i \times U_j$. So its preimage is a countable union of sets $\{x \mid \sigma(xH) \in U_i, x \in U_j\}$. Now $\sigma(xH) \in U_i$ if and only if $x \in \kappa^{-1}(\sigma^{-1}(U_i))$ which is Borel in G. Thus $\{x \mid \sigma(xH) \in U_i, x \in U_j\} = \kappa^{-1}(\sigma^{-1}(U_i)) \cap U_j$ is Borel and so f_0 is Borel. Since $Tf_0 = f$ we see T is a unitary isomorphism of $L^2_\pi(G,\mathcal{H})$ onto $L^2(G/H,\mathcal{H})$. Moreover,

$$\begin{aligned}
T(\pi^G(x)f)(yH) &= \pi^G(x)f(\sigma(yH)) \\
&= \sqrt{\frac{\rho(x^{-1}y)}{\rho(y)}}f(x^{-1}\sigma(yH)) \\
&= \sqrt{\frac{\rho(x^{-1}y)}{\rho(y)}}f(\sigma(x^{-1}yH)\sigma(x^{-1}yH)^{-1}x^{-1}\sigma(yH)) \\
&= \sqrt{\frac{\rho(x^{-1}y)}{\rho(y)}}\pi(\sigma(yH)^{-1}x\sigma(x^{-1}yH))f(\sigma(x^{-1}yH)) \\
&= \sqrt{\frac{\rho(x^{-1}y)}{\rho(y)}}\pi(\sigma(yH)^{-1}x\sigma(x^{-1}yH))Tf(x^{-1}yH) \\
&= L(x)Tf(yH).
\end{aligned}$$

□

4. Let G be a second countable locally compact Hausdorff group acting continuously on a second countable locally compact Hausdorff space X with a quasi-invariant measure μ. Assume $[E]$ is in the measure algebra of μ and $[gE] = [E]$ for all g in G. Show there is a G invariant Borel subset W such that $[W] = [E]$. (**Hint:** Take a probability measure λ equivalent to Haar measure on G and consider $H(x) = \int_G \chi_E(gx)\,d\lambda(g)$.)

Solution:

Let $W = \{(g,x) \mid g\cdot x \in E,\, x \notin E \text{ or } g\cdot x \notin E, x \in E\}$. Since $g\cdot E = E$ in the measure algebra, $W_g = \{x \mid (g,x) \in W\}$ has μ measure 0 for each g. By Fubini's Theorem, $\lambda\times\mu(W) = \int \mu(W_g)\,d\lambda(g) = 0 = \int \lambda(W^x)\,d\mu(x)$ where $W^x = \{g \mid (g,x) \in W\}$. Thus $\lambda\{g \mid g\cdot x \in E,\, x \notin E\} + \lambda\{g \mid x \in E,\, g\cdot x \notin E\} = 0$ for μ a.e. x. So $\lambda\{g \mid x \in E,\, g\cdot x \in E\} = 1$ and $\lambda\{g \mid x \in E^c,\, g\cdot x \in E^c\} = 1$ for μ a.e. x. Consequently $H(x) = 1$ for a.e. x in E and $H(x) = 0$ for a.e. x in E^c. Also $H(gx) = 1$ whenever $H(x) = 1$. Similarly, $H(gx) = 0$ when $H(x) = 0$. Take $U = H^{-1}(1)$ and $V = H^{-1}(0)$. These are invariant sets. Also $\mu(E\cap U^c) = 0$ and $\mu(E^c \cap V^c) = 0$. So $E^c \cap U \subseteq E^c \cap V^c$ has measure 0. Thus $[E] = [U]$. □

5. Let E be a Borel subset of $\mathbb{R}$ and suppose one has $a + E = E$ in the measure algebra given by Lebesgue measure for all $a \in \mathbb{R}$. Show E or its complement has measure 0.

Solution:

This is a direct consequence of the previous problem. Lebesgue measure is invariant under translation and thus is quasi-invariant. Thus if $a + E = E$ in the measure algebra for all a, then there is a Borel subset E' with $E' + a = E'$ for all $a \in \mathbb{R}$ and $[E] = [E']$. If $E' \neq \varnothing$, then $\mathbb{R} = E' + \mathbb{R} \subseteq E'$. So $E' = \varnothing$ or $E' = \mathbb{R}$. Thus $m(E) = 0$ or $m(\mathbb{R} - E) = 0$.

□

6. Let $\phi : \mathbb{R}^n \to \mathbb{R}^n$ be a homeomorphism of $\mathbb{R}^n$ and let m be Lebesgue measure on $\mathbb{R}^n$. Define a Borel measure $\phi_* m$ by $\phi_* m(E) = m(\phi^{-1}(E))$ and a representation π on $L^2(\mathbb{R}^n, \lambda)$ by

$$\pi(x)f(y) = e^{-2\pi i x \cdot \phi(y)} f(y).$$

Assume $\phi_* m$ and m are equivalent measures.

(a) Show π and $\hat{\lambda}$ where λ is the left regular representation of $\mathbb{R}^n$ are unitarily equivalent.

(b) Show $\mathrm{Hom}(\pi, \pi) = \{M_h \mid h \in L^\infty(\mathbb{R}^n)\}$.

Then for $n = 1$, determine π when ϕ is given by:

(c) $\phi(y) = -y$ (d) $\phi(y) = \frac{y}{2\pi}$ (e) $\phi(y) = y^3$.

Solution: Define $T : L^2(\mathbb{R}^n, m) \to L^2(\mathbb{R}^n, m)$ by $Tf(y) = \sqrt{\frac{d\phi_* m}{dm}(y)} f(\phi^{-1}(y))$. Note T is unitary for

$$\begin{aligned}
|Tf|_2^2 &= \int \frac{d\phi_* m}{dm}(y) |f(\phi^{-1}(y))|^2 \, dm(y) \\
&= \int |f(\phi^{-1}(y)|^2 \, d\phi_* m(y) \\
&= \int |f(y)|^2 \, dm(y) .
\end{aligned}$$

Also

$$\begin{aligned}
T\pi(x)f(y) &= \sqrt{\frac{d\phi_* m}{dm}}(y)\pi(x)f(\phi^{-1}(y)) \\
&= \sqrt{\frac{d\phi_* m}{dm}}(y)e^{-2\pi i x \cdot y} f(\phi^{-1}(y)) \\
&= e^{-2\pi i x \cdot y} \sqrt{\frac{d\phi_* m}{d\lambda}}(y) f(\phi^{-1}(y)) \\
&= e^{-2\pi i x \cdot y} Tf(y) \\
&= \hat{\lambda}(x) Tf(y).
\end{aligned}$$

So $T\pi(x)T^{-1} = \hat{\lambda}(x)$ for all $x \in \mathbb{R}^n$. This implies $\mathrm{Hom}(\pi, \pi) = \{T^{-1}AT \mid A \in \mathrm{Hom}(\hat{\lambda}, \hat{\lambda})\}$. But by Exercise 6.8.10, $\mathrm{Hom}(\hat{\lambda}, \hat{\lambda})$ consists of all M_h where

$h \in L^\infty$. Now

$$\begin{aligned}
T^{-1}M_hT(f)(x) &= \sqrt{\frac{dm}{d\phi_*m}}(x)(M_hT(f))(\phi(x)) \\
&= \sqrt{\frac{dm}{d\phi_*m}}(x)h(\phi(x))T(f)(\phi(x)) \\
&= \sqrt{\frac{dm}{d\phi_*m}}(x)h(\phi(x))\sqrt{\frac{d\phi_*m}{dm}}(x)f(\phi^{-1}(\phi(x))) \\
&= h(\phi(x))f(x) \\
&= (M_{h\circ\phi}f)(x).
\end{aligned}$$

This gives $\mathrm{Hom}(\pi,\pi) = \{M_{h\circ\phi} \mid h \in L^\infty(\mathbb{R}^n)\} = \{M_h \mid h \in L^\infty(\mathbb{R}^n)\}$.

Now for (c), if $\phi(y) = -y$, then $\phi_*m = m$ and $\pi(x)f(y) = e^{-2\pi i x\cdot\phi(y)}f(y) = e^{2\pi i x\cdot y}f(y)$ for $f \in L^2(\mathbb{R}^n)$.

If $\phi(y) = \frac{y}{2\pi}$, then $\phi_*m(E) = m(2\pi E) = 2\pi m(E)$ and thus $\pi(x)f(y) = e^{-\pi i x\cdot y}f(y)$ for $f \in L^2(\mathbb{R}^n)$.

Finally if $\phi(y) = y^3$, then using Theorem 2.22,

$$\begin{aligned}
\int h(y)\,d\phi_*m(y) &= \int h(\phi(y))\,dm(y) \\
&= \int h(y^3)\frac{3y^2}{3y^2}\,dm(y) \\
&= \int h(y^3)|\det\phi'(y)|\frac{1}{3\phi(y)^{2/3}}\,dm(y) \\
&= \int h(x)\frac{1}{3x^{2/3}}\,dm(x)
\end{aligned}$$

and we see $\phi_*m(E) = \int_E \frac{1}{3x^{2/3}}\,dm(x)$. This also follow easily by change of variables, $\int h(y^3)\,dy = \int h(x)\,d(x^{1/3}) = \int h(x)\frac{1}{3}x^{-2/3}\,dx$. The representation is given by $\pi(x)f(y) = e^{-2\pi i x\cdot y^3}f(y)$.

□

7. Let π be a unitary representation of a locally compact Hausdorff group G and let π_0 be a subrepresentation on the invariant subspace $P\mathcal{H}$ where P is an orthogonal projection in $\mathrm{Hom}(\pi,\pi)$. Show $\mathrm{Hom}(\pi_0,\pi_0) = P\,\mathrm{Hom}(\pi,\pi)P$.
Solution: Let $A \in \mathrm{Hom}(\pi,\pi)$. Then

$$\begin{aligned}
\pi_0(x)PAP = P\pi_0(x)PAP &= P\pi(x)PAP \\
&= P\pi(x)AP = PAP\pi(x)P = PAP\pi_0(x).
\end{aligned}$$

Now let $B \in \mathrm{Hom}(\pi_0,\pi_0)$. Define A by $A = PBP$. Then $\pi(x)A = \pi(x)PBP = P\pi(x)PPB = \pi_0(x)B = B\pi_0(x) = PBPP\pi(x)P = PBP\pi(x)$. Thus $A \in \mathrm{Hom}(\pi,\pi)$ and $PAP = B$. □

8. Let X be a Borel subset of $\mathbb{R}^n$ with positive Lebesgue measure. Let π be the unitary representation of $\mathbb{R}^n$ on $L^2(X)$ given by $\pi(x)f(y) = e^{ix\cdot y}f(y)$. Show $\operatorname{Hom}(\pi,\pi) = \{M_h \mid h \in L^\infty(X)\}$ where $M_h f = hf$.

Solution:

It is easy to check $M_h \in \operatorname{Hom}(\pi,\pi)$ for $h \in L^\infty(X)$. Conversely, let $A \in \operatorname{Hom}(\pi,\pi)$. If λ is the left regular representation of $\mathbb{R}^n$, then $\hat{\lambda}(x)f(y) = e^{-2\pi i x\cdot y}f(y)$ for $f \in L^2(\mathbb{R}^n)$.

Now by Corollary 6.115, $M_{\chi_X}L^2(\mathbb{R}^n)$ is an invariant subspace of $\hat{\lambda}$ and clearly can be identified with the Hilbert space $L^2(X)$. If $\hat{\lambda}'$ is the representation $\hat{\lambda}$ restricted to $L^2(X)$, then $\pi(x) = \hat{\lambda}'(-x/2\pi)$ for each $x \in \mathbb{R}^n$. This implies $AM_{\chi_X} \in \operatorname{Hom}(\hat{\lambda},\hat{\lambda})$. Exercise 6.8.10 then shows $AM_{\chi_X} = M_g$ for some $g \in L^\infty(\mathbb{R}^n)$. Thus $AM_{\chi_X} = AM_{\chi_X}M_{\chi_X} = M_gM_{\chi_X} = M_{g\chi_X}$. So $Af = g|_X f$ for $f \in L^2(X)$. Setting $h = g|_X$, we have $A = M_h$.

□

9. Show the representation $\widehat{\pi^+}$ given in (6.40) of Example 6.137 of the $ax+b$ group is irreducible.

Solution:

We let $\rho(a,b) = \widehat{\pi^+}(a,b)$. Then $\rho(a,b)f(t) = \sqrt{a}e^{2\pi ibt}f(at)$. Now let $A \in \operatorname{Hom}(\rho,\rho)$. $A\rho(0,b) = \rho(0,b)A$ for all $b \in \mathbb{R}$. The argument given in Exercise 6.10.8 shows $Af(t) = h(t)f(t)$ for some $h \in L^\infty(\mathbb{R}^+)$. But then since $A\rho(a,0)f(t) = h(t)\sqrt{a}f(at)$ and $\rho(a,0)Af(t) = \sqrt{a}Af(at) = \sqrt{a}h(at)f(at)$ a.e. t for each $f \in L^2(\mathbb{R}^+)$ and $a > 0$, we see $h(at) = h(t)$ a.e. t for each $a > 0$. By Fubini's Theorem, one has $h(at) = h(t)$ a.e. $a > 0$ for a.e. $t > 0$. Choose a t_0 with $h(at_0) = h(t_0)$ for a.e. $a > 0$. Now $a \mapsto at_0$ preserving sets of Lebesgue measure 0 implies $h(a) = h(t_0)$ a.e. a. Consequently if $c = h(t_0)$, $Af = cf$ for $f \in L^2(\mathbb{R}^+)$. By Schur's Lemma 6.48, ρ is irreducible.

□

10. Let G be the $ax+b$ group. Let $\widehat{\pi^+}$ and $\widehat{\pi^-}$ be the unitary representation given in Equations (6.40) and (6.41). Show if in Example 6.137, one takes $\pi(1,b) = e^{i\lambda b}$ where $\lambda \in \mathbb{R}$, then π^G is unitarily equivalent to the representation $\widehat{\pi^+}$ if $\lambda > 0$ and is unitarily equivalent to $\widehat{\pi^-}$ when $\lambda < 0$. (**Hint:** Find a formula for π^G and consider Exercise 6.10.6.)

Solution:

Let π be the representation of the subgroup H given by $\pi(1,b) = e^{i\lambda b}$. Following Example 6.137, one sees

$$L(a,ab)f(x) = L((a,0)(1,b))f(x) = e^{\lambda iax^{-1}b}f(a^{-1}x)$$

for $f \in L^2(\mathbb{R}^+, \frac{dx}{x})$. This then implies $L(a,b)f(x) = e^{i\lambda x^{-1}b}f(a^{-1}x)$. Again using the unitary mapping W where $Wf(x) = f(x^{-1})$, we find

$$WL(a,b)W^{-1}f(x) = e^{i\lambda xb}f(ax).$$

As before set $Tf(t) = \frac{1}{\sqrt{t}}f(t)$. Then T is unitary from $L^2(\mathbb{R}^+, \frac{dx}{x})$ onto $L^2(\mathbb{R}^+, dt)$ and $TWL(a,b)W^{-1}T^{-1}f(t) = \sqrt{a}e^{i\lambda bt}f(at)$. Letting this representation be ρ_λ, we finish by showing ρ_λ is unitarily equivalent to $\widehat{\pi^-}$ when $\lambda < 0$. To do this define $S : L^2(\mathbb{R}^+, dt) \to L^2(\mathbb{R}^-, dt)$ by

$$Sf(-t) = \sqrt{-\frac{2\pi}{\lambda}} f(-\frac{2\pi}{\lambda}t).$$

This is unitary and

$$\begin{aligned} S\rho_\lambda(a,b)S^{-1}f(t) &= \sqrt{-\frac{2\pi}{\lambda}}\rho_\lambda(a,b)S^{-1}f(\frac{2\pi}{\lambda}t) \\ &= \sqrt{-\frac{2\pi}{\lambda}}\sqrt{a}e^{i\lambda b(\frac{2\pi}{\lambda}t)}(S^{-1}f)(\frac{2\pi a}{\lambda}at) \\ &= \sqrt{-\frac{2\pi}{\lambda}}\sqrt{a}e^{2\pi ibt}\sqrt{-\frac{\lambda}{2\pi}}f(\frac{\lambda}{2\pi}\frac{2\pi}{\lambda}at) \\ &= \sqrt{a}e^{2\pi ibt}f(at)\,. \end{aligned}$$

Thus for $\lambda < 0$, π^G is unitarily equivalent to $\widehat{\pi^-}$. For $\lambda > 0$, one takes $Sf(t) = \sqrt{\frac{2\pi}{\lambda}} f(\frac{2\pi}{\lambda}t)$ to show π^G is unitarily equivalent to $\widehat{\pi^+}$.

□

11. Let G be the $ax+b$ group and consider the quasi-regular representation ρ given in Equation (6.19) of Example 6.95. Show ρ and $\widehat{\pi^+} \oplus \widehat{\pi^-}$ are unitarily equivalent. (**Hint:** Use a Fourier transform.)

Solution:

We let G be the $ax+b$ group and ρ be the quasi-regular representation L. Then by (6.19),

$$\rho(a,b)f(x) = \frac{1}{\sqrt{a}} f\left(\frac{x+b}{a}\right) \quad \text{where } f \in L^2(\mathbb{R})\,.$$

Let ρ' be the representation $\widehat{\pi^+} \oplus \widehat{\pi^-}$. Since $L^2(\mathbb{R}) = L^2(\mathbb{R}^+) \oplus L^2(\mathbb{R}^-)$, using (6.40) and (6.41) we see ρ' is the representation on $L^2(\mathbb{R})$ given by

$$\rho'(a,b)f(x) = \sqrt{a}e^{2\pi ibx}f(ax)\,.$$

Take W to the Fourier transform on $L^2(\mathbb{R})$ given by $Wf(x) = \int f(t)e^{-2\pi ixt}\,dt$ for $f \in \mathcal{S}(\mathbb{R})$. Now

$$\begin{aligned} W\rho(a,b)f(x) &= \int \frac{1}{\sqrt{a}} f\left(\frac{t+b}{a}\right) e^{-2\pi ixt}\,dt \\ &= \frac{1}{\sqrt{a}}\int f(t/a)e^{-2\pi ix(t-b)}\,dt \\ &= \frac{1}{\sqrt{a}}\int f(t)e^{-2\pi ix(at-b)}\,a\,dt \\ &= \sqrt{a}e^{2\pi ixb}Wf(ax) \\ &= \rho'(a,b)Wf(x) \end{aligned}$$

for Schwartz functions f. By the L^2 density of $\mathcal{S}(\mathbb{R})$, one has $W\rho(a,b)W^{-1} = \rho'(a,b)$ for all $(a,b) \in G$ and we see ρ and ρ' are unitarily equivalent.

$\square$

12. Show the representation $\widehat{\pi^-}$ is the conjugate representation to the representation $\widehat{\pi^+}$ of the $ax+b$ group.
Solution: As seen in Example 6.137, $\widehat{\pi^+}$ and following the same argument $\widehat{\pi^-}$ are unitarily equivalent to representations ρ_+ and ρ_- given by $\rho_\pm(a,b)f(x) = \sqrt{a}e^{\pm 2\pi ixb}f(ax)$ for $f \in L^2(\mathbb{R}^+)$. Now $\overline{L^2(\mathbb{R}^+)}$ is the conjugate space to $L^2(\mathbb{R}^+)$ and consists of all functions $\bar{f}$ where $f \in L^2$. Since $\bar{\rho}_+(a,b)(\bar{f}) = \overline{\rho(a,b)f}$, the mapping $W(f) = \bar{f}$ is a unitary mapping of $\overline{L^2(\mathbb{R}^+)}$ onto $L^2(\mathbb{R}^+)$ and

$$\begin{aligned} W(\bar{\rho}_+(a,b)f)(x) &= \overline{\rho_+(a,b)f(x)} \\ &= \overline{\sqrt{a}e^{2\pi ibx}f(ax)} \\ &= \sqrt{a}e^{-2\pi ibx}\overline{f(ax)} \\ &= \sqrt{a}e^{-2\pi ibx}Wf(ax) \\ &= \rho_-(a,b)Wf(x)\,. \end{aligned}$$

So the conjugate of ρ_+ is unitarily equivalent to ρ_-.

$\square$

13. Show if λ_1 and λ_2 are nonzero real numbers, the unitary representations π_{λ_1} and π_{λ_2} given in Example 6.136 are not equivalent.
Solution:
The representation π_λ is given by

$$\pi_\lambda((x,y),t)f(w) = e^{i\lambda t}\, e^{.5\lambda ix\cdot y}e^{-i\lambda w\cdot y}f(w-x).$$

Thus $U_\lambda((0,0),t) = e^{i\lambda t}I$. Now if U is a unitary equivalence of π_{λ_1} with π_{λ_2}, then $U\pi_{\lambda_1}((0,0),t) = \pi_{\lambda_2}((0,0),t)U$ for $t \in \mathbb{R}$. So $Ue^{i\lambda_1 t}I = e^{i\lambda_2 t}IU$. This gives $e^{i\lambda_1 t} = e^{i\lambda_2 t}$ for all t and thus $\lambda_1 = \lambda_2$.

$\square$

14. Show the unitary representation π_λ of the Heisenberg group given in Example 6.136 is irreducible.
Solution:
The representation π_λ for $\lambda \neq 0$ is given by

$$\pi_\lambda((x,y),t)f(w) = e^{i\lambda t}\, e^{.5\lambda ix\cdot y}e^{-i\lambda w\cdot y}f(w-x).$$

Thus $\pi_\lambda((0,y),0)f(w) = e^{-i\lambda w\cdot y}f(w)$. By (b) of Exercise 6.10.6 with $\phi(y) = \frac{\lambda}{2\pi}y$, we see if $A \in \mathrm{Hom}(\pi_\lambda,\pi_\lambda)$, there is an $h \in L^\infty(\mathbb{R}^n)$ with $A = M_h$. But $A\pi_\lambda((x,0),0) = \pi_\lambda((x,0),0)A$ and thus

$$h(w)f(w-x) = h(w-x)f(w-x)$$

for a.e. w for each x for any $f \in L^2(\mathbb{R})$. Consequently $h(w) = h(w-x)$ a.e. w for each x. Set $E = \{(w,x) \mid h(w) \neq h(w-x)\}$. Then $m(E^x) = m\{w \mid (w,x) \in$

$E\} = 0$ for each x. By Fubini's Theorem, $0 = \int m(E^x)\, dm(x) = m \times m(E) = \int m(E_w)\, dm(w)$ where $E_w = \{x \mid (w, x) \in E\}$. Thus $m(E_w) = 0$ a.e. w. Fix w_0 with $m(E_{w_0}) = 0$. Thus $m\{x \mid h(w_0) \neq h(w_0 - x)\} = 0$. So $h(w_0) = h(w_0 - x)$ a.e. x. Since $x \mapsto w_0 - x$ preserves Lebesgue measure, $h(x) = h(w_0)$ a.e. x. This gives $A = M_h = cI$ where $c = h(w_0)$. By Schur's Lemma, π_λ is irreducible.

□

15. Let X be a locally compact Hausdorff space with a regular measure μ and let $\mathcal{H}$ be a separable Hilbert space. A function $x \mapsto T(x)$ from X into $\mathcal{B}(\mathcal{H})$ is strongly Borel if $x \mapsto T(x)v$ is Borel for each $v \in \mathcal{H}$.

(a) Show T is strongly Borel if and only if $x \mapsto (T(x)v, w)_{\mathcal{H}}$ is Borel for any $v, w \in \mathcal{H}$.

(b) Show $x \mapsto ||T(x)||$ is Borel if T is strongly Borel.

(c) If T is strongly Borel, define $||T||_\infty = \inf\{a \geqslant 0 \mid \mu\{x \mid ||T(x)|| > a\} = 0\}$. Show if $L^\infty(X, \mathcal{B}(\mathcal{H}))$ is the space of all strongly Borel functions T with $||T||_\infty < \infty$, then $L^\infty(X, \mathcal{B}(\mathcal{H}))$ with pointwise addition and multiplication, adjoint $T^*(x) = T(x)^*$, and norm $||\cdot||$ is a C^* algebra with identity.

(d) Define a representation ρ of $L^\infty(X, \mathcal{B}(\mathcal{H}))$ on $L^2(X, \mathcal{H})$ by

$$(\rho(T)f)(x) = T(x)f(x).$$

Show ρ is a representation of the C^* algebra $L^\infty(X, \mathcal{B}(\mathcal{H}))$ and the mapping $T \mapsto \rho(T)$ is an isometric $*$ algebra homomorphism onto its range.

Solution:

For (a), first note if $x \mapsto T(x)v$ is Borel for each v, then since the inner product is continuous, $x \mapsto (T(x)v, w)$ is Borel for each v and w. Conversely, let e_i for $i = 1, 2, \ldots$ be an orthonormal basis. Then $||T(x)v||^2 = \sum_i |(T(x)v, e_i)|^2$ is a pointwise limit of Borel functions and thus is Borel.

For (b), since $\mathcal{H}$ is separable, the unit ball $S = \{v \mid ||v|| = 1\}$ is separable. Being metric spaces, it follows there is a sequence v_n for $n \in \mathbb{N}$ whose elements form a dense subset of S. Thus $||T(x)|| = \sup_n ||T(x)v_n||$. This implies $x \mapsto ||T(x)||$ is Borel.

To see (c), we note first $||\cdot||_\infty$ is a norm. Note we are taking $T = 0$ if and only if $T(x) = 0$ a.e. x. So if $\{x \mid T(x) \neq 0\}$ has positive measure, one has $\mu(\{x \mid ||T(x)|| > 0\}) = \lim_{k\to 0} \mu\{x \mid ||T(x)|| > \frac{1}{k}\} > 0$. This implies there is a k with $\mu\{x \mid ||T(x)|| > \frac{1}{k}\} > 0$. So $||T||_\infty > 0$. One follows the standard argument for L^∞ to show $||aT||_\infty = |a|\, ||T||_\infty$ and $||T + S||_\infty \leqslant ||T||_\infty + ||S||_\infty$. To see it is complete, let T_n be Cauchy. By taking a subsequence we may assume $||T_n - T_m||_\infty < \frac{1}{2^{n+1}}$ for $m \geqslant n$. By definition, the sets $E_{n,m} = \{x \mid ||T_n(x) - T_m(x)|| > \frac{1}{2^n}\}$ have measure 0 for $n \leqslant m$ and $n \geqslant 1$. Thus $E_n = \cup_{m \geqslant n} E_{n,m}$ has measure 0 and one has $||T_n(x) - T_m(x)|| \leqslant \frac{1}{2^n}$ for $x \notin E_n$ when $m \geqslant n$. Now let $E = \cup_n E_n$. It has measure 0 and if $x \notin E$, there is an n_0 with $x \notin E_{n_0}$ and so $||T_n(x) - T_m(x)|| \leqslant ||T_n(x) - T_{n_0}(x)|| + ||T_{n_0}(x) - T_m(x)|| \leqslant \frac{1}{2^{n_0}} + \frac{1}{2^{n_0}} = \frac{1}{2^{n_0-1}}$ for $m, n \geqslant n_0$. Consequently $T_n(x)$ is Cauchy for each $x \notin E$. Since $\mathcal{B}(\mathcal{H})$ is complete, there is a $T(x) \in \mathcal{B}(\mathcal{H})$ with $T_n(x) \to T(x)$ for $x \notin E$. Set $T(x) = 0$ for $x \notin E$. Letting $m \to \infty$ in the above inequality gives $||T_n(x) - T(x)|| \leqslant \frac{1}{2^{n_0-1}}$ for $x \notin E$. So $||T_n - T||_\infty < \frac{1}{2^{n_0-1}}$ for $n \geqslant n_0$. Putting this together we see T_n converges in $L^\infty(X, \mathcal{B}(\mathcal{H}))$ to T. Thus $L^\infty(X, \mathcal{B}(\mathcal{H}))$ is a Banach space.

Since $||T(x)^*|| = ||T(x)||$ for all x, one has $||T^*||_\infty = ||T||_\infty$. Also $||TS||_\infty \leqslant ||T||_\infty ||S||_\infty$ and $||TT^*|| = ||T||^2$ follow from $||T(x)S(x)|| \leqslant ||T(x)||\,||S(x)||$ and $||T(x)T(x)^*|| = ||T(x)||^2$ for all x. Thus $L^\infty(X, \mathcal{B}(\mathcal{H}))$ is a C^* algebra and I defined by $I(x) = I$ for all x is a multiplicative identity.

Next we show ρ is a representation. It is easy to check $T \mapsto \rho(T)$ is linear and multiplicative into $\mathcal{B}(L^2(X, \mathcal{H}))$. Also $\rho(I) = I$. That it is a $*$ representation follows from

$$\begin{aligned} ((\rho(T^*)f), f')_{L^2(X,\mathcal{H})} &= \int (T^*(x)f(x), f'(x))_{\mathcal{H}}\, d\mu(x) \\ &= \int (f(x), T(x)f'(x))_{\mathcal{H}}\, d\mu(x) \\ &= \int (f(x), (\rho(T)f')(x))_{\mathcal{H}}\, d\mu(x) \\ &= (f, \rho(T)f')_{L^2(X,\mathcal{H})}. \end{aligned}$$

Moreover, by Proposition 6.83, we have $||\rho(T)|| \leqslant ||T||_\infty$ although this is easy to show directly. Note if $a > 0$, $\{x \mid ||T(x)|| > ||T||_\infty - a\}$ has positive measure. Consider the dense subset v_n of the unit ball of $\mathcal{H}$ used earlier. Then for some n, the set E_n where $E_n = \{x \mid ||T(x)v_n|| > ||T||_\infty - a\}$ has positive measure. By regularity, there is a compact subset K of E_n having positive measure. Set $f = \chi_K v_n$. Note $||f||_2 = \sqrt{\mu(K)}$ and

$$\begin{aligned} ||\rho(T)f||^2_{L^2(X,\mathcal{H})} &= \int (T(x)f(x), T(x)f(x))_{\mathcal{H}}\, d\mu(x) \\ &= \int_K (T(x)v_n, T(x)v_n)_{\mathcal{H}}\, d\mu(x) \\ &\geqslant \int_K (||T||_\infty - a)^2\, d\mu(x) \\ &= (||T||_\infty - a)^2\, \mu(K)\,. \end{aligned}$$

Thus $||\rho(T)f||_{L^2(X,\mathcal{H})} \geqslant (||T||_\infty - a)\sqrt{\mu(K)} = (||T||_\infty - a)||f||_{L^2(X,\mathcal{H})}$. So $||\rho(T)|| \geqslant ||T||_\infty - a$ for $a > 0$. This gives $||\rho(T)||_\infty \geqslant ||T||_\infty$. So ρ is a $*$ isometric isomorphism.

□

16. Let X be a locally compact Hausdorff space with a regular Borel measure μ. Suppose $\mathcal{H}$ is a separable Hilbert space. Let π the representation of $L^\infty(X, \mu)$ on $L^2(X, \mathcal{H})$ given by $\pi(f)h = fh$. Show if $A \in \mathrm{Hom}(\pi, \pi)$, then there is an essentially bounded strongly Borel function $T : X \to \mathcal{B}(\mathcal{H})$ such that $Ah(x) = T(x)f(x)$ a.e. x for each $f \in L^2(X, \mathcal{H})$. (**Hint:** Take a countable dense $\mathbb{Q} + i\mathbb{Q}$ linear subspace V of $\mathcal{H}$ and show $A_{v,w}(x) : L^2(X) \to L^2(X)$ defined a.e. x on

$V \times V$ by $A_{v,w}f(x) = (A(fv)(x), w)_{\mathcal{H}}$ extends a.e. from $V \times V$ to a bounded sesquilinear form on $\mathcal{H}$. Then apply Exercise 6.5.25.)

Solution: Since $\mathcal{H}$ is separable, there is a countable dense subset $\{v_j\}_{j=1}^{\infty}$. We take $\mathbb{F}$ to be the countable subring of $\mathbb{C}$ consisting of the complex numbers with rational real and complex parts. Let V be the $\mathbb{F}$ linear span of the vectors v_j. Then V is a countable module space over $\mathbb{F}$. For each pair (v, w) where $v, w \in V$, define $A_{v,w} : L^2(X) \to L^2(X)$ by $A_{v,w}f(x) = (A(fv)(x), w)_{\mathcal{H}}$. Note

$$\begin{aligned} |A_{v,w}f|_2^2 &= \int |(A(fv)(x), w)_{\mathcal{H}}|^2 \, d\mu(x) \\ &\leqslant \int ||A(fv)(x)||_{\mathcal{H}}^2 \, ||w||_{\mathcal{H}}^2 \, d\mu(x) \\ &= ||w||_{\mathcal{H}}^2 \, |A(fv)|_2^2 \\ &\leqslant ||A||^2 |fv|_2^2 \, ||w||_2^2 \\ &= ||A||^2 \, |f|_2^2 \, ||v||_{\mathcal{H}}^2 \, ||w||_{\mathcal{H}}^2 \end{aligned}$$

and thus $A_{v,w}$ is bounded. It is clearly linear. We claim it commutes with the representation π_0 of $L^\infty(X)$ on $L^2(X)$ given by $\pi_0(f)h = fh$. Indeed,

$$\begin{aligned} (A_{v,w}\pi_0(f)h)(x) &= (A((\pi_0(f)h)v)(x), w)_{\mathcal{H}} \\ &= (A(\pi(f)(hv))(x), w)_{\mathcal{H}} \\ &= (\pi(f)A(hv)(x), w)_{\mathcal{H}} \\ &= (f(x)A(hv)(x), w)_{\mathcal{H}} \\ &= f(x)(A(hv)(x), w)_{\mathcal{H}} \\ &= \pi_0(f)A_{v,w}(h)(x). \end{aligned}$$

By Exercise 6.5.25, there is a unique Borel function $T_{v,w} \in L^\infty(X)$ such that

$$A_{v,w}h(x) = T_{v,w}(x)h(x) \text{ a.e. } x.$$

Moreover, $T_{v,w}$ may be taken so that $|T_{v,w}(x)| \leqslant ||A|| \, ||v||_{\mathcal{H}} \, ||w||_{\mathcal{H}}$. Since $(v, w) \mapsto A_{v,w}$ is sesquilinear, one has

$$T_{c_1v_1+c_2v_2,w}(x) = c_1T_{v_1,w}(x) + c_2T_{v_2,w}(x) \text{ for } \mu \text{ a.e. } x\,.$$

We obtain $(v, w) \mapsto T_{v,w}(x)$ is bounded and $\mathbb{F}$ sesquilinear on $V \times V$ on a conull Borel subset X_0 of X. Hence for each $x \in X_0$, one has an extension to a bounded sesquilinear functional $(v, w) \mapsto B_{v,w}(x)$ on $\mathcal{H} \times \mathcal{H}$. We have $||B_{v,w}(x)|| \leqslant ||A|| \, ||v||_{\mathcal{H}} \, ||w||_{\mathcal{H}}$. This implies for each $x \in X_0$, there is a bounded linear operator $B(x)$ on $\mathcal{H}$ such that $B_{v,w}(x) = (B(x)v, w)_{\mathcal{H}}$ for $x \in X_0$ and $v, w \in \mathcal{H}$. Also, $||B(x)|| \leqslant ||A||$. Since the simple functions are dense in $L^2(X, \mathcal{H})$, we have $Af(x) = B(x)f(x)$ for $x \in X_0$ and f simple. The result now follows. □

17. Let G be a σ-compact locally compact Hausdorff group and let H be a closed subgroup of G having a unitary representation π on a separable Hilbert space $\mathcal{H}$. Let ρ be a continuous rho function for H and let μ be the corresponding quasi-invariant measure on G/H. Let M be the representation of $L^\infty(G/H,\mu)$ on $L^2_\pi(G,\mathcal{H})$ given by

$$M(h)f = (h \circ \kappa)f \text{ for } h \in L^\infty(G/H) \text{ and } f \in L^2_\pi(G,\mathcal{H}).$$

For each $A \in \text{Hom}(\pi,\pi)$, define $T_A : L^2_\pi(G,\mathcal{H}) \to L^2_\pi(G,\mathcal{H})$ by

$$T_A f(x) = Af(x).$$

Show $A \mapsto T_A$ is a $*$ algebra isomorphism of the C^* algebra $\text{Hom}(\pi,\pi)$ onto the C^* algebra $\text{Hom}(\pi^G,\pi^G) \cap \text{Hom}(M,M)$. Consequently, show π^G is irreducible if and only if π is irreducible and $\text{Hom}(\pi^G,\pi^G) \subseteq \text{Hom}(M,M)$. (**Hint:** Use the argument suggested in the previous exercise to show if $B \in \text{Hom}(M,M)$, then there is a an essentially bounded strongly Borel function $g \mapsto B(g) \in \mathcal{B}(\mathcal{H})$ such that $Bf(g) = B(g)f(g)$ for $f \in L^2_\pi(G,\mathcal{H})$.)

Solution:

Working this problem with the suggestion to modify the previous exercise ran into difficulty and may not work out. So as in the case of second countable locally compact groups, we assume there is a Borel cross section for G/H; that is a Borel subset B of G such that the mapping σ given by $xH \to \{\sigma(xH)\} = B \cap xH$ is a Borel isomorphism of G/H with B. This is the setup in Proposition 6.138. We start by showing the induced representation $L = \pi^G$ has the following behavior:

$$\begin{aligned} L(x)f(\sigma(yH)) &= \sqrt{\frac{\rho(x^{-1}y)}{\rho(y)}} f(x^{-1}\sigma(yH)) \\ &= \sqrt{\frac{\rho(x^{-1}y)}{\rho(y)}} f(\sigma(x^{-1}yH)\sigma(x^{-1}yH)^{-1}x^{-1}\sigma(yH)) \\ &= \sqrt{\frac{\rho(x^{-1}y)}{\rho(y)}} \pi(\sigma(yH)^{-1}x\sigma(x^{-1}yH))f(\sigma(x^{-1}yH)). \end{aligned}$$

This by identifying G/H with B via the Borel isomorphism $xH \leftrightarrow \sigma(xH)$ essentially establishes Proposition 6.138. So the exercise becomes the following. Let L be the representation of G on $L^2(G/H,\mathcal{H})$ given by

$$L(x)f(yH) = \sqrt{\frac{\rho(x^{-1}y)}{\rho(y)}} \pi(\sigma(yH)^{-1}x\sigma(x^{-1}yH))f(x^{-1}yH)$$

and M be the representation of $L^\infty(G/H,\mu)$ on $L^2(G/H,\mu,\mathcal{H})$ defined by $M(h)f = hf$. Then the mapping $A \mapsto T_A$ where $T_Af(xH) = Af(xH)$ is a $*$ algebra isomorphism of $\text{Hom}(\pi,\pi)$ onto $\text{Hom}(L,L) \cap \text{Hom}(M,M)$.

Exercise 6.10.15 implies $A \mapsto T_A$ is a $*$ algebra isometry of $\text{Hom}(\pi,\pi)$ into the space of bounded linear operators on $L^2(G/H,\mathcal{H})$. Furthermore, if $h \in$

$L^\infty(G/H, \mu)$, then

$$\begin{aligned} T_A M(h) f(xH) &= Ah(xH) f(xH) \\ &= h(xH) A f(xH) \\ &= M(h) T_A f(xH) \text{ a.e. } xH \end{aligned}$$

and for $x \in G$,

$$\begin{aligned} T_A L(x) f(yH) &= A L(x) f(yH) \\ &= \sqrt{\frac{\rho(x^{-1}y)}{\rho(y)}} A\pi(\sigma(yH)^{-1} x \sigma(x^{-1}yH)) f(x^{-1}yH) \\ &= \sqrt{\frac{\rho(x^{-1}y)}{\rho(y)}} \pi(\sigma(yH)^{-1} x \sigma(x^{-1}yH)) A f(x^{-1}yH) \\ &= L(x) T_A f(yH) \text{ a.e. } yH. \end{aligned}$$

Hence $T_A \in \operatorname{Hom}(L, L) \cap \operatorname{Hom}(M, M)$.

Now suppose $T \in \operatorname{Hom}(L, L) \cap \operatorname{Hom}(M, M)$. From Exercise 6.10.16, since $T \in \operatorname{Hom}(M, M)$, there is a strongly Borel function $yH \mapsto A(yH) \in \mathcal{B}(\mathcal{H})$ with $||T|| = ||A||_\infty$ and $Tf(yH) = A(yH) f(yH)$ a.e. yH for each $f \in L^2(G/H)$. Since T belongs to $\operatorname{Hom}(L, L)$, one has for each $f \in L^2(G/H)$ and each $x \in G$,

$$A(yH)\sqrt{\frac{\rho(x^{-1}y)}{\rho(y)}} \pi(\sigma(yH)^{-1} x \sigma(x^{-1}yH)) f(x^{-1}yH) = TL(x) f(yH)$$

$$= L(x) T f(yH) = \sqrt{\frac{\rho(x^{-1}y)}{\rho(y)}} \pi(\sigma(yH)^{-1} x \sigma(x^{-1}yH)) A(x^{-1}yH) f(x^{-1}yH)$$

a.e. yH. In particular, we see

$$\begin{aligned} &A(yH)\pi(\sigma(yH)^{-1} x \sigma(x^{-1}yH)) f(x^{-1}yH) \\ &\qquad = \pi(\sigma(yH)^{-1} x \sigma(x^{-1}yH)) A(x^{-1}yH) f(x^{-1}yH) \end{aligned}$$

for a.e. yH. Now let $G/H = \cup_{i=1}^\infty K_i$ where each K_i is a compact subset of G/H and v_j form a countable dense subset of $\mathcal{H}$. Then the set of yH where

$$\begin{aligned} &A(yH)\pi(\sigma(yH)^{-1} x \sigma(x^{-1}yH)) 1_{K_i}(x^{-1}yH) v_j \\ &\qquad \neq \pi(\sigma(yH)^{-1} x \sigma(x^{-1}yH)) A(x^{-1}yH) 1_{K_i}(x^{-1}yH) v_j \end{aligned}$$

has measure zero for each i and j. This by continuity gives

$$A(yH)\pi(\sigma(yH)^{-1} x \sigma(x^{-1}yH)) = \pi(\sigma(yH)^{-1} x \sigma(x^{-1}yH)) A(x^{-1}yH) \tag{A}$$

a.e. yH for each x in G. Set

$$\begin{aligned} L(x, yH) &= \pi(\sigma(yH)^{-1} x \sigma(x^{-1}yH)) \text{ and} \\ L'(yH, x) &= A(yH)^{-1} \pi(\sigma(yH)^{-1} x \sigma(x^{-1}yH)) A(x^{-1}yH). \end{aligned}$$

Note

$$\begin{aligned} &L(x_1, yH)L(x_2, x_1^{-1}yH) \\ &\quad = \pi(\sigma(yH)^{-1}x_1\sigma(x_1^{-1}yH)\sigma(x_1^{-1}yH)^{-1}x_2\sigma(x_2^{-1}x_1^{-1}yH)) \\ &\qquad = \pi(\sigma(yH)^{-1}x_1x_2\sigma(x_2^{-1}x_1^{-1}yH)) = L(x_1x_2, yH) \end{aligned}$$

for all x_1, x_2, and yH. A similar calculation shows

$$L'(x_1, yH)L'(x_2, x_1^{-1}yH) = L'(x_1x_2, yH)$$

also holds for all x_1, x_2, yH. Now let $S = \{(x, yH) \mid L(x, yH) = L'(x, yH)\}$. By Fubini's Theorem, this set is conull in $G \times G/H$. Moreover, if (x_1, yH) and $(x_2, x_1^{-1}yH)$ are in S, then

$$\begin{aligned} L(x_1x_2, yH) &= L(x_1, yH)L(x_2, x_1^{-1}yH) \\ &= L'(x_1, yH)L'(x_2, x_2^{-1}yH) \\ &= L'(x_1x_2, yH) \end{aligned}$$

and we see $(x_1x_2, yH) \in S$. Now let U be the set of all yH such that (x, yH) and $(x^{-1}, x^{-1}yH)$ are in S for a.e. x. Note U is a conull measurable subset of G/H. Assume yH and $x^{-1}yH$ are in U. There is a z in G so that both (xz, yH) and $(z^{-1}, z^{-1}x^{-1}yH)$ are in S. This implies $(x, yH) \in S$. Now choose g_0 with $g_0H \in U$ so that (A) holds for a.e. x. Note $(g_0hg_0^{-1})^{-1}g_0H \in U$. As just seen, $(g_0hg_0^{-1}, g_0H) \in S$ for all $h \in H$. We thus have $L(g_0hg_0^{-1}, g_0H) = L'(g_0hg_0^{-1}, g_0H)$ for all h. This gives

$$\begin{aligned} &\pi(\sigma(g_0H)^{-1}g_0hg_0^{-1}\sigma(g_0h^{-1}g_0^{-1}g_0H)) \\ &\quad = A(g_0H)^{-1}\pi(\sigma(g_0H)^{-1}g_0hg_0^{-1}\sigma(g_0h^{-1}g_0^{-1}g_0H))A(g_0h^{-1}g_0^{-1}g_0H) \end{aligned}$$

or

$$\begin{aligned} &\pi(\sigma(g_0H)^{-1}g_0)\pi(h)\pi(g_0^{-1}\sigma(g_0H)) \\ &\qquad = A(g_0H)^{-1}\pi(\sigma(g_0H)^{-1}g_0)\pi(h)\pi(g_0^{-1}\sigma(g_0H))A(g_0H) \end{aligned}$$

for $h \in H$. Hence $\pi(\sigma(g_0H)^{-1}g_0)^{-1}A(g_0H)\pi(\sigma(g_0H)^{-1}g_0) \in \mathrm{Hom}(\pi, \pi)$. Since $h_0 = \sigma(g_0H)^{-1}g_0 \in H$, we see

$$\pi(h)\pi(h_0^{-1})A(g_0H)\pi(h_0) = \pi(h_0^{-1})A(g_0H)\pi(h_0)\pi(h)$$

which shows $\pi(h_0hh_0^{-1})A(g_0H) = A(g_0H)\pi(h_0hh_0^{-1})$ for all h. We hence have $A(g_0H) \in \mathrm{Hom}(\pi, \pi)$.

Finally since

$$A(g_0H)\pi(\sigma(g_0H)^{-1}x\sigma(x^{-1}g_0H)) = \pi(\sigma(g_0H)^{-1}x\sigma(x^{-1}g_0H))A(x^{-1}g_0H)$$

for a.e. x and $\sigma(g_0H)^{-1}x\sigma(x^{-1}g_0H) \in H$, we see $A(x^{-1}g_0H) = A(g_0H)$ a.e. x. Setting $A_0 = A(g_0H)$, we then obtain $T = T_{A_0}$ and the result is shown.

□

18. Let G be the group $\mathrm{SL}(2,\mathbb{R})$ of 2×2 matrices of determinant 1. Let A be the subgroup of matrices of form $a(t) = \left[\begin{smallmatrix} e^t & 0 \\ 0 & e^{-t} \end{smallmatrix}\right]$ where $t \in \mathbb{R}$ and let N be the subgroup consisting of the matrices $n(x) = \left[\begin{smallmatrix} 1 & x \\ 0 & 1 \end{smallmatrix}\right]$ where $x \in \mathbb{R}$. By Corollary 5.26, the mapping $(k,a,n) \mapsto kan$ from $\mathrm{SO}(2) \times A \times N$ to G given by $(k,a,n) \mapsto kan$ is a homeomorphism. Let M be the two element subgroup consisting of the matrices I and $-I$.

(a) Show $P = MAN$ is a closed subgroup of G.

(b) Find a left Haar measure for P and find the modular function for P.

(c) Find a Haar measure for G and show G is unimodular.

(d) Find a rho function for the subgroup P.

(e) Let π be the unitary representation of P defined by $\pi(\pm I a(t)n(x)) = \epsilon_{\pm}(\pm I)e^{i\lambda t}$ where $\lambda \in \mathbb{R}$ and $\epsilon_+(I) = \epsilon_+(-I) = \epsilon_-(I) = 1$ and $\epsilon_-(-I) = -1$. Find $\rho_{\pm,\lambda} = \pi^G$.

It is known $\rho_{+,\lambda}$ is irreducible for all $\lambda \in \mathbb{R}$ and $\rho_{-,\lambda}$ is irreducible for nonzero $\lambda \in \mathbb{R}$. These representation are known as *the unitary principal series* of $\mathrm{SL}(2,\mathbb{R})$.

Solution:

First, some notation. We let $k(\theta) = \left[\begin{smallmatrix} \cos\theta & -\sin\theta \\ \sin\theta & \cos\theta \end{smallmatrix}\right]$, $a(t) = \left[\begin{smallmatrix} t & 0 \\ 0 & t^{-1} \end{smallmatrix}\right]$ for $t > 0$, and $n(x) = \left[\begin{smallmatrix} 1 & x \\ 0 & 1 \end{smallmatrix}\right]$ and $\bar{n}(x) = \left[\begin{smallmatrix} 1 & 0 \\ x & 1 \end{smallmatrix}\right]$ for $x \in \mathbb{R}$.

We know $M = \{\pm I\}$ is a closed subgroup of $\mathrm{SO}(2)$ and thus $M \times A \times N$ is a closed subset of $\mathrm{SO}(2) \times A \times N$ in the product topology. By Corollary 5.26 the mapping $(k,a,n) \mapsto kan$ is a homeomorphism of $\mathrm{SO}(2) \times A \times N$ onto $\mathrm{SL}(2,\mathbb{R})$. This implies $P = MAN$ is a closed subset of G. Since AN is a subgroup and $\pm I$ are central in G, P is a subgroup.

A direct computation shows

$$I_P(f) = \int_{-\infty}^{\infty}\int_{-\infty}^{\infty} f\left[\left(\begin{smallmatrix} a & x \\ 0 & a^{-1} \end{smallmatrix}\right]\right) \frac{1}{a^2}\, da\, dx$$

is a left invariant integral and thus gives a Haar measure for P. Indeed, note

$$\begin{aligned}
\int_{-\infty}^{\infty}\int_{-\infty}^{\infty} f\left(\left[\begin{smallmatrix} b & y \\ 0 & b^{-1} \end{smallmatrix}\right]\left[\begin{smallmatrix} a & x \\ 0 & a^{-1} \end{smallmatrix}\right]\right)\frac{1}{a^2}\, da\, dx &= \int_{-\infty}^{\infty}\int_{-\infty}^{\infty} f\left(\left[\begin{smallmatrix} ab & bx+a^{-1}y \\ 0 & (ab)^{-1} \end{smallmatrix}\right]\right)\frac{1}{a^2}\, da\, dx \\
&= \int_{-\infty}^{\infty}\int_{-\infty}^{\infty} f\left(\left[\begin{smallmatrix} ab & x+a^{-1}y \\ 0 & (ab)^{-1} \end{smallmatrix}\right]\right)\frac{1}{a^2}\, da\frac{1}{b}\, dx \\
&= \int_{-\infty}^{\infty}\int_{-\infty}^{\infty} f\left(\left[\begin{smallmatrix} ab & x \\ 0 & (ab)^{-1} \end{smallmatrix}\right]\right)\frac{1}{a^2}\, da\frac{1}{b}\, dx \\
&= \int_{-\infty}^{\infty}\int_{-\infty}^{\infty} f\left(\left[\begin{smallmatrix} a & x \\ 0 & a^{-1} \end{smallmatrix}\right]\right)\frac{b^2}{a^2}\frac{1}{b}\, da\frac{1}{b}\, dx \\
&= I_P(f)
\end{aligned}$$

and so I_P is left invariant.

Note this implies one has the formula

$$\begin{aligned} I_P(f) &= \int_{-\infty}^{\infty}\int_{-\infty}^{\infty} f\left(\left[\begin{smallmatrix} a & x \\ 0 & a^{-1}\end{smallmatrix}\right]\right)\frac{1}{a^2}\,da\,dx \\ &= \int_{-\infty}^{\infty}\int_{-\infty}^{\infty} f\left(\left[\begin{smallmatrix} a & 0 \\ 0 & a^{-1}\end{smallmatrix}\right]\left[\begin{smallmatrix} 1 & x/a \\ 0 & 1\end{smallmatrix}\right]\right)\frac{1}{a^2}\,da\,dx \\ &= \int_{-\infty}^{\infty}\int_{-\infty}^{\infty} f\left(\left[\begin{smallmatrix} a & 0 \\ 0 & a^{-1}\end{smallmatrix}\right]\left[\begin{smallmatrix} 1 & x \\ 0 & 1\end{smallmatrix}\right]\right)\frac{1}{|a|}\,da\,dx\,. \end{aligned}$$

As AN is an open subgroup of P, a left invariant integral on AN is given by

$$\begin{aligned} \int f\,dm_{AN} &= \int_{-\infty}^{\infty}\int_{0}^{\infty} f\left(\left[\begin{smallmatrix} a & x \\ 0 & a^{-1}\end{smallmatrix}\right]\right)\frac{1}{a^2}\,da\,dx \\ &= \int_{-\infty}^{\infty}\int_{-\infty}^{\infty} f\left(\left(\begin{smallmatrix} e^t & x \\ 0 & e^{-t}\end{smallmatrix}\right]\right)\frac{1}{e^{2t}}\,(e^t\,dt)\,dx \\ &= \int_{-\infty}^{\infty}\int_{-\infty}^{\infty} f(a(t)n(e^{-t}x))\frac{1}{e^t}\,dt\,dx \\ &= \int_{-\infty}^{\infty}\int_{-\infty}^{\infty} f(a(t)n(x))\,dt\,dx \end{aligned}$$

for $f \in C_c(AN)$.

Now

$$\begin{aligned} \int_{-\infty}^{\infty}\int_{-\infty}^{\infty} f\left(\left[\begin{smallmatrix} b & y \\ 0 & b^{-1}\end{smallmatrix}\right]\left[\begin{smallmatrix} a & x \\ 0 & a^{-1}\end{smallmatrix}\right]\right)\frac{1}{b^2}\,db\,dy &= \int_{-\infty}^{\infty}\int_{-\infty}^{\infty} f\left(\left[\begin{smallmatrix} ab & bx+a^{-1}y \\ 0 & (ab)^{-1}\end{smallmatrix}\right]\right)\frac{1}{b^2}\,db\,dy \\ &= \int_{-\infty}^{\infty}\int_{-\infty}^{\infty} f\left(\left[\begin{smallmatrix} ab & bx+y \\ 0 & (ab)^{-1}\end{smallmatrix}\right]\right)\frac{1}{b^2}\,db\,(a\,dy) \\ &= \int_{-\infty}^{\infty}\int_{-\infty}^{\infty} f\left(\left[\begin{smallmatrix} ab & y \\ 0 & (ab)^{-1}\end{smallmatrix}\right]\right)\frac{a}{b^2}\,db\,dy \\ &= \int_{-\infty}^{\infty}\int_{-\infty}^{\infty} f\left(\left[\begin{smallmatrix} b & y \\ 0 & b^{-1}\end{smallmatrix}\right]\right)\frac{a}{(b/a)^2}\frac{1}{a}\,db\,dy \\ &= a^2 I_P(f)\,. \end{aligned}$$

This implies by (e) of Proposition 6.3 that $\Delta_P\left(\left[\begin{smallmatrix} a & x \\ 0 & a^{-1}\end{smallmatrix}\right]\right) = \frac{1}{a^2}$ for $a \neq 0$ and $x \in \mathbb{R}$.

Next note $\check{I}_P(f) = I_P(\check{f})$ is right invariant and

$$\begin{aligned} \check{I}_P(f) &= \int\int f\left(\left[\begin{smallmatrix} a & x \\ 0 & a^{-1}\end{smallmatrix}\right]^{-1}\right)\frac{1}{a^2}\,da\,dx \\ &= \int\int f\left(\left[\begin{smallmatrix} a^{-1} & -x \\ 0 & a\end{smallmatrix}\right]\right)\frac{1}{a^2}\,da\,dx \\ &= \int\int f\left(\left[\begin{smallmatrix} a & x \\ 0 & a^{-1}\end{smallmatrix}\right]\right)a^2 d(a^{-1})\,dx \\ &= \int\int f\left(\left[\begin{smallmatrix} a & x \\ 0 & a^{-1}\end{smallmatrix}\right]\right)\,da\,dx \end{aligned}$$

and one can check $\check{\Delta}_P\left(\left[\begin{smallmatrix} a & x \\ 0 & a^{-1}\end{smallmatrix}\right]\right) = a^2$ is the modular function; i.e.

$$\check{m}_P\left(\left[\begin{smallmatrix} a & x \\ 0 & a^{-1}\end{smallmatrix}\right] E\right) = a^2 \check{m}_P(E)$$

where m_P is the measure corresponding to the left invariant integral I_P for P.

We show G is unimodular. We first note $\Delta_G(k(\theta)) = 1$ for all θ since the only continuous homomorphism of a connected compact group into $\mathbb{R}^+$ is 1. Also since

$$\left[\begin{smallmatrix} a & 0 \\ 0 & a^{-1}\end{smallmatrix}\right]\left[\begin{smallmatrix} 1 & 1 \\ 0 & 1\end{smallmatrix}\right]\left[\begin{smallmatrix} a^{-1} & 0 \\ 0 & a\end{smallmatrix}\right] = \left[\begin{smallmatrix} 1 & a^2 \\ 0 & 1\end{smallmatrix}\right]$$

and Δ_G is a homomorphism, we have

$$\Delta_G\left(\left[\begin{smallmatrix} 1 & a^2 \\ 0 & 1\end{smallmatrix}\right]\right) = \Delta_G\left(\left[\begin{smallmatrix} 1 & 1 \\ 0 & 1\end{smallmatrix}\right]\right)$$

for all $a > 0$. This implies $\Delta_G\left(\left[\begin{smallmatrix} 1 & x \\ 0 & 1\end{smallmatrix}\right]\right) = 1$ for $x > 0$. Now $\Delta_G\left(\left[\begin{smallmatrix} 1 & -x \\ 0 & 1\end{smallmatrix}\right]\right) = \Delta_G\left(\left[\begin{smallmatrix} 1 & x \\ 0 & 1\end{smallmatrix}\right]\right)^{-1} = 1$ for $x > 0$. So $\Delta_G(n(x)) = 1$ for all x. Similarly, $\Delta_G(\bar{n}(x)) = 1$ for all x.

Now for A. We know by Corollary 5.26, the Iwasawa decomposition, that $\bar{n}(x) = \left[\begin{smallmatrix} 1 & 0 \\ x & 1\end{smallmatrix}\right]$ has form $k(\theta)a(t)n(x)$ for some θ, x, and t. Consider $\mathbb{R}^2$ with the usual inner product. Then $f_1 = \frac{e_1 + xe_2}{\sqrt{1+x^2}}$ and $f_2 = \frac{-xe_1+e_2}{\sqrt{1+x^2}}$ form an orthonormal basis. Moreover, $\begin{bmatrix} \frac{1}{\sqrt{1+x^2}} & \frac{x}{\sqrt{1+x^2}} \\ -\frac{x}{\sqrt{1+x^2}} & \frac{1}{\sqrt{1+x^2}} \end{bmatrix}$ is in $SO(2,\mathbb{R})$ and

$$\begin{aligned}\begin{bmatrix} \frac{1}{\sqrt{1+x^2}} & \frac{x}{\sqrt{1+x^2}} \\ -\frac{x}{\sqrt{1+x^2}} & \frac{1}{\sqrt{1+x^2}} \end{bmatrix}\left[\begin{smallmatrix} 1 & 0 \\ x & 1\end{smallmatrix}\right] &= \begin{bmatrix} \sqrt{1+x^2} & \frac{x}{\sqrt{1+x^2}} \\ 0 & \frac{1}{\sqrt{1+x^2}} \end{bmatrix} \\ &= \begin{bmatrix} \sqrt{1+x^2} & 0 \\ 0 & \frac{1}{\sqrt{1+x^2}} \end{bmatrix}\begin{bmatrix} 1 & \frac{x}{1+x^2} \\ 0 & 1 \end{bmatrix}\end{aligned}$$

and thus $1 = \Delta_G(a(\ln\sqrt{1+x^2}))$. This gives $\Delta_G \equiv 1$ and G is unimodular.

We find a Haar measure on G. By the Iwasawa decomposition, the map $(k, a, n) \mapsto kan$ is a homeomorphism. Since inversion is a homeomorphism, the map $(a, n, k) \mapsto (k^{-1}, a^{-1}, n^{-1}) \mapsto (k^{-1}a^{-1}n^{-1})^{-1}$ is a homeomorphism. So $(a, n, k) \mapsto nak$ is a homeomorphism. Since $na \mapsto an$ is homeomorphism, for $n(x)a(t) = a(t)n(e^{-t}x)$, the map $(a, n, k) \mapsto ank$ is a homeomorphism. We know using Theorem 6.15 and $\Delta_G(k) = \Delta_K(k) = 1$ for $k \in K$ that G/K has a G invariant integral I_K and if m is a left Haar measure for G, the integral I_K is given by

$$I_K(f_K) = \int f(g)\, dm(g) \text{ where } f_K(xK) = \frac{1}{2\pi}\int_0^{2\pi} f(xk(\theta))\, d\theta\,.$$

Let $m_{G/K}$ be the Radon measure (or the regular Borel measure) giving I_K. Now G/K and AN are homeomorphic under the mapping $\phi : ankK \mapsto an$ and under this isomorphism the action of AN is preserved. Thus $I_{AN}(f) = \int_{G/K} f(\phi(gK))\, dm_{G/K}(gK)$ is invariant under AN on the left. Note the Haar integral on AN is I_P for continuous function f with compact support in the open subgroup AN of P. Thus there is a $c > 0$ with

$$I_{AN}(f) = cI_P(f).$$

Hence

$$\begin{aligned}\int f(x)\,dm(x) &= \int_{G/K} \frac{1}{2\pi}\int_0^{2\pi} f(ank(\theta))\,d\theta\,dm_{G/K}(anK)\\ &= \frac{c}{2\pi}\int_{AN}\int_0^{2\pi} f(ank(\theta))\,d\theta\,dm_P(an)\\ &= \frac{c}{2\pi}\int_0^{2\pi}\int_{-\infty}^{\infty}\int_{-\infty}^{\infty} f(a(t)n(x)k(\theta))\,dx\,dt\,d\theta\,.\end{aligned}$$

Changing m to $\frac{m}{c}$, one then has a left Haar measure on G is given by

$$\int f(g)\,dm(g) = \frac{1}{2\pi}\int_0^{2\pi}\int_0^{\infty}\int_{-\infty}^{\infty} f(\left[\begin{smallmatrix} a & 0\\ 0 & a^{-1}\end{smallmatrix}\right]\left[\begin{smallmatrix} 1 & x\\ 0 & 1\end{smallmatrix}\right]\left[\begin{smallmatrix} \cos\theta & -\sin\theta\\ \sin\theta & \cos\theta\end{smallmatrix}\right])\frac{1}{a}\,dx\,da\,d\theta.$$

Since G is unimodular, $\check{m}$ is equal to m, one has by (g) of Proposition 6.3 and the change in variables $\theta \mapsto -\theta$, $x \mapsto -x$, and $t \mapsto \frac{1}{t}$ that

$$\begin{aligned}\int f(g)\,dm(g) &= \int \check{f}(g)\,dm(g)\\ &= \int f(g^{-1})\,dm(g)\\ &= \frac{1}{2\pi}\int_0^{2\pi}\int_{-\infty}^{\infty}\int_{-\infty}^{\infty} f((a(t)n(x)k(\theta))^{-1})\,dx\,dt\,d\theta\\ &= \frac{1}{2\pi}\int_0^{2\pi}\int_{-\infty}^{\infty}\int_{-\infty}^{\infty} f(k(-\theta)n(-x)a(-t))\,dx\,dt\,d\theta\\ &= \frac{1}{2\pi}\int_0^{2\pi}\int_{-\infty}^{\infty}\int_{-\infty}^{\infty} f(k(\theta)n(x)a(t))\,dx\,dt\,d\theta\\ &= \frac{1}{2\pi}\int_0^{2\pi}\int_{-\infty}^{\infty}\int_{-\infty}^{\infty} f(k(\theta)a(t)a(-t)n(x)a(t))\,dx\,dt\,d\theta\\ &= \frac{1}{2\pi}\int_0^{2\pi}\int_{-\infty}^{\infty}\int_{-\infty}^{\infty} f(k(\theta)a(t)n(e^{-t}x))\,dx\,dt\,d\theta\\ &= \frac{1}{2\pi}\int_0^{2\pi}\int_{-\infty}^{\infty}\int_{-\infty}^{\infty} e^t f(k(\theta)a(t)n(x))\,dx\,dt\,d\theta\,.\end{aligned}$$

Let π be the representation $\pi(\pm Ia(t)n(x)) = \epsilon_{\pm}(\pm I)e^{i\lambda t}$ of P. We use Proposition 6.138 to determine π^G. For a rho a rho function we need $\rho(xp) = \rho(x)\Delta_P(p)$ for $x \in G$ and $p \in P$. Take $\rho(k(\theta)a(t)n(x)) = \Delta_P(a(t)n(x))$ where $t \neq 0$ and $x \in \mathbb{R}$. Now $G/P = K/M$ under the mapping $k(\theta)a(t)n(x)G \mapsto k(\theta)M$. Using

Theorem 6.15, the measure μ corresponding to ρ on G/P is given by

$$\begin{aligned}
\int f_P(xP)\,d\mu(xP) &= \int \rho(x)f(x)\,dm(x) \\
&= \frac{1}{2\pi}\int_0^{2\pi}\int_{-\infty}^{\infty}\int_{-\infty}^{\infty} e^t f(k(\theta)a(t)n(x))\Delta_P(a(t)n(x))\,dx\,dt\,d\theta \\
&= \frac{1}{2\pi}\int_0^{2\pi}\int_{-\infty}^{\infty}\int_{-\infty}^{\infty} e^t f(k(\theta)a(t)n(x))e^{-2t}\,dx\,dt\,d\theta \\
&= \frac{1}{2\pi}\int_0^{2\pi}\int_{-\infty}^{\infty}\int_{-\infty}^{\infty} e^{-t} f(k(\theta)a(t)n(x))\,dx\,dt\,d\theta
\end{aligned}$$

where

$$\begin{aligned}
f_P(k(\theta)P) &= \int_P f(k(\theta)p)\,d\mu_P(p) \\
&= \int_{-\infty}^{\infty}\int_{-\infty}^{\infty} f\left(k(\theta)\left[\begin{smallmatrix} a & 0 \\ 0 & a^{-1}\end{smallmatrix}\right]n(x)\right)\frac{1}{a^2}\,da\,dx \\
&= \sum_{\pm}\int_{-\infty}^{\infty}\int_{a>0} f\left(\pm Ik(\theta)\left[\begin{smallmatrix} a & 0 \\ 0 & a^{-1}\end{smallmatrix}\right]n(x)\right)\frac{1}{a^2}\,da\,dx \\
&= \sum_{\pm}\int_{-\infty}^{\infty}\int_{-\infty}^{\infty} f\left(\pm Ik(\theta)\left[\begin{smallmatrix} e^t & 0 \\ 0 & e^{-t}\end{smallmatrix}\right]n(x)\right)\frac{1}{e^{2t}}\,e^t\,dt\,dx \\
&= \sum_{\pm}\int_{-\infty}^{\infty}\int_{-\infty}^{\infty} f\left(\pm Ik(\theta)a(t)n(x)\right)\frac{1}{e^t}\,dt\,dx
\end{aligned}$$

Now let $F \in C(K/M)$ and take $\phi(t) = \chi_{[0,1]}(t)$ and $\psi(x) = \chi_{[0,1]}(x)$. Define $f(k(\theta)a(t)n(x)) = F(k(\theta)M)\phi(t)\psi(x)$. The above formula implies

$$\begin{aligned}
f_P(k(\theta)P) &= \sum_{\pm}\int_0^1\int_0^1 F(\pm Ik(\theta)M)\frac{1}{e^t}\,dt\,dx \\
&= 2\int_0^1\int_0^1 F(k(\theta)M)\frac{1}{e^t}\,dt\,dx \\
&= 2(1-\frac{1}{e})F(k(\theta)M)
\end{aligned}$$

So

$$\begin{aligned}
2(1-\frac{1}{e})\int F(k(\theta)M)\,d\mu(k(\theta)M) &= \frac{1}{2\pi}\int_0^{2\pi}\int\int e^{-t}f(k(\theta)a(t)n(x))\,dx\,dt\,d\theta \\
&= \frac{1}{2\pi}\int_0^{2\pi}\int_0^1\int_0^1 e^{-t}F(k(\theta)M)\,dx\,dt\,d\theta \\
&= \frac{1}{2\pi}(1-\frac{1}{e})\int_0^{2\pi} F(k(\theta)M)\,d\theta \\
&= \frac{1}{\pi}(1-\frac{1}{e})\int_0^{\pi} F(k(\theta)M)\,d\theta.
\end{aligned}$$

This implies up to a scalar, the measure μ on K/M is given by

$$\int F(k(\theta)M)\,d\mu(k(\theta)M) = \frac{1}{\pi}\int_0^\pi F(k(\theta)M)\,d\theta.$$

Now take $\sigma(k(\theta)a(t)n(x)P) = k(\theta)$ for $\theta \in [0,\pi)$. σ is a well defined Borel mapping satisfying $\sigma(gP)P = gP$.

By Proposition 6.138, if $\mathcal{H} = L^2([0,\pi], \frac{d\theta}{\pi})$, π^G is unitarily equivalent to the representation L on $\mathcal{H}$ given by $L(k(\theta))$, $L(a(t))$, $L(n(x))$ where

$$L(g)f(\theta') = \sqrt{\frac{\rho(g^{-1}k(\theta'))}{\rho(k(\theta'))}}\pi(k(\theta')^{-1}g\sigma(g^{-1}k(\theta')P))f(\theta(g^{-1}k(\theta')))$$

and $\theta(g)$ is the unique angle in $[0,\pi)$ with $g = k(\theta(g))p$ for a unique p in P. We leave the details as to the values of $L(k(\theta))$, $L(a(t))$, and $L(n(x))$. These become fairly messy but can be worked out.

We could also use $L^2_\pi(G,\mathbb{C})$ which is the collection of all Borel functions f on G into $\mathbb{C}$ satisfying $f(gp) = \pi(p^{-1})f(g)$ and

$$\int_{G/P} |f(g)|^2\,d\mu(gP) = \frac{1}{\pi}\int_0^\pi |f(k(\theta)P|^2\,d\theta < \infty$$

and $\pi^G(x)f(g) = f(x^{-1}g)$.

□

Exercise Set 6.11

1. Define admissibility with respect to a right Haar measure and show if a vector is admissible with respect to a left Haar measure, then it is admissible with respect to every right Haar measure.

Solution:

Each right Haar measures is a measure of form dg^{-1} where dg is a left Haar measures. Also each right Haar measure on G/Z is a measure of form $d(g^{-1}Z)$ where $d(gZ)$ is a left Haar measures on G/Z. Now a vector v is defined to be admissible with respect to the right Haar measure if $gZ \mapsto |(v,\pi(g)v)|^2$ is integrable with respect to $d(g^{-1}Z)$. So it is admissible if and only if

$$\begin{aligned}\infty &> \int_{G/Z} |(v,\pi(g)v)|^2\,d(g^{-1}Z)\\ &= \int_{G/Z} |(v,\pi(g^{-1})v)|^2\,d(gZ)\\ &= \int_{G/Z} |(\pi(g)v,v)|^2\,d(gZ)\\ &= \int_{G/Z} |(v,\pi(g)v)|^2\,d(gZ).\end{aligned}$$

This implies one has v is admissible using a right Haar measure if and only if it is admissible using a left Haar measure.

□

2. Let π be an irreducible unitary representation of a locally compact Hausdorff group. Show a nonzero vector v is admissible if and only if there is a nonzero vector w such that

$$\int_{G/Z} |(w, \pi(g)v)_{\mathcal{H}}|^2 \, d(gZ) < \infty.$$

Solution: Again consider $\theta w'(g) = (w', \pi(g)v)_{\mathcal{H}}$. θ is a closed densely defined intertwining operator. Indeed, the set of w' for which this is $L^2_\chi(G)$ is a nonzero π invariant subspace. Now just follow the argument in Theorem 6.142 to show θ is bounded, etc. So $v \in D(\theta)$. □

3. Let G be a unimodular group and let π be an irreducible unitary representation of G. Then the linear subspace of admissible vectors is either $\{0\}$ or $\mathcal{H}$, the Hilbert space for π.
Solution: Let v be admissible where $v \neq 0$. Then $(w, \pi(g)v)_{\mathcal{H}}$ is in $L^2_\chi(G)$ for all $w \in \mathcal{H}$. So $(\pi(g^{-1})w, v)_{\mathcal{H}}$ is in $L^2_\chi(G)$. But by unimodularity, $(\pi(g)w, v)_{\mathcal{H}}$ is in $L^2_\chi(G)$. So $w \in \mathcal{H}_a$. □

4. Show if G is unimodular and the irreducible unitary representation π is square integrable, then there is a $d > 0$ such that

$$\frac{1}{d}(v_1, v_2)_{\mathcal{H}}(w_1, w_2)_{\mathcal{H}} = \int_{G/Z} (w_1, \pi(g)v_2)_{\mathcal{H}}(\pi(g)v_1, w_2)_{\mathcal{H}} \, d(gZ)$$

for all $v_1, v_2, w_1, w_2 \in \mathcal{H}$, the Hilbert space for π. The positive constant d is called the formal degree of the representation π.
Solution: By Proposition 6.144 the sesquilinear form B satisfies

$$\begin{aligned} B(v_1, v_2)(w_1, w_2)_{\mathcal{H}} &= \int_{G/Z} (w_1, \pi(g)v_2)_{\mathcal{H}}(\pi(g)v_1, w_2)_{\mathcal{H}} \, d(gZ) \\ &= \int_{G/Z} (v_1, \pi(g^{-1})w_2)_{\mathcal{H}}(\pi(g^{-1})w_1, v_2)_{\mathcal{H}} \, d(gZ) \\ &= \int_{G/Z} (v_1, \pi(g)w_2)_{\mathcal{H}}(\pi(g)w_1, v_2)_{\mathcal{H}} \, d(gZ) \\ &= B(w_1, w_2)(v_1, v_2)_{\mathcal{H}} \end{aligned}$$

Fix $w_1 = w_2 = w$. Then $B(v_1, v_2) = \frac{B(w,w)}{||w||^2}(v_1, v_2)_{\mathcal{H}}$. Take $d = \frac{||w||^2}{B(w,w)}$. So $\frac{1}{d}(v_1, v_2)_{\mathcal{H}}(w_1, w_2)_{\mathcal{H}} = \int_{G/Z}(w_1, \pi(g)v_2)_{\mathcal{H}}(\pi(g)v_1, w_2)_{\mathcal{H}} \, d(gZ)$. □

5. Let π be a finite dimensional unitary square integrable representation of a unimodular group G. Show G/Z is compact and if $m_{G/Z}(G/Z) = 1$, then the formal degree of π is its dimension. (**Hint:** Use $T = \theta^*\theta = \int_{G/Z}(\pi(g)v \otimes \overline{\pi(g)v}) \, d(gZ)$ to calculate its trace.)

Solution: From the proof of Theorem 6.142, we know $T = A^2 I$ where

$$A^2 = ||v||_{\mathcal{H}}^{-2} \int_{G/Z} |(v, \pi(g)v)_{\mathcal{H}}|^2 \, d(gZ).$$

Hence $\mathrm{Tr}(T) = A^2 n$. Now if $e_k, k = 1, 2, \ldots, n$ is a basis of $\mathcal{H}_\pi$, one has

$$\begin{aligned} nA^2 = \mathrm{Tr}(T) &= \sum_{k=1}^{n} (Te_k, e_k) \\ &= \sum_{k=1}^{n} \int (e_k, \pi(g)v)(\pi(g)v, e_k) \, d(gZ) \\ &= \int (\pi(g)v, \pi(g)v) \, d(gZ) \\ &= \int ||v||^2 \, d(gZ) \\ &= m_{G/Z}(G/Z) \, ||v||^2 < \infty \, . \end{aligned}$$

Hence $m_{G/Z}$ is finite and

$$n||v||^{-2} \int_{G/Z} |(v, \pi(g)v)|^2 \, d(gZ) = m_{G/Z}(G/Z)||v||^2 \, .$$

Thus

$$\int_{G/Z} (v, \pi(g)v)_{\mathcal{H}} (\pi(g)v, v)_{\mathcal{H}} \, d(gZ) = \frac{1}{d}(v, v)_{\mathcal{H}}^2 = \frac{m_{G/H}(G/Z)}{n}(v, v)_{\mathcal{H}}^2$$

and we see $d = \frac{n}{m_{G/Z}(G/Z)}$. Note the compactness of G/Z follows from Exercise 6.1.13 and $n = d$ when $m_{G/Z}(G/Z) = 1$. □

6. Show the modular function Δ for G is integrable if and only if G is compact.
Solution: If G is compact, Δ is integrable for it is continuous. Conversely, we start by recalling by Exercise 6.1.13, G is compact if and only if left Haar measure is finite.

We remark if $\check{\Delta}$ is integrable, then G is compact. To see this one has

$$m(G) = \int 1 \, dg = \int \Delta(g^{-1}) \, dg = \int \check{\Delta} \, dg < \infty$$

and thus G is compact.

Assume Δ is integrable. Thus $\alpha = \int \Delta(g) \, dg < \infty$. So $\alpha = \int \Delta(g^{-1}) d(g^{-1}) = \int \Delta(g^{-1})^2 \, dg < \infty$. Thus $|\Delta^{-1}|_2 = |\check{\Delta}|_2 = \sqrt{\alpha}$ and $|\Delta^{1/2}|_2 = \sqrt{\alpha}$. We use Hölder's inequality repeatedly. $\Delta^{-1}, \Delta^{1/2} \in L^2 \implies \Delta^{-1/2} = \Delta^{-1}\Delta^{1/2} \in L^1$ and $|\Delta^{-1/2}|_1 \leqslant |\Delta^{-1}|_2 |\Delta^{1/2}|_2 = \alpha$. So $\Delta^{-1/4} \in L^2$ and $|\Delta^{-1/4}|_2 \leqslant \sqrt{\alpha}$. We now argue inductively that $|\Delta^{(-1)^{n+1}/2^n}|_2 \leqslant \sqrt{\alpha}$ for all n. Assume true up to n. By Hölder's inequality,

$$\Delta^{(-1)^{n+2}/2^n} = \Delta^{(-1)^{n+1}/2^n} \Delta^{(-1)^n/2^{n-1}} \in L^1$$

and

$$|\Delta^{(-1)^{n+2}/2^n}|_1 \leqslant |\Delta^{(-1)^{n+1}/2^n}|_2 |\Delta^{(-1)^n/2^{n-1}}|_2 \leqslant \alpha\,.$$

So $\Delta^{(-1)^{n+2}/2^{n+1}} \in L^2$ and $|\Delta^{(-1)^{n+2}/2^{n+1}}|_2 \leqslant \sqrt{\alpha}$. Thus the inductive step is established. So $|\Delta^{(-1)^{n+1}/2^n}|_2 \leqslant \sqrt{\alpha}$ for all n and one has

$$|\Delta^{(-1)^{n+1}/2^n}|_2^2 = \int_G \Delta^{(-1)^{n+1}/2^{n-1}}(g)\,dg \leqslant \alpha$$

for all n.

Using Fatou's Lemma, one concludes

$$\begin{aligned} m(G) = \int_G 1\,dg &= \int_G \lim_{n\to\infty} \Delta^{(-1)^{n+1}/2^{n-1}}(g)\,dg \\ &\leqslant \liminf_{n\to\infty} \int_G \Delta^{(-1)^{n+1}/2^{n-1}}(g)\,dg \leqslant \alpha\,. \end{aligned}$$

Consequently, left Haar measure is finite and thus G is compact.

□

7. Let π be an irreducible square integrable unitary representation. Recall the formal degree operator is a closed invertible positive operator D on $\mathcal{H}_a$ onto $\mathcal{H}_a$ such that

$$(D^{-1}v_1, v_2)_{\mathcal{H}}(w_1, w_2)_{\mathcal{H}} = \int_{G/Z} (w_1, \pi(g)v_2)_{\mathcal{H}}(\pi(g)v_1, w_2)_{\mathcal{H}}\,d(gZ)$$

for all $v_1, v_2 \in \mathcal{H}_a$ and $w_1, w_2 \in \mathcal{H}$. Show $D\pi(g)v = \Delta(g)\pi(g)Dv$ for $v \in \mathcal{H}_a$.

Solution:

$$\begin{aligned} (D^{-1}\pi(x)v_1, v_2)_{\mathcal{H}}(w_1, w_2)_{\mathcal{H}} &= \int_{G/Z} (w_1, \pi(g)v_2)_{\mathcal{H}}(\pi(g)\pi(x)v_1, w_2)_{\mathcal{H}}\,d(gZ) \\ &= \int_{G/Z} (w_1, \pi(g)v_2)_{\mathcal{H}}(\pi(gx)v_1, w_2)_{\mathcal{H}}\,d(gZ) \\ &= \int_{G/Z} (w_1, \pi(gx^{-1})v_2)_{\mathcal{H}}(\pi(g)v_1, w_2)_{\mathcal{H}}\,d(gZx^{-1}Z) \\ &= \Delta_{G/Z}(x^{-1}) \int_{G/Z} (w_1, \pi(g)\pi(x^{-1})v_2)_{\mathcal{H}}(\pi(g)v_1, w_2)_{\mathcal{H}}\,d(gZ) \\ &= \Delta(x^{-1})(D^{-1}v_1, \pi(x^{-1})v_2)_{\mathcal{H}}(w_1, w_2)_{\mathcal{H}} \\ &= \Delta(x^{-1})(\pi(x)D^{-1}v_1, v_2)_{\mathcal{H}}(w_1, w_2)_{\mathcal{H}}. \end{aligned}$$

for $\Delta(g) = \Delta_G(g) = \Delta_{G/Z}(gZ)$. This implies

$$D^{-1}\pi(x) = \Delta(x^{-1})\pi(x)D^{-1}.$$

Taking inverses gives

$$\Delta(x)D\pi(x^{-1}) = \pi(x^{-1})D$$

or

$$D\pi(x) = \Delta(x)\pi(x)D.$$

□

8. Let $f \in L^1_\chi(G)$ and π be a unitary representation of G with central character χ. Show $\pi_\chi(f)w$ exists weakly for all $w \in \mathcal{H}_\pi$, $\pi_\chi(f)$ is a linear operator, and

$$||\pi_\chi(f)|| \leqslant |f|_1 = \int |f(g)| \, d(gZ).$$

Solution:

Let $f \in L^1_\chi(G)$. Thus $f(gz) = \chi(z)^{-1}f(g)$ for all z a.e. g. For $v \in \mathcal{H}_\pi$, $\pi_\chi(f)v$ is defined weakly by

$$(\pi_\chi(f)v, w) = \int_{G/Z} f(g)(\pi(g)v, w) \, d(gZ)\,.$$

Note $f(g)(\pi(g)v, w)$ is constant on cosets for

$$f(gz)(\pi(gz)v, w) = \chi(z)^{-1}f(g)(\chi(z)\pi(g)v, w) = f(g)(\pi(g)v, w)$$

for all z a.e. g. Moreover, $|f(g)(\pi(g)v, w)| \leqslant |f(g)| \, ||v|| \, ||w||$ implies the function $gZ \mapsto f(g)(\pi(g)v, w)$ is integrable on G/Z for $f \in L^1_\chi(G)$ and

$$|(\pi_\chi(f)v, w)| \leqslant |f|_{L^1_\chi(G)} ||v|| \, ||w||\,.$$

So $\pi_\chi(f)v$ exists weakly and $||\pi_\chi(f)v|| \leqslant |f|_{L^1_\chi(G)} ||v||$. It is easy to check $v \mapsto \pi_\chi(f)v$ is linear.

□

9. Let π be an irreducible unitary representation of G. Suppose there is a nonzero vector v such that $g \mapsto (v, \pi(g)v)_{\mathcal{H}}$ is in $L^2(G)$. Show the center Z of G is compact. Now suppose Z is compact. Show

(a) If v is an admissible vector for π, then $g \mapsto (v, \pi(g)v)_{\mathcal{H}}$ is in $L^2(G)$.

(b) Let χ be a one-dimensional character of Z. Show if $d\xi$ is a Haar measure on Z of total mass 1 and Haar measure on G/Z is given as in (6.45), then the inclusion mapping $L^2_\chi(G) \to L^2(G)$ is an isometry.

(c) Show if $\chi_1 \neq \chi_2$, then $L^2_{\chi_1}(G)$ and $L^2_{\chi_2}(G)$ are orthogonal in $L^2(G)$.

(d) Show if v is a wavelet vector for π, then the properties for W_v given in Theorem 6.150 hold for $L^2(G)$; thus all integrals are over G instead of G/Z and use dg instead of $d(gZ)$.

Solution: By Equation 6.45, we have

$$\begin{aligned}\int_{G/Z} |(v, \pi(g)v)_{\mathcal{H}}|^2 \, dg &= \int_{G/Z} \int |(v, \pi(g\xi)v)_{\mathcal{H}}|^2 \, d\xi \, d(gZ) \\ &= \int_{G/Z} |(v, \pi(g)v)|^2_{\mathcal{H}} \, d(gZ).\end{aligned}$$

This shows both (a) and (b). It also shows $g \mapsto (v, \pi(g)v)$ is in $L^2(G)$ if and only if Z is compact. Indeed, if Z is not compact, then $\int |(v, \pi(g\xi)v)_{\mathcal{H}}|^2 \, d\xi = \infty$ if $(v, \pi(g)v)_{\mathcal{H}} \neq 0$. Now the set of g with $(v, \pi(g)v)_{\mathcal{H}} \neq 0$ is open and thus has positive measure. So the set of gZ where $|(v, \pi(g)v)_{\mathcal{H}}| \neq 0$ has positive measure in G/Z. Thus if Z is noncompact, $\int |(v, \pi(g)v)_{\mathcal{H}}|^2 \, dg = \infty$. For (c), note $\int_G f_1(g)\overline{f_2(g)} \, dg = \int_G f_1(g\xi)\overline{f_2(g\xi)} \, d\xi = \chi_1(\xi^{-1})\chi_2(\xi) \int f_1(g)\overline{f_2(g)} \, dg$. Choose ξ with $\chi_1(\xi) \neq \chi_2(\xi)$. Thus

$$(1 - \chi_1(\xi^{-1})\chi_2(\xi)) \int f_1(g)\overline{f_2(g)} \, dg = 0 \, .$$

So $f_1 \perp f_2$. (d) is immediate because of $\int_{G/Z} f(g) \, d(gZ) = \int f(g) \, dg$ if $f(g\xi) = f(g)$ and f is integrable. □

10. Suppose π is a unitary representation of a group G on a Hilbert space $\mathcal{H}$ and there is a one-dimensional character χ of the center Z of G such that $\pi(\xi) = \chi(\xi)I$ for $\xi \in Z$. Assume the set $\mathcal{H}_a$ of vectors v in $\mathcal{H}$ such that $g \mapsto (w, \pi(g)v)$ is in $L^2_\chi(G)$ for all $w \in \mathcal{H}$ is a dense linear subspace of $\mathcal{H}$. Show if there is a sesquilinear form B on $\mathcal{H}_a$ with

$$\int (w_1, \pi(g)v_1)_{\mathcal{H}} (\pi(g)v_2, w_2)_{\mathcal{H}} \, d(gZ) = B(v_2, v_1)(w_1, w_2)_{\mathcal{H}}$$

for $v_1, v_2 \in \mathcal{H}_a$ and $w_1, w_2 \in \mathcal{H}$, then π is irreducible.

Solution:

It suffices to show if $A \in \mathrm{Hom}(\pi, \pi)$, then $A = cI$ for some constant c. We first note $A\mathcal{H}_a \subseteq \mathcal{H}_a$. In fact, let $w \in \mathcal{H}$. Note $g \mapsto (A^*w, \pi(g)v)$ is in $L^2_\chi(G)$ for each $w \in \mathcal{H}$. But $(A^*w, \pi(g)v) = (w, A\pi(g)v) = (w, \pi(g)Av)$. Thus $g \mapsto (w, \pi(g)Av)$ is in $L^2_\chi(G)$ for all w. This gives $Av \in \mathcal{H}_a$. Since $\mathrm{Hom}(\pi, \pi)$ is a $*$ algebra, one has $A^*v \in \mathcal{H}_a$ for $v \in \mathcal{H}_a$. So

$$\begin{aligned} B(v_2, v_1)(Aw_1, w_2) &= \int (Aw_1, \pi(g)v_1)(\pi(g)v_2, w_2) \, d(gZ) \\ &= \int (w_1, \pi(g)A^*v_1)(\pi(g)v_2, w_2) \, d(gZ) \\ &= B(v_2, A^*v_1)(w_1, w_2) \\ &= (B(v_2, A^*v_1)w_1, w_2) \end{aligned}$$

for all w_1 and w_2 for v_1 and v_2 in $\mathcal{H}_a$. This implies $B(v, v)Aw_1 = B(v, A^*v)w_1$ for all w_1. Note if we choose $v \neq 0$ in $\mathcal{H}_a$, then

$$B(v, v)(v, v) = \int_{G/Z} |(v, \pi(g)v)|^2 \, d(gZ) > 0 \, .$$

So $B(v, v) > 0$. Hence $Aw = cw$ where $c = \frac{B(v, A^*v)}{B(v,v)}$. Now Schur's Lemma 6.48 implies π is irreducible.

□

11. Let G be the group $H \ltimes \mathbb{R}^n$ with multiplication defined in (6.54). Show if dh is a left Haar measure for H and dx is Lebesgue measure, then $d(h \times x) = |\det h^{-1}|\,(dh \times dx)$ is a left Haar measure for G. Then show if Δ is the modular function for H, then the modular function Δ_G for G is given by $\Delta_G(a, y) = \frac{\Delta(a)}{|\det a|}$.

Solution: We have

$$\begin{aligned}
\int f((a,y)(h,x))\, d(h \times x) &= \int\int f(ah, y + ax)|\det h|^{-1}\, dh\, dx \\
&= \int\int f(h, y + ax)|\det(a^{-1}h)|^{-1}\, dh\, dx \\
&= \int\int f(h, y + x)|\det(a^{-1}h)|^{-1}\, d(a^{-1}x)\, dh \\
&= \int\int f(h, y + x)|\det(a^{-1}h)|^{-1}\, |\det(a^{-1})|\, dx\, dh \\
&= \int\int f(h, x)|\,|\det h|^{-1}\, dh\, dx
\end{aligned}$$

and

$$\begin{aligned}
\int f((h,x)(a,y))\, d(h \times x) &= \int\int f(ha, x + hy)|\det h|^{-1}\, dh\, dx \\
&= \int\int f(ha, x)|\det(h)|^{-1}\, dx\, dh \\
&= \int\int f(h, x)|\det(ha^{-1})|^{-1}\, d(ha^{-1})\, dx \\
&= |\det(a)|\Delta(a^{-1}) \int\int f(h, x)|\det(h)|^{-1}\, dx\, dh \\
&= \frac{|\det a|}{\Delta(a)} \int\int f(h, x)|\,|\det h|^{-1}\, dh\, dx.
\end{aligned}$$

□

12. Let G be the group $H \ltimes \mathbb{R}^n$ with multiplication defined by (6.54). Show the *right* quasi-regular representation $\mathrm{ind}_H^G 1$ is unitarily equivalent to the representation ρ defined on $L^2(\mathbb{R}^n)$ by

$$\rho(h, x)f(y) = |\det h|^{-1/2} f(h^{-1}(y + x)).$$

(**Hint:** Follow the argument in Example 6.95).

Solution:

We first note $H\backslash G$ is homeomorphic to $\mathbb{R}^n$ under the mapping

$$H(h, x) = H(h, 0)(1, h^{-1}x) \mapsto h^{-1}x.$$

Indeed, if $h^{-1}x = h_1^{-1}x_1$, then $h_1h^{-1}x = x_1$ and so $(h_1h^{-1}, 0)(h, x) = (h_1, x_1)$. Hence $H(h, x) = H(h_1, x_1)$. Thus the mapping is one-to-one and onto. It is continuous since $(h,x) \mapsto h^{-1}x$ is continuous. Its inverse is the mapping $x \mapsto H(1,x)$ which is also continuous.

The right action of G on $H\backslash G$ is given by $H(1,x)\cdot(h,y) = H(h, x+y) = H(1, h^{-1}(x+y))$ and thus on $\mathbb{R}^n$ is given by

$$x\cdot(h,y) = h^{-1}(x+y)\,.$$

We by the previous exercise know the modular functions for left Haar measure on G is given by $\Delta_G(h,y) = \frac{\Delta(h)}{|\det h|}$ where Δ is the modular function for the left Haar measures on H. Define ϕ by $\phi(h,y) = |\det h^{-1}|$. Note

$$\begin{aligned}
\phi((h,0)(h',y)) &= \phi(hh', hy)\\
&= \frac{1}{|\det hh'|}\\
&= \frac{1}{|\det h'|}\frac{1}{|\det h|}\\
&= \phi(h',y)\frac{\Delta(h)}{|\det h|}\frac{1}{\Delta(h)}\\
&= \phi(h',y)\frac{\Delta_G(h,0)}{\Delta(h)}
\end{aligned}$$

and we see ϕ is a right rho function. Let ν be the corresponding right quasi-invariant measure. We calculate the integral relative to ν. Take $f \in C_c(\mathbb{R}^n)$ and $g \in C_c(H)$ satisfying $\int g(h^{-1})\,dh = 1 = \int g(h)\,d_rh$ where d_rh is right Haar measure on H. Set $F(h,x) = g(h)f(h^{-1}x)$. Then

$$\begin{aligned}
F^H(1,y) &= \int F((h,0)(1,y))\,d_rh\\
&= \int F(h,hy)\,d_rh\\
&= \int F(h^{-1},h^{-1}y)\,dh\\
&= \int g(h^{-1})f(y)\,dh\\
&= f(y)\,.
\end{aligned}$$

Identifying f on $\mathbb{R}^n$ with the function f' on $H\backslash G$ given by

$$f'(H(1,y)) = f(y),$$

we have

$$
\begin{aligned}
\int f(y)\,d\nu(y) &= \int_{H\backslash G} f'(H(1,y))\,d\nu(H(1,y)) \\
&= \int F(h,x)\phi(h,x)\,d_r(h\times x) \\
&= \int F((h,x)^{-1})\phi((h,x)^{-1})\,d(h\times x) \\
&= \int F(h^{-1},-h^{-1}x)\phi(h^{-1},-h^{-1}x)\,d(h\times x) \\
&= \int g(h^{-1})f(-x)|\det h|\,d(h\times x) \\
&= \int\int g(h^{-1})f(-x)|\det h|\,|\det h^{-1}|\,dx\,dh \\
&= \int g(h^{-1})\,dh \int f'(H(0,x))\,dx \\
&= \int f(y)\,dy
\end{aligned}
$$

and so ν is Lebesgue measure on $\mathbb{R}^n$. Now again using the identification $f \mapsto f'$ where $f(y) = f'(H(1,y))$ and Equation 6.18 we have by the identification of $H\backslash G$ with $\mathbb{R}^n$ that

$$
\begin{aligned}
\rho(h,x)f(y) &= \rho(h,x)f'(H(1,y)) \\
&= \left(\frac{\phi((1,y)(h,x))}{\phi(1,y)}\right)^{1/2} f'(H(1,y)(h,x)) \\
&= \left(\frac{\phi(h,y+x)}{\phi(1,y)}\right)^{1/2} f'(H(h,x+y)) \\
&= |\det h|^{-1/2} f'(H(h^{-1},0)(h,x+y)) \\
&= |\det h|^{-1/2} f'(H(1,h^{-1}(x+y)) \\
&= |\det h|^{-1/2} f(h^{-1}(x+y))\,.
\end{aligned}
$$

□

13. Let $\hat{\rho}$ be the representation given in (6.57). Follow the argument in Example 6.139 to show:

(a) Every closed $\hat{\rho}$ invariant nonzero subspace of $L^2(\mathbb{R}^n)$ has form

$$L^2_W(\mathbb{R}^n) = \{f \in L^2(\mathbb{R}^n) \mid f = 0 \text{ off } W\}$$

where W is a Borel subset of $\mathbb{R}^n$ of positive Lebesgue measure satisfying $h^t W = W$ for all $h \in H$.

(b) Let $\hat{\rho}_W$ be $\hat{\rho}$ restricted to $L^2_W(\mathbb{R}^n)$. Show $\hat{\rho}_W$ is irreducible if and only if every Borel subset E of W satisfying $(h^t E \cap E^c) \cup (E \cap (h^t E)^c)$ has

Lebesgue measure 0 for all $h \in H$ has Lebesgue measure 0 or has complement in W with Lebesgue measure 0. (This says the action of H on W is ergodic.)

Solution: By (6.57), $\hat{\rho}(h,x)f(y) = |\det h|^{1/2}e^{2\pi i x\cdot y}f(h^t y)$ for $f \in L^2(\mathbb{R}^n)$. Suppose P is an orthogonal projection onto a closed invariant subspace of $L^2(\mathbb{R}^n)$ for $\hat{\rho}$. Then $P \in \mathrm{Hom}(\hat{\rho},\hat{\rho})$ and we have $P\hat{\rho}(h,x) = \hat{\rho}(h,x)P$ for all pairs (h,x). Thus if $\sigma(x) = \hat{\rho}(-x,0)$, then σ is a representation of $\mathbb{R}^n$ given by $\sigma(x)f(y) = e^{-2\pi i x\cdot y}f(y)$ on $L^2(\mathbb{R}^n,\mu)$ where $\mu = \lambda_n$. Note $P \in \mathrm{Hom}(\sigma,\sigma)$. By Exercise 6.8.10, $P = M_\phi$ where $\phi \in L^\infty(\mathbb{R}^n)$. Since $P^2 = P$, $\phi^2(y) = \phi(y)$ a.e. y. By redefining ϕ to be 1 on the set where $\phi(y)^2 \neq \phi(y)$, we may assume $\phi^2(y) = \phi(y)$ for all y. Hence $\phi = \chi_W$ where $W = \phi^{-1}(1)$. This implies $PL^2(\mathbb{R}^n) = L^2_W(\mathbb{R}^n)$. Now since $L^2_W(\mathbb{R}^n)$ is invariant under $\hat{\rho}$ if and only if it is invariant under all $\hat{\rho}(h,0)$ and all $\hat{\rho}(1,-x)$, we see $L^2_W(\mathbb{R}^n)$ is invariant if and only if it is invariant under $\hat{\rho}(h,0)$ for all h. Using $\phi = \chi_W$, this occurs if and only if $\hat{\rho}(h,0)M_\phi = M_\phi\hat{\rho}(h,0)$ for $h \in H$. Thus

$$|\det h|^{1/2}\chi_W(h^t y)f(h^t y) = |\det h|^{1/2}\chi_W(y)f(h^t y)$$

a.e. y for each $h \in H$ and $f \in L^2$. This implies the closed invariant subspaces of $\hat{\rho}$ are the subspaces $L^2_W(\mathbb{R}^n)$ where $\chi_W(h^t y) = \chi_W(y)$ a.e. y for each $h \in H$. This gives (a).

For (b), let $L^2_W(\mathbb{R}^n)$ be a nonzero invariant subspace for $\hat{\rho}$. Then W has positive Lebesgue measure and $h^t W = W$ in the measure algebra for λ_n for all $h \in H$. Set $\hat{\rho}_W$ to be $\hat{\rho}$ restricted to $L^2_W(\mathbb{R}^n)$. The same argument for (a) shows a subspace of $L^2_W(\mathbb{R}^n)$ is invariant under $\hat{\rho}_W$ if and only if it is an $L^2_E(\mathbb{R}^n)$ where $E \subseteq W$ and $h^t E = E$ for all h. Thus $\hat{\rho}_W$ is irreducible if and only if for all Borel subsets E of W, $h^t E = E$ in the measure algebra of λ_n for all h implies E or $W - E$ has Lebesgue measure 0 and thus the action of H on W is ergodic.

□

14. Let H be the product group $\mathbb{R}^+ \times \mathrm{SO}(n)$ with product topology and $G = H \ltimes \mathbb{R}^n$ where H acts on $\mathbb{R}^n$ by $(a,A)x = aAx$. Show the representation π defined on $L^2(\mathbb{R}^n)$ by

$$\pi(a,A,b)f(x) = a^{-n/2}f(a^{-1}A^{-1}(x-b))$$

is square-integrable and a nonzero function ψ is a wavelet for π if and only if $\int_{\mathbb{R}^n} |\hat{\psi}(x)|^2 \frac{dx}{|x|^n} < \infty$. Let

$$W_\psi(f)(a,A,x) = a^{-n/2}\int f(y)\overline{\psi(a^{-1}A^{-1}(y-x))}\,dy$$

be the wavelet transform. Show if ϕ and ψ are wavelet functions and dA is a Haar measure on $\mathrm{SO}(n)$ satisfying $\int_{\mathrm{SO}(n)} 1\,dA = \sigma(S^{n-1})$, the surface area of S^{n-1}, then

$$f(y) = \frac{1}{B(\phi,\psi)}\int\int\int W_\psi(f)(a,A,x)a^{-n/2}\phi(a^{-1}A^{-1}(y-x))\,\frac{da}{a^{n+1}}\,dA\,dx$$

weakly provided

$$B(\phi,\psi)=\int_{\mathbb{R}^n}\hat{\phi}(x)\overline{\hat{\psi}(x)}\,\frac{dx}{|x|^n}\neq 0.$$

Solution: We apply Corollary 6.158 and the discussion in Remark 6.159. The action $(a,A)x = aAx$ satisfies $(a,A)x = x$ for all x implies $a = 1$ and $A = I$ and $(a,A)x = x$ for all (a,A) if and only if $x = 0$. Also if $y = (1,0,\ldots,0)$, one has $S = H^t y = \mathbb{R}^n - \{0\}$. The representation ρ of G on $L^2(\mathbb{R}^n) = \mathcal{F}^{-1}L^2_S(\mathbb{R}^n)$ in formula (6.56) is square integrable since the stabilizer H_y is the compact group $\{1\}\times \mathrm{SO}(n-1)$. Note a Haar measure on H is $\frac{da}{a}\times dA$ where da is Lebesgue measure and dA is Haar measure on $\mathrm{SO}(n)$. We normalize dA so the measure of the whole group is $\sigma(S^{n-1})$ where σ is Lebesgue surface measure (see Proposition 2.25). In this case we note

$$|\det h|\,d\mu_y(hH_y) = dx$$

under the identification of H/H_y to $\mathbb{R}^n$ by the mapping $(a,A)H_y \mapsto x = aAy$. To see this, one has $H = \mathbb{R}^+\times\mathrm{SO}(n)$ and $H_y = \{1\}\times\mathrm{SO}(n-1)$. Since Lebesgue surface measure is $\mathrm{SO}(n)$ invariant, by uniqueness of invariant measures and the choice of $\sigma(S^{n-1})$ as the total mass of dA , we see the measure $d\mu_y$ on H/H_y is $\frac{da}{a}\times d\sigma$. Hence $dx = a^{n-1}\,da\,d\sigma$ which is just Lebesgue measure in polar form; see the corollary to Proposition 2.25.

Now by Corollary 6.158, the sesquilinear form B is given by

$$\begin{aligned}
B(\phi,\psi) &= \int_H \hat{\phi}(h^t y)\overline{\hat{\psi}(h^t y)}\,dh \\
&= \int_H \hat{\phi}(hy)\overline{\hat{\psi}(hy)}\,dh \\
&= \int_{H/H_y} \hat{\phi}(hy)\overline{\hat{\psi}(hy)}\,d\mu_y(hH_y) \\
&= \int_{H/H_y} \frac{1}{|\det h|}\hat{\phi}(hy)\overline{\hat{\psi}(hy)}\,|\det h|\,d\mu_y(hH_y) \\
&= \int_{\mathbb{R}^n} \frac{1}{|x|^n}\hat{\phi}(x)\overline{\hat{\psi}(x)}\,dx.
\end{aligned}$$

Thus ϕ is a wavelet vector if and only if $\int_{\mathbb{R}^n}|\hat{\phi}(x)|^2\,\frac{dx}{|x|^n}<\infty$ and when $B(\phi,\psi)\neq$

0, the inversion formula (6.60) gives

$$\begin{aligned}
f(y) &= \frac{1}{B(\phi,\psi)} \iint\limits_{H\times\mathbb{R}^n} |\det h|^{-1} W_\psi(f)(h,x)\pi\circ\alpha(h,x)\phi(y)\,(dh\times dx) \\
&= \frac{1}{B(\phi,\psi)} \iint\limits_{H\times\mathbb{R}^n} |\det h|^{-1} W_\psi(f)(h,x)|\det h|^{-1/2}\phi(h^{-1}(y-x))\,(dh\times dx) \\
&= \frac{1}{B(\phi,\psi)} \iint\limits_{H\times\mathbb{R}^n} |\det h|^{-3/2} W_\psi(f)(h,x)\phi(h^{-1}(y-x))\,(dh\times dx) \\
&= \frac{1}{B(\phi,\psi)} \int_0^\infty \int_{\mathrm{SO}(n)} \int_{\mathbb{R}^n} a^{-3n/2} W_\psi(f)(a,A,x)\phi(a^{-1}A^{-1}(y-x))\,\frac{da}{a}\,dA\,dx
\end{aligned}$$

where

$$\begin{aligned}
W_\psi(f)(a,A,x) &= (f,\pi\circ\alpha(a,A,x)\psi)_2 \\
&= a^{-n/2}\int_{\mathbb{R}^n} f(y)\overline{\psi(a^{-1}A^{-1}(y-x))}\,dy
\end{aligned}$$

and left Haar measure dA on $\mathrm{SO}(n)$ is chosen with total mass the surface area of S^{n-1}.

□

15. Let $S = \{(x,y)\in\mathbb{R}^2 \mid x>0 \text{ and } x^2-y^2>0\}$. Define a nonzero function ψ in $\mathcal{F}^{-1}L^2_S(\mathbb{R}^2)$ to be a wavelet function if $B(\psi,\psi) = \iint_S \frac{|\hat{\psi}(x,y)|^2}{x^2-y^2}\,dx\,dy < \infty$. For $t\in\mathbb{R}$, let $A(t)$ be the linear transformation of $\mathbb{R}^2$ given by $A(t)(x_1,x_2) = (x_1,x_2)\begin{pmatrix}\cosh t & \sinh t\\ \sinh t & \cosh t\end{pmatrix} = (\cosh t\,x_1+\sinh t\,x_2, \sinh t\,x_1+\cosh t\,x_2)$. Define the wavelet transform $W_\psi f$ for $f\in\mathcal{F}^{-1}L^2_S(\mathbb{R}^2)$ by

$$W_\psi f(a,t,y) = a\iint\limits_S \hat{f}(\omega)\overline{\hat{\psi}(aA(t)\omega)}e^{2\pi i y\cdot\omega}\,d\omega$$

where $a>0$, $t\in\mathbb{R}$, and $y\in\mathbb{R}^2$. Show one has the weak formula

$$f(x) = \frac{1}{B(\psi,\psi)}\int_{\mathbb{R}^2}\int_{\mathbb{R}}\int_{\mathbb{R}+}\frac{1}{a^4}W_\psi f(a,t,b)\psi(a^{-1}A(-t)(x-b))da\,dt\,db.$$

(**Hint:** Use the Jacobian to show the appropriate measures are equal.)
Solution:

We argue using the Remark 6.159 and Corollary 6.158 with the definitions of ρ and π changed using the automorphism α. That is ρ now is the the rho in formula (6.56) composed with α and π is ρ restricted to an irreducible subspace of form $\mathcal{F}^{-1}L^2_S(\mathbb{R}^n)$.

More specifically, let S be the open subset of $\mathbb{R}^2$ consisting of (x,y) with $x>0$ and $x^2-y^2>0$. We have a transformation group $H=\{(a,t)\mid a>0, t\in\mathbb{R}\}$ acting on S by $(a,t)(x,y) = (a,0)(A(t)(x,y))$ where $(a,0)(x,y) = (ax,ay)$. This action is transitive. Indeed, if $x>0$, then if $a = \frac{1}{\sqrt{x^2-y^2}}$, $a(x,y)\in\{(u,v)\mid$

$u > 0$ and $u^2 - v^2 = 1\}$. Thus $v = \sinh t$ for a unique t and then $u = \cosh t$. Therefore $a^{-1}A(t)(1,0) = a^{-1}(\cosh t, \sinh t) = a^{-1}(u,v) = (x,y)$. Now the representation ρ of $G = H \ltimes \mathbb{R}^2$ on $L^2(\mathbb{R}^2)$ is given by

$$\rho((a,t),b)f(y) = |\det(a,t)|^{-1/2} f((a,t)^{-1} \cdot (y-b)) = \frac{1}{a} f(\frac{1}{a} A(-t)(y-b)).$$

Its Fourier transform is the representation on $L^2(\mathbb{R}^2)$ given by

$$\hat{\rho}((a,t),b)\hat{f}(\omega) = ae^{-2\pi i b \cdot \omega} \hat{f}(aA(t)\omega).$$

Now by Corollary 6.158 since S is the orbit of $(1,0)$ and has positive Lebesgue measure, the irreducible representation π obtained by restricting ρ to $\mathcal{F}^{-1}L^2_S(\mathbb{R}^2)$ is square integrable since the stabilizer of $(1,0)$ under the action of H is the compact subgroup $\{(1,0)\}$. Left Haar measure on H is given by $dh = d(a,t) = \frac{da}{a} \times dt$ and the modular function is $\Delta(a,t) = 1$.

We now consider $\int_H f(h(1,0))\, dh = \int_0^\infty \int_{-\infty}^\infty f((a,0)(0,t) \cdot (1,0)) \frac{1}{a}\, dt\, da$. This integral is

$$\int_0^\infty \int_{-\infty}^\infty f(a\cosh t, a\sinh t)\, dt \frac{da}{a}.$$

$T(a,t) = (a,t) \cdot (1,0) = (a\cosh t, a\sinh t)$ is an invertible diffeomorphism of $\mathbb{R}^+ \times \mathbb{R}$ onto S. By Theorem 2.22,

$$\int_{\mathbb{R}^+ \times \mathbb{R}} f(a\cosh t, a\sinh t) \left|\det \left(\begin{smallmatrix} \cosh t & a\sinh t \\ \sinh t & a\cosh t \end{smallmatrix} \right)\right| da \times dt = \int_S f(x)\, d_2x\,.$$

We thus have

$$\int_{\mathbb{R}^+ \times \mathbb{R}} f((a,t) \cdot (1,0)) a\, da \times dt = \int_S f(x,y)\, dx \times dy.$$

Taking $f(x,y) = F(x,y)/(x^2 - y^2)$, we have

$$\int_H F(h \cdot (1,0))\, dh = \int_{\mathbb{R}^+ \times \mathbb{R}} F(T(a,t)) \frac{1}{a}\, da \times dt = \int_S \frac{F(x,y)}{x^2 - y^2}\, dx \times dy.$$

So again using the corollary, ϕ is a wavelet vector for π if and only if

$$\iint\limits_S \frac{|\hat{\phi}(x,y)|^2}{x^2 - y^2}\, dx\, dy < \infty.$$

We remark if we take $a(x,y)^{-1} = (\sqrt{x^2 - y^2}, \sinh^{-1}(\frac{y}{\sqrt{x^2-y^2}})) \in H$, then $(a(x,y)^{-1})^t \cdot (1,0) = (x,y)$ for $(x,y) \in S$ and thus

$$\frac{|\det(a(x,y))|}{\Delta(a(x,y))} = |\det(a(x,y))| = \frac{1}{x^2 - y^2}$$

and this concurs with the formula $\int_S |\hat{\phi}(x)|^2 \frac{|\det(a(x))|}{\Delta(a(x))}\, dx < \infty$ of Corollary 6.158. The sesquilinear form on pairs of admissible functions is given by

$$\begin{aligned} B(\phi, \psi) &= \int_H \hat{\phi}(h^t(1,0))\overline{\hat{\psi}(h^t(1,0))}\, dh \\ &= \int_H \hat{\phi}(h \cdot (1,0))\overline{\hat{\psi}(h \cdot (1,0))}\, dh \\ &= \int_{-\infty}^{\infty} \int_0^{\infty} \frac{\hat{\phi}(x,y)\overline{\hat{\psi}(x,y)}}{x^2 - y^2}\, dx\, dy\,. \end{aligned}$$

Furthermore, the weak inversion formula for $\hat{f} \in L_S^2(\mathbb{R}^2)$ using the wavelet function ψ is

$$\begin{aligned} f(x) &= \frac{1}{B(\psi,\psi)} \iint\limits_{H\times\mathbb{R}^2} |\det(a,t)|^{-1} W_\psi(f)((a,t),b)\pi((a,t),b)\psi(x)\, d(a,t) \times db \\ &= \frac{1}{B(\psi,\psi)} \int_{\mathbb{R}^2} \left(\int_{-\infty}^{\infty} \int_0^{\infty} \frac{1}{a^2} W_\psi(f)((a,t),b) \frac{1}{a} \psi(a^{-1}A(-t)(x-b)) \frac{1}{a}\, da\, dt \right) db \\ &= \frac{1}{B(\psi,\psi)} \int_{\mathbb{R}^2} \left(\int_{-\infty}^{\infty} \int_0^{\infty} \frac{1}{a^4} W_\psi(f)((a,t),b) \psi(a^{-1}A(-t)(x-b))\, da\, dt \right) db \end{aligned}$$

where

$$\begin{aligned} W_\psi(f)(((a,t),b) &= (f, \pi((a,t),b)\psi) \\ &= (\hat{f}, \hat{\pi}((a,t),b)\hat{\psi})_2 \\ &= \iint\limits_S \hat{f}(\omega)\overline{\hat{\rho}((a,t),b)\hat{\psi}(\omega)}\, d\omega \\ &= \iint\limits_S \hat{f}(\omega)\overline{a e^{-2\pi i b\cdot\omega} \hat{\psi}(aA(t)\omega)}\, d\omega \\ &= a \iint\limits_S \hat{f}(\omega)\overline{\hat{\psi}(aA(t)\omega)} e^{2\pi i b\cdot\omega}\, d\omega\,. \end{aligned}$$

□

16. Let G be the group $H \ltimes \mathbb{R}^n$ with multiplication defined in (6.54). Assume $y \in \mathbb{R}^n$ and H_y is the closed subgroup stabilizing y under the action $h \cdot x = (h^{-1})^t x$. It is known that the orbit $H \cdot y$ is a Borel subset of $\mathbb{R}^n$ and the mapping ϕ defined by $hH_y \mapsto h \cdot y$ is a Borel isomorphism. That is the Borel subsets of H/H_y map under this mapping in a one-to-one manner onto the Borel subsets of $\mathbb{R}^n$ contained in the orbit $H \cdot y$. Let ρ be a rho function for H_y and let μ_y be the corresponding quasi-invariant measure on H/H_y. Let μ be the measure on the Borel subsets of $H \cdot y$ that satisfies

$$\int f(x)\, d\mu(x) = \int_{H/H_y} f(h \cdot y)\, d\mu_y(hH_y)$$

for positive Borel functions f on $H \cdot y$.

(a) Let χ be defined on the closed subgroup $H_y \ltimes \mathbb{R}^n$ of G by $\chi(h,x) = e^{i2\pi x \cdot y}$. Show χ is a unitary representation.

(b) Show $\operatorname{ind}\chi$ is unitarily equivalent to the unitary representation $\hat{\pi}$ defined on $L^2(H \cdot y, \mu)$ by

$$\hat{\pi}(h,x')f(x) = \left(\frac{d(h\cdot\mu)}{d\mu}(x)\right)^{1/2} e^{2\pi i x\cdot x'} f(h^t x).$$

(c) Show $\hat{\pi}$ is irreducible.

(d) Show $\hat{\pi}$ need not be square-integrable even if H_y is compact.

Solution: Note

$$\begin{aligned}\chi(h_1,x)\chi(h_2,x') &= e^{2\pi i x\cdot y}e^{2\pi x'\cdot y} = e^{2\pi i x\cdot y}e^{2\pi i x'\cdot h_1^t y}\\ &= e^{2\pi i x\cdot y}e^{i h_1 x'\cdot y} = e^{2\pi i(x+h_1x')\cdot y}\\ &= \chi(h_1h_2, x+h_1x').\end{aligned}$$

So (a) holds. For (b) it suffices to use Proposition 6.138 and note $G/(H_y \ltimes \mathbb{R}^n) = H/H_y$ and if $\sigma : H/H_y \to H$ is a Borel section, it is also a section of $G/(H_y \ltimes \mathbb{R}^n)$. So

$$\operatorname{ind}\chi(a,x')f(bH_y) = \sqrt{\frac{\rho(a^{-1}b)}{\rho(b)}}\chi(\sigma(bH_y)^{-1}(a,x')\sigma((a,x')^{-1}bH_y)f(a^{-1}bH_y)$$

Thus if $bH_y \mapsto x = (b^{-1})^t y$, then $a^{-1}bH_y \mapsto ((a^{-1}b)^{-1})^t y = a^t(b^{-1})^t y = a^t x$. Also $\sqrt{\frac{\rho(a^{-1}b)}{\rho(b)}} = \sqrt{\frac{d(a\mu_y)}{d\mu_y}(bH_y)} = \sqrt{\frac{d(a\cdot\mu)}{d\mu}}(x)$ can be checked. Indeed, by Theorem 6.15, the measure μ_y satisfies $d\mu_y(h_0hH_y) = \frac{\rho(h_0h)}{\rho(h)}\,d\mu_y(hH_y)$, we have the measure $d\mu$ satisfies $d\mu(a^{-1}\cdot b\cdot y) = \frac{\rho(a^{-1}b)}{\rho(b)}\,d\mu(b\cdot y)$. Finally, we calculate,

$$\begin{aligned}\sigma(bH_y)^{-1}(a,x')\sigma((a,x')^{-1}bH_y) &= (bh_1,0)^{-1}(a,x')(a^{-1}bh_2,0)\\ &= (h_1^{-1}b^{-1},0)(bh_2,x')\\ &= (h_1^{-1}h_2, h_1^{-1}b^{-1}x')\end{aligned}$$

where h_1 and h_2 are in H_y. This gives

$$\begin{aligned}\chi(\sigma(bH_y)^{-1}(a,x')\sigma((a,x')^{-1}bH_y) &= \chi(h_1^{-1}h_2, h_1^{-1}b^{-1}x')\\ &= e^{2\pi i(h_1^{-1}b^{-1}x')\cdot y}\\ &= e^{2\pi i x'\cdot((b^{-1})^t(h_1^{-1})^t y)}\\ &= e^{2\pi i x'\cdot x}.\end{aligned}$$

Thus $\hat{\pi}$ is given by

$$\hat{\pi}(a,x')f(x) = \sqrt{\frac{d(a\cdot\mu)}{d\mu}(x)}e^{2\pi i x'\cdot x}f(a^t x).$$

We next show $\hat{\pi}$ is irreducible. Let $A \in \mathrm{Hom}(\hat{\pi},\hat{\pi})$. Set $\rho(x)f(y) = \hat{\pi}(1,-x)f(y) = e^{-2\pi i x\cdot y}f(y)$ for $f \in L^2(\mu)$. Thus $\rho = \hat{\pi}_\mu$ where $\hat{\pi}_\mu$ is the representation defined in Exercise 6.8.7. Using Exercise 6.8.10, we then know $A = M_\phi$ for some $\phi \in L^\infty(\mu)$. But $A\hat{\pi}(h,0) = \hat{\pi}(h,0)A$ implies

$$\phi(x)\left(\frac{d(h\cdot\mu)}{d\mu}(x)\right)^{1/2} f(h^t x) = \left(\frac{d(h\cdot\mu)}{d\mu}(x)\right)^{1/2}\phi(h^t x)f(h^t x)$$

a.e. x for each $h \in H$ and $f \in L^2(\mu)$. Consequently, one sees $\mathrm{Hom}(\hat{\pi},\hat{\pi})$ consists of all operators $A = M_\phi$ where $\phi \in L^\infty(\mu)$ and $\phi(h^t x) = \phi(x)$ a.e. x for each $h \in H$. Using left Haar measure on H, let $E = \{x \in H^t y \mid \phi(h^t x) = \phi(x) \text{ a.e. } h\}$. We note E is conull and so there is an $x = h_0^t y$ with $\phi(h^t h_0^t y) = \phi(h_0^t y)$ a.e. h. But since we are using the quasi-invariant measure μ_y on H/H_y carried over to $H^t y$, and a Borel subset of H/H_y has μ_y measure 0 if and only if the Haar measure of its preimage in H has measure 0, we have $\phi(x) = c$ for μ a.e. $x \in H^t y$ where $c = \phi(h_0^t y)$. This gives $\mathrm{Hom}(\hat{\pi},\hat{\pi}) = \mathbb{C}I$ and thus by Schur's Lemma, $\hat{\pi}$ is irreducible.

Finally, in order for $\hat{\pi}$ to be square integrable, there needs to be a nonzero function $\psi \in L^2(\mu)$ with $g \mapsto (\psi, \hat{\pi}(g)\psi)$ in $L^2(G)$. Now by Exercise 6.11.11, $dg = d(h,x) = |\det h^{-1}|(dh \times dx)$ is a left Haar measure on $G = H \ltimes \mathbb{R}^n$. Thus one would need a ψ with

$$\int_H |\det h^{-1}| \int_{\mathbb{R}^n} |(\psi, \hat{\pi}(h,x)\psi)_{L^2(\mu)}|^2\, dx\, dh < \infty.$$

We give an example where this is not true. Let $n = 2$ and $H = \mathbb{R}^+$ act on $\mathbb{R}^2$ by

$$a(x,y) = (ax, ay).$$

Then $H\cdot(1,0) = \mathbb{R}^+ \times \{0\}$ and if $p = (1,0)$, then $H_p = \{1\}$. Identifying $\mathbb{R}^+$ with $\mathbb{R}^+ \times \{0\}$, the representation $\hat{\pi}$ is then given by

$$\hat{\pi}(a,(x,y))f(s,0) = a^{1/2}e^{2\pi i (x,y)\cdot(s,0)}f(as) = \sqrt{a}e^{2\pi i s x}f(as).$$

Note this is essentially the formula (6.40) and this representation cannot be square integrable for

$$\begin{aligned}(\psi, \hat{\pi}(a,(x,y))\psi)_{L^2(\mathbb{R}^+)} &= \sqrt{a}\int_0^\infty \psi(s)\overline{e^{2\pi i s x}\psi(as)}\, ds \\ &= \sqrt{a}\int_0^\infty e^{-2\pi i s x}\psi(s)\bar{\psi}(as)\, ds\end{aligned}$$

and thus one would have

$$\int_H |\det h^{-1}| \int_{\mathbb{R}^2} |(\psi, \hat{\pi}(h,(x,y))\psi)_{L^2(\mu)}|^2 \, d(x,y)\, dh$$
$$= \int_0^\infty \frac{1}{a^2} \int_{-\infty}^\infty \int_{-\infty}^\infty a \left| \int_0^\infty e^{-2\pi i s x} \psi(s)\bar{\psi}(as)\, ds \right|^2 dx\, dy\, \frac{da}{a}$$
$$= \int_0^\infty \frac{1}{a^2} \int_{-\infty}^\infty \int_{-\infty}^\infty \left| \int_0^\infty e^{-2\pi i s x} \psi(s)\bar{\psi}(as)\, ds \right|^2 dx\, dy\, da$$
$$= \int_{-\infty}^\infty \int_0^\infty \frac{1}{a^2} \int_{-\infty}^\infty \left| \int_0^\infty e^{-2\pi i s x} \psi(s)\bar{\psi}(as)\, ds \right|^2 dx\; da\, dy$$
$$= \int_{-\infty}^\infty \int_0^\infty \frac{1}{a^2} F(a)\, da\, dy$$

where $F(a) = \int_{-\infty}^\infty \left| \int_0^\infty e^{-2\pi i s x} \psi(s)\bar{\psi}(as)\, ds \right|^2 dx$. So

$$\int_H |\det h^{-1}| \int_{\mathbb{R}^2} |(\psi, \hat{\pi}(h,(x,y))\psi)_{L^2(\mu)}|^2 \, d(x,y)\, dh$$

is either 0 or ∞ and there are no admissible functions.

□

Chapter 7

The Heisenberg Group

Exercise Set 7.1

1. Let $\langle \cdot, \cdot \rangle$ be a symplectic form on a real finite dimensional vector space V. Show that V may be turned into a complex vector space with a complex inner product $(\cdot, \cdot)$ satisfying

$$\langle v, w \rangle = -\text{Im}(v, w) \text{ for all } v, w \in V.$$

Solution: Let $e_1, e_2, \cdots, e_n, f_1, f_2, \cdots, f_n$ be a symplectic basis. Define a real linear transformation $i : V \to V$ by $ie_j = f_j$ and $if_j = -e_j$. It is clear that $i^2 v = -v$ for all v and thus $(a + ib)(c + id)v = (ac - bd + (bc + ad)i)v$ for all v and $a, b, c, d \in \mathbb{R}$. Other properties to be a complex vector space are similarly treated.

Clearly $e_1, e_2, \cdots, e_n$ form a complex basis of V. Define a complex inner product $(\cdot, \cdot)$ on V by

$$(\sum c_j e_j, \sum d_j e_j) = \sum c_j \bar{d}_j.$$

Then $e_1, e_2, \cdots, e_n$ is an orthonormal basis. Also

$$\begin{aligned} \langle \sum (a_j e_j + b_j f_j), \sum (a'_j e_j + b'_j f_j) \rangle &= \sum_j (b_j a'_j \langle f_j, e_j \rangle + a_j b'_j \langle e_j, f_j \rangle) \\ &= \sum_j (a_j b'_j - b_j a'_j) \\ &= \sum_j -\text{Im}((a_j + b_j i)e_j, (a'_j + b'_j i)e_j) \\ &= -\text{Im}(\sum_j (a_j + b_j i)e_j, \sum_j (a'_j + b'_j i)e_j). \end{aligned}$$

□

2. Let H be the collection of all matrices of form

$$\begin{bmatrix} 1 & x & t \\ 0 & I_n & y^t \\ 0 & 0 & 1 \end{bmatrix}$$

where $x, y \in \mathbb{R}^n$, I_n is the $n \times n$ identity matrix and $t \in \mathbb{R}$. Show H is a group and find an isomorphism of H_n onto H.

Solution: Multiplication in H can be described as

$$(x, y, t)(x', y', t') = (x + x', y + y', t + t' + x \cdot y').$$

For H_n we use $\mathbb{R}^n \times \mathbb{R}^n \times \mathbb{R}$ with multiplication

$$(x, y, t)(x', y', t') = (x + x', y + y' + \frac{1}{2}(x \cdot y' - y \cdot x')).$$

Define

$$\Phi(x, y, t) = (x, y, t + \frac{1}{2}x \cdot y).$$

Clearly, Φ is one-to-one from H_n onto H.

Next note

$$\begin{aligned}
\Phi((x, y, t)(x', y', t')) &= \Phi(x + x', y + y', t + t' + \frac{1}{2}(x \cdot y' - y \cdot x')) \\
&= (x + x', y + y', t + t' + \frac{1}{2}(x \cdot y' - y \cdot x') + \frac{1}{2}(x + x') \cdot (y + y')) \\
&= (x + x', y + y', t + t' + \frac{1}{2}x \cdot y + \frac{1}{2}x' \cdot y' + x \cdot y') \\
&= (x, y, t + \frac{1}{2}x \cdot y)(x', y', t' + \frac{1}{2}x' \cdot y') \\
&= \Phi(x, y, t)\Phi(x', y', t').
\end{aligned}$$

Note also that

$$\Phi((x, y, t)^{-1}) = \Phi(-x, -y, -t) = (-x, -y, -t + \frac{1}{2}x \cdot y)$$

and thus $\Phi((x, y, t)^{-1})\Phi(x, y, t) = (-x, -y, -t + \frac{1}{2}x \cdot y)(x, y, t + \frac{1}{2}x \cdot y) = (0, 0, 0)$. Thus $\Phi((x, y, t)^{-1}) = \Phi(x, y, t)^{-1}$. □

3. A C^∞ vector field on $\mathbb{R}^n$ is a differential operator of form

$$X = \sum_{j=1}^{n} F_j \frac{\partial}{\partial x_j}$$

where $F_j \in C^\infty(\mathbb{R}^n)$.

(a) Suppose $X = \sum F_j \frac{\partial}{\partial x_j}$ and $X' = \sum F'_j \frac{\partial}{\partial x_j}$ are C^∞ vector fields on $\mathbb{R}^n$. Show the bracket $[X, X'] = XX' - X'X$ is the C^∞ vector field

$$\sum H_k \frac{\partial}{\partial x_k}$$

where $H_k = \sum_j \left(F_j \frac{\partial F'_k}{\partial x_j} - F'_j \frac{\partial F_k}{\partial x_j} \right)$.

(b) Show $(X, X') \mapsto [X, X']$ is real bilinear on the vector space of C^∞ vector fields.

(c) Show $[X, X'] = -[X', X]$ if X and X' are C^∞ vector fields.

(d) Show $[X,[X',X'']]+[X',[X'',X]]+[X'',[X,X']] = 0$ for C^∞ vector fields X, X', and X.

Solution:

One has

$$\begin{aligned} XX'(\phi) &= \sum_{i=1}^{n} F_i \partial_i (\sum_{j=1}^{n} F'_j \partial_j \phi) \\ &= \sum_{i=1}^{n} F_i \sum_{j=1}^{n} ((\partial_i F'_j)\partial_j \phi + F'_j \partial_i \partial_j \phi) \\ &= \sum_j \left(\sum_i F_i \partial_i F'_j \right) \partial_j \phi + \sum_{i,j} F_i F'_j \partial_i \partial_j \phi \end{aligned}$$

and similarly

$$X'X(\phi) = \sum_j \left(\sum_i F'_i \partial_i F_j \right) \partial_j \phi + \sum_{i,j} F_i F'_j \partial_j \partial_i \phi.$$

Taking $\phi \in C^2(\mathbb{R}^n)$, one has $\partial_i \partial_j \phi = \partial_j \partial_i \phi$ and thus

$$[X, X']\phi = XX'\phi - X'X\phi = \sum_j H_j \partial_j \phi$$

where

$$H_j = \sum_i (F_i \partial_i F'_j - F'_i \partial_i F_j)$$

and we see (a) holds.

That $(X, X') \mapsto [X, X']$ is bilinear follows from

$$\sum_i a_i X_i(\phi) = (\sum a_i X_i)\phi$$

and clearly $[X', X] = X'X - XX' = -(XX' - X'X) = -[X, X']$.

Finally note

$$\begin{aligned} [X,[X',X'']] &= X(X'X'' - X''X') - (X'X'' - X''X')X \\ &= XX'X'' - XX''X' - X'X''X + X''X'X \end{aligned}$$

and similarly

$$\begin{aligned} [X',[X'',X]] &= X'X''X - X'XX'' - X''XX' + XX''X' \\ [X'',[X,X']] &= X''XX' - X''X'X - XX'X'' + X'XX''. \end{aligned}$$

Adding these gives 0 and we are done.

□

4. A one parameter group in H_n is a continuous homomorphism from $(\mathbb{R},+)$ into H_n. Show a one parameter group ϕ satisfies

$$\phi(t) = t\phi(1) = (tv, ts)$$

where $\phi(1) = (v, s)$.

Solution: Let $(v, s) = \phi(1)$. Let $m, k \in \mathbb{N}$. Set $(w, r) = \phi(\frac{m}{k})$. Then $(kw, kr) = \phi(\frac{m}{k})^k = \phi(m) = \phi(1)^m = (mv, ms)$. Thus $(w, r) = (\frac{m}{k}v, \frac{m}{k}s)$. Thus $\phi(q) = q\phi(1)$ for all rational numbers $q \geqslant 0$.

Since $\phi(-\frac{k}{m}) = -\phi(\frac{k}{m})$, we see $\phi(q) = q\phi(1)$ for all $q \in \mathbb{Q}$. By continuity, we have $\phi(t) = t\phi(1) = (tv, ts)$ for all t. □

5. Show a vector field D on H_n has form

$$D = \sum_j (a_j(\frac{\partial}{\partial x_j} - \frac{1}{2}y_j\frac{\partial}{\partial t}) + b_j(\frac{\partial}{\partial y_j} + \frac{1}{2}x_j\frac{\partial}{\partial t})) + c\frac{\partial}{\partial t}$$

if and only if it is left invariant on H_n; i.e.,

$$L_h(Df) = D(L_h f)$$

for all $h \in H_n$ and $f \in C^\infty(H_n)$.

Solution: A vector field has form $D = \sum_j \left(F_j\frac{\partial}{\partial x_j} + G_j\frac{\partial}{\partial y_j}\right) + H\frac{\partial}{\partial t}$ where F_j, G_j, and H are functions on H_n. Suppose it is left invariant. Set $a_j = F_j(0,0,0)$, $b_j = G_j(0,0,0)$ and $c = H(0,0,0)$. Now the chain rule yields

$$\begin{aligned}
Df(x,y,t) &= L_{(-x,-y,-t)}Df(0) \\
&= (DL_{(-x,-y,-t)}f)(0) \\
&= \left\{\sum_j \left(a_j\frac{\partial}{\partial x_j} + b_j\frac{\partial}{\partial y_j}\right) + c\frac{\partial}{\partial t}\right\} L_{(-x,-y,-t)}f(0) \\
&= \frac{d}{ds}|_{s=0}L_{(-x,-y,-t)}f(sa, sb, st) \\
&= \frac{d}{ds}|_{s=0}f((x,y,t)(sa,sb,sc)) \\
&= \frac{d}{ds}|_{s=0}f((x+sa, y+sb, t+sc+\frac{1}{2}(x\cdot sb - y\cdot sa)) \\
&= \sum_{j=1}^n a_j\frac{\partial}{\partial x_j}f(x,y,t) + b_j\frac{\partial}{\partial y_j}f(x,y,t) + (c+\frac{1}{2}(x\cdot b - y\cdot a))\frac{\partial}{\partial t}f(x,y,t) \\
&= \left(\sum_{j=1}^n (a_j\frac{\partial}{\partial x_j} - \frac{1}{2}a_jy_j\frac{\partial}{\partial t}) + \sum_{j=1}^n (b_j\frac{\partial}{\partial y_j} + \frac{1}{2}b_jx_j\frac{\partial}{\partial t}) + c\frac{\partial}{\partial t}\right) f(x,y,t) \\
&= (X(a) + Y(b) + cZ)f(x,y,t).
\end{aligned}$$

□

6. If D is a C^∞ vector field on H_n, then $\check{D}$ is the vector field defined on H_n by $\check{D}(f) = \widecheck{D(\check{f})}$ for $f \in C^\infty(H_n)$.

(a) Show $\check{D}$ is a C^∞ vector field.

(b) Show $D \mapsto \check{D}$ is linear and $\check{\check{D}} = D$.

(c) Show D is right invariant if and only if the the vector field $\check{D}$ is left invariant.

(d) Show $\widecheck{[D_1, D_2]}$ is $[\check{D}_1, \check{D}_2]$ for any pair D_1, D_2 of C^∞ vector fields on H_n.

(e) Find $\check{Z}$, and $\check{X}_j$ and $\check{Y}_j$ for $j = 1, \ldots, n$.

(f) Show $[\check{X}(a) + \check{Y}(b) + c\check{Z}, \check{X}(a') + \check{Y}(b') + c'\check{Z}] = \langle (a, b), (a', b')\rangle \check{Z}$.

(g) Show if $D = X(a) + Y(b) + cZ$, then

$$\check{D}f(x, y, t) = \frac{d}{ds}|_{s=0} f((-sa, -sb, -sc)(x, y, t)).$$

Solution:

Take $D = \sum_{j=1}^n (F_j \frac{\partial}{\partial x_j} + G_j \frac{\partial}{\partial y_j}) + H\frac{\partial}{\partial t}$. Now if $g = (x, y, t)$, $g^{-1} = (x, y, t)^{-1} = (-x, -y, -t)$. Thus since

$$\begin{aligned} \frac{\partial}{\partial x_j}\check{f}(g) &= \frac{d}{ds}|_{s=0}\check{f}(x + se_j, y, t) \\ &= \frac{d}{ds}|_{s=0} f(-x - se_j, -y, -t) \\ &= -\frac{\partial}{\partial x_j} f(g^{-1}) \end{aligned}$$

and similarly, $\frac{\partial}{\partial y_j}\check{f}(g) = -\frac{\partial}{\partial y_j} f(g^{-1})$ and $\frac{\partial}{\partial t}\check{f}(g) = -\frac{\partial}{\partial t} f(g^{-1})$, we see

$$\begin{aligned} D\check{f}(g) &= \sum_{j=1}^n (F_j(g)\frac{\partial}{\partial x_j}\check{f}(g) + G_j(g)\frac{\partial}{\partial y_j}\check{f}(g)) + H(g)\frac{\partial}{\partial t}\check{f}(g) \\ &= -\sum_{j=1}^n (F_j(g)\frac{\partial}{\partial x_j} f(g^{-1}) + G_j(g)\frac{\partial}{\partial y_j} f(g^{-1})) - H(g)\frac{\partial}{\partial t} f(g^{-1}). \end{aligned}$$

This implies

$$\check{D}(f) = \widecheck{D\check{f}} = -\sum_{j=1}^n (\check{F}_j \frac{\partial}{\partial x_j} f + \check{G}_j \frac{\partial}{\partial y_j} f) - \check{H}\frac{\partial}{\partial t} f$$

and thus $\check{D}$ is a C^∞ vector field on H_n. One also sees $\check{\check{D}} = D$. Note $D \mapsto \check{D}$ is linear since $F \mapsto \check{F}$ is linear.

To see D is left invariant if and only if $\check{D}$ is right invariant, note $\widecheck{R_h f}(x) = f(x^{-1}h) = \check{f}(h^{-1}x) = L(h)\check{f}(x)$. Thus $\check{D}(R_h f) = R_h \check{D}(f)$ for all $f \in C^\infty(H_n)$ if and only if $D(\widecheck{R_h f})(g^{-1}) = D(\check{f})((gh)^{-1})$ for all $g \in H_n$ and all f if and only if $D(L_h\check{f})(g^{-1}) = D(\check{f})(h^{-1}g^{-1}) = L_hD(\check{f})(g^{-1})$ for all g for all f if and only if $DL_h = L_hD$.

Using $\check{D}(f) = \widecheck{D\check{f}} = -\sum_{j=1}^n(\check{F}_j\frac{\partial}{\partial x_j}f + \check{G}_j\frac{\partial}{\partial y_j}f) - \check{H}\frac{\partial}{\partial t}f$, we see since $X(a) = \sum_{j=1}^n a_j(\frac{\partial}{\partial x_j} - \frac{1}{2}y_j\frac{\partial}{\partial t})$, that $\check{X}(a) = -\sum_{j=1}^n a_j(\frac{\partial}{\partial x_j} + \frac{1}{2}y_j\frac{\partial}{\partial t})$ and similarly $\check{Y}(b) = -\sum_{j=1}^n b_j(\frac{\partial}{\partial y_j} - \frac{1}{2}x_j\frac{\partial}{\partial t})$ and $\check{Z} = -Z$. Thus $\check{X}_j = -\frac{\partial}{\partial x_j} - \frac{1}{2}y_j\frac{\partial}{\partial t}$, $\check{Y}_j = -\frac{\partial}{\partial y_j} + \frac{1}{2}x_j\frac{\partial}{\partial t}$, and $\check{Z} = -\frac{\partial}{\partial t}$.

Note

$$\begin{aligned}
[\check{D}_1, \check{D}_2](f) &= (\check{D}_1\check{D}_2 - \check{D}_2\check{D}_1)f \\
&= \check{D}_1\left((D_2\check{f})\check{}\,\right) - \check{D}_2\left((D_1\check{f})\check{}\,\right) \\
&= (D_1D_2\check{f})\check{} - (D_2D_1\check{f})\check{} \\
&= ((D_1D_2)\check{f} - (D_2D_1)\check{f})\check{} \\
&= ([D_1, D_2]\check{f})\check{} \\
&= \widecheck{[D_1, D_2]}(f)\,.
\end{aligned}$$

Now by the last part of Proposition 7.5, if $D = X(a) + X(b) + cZ$ and $D' = X(a') + Y(b') + c'Z$, then $[\check{D}, \check{D}'] = \widecheck{[D, D']} = (\langle(a,b),(a',b')\rangle Z)\check{} = \langle(a,b),(a',b')\rangle\check{Z}$.

Finally, if $D = X(a) + Y(b) + cZ$, then

$$\begin{aligned}
D\check{f}(x,y,t) &= \frac{d}{ds}|_{s=0}\check{f}((x,y,t)(sa,sb,sc)) \\
&= \frac{d}{ds}|_{s=0}\check{f}(x+sa, y+sb, t+sc+\frac{1}{2}s(x\cdot b - y\cdot a)) \\
&= \frac{d}{ds}|_{s=0}f(-x-sa, -y-sb, -t-sc-\frac{1}{2}s(x\cdot b - y\cdot a))
\end{aligned}$$

and so

$$\begin{aligned}
\check{D}f(x,y,t) &= D\check{f}(-x,-y,-t) \\
&= \frac{d}{ds}|_{s=0}f(x-sa, y-sb, t-sc-\frac{1}{2}s(-x\cdot b + y\cdot a)) \\
&= \frac{d}{ds}|_{s=0}f((-sa,-sb,-sc)(x,y,t))\,.
\end{aligned}$$

□

7. Let $\mathfrak{h}_n = \mathbb{R}^{2n} \times \mathbb{R}$ be the Lie algebra of the Heisenberg group H_n.

(a) Show all triple Lie brackets are 0 in $\mathfrak{h}_n$.

(b) Suppose N is a finite dimensional Lie algebra having the property that all triple Lie brackets are 0. Define a multiplication in N by $x \cdot y = x + y + \frac{1}{2}[x,y]$. Show N is a group.

Solution:

Note (a) follows directly from Proposition 7.5. For (b), let N be a finite dimensional Lie algebra with triple brackets equal to 0. We have $x \cdot y = x + y + \frac{1}{2}[x, y]$ for $x, y \in N$. Note $0 \cdot x = x \cdot 0 = x$ and so 0 works as a multiplicative identity. Also $x \cdot (-x) = (-x) \cdot x = 0$ for $[x, -x] = [-x, x] = 0$ and thus $-x$ is a multiplicative inverse. We thus only need to show associativity. We have

$$\begin{aligned} x \cdot (y \cdot z) &= x \cdot (y + z + \frac{1}{2}[y, z]) \\ &= x + y + z + \frac{1}{2}[y, z] + \frac{1}{2}[x, y + z + \frac{1}{2}[y, z]] \\ &= x + y + \frac{1}{2}[x, y] + z + \frac{1}{2}[y, z] + \frac{1}{2}[x, z] + \frac{1}{4}[[x, y], z] \\ &= x + y + \frac{1}{2}[x, y] + z + \frac{1}{2}[x + y + \frac{1}{2}[x, y], z] \\ &= x \cdot y + z + \frac{1}{2}[x \cdot y, z] \\ &= (x \cdot y) \cdot z\,. \end{aligned}$$

□

8. Consider the vector fields $X' = \frac{\partial}{\partial x}$, $Y' = \frac{\partial}{\partial y} + x\frac{\partial}{\partial t}$, and $Z' = \frac{\partial}{\partial t}$ on $\mathbb{R}^3 = \{(x, y, t) \mid x, y, t \in \mathbb{R}\}$. Show X', Y', Z' satisfy the same brackets as $X = \frac{\partial}{\partial x} - \frac{1}{2}y\frac{\partial}{\partial t}$, $Y = \frac{\partial}{\partial y} + \frac{1}{2}x\frac{\partial}{\partial t}$, $Z = \frac{\partial}{\partial t}$, a basis for $\mathfrak{h}_1$.

Determine a multiplication $\cdot$ on $\mathbb{R}^3$ such that

(a) $X'f(x, y, t) = \frac{d}{ds}f((x, y, t) \cdot (s, 0, 0))|_{s=0}$

(b) $Y'f(x, y, t) = \frac{d}{ds}f((x, y, t) \cdot (0, s, 0))|_{s=0}$

(c) $Z'f(x, y, t) = \frac{d}{ds}f((x, y, t) \cdot (0, 0, s))|_{s=0}$.

Then show this group is isomorphic to the group H_1. How does this relate to Exercise 7.1.2?

Solution:

Note one can easily check $[X', Y'] = Z'$ and the brackets with Z' are 0. So one has the same brackets as X, Y and Z. Assume we have a group multiplication $\cdot$ on $\mathbb{R}^3$ so that (a), (b), and (c) hold with $(0, 0, 0)$ as the identity. In particular, $(x, y, t) \cdot (0, s, 0) = (\phi_1(s), \phi_2(s), \phi_3(s))$ with $\phi_1(0) = x$, $\phi_2(0) = y$, and $\phi_3(0) = t$. Let f_j denote the partial derivative $D_j f$. By (b) and the chain rule, we have

$$f_2(x, y, t) + x f_3(x, y, t) = f_1(x, y, t)\phi_1'(0) + f_2(x, y, t)\phi_2'(0) + f_3(x, y, t)\phi_3'(0).$$

This gives $\phi_1' = 0$, $\phi_2' = 1$, and $\phi_3' = x$. We thus obtain $\phi_1(s) = x$, $\phi_2(s) = y + s$, and $\phi_3(s) = t + xs$. So $(x, y, t) \cdot (0, s, 0) = (x, y + s, t + xs)$. Similarly $(x, y, t) \cdot (s, 0, 0) = (x + s, y, t)$ and $(x, y, t) \cdot (0, 0, s) = (x, y, t + s)$. So $(x, y, z) \cdot (x', y', z') = (x + x', y + y', z + z' + xy')$.

To finish showing this is a group, it suffices to find a diffeomorphism ϕ of $\mathbb{R}^3$ that carries this new multiplication to the multiplication on H_1. We let $\phi(x,y,t) = (x,y,t-\frac{1}{2}xy)$. Note

$$\begin{aligned}\phi((x,y,t)\cdot(x',y',t')) &= \phi(x+x',y+y',t+t'+xy')\\ &= (x+x',y+y',t+t'+xy'-\frac{1}{2}(x+x')(y+y'))\\ &= (x+x',y+y',t+t'-\frac{1}{2}xy-\frac{1}{2}x'y'+\frac{1}{2}(xy'-x'y))\\ &= (x,y,t-\frac{1}{2}xy)(x',y',t'-\frac{1}{2}x'y')\\ &= \phi(x,y,t)\phi(x',y',t')\,.\end{aligned}$$

So these groups are isomorphic. Note ϕ is the inverse of Φ in Exercise 7.1.2.

□

9. Let $\mathbb{R}^4 = \{(w,x,y,z) \mid w,x,y,z \in \mathbb{R}\}$. Then consider the C^∞ vector fields $W = \frac{\partial}{\partial w}$, $X = \frac{\partial}{\partial x} + w\frac{\partial}{\partial y} + wx\frac{\partial}{\partial z}$, $Y = \frac{\partial}{\partial y} + x\frac{\partial}{\partial z}$, and $Z = \frac{\partial}{\partial z}$. Show $[W,X] = Y$, $[X,Y] = Z$, $[W,Y] = 0$, $[W,Z] = 0$, $[X,Z] = 0$, $[Y,Z] = 0$. Thus the linear span of W, X, Y, Z form a four dimensional Lie algebra of C^∞ vector fields on $\mathbb{R}^4 = \{(w,x,y,z) \mid w,x,y,z \in \mathbb{R}\}$. Find a group multiplication on $\mathbb{R}^4$ such that for $f \in C^\infty(\mathbb{R}^4)$ one has

$$\frac{d}{ds}f((w,x,y,z)\cdot se_k) = \begin{cases} Wf((w,x,y,z)\cdot se_k) & \text{when } k=1\\ Xf((w,x,y,z)\cdot se_k) & \text{when } k=2\\ Yf((w,x,y,z)\cdot se_k) & \text{when } k=3\\ Zf((w,x,y,z)\cdot se_k) & \text{when } k=4.\end{cases}$$

Solution:

First we calculate the brackets: for $f \in C^\infty$,

$$\begin{aligned}[W,X]f &= \frac{\partial}{\partial w}\left(\frac{\partial}{\partial x}f + w\frac{\partial}{\partial y}f + wx\frac{\partial}{\partial z}f\right) - \left(\frac{\partial}{\partial x} + w\frac{\partial}{\partial y} + wx\frac{\partial}{\partial z}\right)\frac{\partial}{\partial w}f\\ &= D_{1,2}f - D_{2,1}f + wD_{1,3}f - wD_{3,1}f + wxD_{1,4}f - wxD_{4,1}f\\ &\quad + \frac{\partial}{\partial y}f + x\frac{\partial}{\partial z}f\\ &= \left(\frac{\partial}{\partial y} + x\frac{\partial}{\partial x}\right)f\\ &= Yf\end{aligned}$$

and

$$\begin{aligned}[X,Y]f &= \left(\frac{\partial}{\partial x} + w\frac{\partial}{\partial y} + wx\frac{\partial}{\partial z}\right)\left(\frac{\partial}{\partial y}f + x\frac{\partial}{\partial z}f\right) \\ &\quad - \left(\frac{\partial}{\partial y} + x\frac{\partial}{\partial z}\right)\left(\frac{\partial}{\partial x}f + w\frac{\partial}{\partial y}f + wx\frac{\partial}{\partial z}f\right) \\ &= D_{2,3}f - D_{3,2}f + xD_{2,4}f - xD_{4,2}f + wD_{3,3}f - wD_{3,3}f \\ &\quad + wxD_{3,4}f - wxD_{4,3}f + wxD_{4,3}f - wxD_{3,4}f \\ &\quad + wx^2D_{4,4}f - wx^2D_{4,4}f + \frac{\partial}{\partial z}f \\ &= \frac{\partial}{\partial z}f = Zf\,.\end{aligned}$$

In order to determine a group multiplication with these vector fields as the left invariant vector fields one would look for a multiplication satisfying

$$Wf((w,x,y,z)\cdot(s,0,0,0)) = \frac{d}{ds}f((w,x,y,z)\cdot(s,0,0,0))\,.$$

We let $\Phi(s) = (\phi_1(s), \phi_2(s), \phi_3(s), \phi_4(s)) = (w,x,y,z)\cdot(s,0,0,0)$. Using the notation $f_j = D_jf$, $f_1(\Phi(s)) = f_1(\Phi(s))\phi_1'(s) + f_2(\Phi(s))\phi_2'(s) + f_3(\Phi(s))\phi_3'(s) + f_4(\Phi(s))\phi_4'(s)$ and $\Phi(0) = (w,x,y,z)$. This being true for all f which are C^∞ implies $\phi_1'(s) = 1$ and ϕ_2, ϕ_3, and ϕ_4 are constant. So $\phi_1(s) = w + s$ and $\phi_2(s) = x$, $\phi_3(s) = y$ and $\phi_3(s) = z$. Thus we take

$$(w,x,y,z)\cdot(s,0,0,0) = (w+s,x,y,z).$$

We repeat this for X, Y and Z. For X we have if $(w,x,y,z)\cdot(0,s,0,0) = (\phi_1(s), \phi_2(s), \phi_3(s), \phi_4(s))$, then

$$\begin{aligned}&f_2(\Phi(s)) + \phi_1(s)f_3(\Phi(s)) + \phi_1(s)\phi_2(s)f_4(\Phi(s)) \\ &\qquad = f_1(\Phi(s))\phi_1'(s) + f_2(\Phi(s))\phi_2'(s) + f_3(\Phi(s))\phi_3'(s) + f_4(\Phi(s))\phi_4'(s)\end{aligned}$$

and so we take $\phi_1'(s) = 0$, $\phi_2'(s) = 1$, $\phi_3'(s) = \phi_1(s)$ and $\phi_4'(s) = \phi_1(s)\phi_2(s)$. With $\Phi(0) = (w,x,y,z)$, we see $\phi_1(s) = w$, $\phi_2(s) = x+s$, $\phi_3(s) = y + ws$, and $\phi_4'(s) = w(x+s)$ and so $\phi_4(s) = z + wxs + \frac{1}{2}ws^2$.

$$(w,x,y,z)\cdot(0,s,0,0) = (w, x+s, y+ws, z+wxs+\frac{1}{2}ws^2).$$

For Y we have

$$\begin{aligned}&f_3(\Phi(s)) + \phi_2(s)f_4(\Phi(s)) \\ &\qquad = f_1(\Phi(s))\phi_1'(s) + f_2(\Phi(s))\phi_2'(s) + f_3(\Phi(s))\phi_3'(s) + f_4(\Phi(s))\phi_4'(s)\end{aligned}$$

and this gives $\phi_1'(s) = 0$, $\phi_2'(s) = 0$, $\phi_3'(s) = 1$ and $\phi_4'(s) = \phi_2(s)$. Thus $\phi_1(s) = w$, $\phi_2(s) = x$, $\phi_3(s) = y+s$, and $\phi_4(s) = z + xs$. So

$$(w,x,y,z)\cdot(0,0,s,0) = (w,x,y+s,z+xs).$$

Finally, for Z one has

$$f_4(\Phi(s)) = f_1(\Phi(s))\phi_1'(s) + f_2(\Phi(s))\phi_2'(s) + f_3(\Phi(s))\phi_3'(s) + f_4(\Phi(s))\phi_4'(s)$$

and thus $\phi_1' = \phi_2' = \phi_3' = 0$ and $\phi_4' = 1$. So

$$(w,x,y,z)\cdot(0,0,0,s) = (w,x,y,z+s)\,.$$

Putting all together we have formulas

$$\begin{aligned}
(w,x,y,z)\cdot(w',0,0,0) &= (w+w',x,y,z)\\
(w,x,y,z)\cdot(0,x',0,0) &= (w,x+x',y+wx',z+wxx'+\frac{1}{2}wx'^2)\\
(w,x,y,z)\cdot(0,0,y',0) &= (w,x,y+y',z+xy')\\
(w,x,y,z)\cdot(0,0,0,z') &= (w,x,y,z+z')\,.
\end{aligned}$$

So

$$\begin{aligned}
(w,x,y,z) &= (0,x,y,z)\cdot(w,0,0,0)\\
&= (0,x,y,0)\cdot(0,0,0,z)\cdot(w,0,0,0)\\
&= (0,0,y,0)\cdot(0,x,0,0)\cdot(0,0,0,z)\cdot(w,0,0,0)\,.
\end{aligned}$$

Thus

$$\begin{aligned}
(w,x,y,z)\cdot(w',x',y',z') &= (w,x,y,z)\cdot(0,0,y',0)\cdot(0,x',0,0)\cdot(0,0,0,z')\cdot(w',0,0,0)\\
&= (w,x,y+y',z+xy')\cdot(0,x',0,0)\cdot(0,0,0,z')\cdot(w',0,0,0)\\
&= (w,x+x',y+y'+wx',z+xy'+wxx'+\frac{1}{2}wx'^2)\cdot(0,0,0,z')\cdot(w',0,0,0)\\
&= (w,x+x',y+y'+wx',z+z'+xy'+wxx'+\frac{1}{2}wx'^2)\cdot(w',0,0,0)\\
&= (w+w',x+x',y+y'+wx',z+z'+xy'+wxx'+\frac{1}{2}wx'^2)
\end{aligned}$$

These formulas imply $(0,0,0,z)$ is in the center for any z. Note $(0,0,0,0)$ is the multiplicative identity and

$$(w,x,y,z)^{-1} = (-w,-x,-y+wx,-z+xy-\frac{1}{2}wx^2)$$

for

$$\begin{aligned}
&(w,x,y,z)\cdot(-w,-x,-y+wx,-z+xy-\frac{1}{2}wx^2)\\
&= (0,0,0,xy-\frac{1}{2}wx^2+wx(-x)+x(-y+wx)+\frac{1}{2}wx^2) = (0,0,0,0) \text{ and}
\end{aligned}$$

$$\begin{aligned}
&(-w,-x,-y+wx,-z+xy-\frac{1}{2}wx^2)\cdot(w,x,y,z)\\
&= (0,0,0,xy-\frac{1}{2}wx^2+(-x)y+(-w)(-x)x+\frac{1}{2}(-w)x^2) = (0,0,0,0)\,.
\end{aligned}$$

To finish, one needs to show associativity.

$$((w,x,y,z)\cdot(w',x',y',z'))\cdot(w'',x'',y'',z'') =$$
$$(w+w',x+x',y+y'+wx',z+z'+xy'+wxx'+\frac{1}{2}wx'^2)\cdot(w'',x'',y'',z'')$$
$$=(w+w'+w'',x+x'+x'',y+y'+wx'+y''+(w+w')x'',$$
$$z+z'+xy'+wxx'+\frac{1}{2}wx'^2+z''+(x+x')y''+(w+w')(x+x')x''+\frac{1}{2}(w+w')x''^2)$$
$$=(w+w'+w'',x+x'+x'',y+y'+y''+w(x'+x'')+w'x'',$$
$$z+z'+z''+xy'+wxx'+\frac{1}{2}wx'^2+(x+x')y''+(w+w')(x+x')x''+\frac{1}{2}(w+w')x''^2)$$

while

$$(w,x,y,z)\cdot((w',x',y',z')\cdot(w'',x'',y'',z'')) =$$
$$(w,x,y,z)\cdot(w'+w'',x'+x'',y'+y''+w'x'',z'+z''+x'y''+w'x'x''+\frac{1}{2}w'x''^2)$$
$$=(w+w'+w'',x+x'+x'',y+y'+y''+w'x''+w(x'+x''),$$
$$z+z'+z''+x'y''+w'x'x''+\frac{1}{2}w'x''^2+x(y'+y''+w'x'')+wx(x'+x'')+\frac{1}{2}w(x'+x'')^2).$$

We note the first and second entries of these triple products are equal. Also for the third we have $y+y'+y''+wx'+wx''+w'x''$ in both triple products. Thus we have the associative law if

$$xy'+wxx'+\frac{1}{2}wx'^2+(x+x')y''+(w+w')(x+x')x''+\frac{1}{2}(w+w')x''^2$$

and

$$x'y''+w'x'x''+\frac{1}{2}w'x''^2+x(y'+y''+w'x'')+wx(x'+x'')+\frac{1}{2}w(x'+x'')^2$$

are equal. But

$$x'y''+w'x'x''+\frac{1}{2}w'x''^2+x(y'+y''+w'x'')+wx(x'+x'')+\frac{1}{2}w(x'+x'')^2 =$$
$$x'y''+w'x'x''+x(y'+y''+w'x'')+wx(x'+x'')+\frac{1}{2}wx'^2+wx'x''+\frac{1}{2}(w+w')x''^2$$
$$=xy'+wxx'+\frac{1}{2}wx'^2+(x+x')y''+w'x'x''+w'xx''+wxx''+wx'x''+\frac{1}{2}(w+w')x''^2$$
$$=xy'+wxx'+\frac{1}{2}wx'^2+(x+x')y''+(w+w')(x+x')x''+\frac{1}{2}(w+w')x''^2$$

and so we see associativity and we have a group with the needed properties.

□

Exercise Set 7.2

1. Show the representation π_λ where $\lambda \neq 0$ can be obtained from the representation π_1 by $\pi_\lambda(g) = \pi_1(\phi_\lambda(g))$ where ϕ_λ is the group automorphism of H_n defined by

$$\phi_\lambda(x + iy, t) = (x + i\lambda y, \lambda t).$$

Solution:

$$\begin{aligned}\phi_\lambda((x+iy,t)(x'+iy',t')) &= \phi(x+x'+i(y+y'), t+t'+\frac{1}{2}(x\cdot y' - y\cdot x'))\\ &= (x+x'+i(\lambda y+\lambda y'), \lambda t+\lambda t'+\frac{1}{2}(x\cdot\lambda y' - \lambda y\cdot x'))\\ &= (x+i\lambda y, \lambda t)(x'+i\lambda y', \lambda t')\\ &= \phi_\lambda(x+iy,t)\phi_\lambda(x'+iy',t').\end{aligned}$$

The result now follows from the formulas for π_λ given in Proposition 7.6 and its proof. □

2. In Remark 4.81, it was stated that the windowed Fourier transform is a matrix coefficient of a representation of the Heisenberg group. Indeed, let ρ be the representation π_λ when $\lambda = -2\pi$. Show if ψ and f are in $L^2(\mathbb{R}^n)$, then the windowed Fourier transform given in Definition 4.22 satisfies

$$S_\psi(f)(u,\omega) = \frac{e^{\pi i(u\cdot\omega)}}{(2\pi)^{n/2}}(f, \rho(u+i\omega,0)\psi)_2.$$

Then show the orthogonality relation given in Theorem 7.9 is the Plancherel formula given in Theorem 4.80.

Solution: We have

$$\begin{aligned}(f,\rho(u+i\omega,0)\psi)_2 &= \int f(x)\overline{\rho(u+i\omega,0)\psi(x)}\,d_nx\\ &= \int f(x)e^{\pi i(u,\omega)}e^{-2\pi i(x,\omega)}\overline{\psi(x-u)}\,d_nx\\ &= \frac{e^{\pi i(u\cdot\omega)}}{(2\pi)^{n/2}}\int f(x)e^{-2\pi i x\cdot\omega}\overline{\psi(x-u)}\,dx\\ &= \frac{e^{\pi i(u\cdot\omega)}}{(2\pi)^{n/2}}S_\psi(f)(u,\omega).\end{aligned}$$

From the orthogonality relation, we obtain

$$\begin{aligned}\frac{1}{(2\pi)^n}(f,g)_2(\phi,\psi)_2 &= \iint (f,\rho(u+i\omega,0)\psi)\overline{(g,\rho(u+i\omega,0)\phi)}\,d_nu\,d_n\omega\\ &= \frac{1}{(2\pi)^{2n}}\iint S_\psi(f)(u,\omega)\overline{S_\phi(g)(u,\omega)}\,du\,d\omega.\end{aligned}$$

Consequently since $(f,h)_2 = \frac{1}{(2\pi)^{n/2}} \int f(x)\overline{g(x)}\,dx$, we see

$$\iint S_\psi(f)(u,\omega)\overline{S_\phi(g)(u,\omega)}\,du\,d\omega = \left(\int f(x)\overline{g(x)}\,dx\right)\left(\int \phi(x)\overline{\psi(x)}\,dx\right).$$

□

3. Show if $F \in L^2(\mathbb{R}^n \times \mathbb{R}^n)$, the Weyl transform $\pi_\lambda(F)$ as a distribution is the distribution given by the L^2 function $\mathcal{F}_2(F) \circ \Phi$.
Solution: From the definition $\pi_\lambda(U) = \frac{1}{|\lambda|^n}(\mathcal{F}_2 U) \circ (\Phi')^{-1}$. Thus if $H \in \mathcal{S}(\mathbb{R}^n \times \mathbb{R}^n)$,

$$\begin{aligned}
\pi_\lambda(F)(H) &= \frac{1}{|\lambda|^n}(\mathcal{F}_2 F) \circ (\Phi^{-1})'(H) \\
&= \frac{1}{|\lambda|^n}\mathcal{F}_2 F(H \circ \Phi^{-1}) \\
&= \frac{1}{|\lambda|^n} F(\mathcal{F}_2(H \circ \Phi^{-1})) \\
&= \frac{1}{|\lambda|^n} \iint F(x,y)\mathcal{F}_2(H \circ \Phi^{-1})(x,y)\,d_n x\,d_n y \\
&= \frac{1}{|\lambda|^n} \iint \mathcal{F}_2 F(x,y)\,H(\frac{x}{2} + \frac{y}{\lambda}, -\frac{x}{2} + \frac{y}{\lambda})\,d_n x\,d_n y
\end{aligned}$$

Changing variables yields

$$\begin{aligned}
\pi_\lambda(F)(H) &= \iint \mathcal{F}_2 F(x,\lambda y)H(\frac{x}{2} + y, -\frac{x}{2} + y)\,d_n x\,d_n y \\
&= \iint \mathcal{F}_2 F(x, \lambda(y + \frac{x}{2}))H(x+y,y)\,d_n x\,d_n y \\
&= \iint \mathcal{F}_2 F(x-y, \frac{\lambda}{2}(x+y))H(x,y)\,d_n x\,d_n y \\
&= \iint (\mathcal{F}_2 F) \circ \Phi(x,y)\,H(x,y)\,d_n x\,d_n y.
\end{aligned}$$

□

4. Let $F \in L^2(\mathbb{R}^n \times \mathbb{R}^n)$. Show $W_\lambda(F)$ as a distribution is given by the function $\frac{1}{|\lambda|^n}\mathcal{F}_2^{-1}(F \circ \Phi^{-1})$.
Solution: Let $H \in \mathcal{S}(\mathbb{R}^n \times \mathbb{R}^n)$. Then

$$\begin{aligned}
W_\lambda(F)(H) &= \mathcal{F}_2^{-1}(F \circ \Phi')(H) \\
&= (F \circ \Phi')(\mathcal{F}_2^{-1} H) \\
&= F((\mathcal{F}_2^{-1} H) \circ \Phi) \\
&= \iint F(x,y)(\mathcal{F}_2^{-1} H)(x-y, \frac{\lambda}{2}(x+y))\,d_n x\,d_n y \\
&= \iint F(x+y,y)\mathcal{F}_2^{-1} H(x, \frac{\lambda}{2}x + \lambda y)\,d_n x\,d_n y
\end{aligned}$$

$$
\begin{aligned}
&= \frac{1}{|\lambda|^n} \iint F(x + \frac{y}{\lambda}, \frac{y}{\lambda}) \mathcal{F}_2^{-1} H(x, \frac{\lambda}{2} x + y) \, d_n x \, d_n y \\
&= \frac{1}{|\lambda|^n} \iint F(\frac{x}{2} + \frac{y}{\lambda}, -\frac{x}{2} + \frac{y}{\lambda}) \mathcal{F}_2^{-1} H(x, y) \, d_n x \, d_n y \\
&= \frac{1}{|\lambda|^n} \iint (F \circ \Phi^{-1})(x, y) \mathcal{F}_2^{-1} H(x, y) \, d_n x \, d_n y \\
&= \frac{1}{|\lambda|^n} \iint \mathcal{F}_2^{-1} (F \circ \Phi^{-1})(x, y) H(x, y) \, d_n x \, d_n y.
\end{aligned}
$$

□

5. Show the mappings $(f, h) \mapsto f \otimes_\lambda \bar{h}$ and $(f, h) \mapsto f \otimes_{-\lambda} h$ are continuous bilinear mappings of $\mathcal{S}(\mathbb{R}^n) \times \mathcal{S}(\mathbb{R}^n)$ into $\mathcal{S}(\mathbb{R}^n \times \mathbb{R}^n)$ whose ranges span Schwartz dense linear subspaces.

Solution:

First we note for $\lambda \neq 0$ that the mapping $F \mapsto F \circ \Phi^{-1}$ on $\mathcal{S}(\mathbb{R}^{2n})$ where $\Phi_\lambda(x, y) = (x - y, \frac{\lambda}{2}(x + y))$ and $\Phi_\lambda^{-1}(x, y) = (\frac{x}{2} + \frac{y}{\lambda}, -\frac{x}{2} + \frac{y}{\lambda})$ is a linear topological isomorphism of $\mathcal{S}(\mathbb{R}^{2n})$. Indeed, Φ_λ is an invertible linear transformation of $\mathbb{R}^{2n}$ and thus this follows from Proposition 2.62. By Theorem 3.10, we know $\mathcal{F}_2$ is a linear topological isomorphism of $\mathcal{S}(\mathbb{R}^{2n})$.

We have by Lemma 7.7 that $f \otimes_\lambda \bar{h}(x, y) = \frac{1}{|\lambda|^n} \mathcal{F}_2^{-1}((f \otimes \bar{h}) \circ \Phi_\lambda^{-1})(x, y)$ and $f \otimes_{-\lambda} h(x, y) = \frac{1}{|\lambda|^n} \mathcal{F}_2^{-1}(f \otimes h) \circ \Phi_{-\lambda}^{-1})(x, y)$. Thus to see the linear span of the functions $f \otimes_\lambda \bar{h}$ and $f \otimes_{-\lambda} h$ are dense in $\mathcal{S}(\mathbb{R}^{2n})$, it suffices to note the linear spans of the tensors $f \otimes \bar{h}$ and $f \otimes h$ for $f, h \in \mathcal{S}(\mathbb{R}^n)$ are dense in $\mathcal{S}(\mathbb{R}^{2n})$. This however follows from Theorem 2.85.

□

6. Show if U is a tempered distribution on $\mathbb{R}^n \times \mathbb{R}^n$, then the Weyl transform $\pi_\lambda(U)$ satisfies

$$
\pi_\lambda(U)(f \otimes h) = \iint \left(\int_{\mathbb{R}^n} f(w) \pi_\lambda(x + iy, 0) h(w) \, d_n w \right) dU(x, y).
$$

Solution:

$$\begin{aligned}
\pi_\lambda(U)(f\otimes h) &= \frac{1}{|\lambda|^n} U(\mathcal{F}_2((f\otimes h)\circ\Phi^{-1})) \\
&= \frac{1}{|\lambda|^n}\iint \mathcal{F}_2((f\otimes h)\circ\Phi^{-1})(x,y)\,dU(x,y) \\
&= \frac{1}{|\lambda|^n}\iint \left(\int (f\otimes h)\circ\Phi^{-1}(x,w)e^{-i(y,w)}\,d_n w\right)\,dU(x,y) \\
&= \frac{1}{|\lambda|^n}\iint \left(\int (f\otimes h)(\frac{x}{2}+\frac{w}{\lambda},-\frac{x}{2}+\frac{w}{\lambda})e^{-i(y,w)}\,d_n w\right)\,dU(x,y) \\
&= \iint \left(\int (f\otimes h)(\frac{x}{2}+w,-\frac{x}{2}+w)e^{-i\lambda(y,w)}\,d_n w\right)\,dU(x,y) \\
&= \iint \left(\int (f\otimes h)(w,w-x)e^{\frac{i\lambda(x,y)}{2}}e^{-i\lambda(y,w)}\,d_n w\right)\,dU(x,y) \\
&= \iint \left(\int f(w)\,e^{\frac{i\lambda(x,y)}{2}}e^{-i\lambda(y,w)}h(w-x)\,d_n w\right)\,dU(x,y) \\
&= \iint \left(\int f(w)\pi_\lambda(x+iy,0)h(w)\,d_n w\right)\,dU(x,y)
\end{aligned}$$

□

Exercise Set 7.3

1. Let F and H be in $L^2(\mathbb{R}^n\times\mathbb{R}^n)$. Show $F\#_\lambda H$ is a continuous function on $\mathbb{R}^n\times\mathbb{R}^n$. Moreover, show if F_k and H_k are sequences in $L^2(\mathbb{R}^n\times\mathbb{R}^n)$ converging in norm to F and H, then $F_k\#_\lambda H_k$ converges uniformly to $F\#_\lambda H$ as $k\to\infty$.
Solution:

$$F\#_\lambda H(w) = \int e^{i\frac{\lambda}{2}\langle z,w\rangle}F(z)H(w-z)\,d_{2n}z\,.$$

Set $F_w(z) = e^{i\frac{\lambda}{2}\langle z,w\rangle}F(z)$ and $H^w(z) = H(w-z) = \check{H}(z-w)$. First note $|F_w|_2 = |F|_2$ and $|H^w|_2 = |H|_2$ and thus $F\mapsto F_w$ and $H\mapsto H^w$ are unitary mappings on $L^2(\mathbb{C}^n)$.

For fixed H, $w\mapsto H^w$ is continuous into $L^2(\mathbb{C}^n)$ by Lemma 2.74. Also $w\mapsto F_w$ is continuous into $L^2(\mathbb{C}^n)$ for any $F\in L^2(\mathbb{C}^n)$. Indeed,

$$|F_w - F_{w_0}|_2^2 = \int |e^{i\frac{\lambda}{2}\langle z,w\rangle} - e^{i\frac{\lambda}{2}\langle z,w_0\rangle}|^2|F(z)|^2\,d_{2n}z \to 0$$

as $w\to w_0$ by the Lebesgue Dominated Convergence Theorem since $|e^{i\frac{\lambda}{2}\langle z,w\rangle} - e^{i\frac{\lambda}{2}\langle z,w_0\rangle}|^2|F(z)|^2 \to 0$ pointwise in z as $w\to w_0$ and

$$z\mapsto |e^{i\frac{\lambda}{2}\langle z,w\rangle} - e^{i\frac{\lambda}{2}\langle z,w_0\rangle}|^2|F(z)|^2 \leqslant 4|F(z)|^2$$

and thus is dominated by $4|F|^2 \in L^1(\mathbb{C}_n)$. Since the inner product is continuous on $L^2(\mathbb{C}^n)$, we have $w \mapsto F\#_\lambda H(w) = (F_w, H^w)_2$ is a continuous function. Furthermore if $F_k \to F$ and $H_k \to H$ in $L^2(\mathbb{C}^n)$, then $(F_k)_w$ and $(H_k)^w$ converge uniformly in w to F_w and H^w in $L^2(\mathbb{C}^n)$. This follows from

$$|(F_k)_w - F_w|_2 = |F_k - F|_2 \text{ and } |(H_k)^w - H^w|_2 = |H_k - H|_2 .$$

By the Cauchy-Schwarz inequality,

$$\begin{aligned}
|F_k\#_\lambda H_k(w) - F\#_\lambda H(w)| &= |((F_k)_w, (H_k)^w)_2 - (F_w, H^w)_2| \\
&\leqslant |((F_k)_w, (H_k)^w)_2 - (F_w, (H_k)^w)_2| + |(F_w, (H_k)^w)_2 - (F_w, H^w)_2| \\
&= |((F_k)_w - F_w, (H_k)^w)_2| + |(F_w, (H_k)^w - H^w)_2| \\
&\leqslant |(F_k)_w - F_w|_2 |(H_k)^w|_2 + |F_w|_2 |(H_k)^w - H^w|_2 \\
&= |(F_k - F)_w|_2 |(H_k)^w|_2 + |F_w|_2 |(H_k - H)^w|_2 \\
&= |F_k - F|_2 |H_k|_2 + |F|_2 |H_k - H|_2 \to 0
\end{aligned}$$

uniformly in w as $k \to \infty$.

□

2. Let $F \in L^1(\mathbb{R}^n \times \mathbb{R}^n)$. Show if H is Schwartz, then

$$\iint\int |H(x,y)F(x,v)|\, d_n x\, d_n y\, d_n v < \infty.$$

Solution: Choose $N > n/2$. Since H is Schwartz, there is a K with $|H(x,y)| \leqslant K(1+|x|^2+|y|^2)^{-2N}$. Now $(1+a+b)^2 \geqslant (1+a)(1+b)$ for $a, b \geqslant 0$. So $(1+|x|^2+|y|^2)^{-2N} \leqslant (1+|y|^2)^{-N}$. By Remark 2.27, $(1+|y|^2)^{-N}$ is integrable. Thus

$$\iint\int |H(x,y)F(x,v)|\, d_n x\, d_n y\, d_n v \leqslant$$
$$K \iint\int (1+|y|^2)^{-N} |F(x,v)|\, d_n v\, d_n y\, d_n x < \infty.$$

□

3. Show if F is in $L^1(\mathbb{R}^n \times \mathbb{R}^n)$ and $\lambda \neq 0$, then the Weyl transform $\pi_\lambda(F)$ as a distribution is given by the function $\mathcal{F}_2(F) \circ \Phi$.

Solution: We first determine the distribution $\pi_\lambda(F)$. Recall from Equation 7.33 that if $H \in \mathcal{S}(\mathbb{R}^n \times \mathbb{R}^n)$, then

$$\begin{aligned}
\pi_\lambda(F)(H) &= \frac{1}{|\lambda|^n} \iint \mathcal{F}_2(H \circ \Phi^{-1})(x,y)\, F(x,y)\, d_n x\, d_n y \\
&= \frac{1}{|\lambda|^n} \iint (H \circ \Phi^{-1})(x,y)\, \mathcal{F}_2 F(x,y)\, d_n x\, d_n y.
\end{aligned}$$

Now $H \circ \Phi^{-1}$ is Schwartz and

$$\iint |H \circ \Phi^{-1}(x,y)\mathcal{F}_2 F(x,y)|\, d_n x\, d_n y =$$
$$\iint |H \circ \Phi^{-1}(x,y) \int F(x,v) e^{-i(y,v)}\, d_n v|\, d_n x\, d_n y$$
$$\leqslant \iint \int |H \circ \Phi^{-1}(x,y) F(x,v)|\, d_n v\, d_n x\, d_n y < \infty$$

by Exercise 7.3.2. Since Φ preserves integrability, we have by Equation 7.20

$$H(x,y)(\mathcal{F}_2 F) \circ \Phi$$

is integrable for each Schwartz function and

$$\pi_\lambda(F)(H) = \iint H(x,y)(\mathcal{F}_2 F) \circ \Phi(x,y)\, d_n x\, d_n y.$$

Thus the distribution $\pi_\lambda(F)$ is given by the function $(\mathcal{F}_2 F) \circ \Phi$. □

4. Show if F is in $L^1(\mathbb{R}^n \times \mathbb{R}^n)$ and $\lambda \neq 0$, then the bounded linear operator $\pi_\lambda(F)$ on $L^2(\mathbb{R}^n)$ is given on Schwartz functions f by

$$\pi_\lambda(F)(f)(x) = \int (\mathcal{F}_2 F) \circ \Phi(x,y) f(y)\, d_n y$$

a.e. x. (**Hint:** Show $(\mathcal{F}_2 F) \circ \Phi(x,y) f(y)$ is integrable in y for a.e. x.)

Solution:

We have by Definition 7.18 that $\pi_\lambda(F)$ for an integrable function F is the operator given weakly by

$$(\pi_\lambda(F)f, h)_2 = \int_{\mathbb{R}^n} \int_{\mathbb{R}^n} F(u,v)(\pi_\lambda((u,v),0))f, h)_2\, d_n u\, d_n v\,.$$

Let f be Schwartz. Since F is integrable, the function

$$W_x : (u,v) \mapsto F(u,v) e^{\frac{1}{2} i\lambda(u,v)} e^{-i\lambda(v,x)} f(x-u)$$

is integrable for each x for

$$\int\int |F(u,v) e^{\frac{1}{2} i\lambda(u,v)} e^{-i\lambda(v,x)} f(x-u)|\, d_n u\, d_n v = \int\int |F(u,v) f(x-u)|\, d_n u\, d_n v$$
$$\leqslant |f|_\infty \int\int |F(u,v)|\, d_n u\, d_n v < \infty\,.$$

So if h is bounded and integrable, we see by Fubini

$$
\begin{aligned}
(\pi_\lambda(F)f,h)_2 &= \int\int F(u,v)\int e^{\frac{1}{2}i\lambda(u,v)}e^{-i\lambda(v,x)}f(x-u)\bar{h}(x)\,d_nu\,d_nv\,d_nx \\
&= \int\int\int F(x+u,v)e^{\frac{1}{2}i\lambda(u+x,v)}e^{-i\lambda(v,x)}f(-u)\bar{h}(x)\,d_nu\,d_nv\,d_nx \\
&= \int\int\int F(x-u,v)e^{\frac{1}{2}i\lambda(-u+x,v)}e^{-i\lambda(v,x)}f(u)\bar{h}(x)\,d_nu\,d_nv\,d_nx \\
&= \int\int\int F(x-u,v)e^{\frac{1}{2}i\lambda(-x-u,v)}\,d_nv\,f(u)\bar{h}(x)\,d_nu\,d_nx \\
&= \int\int \mathcal{F}_2F(x-u,\frac{1}{2}\lambda(x+u))f(u)\bar{h}(x)\,d_nu\,d_nx \\
&= \int\left(\int \mathcal{F}_2F\circ\Phi(x,u)f(u)\,d_nu\right)\bar{h}(x)\,d_nx
\end{aligned}
$$

where the integral $\int\mathcal{F}_2F\circ\Phi(x,u)f(u)\,d_nu$ exists a.e. x and is integrable in x. Indeed,

$$
\begin{aligned}
\int\left|\int \mathcal{F}_2(F\circ\Phi(x,u)f(u)\,d_nu\right|\,dx &= \int\left|\int\int F\circ\Phi(x,\omega)e^{-i(u,\omega)}\,f(u)d_n\omega\,d_nu\right|\,d_nx \\
&= \int\left|\int F\circ\Phi(x,\omega)\int f(u)e^{-i(u,\omega)}\,d_nu\,d_n\omega\right|\,d_nx \\
&= \int\left|\int F\circ\Phi(x,\omega)\mathcal{F}(f)(\omega)\,d_n\omega\right|\,d_nu \\
&\leqslant \int\int |F\circ\Phi(x,\omega)\mathcal{F}(f)(\omega)|\,d_n\omega\,d_nu \\
&\leqslant |F\circ\Phi|_1|\mathcal{F}(f)|_\infty < \infty\,.
\end{aligned}
$$

Putting these together implies the L^2 function $\pi_\lambda(F)f$ is equal a.e. to the function $x\mapsto\int\mathcal{F}_2F\circ\Phi(x,u)f(u)\,d_nu$.

□

5. Show $L^1(\mathbb{R}^n\times\mathbb{R}^n)$ is a noncommutative Banach $*$ algebra. That is show twisted convolution is associative, $||F\#_\lambda H||_1\leqslant||F||_1||H||_1$, $F\mapsto F^*$ is a norm preserving conjugate linear isomorphism and $(F\#_\lambda H)^*=H^*\#_\lambda F^*$.

Solution: The following arguments use the Fubini Theorem and translation invariance of Lebesgue measure. We show twisted convolution is associative.

$$\begin{aligned}
(F\#G)\#H(z') &= \int e^{\frac{1}{2}i\lambda\langle z,z'\rangle} F\#G(z)H(z'-z)\,d_{2n}z \\
&= \int e^{\frac{1}{2}i\lambda\langle z,z'\rangle} \int e^{\frac{1}{2}i\lambda\langle w,z\rangle} F(w)G(z-w)H(z'-z)\,d_{2n}w\,d_{2n}z \\
&= \int e^{\frac{1}{2}i\lambda\langle z+w,z'\rangle} \int e^{\frac{1}{2}i\lambda\langle w,z+w\rangle} F(w)G(z)H(z'-w-z)\,d_{2n}z\,d_{2n}w \\
&= \int e^{\frac{1}{2}i\lambda\langle w,z'\rangle} F(w) \int e^{\frac{1}{2}i\lambda\langle z,z'-w\rangle} G(z)H(z'-w-z)\,d_{2n}z\,d_{2n}w \\
&= \int e^{\frac{1}{2}i\lambda\langle w,z'\rangle} F(w)G\#H(z'-w)\,d_{2n}w \\
&= F\#(G\#H)(z').
\end{aligned}$$

Thus # is associative.

$$\begin{aligned}
||F\#H||_1 &= \int |F\#H(w)|\,d_{2n}w \\
&= \int \left| \int e^{\frac{1}{2}i\lambda\langle z,w\rangle} F(z)H(w-z)\,d_{2n}z \right| d_{2n}w \\
&\leqslant \int\int |F(z)H(w-z)|\,d_{2n}z\,d_{2n}w \\
&\leqslant \int\int |F(z)H(w)|\,d_{2n}z\,d_{2n}w \\
&= ||F||_1||H||_1.
\end{aligned}$$

Clearly $||F^*||_1 = \int |\bar{F}(-z)|\,d_{2n}z = \int |F(z)|\,d_{2n}z = ||F||_1$.
Finally note

$$\begin{aligned}
(F\#H)^*(z) &= \overline{F\#H(-z)} \\
&= \overline{\int e^{\frac{1}{2}i\lambda\langle w,-z\rangle} F(w)H(-z-w)\,d_{2n}w} \\
&= \int e^{\frac{1}{2}i\lambda\langle w,z\rangle} \bar{H}(-z-w)\bar{F}(w)\,d_{2n}w \\
&= \int e^{\frac{1}{2}i\lambda\langle w-z,z\rangle} \bar{H}(-z-w+z)\bar{F}(w-z)\,d_{2n}w \\
&= \int e^{\frac{1}{2}i\lambda\langle w,z\rangle} H^*(w)F^*(z-w)\,d_{2n}w \\
&= (H^*\#F^*)(z).
\end{aligned}$$

Note for $\lambda \neq 0$, Theorem 7.20 shows the mapping $F \mapsto \pi_\lambda(F)$ extends from $L^1 \cap L^2$ to an algebra isomorphism of $L^2(\mathbb{R}^n \times \mathbb{R}^n)$ onto the space of Hilbert-Schmidt operators on $L^2(\mathbb{R}^n)$. Thus if twisted convolution were commutative on $L^1(\mathbb{R}^n \times \mathbb{R}^n)$, all Hilbert-Schmidt operators would commute. □

6. Show a unitary representation π of H_n on a Hilbert space $\mathcal{H}$ with $\pi(0,t) = e^{i\lambda t}I$ for all t where λ is a nonzero real number is irreducible if and only if the only closed invariant subspaces under all the operators $\pi(F) = \iint F(x,y)\pi(x + iy, 0)\, d_n x\, d_n y$ for $F \in L^1(\mathbb{R}^n \times \mathbb{R}^n)$ are $\{0\}$ and $\mathcal{H}$.

Solution:

By Corollary 6.110, we know π is irreducible if and only if the only invariant subspaces of $\pi(H)$ where $H \in L^1(\mathbb{R}^n \times \mathbb{R}^n \times \mathbb{R})$ are $\{0\}$ and $\mathcal{H}$. Since the linear span of the function $F \times f$ where $F \in L^1(\mathbb{R}^n \times \mathbb{R}^n)$ and $f \in L^1(\mathbb{R}^n)$ and $F \times f(x,y,t) = F(x,y)f(t)$ is dense in $L^1(\mathbb{R}^n \times \mathbb{R}^n \times \mathbb{R})$ and $||\pi(H)|| \leqslant |H|_1$, we see π is irreducible if and only if the only closed invariant subspaces under the operators $\pi(F \times f)$ for $F \in L^1(\mathbb{R}^n \times \mathbb{R}^n)$ and $f \in L^1(\mathbb{R})$ are $\{0\}$ and $\mathcal{H}$. But since $\pi(0,0,t) = e^{i\lambda t}I$, one can check $\pi(F \times f) = \hat{f}(-\lambda)\pi(F)$ where $\hat{f}(\lambda) = \int f(t)e^{-i\lambda t} d_1 t$. This implies π is irreducible if and only if the only closed invariant subspaces under the operators $\pi(F)$ for $F \in L^1(\mathbb{R}^n \times \mathbb{R}^n)$ are $\{0\}$ and $\mathcal{H}$.

□

7. As seen in Exercise 2.5.16, the space of Hilbert-Schmidt operators is an example of a Hilbert algebra. Let $d_\lambda z$ be the measure on $\mathbb{R}^n \times \mathbb{R}^n$ given by $d_\lambda z = |\lambda|^{-n}\, d_{2n} z = |\lambda|^{-n}\, d_n x \times d_n y$. Show the Weyl transform π_λ is a Hilbert algebra isometry of $L^2(\mathbb{R}^n \times \mathbb{R}^n, d_\lambda z)$ under λ-twisted convolution and function adjoints onto the Hilbert algebra of Hilbert-Schmidt operators on $L^2(\mathbb{R}^n)$ under composition and operator adjoints.

Solution:

By Theorem 7.20, the mapping π_λ maps $L^2(\mathbb{R}^n \times \mathbb{R}^n)$ linearly onto the Hilbert algebra of Hilbert-Schmidt operators on $L^2(\mathbb{R}^n)$. Moreover,

$$\pi_\lambda(F \#_\lambda H) = \pi_\lambda(F)\pi_\lambda(H), \quad \pi_\lambda(F^*) = \pi_\lambda(F)^*,$$

and

$$\begin{aligned} ||\pi_\lambda(F)|| &= \frac{1}{|\lambda|^{n/2}} \left(\iint |F(x,y)|^2\, d_n x\, d_n y \right)^{1/2} \\ &= \left(\iint |F(z)|^2 |\lambda|^{-n}\, d_{2n} z \right)^{1/2} \\ &= |F|_{L^2(\mathbb{R}^n \times \mathbb{R}^n,\, |\lambda|^{-n} d_n x \times dy_n)}. \end{aligned}$$

This gives the result.

□

8. Suppose λ is a nonzero real number. Let π be a unitary representation of the Heisenberg group H_n on a Hilbert space $\mathcal{H}$ satisfying $\pi(0,t) = e^{i\lambda t}I$ for all t. Define π' on H_n by $\pi'(x + iy, t) = \pi(x + \frac{iy}{\lambda}, \frac{t}{\lambda})$. Show π' is a unitary representation of H_n satisfying $\pi'(0,t) = e^{it}I$ for all t.

Solution: It suffices to show $\phi : (x + iy, t) \mapsto (x + \frac{iy}{\lambda}, \frac{t}{\lambda})$ is an automorphism of the group H_n and then note $\pi'(0,t) = \pi(0, \frac{t}{\lambda}) = e^{i\lambda(t/\lambda)}I = e^{it}I$. But

$$\begin{aligned}\phi((x+iy,t)^{-1}) &= \phi(-x-iy,-t) = (-x - \frac{iy}{\lambda}, -\frac{t}{\lambda}) \\ &= (x + \frac{iy}{\lambda}, \frac{t}{\lambda})^{-1} = \phi(x+iy,t)^{-1}\end{aligned}$$

and

$$\begin{aligned}\phi((x_1+iy_1,t_1)(x_2+iy_2,t_2)) &= \phi(x_1+x_2+i(y_1+y_2), t_1+t_2+\frac{1}{2}(x_1\cdot y_2 - x_2\cdot y_1)) \\ &= (x_1+x_2+i\frac{y_1+y_2}{\lambda}, \frac{1}{\lambda}(t_1+t_2+\frac{1}{2}(x_1\cdot y_2 - x_2\cdot y_1))) \\ &= (x_1+i\frac{y_1}{\lambda}, \frac{t_1}{\lambda})(x_2+i\frac{y_2}{\lambda}, \frac{t_2}{\lambda}) \\ &= \phi(x_1+iy_1,t_1)\phi(x_2+iy_2,t_2)\,.\end{aligned}$$

□

9. Let U be an $n \times n$ unitary matrix. Let $\lambda \neq 0$. Show the representations π_λ and $(z,t) \mapsto \pi_\lambda(zU,t)$ are unitarily equivalent.

Solution:

We first note $(z,t) \mapsto (Uz,t)$ is an automorphism of H_n for $\langle Uz, Uz'\rangle = -\mathrm{Im}(Uz,Uz') = -\mathrm{Im}(z,z') = \langle z,z'\rangle$ and we see

$$\begin{aligned}(z_1,t_1)(z_2,t_2)^{-1} &= (z_1 - z_2, t_1 - t_2 + \frac{1}{2}\langle z_1, -z_2\rangle) \\ &\mapsto (U(z_1-z_2), t_1-t_2+\frac{1}{2}\langle Uz_1, -Uz_2\rangle) \\ &= (Uz_1 - Uz_2, t_1-t_2+\frac{1}{2}\langle Uz_1,-Uz_2\rangle) \\ &= (Uz_1,t_1)(-Uz_2,-t_2) \\ &= (Uz_1,t_1)(Uz_2,t_2)^{-1}\,.\end{aligned}$$

It follows that $\pi' : (x,t) \mapsto \pi_\lambda(Uz,t)$ is an irreducible representation with $\pi'(0,t) = \pi_\lambda(0,t) = e^{i\lambda t}I$. By Theorem 7.26, π' and π_λ are equivalent.

□

10. Let $\tilde{\pi}_\lambda(z,t) = \pi_\lambda(iz,t)$. Show the Fourier transform $\mathcal{F} : L^2(\mathbb{R}^n) \to L^2(\mathbb{R}^n)$ is a unitary equivalence of π_1 with $\tilde{\pi}_1$.

Solution:

$$\begin{aligned}
F\pi_1(x+iy,0)f(w) &= \int \pi_1(x+iy,0)f(u)e^{-iw\cdot u}\,d_nu \\
&= \int e^{\frac{1}{2}ix\cdot y}e^{-y\cdot u}f(u-x)e^{-iw\cdot u}\,d_nu \\
&= \int e^{\frac{1}{2}ix\cdot y}e^{-iy\cdot(u+x)}f(u)e^{-iw\cdot(u+x)}\,d_nu \\
&= e^{-\frac{1}{2}iy\cdot x}\int e^{-ix\cdot w}e^{-iw\cdot u}e^{-iy\cdot u}f(u)\,d_nu \\
&= e^{-\frac{1}{2}iy\cdot x}e^{-ix\cdot w}\int e^{-i(w+y)\cdot u}f(u)\,d_nu \\
&= e^{-\frac{1}{2}iy\cdot x}e^{-ix\cdot w}Ff(w+y) \\
&= \pi_1(-y+ix,0)Ff(w) \\
&= \pi_1(i(x+iy),0)Ff(w) \\
&= \tilde{\pi}_1(x+iy,0)Ff(w).
\end{aligned}$$

□

11. Show $\int_{\mathbb{R}^n} e^{-i(z,x)}e^{-\frac{1}{2}|x|^2}\,d_nx = e^{-\frac{1}{2}z^2}$ where $z^2 = \sum z_j^2$. (Hint: It suffices to do the one-dimensional case.)

Solution:

We start with the case $n = 1$. Using the Cauchy integral formula,

$$\begin{aligned}
\int_{-\infty}^{\infty} e^{-izx}e^{-\frac{1}{2}x^2}\,d_1x &= e^{-\frac{1}{2}z^2}\int_{-\infty}^{\infty} e^{-\frac{1}{2}(iz+x)^2}\,d_1x \\
&= \frac{1}{\sqrt{2\pi}}e^{-\frac{1}{2}z^2}\int_{-\infty}^{\infty} e^{-\frac{1}{2}(iz+x)^2}\,dx \\
&= \frac{1}{\sqrt{2\pi}}e^{-\frac{1}{2}z^2}\int_{-\infty}^{\infty} e^{-\frac{1}{2}x^2}\,dx \\
&= e^{-\frac{1}{2}z^2}
\end{aligned}$$

for $\int_{-\infty}^{\infty} e^{-\frac{1}{2}x^2}\,dx = \sqrt{2\pi}$ since by Exercise 2.3.3, $\int_{\mathbb{R}} e^{-\pi y^2}\,dy = 1$; just make the change in variables $y = \frac{1}{\sqrt{2\pi}}x$. The general result is an immediate application of Fubini for $e^{-i(z,x)}e^{-\frac{1}{2}|x|^2} = \prod_{j=1}^n e^{-iz_jx_j}e^{-\frac{1}{2}x_j^2}$ and one is dealing with integrable functions.

□

Exercise Set 7.4

1. Let f be a Schwartz function on H_n. Show $f * f^*(0,0) = |f|_2^2$. Use the inversion formula $\phi(0,0) = \int \operatorname{Tr}(\hat{\phi}(\lambda))\,|\lambda|^n\,d_1\lambda$ to show

$$|f|_2^2 = \int |\hat{f}(\lambda)|_2^2\,|\lambda|^n\,d_1\lambda.$$

Solution:

Let f be Schwartz. Then

$$\begin{aligned} f * f^*(0,0) &= \int\int f(z,t) f^*((z,t)^{-1}(0,0))\, d_{2n}z\, d_1 t \\ &= \int\int f(z,t) f^*((z,t)^{-1})\, d_{2n}z\, d_1 t \\ &= \int\int f(z,t) \overline{f(((z,t)^{-1})^{-1})}\, d_{2n}z\, d_1 t \\ &= \int\int f(z,t) \overline{f(z,t)}\, d_{2n}z\, d_1 t \\ &= |f|_2^2 . \end{aligned}$$

Moreover, $f, f^* \in L^1(H_n) \cap L^2(H_n)$ and thus by the Plancherel Theorem 7.28,

$$f * f^*(z,t) = \int \mathrm{Tr}(\pi_\lambda(-z,-t)\pi_\lambda(f * f^*))|\lambda|^n\, d_1\lambda$$

and we see

$$|f|_2^2 = \int \mathrm{Tr}(\pi_\lambda(f)\pi_\lambda(f^*))|\lambda|^n\, d_1\lambda = \int |\hat{f}(\lambda)|_2^2 |\lambda|^n\, d_1\lambda .$$

□

2. Let T, T_1 and T_2 be strongly measurable functions from $\mathbb{R}^n$ into $\mathcal{H}_2$, the Hilbert space of Hilbert-Schmidt operators on $L^2(\mathbb{R}^n)$. Show:

(a) $x \mapsto (T_1(x)f_1, T_2(x)f_2)_2$ is measurable for each pair of functions f_1 and f_2 in $L^2(\mathbb{R}^n, d_n x)$.

(b) Show $x \mapsto ||T(x)||_2$ is measurable.

(c) Show $x \mapsto (T_1(x), T_2(x))_2$ is measurable.

(d) Show $x \mapsto T(x)$ is strongly measurable if and only if it is measurable; i.e., $T^{-1}(V)$ is measurable in $\mathbb{R}^n$ for any open subset V of $\mathcal{H}_2$.

(e) Define $L^2(\mathbb{R}^n, \mathcal{H}_2)$ to be the vector space of all measurable functions $T : \mathbb{R}^n \to \mathcal{H}_2$ satisfying

$$\int_{\mathbb{R}^n} ||T(x)||_2^2\, d_n x < \infty$$

identifying them if they are equal almost everywhere. Show

$$(T_1, T_2) = \int (T_1(x), T_2(x))_2\, d_n x.$$

defines an inner product on $L^2(\mathbb{R}^n, \mathcal{H}_2)$ which makes it a Hilbert space.

Solution:

We know $x \mapsto T_j(x)f_j$ is measurable from X into $L^2(\mathbb{R}^n)$. $L^2(\mathbb{R}^n)$ is separable and consequently $L^2(\mathbb{R}^n) \times L^2(\mathbb{R}^n)$ is separable. This implies every open subset of this space is a countable union of sets of form $U_1 \times U_2$ where U_1 and U_2 are open in $L^2(\mathbb{R}^n)$. But the preimage of $U_1 \times U_2$ in X under the mapping $x \mapsto (T_1(x)f_1, T_2(x)f_2)$ is $A_1 \cap A_2$ where $A_j = \{x \mid T_j(x)f_j \in U_j\}$ are measurable. We therefore see the mapping $x \mapsto (T_1(x)f_1, T_2(x)f_2(x))$ is measurable into $L^2(\mathbb{R}^n) \times L^2(\mathbb{R}^n)$. The inner product $(f, h) \mapsto \int f(x)\bar{h}(x)\, d_n x$ being continuous on $L^2(\mathbb{R}^n) \times L^2(\mathbb{R}^n)$ then gives $x \mapsto (T_1(x)f_1, T_2(x)f_2)_2$ is measurable. So (a) holds.

Let f_k for $k = 1, 2, \ldots$ be an orthonormal basis of $L^2(\mathbb{R}^n)$. Definition 2.31 gives $||T(x)||_2^2 = \sum |T(x)f_k|_2^2 = \sum (T(x)f_k, T(x)f_k)_2$ which is measurable by part (a). So (b) holds.

For (c), note by Proposition 2.32, if f_k for $k \in \mathbb{N}$ is an orthonormal basis of $L^2(\mathbb{R}^n)$, then $(T_1(x), T_2(x))_2 = \sum_k (T_1(x)f_k, T_2(x)f_k)_{L^2(\mathbb{R}^n)}$ which is measurable by (a).

Let $A \in \mathcal{H}_2$ and $\epsilon > 0$. Note $x \mapsto T(x) - A$ is measurable or strongly measurable if and only if $x \mapsto T(x)$ is measurable or strongly measurable. Setting $V = \{B \in \mathcal{H}_2 \mid ||B - A||_2 < \epsilon\}$, we then see $T(x) - A \in V$ if and only if $||T(x) - A||_2 < \epsilon$. So by (b), $x \mapsto T(x) - A$ is measurable if it is strongly measurable. Now $x \mapsto T(x)$ measurable implies $x \mapsto T(x)f$ is measurable for any $f \in L^2(\mathbb{R}^n)$ for $A \mapsto Af$ is continuous on $\mathcal{H}_2$. Hence (d) holds.

Finally for (e), it is easy to check $(\,\cdot\,,\,\cdot\,)$ is an inner product. To see this gives a Hilbert space, let T_k be a Cauchy sequence. By taking a subsequence, we may assume $||T_k - T_m|| < \frac{1}{2^{2m}}$ if $k \geqslant m$. Let $E_{k,m} = \{x \mid ||T_k(x) - T_m(x)||_2 > \frac{1}{2^m}\}$. Note

$$\begin{aligned}\frac{1}{2^{4m}} > ||T_k - T_m||^2 &= \int ||T_k(x) - T_m(x)||_2^2 \, d_n x \\ &\geqslant \frac{1}{2^{2m}} m(E_{k,m})\,.\end{aligned}$$

This give $m(E_{k,m}) < \frac{1}{2^{2m}}$. Set $E_m = \cup_{k=m+1}^{\infty} E_{k+1,k}$. Then $m(E_m) < \frac{1}{2^{2m}}$. So if $E = \cap E_m$, $m(E) = 0$. Let $x \notin E$. Then $x \notin E_m$ for some m. Thus $x \notin E_{k+1,k}$ for all $k \geqslant m+1$ and so $||T_{k+1}(x) - T_k(x)||_2 \leqslant \frac{1}{2^k}$ for $k \geqslant m+1$. Thus for $l > k > m$, one has

$$\begin{aligned}||T_l(x) - T_k(x)||_2 &\leqslant \sum_{j=k}^{l-1} ||T_{j+1}(x) - T_j(x)||_2 \\ &\leqslant \sum_{j=k}^{l-1} \left(\frac{1}{2}\right)^j < \frac{1}{2^{k-1}}\,.\end{aligned}$$

This implies $T_k(x)$ is Cauchy in $\mathcal{H}_2$ and by the completeness of $\mathcal{H}_2$, there is a $T(x) \in \mathcal{H}_2$ with $T_k(x) \to T(x)$ as $k \to \infty$. Set $T(x) = 0$ if $x \notin E$. Then

$x \mapsto T(x) \in \mathcal{H}_2$ is measurable and by Fatou's Lemma

$$\int ||T(x) - T_m(x)||_2^2 \, d_n x \leqslant \liminf_k \int ||T_k(x) - T_m(x)||^2 \, d_n \leqslant \frac{1}{2^{2m}}.$$

Thus $T_m \to T$ in $L^2(\mathbb{R}^n, \mathcal{H}_2)$.

□

3. Let X be a measurable space. Show a mapping $T : X \to \mathcal{H}_2$ is strongly measurable if and only if $x \mapsto (h_1, T(x)h_2)_2$ is measurable for each pair h_1 and h_2 of Schwartz functions on $\mathbb{R}^n$.

Solution:

If $x \mapsto T(x)$ is strongly measurable, then $x \mapsto T(x)h_2$ for a Schwartz function h_2 is measurable. Consequently the mapping $x \mapsto (h_1, T(x)h_2)$ of X into $L^2(\mathbb{R}^n) \times L^2(\mathbb{R}^n)$ is measurable. The continuity of the inner product on L^2 then implies $x \mapsto (h_1, T(x)h_2)_{L^2}$ is measurable for it is a composition of a continuous function with a measurable function.

Conversely, suppose $x \mapsto (h_1, T(x)h_2)_2$ is measurable for all Schwartz functions h_1 and h_2. Fix a Schwartz function h_2. Then if f_k, $k = 1, 2, \ldots$ is an orthonormal basis of $L^2(\mathbb{R}^n)$ consisting of Schwartz functions, one has $T(x)h_2 = \sum_k (T(x)h_2, f_k)_2 f_k$. Since $x \mapsto (f_k, T(x)h_2)_2$ is measurable for each k, $x \mapsto (T(x)h_2, f_k)_2 f_k$ is measurable from X into $L^2(\mathbb{R}^n)$ for each k. Indeed, the mapping $(c, f) \mapsto cf$ is measurable on $\mathbb{C} \times L^2(\mathbb{R}^n)$ for it is continuous and so again $x \mapsto (T(x)h_2, f_k)f_k$ is a composition of a continuous mapping with a measurable mapping. Since $x \mapsto T(x)h_2$ is a limit of measurable functions, it is a measurable function. Now if $f \in L^2(\mathbb{R}^n)$, one can choose a sequence $\{h_k\}$ of Schwartz functions such that $h_k \mapsto f$ in $L^2(\mathbb{R}^n)$. Then $T(x)f = \lim_k T(x)h_k$ for each x. We thus have $x \mapsto T(x)f$ is measurable for each $f \in L^2(\mathbb{R}^n)$. So T is strongly measurable. □

4. Suppose $\lambda \mapsto F(\lambda)$ is a strongly measurable function from $\mathbb{R}$ into $L^2(\mathbb{R}^n \times \mathbb{R}^n)$ satisfying $\int |F(\lambda)|_2^2 |\lambda|^n \, d_1\lambda < \infty$. Show $\lambda \mapsto F(\lambda) \circ \Phi^{-1}$ is a measurable function from $\mathbb{R}$ into $L^2(\mathbb{R}^n \times \mathbb{R}^n)$ satisfying

$$\int |F(\lambda) \circ \Phi^{-1}|_2^2 \, d_1\lambda = \int |F(\lambda)|_2^2 |\lambda|^n \, d_1\lambda.$$

Solution: By Exercise 7.4.3, we need only show

$$\lambda \mapsto \iint F(\lambda) \circ \Phi^{-1}(x, y) h(x, y) \, d_n x \, d_n y$$

is measurable for Schwartz functions h on $\mathbb{R}^n \times \mathbb{R}^n$. But this follows since

$$\lambda \mapsto \iint F(\lambda)(x, y) h \circ \Phi(x, y) \, d\Phi(x, y) = |\lambda|^n \iint F(\lambda)(x, y) h \circ \Phi(x, y) \, d_n x \, d_n y$$

is measurable for $h \circ \Phi$ is again Schwartz. Note $d\Phi(x, y) = |\lambda|^n \, d_n x \, d_n y$ implies $|F(\lambda) \circ \Phi^{-1}|_2^2 = |F(\lambda)|_2^2 |\lambda|^n$ which yields the equality of the integrals. □

5. Show if $\lambda \mapsto F(\lambda)$ is measurable from $\mathbb{R}$ into $\mathcal{H}_2 = L^2(\mathbb{R}^n \times \mathbb{R}^n)$ and $\int |F(\lambda)|_2^2 \, d_1\lambda < \infty$, then there is a function $\widetilde{F} \in L^2(\mathbb{R}^n \times \mathbb{R}^n \times \mathbb{R})$ satisfying

$$F(\lambda)(x,y) = \widetilde{F}(x,y,\lambda)$$

a.e. (x,y) for a.e. λ.
Solution:
Pick an orthonormal basis $H_1, H_2, \ldots$ of Schwartz functions for $L^2(\mathbb{R}^n \times \mathbb{R}^n)$. We know $\lambda \mapsto (F(\lambda), H_k)$ is measurable for each k and

$$\sum_k |(F(\lambda), H_k)_2|^2 = |F(\lambda)|_2^2 \text{ for a.e. } \lambda\,.$$

Set $a_k(\lambda) = (F(\lambda), H_k)_2$. Define a measurable function $\widetilde{F}$ by $\widetilde{F}(x,y,\lambda) = \sum_k a_k(\lambda) H_k(x,y)$. Note this function converges in L^2 for

$$\begin{aligned}\sum_k \int |a_k(\lambda) H_k(x,y)|^2 \, d_n x \, d_n y \, d_1\lambda &= \sum_k \int |a_k(\lambda)|^2 \, d_1\lambda \\ &= \int \sum_k |(F(\lambda), H_k)_2|^2 \, d\lambda \\ &= \int |F(\lambda)|_2^2 \, d_1\lambda < \infty\,.\end{aligned}$$

□

6. Let U be the unitary mapping from $L^2(H_n)$ onto $L^2(\mathbb{R}, \mathcal{H}_2, |\lambda|^n d_1\lambda)$ given by $U(f)(\lambda) = \hat{f}(\lambda)$. Let B be the biregular representation of H_n. Show

$$U(B(g_1,g_2)f)(\lambda) = \pi_\lambda(g_1) U f(\lambda) \pi_\lambda(g_2)^{-1} = \pi_\lambda \times \bar{\pi}_\lambda(g_1,g_2)(U(f)(\lambda))$$

a.e. λ for each $f \in L^2(H_n)$ and any g_1 and g_2 in H_n.
Solution:
Note

$$\begin{aligned}U(B(g_1,g_2)f)(\lambda)\phi &= \pi_\lambda(B(g_1,g_2)f)\phi \\ &= \int_{H_n} (B(g_1,g_2)f(g)) \pi_\lambda(g)\phi \, dg \\ &= \int_{H_n} f(g_1^{-1} g g_2) \pi_\lambda(g)\phi \, dg \\ &= \int_{H_n} f(g) \pi_\lambda(g_1 g g_2^{-1})\phi \, dg \\ &= \int_{H_n} f(g) \pi_\lambda(g_1) \pi_\lambda(g) \pi_\lambda(g_2^{-1})\phi \, dg \\ &= \pi_\lambda(g_1) \int_{H_n} f(g) \pi_\lambda(g) \, dg \pi_\lambda(g_2)^{-1}\phi \\ &= \pi_\lambda(g_1) \pi_\lambda(f) \pi_\lambda(g_2)^{-1}\phi\end{aligned}$$

for functions $f \in L^1(H_n) \cap L^2(H_n)$ and $\phi \in L^2(\mathbb{R}^n)$. Since π_λ, B and U are unitary, this implies $UB(g_1, g_2)f(\lambda) = \pi_\lambda(g_1)Uf(\lambda)\pi_\lambda(g_2)^{-1}$ a.e. λ for each $f \in L^2(H_n)$ and g_1 and g_2.

But by Definitions 6.69 and 6.71, the tensor product $\pi_\lambda \times \bar{\pi}_\lambda$ is the representation on $\mathcal{H}_2 = L^2(\mathbb{R}^n) \otimes \overline{L^2(\mathbb{R}^n)} = \mathcal{B}_2(L^2(\mathbb{R}^n), L^2(\mathbb{R}^n))$ given by $\pi_\lambda(g_1) \otimes \bar{\pi}_\lambda(g_2)(T) = \pi_\lambda(g_1)T\pi_\lambda(g_2)^* = \pi_\lambda(g_1)T\pi_\lambda(g_2)^{-1}$.

So $UB(g_1, g_2)f(\lambda) = \pi_\lambda \times \bar{\pi}_\lambda(g_1, g_2)Uf(\lambda)$ a.e. λ for each $f \in L^2(H_n)$ and g_1 and g_2 in H_n.

□

7. Use the Plancherel Theorem to show if f_1 and f_2 are in $L^2(H_n)$, then the continuous function $f_1 * f_2^*$ on H_n is given at each point by

$$f_1 * f_2^*(z, t) = \int \mathrm{Tr}(\pi_\lambda(-z, -t)\pi_\lambda(f_1)\pi_\lambda(f_2)^*)\, |\lambda|^n \, d_1\lambda.$$

Solution: We first note this was already part of Theorem 7.28. We give a slightly different argument here.

Take $f_1, f_2 \in L^2(H_n)$. Then as seen in the proof of the Plancherel Theorem,

$$f_1 * f_2^*(g) = (f_1, L(g)f_2)_{L^2(H_n)}\,.$$

Also the mapping $U : L^2(H_n) \to L^2(\mathbb{R}, \mathcal{H}_2, |\lambda|^n d_1\lambda)$ is given for a.e. λ by

$$Uf(\lambda) = \hat{f}(\lambda) = \pi_\lambda(F_\lambda) = \pi_\lambda(\mathcal{F}_c^{-1} f(\cdot, \lambda))\,.$$

Indeed, this formula holds for the λ for which $\mathcal{F}_c^{-1} f(\cdot, \lambda) \in L^2(\mathbb{R}^n \times \mathbb{R}^n)$ and this is for a.e. λ by Fubini's Theorem, since

$$\int_{\mathbb{R}} \iint_{\mathbb{R}^n \times \mathbb{R}^n} |f(z, t)|^2 \, d_{2n}z \, d_1 t = \int_{\mathbb{R}} \iint_{\mathbb{R}^n \times \mathbb{R}^n} |\mathcal{F}_c^{-1} f(z, \lambda)|^2 \, d_{2n}z \, d_1\lambda\,.$$

Also we claim

$$U(L(g)f)(\lambda) = \pi_\lambda(g)Uf(\lambda) \text{ a.e. } \lambda$$

for any $f \in L^2(H_n)$. Indeed, we know this holds for $f \in L^1 \cap L^2$. Thus take a sequence f_k in $L^1 \cap L^2$ converging in L^2 to f. Since U is unitary, $U(L(g)f_k))$ converges in $L^2(\mathbb{R}, \mathcal{H}_2, |\lambda|^n \, d_1\lambda)$ to $U(L(g)f)$. By taking a subsequence we may assume this converges pointwise almost everywhere.

Finally, since U is unitary, one has

$$\begin{aligned}
f_1 * f_2^*(g) &= (Uf_1, U(L(g)f_2))_{L^2(\mathbb{R},\mathcal{H}_2,|\lambda|^n d_1\lambda)} \\
&= \int (Uf_1(\lambda), U(L(g)f_2)(\lambda))_2 |\lambda|^n \, d_1\lambda \\
&= \int (Uf_1(\lambda), \pi_\lambda(g) Uf_2(\lambda))_2 |\lambda|^n \, d_1\lambda \\
&= \int \mathrm{Tr}\left(\hat{f}_1(\lambda)(\pi_\lambda(g)\hat{f}_2(\lambda))^*\right) |\lambda|^n d_1\lambda \\
&= \int \mathrm{Tr}\left(\hat{f}_1(\lambda)\hat{f}_2(\lambda)^* \pi_\lambda(g)^*\right) |\lambda|^n \, d_1\lambda \\
&= \int \mathrm{Tr}\left(\hat{f}_1(\lambda)\hat{f}_2(\lambda)^* \pi_\lambda(g^{-1})\right) |\lambda|^n \, d_1\lambda \\
&= \int \mathrm{Tr}\left(\pi_\lambda(g^{-1})\hat{f}_1(\lambda)\hat{f}_2(\lambda)^*\right) |\lambda|^n \, d_1\lambda \, .
\end{aligned}$$

□

Chapter 8

Compact Groups

Exercise Set 8.1

1. Let G be a compact Hausdorff group with Haar measure m normalized so that $m(G) = 1$.

(a) Show $|f|_p \leqslant |f|_q$ if $1 \leqslant p \leqslant q \leqslant \infty$ for measurable functions f.

(b) Show $L^1(G) \supseteq L^p(G) \supseteq L^q(G) \supseteq L^\infty(G) \supseteq C(G)$ if $1 < p < q < \infty$.

Solution: Let $1 \leqslant p \leqslant q \leqslant \infty$. Note $1 \in L^s(G)$ for $1 \leqslant s \leqslant \infty$ and $|1|_s = 1$. Let $1 \leqslant p < q < \infty$. Then if $\frac{1}{s} + \frac{p}{q} = 1$, we have

$$\begin{aligned}\int |f|^p \, dm &\leqslant \left(\int (|f|^p)^{\frac{q}{p}} \, dm \right)^{\frac{p}{q}} |1|_s \\ &= \left(\int |f|^q \, dm \right)^{\frac{p}{q}}.\end{aligned}$$

Thus $|f|_p^p \leqslant |f|_q^p$ and we see $|f|_p \leqslant |f|_q$. Clearly $|f|_q \leqslant |f|_\infty$. Hence we have the inclusions. □

2. Let G be a compact Hausdorff group. Let π and π' be unitary representations of G on $\mathcal{H}$ and $\mathcal{H}'$. Define P on $\mathcal{B}(\mathcal{H}, \mathcal{H}')$ by

$$P(T) = \int_G \pi'(g) T \pi(g^{-1}) \, dg.$$

(a) Show $||P|| \leqslant 1$ and $P^2 = P$.

(b) Show P is a projection of $\mathcal{B}(\mathcal{H}, \mathcal{H}')$ onto $\mathrm{Hom}_G(\pi, \pi')$.

(c) Show P restricted to $\mathcal{B}_2(\mathcal{H}, \mathcal{H}')$ is the orthogonal projection of $\mathcal{H}' \otimes \bar{\mathcal{H}}$ onto $\mathrm{Hom}_G(\pi, \pi')_2$.

Solution:

We will use this in a strong sense and work only formally. First

$$\begin{aligned}
P^2(T)v &= \int \pi'(g)P(T)\pi(g^{-1})v\,dg \\
&= \int \pi'(g)\int \pi'(x)T\pi(x^{-1})\pi(g^{-1})v\,dx\,dg \\
&= \iint \pi'(gx)T\pi((gx)^{-1})v\,dx\,dg \\
&= \iint \pi'(x)T\pi(x^{-1})v\,dx\,dg \\
&= \int P(T)v\,dg \\
&= P(T)v
\end{aligned}$$

since we are taking Haar measure on G to have total mass 1.

Since $||P(T)v||_{\mathcal{H}'} \leqslant ||T||\,||v||_{\mathcal{H}}$, we see $||P|| \leqslant 1$. Next note if $T \in \mathrm{Hom}(\pi, \pi')$, then $P(T) = T$ for $\pi'(x)T\pi(x^{-1}) = T$ for all x. Also

$$P(T) \in \mathrm{Hom}(\pi, \pi')$$

for each T since

$$\begin{aligned}
P(T)\pi(x)v &= \int \pi'(g)T\pi(g^{-1}x)v\,dg \\
&= \int \pi'(xg)T\pi(g^{-1})v\,dg \\
&= \pi'(x)\int \pi'(g)T\pi(g^{-1})v\,dg \\
&= \pi'(x)P(T)v\,.
\end{aligned}$$

Now if T is Hilbert-Schmidt, $P(T)$ is Hilbert-Schmidt for if $\{e_\alpha\}_{\alpha\in A}$ is an orthonormal basis of $\mathcal{H}$, then the vectors $\pi(x^{-1})e_\alpha$ form an orthonormal basis and we see if F is a finite subset of A, then

$$\begin{aligned}
\sum_{\alpha\in F} ||P(T)e_\alpha||^2_{\mathcal{H}'} &\leqslant \sum_{\alpha\in F}\int ||\pi'(x)T\pi(x^{-1})e_\alpha||^2_{\mathcal{H}'}\,dx \\
&= \int \sum_{\alpha\in F} ||\pi'(x)T\pi(x^{-1})e_\alpha||^2\,dx \\
&= \int \sum_{\alpha\in F} ||T\pi(x^{-1})e_\alpha||^2_{\mathcal{H}'}\,dx \\
&\leqslant \int ||T||_2^2\,dx \\
&= ||T||_2^2\,.
\end{aligned}$$

Taking $F \nearrow A$, we see $P(T)$ is Hilbert-Schmidt if T is Hilbert-Schmidt and also $||P(T)||_2 \leqslant ||T||_2$. So P is a projection of $\mathcal{B}_2(\mathcal{H}, \mathcal{H}')$ onto $\mathrm{Hom}(\pi, \pi')_2$. We claim it is orthogonal. It suffices to show $P^* = P$. Let T and S be Hilbert-Schmidt operators from $\mathcal{H}$ into $\mathcal{H}'$ and let $\{e_\alpha\}_{\alpha \in A}$ be an orthonormal basis of $\mathcal{H}$. Then the set C consisting of the α where $Te_\alpha \neq 0$ or $Se_\alpha \neq 0$ is countable. Again note for each x, the collection $\{\pi(x)e_\alpha\}$ is an orthonormal basis. Using Proposition 2.32, we see

$$\begin{aligned}
(P(T), S)_2 &= \sum_{\alpha \in A} (P(T)e_\alpha, Se_\alpha)_{\mathcal{H}'} \\
&= \sum_{\alpha \in C} (\int \pi'(x)T\pi(x^{-1})e_\alpha \, dx, Se_\alpha)_{\mathcal{H}'} \\
&= \int \sum_{\alpha \in C} (\pi'(x)T\pi(x^{-1})e_\alpha, Se_\alpha)_{\mathcal{H}'} \, dx \\
&= \int \sum_{\alpha \in C} (\pi'(x)T\pi(x^{-1})\pi(x)e_\alpha, S\pi(x)e_\alpha)_{\mathcal{H}'} \, dx \\
&= \int \sum_{\alpha \in C} (Te_\alpha, \pi'(x^{-1})S\pi(x)e_\alpha)_{\mathcal{H}'} \, dx \\
&= \sum_{\alpha \in C} (Te_\alpha, \int \pi'(x^{-1})S\pi(x)e_\alpha \, dx)_{\mathcal{H}'} \\
&= \sum_{\alpha \in A} (Te_\alpha, \int \pi'(x)S\pi(x^{-1})e_\alpha \, dx)_{\mathcal{H}'} \\
&= (T, P(S))_2 \,.
\end{aligned}$$

Note we can interchange summation and integration since

$$\begin{aligned}
\sum_{\alpha \in C} |(\pi'(x)T\pi(x^{-1})e_\alpha, Se_\alpha)| &\leqslant \sum_{\alpha \in C} ||\pi'(x)T\pi(x^{-1})e_\alpha||_{\mathcal{H}'} ||Se_\alpha||_{\mathcal{H}'} \\
&\leqslant \left(\sum_{\alpha \in C} ||\pi(x)T\pi(x^{-1})e_\alpha||^2 \right)^{1/2} \left(\sum_{\alpha \in C} ||Se_\alpha||^2_{\mathcal{H}'} \right)^{1/2} \\
&\leqslant \left(\sum_{\alpha} ||\pi(x)Te_\alpha||^2 \right)^{1/2} \left(\sum_{\alpha} ||Se_\alpha||^2_{\mathcal{H}'} \right)^{1/2} \\
&= ||T||_2 ||S||_2
\end{aligned}$$

and so we have domination by the L^1 function $x \mapsto ||T||_2||S||_2$.

□

3. Let π and π' be unitary representations of a compact Hausdorff group G. Show $\mathrm{Hom}_G(\pi, \pi') = \{0\}$ if and only if the matrix coefficients of π are orthogonal to the matrix coefficients of π'.

Solution:

Assume $\mathrm{Hom}(\pi, \pi') = \{0\}$. By Exercise 8.1.2, we have $P(T) = 0$ for all $T \in \mathcal{B}(\mathcal{H}, \mathcal{H}')$. Thus $P(v' \otimes \bar{v}) = 0$ for $v' \in \mathcal{H}'$ and $v \in \mathcal{H}$. Thus $(P(v' \otimes \bar{v})w, w') = 0$

for any $w \in \mathcal{H}$ and $w' \in \mathcal{H}'$. But this is $\int(\pi'(x)(v' \otimes \bar{v})(\pi(x^{-1})w), w')_{\mathcal{H}'}\,dx = \int((\pi(x^{-1})w, v)_{\mathcal{H}}\pi'(x)v', w')_{\mathcal{H}'}\,dx$. Thus we see $\int(w, \pi(x)v)_{\mathcal{H}}(\pi'(x)v', w')_{\mathcal{H}'}\,dx = 0$ for all $w, v \in \mathcal{H}$ and $w', v' \in \mathcal{H}'$ and thus the matrix coefficients are orthogonal. This is also an immediate consequence of Theorem 8.12.

Conversely, let $T \in \mathcal{B}(\mathcal{H}, \mathcal{H}')$ and $v, w_1, w_2 \in \mathcal{H}$ and $v' \in \mathcal{H}'$. Again using the previous exercise, we see

$$\begin{aligned}
(P(T(w_1 \otimes \bar{w}_2))v, v') &= \int(\pi'(x)T(w_1 \otimes \bar{w}_2)\pi(x^{-1})v, v')_{\mathcal{H}'}\,dx \\
&= \int(\pi(x^{-1})v, w_2)_{\mathcal{H}}(\pi'(x)Tw_1, v')_{\mathcal{H}'}\,dx \\
&= \int(v, \pi(x)w_2)_{\mathcal{H}}(\pi'(x)Tw_1, v')_{\mathcal{H}'}\,dx \\
&= \int(v, \pi(x)w_2)_{\mathcal{H}}\overline{(v', \pi'(x)Tw_1)_{\mathcal{H}'}}\,dx \\
&= 0\,.
\end{aligned}$$

So $P(T(w_1 \otimes \bar{w}_2)) = 0$ for all $w_1, w_2 \in \mathcal{H}$. Taking an orthonormal basis $\{e_\alpha\}$ of $\mathcal{H}$ and using $I = \sum_\alpha e_\alpha \otimes \bar{e}_\alpha$ and the Lebesgue Dominated Convergence Theorem, one has

$$\begin{aligned}
(P(T)v, w')_{\mathcal{H}'} &= \left(\left(\int \pi'(x)T\left(\sum_\alpha e_\alpha \otimes \bar{e}_\alpha(\pi(x^{-1})v)\right)\,dx, w'\right)\right)_{\mathcal{H}'} \\
&= \sum_\alpha \left(\left(\int \pi'(x)T(e_\alpha \otimes \bar{e}_\alpha)\pi(x^{-1})v\,dx, w'\right)\right)_{\mathcal{H}'} \\
&= \sum_\alpha (P(T(e_\alpha \otimes \bar{e}_\alpha))v, w')_{\mathcal{H}'} \\
&= 0
\end{aligned}$$

for the sum is dominated. Indeed,

$$\begin{aligned}
|\sum_\alpha(\pi'(x)T(e_\alpha \otimes \bar{e}_\alpha)\pi(x^{-1})v, w')_{\mathcal{H}'}| &\leqslant \sum_\alpha |(\pi'(x)T(e_\alpha \otimes \bar{e}_\alpha)\pi(x^{-1})v, w')_{\mathcal{H}'}| \\
&\leqslant \sum_\alpha |(\pi'(x)Te_\alpha, w')_{\mathcal{H}'}(\pi(x^{-1})v, e_\alpha)_{\mathcal{H}}| \\
&\leqslant \left(\sum_\alpha |(\pi'(x)Te_\alpha, w')_{\mathcal{H}'}|^2\right)^{1/2} \left(\sum_\alpha |(\pi(x^{-1})v, e_\alpha)_{\mathcal{H}}|^2\right)^{1/2} \\
&\leqslant \left(\sum_\alpha |(e_\alpha, T^*\pi'(x^{-1})w')^2_{\mathcal{H}}\right)^{1/2} ||\pi(x^{-1})v||^2_{\mathcal{H}} \\
&= ||T^*\pi'(x^{-1})w'||_{\mathcal{H}}||v||_{\mathcal{H}} \\
&\leqslant ||T^*||\,||w'||_{\mathcal{H}}\,||v||
\end{aligned}$$

which is integrable on G. One can note or show this sum is actually a countable sum. So $P(T) = 0$ for any T. But the range of P is $\mathrm{Hom}(\pi, \pi')$.

Another argument. Again assume the matrix coefficients are orthogonal. Let $T \in \mathrm{Hom}(\pi, \pi')$ and let Q be an orthogonal projection onto a π invariant finite dimensional subspace of $\mathcal{H}$. Then $TQ \in \mathrm{Hom}(\pi, \pi')$ has finite dimensional range. Let $e_1', e_2', \ldots, e_m'$ be an orthonormal basis of the range of TQ. Then $TQv = \sum_{i=1}^m (TQv, e_i')_{\mathcal{H}'} e_i' = \sum_{i=1}^m (v, QT^* e_i')_{\mathcal{H}} e_i' = \sum_{i=1}^m \left(e_i' \otimes \overline{QT^* e_i'} \right)(v)$. But

$$\begin{aligned}
(TQv, v')_{\mathcal{H}'} &= \int (\pi'(x) TQ \pi(x^{-1}) v, v')\, dx \\
&= \int \sum_{i=1}^m (\pi'(x)(\pi(x^{-1})v, QT^* e_i')_{\mathcal{H}} e_i', v')_{\mathcal{H}'}\, dx \\
&= \sum_{i=1}^m \int (v, \pi(x) QT^* e_i')_{\mathcal{H}} (\pi'(x) e_i', v')'_{\mathcal{H}}\, dx \\
&= \sum_{i=1}^m \int (v \otimes_\pi \overline{QT^* e_i'})(x) \overline{(v' \otimes_{\pi'} \overline{e_i'})(x)}\, dx \\
&= 0 .
\end{aligned}$$

So $TQ = 0$ for any such Q. By Theorem 8.5, the sum of a family of such Q's is the identity operator on $\mathcal{H}$. So $T = 0$ and thus $\mathrm{Hom}(\pi, \pi') = \{0\}$.

□

4. Let G be a compact Hausdorff group and let π be a unitary representation of G with cyclic vector w. For $u, v \in \mathcal{H}$, the Hilbert space of π, define

$$B(u, v) = \int_G (w, \pi(g)v)_{\mathcal{H}} (\pi(g)u, w)_{\mathcal{H}}\, dg.$$

(a) Show B is a bounded sesquilinear form on $\mathcal{H} \times \mathcal{H}$.

(b) Show there is a positive continuous linear one-to-one operator T on $\mathcal{H}$ with

$$B(u, v) = (Tu, v)_{\mathcal{H}} \text{ for all } u, v \in \mathcal{H}.$$

(c) Show $T \in \mathrm{Hom}_G(\pi, \pi)$.

(d) Show if v_n is bounded and converges weakly to a vector v, then

$$\lim_n ||Tv_n||^2 = \int_G (w, \pi(g)Tv)_{\mathcal{H}} (\pi(g)v, w)_{\mathcal{H}}\, dg = ||Tv||^2.$$

(e) Use Exercise 6.2.4 and the weak compactness of the closed unit ball in $\mathcal{H}$ to show T is a compact operator.

(f) Using the spectral theorem for compact linear operators, show π has a nonzero finite dimensional irreducible subrepresentation.

Solution:

Sesquilinearity follows from the linearity of $u \mapsto (\pi(g)u, w)_{\mathcal{H}}$ and the conjugate linearity of $v \mapsto (w, \pi(g)v)_{\mathcal{H}}$. To see B is bounded, note

$$\begin{aligned}|B(u,v)| &= \left|\int (w, \pi(g)v)_{\mathcal{H}}(\pi(g)u, w)_{\mathcal{H}}\, dg\right| \\ &\leqslant \int |(w, \pi(g)v)_{\mathcal{H}}(\pi(g)u, w)_{\mathcal{H}}|\, dg \\ &\leqslant \int ||w||_{\mathcal{H}}||\pi(g)v||_{\mathcal{H}}||\pi(g)u||_{\mathcal{H}}||w||_{\mathcal{H}}\, dg \\ &= ||w||^2_{\mathcal{H}} \int ||v||_{\mathcal{H}}||u||_{\mathcal{H}}\, dg \\ &= ||w||^2_{\mathcal{H}}||u||_{\mathcal{H}}||v||_{\mathcal{H}}\,.\end{aligned}$$

One can easily check $B(u, v) = \overline{B(v, u)}$. Proposition 5.28 implies there is a bounded self adjoint linear operator T with $B(u, v) = (Tu, v)_{\mathcal{H}}$ for all u, v. Let $v \neq 0$. Since w is a cyclic vector, $(w, \pi(g)v)_{\mathcal{H}} \neq 0$ for g in a nonempty open set for otherwise $v \perp \pi(g^{-1})w$ for all g. So $(Tv, v)_{\mathcal{H}} = \int(w, \pi(g)v)_{\mathcal{H}}(\pi(g)v, w)_{\mathcal{H}}\, dg = \int|(w, \pi(g)v)_{\mathcal{H}}|^2\, dg > 0$. Thus T is positive and one-to-one.

That T is in $\text{Hom}(\pi, \pi)$ follows from

$$\begin{aligned}(T\pi(x)u, v)_{\mathcal{H}} &= B(\pi(x)u, v) \\ &= \int (w, \pi(g)v)_{\mathcal{H}}(\pi(g)\pi(x)u, w)_{\mathcal{H}}\, dg \\ &= \int (w, \pi(g)v)_{\mathcal{H}}(\pi(gx)u, w)_{\mathcal{H}}\, dg \\ &= \int (w, \pi(gx^{-1})v)_{\mathcal{H}}(\pi(g)u, w)_{\mathcal{H}}\, dg \\ &= \int (w, \pi(g)\pi(x^{-1})v)_{\mathcal{H}}(\pi(g)u, w)_{\mathcal{H}}\, dg \\ &= B(u, \pi(x^{-1})v) \\ &= (Tu, \pi(x)^{-1}v)_{\mathcal{H}} \\ &= (\pi(x)Tu, v)_{\mathcal{H}}\end{aligned}$$

for all $u, v \in \mathcal{H}$ and $x \in G$.

For (d), if v_n converges weakly to v and v_n is bounded, then $(w, \pi(g)Tv_n)_{\mathcal{H}} = (T^*\pi(g^{-1})w, v_n)_{\mathcal{H}} \to (T^*\pi(g^{-1})w, v)_{\mathcal{H}} = (w, \pi(g)Tv)_{\mathcal{H}}$ for each $g \in G$. Similarly, $(\pi(g)v_n, w)_{\mathcal{H}} \to (\pi(g)v, w)_{\mathcal{H}}$ for all g. Since

$$|(w, \pi(g)Tv_n)_{\mathcal{H}}| = |(\pi(g)Tv_n, w)_{\mathcal{H}}| \leqslant ||w||_{\mathcal{H}}||\pi(g)Tv_n||_{\mathcal{H}} \leqslant ||w||_{\mathcal{H}}||T||\,||v_n||_{\mathcal{H}}\,,$$

we see these functions are bounded by some positive constant. Haar measure on G being finite makes constants integrable and thus by the Lebesgue Dominated

Convergence Theorem,

$$\begin{aligned}(Tv,Tv)_{\mathcal{H}} &= B(v,Tv)\\ &= \int (w,\pi(g)Tv)_{\mathcal{H}}(\pi(g)v,w)_{\mathcal{H}}\,dg\\ &= \lim_n \int (w,\pi(g)Tv_n)_{\mathcal{H}}(\pi(g)v_n,w)_{\mathcal{H}}\,dg\\ &= \lim_n B(v_n,Tv_n)\\ &= \lim_n (Tv_n,Tv_n)_{\mathcal{H}}\,.\end{aligned}$$

To see T is a compact operator, it suffices to show if v_n is a bounded sequence, then Tv_n has a subsequence which converges in $\mathcal{H}$. But if v_n is bounded, since bounded spheres are weakly compact, there is a subsequence $\{w_k\}$ of $\{v_n\}$ such that w_k converges weakly to some vector w. Using (d) one has $||Tw_k|| \to ||Tw||$ and thus by Exercise 6.2.4, $Tv_k \to Tw$ in $\mathcal{H}$. Hence $\{Tv_n\}$ has a convergent subsequence. So (e) holds.

The Spectral Theorem (2.51) shows $\lambda = ||T|| > 0$ is an eigenvalue of T and $\mathcal{H}_\lambda = \{u \in \mathcal{H} \mid Tu = \lambda u\}$ is finite dimensional. This subspace is invariant under π for $T\pi(x)u = \pi(x)Tu = \lambda\pi(x)u$ if $u \in \mathcal{H}_\lambda$.

□

5. Let π and π' be finite dimensional inequivalent unitary representations of a compact group G on Hilbert spaces $\mathcal{H}$ and $\mathcal{H}'$. For $T \in \mathcal{L}(\mathcal{H},\mathcal{H}')$, let A be the operator given by

$$Av = \int_G \pi'(g)T\pi(g)^{-1}v\,dg.$$

(a) If $T = v' \otimes \bar{v}$, show $(Aw,w')_{\mathcal{H}'} = 0$ yields the orthogonality of the matrix coefficients $w \otimes_\pi \bar{v}$ and $w' \otimes_{\pi'} \bar{v}'$ in $L^2(G)$.

(b) In the case where $\pi' = \pi$ and π is irreducible, show Schur's Lemma implies that there is a $c > 0$ so that

$$\int (w \otimes_\pi \bar{v})(g)\overline{(w' \otimes_\pi \bar{v}')(g)}\,dg = c(w,w')_{\mathcal{H}}(v',v)_{\mathcal{H}}.$$

Solution:

For (a), note if $T = v' \otimes \bar{v}$, then

$$Aw = \int \pi'(g)(v' \otimes \bar{v})\pi(g)^{-1}w\,dg = \int (\pi(g)^{-1}w,v)_{\mathcal{H}}\pi'(g)v'\,dg.$$

Thus if $(Aw,w') = 0$, we see

$$\begin{aligned}\int (w,\pi(g)v)_{\mathcal{H}}\overline{(w',\pi'(g)v')_{\mathcal{H}'}}\,dg &= \int (\pi(g)^{-1}w,v)_{\mathcal{H}}(\pi'(g)v',w')_{\mathcal{H}'}\,dg\\ &= (Aw,w')_{\mathcal{H}'} = 0\,.\end{aligned}$$

So $w \otimes_\pi \bar{v} \perp w' \otimes_{\pi'} \bar{v}'$ in $L^2(G)$. We remark this argument was also given in the solution of Exercise **??**.

Now suppose $\pi = \pi'$ and π is irreducible. Again take A for the case when $T = v' \otimes \bar{v}$. By Lemma 8.2, $A \in \mathrm{Hom}(\pi, \pi)$ and thus by Schur's Lemma, there is a $C(v', v) \in \mathbb{C}$ with $A = C(v', v)I$. But then from above we see

$$\begin{aligned} C(v', v)(w, w')_{\mathcal{H}} &= (Aw, w')_{\mathcal{H}} \\ &= \int (w \otimes_\pi \bar{v})(g)\overline{(w' \otimes_\pi \bar{v}')(g)}\, dg. \end{aligned}$$

Note C is a sesquilinear form on a finite dimensional Hilbert space. Consequently, there is a linear transformation B on $\mathcal{H}$ with $C(v, v') = (Bv, v')_{\mathcal{H}}$ for all $v, v' \in \mathcal{H}$. So

$$\begin{aligned} (B\pi(x)v', v)_{\mathcal{H}}(w, w')_{\mathcal{H}} &= \int (w, \pi(g)v)_{\mathcal{H}}(\pi(g)\pi(x)v', w')_{\mathcal{H}}\, dg \\ &= \int (w, \pi(g)v)_{\mathcal{H}}(\pi(gx)v', w')_{\mathcal{H}}\, dg \\ &= \int (w, \pi(gx^{-1})v)_{\mathcal{H}}(\pi(g)v', w')_{\mathcal{H}}\, dg \\ &= \int (w, \pi(g)\pi(x^{-1})v)_{\mathcal{H}}(\pi(g)v', w')_{\mathcal{H}}\, dg \\ &= (Bv', \pi(x^{-1})v)_{\mathcal{H}}(w, w')_{\mathcal{H}} \end{aligned}$$

for all vectors $v, w, v', w' \in \mathcal{H}$. Thus

$$(B\pi(x)v', v')_{\mathcal{H}} = (Bv', \pi(x^{-1})v)_{\mathcal{H}} = (\pi(x)Bv', v)_{\mathcal{H}}$$

for all v and v' and any $x \in G$. This gives $B \in \mathrm{Hom}(\pi, \pi)$ and by Schur's Lemma there is a $c \in \mathbb{C}$ with $B = cI$. Consequently

$$\int (w \otimes_\pi \bar{v})(g)\overline{(w' \otimes_\pi \bar{v}')(g)}\, dg = c(v', v)_{\mathcal{H}}(w, w')_{\mathcal{H}}.$$

Note $c > 0$ follows from taking $w = v = v' = w' \neq 0$ for then $v \otimes_\pi \bar{v}$ is a nonzero function in $L^2(G)$.

□

6. Let π be an irreducible unitary representation of a compact Hausdorff group G on a Hilbert space $\mathcal{H}$ of dimension d. Let v and w be in $\mathcal{H}$. Define $T = \int \pi(g)v \otimes \bar{\pi}(g)\bar{w}\, dg$. Show $T = \frac{1}{d}(v, w)I$ and use this to show

$$(v \otimes_\pi \bar{w}, v' \otimes_\pi \bar{w}')_{L^2(G)} = \frac{1}{d}(v, v')_{\mathcal{H}}(w', w)_{\mathcal{H}} \text{ for all } v, w, v', w' \in \mathcal{H}.$$

Solution:

We first note $\int \pi(g)v \otimes \bar{\pi}(g)\bar{w}\, dg = \int \pi(g)(v \otimes \bar{w})\pi(g^{-1})\, dg$ for

$$\begin{aligned}\pi(g)(v \otimes \bar{w})\pi(g^{-1})u &= (\pi(g^{-1})u, w)\pi(g)v \\ &= (u, \pi(g)w)\pi(g)v \\ &= (\pi(g)v \otimes \overline{\pi(g)w})(u) \\ &= (\pi(g)v \otimes \bar{\pi}(g)\bar{w})(u)\,.\end{aligned}$$

Thus by Lemma 8.2, $T \in \mathrm{Hom}(\pi, \pi)$. Since π is irreducible, Schur's Lemma give $T = C(v, w)I$. Now by Definition 2.38,

$$\begin{aligned}dC(v, w) &= \mathrm{Tr}(T) \\ &= \int \mathrm{Tr}(\pi(g)(v \otimes \bar{w})\pi(g^{-1}))\, dg \\ &= \int \mathrm{Tr}(v \otimes \bar{w})\, dg \\ &= \mathrm{Tr}(v \otimes \bar{w}) \\ &= (v, w)\,.\end{aligned}$$

So $C(v, w) = \frac{1}{d}(v, w)$ and thus $T = \frac{1}{d}(v, w)I$. So from this we see

$$\begin{aligned}\frac{1}{d}(v, w)(w', v') &= \int (\pi(g)(v \otimes \bar{w})\pi(g^{-1})w', v')\, dg \\ &= \int ((\pi(g^{-1})w', w)\pi(g)v, v')\, dg \\ &= \int (w', \pi(g)w)(\pi(g)v, v')\, dg \\ &= \int (w' \otimes_\pi \bar{w})(g)\overline{(v' \otimes_\pi \bar{v})(g)}\, dg \\ &= (w' \otimes_\pi \bar{w}, v' \otimes_\pi \bar{v})_{L^2(G)}\,.\end{aligned}$$

□

7. Let π be a nontrivial irreducible unitary representation of a compact Hausdorff group G. Show

$$\iint \pi(x)A\pi(y^{-1})\, dx\, dy = 0$$

for all linear transformations A of $\mathcal{H}_\pi$.

Solution: Recall the outer tensor product $\pi \times \bar{\pi}$ is the representation on $\mathcal{B}(\mathcal{H}) = \mathcal{H} \otimes \bar{\mathcal{H}}$ that satisfies $(\pi \times \bar{\pi})(g_1, g_2)(v_1 \otimes \bar{v}_2) = \pi(g_1)v_1 \otimes \bar{\pi}(g_2)\bar{v}_2$. Note by Definition 6.69, if $A \in \mathcal{B}(\mathcal{H})$, then $\pi(g_1) \otimes \bar{\pi}(g_2)(A) = \pi(g_1)A\bar{\bar{\pi}}(g_2)^* = \pi(g_1)A\pi(g_2^{-1})$. Define $T(A) = \iint \pi(x)A\pi(y^{-1})\, dx\, dy$ for $A \in \mathcal{B}(\mathcal{H})$. Using

Fubini's Theorem and left and right translation invariance, one sees

$$\begin{aligned}(\pi \times \bar{\pi})(g_1, g_2)T(A) &= \iint \pi(g_1)\pi(x)A\pi(y^{-1})\pi(g_2^{-1})\,dx\,dy \\ &= \iint \pi(g_1x)A\pi(y^{-1}g_2^{-1})\,dx\,dy \\ &= \iint \pi(x)A\pi(y^{-1})\,dy\,dx \\ &= \iint \pi(xg_1)A\pi(g_2^{-1}y^{-1})\,dx\,dy \\ &= \iint \pi(x)(\pi \times \bar{\pi})(g_1, g_2)(A)\pi(y^{-1})\,dy\,dx \\ &= T((\pi \times \bar{\pi})(g_1, g_2)(A))\end{aligned}$$

and thus T commutes with $\pi \times \bar{\pi}$. Proposition 6.75 shows $\pi \times \bar{\pi}$ is irreducible. By Schur's Lemma, there is a scalar c such that $T = cI_{\mathcal{B}(\mathcal{H})}$ and thus $T(A) = cA$ for each $A \in \mathcal{B}(\mathcal{H})$. By taking $A = I$ and using traces and Theorem 8.9, one sees

$$\begin{aligned}cd &= \mathrm{Tr}(T(I)) \\ &= \iint \mathrm{Tr}(\pi(x)I\pi(y^{-1}))\,dx\,dy \\ &= \iint \mathrm{Tr}(xy^{-1})\,dx\,dy \\ &= \iint \mathrm{Tr}(y^{-1})\,dx\,dy \\ &= \int I_\pi(I)(y)dy \\ &= (I_\pi(I), I_1(1))_2\,.\end{aligned}$$

Theorem 8.12 and its corollary show $I_\pi(\mathcal{B}(\mathcal{H}))$ and $I_1(\mathbb{C})$ are orthogonal. Thus $cd = 0$ and so $T(A) = 0$ for all A. We remark $I_\pi(I) = \bar{\chi}_\pi$, the conjugate of the character of π.

□

8. Show a continuous function f on a compact Hausdorff group G is central if and only if $f * g = g * f$ for all functions $g \in L^1(G)$.

Solution:

Let f be a continuous function on G with $f * g = g * f$ for all $g \in L^1(G)$. By Exercise 6.7.4, $f * g$ and $g * f$ are continuous functions. Thus $\int f(y)g(y^{-1}x)\,dy = \int g(y)f(y^{-1}x)\,dy$ for each $g \in L^1(G)$ and all x. This implies $\int f(xy)g(y^{-1})\,dy = \int g(y^{-1})f(yx)\,dy$. Replacing g by $\check{g}$ and using $\check{g}$ is integrable if and only if g is integrable, one has $\int (f(xy) - f(yx))\,g(y)\,dy = 0$ for all $g \in L^1(G)$ and any x. Set $F_x(y) = f(xy) - f(yx)$. Note $F_x \in L^\infty(G)$ and $\int F_x(y)g(y)\,dy = 0$ for all $g \in L^1(G)$. Since the dual of L^1 is L^∞, this implies $F_x(y) = 0$ a.e. y. But F_x

is continuous and so one must have $F_x(y) = 0$ for all y. Hence $f(xy) = f(yx)$ for all x, y and f is central.

□

9. Let G be a compact Hausdorff group with left regular representation λ. Suppose $f, h \in L^1(G)$. Show $\lambda(f) = \lambda(h)$ if and only if $f = h$.

Solution:

Since λ is linear it suffices to show $f = 0$ if $\lambda(f) = 0$. So assume $f \in L^1(G)$ and $\lambda(f) = 0$. By Example 6.113, $\lambda(f)g = f * g$ for $g \in L^2(G)$. Thus $f * g = 0$ for all $g \in L^2$. Take $g = \chi_{E^{-1}}$ where E is measurable. Then

$$f * g(x) = \int f(y)\chi_{E^{-1}}(y^{-1}x)\,dy = \int f(y)\chi_{xE}(y)\,dy\,.$$

Since g is bounded, Exercise 6.7.4 shows $f * g$ is continuous. So $f * g = 0$ in L^2 implies $f * g(x) = 0$ for all x. Thus $f * g(e) = 0$ and we see $\int_E f(y)\,dy = 0$ for all measurable subsets E. This implies $f(y) = 0$ a.e. y and $f = 0$ in $L^1(G)$.

□

10. Show an L^1 function f on a compact Hausdorff group G is central if and only if $\lambda(f) \in \mathrm{Hom}_G(\lambda, \lambda)$.

Solution: Assume $\lambda(f)$ commutes with all $\lambda(x)$. Then $\lambda(f)\lambda(h) = \lambda(h)\lambda(f)$ for all $h \in L^1(G)$. So $\lambda(f * h) = \lambda(h * f)$ for all $h \in L^1(G)$. By Exercise 8.1.9, $f * h = h * f$ for all $h \in L^1$. Thus $\int f(y)h(y^{-1}x)\,dy = \int h(y)f(y^{-1}x)\,dy$. Translating and replacing y by y^{-1} gives $\int f(xy^{-1})h(y)\,dy = \int f(y^{-1}x)h(y)\,dy$ for all $h \in L^\infty(G)$. Thus $f(xy^{-1}) = f(y^{-1}x)$ a.e. y for each x. Now apply Lemma 6.118. □

Exercise Set 8.2

1. Let π be an irreducible unitary representation of a compact Hausdorff group G and let λ and ρ be the left and right regular unitary representations of G. Show $\lambda|_{P_\pi L^2(G)}$ is unitarily equivalent to $d(\pi)\pi$ and $\rho|_{P_\pi L^2(G)}$ is unitarily equivalent to $d(\pi)\bar{\pi}$.

Solution:

By the Plancherel Theorem (8.20) we know the mapping $f \mapsto \sqrt{d(\pi)}\pi(f)$ from $P_\pi(L^2(G))$ into $\mathcal{H}_\pi \otimes \bar{\mathcal{H}}_\pi$ is a unitary equivalence of B on $P_\pi(L^2(G))$ with $\pi \times \bar{\pi}$. Indeed, if $f \in P_\pi(L^2(G))$ and $\pi' \in \hat{G}_c$ with $\pi' \neq \pi$, then $P_{\pi'}(f) = P_{\pi'}P_\pi(f) = 0$. So $f = P_\pi f$. Consequently, the biregular representation B on $P_\pi(L^2(G))$ is unitarily equivalent to $\pi \times \bar{\pi}$ under the mapping $U_\pi(f) = \sqrt{d(\pi)}\pi(f)$. Now the biregular representation restricted to $G \times \{e\}$ is λ and restricted to $\{e\} \times G$ is ρ. Since $\pi \times \bar{\pi}(g, e) = \pi(g) \otimes \bar{I}_\pi$ and $\pi \times \bar{\pi}(e, g) = I_\pi \otimes \bar{\pi}(g)$, we know by Exercise 6.4.9, that λ restricted to $P_\pi(L^2(G))$ under the unitary mapping U_π is unitarily equivalent to $d(\pi)\pi$ and ρ restricted to this space is unitarily equivalent to $d(\pi)\bar{\pi}$.

□

2. Let G be a compact Hausdorff group. Show a function $f \in L^1(G)$ is central if and only if for every irreducible unitary representation π of G, $\pi(f) \in \mathrm{Hom}_G(\pi, \pi)$.

Solution:

First we note if $f \in L^1(G)$ is central, Lemma 6.119 implies $\pi(f) \in \mathrm{Hom}_G(\pi, \pi)$ for each irreducible unitary representation π. Now suppose $\pi(f) \in \mathrm{Hom}_G(\pi, \pi)$ for any irreducible unitary representation π. Let Γ be a unitary representation of G on a Hilbert space $\mathcal{H}$. By Theorem 8.5, $\mathcal{H}$ decomposes into an orthogonal internal direct sum of invariant subspaces $\mathcal{H}_i$ such that $\pi_i = \Gamma|_{\mathcal{H}_i}$ is an irreducible representation. Since $\Gamma(f) = \oplus \pi_i(f)$ for $f \in L^1(G)$, we see $\Gamma(f) \in \mathrm{Hom}_G(\Gamma, \Gamma)$. A reapplication of Lemma 6.119 gives the conclusion.

□

3. Let G be a compact Hausdorff group with concrete dual $\hat{G}_c$. For $\pi \in \hat{G}_c$, let $e_1, e_2, \ldots, e_{d(\pi)}$ be an orthonormal basis of the Hilbert space for π. Let $\pi_{i,j} = e_i \otimes_\pi \bar{e}_j$ be the matrix coefficients of π obtained from the basis $\{e_i\}_{i=1}^{d(\pi)}$.

(a) Show $\{\sqrt{d(\pi)}\pi_{i,j} \mid \pi \in \hat{G}_c,\ 1 \leqslant i, j \leqslant d(\pi)\}$ form an orthonormal basis of $L^2(G)$.

(b) Show for $f \in L^2(G)$, the series

$$\sum_{\pi \in \hat{G}_c} \sum_{1 \leqslant i,j \leqslant d(\pi)} d(\pi)(f, \pi_{i,j})_2 \pi_{i,j}$$

converges in $L^2(G)$ to f.

(c) Show if $f \in \langle L^2(G) * L^2(G) \rangle$, then the series in (b) converges uniformly on G to f.

Solution:

Note we have $\pi_{i,j}(g) = e_i \otimes_\pi \bar{e}_j(g) = (e_i, \pi(g)e_j)_{\mathcal{H}_\pi}$. The orthonormality of the functions $\sqrt{d(\pi)}\pi_{i,j}$ follows directly from Corollary 8.10 and the orthonormality of the vectors e_i. The linear span of these functions is the same as the linear span of all matrix coefficients of π for $v \otimes_\pi w = \sum_{i,j} (v, e_i)\overline{(w, e_j)} e_i \otimes_\pi \bar{e}_j$. Thus the orthogonal projection Q_π of $L^2(G)$ onto the linear span of of the matrix coefficients of π is given by $Q_\pi f = \sum_{i,j} (f, \sqrt{d(\pi)}\pi_{i,j})\sqrt{d(\pi)}\pi_{i,j} = d(\pi) \sum_{i,j \leqslant d(\pi)} (f, \pi_{i,j})_{L^2(G)} \pi_{i,j}$. By Lemma 8.15, we see $Q_\pi = P_\pi$. So (b) and (c) follow from Theorem 8.19 and Corollary 8.21. □

4. Let G be a compact Hausdorff group and let π be an irreducible unitary representation of G. Show if f is in the π-primary subspace $P_\pi L^2(G)$ for the left regular representation λ, then f^* is also in this subspace.

Solution:

As seen in the previous exercise, the range of P_π is the linear span of the matrix coefficients of π. Thus to show f^* is range of P_π if f is in the range of

P_π, it suffices to show $(v \otimes_\pi \bar{w})^* = w \otimes_\pi \bar{v}$. But if $f = v \otimes_\pi \bar{w}$, then

$$\begin{aligned} f^*(g) &= \overline{f(g^{-1})} \\ &= \overline{v \otimes_\pi \bar{w}(g^{-1})} \\ &= \overline{(v, \pi(g^{-1})w)_{\mathcal{H}_\pi}} \\ &= (\pi(g^{-1})w, v)_{\mathcal{H}_\pi} \\ &= (w, \pi(g)v)_{\mathcal{H}_\pi} \\ &= w \otimes_\pi \bar{v}(g) . \end{aligned}$$

□

5. Show if $\mathcal{H}$ is a finite dimensional Hilbert space and there is unitary isomorphism from Hilbert space $m\mathcal{H}$ onto $n\mathcal{H}$ where m and n are cardinals, then $m = n$.

Solution:

We use the fact that any two orthonormal bases of a Hilbert space have the same cardinality. Let $e_1, e_2, \ldots, e_d$ be an orthonormal basis of $\mathcal{H}$. Suppose A has cardinality n. Then $n\mathcal{H}$ is $\{(v_a) \mid v_a \in \mathcal{H}, \sum_a ||v_a||^2 < \infty\}$ with inner product

$$(v, w) = \sum_a (v_a, w_a)_{\mathcal{H}} .$$

This has orthonormal basis $\epsilon(a, i)$ where $\epsilon(a, i)_b = \delta_{a,b} e_i$. So the mapping $(a, i) \mapsto \epsilon(a, i)$ is a one-to-one mapping of $A \times \{1, 2, \ldots, d\}$ onto an orthonormal basis of $n\mathcal{H}$. But this basis of $n\mathcal{H}$ has cardinality nd. Thus if there is a unitary isomorphism of $m\mathcal{H}$ onto $n\mathcal{H}$, one would have $md = nd$ as cardinals. Now $md = m$ for any infinite cardinal. So m is infinite if and only if n is infinite and m is finite if and only if n is finite, and in either case $m = n$.

□

6. Show if π' is a unitary representation on a Hilbert space $\mathcal{H}'$ and I is the identity representation on Hilbert space $\mathcal{H}$, then the representation $I \otimes \pi'$ is unitarily equivalent to $n\pi'$ where n is the cardinality of an orthonormal basis of $\mathcal{H}$.

Solution: This has already been done in Exercise 6.4.9 □

Exercise Set 8.3

1. When $\mathcal{H}'$ is finite dimensional, then the $\mathcal{K}$-trace is actually a contraction; i.e., show there is a bounded linear transformation C of $(\mathcal{K} \otimes \bar{\mathcal{H}}') \otimes \mathcal{H}'$ into $\mathcal{K}$ satisfying $C(T \otimes v') = T(v')$. In particular, $C((w \otimes \bar{v}') \otimes v) = (v, v')w$. Then show

$$C \circ \Phi^{-1} = \mathrm{Tr}_{\mathcal{K}}$$

where Φ is the unitary isomorphism given by (8.6).

Solution:

Define a sesquilinear mapping B on $(\mathcal{K} \otimes \bar{\mathcal{H}}') \times \bar{\mathcal{H}}'$ by $B(T, \bar{v}') = T(v')$. Note B is bounded. Now if A is a linear mapping from $\bar{\mathcal{H}}'$ into $\mathcal{K} \otimes \bar{\mathcal{H}}'$, then A is bounded and trace class for if $\bar{e}'_1, \bar{e}'_2, \ldots \bar{e}'_{d'}$ is an orthonormal basis of $\bar{\mathcal{H}}'$, then $A = \sum_{i=1}^{d'} A\bar{e}'_i \otimes e'_i$. Thus we can use Definition 8.33 and take $\mathrm{Tr}_B(A) = \sum_{i=1}^{d'} B(A\bar{e}'_i, \bar{e}'_i) = \sum_{i=1}^{d'} A(\bar{e}'_i)(e'_i) \in \mathcal{K}$. So $C(A) = \mathrm{Tr}_B(A)$ is a linear mapping of $(K \otimes \bar{\mathcal{H}}') \otimes \mathcal{H}'$ into $\mathcal{K}$. It satisfies

$$\begin{aligned} C(T \otimes v') &= \sum_{i=1}^{d'} (T \otimes v')(\bar{e}'_i)(e'_i) = \sum_{i=1}^{d'} ((\bar{e}'_i, \bar{v}')T)(e'_i) \\ &= \sum_{i=1}^{d'} (v', e'_i)T(e'_i) = T\left(\sum_{i=1}^{d'} (v', e'_i)e'_i\right) = T(v'). \end{aligned}$$

C is also bounded for

$$||C(A)||_{\mathcal{K}} = ||\mathrm{Tr}_B(A)||_{\mathcal{K}} \leqslant \sum_{i=1}^{d'} ||B|| \, ||A(\bar{e}'_i)|| \, ||\bar{e}'_i|| \leqslant d' ||B|| \, ||A||.$$

Now by (8.6), the unitary isomorphism Φ satisfies

$$\Phi^{-1}((v'_1 \otimes \bar{v}'_2) \otimes w) = (w \otimes \bar{v}'_2) \otimes v'_1.$$

Using (8.5) one sees

$$\begin{aligned} C(\Phi^{-1}((v'_1 \otimes \bar{v}'_2) \otimes w)) &= C((w \otimes \bar{v}'_2) \otimes v'_1) \\ &= (w \otimes \bar{v}'_2)(v'_1) \\ &= (v'_1, v'_2)w \\ &= \mathrm{Tr}_{\mathcal{K}}((v'_1 \otimes \bar{v}'_2) \otimes w) \end{aligned}$$

for $v'_1, v'_2 \in \mathcal{H}'$ and $w \in \mathcal{K}$. This implies $C \circ \Phi^{-1} = \mathrm{Tr}_{\mathcal{K}}$.

□

2. Let $T \in \mathcal{H}_1 \otimes \bar{\mathcal{H}}_2$ be a trace class operator.

(a) Show there is a constant $M > 0$ such that if A is an index set and e_α and f_α for $\alpha \in A$ are orthonormal sets in $\mathcal{H}_1$ and $\mathcal{H}_2$, then $\sum_\alpha |(e_\alpha, Tf_\alpha)| < M$.

(b) Let $|T|_1$ be the infinimum of all such M in (a). Show $|\cdot|_1$ is a norm on the linear space of trace class operators from $\mathcal{H}_2$ to $\mathcal{H}_1$.

(c) Show $||T|| \leqslant |T|_1$.

(d) Show the trace class operators with this norm is a Banach space.

Solution: Let $T = \sum v_i \otimes \bar{w}_i$ where $\sum ||v_i||^2 < \infty$ and $\sum ||w_i||^2 < \infty$. Then

$$\begin{aligned}\sum |(e_\alpha, Tf_\alpha)| &= \sum |(e_\alpha, \sum_i (f_\alpha, w_i)v_i)| \\ &\leqslant \sum_i \sum_\alpha |(e_\alpha, v_i)|\,|(f_\alpha, w_i)| \\ &= \left(\sum_i \sum_\alpha |(e_\alpha, v_i)|^2\right)^{\frac{1}{2}} \left(\sum_i \sum_\beta |(f_\beta, w_i)|^2\right)^{\frac{1}{2}} \\ &\leqslant \left(\sum_i ||v_i||^2\right)^{\frac{1}{2}} \left(\sum_i ||w_i||^2\right)^{\frac{1}{2}} \\ &< \infty.\end{aligned}$$

Thus the supremum over all such sums is finite and is called $|T|_1$.

Clearly $|T|_1 > 0$ if $T \neq 0$. Also $|aT|_1 = |a|\,|T|_1$ and $|T + S|_1 \leqslant |T|_1 + |S|_1$. Note $||T|| \leqslant |T|_1$. Indeed, if $Tv \neq 0$, then

$$||v||^{-1}||Tv|| = |(||Tv||^{-1}Tv, T(||v||^{-1}v))| \leqslant |T|_1$$

and so $||Tv|| \leqslant |T|_1||v||$.

We need completeness. Suppose T_n is Cauchy. Then since $||T_n - T_m|| \leqslant |T_n - T_m|_1$, T_n is Cauchy in operator norm. Thus there is a limit operator T. Since convergence is in norm and each T_n is compact, T is compact. By the spectral theorem for compact operators, Theorem 2.52, $T = \sum \lambda_i e_i \otimes \bar{f}_i$ where the sequences e_i and f_i are orthonormal and $\lambda_i \searrow 0$ (if infinite).

We have an $M > 0$ such that for any e_α, f_α we have

$$\sum |(e_\alpha, T_n f_\alpha)| \leqslant M.$$

This implies we are dealing with countable sums. Using Fatou's Lemma, we obtain $\sum |(e_\alpha, Tf_\alpha)| \leqslant M$ for any sets e_α and f_α. We now show this implies T is trace class. Indeed, $\sum \lambda_i = \sum |(Tf_i, e_i)| < M$. Take $v_i = \sqrt{\lambda_i} e_i$ and $w_i = \sqrt{\lambda_i} f_i$. Then $\sum ||v_i||^2 = \sum \lambda_i \leqslant M$ and $\sum ||w_i||^2 = \sum \lambda_i \leqslant M$. Also $T = \sum_i v_i \otimes \bar{w}_i$. So T is trace class. To see $|T_n - T|_1 \to 0$, note if $\epsilon > 0$, then there is an N such that $n, m \geqslant N$ implies $\sum |(e_\alpha, (T_n - T_m)f_\alpha)| \leqslant \epsilon$ for orthonormal sets $\{e_\alpha\}$ in $\mathcal{H}_1$ and $\{f_\alpha\}$ in $\mathcal{H}_2$. So again an application of Fatou's Lemma yields $\sum |(e_\alpha, (T_n - T)f_\alpha)| \leqslant \epsilon$ for any such sets for $n \geqslant N$. Thus $|T_n - T|_1 \leqslant \epsilon$ for $n \geqslant N$ and we have completeness. □

3. Show the inverse Φ^{-1} of the transformation Φ given in Definition 8.6 is defined by

$$\left(\Phi^{-1}(S)(\bar{v}_1')\right)(v_2') = \overline{S^*(v_1' \otimes \bar{v}_2')}$$

for $S \in (\mathcal{H}' \otimes \bar{\mathcal{H}}') \otimes \mathcal{K}$.

Solution: Define $\Psi(S) = \overline{S^*(v_1' \otimes \bar{v}_2')}$. We need to show

$$\Psi((v_1' \otimes \bar{v}_2') \otimes w) = (w \otimes \bar{v}_2') \otimes v_1'.$$

But

$$\begin{aligned}
\left(\Psi((v_1' \otimes \bar{v}_2') \otimes w)(\bar{v}_3')\right)(v_4') &= \overline{((v_1' \otimes \bar{v}_2') \otimes w)^* \, (v_3' \otimes \bar{v}_4')} \\
&= \overline{\left(\bar{w} \otimes \overline{v_1' \otimes \bar{v}_2'}\right)(v_3' \otimes \bar{v}_4')} \\
&= \overline{(v_3' \otimes \bar{v}_4', v_1' \otimes \bar{v}_2')\bar{w}} \\
&= \overline{(v_3', v_1')_{\mathcal{H}'}(\bar{v}_4', \bar{v}_2')_{\bar{\mathcal{H}}'} w} \\
&= (\bar{v}_3', \bar{v}_1')_{\bar{\mathcal{H}}'}(v_4', v_2')_{\mathcal{H}'} w \\
&= (\bar{v}_3', \bar{v}_1')_{\bar{\mathcal{H}}'}(w \otimes \bar{v}_2')(v_4') \\
&= \left(((w \otimes \bar{v}_2') \otimes v_1')(\bar{v}_3)\right)(v_4').
\end{aligned}$$

Therefore,

$$\Psi((v_1' \otimes \bar{v}_2') \otimes w) = (w \otimes \bar{v}_2') \otimes v_1$$

for $v_1', v_2' \in \mathcal{H}'$ and $w \in \mathcal{K}$. □

4. Let G be a compact Hausdorff group with closed subgroup K having one dimensional character χ. Show the orthogonal projection of $L^2(G)$ onto $L^2_K(G, \chi)$ is given by

$$Qf(x) = \int_K \chi(k) f(xk)\, dk.$$

Solution:

Note using Fubini's Theorem and $\int_K 1\, dk = 1$, we have

$$\int |f(g)|^2\, dg = \int_K \int_G |f(gk)|^2\, dg\, dk = \int_G \int_K |\chi(k) f(gk)|^2\, dk\, dg\,.$$

Thus using Cauchy-Schwarz,

$$\begin{aligned}
|Qf|_2^2 &= \int \left| \int \chi(k) f(xk)\, dk \right|^2 dx \\
&\leqslant \int \left(\int |\chi(k) f(xk)|\, dk \right)^2 dx \\
&\leqslant \int \left(\int |f(xk)|^2\, dk \right) \left(\int |\chi(k)|^2\, dk \right) dx \\
&= \int |f(x)|^2\, dx\,,
\end{aligned}$$

we see $Qf \in L^2(G)$ and $|Qf|_2 \leqslant |f|_2$. Note $Qf(xk') = \int \chi(k) f(xk'k)\, dk = \int \chi(k')^{-1} \chi(k) f(xk)\, dk = \chi(k')^{-1} Qf(x)$ for all k' for a.e. x. Thus $Qf \in L^2_K(G, \chi)$. Also clearly if $f \in L^2_K(G, \chi)$, then

$$Qf(x) = \int \chi(k) f(xk)\, dk = \int \chi(k) \chi(k^{-1}) f(x)\, dk = f(x)$$

for a.e. x. Thus Q is a projection onto $L_K(G,\chi)$. To see it is orthogonal, we show it is self adjoint. Specifically,

$$\begin{aligned}
(Qf_1, f_2) &= \int Qf_1(x)\overline{f_2(x)}\,dx \\
&= \int\int \chi(k) f_1(xk)\,dk\,\overline{f_2(x)}\,dx \\
&= \int\int \chi(k) f_1(xk)\ \overline{f_2(x)}\,dx\,dk \\
&= \int\int \chi(k) f_1(x)\ \overline{f_2(xk^{-1})}\,dx\,dk \\
&= \int\int f_1(x)\,\overline{\chi(k^{-1}) f_2(xk^{-1})}\,dk\,dx \\
&= \int f_1(x)\left(\overline{\int \chi(k^{-1}) f_2(xk^{-1})\,dk}\right)\,dk\,dx \\
&= \int f_1(x)\left(\overline{\int \chi(k) f_2(xk)\,dk}\right)\,dk\,dx \\
&= (f_1, Qf_2)
\end{aligned}$$

for $f_1, f_2 \in L^2(G)$.

□

5. Let G be a compact Hausdorff group and suppose π is an irreducible unitary representation of G. Let χ be a one-dimensional character of a closed subgroup K of G. Let f be a function in $L^2(G)$ orthogonal to the subspace $L^2_K(G,\chi)$. Show $\pi(f)w = 0$ for all $w \in E(\pi,\chi)$.

Solution:

We use Exercise 8.3.4. From it we know $Qf = 0$ and thus $\int \chi(k) f(xk)\,dk = 0$ a.e. x. For $w \in E(\pi,\chi)$ one has $\pi(k)w = \chi(k)w$ for all k. Using the unimodularity of G, $\int 1\,dk = 1$, and Fubini's Theorem,

$$\begin{aligned}
\pi(f)w &= \pi(f)w - \pi(Qf)w \\
&= \int f(g)\pi(g)w\,dg - \int Qf(g)\pi(g)w\,dg \\
&= \int_K \chi(k)\int_G f(g)\pi(gk^{-1})w\,dg\,dk - \int_G Qf(g)\pi(g)w\,dg \\
&= \int_K \chi(k)\int_G f(gk)\pi(g)w\,dg\,dk - \int_G Qf(g)\pi(g)w\,dg \\
&= \int_G \left(\int_K \chi(k) f(gk)\,dk\right)\pi(g)w\,dg - \int_G Qf(g)\pi(g)w\,dg \\
&= \int_G (Qf(g) - Qf(g))\,dg \\
&= 0\,.
\end{aligned}$$

□

6. Let G be a compact Hausdorff group and suppose π and π' are irreducible unitary representations of G. Let χ be a one-dimensional character of a closed subgroup K of G. Let $w \in E(\pi, \chi)^{\perp}$.

(a) Show $\pi(w' \otimes_{\pi'} \bar{v}')(w) = 0$ if $v' \in E(\pi', \chi)$ and $w' \in \mathcal{H}_{\pi'}$.

(b) Use Theorem 8.41 and the description of the primary subspaces given in Equation (8.9) to show $\pi(f)w = 0$ for all $f \in L^2_K(G, \chi)$.

This gives an alternative proof to Lemma 8.44.

Solution:

Note if π and π' are not unitarily equivalent, then by Theorem 8.12, $\pi(w' \otimes_{\pi'} \bar{v}') = 0$ and so (a) holds.

Now suppose $\pi = \pi'$, $v' \in E(\pi, \chi)$ and $w \in E(\pi, \chi)^{\perp}$. Let $v \in \mathcal{H}_{\pi}$. Then by Corollary 8.10,

$$\begin{aligned}
(\pi(w' \otimes_{\pi} \bar{v}')w, v)_{\mathcal{H}_{\pi}} &= \int_G (w' \otimes_{\pi} \bar{v}')(g)\, (\pi(g)w, v)_{\mathcal{H}_{\pi}}\, dg \\
&= \int_G (w' \otimes_{\pi} \bar{v}')(g)\, \overline{(v, \pi(g)w)_{\mathcal{H}_{\pi}}}\, dg \\
&= \int_G (w' \otimes_{\pi} \bar{v}')(g)\, \overline{(v \otimes_{\pi} \bar{w})(g)}\, dg \\
&= (w' \otimes_{\pi} \bar{v}', v \otimes_{\pi} \bar{w})_{L^2(G)} \\
&= \frac{1}{d_{\pi}} (w', v)_{\mathcal{H}_{\pi}} (w, v')_{\mathcal{H}_{\pi}} \\
&= 0\,.
\end{aligned}$$

So (a) also holds if π and π' are equivalent.

Using (a) and Equation (8.9) we see $\pi'(f)w = 0$ for all $f \in P(\pi')(L^2_K(G, \chi))$ and any $\pi' \in \hat{G}_c$. By Theorem 8.41, $L^2_K(G, \chi) = \bigoplus_{\pi' \in \hat{G}_c} \mathcal{H}(\pi')$ and since $|f|_1 \leqslant |f|_2$, we have $f = \sum_{\pi' \in \hat{G}_c} P(\pi')f$ in $L^1(G)$. So $\pi(f) = \sum \pi(P(\pi')f)$ in $\mathcal{B}(\mathcal{H}_{\pi})$ for $||\pi(f)|| \leqslant |f|_1$. This implies $\pi(f)w = \sum \pi'(P(\pi')f)w = 0$ in $\mathcal{H}_{\pi}$.

□

7. Show directly that if U is the unitary mapping from $L^2(G)$ onto $L^2_{G_d}(G \times G, 1)$ given by

$$Uf(x, y) = f(xy^{-1}),$$

then

$$d(\pi \times \bar{\pi})\mathrm{Tr}((\pi \times \bar{\pi})(x, y)^{-1}(\pi \times \bar{\pi})(Uf)) = UP(\pi)f(x, y)$$

where

$$P(\pi)f(x) = d(\pi)\mathrm{Tr}(\pi(x^{-1})\pi(f)).$$

Solution: We simplify $d(\pi \times \bar{\pi})\mathrm{Tr}((\pi \times \bar{\pi})(x, y)^{-1}(\pi \times \bar{\pi})(Uf))$. Using the orthogonality relations given in Corollary 8.10, we see:

$$
\begin{aligned}
&\operatorname{Tr}\left((\pi\times\bar{\pi})(x,y)^{-1}(\pi\times\bar{\pi})(Uf)\right)\\
&=\operatorname{Tr}\left(\pi(x^{-1})\times\bar{\pi}(y^{-1})\int_{G\times G} f(ab^{-1})(\pi\times\bar{\pi})(a,b)\,d(a,b)\right)\\
&=\sum_{i,j}(\pi(x^{-1})\times\bar{\pi}(y^{-1})\int_{G\times G} f(ab^{-1})(\pi\times\bar{\pi})(a,b)\,d(a,b))e_i\otimes\bar{e}_j, e_i\otimes\bar{e}_j)\\
&=\sum_{i,j}\iint(\pi(x^{-1})\times\bar{\pi}(y^{-1})f(ab^{-1})\pi(a)e_i\times\bar{\pi}(b)\bar{e}_j, e_i\otimes\bar{e}_j)\,da\,db\\
&=\sum_{i,j}\iint f(ab^{-1})(\pi(x^{-1}a)e_i\times\bar{\pi}(y^{-1}b)\bar{e}_j, e_i\otimes\bar{e}_j)\,da\,db\\
&=\sum_{i,j}\iint f(ab^{-1})(\pi(x^{-1}a)e_i,e_i)(e_j,\pi(y^{-1}b)e_j)\,da\,db\\
&=\sum_{i,j}\iint f(a)(\pi(x^{-1}ab)e_i,e_i)(e_j,\pi(y^{-1}b)e_j)\,da\,db\\
&=\sum_{i,j}\iint f(a)(\pi(y)e_j,\pi(b)e_j)\overline{(\pi(a^{-1}x)e_i,\pi(b)e_i)}\,db\,da\\
&=\sum_{i,j}\int f(a)(\pi(y)e_j\otimes_\pi e_j,\pi(a^{-1}x)e_i\otimes_\pi e_i)_2\,da\\
&=\frac{1}{d(\pi)}\sum_{i,j}\int f(a)(\pi(y)e_j,\pi(a^{-1}x)e_i)(e_i,e_j)\,da\\
&=\frac{1}{d(\pi)}\sum_{i}\int f(a)(\pi(x^{-1}ay)e_i,e_i)\,da\\
&=\frac{1}{d(\pi)}\sum_{i}(\pi(x^{-1})\pi(f)\pi(y)e_i,e_i)\,da\\
&=\frac{1}{d(\pi)}\operatorname{Tr}(\pi(x^{-1})\pi(f)\pi(y)).
\end{aligned}
$$

Thus

$$
\begin{aligned}
d(\pi\times\bar{\pi})\operatorname{Tr}\left((\pi\times\bar{\pi})(x,y)^{-1}(\pi\times\bar{\pi})(Uf)\right) &= d(\pi)\operatorname{Tr}\left(\pi(x^{-1})\pi(f)\pi(y)\right)\\
&= d(\pi)\operatorname{Tr}\left(\pi(yx^{-1})\pi(f)\right)\\
&= P(\pi)f(xy^{-1})\\
&= UP(\pi)f(x,y).
\end{aligned}
$$

□

8. Use Frobenius Reciprocity Theorem 8.25 to show if G is a compact Hausdorff group with closed subgroups K and H where $K\subseteq H$ and π is a unitary representation of K, then $\operatorname{ind}_H^G(\operatorname{ind}_K^H\pi)$ and $\operatorname{ind}_H^G\pi$ are unitarily equivalent.

Solution:

We assume all groups are separable and all Hilbert spaces are separable. We let $\hat{G}_c$ be a concrete dual for G and $\hat{H}_c$ be a concrete dual for H. It suffices to show the multiplicity of any $\pi' \in \hat{G}_c$ in $\mathrm{ind}_K^G \pi$ is the same as the multiplicity of π' in $\mathrm{ind}_H^G(\mathrm{ind}_K^H \pi)$. Take $\rho = \mathrm{ind}_K^H \pi$ and use Theorem 8.25. Note

$$\begin{aligned} m(\pi', \mathrm{ind}_H^G \rho) &= \dim \mathrm{Hom}_H(\pi'|_H, \rho) \\ &= \dim \mathrm{Hom}_H(\pi'|_H, \mathrm{ind}_K^H \pi) \\ &= \dim \mathrm{Hom}_K(\pi'|_H|_K, \pi) \\ &= \dim \mathrm{Hom}_K(\pi'|_K, \pi) \\ &= \dim \mathrm{Hom}_G(\pi', \mathrm{ind}_K^G \pi) \\ &= m(\pi', \mathrm{ind}_K^G \pi)\,. \end{aligned}$$

We remark in the proof of Theorem 6.53 that the cardinality of the multiplicity of π' in a unitary representation ρ was shown to be equal to the cardinality of the collection of orthogonal projections P_α which was equal to the cardinality of a strong basis of $\mathrm{Hom}_G(\pi', \rho)$. Thus instead of using vector space dimensions, one uses $\dim_s \mathrm{Hom}_G(\pi', \rho)$ to be the cardinality of a strong operator basis of the Banach space $\mathrm{Hom}_G(\pi', \rho)$, one could use $\dim_s$ instead of dim in the above argument and obtain the result in the nonseparable case. □

9. Let G be a compact Hausdorff group with closed subgroup K having one dimensional character χ. Recall we know $L^2_K(G, \chi)$, the Hilbert space for $\mathrm{ind}_K^G \chi$, is a subspace of $L^2(G)$ and the induced representation $\mathrm{ind}_K^G \chi$ is just the restriction of the regular representation λ to this subspace. Show if π is an irreducible unitary representation of G, then the orthogonal projection of $L^2(G)$ onto the π-primary subspace for $\mathrm{ind}_K^G \chi$ is $P_\pi Q$ where Q is the orthogonal projection of $L^2(G)$ onto $L^2_K(G, \chi)$.

Solution:

We know by Corollary 8.45 that $P(\pi)f(x) = d(\pi)\mathrm{Tr}(\pi(x^{-1})\pi(f))$ for $f \in L^2_K(G, \chi)$. Also by Theorem 8.19, $P_\pi f(x) = d(\pi)\mathrm{Tr}(\pi(x^{-1})\pi(f))$ for $f \in L^2(G)$. Exercise 8.3.4 shows $Qf(x) = \int \chi(k) f(xk)\, dk$. Since Q is the orthogonal projection of $L^2(G)$ onto $L^2_K(G, \chi)$ and P_π is the orthogonal projection of $L^2_K(G, \chi)$ onto the π-primary subspace of $L^2_K(G, \chi)$, $P_\pi Q$ is the orthogonal projection of $L^2(G)$ onto the π-primary subspace of $L^2_K(G, \chi)$ if P_π and Q commute. But using Fubini's Theorem and the linearity of trace and $\mathrm{Tr}(AB) = \mathrm{Tr}(BA)$, one sees

$$
\begin{aligned}
P_\pi(Qf)(x) &= d(\pi)\mathrm{Tr}(\pi(x^{-1})\pi(Qf)) \\
&= d(\pi)\mathrm{Tr}\left(\pi(x^{-1})\int Qf(g)\pi(g)\,dg\right) \\
&= d(\pi)\mathrm{Tr}\left(\pi(x^{-1})\int_G\int_K \chi(k)f(gk)\,dk\,\pi(g)\,dg\right) \\
&= d(\pi)\mathrm{Tr}\left(\pi(x^{-1})\int_G\int_K \chi(k)f(g)\,dk\,\pi(g)\pi(k^{-1})\,dg\right) \\
&= d(\pi)\mathrm{Tr}\left(\int_K \pi(x^{-1})\chi(k)\int_G f(g)\pi(g)\;dg\,\pi(k^{-1})\,dk\right) \\
&= d(\pi)\mathrm{Tr}\left(\int_K \pi(x^{-1})\chi(k)\pi(f)\pi(k^{-1})\,dk\right) \\
&= \int_K \chi(k)d(\pi)\mathrm{Tr}\left(\pi(k^{-1}x^{-1})\pi(f)\right)\,dk \\
&= \int \chi(k)P_\pi(f)(xk)\,dk \\
&= QP_\pi f(x)\,.
\end{aligned}
$$

We remark an easier way to see this is that Q commutes with the regular representation and P_π is a central projection in $\mathrm{Hom}(\lambda,\lambda)$.

Indeed, we know P_π is in $\mathrm{Hom}(\lambda,\lambda)$ and if $A \in \mathrm{Hom}(\lambda,\lambda)$ and $B \in \mathrm{Hom}(\pi,\lambda)$, then $AB \in \mathrm{Hom}(\pi,\lambda)$ and thus by Corollary 6.54, $P_\pi AB = AB = AP_\pi B$ for all such B. Thus $P_\pi A = AP_\pi$ on the range of every B in $\mathrm{Hom}(\pi,\lambda)$ and thus P_π commutes with A. □

Exercise Set 8.4

1. Let G be a compact Hausdorff group. Let χ be a 1-dimensional representation of G. Show χ is unitary; i.e., $|\chi(g)| = 1$ for all g.

Solution:

Let χ be a 1-dimensional representation of a compact Hausdorff group G. Then $\chi(G)$ is a compact subgroup of the nonzero complex numbers under complex multiplication. It thus is a bounded subset of $\mathbb{C}$. Now if $|\chi(g)| \neq 1$, then either $|\chi(g)|$ or $|\chi(g^{-1})|$ is strictly greater than 1. Thus the set $\{\chi(g^n) \mid n \in \mathbb{N}\} \cup \{\chi((g^{-1})^n) \mid n \in \mathbb{N}\}$ is unbounded. This being impossible implies $|\chi(g)| = 1$ and $\chi : G \to \mathbb{T}$ and thus is a unitary representation.

□

2. Let G be a compact group. For $\pi \in \hat{G}$, show by a direct calculation that if $f \in L^2(G)$, one has

$$\bar{\xi}_\pi * f(x) = d(\pi)\mathrm{Tr}(\pi(\lambda(x^{-1})f)) = d(\pi)\mathrm{Tr}(\pi(x^{-1})\pi(f)) \text{ for all } x.$$

Solution:

Note

$$\begin{aligned}
\bar{\xi}_\pi * f(x) &= \int d(\pi) \sum_{j=1}^{d(\pi)} \overline{(\pi(y)e_j, e_j)} f(y^{-1}x)\, dy \\
&= \int d(\pi) \sum_{j=1}^{d(\pi)} (e_j, \pi(y)e_j) f(y^{-1}x)\, dy \\
&= d(\pi) \sum_{j=1}^{d(\pi)} \int (e_j, \pi(xy)e_j) f(y^{-1})\, dy \\
&= d(\pi) \sum_{j=1}^{d(\pi)} \int (e_j, \pi(xy^{-1})e_j) f(y)\, dy \\
&= d(\pi) \sum_{j=1}^{d(\pi)} \int (f(y)\pi(yx^{-1})e_j, e_j)\, dy \\
&= d(\pi) \sum_{j=1}^{d(\pi)} (\int f(y)\pi(y)\pi(x^{-1})e_j\, dy, e_j) \\
&= d(\pi) \sum_{j=1}^{n} (\pi(f)\pi(x^{-1})e_j, e_j) \\
&= d(\pi)\mathrm{Tr}(\pi(f)\pi(x^{-1})) \\
&= d(\pi)\mathrm{Tr}(\pi(x^{-1})\pi(f))
\end{aligned}$$

and

$$\begin{aligned}
\pi(x^{-1})\pi(f) &= \int \pi(x^{-1}) f(y)\pi(y)\, dy \\
&= \int f(y)\pi(x^{-1})\pi(y)\, dy \\
&= \int f(y)\pi(x^{-1}y)\, dy \\
&= \int f(xy)\pi(y)\, dy \\
&= \int (\lambda(x^{-1})f)(y)\pi(y)\, dy \\
&= \pi(\lambda(x^{-1})f)
\end{aligned}$$

for $f \in L^2(G)$ give the result.

□

3. Let π be a finite dimensional unitary representation of compact Hausdorff group G. Let f be in $L^1(G)$. Show

$$\mathrm{Tr}(\pi(x^{-1})\pi(f)) = \bar{\chi}_\pi * f(x).$$

Solution: Using $\pi_{i,i}$ for the matrix coefficient $e_i \otimes_\pi \bar{e}_i$, we have:

$$\begin{aligned}
\mathrm{Tr}(\pi(x^{-1})\pi(f)) &= \mathrm{Tr}(\pi(f)\pi(x^{-1})) \\
&= \sum_{i=1}^{d} (\pi(f)\pi(x^{-1})e_i, e_i) \\
&= \sum_{i=1}^{d} \int f(y^{-1})(\pi(y^{-1})\pi(x^{-1})e_i, e_i)\, dy \\
&= \sum_{i=1}^{d} \int (e_i, \pi(xy)e_i) f(y^{-1})\, dy \\
&= \sum_{i=1}^{d} \int (e_i, \pi(y)e_i) f(y^{-1}x)\, dy \\
&= \sum_{i=1}^{d} \pi_{i,i} * f(x) \\
&= \bar{\chi}_\pi * f(x).
\end{aligned}$$

□

4. Let λ, ρ, and B be the left regular, the right regular, and the biregular representations of a compact Hausdorff group G. Let π be an irreducible unitary representation of G. Show from direct calculations of the definitions of ξ_π and $\xi_{\pi\times\bar{\pi}}$ that for $f \in L^2(G)$, one has:

(a) $\lambda(\bar{\xi_\pi})f(x) = d(\pi)\mathrm{Tr}(\pi(x^{-1})\pi(f))$,

(b) $\rho(\bar{\xi_{\bar{\pi}}})f(x) = d(\pi)\mathrm{Tr}(\pi(x^{-1})\pi(f))$, and

(c) $B(\bar{\xi}_{\pi\times\bar{\pi}})f(x) = d(\pi)\mathrm{Tr}(\pi(x^{-1})\pi(f))$.

Solution: Take a basis $\{e_j\}_{j=1}^{d(\pi)}$ of $\mathcal{H}_\pi$. Using properties of trace and the unimodularity of G, one sees

$$\begin{aligned}
\lambda(\bar{\xi_\pi})f(x) &= \int \bar{\xi}_\pi(g)\lambda(g)f(x)\; dg \\
&= \int \bar{\xi}_\pi(g)f(g^{-1}x)\; dg \\
&= d(\pi)\int \sum_{j=1}^{d(\pi)} \overline{(\pi(g)e_j, e_j)}\, f(g^{-1}x)\, dg \\
&= d(\pi)\int \sum_{j=1}^{d(\pi)} (e_j, \pi(g)e_j)\, f(g^{-1}x)\, dg \\
&= d(\pi)\int \sum_{j=1}^{d(\pi)} (e_j, \pi(g^{-1})e_j)\, f(gx)\, dg
\end{aligned}$$

$$
\begin{aligned}
&= d(\pi) \int \sum_{j=1}^{d(\pi)} (\pi(g)e_j, \pi(g)\pi(g^{-1})e_j)\, f(gx)\, dg \\
&= d(\pi) \int \sum_{j=1}^{d(\pi)} (\pi(g)e_j, e_j)\, f(gx)\, dg \\
&= d(\pi) \int \mathrm{Tr}(f(gx)\pi(g))\ dg \\
&= d(\pi) \int \mathrm{Tr}(f(g)\pi(g)\pi(x^{-1}))\ dg \\
&= d(\pi)\mathrm{Tr}\left[\left(\int f(g)\pi(g)\, dg\right) \pi(x^{-1})\right] \\
&= d(\pi)\mathrm{Tr}(\pi(f)\pi(x^{-1})) \\
&= d(\pi)\mathrm{Tr}(\pi(x^{-1})\pi(f))\,.
\end{aligned}
$$

Similarly for $\rho(\bar{\xi}_{\bar{\pi}})$,

$$
\begin{aligned}
\rho(\bar{\xi}_{\bar{\pi}})f(x) &= \int \bar{\xi}_{\bar{\pi}}(g)\rho(g)f(x)\, dg \\
&= \int \bar{\xi}_{\bar{\pi}}(g) f(xg)\, dg \\
&= d(\pi) \int \sum_{j=1}^{d(\pi)} (\bar{e}_j, \bar{\pi}(g)\bar{e}_j)_{\bar{\mathcal{H}}_\pi} f(xg)\, dg \\
&= d(\pi) \int \sum_{j=1}^{d(\pi)} \overline{(e_j, \pi(g)e_j)_{\mathcal{H}_\pi}} f(xg)\, dg \\
&= d(\pi) \int \sum_{j=1}^{d(\pi)} (\pi(g)e_j, e_j)_{\mathcal{H}_\pi} f(xg)\, dg \\
&= d(\pi) \int \sum_{j=1}^{d(\pi)} (\pi(x^{-1}g)e_j, e_j)_{\mathcal{H}_\pi} f(g)\, dg \\
&= d(\pi) \sum_{j=1}^{d(\pi)} (\pi(x^{-1}) \int f(g)\pi(g)e_j\, dg, e_j)_{\mathcal{H}_\pi} \\
&= d(\pi) \sum_{j=1}^{d(\pi)} (\pi(x^{-1})\pi(f)e_j, e_j)_{\mathcal{H}_\pi} \\
&= d(\pi)\mathrm{Tr}(\pi(x^{-1})\pi(f))\,.
\end{aligned}
$$

Finally, one has basis $e_j \otimes \bar{e}_k$ where $j, k = 1, 2, \ldots, d(\pi)$ for $\mathcal{H}_\pi \otimes \bar{\mathcal{H}}_\pi$ and thus by the orthogonality formula (Corollary 8.10),

$$
\begin{aligned}
&B(\bar{\xi}_{\pi\times\bar{\pi}})f(x)\\
&= d(\pi)^2 \iint \sum_{j=1}^{d(\pi)} \sum_{k=1}^{d(\pi)} (e_j \otimes \bar{e}_k, \pi(g_1)e_j \otimes \bar{\pi}(g_2)\bar{e}_k)_{\mathcal{H}_\pi \otimes \bar{\mathcal{H}}_\pi} B(g_1,g_2)f(x)\, dg_1\, dg_2\\
&= d(\pi)^2 \iint \sum_{j=1}^{d(\pi)} \sum_{k=1}^{d(\pi)} (e_j, \pi(g_1)e_j)_{\mathcal{H}_\pi} (\bar{e}_k, \bar{\pi}(g_2)\bar{e}_k)_{\bar{\mathcal{H}}_\pi} f(g_1^{-1}xg_2)\, dg_1\, dg_2\\
&= d(\pi)^2 \iint \sum_{j=1}^{d(\pi)} \sum_{k=1}^{d(\pi)} (e_j, \pi(g_1)e_j)_{\mathcal{H}_\pi} (\pi(x^{-1}g_1g_2)e_k, e_k)_{\mathcal{H}_\pi} f(g_2)\, dg_2\, dg_1\\
&= d(\pi)^2 \iint \sum_{j=1}^{d(\pi)} \sum_{k=1}^{d(\pi)} (e_j, \pi(g_1)e_j)_{\mathcal{H}_\pi} (\pi(g_1)\pi(g_2)e_k, \pi(x)e_k)_{\mathcal{H}_\pi} f(g_2)\, dg_2\, dg_1\\
&= d(\pi)^2 \int \sum_{j=1}^{d(\pi)} \sum_{k=1}^{d(\pi)} \left(\int (e_j, \pi(g_1)e_j)_{\mathcal{H}_\pi} (\pi(g_1)\pi(g_2)e_k, \pi(x)e_k)_{\mathcal{H}_\pi}\, dg_1\right) f(g_2)\, dg_2\, dg_1\\
&= d(\pi)^2 \int \sum_{j=1}^{d(\pi)} \sum_{k=1}^{d(\pi)} \left(\int (e_j \otimes_\pi \bar{e}_j)(g_1)\overline{(\pi(x)e_k \otimes_\pi \overline{\pi(g_2)e_k})(g_1)}\, dg_1\right) f(g_2)\, dg_2\\
&= d(\pi)^2 \int \sum_{j=1}^{d(\pi)} \sum_{k=1}^{d(\pi)} \frac{1}{d(\pi)} (e_j, \pi(x)e_k)_{\mathcal{H}_\pi} (\pi(g_2)e_k, e_j)_{\mathcal{H}_\pi} f(g_2)\, dg_2\\
&= d(\pi) \int \sum_{k=1}^{d(\pi)} \Big(\sum_{j=1}^{d(\pi)} (\pi(g_2)e_k, e_j)_{\mathcal{H}_\pi} e_j, \pi(x)e_k)_{\mathcal{H}_\pi} f(g_2)\, dg_2\\
&= d(\pi) \int \sum_{k=1}^{d(\pi)} (\pi(g_2)e_k, \pi(x)e_k)_{\mathcal{H}_\pi} f(g_2)\, dg_2\\
&= d(\pi) \int \sum_{k=1}^{d(\pi)} (\pi(x^{-1}g_2)e_k, e_k)_{\mathcal{H}_\pi} f(g_2)\, dg_2\\
&= d(\pi) \int \sum_{k=1}^{d(\pi)} (f(g_2)\pi(x^{-1})\pi(g_2)e_k, e_k)_{\mathcal{H}_\pi}\, dg_2\\
&= d(\pi) \sum_{k=1}^{d(\pi)} (\pi(x^{-1})\pi(f)e_k, e_k)_{\mathcal{H}_\pi}\\
&= d(\pi)\mathrm{Tr}(\pi(x^{-1})\pi(f)).
\end{aligned}
$$

□

5. For each $\pi \in \hat{G}_c$, let $\mathcal{H}_\pi \otimes \bar{\mathcal{H}}_\pi$ be the Hilbert space of Hilbert-Schmidt operators on $\mathcal{H}_\pi$ with inner product $\langle R_\pi, S_\pi \rangle_\pi = d(\pi)(R_\pi, S_\pi)_{\mathcal{H}\otimes\bar{\mathcal{H}}_\pi} = d(\pi)\mathrm{Tr}(R_\pi S_\pi^*)$ and take $\mathcal{A}$ to be the Hilbert space external direct sum $\bigoplus_{\pi\in\hat{G}_c} \left(\mathcal{H}_\pi \otimes \bar{\mathcal{H}}_\pi\right)$ of these Hilbert spaces. Define a multiplication on $\mathcal{A}$ by $(T_\pi)\cdot(S_\pi) = (T_\pi S_\pi)$ and an adjoint by $(T_\pi)^* = (T_\pi^*)$. Let $\langle \cdot\,, \cdot \rangle$ denotes the inner product on $\mathcal{A}$.

(a) Show the adjoint is an isometry.

(b) Show $\langle RS, T\rangle = \langle S, R^*T\rangle$.

(c) Show $\langle R, S\rangle = \langle S^*, R^*\rangle$.

(d) Show $L_R : \mathcal{A} \to \mathcal{A}$ defined by $L_R S = RS$ is bounded and has adjoint L_{R^*}.

(e) Show $\mathcal{A}^2$ where $\mathcal{A}^2$ is the linear span of all products RS where $R, S \in \mathcal{A}$ is dense in $\mathcal{A}$.

This shows $\mathcal{A}$ is a left Hilbert algebra. Define $\mathcal{F} : L^2(G) \to \mathcal{A}$ by $\mathcal{F}(f)(\pi) = \pi(f)$. Show $\mathcal{F}$ is a unitary isomorphism of $L^2(G)$ onto $\mathcal{A}$ satisfying $\mathcal{F}(f * h) = \mathcal{F}(f)\mathcal{F}(h)$ for $f, h \in L^2(G)$ and $\mathcal{F}(f^*) = \mathcal{F}(f)^*$. In particular, $L^2(G)$ is a left Hilbert algebra. The unitary transformation $\mathcal{F}$ is called the Fourier transform.
Solution: To see the adjoint is an isometry it suffices to note on each $\mathcal{H}_\pi \otimes \bar{\mathcal{H}}_\pi$, the mapping $T \mapsto T^*$ is an isometry of the Hilbert space $\mathcal{B}_2(\mathcal{H}_\pi, \mathcal{H}_\pi)$ of Hilbert-Schmidt operators. But this follows from (2.2).

Next note

$$\begin{aligned}\langle RS, T\rangle &= \sum_{\pi\in\hat{G}_c} d(\pi)(R_\pi S_\pi, T_\pi)_{\mathcal{H}_\pi\otimes\bar{\mathcal{H}}_\pi}\\ &= \sum_{\pi\in\hat{G}_c} d(\pi)\mathrm{Tr}(R_\pi S_\pi T_\pi^*)\\ &= \sum_{\pi\in\hat{G}_c} d(\pi)\mathrm{Tr}(S_\pi T_\pi^* R_\pi)\\ &= \sum_{\pi\in\hat{G}_c} d(\pi)\mathrm{Tr}(S_\pi (R_\pi^* T_\pi)^*)\\ &= \sum_{\pi\in\hat{G}_c} d(\pi)(S_\pi, R_\pi^* T_\pi)_{\mathcal{H}_\pi\otimes\bar{\mathcal{H}}_\pi}\\ &= \langle S, R^*T\rangle .\end{aligned}$$

Similarly,

$$\begin{aligned}\langle R, S\rangle &= \sum_\pi d(\pi)(R_\pi, S_\pi)_{\mathcal{H}_\pi\otimes\bar{\mathcal{H}}_\pi}\\ &= \sum_\pi d(\pi)\mathrm{Tr}(R_\pi S_\pi^*)\\ &= \sum_\pi d(\pi)\mathrm{Tr}((S_\pi^*)(R_\pi^*)^*)\\ &= \langle S^*, R^*\rangle .\end{aligned}$$

Next note $L_R(S) = (R_\pi S_\pi)_\pi$. Since $\sum d(\pi)||R_\pi||_2^2 < \infty$, there is a constant $M > 0$ with $||R_\pi||_2 \leqslant M$ for all $\pi \in \hat{G}_c$. From (2.3), $||R_\pi|| \leqslant ||R_\pi||_2 \leqslant M$ and from Proposition 2.32, $||R_\pi S_\pi||_2 \leqslant ||R_\pi|| \, ||S_\pi||_2 \leqslant M||S_\pi||_2$ and thus

$$\begin{aligned} ||L_R(S)||_{\mathcal{A}}^2 &= \sum_\pi d(\pi)||R_\pi S_\pi||_2^2 \\ &\leqslant \sum_\pi d(\pi)M^2||S_\pi||_2^2 \\ &= M^2||S||_{\mathcal{A}}^2 . \end{aligned}$$

So L_R is bounded. Lastly, $\mathcal{A}^2$ is dense in $\mathcal{A}$. In fact one only needs to note the linear span of the operators $R_\pi S_\pi$ where R_π and S_π are Hilbert-Schmidt on $\mathcal{H}_\pi$ is a dense subspace of $\mathcal{H}_\pi \otimes \bar{\mathcal{H}}_\pi$. But this follows from Exercise 2.5.16. Consequently (a), (b), (c), (d), and (e) hold and $\mathcal{A}$ is a left Hilbert algebra.

The fact that $\mathcal{F}$ is a unitary linear transformation of $L^2(G)$ onto $\mathcal{A}$ is the Plancherel Theorem 8.20. $\mathcal{F}(f * h) = \mathcal{F}(f)\mathcal{F}(h)$ and $\mathcal{F}(f^*) = \mathcal{F}(f)^*$ follow from $\pi(f * h) = \pi(f)\pi(h)$ and $\pi(f^*) = \pi(f)^*$ for $f, h \in L^2(G) \subseteq L^1(G)$. □

6. Let G be a compact Hausdorff group. Let π be an irreducible unitary representation of G. Show

$$d(\pi)\mathrm{Tr}(\pi(f)) = (f, \bar{\xi}_\pi) \text{ for } f \in L^2(G).$$

Solution:

$$\begin{aligned} (f, \bar{\xi}_\pi) &= (f, P_\pi \bar{\xi}_\pi) \\ &= (P_\pi f, \bar{\xi}_\pi) \\ &= d(\pi)(I_\pi I_\pi^* f, \bar{\xi}_\pi) \\ &= d(\pi)(\pi(f), \pi(\bar{\xi}_\pi))_2 \\ &= d(\pi)(\pi(f), I)_2 \\ &= d(\pi)\mathrm{Tr}(\pi(f)). \end{aligned}$$

□

7. Let ρ be an irreducible unitary representation of a compact Hausdorff group G with orthonormal basis $e_1, e_2, \ldots, e_d$. Set $\rho_{i,j} = e_i \otimes_\rho \bar{e}_j$. Using Corollaries 8.11 and 8.13 and Exercise 8.2.3, show $\sum_{i,j} a_{i,j}\rho_{i,j}$ is central if and only if $a_{i,j} = c\delta_{i,j}$ for some constant c. Then conclude Proposition 8.53.

Solution:

Suppose f is a central function in $L^2(G)$ is central. Exercise 8.2.3 shows $f = \sum_{\pi\in\hat{G}_c} \sum_{1\leqslant i,j\leqslant d(\pi)} a_{i,j}^\pi \pi_{i,j}$ in $L^2(G)$ where $a_{i,j}^\pi = d(\pi)(f, \pi_{i,j})$. By Exercise 8.1.10,

$$\lambda(\pi_{k,l} * f) = \lambda(\pi_{k,l})\lambda(f) = \lambda(f)\lambda(\pi_{k,l}) = \lambda(f * \pi_{k,l})$$

if $\pi \in \hat{G}_c$. Exercise 8.1.9 then implies $\pi_{k,l} * f = f * \pi_{k,l}$ for $1 \leqslant k, l \leqslant d(\pi)$. Now fix a ρ in $\hat{G}_c$. Using Corollaries 8.11 and 8.13, we have $\pi_{k,l} * \rho_{i,j} = \rho_{i,j} * \pi_{k,l} = 0$ if $\rho \neq \pi$ and $\rho_{k,l} * \rho_{s,t} = \frac{1}{d(\rho)}\delta_{l,s}\rho_{k,t}$ for all k, l, s, t between 1 and $d(\rho)$. Continuity of convolution, i.e., $|f * g|_2 \leqslant |f|_1|g|_2 \leqslant |f|_2|g|_2$ then implies

$$\sum_\pi \sum_{1\leqslant i,j\leqslant d(\pi)} a_{i,j}^\pi \rho_{k,l} * \pi_{i,j} = \sum_\pi \sum_{1\leqslant i,j\leqslant d(\pi)} a_{i,j}^\pi \pi_{i,j} * \rho_{k,l}$$

for each pair $k\,,l$. Consequently,

$$\sum_{1\leqslant i,j\leqslant d(\rho)} \frac{1}{d(\rho)} a^{\rho}_{i,j}\delta_{l,i}\rho_{k,j} = \sum_{1\leqslant i,j\leqslant d(\rho)} \frac{1}{d(\rho)} a^{\rho}_{i,j}\delta_{j,k}\rho_{i,l}.$$

So we see

$$\sum_j a^{\rho}_{l,j}\rho_{k,j} = \sum_i a^{\rho}_{i,k}\rho_{i,l} \text{ for each pair } k,\, l\,.$$

The linear independence of the functions $\rho_{i,j}$ then implies $a_{l,j} = 0$ if $j \neq l$ and $a_{l,l} = a_{k,k}$ for $k \neq l$. Let c_ρ be the common value of the $a^{\rho}_{k,k}$. Thus in $L^2(G)$, $f = \sum_{\pi\in\hat{G}_c}\sum_{i=1}^{d(\pi)} c_\pi \pi_{i,i} = \sum_{\pi\in\hat{G}_c} c_\pi \bar{\chi}_\pi$.

□

8. Let G be a compact Hausdorff group and let B be the biregular representation. Suppose π is an irreducible unitary representation of G. Let $I_\pi : \mathcal{H}_\pi \otimes \bar{\mathcal{H}}_\pi \to L^2(G)$ be the intertwining operator between $\pi \times \bar{\pi}$ of $G \times G$ with B given in Theorem 8.9 whose range is $P_\pi L^2(G)$. Show the subspace of $P_\pi L^2(G)$ invariant under all $B(g,g)$ for $g \in G$ is $I_\pi(\mathbb{C}I)$ where $I : \mathcal{H}_\pi \to \mathcal{H}_\pi$ is the identity operator. Use this to show the space of central functions in $P_\pi L^2(G)$ is $\mathbb{C}\bar{\chi}_\pi$ and obtain another proof of Proposition 8.53.

Solution:

We know I_π is an intertwining operator from the representation $\pi\times\bar{\pi}$ of $G\times G$ with B on $L^2(G)$. Moreover, $\sqrt{d(\pi)}I_\pi$ is an isometry. From Definitions 6.69 and 6.71, one sees $\pi\times\bar{\pi}(g_1,g_2)(A) = \pi(g_1)A\pi(g_2)^{-1}$ for $A \in \mathcal{H}_\pi\otimes\bar{\mathcal{H}}_\pi = \mathcal{B}_2(\mathcal{H}_\pi,\mathcal{H}_\pi)$. Now the subspace of $\mathcal{H}_\pi \otimes \bar{\mathcal{H}}_\pi$ consisting of the A with $\pi \times \bar{\pi}(g,g)(A) = A$ is the collection of all A with $\pi(g)A\pi(g)^{-1} = A$. By Schur's Lemma, this consists of all multiples cI where I is the identity operator. Thus the range $\{I_\pi(cI) \mid c \in \mathbb{C}\}$ is the subspace of all $f \in P_\pi L^2(G)$ satisfying $B(g,g)f = f$ for all g; i.e., the subspace of central functions in $P_\pi L^2(G)$. But $I_\pi(cI)(x) = \mathrm{Tr}(\pi(x^{-1})cI) = c\mathrm{Tr}(\pi(x^{-1})I) = c\bar{\chi}_\pi(x)$. Thus the central functions in $P_\pi L^2(G)$ are the multiples of $\bar{\chi}_\pi$.

Finally for a central function f in $L^2(G)$, we note each $P_\pi f$ for $\pi \in \hat{G}_c$ is central since $B(g,g)P_\pi f = P_\pi B(g,g)f = P_\pi f$ for all $g \in G$. Indeed, $P_\pi \in \mathrm{Hom}_B(\lambda,\lambda)$. Consequently, $f = \sum_{\pi\in\hat{G}_c} P_\pi f = \sum_{\pi\in\hat{G}_c} c_\pi\bar{\chi}_\pi$ for appropriate constants c_π. But since this is an orthogonal direct sum, Corollary 8.54 then implies $c_\pi = (f,\bar{\chi}_\pi)_2$.

□

9. Let G be a compact Hausdorff group and let π be an irreducible unitary representation of G. Let f be a central L^1 function on G. Show $\pi(f) = \frac{1}{d(\pi)} f * \chi_\pi(e)I$.

Solution: By Lemma 6.119, $\pi(g)\pi(f) = \pi(f)\pi(g)$ for all g. Hence by Schur's Lemma, $\pi(f) = cI$ for some c. So $\mathrm{Tr}(\int f(x)\pi(x)\,dx) = cd(\pi)$. Hence $(f,\bar{\chi}_\pi) = cd(\pi)$. This gives $c = \frac{1}{d(\pi)}(f,\bar{\chi}_\pi) = \frac{1}{d(\pi)} f * \chi_\pi(e)$. □

10. Let σ be a finite dimensional unitary representation of a compact Hausdorff group G. Let $\chi_\sigma(g) = \mathrm{Tr}(\sigma(g))$ for $g \in G$. Show the following are equivalent.

(a) σ is irreducible.

(b) $\chi_\sigma * \chi_\sigma = \frac{1}{d(\sigma)}\chi_\sigma$.

(c) $(\chi_\sigma, \chi_\sigma)_2 = 1$.

Solution:

We know σ is unitary equivalent to $\bigoplus_{\pi\in\hat{G}_c} n_\pi\pi$ where $n_\pi < \infty$ and nonzero for finitely many π. Therefore we may assume $\rho = \sum_{k=1}^m n_k\pi_k$ where each n_k is finite and each π_k is an irreducible unitary representation of G. Set $\chi_k = \chi_{\pi_k}$ and $d_k = \dim(\pi_k)$. Then $\chi_\rho = \sum_{k=1}^m n_k\chi_k$. Now ρ is irreducible if and only if $\sum n_k = 1$. From Proposition 8.47 and $\overline{f * g} = \bar{f} * \bar{g}$, we see $\chi_j * \chi_k = 0$ for $j \neq k$ and $\chi_k * \chi_k = \frac{1}{d_k}\chi_k$ for each k. Hence

$$\chi_\rho * \chi_\rho = \sum_{j,k}(n_j\chi_j) * (n_k\chi_k) = \sum \frac{n_k^2}{d_k}\chi_k = \frac{1}{d(\rho)}\sum_k n_k\chi_k = \frac{1}{d(\rho)}\chi_\rho$$

if and only if $\sum n_k = 1$.

Using Corollary 8.54 we have

$$(\chi_\rho, \chi_\rho)_2 = \sum_{j,k}(n_j\chi_j, n_k\chi_k)_2 = \sum_k n_k^2.$$

So $\sum n_k^2 = 1$ is equivalent to $(\chi_\rho, \chi_\rho)_2 = 1$.

□

11. Let G be a compact Hausdorff group with left regular representation λ and suppose $\pi_1, \pi_2, \ldots, \pi_s$ are inequivalent irreducible unitary representations of G. For each r, let $e_{r,1}, e_{r,2}, \ldots, e_{r,d_r}$ be an orthonormal basis of $\mathcal{H}_{\pi_r}$ and let $\pi_r(i,j) = e_{r,1} \otimes_{\pi_r} \bar{e}_{r,j}$ be the collection of corresponding matrix coefficients. Suppose $n_r \leqslant d_r$ for $r = 1, 2, \ldots, s$. Define f by

$$f = \sum_{r=1}^{s}\sum_{j=1}^{n_r} \pi_r(j,j).$$

Show the smallest invariant subspace of $L^2(G)$ containing f gives a unitary representation unitarily equivalent to

$$\bigoplus_{r=1}^{s} n_r\pi_r.$$

In particular, this representation is cyclic.

Solution:

By Corollary 6.111 and Example 6.113, the smallest closed invariant subspace containing f is the smallest subspace W containing f and satisfying $h*k \in W$ if $h \in L^1(G)$ and $k \in W$. By Corollaries 8.11 and 8.13, $\pi_r(i,j) * \pi_t(k,l) = \frac{1}{d_r}\delta_{r,t}\delta_{j,k}\pi_r(i,l)$. Since $f = \sum_{r=1}^{s}\sum_{j=1}^{n_r}\pi_r(j,j) \in W$ and $\pi_t(k,l) * f = \frac{1}{d_r}\pi_t(k,j)$ if $t \leqslant s$ and $l \leqslant n_r \leqslant d(\pi_r)$, we see $\pi_r(k,j) \in W$ if $1 \leqslant k \leqslant d_r$ and

$j = 1, 2, \ldots, n_r$. So W contains the linear span of the matrix coefficients $\pi_r(i,j)$ for $1 \leqslant i \leqslant d_r$ and $j = 1, 2, \ldots, n_r$. Set W' to be the linear span of these matrix coefficients. This space is finite dimensional and closed. It clearly contains f. Set $W'_{r,j}$ to be the linear span of the matrix coefficients $\pi_r(i,j)$ where $i = 1, 2, \ldots, d_r$. By Corollaries 8.10 and 8.13, $W'_{r,j} \perp W'_{r',j'}$ if $r \neq r'$ or $j \neq j'$ and thus $W' = \bigoplus_{r=1}^{s} \bigoplus_{j \leqslant n_r} W'_{r,j}$. We note λ restricted to $W'_{r,j}$ is unitarily equivalent to π_r. In fact by Corollary 8.10, U defined by $Uv(x) = \sqrt{d_r}\,(v, \pi_r(x)e_{r,j})_{\mathcal{H}_{\pi_r}} = \sqrt{d_r}\,(v \otimes_{\pi_r} \bar{e}_{r,j})(x)$ is a unitary mapping onto $W'_{r,j}$ and

$$\begin{aligned} U(\pi_r(g)v)(x) &= \sqrt{d_r}\,(\pi_r(g)v, \pi_r(x)e_{r,j})_{\mathcal{H}_{\pi_r}} \\ &= \sqrt{d_r}\,(v, \pi_r(g^{-1}x)e_{r,j})_{\mathcal{H}_{\pi_r}} \\ &= \lambda(g)\left(\sqrt{d_r}\, v \otimes_{\pi_r} \bar{e}_{r,j}\right)(x) \\ &= \lambda(g)(Uv)(x)\,. \end{aligned}$$

So W' is λ invariant and we see $W = W'$. Moreover, since $\lambda|_{W'_{r,j}}$ is unitarily equivalent to π_r if $1 \leqslant r \leqslant s$ and $1 \leqslant j \leqslant n_r$, we have the unitary equivalence

$$\lambda|_{W'} \cong \bigoplus_{r=1}^{s} n_r \pi_r.$$

Note: Recall we know $\bigoplus_{r=1}^{s} n_r\pi_r$ is cyclic by Exercise 6.4.24. □

Exercise Set 8.5

1. Let ϕ be a positive definite χ spherical function on a topological Hausdorff group G with closed compact subgroup K having one dimensional character χ. Let S be the linear span of the functions $\lambda(a)\phi$ where $\lambda(a)f(x) = f(a^{-1}x)$ for functions f. Define a sesquilinear form $\langle\cdot,\cdot\rangle$ on S by

$$\langle \sum_{i=1}^{m} a_i\lambda(x_i)\phi, \sum_{j=1}^{n} b_j\lambda(y_j)\phi\rangle = \sum_{i,j} a_i\bar{b}_j\phi(x_i^{-1}y_j).$$

(a) Show

$$|\langle \sum_{i=1}^{m} a_i\lambda(x_i)\phi, \sum_{i=1}^{n} b_j\lambda(y_j)\phi\rangle|^2 \leqslant \sum_{i,j} a_i\bar{a}_j\phi(x_i^{-1}x_j)\sum_{i,j} b_i\bar{b}_j\phi(y_i^{-1}y_j).$$

(b) Show $\langle\cdot,\cdot\rangle$ is well defined.

(c) Show $\langle f, f\rangle \geqslant 0$ for $f \in S$.

(d) Let $N_\phi = \{f \in S \mid \langle f, f\rangle = 0\}$. Show $\lambda(a)N_\phi \subseteq N_\phi$.

(e) Define an inner product on S/N_ϕ by

$$\langle f + N_\phi, f' + N_\phi \rangle = \langle f, f' \rangle.$$

Show $g \mapsto \lambda(g)f + N_\phi$ is continuous from G into the complex inner product space S/N_ϕ.

(f) Set $\lambda'(g)(f + N_\phi) = \lambda(g)f + N_\phi$. Show $\lambda'(g)$ preserves the inner product on S/N_ϕ and extends to a unitary representation of π of G on the completion $\mathcal{H}_\pi$ of S/N_ϕ.

(g) Show $\phi + N_\phi$ is a cyclic vector for π and $\langle \phi + N_\phi, \pi(g)(\phi + N_\phi) \rangle = \phi(g)$.

Solution:

This exercise is essentially the same as Exercise 6.8.12. Since ϕ is continuous and positive definite, (a), (b), (c), (d), (e) and (f) follow immediately from this exercise and the Cauchy-Schwarz inequality for positive semidefinite sesquilinear forms. Note (g) holds for S is the linear span of the vectors $\lambda(x)\phi$ for $x \in G$, and since $\mathcal{H}_\pi$ is the completion of S/N_ϕ, $S + N_\phi$ is dense in $\mathcal{H}_\pi$. Thus $\phi + N_\phi$ is a cyclic vector for π and $\langle \phi + N_\phi, \pi(g)(\phi + N_\phi) \rangle = \langle \lambda(e)\phi, \lambda(g)\phi \rangle = \phi(e^{-1}g) = \phi(g)$.

□

2. Assume G is as in Exercise 8.5.1. Show π is irreducible. (**Hint:** Show the χ-primary projection for the representation $\pi|_K$ has range the one dimensional space consisting of multiples of $\phi + N_\phi$.)

Solution:

Using (d) of Theorem 8.49, we know the χ-primary projection P for $\pi|_K$ is given by $P = \pi(\bar{\chi})$. Thus $P\pi(g)v = \int_K \bar{\chi}(k)\pi(kg)v\,dk$. This implies

$$\begin{aligned} P(\pi(g)(\phi + N_\phi)) &= \int_K \bar{\chi}(k)\pi(kg)(\phi + N_\phi)\,dk \\ &= \int_K \bar{\chi}(k)(\lambda(kg)\phi + N_\phi)\,dk \\ &= \int_K \bar{\chi}(k)\lambda(kg)\phi\,dk + N_\phi\,. \end{aligned}$$

But by Definition 8.59,

$$\begin{aligned} \int_K \bar{\chi}(k)\lambda(kg)\phi(x)\,dk &= \int_K \chi(k^{-1})\phi(g^{-1}k^{-1}x)\,dk \\ &= \int_K \chi(k)\phi(g^{-1}kx)\,dk \\ &= \phi(g^{-1})\phi(x)\,. \end{aligned}$$

Putting these together gives $P(\pi(g)(\phi + N_\phi)) = \phi(g^{-1})(\phi + N_\phi)$. Since $\phi + N_\phi$ is a cyclic vector, the linear span of the vectors $\pi(g)(\phi + N_\phi)$ is dense in $\mathcal{H}_\pi$ and by the continuity of P, one has the range of P is the one dimensional subspace of $\mathcal{H}_\pi$ spanned by ϕ.

Now let S be a nonzero closed π-invariant subspace of $\mathcal{H}_\pi$. If Q is the orthogonal projection of $\mathcal{H}_\pi$ onto S, then $Q\pi(g) = \pi(g)Q$ for all g. Moreover $Q\phi \neq 0$ for otherwise $Q\pi(g)(\phi + N_\phi) = \pi(g)Q(\phi + N_\phi) = 0$ for all g would imply $S = \{0\}$ for $\phi + N_\phi$ is a cyclic vector. Since $\pi(k)$ commutes with Q for all $k \in K$, Exercise 6.5.2 gives $PQ = QP$. Consequently, $PQ(\phi + N_\phi) = QP(\phi + N_\phi) = Q(\phi + N_\phi) \neq 0$. Thus the range of Q contains a nonzero multiple of $\phi + N_\phi$. So $\phi + N_\phi \in S$ and since $\phi + N_\phi$ is cyclic, $S = \mathcal{H}_\pi$.

□

Exercise Set 8.6

1. Suppose G_1 and G_2 are Hausdorff locally compact abelian groups. Show the dual group of $G_1 \times G_2$ is topologically isomorphic to $\hat{G}_1 \times \hat{G}_2$.

Solution: One has $\chi(g_1, g_2) = \chi_1(g_1)\chi_2(g_2)$ where $\chi_1(g_1) = \chi(g_1, e_2)$ and $\chi_2(g_2) = \chi(e_1, g_2)$. Thus every character of $G_1 \times G_2$ has form $\chi(g_1, g_2) = \chi_1(g_1)\chi_2(g_2)$ where χ_1 and χ_2 are characters for G_1 and G_2. It is then easy to check the mapping ϕ from the dual group of $\hat{G}_1 \times \hat{G}_2$ into the dual of $G_1 \times G_2$ defined by $\phi(\chi_1, \chi_2)(g_1, g_2) = \chi_1(g_1)\chi_2(g_2)$ is a group isomorphism. We need only check it is a homeomorphism. To do this it suffices to show the continuity of ϕ and ϕ^{-1} at the identities.

Consider the neighborhood of 1 in $\widehat{G_1 \times G_2}$ given by $N = \{\chi \mid |\chi(g_1, g_2) - 1| < \epsilon \text{ for } (g_1, g_2) \in K\}$ where K is a compact subset of $G_1 \times G_2$ and $\epsilon > 0$. Then the images K_1 and K_2 of K under the mappings $(g_1, g_2) \mapsto g_1$ and $(g_1, g_2) \mapsto g_2$ are compact in G_1 and G_2 and $K \subseteq K_1 \times K_2$. The sets $N_j = \{\chi_j \in \hat{G}_j \mid |\chi_j(g_j) - 1| < \frac{\epsilon}{2} \text{ for } g_j \in K_j\}$ for $j = 1, 2$ are neighborhoods of 1 in $\hat{G}_j$ and if $(\chi_1, \chi_2) \in N_1 \times N_2$, and

$$\begin{aligned} |\chi_1(g_1)\chi_2(g_2) - 1| &\leqslant |\chi_1(g_1)\chi_2(g_2) - \chi_2(g_2)| + |\chi_2(g_2) - 1| \\ &\leqslant |\chi_1(g_1) - 1| + |\chi_2(g_2) - 1| \\ &< \epsilon \end{aligned}$$

and thus $\phi(\chi_1, \chi_2) \in N$. So ϕ is continuous at $(1, 1)$.

For ϕ^{-1} consider neighborhoods of 1 in $\hat{G}_1$ and $\hat{G}_2$ given by $N_j = \{\chi_j \in \hat{G}_j \mid |\chi_j(g_j) - 1| < \epsilon \text{ for } g_j \in K_j\}$ where $\epsilon > 0$ and K_j is a compact subset of G_j for $j = 1, 2$. Set $K = (K_1 \times \{e_2\}) \cup (\{e_1\} \times K_2)$ and take $N = \{\chi \in \widehat{G_1 \times G_2} \mid |\chi(g_1, g_2) - 1| < \epsilon \text{ for } (g_1, g_2) \in K\}$. Then N is a neighborhood of 1 in $\widehat{G_1 \times G_2}$ and $\phi^{-1}(N) \subseteq N_1 \times N_2$. Indeed, $\phi^{-1}(\chi) = (\chi_1, \chi_2)$ where $\chi_1(g_1) = \chi(g_1, e_2)$ and $\chi_2(g_2) = \chi(e_1, g_2)$. Thus ϕ^{-1} is continuous at 1.

□

2. Let A be an index set. Show $\prod_{a \in A} \mathbb{T}$ with the product topology is a compact Hausdorff group.

Solution: Compactness is by the Tychonoff Theorem. We also note the space is Hausdorff. We show multiplication is continuous. Recall a neighborhood in

$\prod_\alpha \mathbb{T}$ of a point x is a set which a contains a subset of form $\prod N_\alpha$ where each N_α is an open neighborhood of $x(\alpha)$ and all but finitely many N_α are $\mathbb{T}$. Let x_0 and y_0 be in $\prod_\alpha \mathbb{T}$ and let $N = \prod N_\alpha$ be as above for the point $x_0 y_0^{-1}$. By continuity of multiplication on $\mathbb{T}$ there are open neighborhoods U_α and V_α of $x_0(\alpha)$ and $y_0(\alpha)$ such that all but finitely many of these are $\mathbb{T}$ and $z_1 z_2^{-1} \in N_\alpha$ if $z_1 \in U_\alpha$ and $z_2 \in V_\alpha$. Set $U = \prod_\alpha U_\alpha$ and $V = \prod_\alpha V_\alpha$. These are open neighborhoods of x_0 and y_0 and $xy^{-1} \in N$ if $x \in U$ and $y \in V$. By Exercise 5.1.1, $\prod_{\alpha\in A} \mathbb{T}$ is a topological group.

□

3. Let G be a locally compact Hausdorff group and suppose χ is a one-dimensional character of G. For $f \in L^1(G)$, show $f * \chi = \hat{f}(\chi)\chi$.

Solution:

One has

$$\begin{aligned} f * \chi(g) &= \int f(x)\chi(x^{-1}g)\,dx \\ &= \int f(x)\chi(x^{-1})\chi(g)\,dx \\ &= \left(\int f(x)\overline{\chi(x)}\,dx\right)\chi(g) \\ &= \hat{f}(\chi)\chi(g)\,. \end{aligned}$$

□

4. Let G be a locally compact Hausdorff group. Give $\widehat{G}$ the topology in (8.17). Show this makes $\widehat{G}$ into a group which is Hausdorff and abelian.

Solution: It is clear $\widehat{G}$ is a group. We show multiplication is continuous. Indeed, consider $N_{K,\epsilon}(\chi_0\chi_0')$. If $\chi \in N_{K,\epsilon/2}(\chi_0)$ and $\chi' \in N_{K,\epsilon/2}(\chi_0')$, then

$$\begin{aligned} |\chi\chi'(k) - \chi_0\chi_0'(k)| &\leqslant |\chi(k)\chi'(k) - \chi_0(k)\chi'(k)| + |\chi_0(k)\chi'(k) - \chi_0(k)\chi_0'(k)| \\ &= |\chi(k) - \chi_0(k)| + |\chi'(k) - \chi_0(k)| \\ &< \frac{\epsilon}{2} + \frac{\epsilon}{2} = \epsilon \end{aligned}$$

for $k \in K$. So multiplication is continuous. For inversion note if $\chi_0 \in \hat{G}$ and $\epsilon > 0$ and K is a compact subset of G, then $\chi^{-1} \in N_{K,\epsilon}(\chi_0^{-1})$ if $|\bar{\chi}(k) - \bar{\chi}_0(k)| < \epsilon$ for $k \in K$ which occurs if $\chi \in N_{K,\epsilon}(\chi_0)$.

To see $\widehat{G}$ is Hausdorff, note if $\chi \neq 1$, then there is a g with $\chi(g) \neq 1$. Let $\epsilon = \frac{1}{2}|\chi(g) - 1|$. Then $N_{\{g\},\epsilon}(\chi)$ and $N_{\{e\},\epsilon}(1)$ are neighborhoods of χ and 1 which are disjoint. □

5. Let G be a locally compact Hausdorff group and let Δ be the collection all one dimensional nonzero * representations of the Banach * algebra $L^1(G)$. By Corollary 6.108, there is a one-to-one correspondence $\chi \to \pi_\chi$ between the collection of one dimensional characters of G and Δ where

$$\pi_\chi(f) = \int f(g)\chi(g)\,dg.$$

Show the collection $\Delta \cup \{0\}$ is a closed subset in the weak $*$ topology of the unit ball of the dual space of $L^1(G)$ and thus with the weak $*$ topology is a compact Hausdorff space. In particular, Δ with the relative weak $*$ topology is a locally compact Hausdorff space. When G is abelian, the correspondence $\chi \to \pi_\chi$ gives a locally compact topology on $\widehat{G}$.

Solution: One can show this using neighborhoods or nets. Note if $\pi_i = \pi_{\chi_i}$ is a net converging in the weak $*$ topology to a π in the dual of $L^1(G)$, then $\pi_i(f * h) = \pi_i(f)\pi_i(h) \to \pi(f)\pi(h)$. But $\pi_i(f * h) \to \pi(f * h)$. So $\pi(f * h) = \pi(f) * \pi(h)$. Similarly, $\pi(f^*) = \overline{\pi(f)}$. Thus π is a representation. Also $|\pi(f)| = \lim |\pi_i(f)| \leqslant |f|_1$. So π is in the unit ball. So π is either 0 or in Δ. Thus $\Delta \cup \{0\}$ is closed in the unit ball. Since the unit ball is compact, this is a compact Hausdorff space in the relative topology. Also Δ would then be locally compact. □

6. Let G be an abelian compact Hausdorff group.

(a) Show $(g, \chi) \mapsto \chi(g)$ is continuous from $G \times \widehat{G}$ into $\mathbb{T}$ where $\widehat{G}$ has the relative weak $*$ topology of Exercise 8.6.5. (**Hint:** Consider $(g, \chi) \mapsto \pi_\chi(\lambda(g)f)$.)

(b) Show if K is a compact subset of G and $\epsilon > 0$, then

$$\{\chi \in \widehat{G} \mid |\chi(k) - 1| < \epsilon \text{ for } k \in K\}$$

is open in the topology on $\widehat{G}$ given in Exercise 8.6.5.

(c) Show if $f \in L^1(G)$ and $\chi \in \widehat{G}$, then the set

$$\{\chi' \in \widehat{G} \mid |\pi_{\chi'}(f) - \pi_\chi(f)| < \epsilon\}$$

is open in the topology on $\widehat{G}$ given by the neighborhood bases of (8.17).

In particular, the topology on $\widehat{G}$ is the weak $*$ topology described in Exercise 8.6.5.

Solution: Let $f \in L^1(G)$. Then

$$\pi_\chi(\lambda(g)f) = \int f(g^{-1}x)\chi(x)\,dx = \int f(x)\chi(gx)\,dx = \chi(g)\pi_\chi(f).$$

Thus if $(g, \chi) \mapsto \pi_\chi(\lambda(g)f)$ is continuous for each $f \in L^1(G)$, then since $\pi_\chi(f)$ is continuous in χ, we would see $(g, \chi) \mapsto \chi(g)$ is continuous on the open set $G \times \{\chi \mid \pi_\chi(f) \neq 0\}$. But by varying f, these sets cover $G \times \hat{G}$.

Fix g_0 and $\epsilon > 0$. By Exercise 6.7.8, there is a neighborhood N of g_0 such that $|\lambda(g)f - \lambda(g_0)f|_1 < \frac{\epsilon}{2}$ if $g \in N$. Thus

$$\begin{aligned}|\pi_{\chi'}(\lambda(g)f) - \pi_\chi(\lambda(g_0)f)| &\leqslant |\pi_{\chi'}(\lambda(g)f) - \pi_{\chi'}(\lambda(g_0)f)| + |\pi_{\chi'}(\lambda(g_0)f) - \pi_\chi(\lambda(g_0)f)| \\ &\leqslant |\lambda(g)f - \lambda(g_0)f|_1 + |\pi_{\chi'}(\lambda(g_0)f) - \pi_\chi(\lambda(g_0)f)| \\ &< \frac{\epsilon}{2} + \frac{\epsilon}{2}\end{aligned}$$

if g is in N and χ' is in the neighborhood $\{\chi' \mid |\pi_{\chi'}(\lambda(g_0)f) - \pi_\chi(\lambda(g_0)f)| < \frac{\epsilon}{2}\}$ of χ.

(b) Set $S = \{\chi \in \widehat{G} \mid |\chi(k) - 1| < \epsilon \text{ for } k \in K\}$. Let χ_0 be in this set. So $|\chi_0(k) - 1| < \epsilon$ for $k \in K$. Let M be the maximum of $|\chi_0(k) - 1|$ for $k \in K$. Then $M < \epsilon$. By part (a), $(g, \chi) \mapsto \chi(g)$ is continuous. Thus for each k in K, there is an open neighborhood V_k of k in G and a weak $*$ neighborhood U_k of χ_0 such that $|\chi(g) - \chi_0(k)| < \epsilon - M$ if $g \in V_k$ and $\chi \in U_k$. Since K is compact, there is a finite subcover $V_{k_1}, V_{k_2}, \ldots, V_{k_n}$ of K. Set $U = \cap U_{k_i}$. U is a neighborhood of χ_0 in the weak $*$ topology. Now if $\chi \in U$ and $k \in K$, then $k \in V_{k_i}$ for some i and thus $|\chi(k) - \chi_0(k_i)| < \epsilon - M$. Consequently,

$$\begin{aligned} |\chi(k) - 1| &\leqslant |\chi(k) - \chi_0(k_i)| + |\chi_0(k_i) - 1| \\ &< \epsilon - M + M = \epsilon \end{aligned}$$

for $\chi \in U$. Thus S is open.

For (c) we may assume $|f|_1 > 0$. Let ξ be in the set $\{\chi' \in \widehat{G} \mid |\pi_{\chi'}(f) - \pi_\chi(f)| < \epsilon\}$. Set $\delta = \epsilon - |\pi_\xi(f) - \pi_\chi(f)| > 0$.

Choose a compact set K such that $\int_{G-K} |f(g)|\,dg < \frac{\delta}{4}$. Then if χ' satisfies $|\chi'(k) - \xi(k)| < \frac{\delta}{2|f|_1}$ for $k \in K$, we see

$$\begin{aligned} |\pi_{\chi'}(f) - \pi_\chi(f)| &\leqslant |\pi_{\chi'}(f) - \pi_\xi(f)| + |\pi_\xi(f) - \pi_\chi(f)| \\ &= \left| \int f(g)(\chi'(g) - \xi(g))\,dg \right| + |\pi_\xi(f) - \pi_\chi(f)| \\ &< \int |f(g)|\,|\chi'(g) - \xi(g)|\,dg + |\pi_\xi(f) - \pi_\chi(f)| \\ &= \int_{G-K} 2|f(g)|\,dg + \int_K |f(g)|\,|\chi'(g) - \xi(g)|\,dg + |\pi_\xi(f) - \pi_\chi(f)| \\ &< \frac{\delta}{2} + \frac{\delta}{2|f|_1} \int_K |f(g)|\,dg + |\pi_\xi(f) - \pi_\chi(f)| \\ &\leqslant \frac{\delta}{2} + \frac{\delta}{2} + |\pi_\xi(f) - \pi_\chi(f)| \\ &< \delta + \epsilon - \delta = \epsilon. \end{aligned}$$

Thus this set is open in the topology given by (8.17). □

7. Assume G is an abelian compact Hausdorff group. Show the topology defined by the neighborhood bases given in (8.17) is discrete.

Solution: Consider the neighborhood $U = \{\chi \mid |\chi(g) - 1| < 1 \text{ for } g \in G\}$. We show $U = \{1\}$. First note $\chi = 1$ is in U. Now suppose $\chi \neq 1$ and choose $g \in G$ with $\chi(g) \neq 1$. If $\mathrm{Re}(\chi(g)) \leqslant 0$, then $|\chi(g) - 1| \geqslant \sqrt{(\mathrm{Re}(\chi(g)) - 1)^2} \geqslant 1$ and thus $\chi \notin U$. So we may suppose $\chi(g) = e^{i\theta}$ where $0 < |\theta| < \frac{\pi}{2}$. Choose n where $\frac{\pi}{2} \leqslant |n\theta| < \pi$. Then $\mathrm{Re}(\chi(g^n)) \leqslant 0$ and again one has $\chi \notin U$. So $U = \{e\}$ and G is discrete. □

8. Assume G is a discrete abelian group. Show the topology on $\widehat{G}$ given by the neighborhood bases given in (8.17) is compact. (**Hint:** You may use Exercises 8.6.5 and 8.6.6.)

Solution:

By Exercise 8.6.6, the mapping $\chi \mapsto \pi_\chi$ from $\hat{G}$ into Δ is a homeomorphism of $\hat{G}$ onto the collection of one-dimensional nonzero $*$ representations of the Banach algebra $L^1(G)$ where Δ has the relative weak $*$ topology. Moreover, by Exercise 8.6.5, $\Delta \cup \{0\}$ is a compact subset in the weak $*$ topology. To show $\hat{G}$ is a compact when G is discrete, it suffices to show $\{0\}$ is open in $\Delta \cup \{0\}$ in the relative weak $*$ topology. Consider the neighborhood at 0 in the weak $*$ topology given by $N_\epsilon = \{L \in L^1(G)^* \mid |L(1_{\{e\}})| < \epsilon\}$. Then $N_\epsilon \cap (\Delta \cup \{0\}) = \{0\} \cup \{\pi_\chi \mid \chi \in \hat{G}, |\pi_\chi(1_{\{e\}})| < \epsilon\}$. But $\pi_\chi(1_{\{e\}}) = \int \chi(g) 1_{\{e\}}\, dg = \chi(e) = 1$. So $N_{1/2} \cap (\Delta \cup \{0\}) = \{0\}$ and we see $\{0\}$ is open in $\Delta \cup \{0\}$.

□

9. Let G be a compact abelian Hausdorff group. Let $A(G)$ to be the space $\mathcal{F}^{-1}(L^1(\widehat{G}))$. Thus

$$A(\mathrm{G}) = \{\mathcal{F}^{-1}(f) \mid f \in L^1(\widehat{G})\}$$

where

$$\mathcal{F}^{-1}(f)(g) = \sum_{\chi \in \widehat{G}} f(\chi)\chi(g).$$

Show $A(G)$ is a dense $*$ subalgebra of the Banach $*$ algebra $C(G)$.

Solution: If $f \in L^1(\hat{G})$, then since $\sum_\chi f(\chi) < \infty$, the series $\sum_\chi f(\chi)\chi$ converges uniformly and hence $A(G) \subset C(G)$. Note $\mathcal{F}^{-1}(f_1 * f_2) = \mathcal{F}^{-1}(f_1)\mathcal{F}^{-1}(f_2)$ and $\mathcal{F}^{-1}(f^*) = \overline{\mathcal{F}^{-1}(f)}$ implies $A(G)$ is a complex algebra closed under conjugation. Since $\mathcal{F}^{-1}(\delta_1) = 1$, we see $A(G)$ contains the constants.

We claim $A(G)$ separates points. Suppose $g_1 \neq g_2$. Then there is a χ with $\chi(g_1) \neq \chi(g_2)$ for the regular representation separates points. Take $f = \delta_\chi$. Then $\mathcal{F}^{-1}(f) = \chi$ and we see $A(G)$ separates points. By the Stone-Weierstrass Theorem, $A(G)$ is uniformly dense in $C(G)$. □

10. Let G be a compact abelian Hausdorff group and let $\xi \in \widehat{\widehat{G}}$. Define π_0 on $A(G)$ by $\pi_0(\mathcal{F}^{-1}(f)) = \sum_{\chi \in \widehat{G}} f(\chi)\xi(\chi)$ for $f \in L^1(\widehat{G})$.

(a) Show π_0 is a $*$ representation of the subalgebra $A(G)$ of $C(G)$.

(b) Assuming π_0 is continuous; i.e. $||\pi(h)|| \leqslant ||h||_\infty$ for h in $A(G)$, show there is a $g \in G$ such that

$$\pi_0(h) = \sum_{\chi \in \widehat{G}} \widehat{h}(\chi)\chi(g).$$

(c) Assuming π_0 is continuous, show $\xi(\chi) = \chi(g)$ for $\chi \in \widehat{G}$.

Solution: (a) By Corollary 6.108, π defined by $\pi(f) = \sum_\chi f(\chi)\xi(\chi)$ is a nondegenerate $*$ representation of the Banach algebra $L^1(\widehat{G})$. Now π_0 on the algebra $A(G)$ is given by

$$\pi_0(\mathcal{F}^{-1}(f)) = \pi(f).$$

We first note π_0 is well defined. Indeed, if $\mathcal{F}^{-1}(f) = 0$, then

$$\sum_{\chi} f(\chi)\chi = 0.$$

Since this sum is uniformly convergent, we have

$$\sum \int f(\chi)\chi(g)\overline{\chi_0(g)}\, dg = 0$$

for each $\chi_0 \in \widehat{G}$. Since the χ_0 in $\widehat{G}$ are an orthonormal basis, we see $f(\chi_0) = 0$ for each χ_0. Thus $f = 0$ and so $\pi(f) = 0$.

We show (b) and (c). Assume π_0 is norm continuous. It therefore by the density of $A(G)$ in $C(G)$ extends to a $*$ representation of $C(G)$. By Theorem 6.84, there is a g with $\pi_0(\mathcal{F}^{-1}(f)) = F^{-1}(f)(g)$ for all $f \in L^1(\widehat{G})$. But $\mathcal{F}^{-1}(f)(g) = \sum_\chi f(\chi)\chi(g)$. This gives

$$\sum f(\chi)\chi(g) = \sum f(\chi)\xi(\chi)$$

for all $f \in L^1(\widehat{G})$. Take $f = \delta_{\chi_0}$. Then $\chi_0(g) = \xi(\chi_0)$ for all $\chi_0 \in \widehat{G}$; i.e., $\xi = \Phi(g)$. □

Exercise Set 8.7

1. Let $\mathcal{R}$ be a finitely generated algebra over the rational numbers $\mathbb{Q}$. Show every number $w \in \mathcal{R}$ is algebraic.

Solution: Let $w \in \mathcal{R}$ with $w \neq 0$. Then there exists a $k \times k$ matrix $A = [a_{i,j}]$ with rational entries such that

$$ww_i = a_{i,1}w_1 + a_{i,2}w_2 + \cdots + a_{i,k}w_k$$

for $i = 1, 2, \ldots, k$. Hence

$$(wI - A)\begin{pmatrix} w_1 \\ \vdots \\ w_k \end{pmatrix} = \begin{pmatrix} 0 \\ \vdots \\ 0 \end{pmatrix}.$$

This implies the matrix $wI - A$ is singular. Hence $\det(wI - A) = 0$. So w is a root of the characteristic polynomial $p(x) = \det(xI - A)$. Since this polynomial has rational coefficients, multiplying by the least common multiple of their denominators produces a polynomial with integer coefficients having w as a root. □

2. Show the collection of algebraic numbers is a field.

Solution: First note if $w_1, \ldots, w_k$ are rational numbers and θ satisfies $\theta w_j = \sum q_{i,j} w_i$ for rational $q_{i,j}$, then θ is algebraic for $(\theta I - Q)w = 0$ gives θ is a root of the polynomial $p(x) = \det(xI - Q)$. Thus the same argument as for the algebraic integers given in the proof of Theorem 8.84 shows the algebraic numbers form a ring. Now suppose $\theta \neq 0$ is algebraic. So $p(\theta) = 0$ where $p(x) = \sum_{k=0}^{n} a_k x^k$ and $a_k \in \mathbb{Z}$. So $\sum a_k \theta^k = 0$. Thus $\sum a_k \theta^{k-n} = 0$. So $\sum_{k=0}^{n} a_k (\theta^{-1})^{n-k} = 0$. So θ^{-1} is an algebraic number. □

3. Let A_4 be the alternating subgroup of the symmetric group S_4 consisting of even permutations.

(a) Show there are 4 conjugacy classes.

(b) Determine the dimensions of the irreducible finite dimensional representations of A_4.

Solution:

A permutation is a 1-1 function from $\{1, 2, \ldots, n\}$ into itself. It can be written as an n-tuple list $a_1, a_2, \ldots, a_n$ of this set. So $\sigma \in S_n$ corresponding to this would be given by $\sigma(i) = a_i$ or more succinctly by $a(i) = a_i$ for $i = 1$ to n. A cycle of order k is a permutation σ where there is a subset $\{a_1, a_2, \ldots, a_k\}$ of $\{1, 2, \ldots, n\}$ such that $\sigma(a_i) = a_{i+1}$ for $i = 1, \ldots, k-1$ and $\sigma(a_k) = a_1$ and $\sigma(j) = j$ for $j \notin \{a_1, a_2, \ldots, a_k\}$. A transposition is a cycle of order 2. For convenience we write $\sigma = (a_1, a_2, \ldots, a_k)$. Two cycles $(a_1, a_2, \ldots, a_k)$ and $(b_1, b_2, \ldots, b_l)$ are disjoint if $a_i \neq b_j$ for all i and j. Every permutation is a product (composition) of disjoint cycles. A cycle $(a_1, a_2, \ldots, a_k)$ can be written as $(a_{k-1}, a_k) \cdots (a_3, a_k)(a_2, a_k)(a_1, a_k)$; e.g..

$$(1, 2, 3, 4, 5) = (4, 5)(3, 5)(2, 5)(1, 5).$$

Consequently every permutation is a product of transpositions. If an even number is used, the permutation is even, otherwise it is odd.

In S_n, two cycles $a = (a_1, a_2, \ldots, a_k)$ and $b = (b_1, b_2, \ldots, b_k)$ of the same length are conjugate for there is a $\sigma \in S_n$ such that $\sigma a_j = b_j$ for each j and thus $\sigma a \sigma^{-1}(b_i) = \sigma a_{i+1} = b_{i+1}$ and if $l \notin a$, then $\sigma l \notin b$ for otherwise $\sigma^{-1}\sigma l = l \in a$. More generally two permutations in S_n are conjugate if and only if they have same number of disjoint cycles with the same lengths.

We call a cycle a monotone if $a_1 < a_2 < \cdots < a_k$ and note two permutations are conjugate under A_n if and only if they have the same number of disjoint cycles with the same lengths and the total number of transpositions needed to make all these cycles monotone is even. This is another way of saying they have the same parity. We leave the details. Using this information one sees A_4 has 4 conjugacy classes:

$$\{1\},$$
$$\{(1, 2)(3, 4), (1, 3)(2, 4), (1, 4)(2, 3)\},$$
$$\{(1, 2, 3), (1, 3, 4), (2, 3, 4), (1, 2, 4)\} \text{ and} \{(2, 1, 3), (3, 1, 4), (3, 2, 4), (2, 1, 4)\}.$$

(b) Thus there are 4 irreducible characters or 4 equivalence classes of finite dimensional irreducible representations of A_4. We let $\chi_1 = 1$ be the character of the 1-rep. So $n_1 = \dim \chi_1 = 1$. Let n_2, n_3, n_4 be their degrees. Then

$1^2 + n_2^2 + n_3^2 + n_4^2 = 12$ and $n_i | 12$ for $i = 2, 3, 4$. So $n_4 \in \{1, 2, 3\}$. If n_2, n_3, $n_4 \in \{1, 2\}$, then the sum can't be 12. So we may take $n_4 = 3$. Thus $n_2^2 + n_3^2 = 2$. Hence $n_1 = n_2 = n_3 = 1$ and $n_4 = 3$. □

4. Find 4 inequivalent irreducible finite dimensional representations of A_4.

Solution: Take $\pi_1 = 1$. Note $N = \{(1), (1,2)(3,4), (1,3)(2,4), (1,4)(2,3)\}$ is a normal subgroup of A_4. Indeed

$$(1,2)(3,4)(1,3)(2,4) = (1,4)(2,3), \ (1,2)(3,4)(1,4)(2,3) = (1,3)(2,4), \ \text{etc.}$$

It clearly is normal. Thus A_4/N has 3 elements and so is abelian. It has 3 characters (one-dimension representations); namely, $\bar{\pi}_1 = 1$, $\bar{\pi}_2$ and $\bar{\pi}_3$ given by $\bar{\pi}_2((1,2,3)N) = e^{2\pi i/3}$ and $\bar{\pi}_3((1,2,3)N) = e^{4\pi i/3}$. If $p : A_4 \mapsto A_4/N$ is the natural projection, then $\pi_1 = \bar{\pi}_1 \circ p$, $\pi_2 = \bar{\pi}_2 \circ p$, and $\pi_3 = \bar{\pi}_3 \circ p$ are 3 inequivalent 1-dimensional representations of A_4. To obtain π_4, let ρ be the permutation representation of S_4. So $\rho(\sigma)(e_i) = e_{\sigma i}$ where e_1, e_2, e_3, e_4 is basis. Note $e_1 + e_2 + e_3 + e_4$ is invariant. Take $V = [e_1, e_2, e_3, e_4]/[e_1 + e_2 + e_3 + e_4]$ and let π_4 be the quotient representation of ρ on V. Then π_4 has dimension 3. It is irreducible since being equivalent to a unitary representation, it is completely reducible and thus is equivalent to a direct sum of irreducible representations. By the previous exercise, the only possible dimensions of these are 1 and 3. So if it were reducible, it would be a sum of three one-dimensional representations. □

5. Determine the degrees of the characters of S_4.

Solution: There are 5 conjugacy classes; namely

$$\begin{aligned}
C_1 &= \{1\} \\
C_2 &= \{(1,2), (1,3), (1,4), (2,3), (2,4), (3,4)\} \\
C_3 &= \{(1,2)(3,4), (1,3)(2,4), (1,4)(2,3)\} \\
C_4 &= \{(1,2,3), (1,3,2), (1,4,2), (1,2,4), (1,3,4), (1,4,3), (2,3,4), (2,4,3)\} \\
C_5 &= \{(1,2,3,4), (1,2,4,3), (1,3,2,4), (1,3,4,2), (1,4,2,3), (1,4,3,2)\}.
\end{aligned}$$

Thus there are five equivalence classes of irreducible representations with degrees $n_1 \leqslant n_2 \leqslant \cdots \leqslant n_5$. Again we take $n_1 = 1$, the degree of 1. We have $n_1^2 + n_2^2 + n_3^2 + n_4^2 + n_5^2 = 24$. Since $n_5 | 24$, the biggest n_5 can be is 4. If $n_5 = 4$, then $1 + n_2^2 + n_3^2 + n_4^2 = 8$. Since $n_4 | 24$, n_4 can't be 2 for the best we can get is 7. So $n_5 = 4$ is impossible. Thus $n_5 = 3$ and this is the representation π_5 obtained by taking the quotient of the permutation representation ρ on $V = [e_1, e_2, e_3, e_4]/[e_1 + e_2 + e_3 + e_4]$. Finally we need $1 + n_2^2 + n_3^2 + n_4^2 = 15$. This forces n_4 to be 3. This then makes $n_3 = 2$ and $n_2 = 1$. □

6. Find the dual of S_4; i.e. one irreducible representation for each distinct irreducible character of S_4. (**Hint:** Consider the parity representation, the three dimensional representations of S_4 which extend the irreducible 3 dimensional representation of A_4, and representations of S_4 induced from one-dimensional representations of A_4.)

Solution:

We take $\rho_1 = 1$ and $\rho_2(\sigma) = (-1)^{\text{parity}(\sigma)}$. These are the one dimensional irreducible representations of S_4. In the previous exercise, we saw $n_1 = n_2 = 1$, $n_3 = 2$, $n_4 = 3$, $n_5 = 3$ are the degrees of the distinct irreducible characters for S_4. The elements of S_4 are the products $(1,2)\sigma$ where $\sigma \in A_4$. Consider the 3-dimensional representation ρ_4 of S_4 given by the permutation representation ρ of S_4 factored to $V = [e_1, e_2, e_3, e_4]/[e_1+e_2+e_3+e_4]$. It is thus the extension of π_4 of Exercise 8.7.4 to S_4. Since it was irreducible on A_4, it is irreducible on S_4. Set $\rho_5(\sigma) = (-1)^{\text{parity}(\sigma)}\rho_4(\sigma)$. Note $\rho_5(\sigma_1\sigma_2) = (-1)^{\text{parity}(\sigma_1\sigma_2)}\rho_4(\sigma_1)\rho_4(\sigma_2) = (-1)^{\text{parity}(\sigma_1)}\rho_4(\sigma_1)(-1)^{\text{parity}(\sigma_2)}\rho_4(\sigma_2) = \rho_5(\sigma_1)\rho_5(\sigma_2)$ and thus ρ_5 is an irreducible representation.

It remains to find an irreducible representation ρ_3 of degree 2. We consider $\rho_3 = \text{ind}_{A_4}^{S_4}\pi_2$ where π_2 is the one-dimensional representation of A_4 given in the solution of Exercise 8.7.4. It satisfies $\pi_2(1,2,3) = e^{2\pi i/3}$ and $\pi_2 = 1$ on N. Set $\rho_3 = \text{ind}_{A_4}^{S_4}\pi_2$. It is a 2 dimensional representation. We claim it is irreducible. If reducible, it would be a sum of two 1 dimensional representations and thus the range of ρ_3 would consist of commuting operators. Hence we need only show $\rho_3(1,2)$ does not commute with $\rho_3(1,2,3)$. Let us recall its construction. It is the representation on the space of complex functions f satisfying $f(xk) = \pi_2(k^{-1})f(x)$ for $k \in A_4$ and $x \in S_4$ given by $\rho_3(g)f(x) = f(g^{-1}x)$. Each f is determined by it values on 1 and $(1,2)$ for $S_4 = A_4 \cup (1,2)A_4$. We take f to be a function with value 1 at $(1,2)$. Note

$$\begin{aligned}
\rho_3(1,2,3)\rho_3(1,2)f(1) &= \rho_3(1,3)f(1)\\
&= f((1,3))\\
&= f((1,2)(1,3,2))\\
&= \pi_2(1,3,2)^{-1}f((1,2))\\
&= \pi_2(1,2,3)\\
&= e^{2\pi i/3}
\end{aligned}$$

while

$$\begin{aligned}
\rho_3(1,2)\rho_3(1,2,3)f(1) &= \rho_3(2,3)f(1)\\
&= f((2,3))\\
&= f((1,2)(1,2,3))\\
&= \pi_2(1,2,3)^{-1}f((1,2))\\
&= \pi_2(1,2,3)^{-1}\\
&= e^{-2\pi i/3}
\end{aligned}$$

and so they don't commute.

□

7. Let D_4 be the dihedral group of order 8. It has generators a and b satisfying $a^4 = b^2 = 1$ and $bab = a^3$. Determine the conjugacy classes of D_4, the degrees of each irreducible character of D_4, and a representation for each of these characters.

Solution:

The elements of D_4 are 1, a, a^2, a^3, b, ab, a^2b, a^3b. The conjugacy classes are $C_1 = \{1\}$, $C_2 = \{a, a^3 = bab\}$, $C_3 = \{a^2\}$, $C_4 = \{b, a^2b\}$, $C_5 = \{ab, a^3b\}$. Thus there are 5 distinct irreducible complex representations of D_4 and their degrees satisfy $n_1^2 + n_2^2 + n_3^2 + n_4^2 + n_5^2 = 8$ and n_j divides 8 for each j. Taking $1 \leqslant n_1 \leqslant n_2 \leqslant n_3 \leqslant n_4 \leqslant n_5$, we see $3 > n_5 > 1$ and thus $n_5 = 2$ and $n_1 = n_2 = n_3 = n_4 = 1$. We first find 4 one dimensional inequivalent representations. Namely, $\pi_1 = 1$, $\pi_2(a) = 1$ and $\pi_2(b) = -1$, $\pi_3(a) = 1$ and $\pi_3(b) = -1$, $\pi_4(a) = -1$ and $\pi_4(b) = -1$. That these give representation follows since each satisfy the defining relations, i.e. $\pi(a^4) = \pi(b^2) = 1$ and $\pi(bab) = \pi(a^3)$. We finally need a two dimensional representation. Take it to be its representation as motions on $\mathbb{C}^2$. So $\pi_5(a) = \left(\begin{smallmatrix} 0 & 1 \\ -1 & 0 \end{smallmatrix}\right)$ and $\pi_5(b) = \left(\begin{smallmatrix} -1 & 0 \\ 0 & 1 \end{smallmatrix}\right)$. Note $\pi_5(a)^4 = I$, $\pi_5(b)^2 = I$, and $\pi_5(bab) = \pi_5(a^3) = \left(\begin{smallmatrix} 0 & -1 \\ 1 & 0 \end{smallmatrix}\right) = \pi_5(b)\pi_5(a)\pi_5(b)$. Note this representation is irreducible for it cannot be the direct sum of two invariant subspaces for then $\pi_5(a)$ and $\pi_5(b)$ would commute.

□

8. Let G be a finite group. Show two finite dimensional complex representations of G are equivalent if and only if they have equal characters.

Solution:

We let π be a finite dimensional complex representation of the finite group G. Let $\hat{G}_c$ be a concrete dual. We know π is equivalent in a unique way to a representation $\bigoplus_{\rho \in \hat{G}_c} n_\rho \rho$ where all but finitely many n_ρ are nonzero. Thus $\chi_\pi = \sum n_\rho \chi_\rho$. But since $\bar{\chi}_\rho = \sum_{j=1}^{d(\rho)} e_j \otimes_\rho \bar{e}_j$, one has by the orthogonality of matrix coefficients (Corollary 8.13) and Exercise 8.4.10 that $(\bar{\chi}_\rho, \bar{\chi}_\pi)_2 = (\chi_\pi, \chi_\rho)_2 = n_\rho \delta_{\rho,\pi}$. Thus $n_\rho = (\chi_\pi, \chi_\rho)_2$ for each $\rho \in \hat{G}_c$. This implies if the characters of two finite dimensional representations are equal, they are equivalent to the same representation and hence equivalent to each other.

One can also just apply Theorem 6.123.

□

9. Let G be a finite group with center Z. Show if χ is an irreducible character, then

$$(\deg\chi)^2 \leqslant [G : Z] = \frac{|G|}{|Z|}.$$

(**Hint:** Let ρ be a representation with character χ. Use Schur's Lemma and Frobenius reciprocity.)

Solution: Let ρ be an irreducible representation with character χ. Then by Schur's Lemma, ρ on Z is given by

$$\rho(z) = c(z)I$$

where c is a one-dimensional representation of Z. Now the dimension of $\mathrm{ind}_Z^G c$ is $\frac{|G|}{|Z|}$ and the multiplicity of ρ in this induced representation is the multiplicity of c in $\rho|_Z$ which is $\dim \rho$. Hence if $n = \deg \chi = \dim \rho$, we see there are n copies of ρ in $\mathrm{ind}_Z^G c$. Thus these n copies have dimension n^2. So $n^2 \leqslant \frac{|G|}{|Z|} = \dim\left(\mathrm{ind}_Z^G c\right)$. □

10. Let G be a finite group and let G^1 be the commutator subgroup of G. Thus G^1 is the smallest group containing all $aba^{-1}b^{-1}$ for $a, b \in G$. Note G/G^1 is an abelian group. Show the number of distinct one dimensional representations of G is the order of the group G/G^1.

Solution: Note we know G/G^1 since abelian has dual equal to itself. Thus the number of 1 dimensional representations of G is bounded below by $|G/G^1|$. Let $\rho_1, \rho_2, \ldots, \rho_s$ be the irreducible 1 dimensional representations of G/G^1. Set $\pi_i = \rho_i \circ p$ where $p : G \to G/G^1$ is the natural projection. Assume π is a 1 dimensional representation not equal to any of the π_i. Clearly π vanishes on G^1. So π factors to a representation ρ of G/G^1. So $\pi = \rho \circ p$. □

11. Let G be a finite group with subgroup H. Find the character of the quasi-regular representation of G on $L^2(G/H)$.

Solution:

We use Theorem 8.90. Since the left quasi-regular representation on $L^2(G/H)$ is $\mathrm{ind}_H^G 1$, we have $\chi(x) = \frac{1}{|H|}\sum_{g\in G} 1_H(g^{-1}xg) = \frac{|\{g|g^{-1}xg\in H\}|}{|H|}$. But $g_1^{-1}xg_1 = g_2^{-1}xg_2$ if and only if $g_2g_1^{-1}x = xg_2g_1^{-1}$ if and only if $g_2g_1^{-1} \in Z_x$, the subgroup of G centralizing x. Let C_x be the conjugacy class of x in G. We claim $|C_x \cap H|\,|Z_x|$ is $|\{g \in G \mid g^{-1}xg \in H\}|$. To see this let $C_x \cap H = \{x_1, x_2, \ldots, x_k\}$ where the x_j are distinct. Fix g_j with $g_j^{-1}xg_j = x_j$. Now suppose $g^{-1}xg = x_j$. So $g^{-1}xg = g_j^{-1}xg_j$. Thus $gg_j^{-1} \in Z_x$. This implies $g = z_xg_j$ for a unique $z_x \in Z_x$ and g_j. Thus the character χ of the left quasi-regular representation on $L^2(G/H)$ is given by

$$\chi(x) = \frac{|C_x \cap H|\,|Z_x|}{|H|}.$$

□

12. Let $G = S_3$ and $X = \{1, 2, 3\}$. Show that $L^2(X)$ decomposes into two irreducible representations, one of them, the trivial representation being one dimensional, and the other two dimensional.

Solution:

The representation on $L^2(X)$ which we denote by ρ is given by $\rho(\sigma)f(j) = f(\sigma^{-1}j)$ for $j = 1, 2, 3$. Note $\rho(\sigma)1 = 1$ for $\sigma \in S_3$. Thus $\langle 1 \rangle = \mathbb{C}1$ is an invariant subspace and the representation ρ on this subspace is trivial.

Since ρ is unitary, $W = \{1\}^\perp$ is a two dimensional invariant subspace. It consists of all functions of form $c_1 1_{\{1\}} + c_2 1_{\{2\}} + c_3 1_{\{3\}}$ where $c_1 + c_2 + c_3 = 0$. Thus $f_1 = \frac{1}{\sqrt{2}}1_{\{1\}} - \frac{1}{\sqrt{2}}1_{\{2\}}$ and $f_2 = \frac{1}{\sqrt{6}}1_{\{1\}} + \frac{1}{\sqrt{6}}1_{\{2\}} - \frac{2}{\sqrt{6}}1_{\{3\}}$ form an orthonormal basis of W.

Note $\sigma(1_{\{j\}})(k) = 1_{\{j\}}(\sigma^{-1}k) = \delta_{j,\sigma^{-1}k} = \delta_{\sigma j,k} = 1_{\{\sigma j\}}(k)$. Thus $\rho(\sigma)f_1 = \frac{1}{\sqrt{2}}1_{\{\sigma 1\}} - \frac{1}{\sqrt{2}}1_{\{\sigma 2\}}$ and $\rho(\sigma)f_2 = \frac{1}{\sqrt{6}}1_{\{\sigma 1\}} + \frac{1}{\sqrt{6}}1_{\{\sigma 2\}} - \frac{2}{\sqrt{6}}1_{\{\sigma 3\}}$. So $\rho(1) = I$ and $\rho(1,2)f_1 = -f_1$, $\rho(1,2)f_2 = f_2$,

For $\sigma = (1,3)$, we first note

$$
\begin{aligned}
1_{\{3\}} - 1_{\{2\}} &= \frac{1}{2}(1_{\{1\}} - 1_{\{2\}}) - \frac{1}{2}(1_{\{1\}} + 1_{\{2\}} - 2\,1_{\{3\}}) \\
1_{\{3\}} + 1_{\{2\}} - 2\,1_{\{1\}} &= \frac{-3}{2}(1_{\{1\}} - 1_{\{2\}}) - \frac{1}{2}(1_{\{1\}} + 1_{\{2\}} - 2\,1_{\{3\}})
\end{aligned}
$$

and thus $\rho(1,3)f_1 = \frac{1}{\sqrt{2}}1_{\{3\}} - \frac{1}{\sqrt{2}}1_{\{2\}} = \frac{1}{2}f_1 - \frac{\sqrt{3}}{2}f_2$ and $\rho(1,3)f_2 = -\frac{\sqrt{3}}{2}f_1 - \frac{1}{2}f_2$. Similarly for $\sigma = (2,3)$, we note

$$
\begin{aligned}
1_{\{1\}} - 1_{\{3\}} &= \frac{1}{2}(1_{\{1\}} - 1_{\{2\}}) + \frac{1}{2}(1_{\{1\}} + 1_{\{2\}} - 2\,1_{\{3\}}) \\
1_{\{1\}} + 1_{\{3\}} - 2\,1_{\{2\}} &= \frac{3}{2}(1_{\{1\}} - 1_{\{2\}}) - \frac{1}{2}(1_{\{1\}} + 1_{\{2\}} - 2\,1_{\{3\}})
\end{aligned}
$$

and so $\rho(2,3)f_1 = \frac{1}{2}f_1 + \frac{\sqrt{3}}{2}f_2$ and $\rho(2,3)f_2 = \frac{\sqrt{3}}{2}f_1 - \frac{1}{2}f_2$.

Thus the corresponding matrix representation satisfies:

$$
\rho(1) = \left(\begin{smallmatrix} 1 & 0 \\ 0 & 1 \end{smallmatrix}\right),\ \rho(1,2) = \left(\begin{smallmatrix} -1 & 0 \\ 0 & 1 \end{smallmatrix}\right),\ \rho(1,3) = \begin{pmatrix} \frac{1}{2} & -\frac{\sqrt{3}}{2} \\ -\frac{\sqrt{3}}{2} & -\frac{1}{2} \end{pmatrix},\ \rho(2,3) = \begin{pmatrix} \frac{1}{2} & \frac{\sqrt{3}}{2} \\ \frac{\sqrt{3}}{2} & -\frac{1}{2} \end{pmatrix}.
$$

Consequently,

$$
\begin{aligned}
\rho(1,2,3) &= \rho(1,2)\rho(2,3) = \left(\begin{smallmatrix} -1 & 0 \\ 0 & 1 \end{smallmatrix}\right)\begin{pmatrix} \frac{1}{2} & \frac{\sqrt{3}}{2} \\ \frac{\sqrt{3}}{2} & -\frac{1}{2} \end{pmatrix} = \begin{pmatrix} -\frac{1}{2} & -\frac{\sqrt{3}}{2} \\ \frac{\sqrt{3}}{2} & -\frac{1}{2} \end{pmatrix} \text{ and} \\
\rho(1,3,2) &= \rho(1,3)\rho(2,3) = \begin{pmatrix} \frac{1}{2} & -\frac{\sqrt{3}}{2} \\ -\frac{\sqrt{3}}{2} & -\frac{1}{2} \end{pmatrix}\begin{pmatrix} \frac{1}{2} & \frac{\sqrt{3}}{2} \\ \frac{\sqrt{3}}{2} & -\frac{1}{2} \end{pmatrix} = \begin{pmatrix} -\frac{1}{2} & \frac{\sqrt{3}}{2} \\ -\frac{\sqrt{3}}{2} & -\frac{1}{2} \end{pmatrix}
\end{aligned}
$$

and thus we have ρ on all 6 elements of S_3.

□

Index

Made in the USA
Middletown, DE
04 November 2026

36537969R00249

Employment of Functionalized Natant Bio-materials as Potential Adsorbents for Industrial Liquid Litters

Scaleup of Plausible Solutions through Greener Approach